MODERN
POWER STATION PRACTICE

Third Edition
(in 12 volumes)

Incorporating Modern Power System Practice

Books are to b

MODERN
POWER STATION PRACTICE

Third Edition

Incorporating Modern Power System Practice

British Electricity International, London

Volume E
Chemistry and Metallurgy

PERGAMON PRESS

OXFORD · NEW YORK · SEOUL · TOKYO

U.K.	Pergamon Press plc, Headington Hill Hall, Oxford OX3 0BW, England
U.S.A.	Pergamon Press, Inc., 660 White Plains Road, Tarrytown, New York 10591–5153, U.S.A.
KOREA	Pergamon Press Korea, KPO Box 315, Seoul 110-603, Korea
JAPAN	Pergamon Press Japan, Tsunashima Building Annex, 3-20-12 Yushima, Bunkyo-ku, Tokyo, Japan

First edition 1963

Second edition 1971

Third edition 1992

Library of Congress Cataloging in Publication Data

Modern power station practice: incorporating modern power system practice/British Electricity International.— 3rd ed. p. cm.
Includes index.
1. Electric power-plants. I. British Electricity International.
TK1191.M49   1990
62.31'21 — dc20   90-43748

British Library Cataloguing in Publication Data

British Electricity International
Modern power station practice.—3rd. ed.
1. Electric power-plants. Design and construction
I. Title II. Central Electricity Generating Board
621.3121.

ISBN 0-08-040510-X (12 Volume Set)
ISBN 0-08-040515-0 (Volume E)
ISBN 0-08-042245-4 (Flexicover)

Printed in the Republic of Singapore
by Singapore National Printers Ltd

Contents

Colour Plates

(between pp 208 and 209)

Foreword

G. A. W. Blackman, CBE, FEng
Chairman, Central Electricity Generating Board
and Chairman, British Electricity International Ltd

FOR OVER THIRTY YEARS, since its formation in 1958, the Central Electricity Generating Board (CEGB) has been at the forefront of technological advances in the design, construction, operation, and maintenance of power plant and transmission systems. During this time capacity increased almost fivefold, involving the introduction of thermal and nuclear generating units of 500 MW and 660 MW, to supply one of the largest integrated power systems in the world. In fulfilling its statutory responsibility to ensure continuity of a safe and economic supply of electricity, the CEGB built up a powerful engineering and scientific capability, and accumulated a wealth of experience in the operation and maintenance of power plant and systems. With the privatisation of the CEGB this experience and capability is being carried forward by its four successor companies — National Power, PowerGen, Nuclear Electric and National Grid.

At the heart of the CEGB's success has been an awareness of the need to sustain and improve the skills and knowledge of its engineering and technical staff. This was achieved through formal and on-job training, aided by a series of textbooks covering the theory and practice for the whole range of technology to be found on a modern power station. A second edition of the series, known as Modern Power Station Practice, was produced in the early 1970s, and it was sold throughout the world to provide electricity undertakings, engineers and students with an account of the CEGB's practices and hard-won experience. The edition had substantial worldwide sales and achieved recognition as the authoritative reference work on power generation.

A completely revised and enlarged (third) edition has now been produced which updates the relevant information in the earlier edition together with a comprehensive account of the solutions to the many engineering and environmental challenges encountered, and which puts on record the achievements of the CEGB during its lifetime as one of the world's leading public electricity utilities.

In producing this third edition, the opportunity has been taken to restructure the information in the original eight volumes to provide a more logical and detailed exposition of the technical content. The series has also been extended to include three new volumes on 'Station Commissioning', 'EHV Transmission' and 'System Operation'. Each of the eleven subject volumes had an Advisory Editor for the technical validation of the many contributions by individual authors, all of whom are recognised as authorities in their particular field of technology.

All subject volumes carry their own index and a twelfth volume provides a consolidated index for the series overall. Particular attention has been paid to the production of draft material, with text refined through a number of technical and language editorial stages and complemented by a large number of high quality illustrations. The result is a high standard of presentation designed to appeal to a wide international readership.

It is with much pleasure therefore that I introduce this new series, which has been attributed to British Electricity International on behalf of the CEGB and its successor companies. I have been closely associated with its production and have no doubt that it will be invaluable to engineers worldwide who are engaged in the design, construction, commissioning, operation and maintenance of modern power stations and systems.

March 1990

Preface

Volumes B and C of Modern Power Station Practice deal with the main items of power station plant — the boilers, turbines and generators — and the ancillary plant which goes with them. Volume E is complementary to them, in that it covers the pre-combustion treatment of the fossil fuels which fire the boilers and the control of the feed and boiler water which constitute the working fluid. The control of cooling water quality and the techniques of plant cleaning, both before commissioning and during overhauls, are further subjects covered in this volume and all of these topics are basically within the province of chemistry.

Much of the development work described in the first four chapters of the volume has been undertaken by station chemists, staff of the Regional Scientific Services Departments or those in the chemistry division of the CEGB Research Laboratory at Leatherhead. The information given in the previous editions of MPSP has been broadened to cover the much greater operational experience gained since that edition was published. There is much new information on the experience gained with water chemical treatment at the CEGB's existing nuclear power stations and the subject of the treatment required at the forthcoming PWR stations is also covered.

The previous edition of Volume E contained a relatively short section on metallurgy but the present volume deals with the subject at much greater length, particularly in covering defect analysis and life assessment; this latter work has already proved invaluable to the operators in deciding on the prolongation of the life of plant items and the safety factors involved in doing so. Again, there has been considerable research and development work carried out by the Regional Scientific Services Departments and all the main Central Research Laboratories, whilst the Materials Engineer in the Design and Construction Division has acted as a focal point in advising stations on practices to be followed.

As a former Director-General of the Research Division, I should like to take this opportunity to thank all those research staff who have made contributions in the fields of chemistry and metallurgy or their excellent work. I should also like to thank all the authors who have contributed to this volume and the three advisory editors, John Brown, John Ray and especially Matt Gemmill for ensuring that the presentation of the technical material is pellucid.

As Chairman of the Main Editorial Panel, I should also like to thank my fellow panel members for their contributions, particularly the late Professor John Davies who helped with the detailed editing of a number of volumes. Finally, I must congratulate Peter Reynolds, the Managing Editor, whose consistent hard work has ensured that the publication is of such a high standard.

D. J. LITTLER
Consulting Editor — Volume E

Contents of All Volumes

Volume A — Station Planning and Design
Power station siting and site layout
Station design and layout
Civil engineering and building works

Volume B — Boilers and Ancillary Plant
Furnace design, gas side characteristics and combustion equipment
Boiler unit — thermal and pressure parts design
Ancillary plant and fittings
Dust extraction, draught systems and flue gas desulphurisation

Volume C — Turbines, Generators and Associated Plant
The steam turbine
Turbine plant systems
Feedwater heating systems
Condensers, pumps and cooling water systems
Hydraulic turbines
The generator

Volume D — Electrical Systems and Equipment
Electrical system design
Electrical system analysis
Transformers
Generator main connections
Switchgear and control gear
Cabling
Motors
Telecommunications
Emergency supply equipment
Mechanical plant electrical services
Protection
Synchronising

Volume E — Chemistry and Metallurgy
Chemistry
Fuel and oil
Corrosion: feed and boiler water chemistry
Water treatment plant and cooling water systems
Plant cleaning and inspection
Metallurgy
Introduction to metallurgy
Materials behaviour
Non-ferrous metals and alloys
Non-metallic materials
Materials selection

Volume L — System Operation

System operation in England and Wales
Operational planning — demand and generation
Operational planning — power system
Operational procedures — philosophy, principles and outline contents
Control in real time
System control structure, facilities, supporting services and staffing

Volume M — Index

Complete contents of all volumes
Cumulative index

Volume E
Chemistry

CHAPTER 1

Fuel and oil

1 Introduction to power industry chemistry

1.1 Historical

Any new technology concentrates in its early years on proving what can initially be achieved and finding better routes to follow in order to reach its ultimate goals. The electricity industry went through this pioneering phase between 1880 and 1920. By then the steam turbine had been demonstrated and reliable electrical machines were available but the efficiency with which the energy in fuel was converted to electricity was still well short of what was theoretically possible. Chemical services to stations were almost non-existent and it was recognised that further progress would require the application of scientific knowledge to the solution of problems of safety, economic operation, reliability of plant and improved efficiency. The insights gained from understanding and controlling fuels, combustion, water treatment and corrosion allowed steady progress to be made.

From 1920 onwards station chemists began to be appointed to work with their electrical and mechanical engineering colleagues. The industry was then made up of many small municipal undertakings and the more enterprising ones led the way. Among the early pioneers in power station chemistry were G. W. Hewson, W. S. Coates, A. B. Owles and J. Dunn of the London, Newcastle, Manchester and Leicester companies respectively, and their laboratories were the first to be established in Britain to provide a chemical service for the generation and transmission of electricity.

After 1945 it was clear that the country's economic development would be hindered if the size and efficiency of the electricity industry was not increased. During the early years of the British Electricity Authority divisional scientific laboratories were set up around the country and chemists were generally appointed to stations. In 1958 the Central Electricity Generating Board was formed and a five year development plan

was implemented by the new CEGB Research Department. Central laboratories were set up at Leatherhead, Nuclear Laboratories at Berkeley, and by 1963 Engineering Laboratories at Marchwood. These three laboratories, together with a small London headquarters group worked closely with the, by now, five Regional Research organisations to meet the needs of the Board as a whole for basic, operational and applied research; thus providing the information required for the design of new plant and for the solving of operational problems.

Chemists were now working in laboratories alongside scientists of other disciplines, metallurgists physicists and biologists, and in addition maintained their links with those chemists assigned the day-to-day tasks of supporting operations at the power station sites. This organisation was effective in supporting the large power station building programme through the 1960s to early 1970s, and the five CEGB Regional Research organisations formed from the earlier Divisional Laboratories brought their facilities together during this period at five Scientific Services Centres, to support the major commissioning programme required.

With the improvements in design, combustion, materials of construction and improved control of water chemistry achieved, efficiencies now approach the theoretical limit and about twice as much electricity was produced from a tonne of fuel than was obtained before 1939.

New challenges await the power industry's chemists and laboratory organisation will evolve to meet their needs. The underlying requirements to control fuel, combustion, water treatment and corrosion will continue, but as nuclear power develops and natural gas is burnt in new combined cycle plant, the industry will become more diverse. Older stations will close and the National Grid will need renewal. Chemical monitoring of operating equipment will continue and environmental control, including measurements of conditions around decommissioning plant, will become increasingly important.

1.2 The work of the power station chemist

In the early development of chemical services the station chemist either stood alone or received support from a divisional chemist. He was responsible for sampling and analysis of fuel delivered to the station, the resulting data being needed for both fuel pricing and calculating station efficiency. Tests of lubricating and insulating oils were carried out and the quality control of raw, treated, feed and boiler waters was undertaken, along with technical control of the processes of water softening, evaporation and chlorination. Staff were also available to carry out plant investigations, monitor and control corrosion and check that working conditions were safe.

As the Research Department and Regional Scientific Services evolved in the 1960s the support available to the station chemist was greatly strengthened, but to an extent his autonomy became limited, as 'best operational practices' were identified and formalised in generation operations memoranda. Centralised services were established as analytical techniques improved in accuracy and speed, and instrumental complexity increased. Plant investigations were often more efficiently mounted by specialists from the regional scientific services who had greater resources than the station could justify for occasional use.

At present the sampling of fuel is still a most important part of the station chemists' duties but preparation and analysis of the samples is usually carried out at a central services laboratory. Most water and steam analysis is carried out on station, as samples deteriorate during transport. Station laboratories will often have atomic absorption spectrometers so that the best modern standards of analysis may be achieved. For more detailed analysis, however, the scientific services have specialist equipment.

The resources of station chemists are sufficient for day-to-day operational needs and also enable them to plan and manage the acid cleaning of the water/steam circuit of the station, to deal with local water authorities regarding consents for effluent disposal, to manage the storage and handling of large amounts of chemicals (sulphuric acid and methanol for example), and to run chemical plants capable of producing hydrogen for generator cooling or sodium hypochlorite for water treatment. In some other work areas the chemist is able to call on the scientific services departments for help.

Scientific services staff who concentrate, for example, on combustion monitoring and control, corrosion, ion exchange resin testing, cooling water pump calibration or lubrication problems are available to assist the chemist when required and as a result station chemistry staff numbers are less than they were, but the functional support available to the station management is comprehensive.

Improvements in communications and data retrieval will probably effect the most significant gains in efficiency of operation in the next few years. When a problem arises, the results of previous similar incidents may be readily retrieved to help decide what action to take, or trends in monitored parameters may be followed and alarm levels set, to alert staff that the plant might be about to operate out of specification. Such developments may change the chemist's role yet again, but will have the advantage of giving him or her a much better overview.

The role of the chemist in a nuclear station covers some of the same tasks as his counterpart in a conventional station but obviously fuel sampling and combustion control are replaced by other tasks involving monitoring of the reactor functions. Control of the gas chemistry in gas-cooled reactors and provision of radio-chemical services are not described in this volume, but are to be found in Volume J. The station chemist of today then, provides a skilled scientific service in order to improve availability, efficiency and the economics of generation, by controlling the raw materials input to the station and minimising chemical damage to the very high cost capital plant. He applies specialised knowledge and experience to the many problems arising in daily operation inside the station and to the increasing problems arising outside the station in the fields of atmospheric pollution, marine fouling and river pollution. To help in his task he has access to resources and specialists both at the Scientific Services Centres and the three CEGB Research Division laboratories at Leatherhead, Berkeley and Marchwood.

1.3 Central services

Operational support to the station chemist was first offered by a divisional chemist and staff; then a Regional Research Organisation evolved, eventually centring its facilities at five Regional Scientific Services centres and latterly in the CEGB these services were combined with engineering support into one Technology and Engineering Division. The private companies formed from the CEGB do not operate in the same way, but some form of centralised support for the chemist continues.

Those functions which can be carried out by specialists, or large instrumental techniques, are provided as a service to the chemist and/or his management. To take the original four chemist's objectives of control of fuel, combustion, water treatment and corrosion; central support is provided for each.

Fuel sampling is carried out at stations, but analysis and sample preparation is undertaken by a central service laboratory who pass the data to a central computer for pricing and control purposes. *Combustion control* may require sampling and investigation of fuel at the mill or classifier outlets or analysis of gas compositions at points in the furnace, or from the back end of the boiler. Additionally, to monitor pollution, samples may be required from the stack or from the field and these tasks will be carried out by specialists. *Water treatment* involves ion-exchange and the resin performance is monitored by a centralised service. *Corrosion control* involves detailed analysis of oxide growth and deposits. Large instruments such as scanning electron microscopes, X-ray spectrometers, and emission spectrometers are available for metallurgists and chemists to use to determine why corrosion is occurring and whether it may be due to the wrong use of materials.

In further support of *environmental studies* plasma emission spectrometers are available, capable of rapid multi-element analysis of solutions such as rainwater or chemical cleaning liquors which might be awaiting disposal.

Both gas and high performance liquid chromatography are available to support the *monitoring of transformers*, a task not necessarily involving station chemists; this is often a service to transmission engineers.

Spectrometers are also available for identification of non-metallic materials, plastics and asbestos, for example, and there are supporting specialists to advise on paints and coatings and the uses and disposal of ash from power stations. The above examples are not exhaustive, but give an indication of the wide chemical support now available to Generation Divisions and the station chemists in particular.

From time to time operational problems of a new kind arise and the scientific services specialists may themselves require support. In the CEGB this was obtained from Research Division colleagues at Leatherhead, Berkeley or Marchwood who undertook research or development work to help in understanding the processes involved. Chemists in the research laboratories also had a key role to play in the CEGB's forward strategies. The CEGB successor companies have a more compact scientific support structure and when research is required it may be contracted outside the company.

2 Coal

2.1 Introduction

Coal is a compacted mass of fossilised plant debris mixed with smaller amounts of inorganic matter and covered by sedimentary rocks. Its chemical properties depend upon the proportions of the different chemical components present in the parent plant material; the nature and extent of the changes which these components have undergone since their deposition; and the nature and quantity of the inorganic matter present.

Plant-material decayed first to a material resembling peat, due to bacterial action. The peaty deposit was then covered by an impervious sedimentary layer and under the influence of temperature, time and finally pressure, from the increasing depth of burial, coal was formed. The process proceeds from peat to lignite (brown coal) through bituminous coals and finally to anthracites; as the process progresses, the coal rank increases. There have been several coal-forming periods in the earth's history and these are summarised

in Table 1.1. The corresponding chemical changes are noted in Table 1.2.

From the point of view of increasing fuel quality the progression takes wood, calorific value on a *dry* basis of 16 000 kJ/kg, through to coals with dry mineral matter-free calorific values in excess of 32 500 kJ/kg. This is achieved by progressive loss of carbon dioxide, water and methane from the original cellulose material. The protein material, also present in the wood, progressively loses its nitrogen and sulphur content, in addition to its oxygen content. The progression is illustrated in Table 1.3.

TABLE 1.1

Age of coal measures

Geological system		Approx. mean age, years $\times 10^6$	Rank of coals formed
Era	Period		
Upper Palaeozoic	Carboniferous	250	Anthracites
	Permian	210	Carbonaceous and bituminous coals
Mesozoic	Triassic	180	Bituminous
	Jurassic	150	Bituminous
	Cretaceous	100	Bituminous and sub-bituminous
	Eocene	60	Sub-bituminous and lignites
Tertiary	Oligocene	40	Lignites
	Miocene	20	Lignites
Quarternary	Pleistocene	1	Peat only

TABLE 1.2

Approximate composition of the ranks of coal listed in Table 1.1

	Carbon (%)	Hydrogen (%)
Anthracites	93 – 95	3.8 – 2.8
Carbonaceous	91 – 93	4.25 – 3.8
Bituminous	80 – 91	5.6 – 4.25
Sub-bituminous	75 – 80	5.6 – 5.1
Lignites	60 – 75	5.7 – 5.0

The moisture and oxygen contents of the fuels fall as the rank increases, and the calorific value rises, with the slight exception of anthracites. Loss of methane in the final stage of coal formation increases the carbon content of anthracite but reduces its hydrogen content, and there is a slight fall in the inherent calorific value. Anthracite coal seams are noted for their methane content (fire-damp) because of this final step in the geological process.

2.2 British coal resources

The known coal reserves in the United Kingdom are massive. About 100 000 million tonnes of coal are in

TABLE 1.3

Changes in composition during coal formation

Fuel	Ultimate analysis dry ash free basis			Proximate air dry basis	Calorific value air dried
	% C	% H	% O	Moisture %	kJ/kg
Cellulose	44.4	6.2	49.4	50	9 000
Peat	60	6	34	20	14 650
Lignite	70	8	22	15	20 900
Sub-bituminous	75–82	6–5	20–12	10	23 000
Bituminous	82–90	6–4.5	12–3	2	32 400
Carbonaceous	91–93	4	4	1	33 500
Anthracites	94	3	2	1	32 900

known coalfields (see Fig 1.1) and a further 70 000 million tonnes could exist elsewhere on the basis of current geological knowledge. Recoverable reserves, using present technology, are in excess of 45 000 million tonnes, sufficient for 360 years at present production rates. However, reserves in existing collieries are only 4000 million tonnes, so that British Coal (BC) has to have a rolling plan to create access to new reserves; presently Selby, the Vale of Belvoir and the area south west of Coventry are being opened up. The quality of British coal is good and is broadly classified by international standards as 'hard' (see Section 3.1 of this chapter). There are no brown coal deposits in the UK and only insignificant amounts of lignite. Although deposits of peat are available in Scotland, this fuel has not proved economical to use on a large scale.

2.3 British coal production and utilisation by the CEGB

Peak coal production in Britain reached 287 million tonnes during the first quarter of the century but as oil has become more widely used for transport, the steel industry has contracted, and the efficiency of energy useage has improved, so the requirement for coal has fallen. In a single year (1982/83) the output of about 124 million tonnes was made up of ~105 million tonnes of deep-mined BC production, some 1.5 million tonnes from privately operated licensed sites, ~15 million tonnes of BC open cast production and about 3 million tonnes of 'non-vested' coal, including slurry and recovered fines from various sources. Figures vary somewhat from year to year and the short term trend may be for a further reduction in output until new pits reduce production costs in real terms, but the long term future is assured, since world reserves of oil are small compared with the quantities of coal available, and a general shift back to coal as a fuel and chemical feedstock should eventually occur everywhere.

The CEGB in 1982/83 burned some 75 million tonnes of coal in its power stations. The industry is by far the biggest customer for UK coal. British Coal typically supplies 89% of the electricity industry's regular requirements from mines and a further 9% from open cast sites, with the private sector supplying only 2%.

The locations of the coalfields are shown in Fig 1.1 and the geographical distribution of the supply is shown in Fig 1.2 for coal received in 1982/83. In this year although the coal consumed was 75 million tonnes, as normal, some 82 million tonnes was purchased and 0.8 million tonnes was imported so that stock levels rose. The importance of supplies from Yorkshire, North Midlands and Western Areas is clear from Fig 1.2 and this, coupled with the availability of cooling water from the Aire and Calder, Trent and Severn rivers, explains why the majority of power stations are located in the Midlands and Yorkshire (Fig 1.3).

The major part of the coal received is 'smalls' and the average qualities of the various categories produced are indicated in Table 1.4. The quality also varies with the coal's geographical origin and this is shown in Table 1.5. There are significant variations in quality from Area to Area, due to the different geological nature of the coal, and power stations have to accommodate this. The net calorific value of the inherently wet South Midlands coals is low, for example, compared with the drier North Eastern coals and more coal has to be milled and burned to obtain sufficient heat input to the boiler. Chlorine also causes problems; corrosion of furnace wall tubes at certain stations receiving Western Area coals. Finally the low volatile South Wales coals require careful combustion and are directed to stations designed specifically to cope with them. These and other factors will be discussed in more detail in Section 17 of this chapter.

3 Classification of coal

3.1 Classification as a mineral by masceral type

Coal can be classified by its general appearance and by its microscopic structure. Its general appearance relates to how well a coal will perform and engineers soon learned to recognise a good fuel or coking coal,

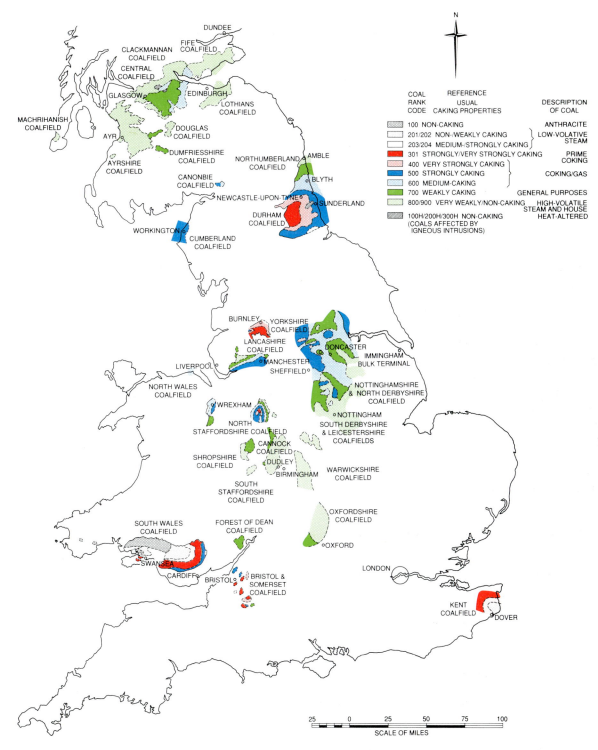

Fig. 1.1 The coal fields of Great Britain

for example. The microscopic nature of coal is of more interest to the geologist and for the purposes of this volume only a brief description will be given to illustrate those aspects relevant to its behaviour as a fuel.

3.1.1 The general appearance of coal

Stopes and Wheeler [1] were the first to publish a comprehensive account of the relationship between the appearance and properties of the banded components of bituminous coals. In modern terminology they recognised four main lithotypes, each ending with 'ain':

Vitrain Narrow bands of brilliant, or vitreous, uniform appearance best described as black glass. It is brittle material and fractures conchoidally. It contains no obvious plant structure and seems to have been derived from bark.

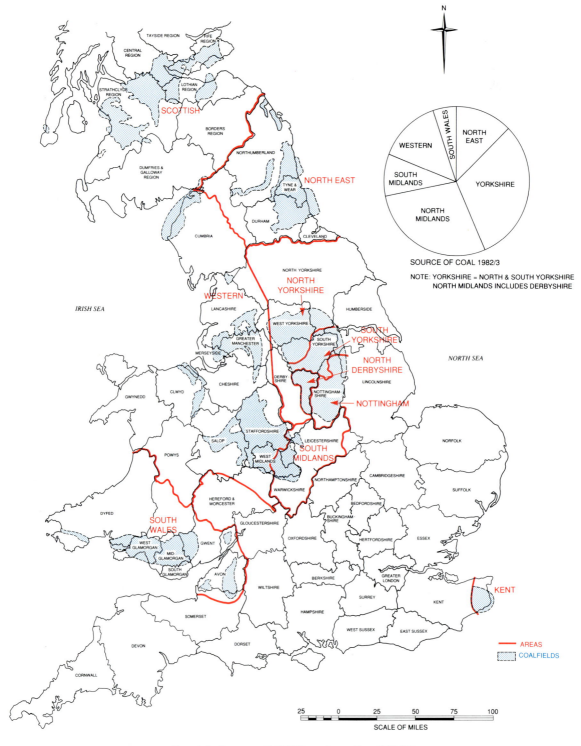

FIG. 1.2 Sources of CEGB coal supplies 1982/83

Clairain Bands of varying thickness, with a pronounced gloss or shine but not so brilliant as vitrain. It has an irregular fracture and can be seen to be made up of layers. It contains more plant remains (spores, etc.) than vitrain and is commonest of the four types of coal substance.

Durain Dull grey hard bands of varying thickness with a matt or granular texture. It also cracks in an

irregular fashion. It is full of durable plant remains (spores, cuticle, etc.) and it is thought that durain was formed from silts or muds composed of small particles of robust vegetable matter. Cannel coal, found in some parts of the country, may be an extreme form of durain, with a high clay content.

Fusain Bands, patches or wedges of soft, fibrous material resembling charcoal. It is the component of coal

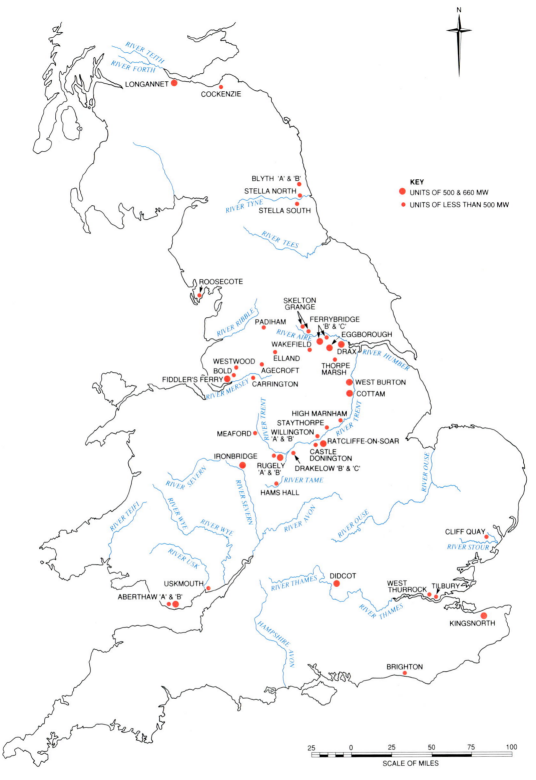

FIG. 1.3 Coal-fired power stations in England and Wales

which dirties your hands. It only occurs in thin seams between bands of the other types, and seems to have formed differently from them. As it is weak, large pieces of coal fracture through the fusain bands.

About half of a coal seam may be clairain, a sixth to a third durain, and vitrain may rise to 10 to 15 per cent, while fusain is only present up to 1 or 2 per cent.

The chemical properties of the four types differ as shown in Table 1.6.

In coal preparation durain and clairain may segregate as 'hards' and 'brights' respectively. These terms have come into commercial use because they accurately

9

TABLE 1.4

Proximate analysis and calorific value of coals delivered to the CEGB (1982/83)

Type/treatment	Delivered Tonne × 1000	Proximate analysis			Calorific value kJ/kg
		Moisture %	Ash %	Volatile matter %	
BC mined Clean coals:					
Sized	88	10.2	6.5	32.9	27 330
Smalls	2 348	10.6	7.0	27.1	27 753
Sub Total	2 436	10.6	6.9	27.3	27 737
General coals:					
Run of mine	28	9.8	27.0	10.3	20 337
Basic smalls	65 111	11.6	16.3	27.3	23 281
Sub Total	65 139	11.6	16.3	27.4	23 280
Dirty coals:					
Smalls	4 618	10.6	22.4	25.7	21 552
Fine coals:					
Dry fines	—	—	—	—	—
Wet fines	103	22.1	16.4	22.0	19 047
Slurry	346	22.0	27.4	17.4	14 267
Sub Total	449	22.0	24.9	18.5	15 364
Total	72 642	11.5	16.4	27.2	23 288
BC open cast	7 563	13.0	10.8	29.9	24 154
Licensed mined	1 797	11.2	15.2	20.6	24 105
Total British coal invoiced 1982/83	82 002	11.7	15.9	27.3	23 386

TABLE 1.5

Variation of coal quality received by the CEGB in 1982/83

Area	Tonnes delivered (1000)	Coal quality					
		Moisture %	Ash %	Volatile matter %	Net calorific value kJ/kg	Sulphur %	Chlorine %
Scotland	114	15.0	13.1	28.9	23 039	1.0	—
North Eastern	10 210	10.3	14.4	28.4	24 656	1.6	0.19
Yorkshire	25 279	10.8	17.5	27.5	23 210	1.7	0.23
North Midlands	23 309	12.4	16.5	27.9	22 868	1.5	0.33
South Midlands	7 525	14.8	14.2	28.3	21 985	1.7	0.15
Western	11 255	11.8	13.7	29.5	23 939	1.6	0.40
South Wales	3 994	9.9	16.3	11.8	25 238	1.2	0.05
Unassigned	317	14.0	7.6	15.9	25 527	1.2	0.21
Total	82 002	11.6	15.9	27.3	23 386	1.6	0.27

describe coal behaviour. Dull coals, containing considerable durain, are hard and strong and are easier to transport than the bright coals. As durain has a lower volatile content than the two bright components, vitrain and clairain (see Table 1.6) it is less likely to coke on a furnace grate. As a result hard coals have been traditionally used for steam raising. The bright coals, containing clairain and vitrain in larger amounts and therefore having a high volatile content, have traditionally been used as coking coals.

3.1.2 The macroscopic nature of coal

If coal is examined with a microscope details of plant remains and crystals of minerals such as pyrites (iron sulphide) are readily discerned. This allows geologists

TABLE 1.6

Composition of the components of a mid-ranking bituminous coal [2]

Analysis	Vitrain	Clairain	Durain	Fusain
Proximate:				
Moisture, air dry %	1.7	1.4	1.2	0.9
Volatile matter	34.6	37.6	32.2	19.1
Ash	0.6	3.5	4.6	9.6
Ultimate:				
Carbon dry ash free %	84.4	82.2	85.8	88.7
Hydrogen dry ash free %	5.4	5.7	5.3	4.0
Sulphur dry ash free %	1.0	2.3	0.9	1.0
Nitrogen dry ash free %	1.5	1.9	1.4	0.7
Oxygen dry ash free %	7.7	7.9	6.6	5.6
Calorific value kJ/kg	31 600	31 600	32 300	31 700

to characterise coal seams accurately. The various components of a coal at the microscopic level are termed 'mascerals' and are denoted by names terminating in 'inite', for example vitrinite, exinite, resinite and inertinite with the latter further subdivided to micrinite and fusinite (Francis [3]). Simple vitrain is composed of vitrinite when examined under the microscope, fusain is made up of inertinite but clairain and durain are complex being composed of vitrinite and exinite, and inertinite and exinite respectively.

From the point of view of a power station engineer, the information on the microscopic structure of coal is not so directly helpful as the observation of the lithotypes. Nevertheless specialist staff in both the CEGB and BC, by studying masceral types, are able to improve methods of coal cleaning and preparation and to understand combustion processes better. Power station operation directly benefits as a result.

3.2 Classification by composition — Seyler's method

It was realised at an early date that coals varied in the amount of carbon (C), hydrogen (H) and oxygen (O) which they contained, i.e., their ultimate analysis. The first satisfactory classification of coals based on C, H, O was proposed in 1837 by Regnault [4] and this was further developed in 1874 by Gruner [5]. Similar work, based mainly on coals from South Wales, led eventually to the classification of Seyler [6] and this has become the most usual way of representing coal classification in the UK from the point of view of elemental composition.

Seyler's approach was first to classify 'pure' coals, i.e., on a dry mineral-matter-free basis, by their carbon content and then to subdivide the classes according to the coal's hydrogen content. The representation assumes C + H + O = 100 so the data is on a sulphur-free basis. A summary of his classification is given in Table 1.7.

The ASTM classification adopted for North American coals in 1936 derived from the work of Stansfield,

Thorn and Parr. A good review of both Seyler's work and the work in America is given by Mott [7].

It is perhaps worth stressing that the data in Table 1.7 is on the basis of sulphur-free values using the Parr formula, or a modified updated form of this involving corrections for carbonate and chlorine (Mott and Spooner [8]).

Parr's early formula is:

Total inorganic matter =

$$\text{Moisture} + 1.08 \text{ Ash} + 0.55 \text{ Sulphur}$$

Thus to convert to a dry mineral-matter-free basis before using Seyler's chart,

$$\text{Calorific value} = \frac{\text{CV of dry coal} - 5000 \text{ Sulphur}}{1 - (1.08 \text{ Ash} - 0.55 \text{ Sulphur})}$$

$$\text{Volatile matter} = \frac{\text{Volatile matter of dry coal}}{1 - (1.08 \text{ Ash} - 0.55 \text{ Sulphur})}$$

$$\text{Carbon} = \frac{\text{Carbon content of dry coal}}{1 - (1.08 \text{ Ash} - 0.55 \text{ Sulphur})}$$

$$\text{Hydrogen} = \frac{\text{Hydrogen content of dry coal}}{1 - (1.08 \text{ Ash} - 0.55 \text{ Sulphur})}$$

Seyler's classification is usually presented in graphical form Fig 1.4. By plotting coal carbon contents against their hydrogen contents most coals are found to lie within a fairly narrow band. His big contribution was in recognising that a second set of axes, relating lines of equal calorific value 'iso-cals' to lines of equal volatile matter 'iso-vols' could be drawn, except for the anthracites at the limit of the band. Now if the carbon and hydrogen content of a coal are known, the volatile matter and calorific value content may be read off the second set of axes on the chart, or vice versa. A third set of axes gives the oxygen content and the curved lines give the British Standard swelling numbers.

It is found that all normal high-rank coals lie within a band of carbon content from 75% to 95% and hydrogen content of 3.0–5.7%, with a maximum width equivalent to 0.7% hydrogen. Any coals lying above this band Seyler referred to as perhydrous and any lying below the band as subhydrous, but most coals encountered lie within the chart and it provides a valuable means of estimating data for coals.

3.3 Classification by volatile matter and coking behaviour — the British Coal ranking

During the Second World War, the Fuel Efficiency Committee of the Ministry of Fuel and Power investigated methods for the classification of commercial coals, and suggested a simple scheme utilising only two pro-

TABLE 1.7

Seyler's classification

Coal rank	Carbon range (%)	Hydrogen (%)	Volatile matter (%)	Calorific value kJ/kg	BS Swelling Index
Anthracite	> 93.3	3.0–3.8	5–10	32 900	1
Carbonaceous					
Semi-anthracite		3.8–4.4	10–14	37 400	1
Semi-bituminous	93.3–91.2	4.4–5.0	14–20	33 700	3½
Bituminous					
Meta —	91.2–89.0	4.4–5.4	20–28	33 500	9
Ortho —	89.0–87.0	4.7–5.6	28–31	33 100	9
Para —	87.0–84.0	4.9–5.7	31–36	32 000	6
Lignitious					
Meta —	84.0–80.0	5.0–5.7	36–42	30 500	2
Ortho —	80.0–75.0	5.0–5.7	42–49	28 400	1
Lignite	< 75.0	5.0–5.7	49–59	25 000	1

perties of coal: volatile matter expressed on the dry-mineral-matter-free basis (DMMF) and the coking power (see Section 14 of this chapter) of the clean coal, i.e., using proximate analysis and a practical test rather than the ultimate analysis used by Seyler. A clean coal for classification purposes is defined as a coal with 10% or less ash content, and preparation of a suitable test sample will be necessary if the seam contains sufficient free dirt to exceed this 10% limit. The then National Coal Board, which became British Coal in 1986, adopted this system for survey and marketing purposes, and for convenience substituted code numbers to describe the various classes of coal, instead of the former names designated by older methods of classification. One of the many advantages of the BC Classification is that it recognises the wide variation of coal types possible in coals of similar volatile matter content. The classification is given graphically in Fig 1.5.

The coal rank Code System was first published in Survey Paper No 58 by the Fuel Research Division of the Department of Scientific and Industrial Research in 1946; since then it has been modified in 1952, 1956 and again in 1964. Because the system was first used in an investigation requiring mechanical analysis of the data, numbers instead of names were used to designate the different classes; apart from this particular application the numerical system provides a useful form of shorthand description, especially suitable for use in tables of analytical data.

The classification is based upon the volatile contents (expressed on a dry mineral-matter-free basis) and the caking properties of the *clean* coals. Using the criterion of volatile matter alone, a first division into the following groups is obtained:

	Volatile matter	Code number
Anthracites	up to and including 9.0 per cent	100
Low-volatile steam coals	9.1–19.5 per cent	200
Medium-volatile coals	19.6–32.0 per cent	301*
High-volatile coals	over 32.0 per cent	(see below)

In the first three groups, i.e., in coals of volatile matter up to 32 per cent, there is a close relationship between volatile matter content and caking properties. Consequently, the effect of subdividing into progressive ranges of volatile matter content is also to produce classes with progressive ranges of caking power.

In the fourth group, i.e., in coals with more than 32% of volatile matter, there is a wide range of caking properties at any given volatile matter content and subdivision has been made on the basis of caking power, as indicated by the type of coke produced in the Gray-King assay. Six ranges of caking properties are recognised:

	Gray-King coke type	Code number
Very strongly caking	G_9, G_{10} and higher	400
Strongly caking	G_5, G_6, G_7 and G_8	500
Medium caking	G_1, G_2, G_3 and G_4	600
Weakly caking	E, F and G	700
Very weakly caking	C and D	800
Non-caking	A and B	900

Each of the classes 400–900 can be further subdivided according to volatile matter content: a 1 in the third figure of the code number indicates that the volatile matter content of a coal lies between 32.1 and 36.0% and a 2 that it is over 36%.

When Gray-King coke types are plotted against dry mineral-matter-free volatile contents for British deep mined coals, the main bulk lie within the band illustrated above (see Fig 1.6).

The average volatile matter content of all coal delivered to the CEGB recently on a dry mineral-matter

---

* Certain coals have been affected by the heat from nearby igneous intrusions, with the result that their caking properties are generally subnormal compared with those of the coals of similar volatile content. These affected coals are distinguished by the code numbers 302 and 303, and by adding H after the code number when they fall in the 100 and 200 classes. They occur mainly in Scotland but some are also found in Durham.

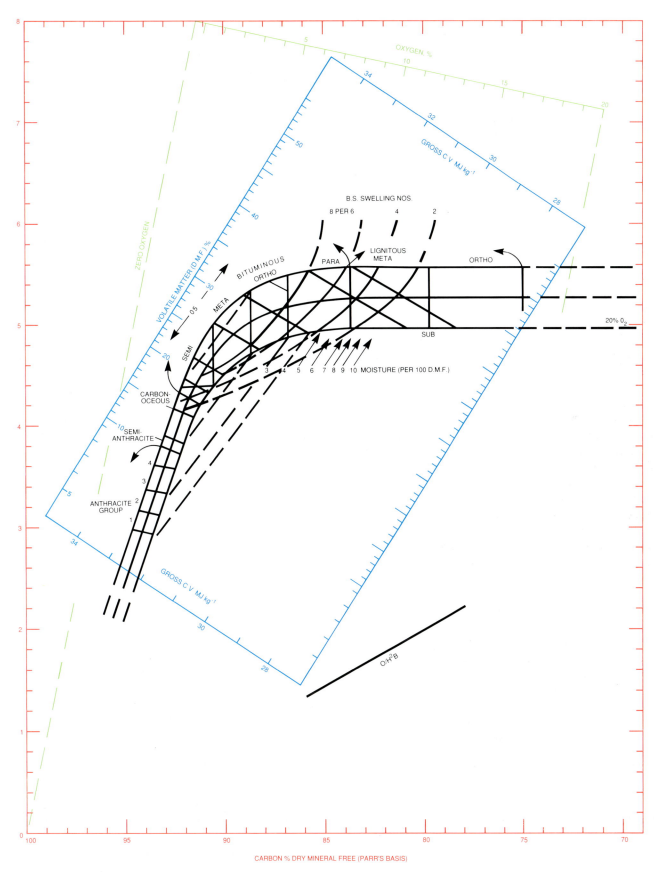

Fig. 1.4 Seyler's classification and fuel chart

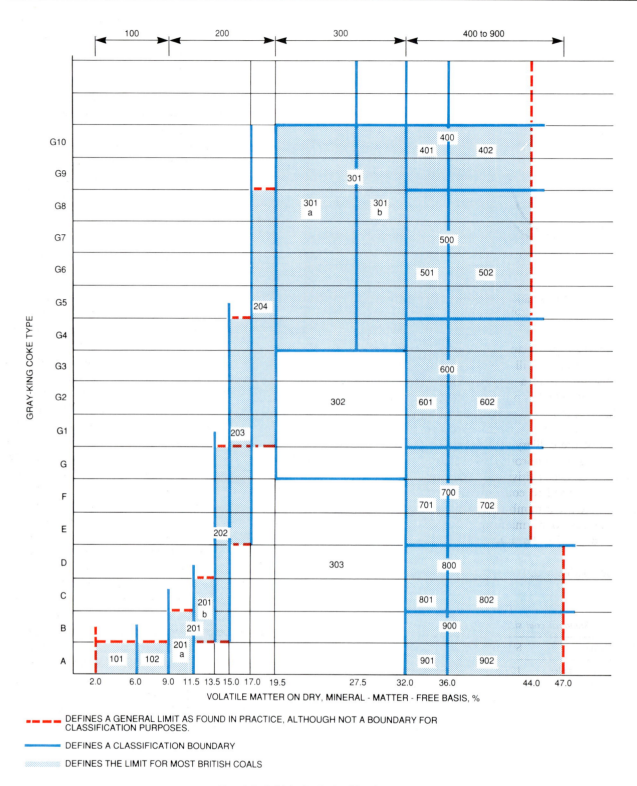

Fig. 1.5 British Coal classification system

free basis is about 40%, and BC ranks 700, 800 and 900 account for about half the total tonnage. These weakly caking or non-caking coals are suitable for power station use and whereas modern power stations with pulverised fuel firing can cope with most ranks of coal, the other types have more beneficial uses. The types 300 and 400, for example, are prime coking coals and rank 100 (anthracite) is mined for use in heating plant and appliances and to make smokeless fuel, although excess anthracite material, fines, can be burned in suitably designed power station boilers.

3.4 Classification by size

Users of coal have an interest in fuels of standard size specification, so that they can be sure that purchases

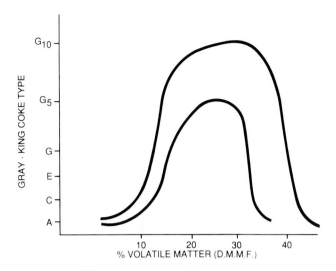

FIG. 1.6 Relationship of caking power to volatile matter

will suit their plant and equipment. Each individual colliery, however, will have an interest in preparing products for sale in a way which ensures the disposal of all the coal which is produced. 'Unnecessary' crushing to achieve a standard size specification may generate excessive fines and create problems at the mine. The interests of the colliery and consumer may conflict, to an extent, and compromises have to be reached. As a result there are British Standard specifications for coal screens used in coal preparation (BS6620: 1985), but as there are a number of standard sizes this allows some choices to be made in the specification of the coal produced by the mine.

In the United Kingdom there are standardised coal sizes, see Table 1.8.

TABLE 1.8
Standard coal sizes in the United Kingdom

Name	Size (mm)	Use
Large cobbles	150 × 75	Household
Cobbles	100 × 50	Household
Trebles	75 × 50	Vertical gas retorts
Doubles	50 × 25	Vertical gas retorts
Singles	25 × 10	Mechanical stokers
Peas	10 × 5	Mechanical stokers
Smalls	25 to 0	Industrial furnaces
	or	
	10 to 0	

The majority of British coal-fired power stations now use pulverised fuel and small coal, or even fines and slurries can be satisfactorily handled. The need to purchase singles or peas has declined. They are relatively expensive, but occasionally they may be needed to help the coal to flow better and they are kept as a strategic stock for cold or wet weather. The distribution of purchases by size specification is illustrated in Fig 1.7 for a typical year (1982/3).

The majority of the tonnage is of a topsize between 25 and 50 mm and most of this is untreated or only partly treated. It contains fines, i.e., it has no lower size limit, and is described as basic smalls. Clean coals with an ash content less than 12% comprise only a small percentage of the total. The coal of 12.5 to 25 mm is almost all untreated and of the coal under 9.3 mm about half is wet fines or slurry. The amount of this wet coal received is less than 1% of the total. It is a cheap source of heat but presents handling problems in cold or wet weather.

4 Fuel behaviour of the different ranks of coal

All large modern coal-fired power stations grind their fuel before combustion to a fineness approaching that of face powder. This pulverised fuel (PF) is blown to the burners, where the flame characteristics are similar to a gas flame. In older power stations the coal, as delivered, was spread onto a grate constructed of an endless belt of linked chains. Combustion commenced at one end of the grate and its movement was adjusted so that 'burnt-out' ash was rejected at the opposite end. The two methods of coal combustion require rather different fuel characteristics. As examples, milling behaviour is extremely important for pulverised fuel firing whereas caking behaviour is an important parameter for chain grates. In the latter case if the coal swells to an extent during combustion it helps to stop fines blowing out of the bed before they are fully burnt.

Table 1.9 gives the average properties of British coals for the main BC ranks from 100 to 900. In order to illustrate the trends in properties as the rank changes, a constant ash content of 8.0% has been chosen. This is about right for the 100 grades, but low for the 700 to 900 grades commonly burnt. The average value of ash in the latter is presently about 16.0%, but the main purpose of the table is to illustrate trends on a comparative basis. Details of the methods of calculation are given in Section 17 of this chapter.

The advantages and disadvantages of each BC rank as a power station fuel are briefly described below. Comments are made for each method of combustion, i.e., chain grate or pulverised fuel. No account has been taken of the effect of the quantity or nature of ash, e.g., fusion temperature or hardness of mineral matter. The size range and distribution, and free moisture content have also been ignored. Although all these factors are relevant to combustion, it is the behaviour resulting from the nature of the clean coal substance which is sought to emphasise here.

BC Rank No	*Remarks*
100	On chain grates this group is very difficult to ignite, and slow in burning.

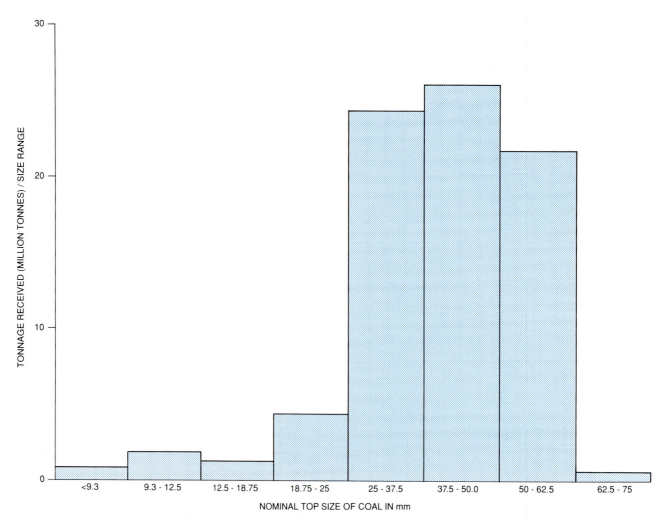

FIG. 1.7 Distribution of coal received from NCB by top size 1982/83

BC Rank No	Remarks
100 (cont'd)	Burns without flame and produces a heavy 'carry-over' of high carbon grit. Grates run hot as almost all the heat release is in the grate.

It can be fired as PF in special combustion chambers which allow a high residence time for the fuel to 'burn-out', but a high mill power is required in order to secure a sufficiently fine product. Carbon in ash is high and is likely to be approaching 20%, even with special plant.

201   On chain grates the main difficulty with this group is due to 'carry-over' of high carbon grit. The burn-off is good but rather slow, and a low ash content gives rise to overheated grates. Small graded sizes are expensive, but can be used very efficiently.

202/203/204   On chain grates, groups 202 and 203 are extremely satisfactory provided that fines in smalls are not excessive. Caking power is just sufficient to bind small particles satisfactorily, without forming large coke masses which prevent a good burn-off. The rapidly increasing caking property causes some 204 coals to give a high carbon in ash loss due to swollen coke formation.

As PF, these coals are more satisfactory, being softer but still requiring fine milling. Carbon in dust is usually 5–12%.

301   On chain grates these coals are moderately satisfactory. The carbon loss due to swollen cokes is balanced by reduced grit loss and the higher proportion of heat released in the form of volatile constituents.

These soft coals are quite satisfactory as PF and the carbon in the pulverised fuel ash (PFA) continues to decrease as the rank decreases. Carbon in PFA is usually 3–10%.

TABLE 1.9

Average properties of a representative range of British coals

Certain basic characteristics of British coals are given below, together with notional examples of washed smalls representing the various types of coal. The moisture and ash content of commercial grades vary with the rank of coal, seam characteristics, mining conditions, and the degree of preparation.

The moisture content of large and graded coals is usually a little higher than that shown for air-dried coal; for untreated smalls the moisture content will be about 2% higher (except where this is affected by water infusion during mining).

The moisture content of washed smalls is commonly 6 to 8% above the moisture content for air-dried coal.

Values of ash content commonly found commercially are:

Large and graded coal	3 to 6%
Washed smalls	5 to 10%
Untreated smalls	12 to 20%

The analysis of a given commercial product of known coal rank code number can be estimated very approximately by conversion of the appropriate dry-mineral-free analysis to the basis of the actual moisture and ash contents of the sample, as follows:-
If 'X' is the value of the parameter on the dry-mineral-free basis, then the value of that parameter on a basis of M% moisture and A% ash is:-

$$\text{Approximately 'X'} \times \frac{100 - (M + 1.15A)}{100}$$

	Anthracite		Dry Steam	Coking Steam		Medium Volatile Coking Coals		High Volatile Coking Coals — Very Strongly Caking	High Volatile Coking Coals — Strongly Caking		High Volatile Coking Coals — Medium Caking		General Purpose Coals — Weakly Caking		General Purpose Coals — Very Weakly Caking	General Purpose Coals — Non-Caking	Heat Altered Coals — Low Volatile	Heat Altered Coals — Medium Volatile	
British Coal - Rank Code Number	101	102	201	202	204	301a	301b	401	501	502	601	602	701	702	802	902	101H to 204H	302	303
Moisture content of air-dried coal	2.0	1.0	1.0	1.0	1.0	1.0	1.0	2.0	2.0	3.0	4.0	4.0	5.0	5.0	8.0	10.0	2.0	1.0	2.0
Moisture content at 96% RH and 30°C	4.0	2.0	1.0	1.0	1.0	1.0	1.0	2.0	3.0	4.0	6.0	5.0	6.0	7.0	11.0	13.0	3.0	2.0	3.0
Analysis of Dry-Mineral-Free Coal																			
Volatile Matter	5.0	7.5	11.5	14.0	18.0	23.0	30.0	35.0	35.0	37.5	35.0	37.5	35.0	38.5	39.0	40.0	15.0	25.0	28.0
Calorific Value, kJ/kg	35 820	36 285	36 520	36 630	36 750	36 750	36 400	36 050	35 700	35 470	34 890	34 890	34 425	34 300	33 730	32 800	36 285	36 285	35 435
Carbon	94.4	93.0	92.4	92.0	91.6	90.4	89.3	87.5	86.8	85.2	85.0	85.5	84.5	83.5	82.0	81.0	90.6	88.8	87.8
Hydrogen	2.9	3.7	4.0	4.2	4.5	4.9	5.0	5.3	5.3	5.4	5.3	5.5	5.2	5.4	5.3	5.1	4.3	5.0	5.1
Nitrogen	1.1	0.7	1.4	1.4	1.4	1.5	1.5	1.9	1.7	1.8	1.8	1.8	1.8	1.7	1.7	1.6	2.0	2.0	1.7
Sulphur	0.7	0.7	0.9	0.7	0.7	0.6	0.8	1.0	1.0	0.9	0.8	0.9	0.8	1.0	1.0	0.7	0.9	0.8	1.0
Oxygen	0.9	1.3	1.3	1.7	1.8	2.6	3.4	4.3	5.2	6.7	7.1	7.3	7.7	8.4	10.0	11.6	2.1	3.4	4.4
Caking Properties																			
BS Swelling Index Number	0	0	1	3	9	8	8	9	8	7.5	6	6	3.5	3.5	1.5	1	0.5	8.5	6
Gray-King Coke Type (600°C)	A	A	B	E	G3	G6	G7	G10	G7	G6	G2	G3	F	F	C-D	B	A	G2	D/E
Maximum Dilation (Audibert-Arnu)	-	-	-	-	25	140	256	280	133	88	20	21	-12	-19	-20	-	-	-	-28
Principal Coalfields of Origin	South Wales, Scotland		South Wales	South Wales, Kent		South Wales, Durham, Kent		Durham, Yorks, Northumberland	Yorkshire, Durham, Northumberland, Lancs, Notts and North Derbyshire		Yorkshire, Notts and North Derbyshire, Lancs, Northumberland, Scotland		Yorkshire, Notts and North Derbyshire, Lancs North, Staffordshire, Scotland		Notts and North Derbyshire, Yorkshire, Cannock Chase, Warwicks, Lancs, Scotland	Notts and North Derbyshire, Leics, Scotland, Warwicks, South Derbyshire	Durham, Scotland	Durham, Scotland	
Analysis of Notional Washed Smalls (as fired)																			
Moisture	8.0	7.0	7.0	7.0	7.0	7.0	7.0	9.0	9.0	10.0	11.0	11.0	13.0	13.0	16.0	18.0	8.0	7.0	8.0
Ash	8.0	8.0	8.0	8.0	8.0	8.0	8.0	8.0	8.0	8.0	8.0	8.0	8.0	8.0	8.0	8.0	8.0	8.0	8.0
Carbon	78.2	77.9	77.4	77.1	76.8	75.8	74.8	71.6	71.0	68.8	67.8	67.4	65.7	65.0	61.3	59.0	75.0	74.4	72.7
Hydrogen	2.4	3.1	3.4	3.5	3.8	4.1	4.2	4.3	4.3	4.4	4.2	4.4	4.0	4.2	4.0	3.7	3.6	4.2	4.2
Nitrogen	0.9	1.1	1.2	1.2	1.2	1.3	1.3	1.6	1.4	1.5	1.4	1.4	1.4	1.3	1.3	1.2	1.7	1.7	1.4
Sulphur	1.0	1.0	1.0	1.0	1.0	1.2	1.2	1.7	1.7	1.7	1.7	1.7	1.7	1.7	1.7	1.7	1.2	1.2	1.2
Oxygen	1.5	1.9	2.0	2.2	2.2	2.6	3.5	3.8	4.6	5.6	5.9	6.1	6.2	6.8	7.7	8.4	2.5	3.5	4.5
Calorific Value, kJ/kg Gross	29 655	30 355	30 585	30 705	30 820	30 820	30 470	29 540	29 190	28 610	27 795	27 795	26 750	26 750	25 235	23 840	30 005	30 355	29 775
Net	28 935	29 495	29 655	29 750	29 820	29 750	29 380	28 375	28 030	27 400	26 610	26 565	25 540	25 495	23 960	22 585	29 005	29 260	28 655
Theoretical Air Requirements per kg of fuel: kg of Air	9.84	10.05	10.10	10.09	10.16	10.15	10.03	9.74	9.61	9.37	9.16	9.17	8.84	8.79	8.26	7.85	9.93	10.03	9.76
m³ of Air at 0°C and 760mm	7.61	7.77	7.81	7.80	7.86	7.85	7.76	7.53	7.44	7.24	7.08	7.08	6.84	6.80	6.39	6.07	7.68	7.75	7.55
Waste gases per kg of fuel: m³ of Wet Waste Gas at 0°C and 760mm	7.84	8.03	8.09	8.08	8.16	8.17	8.09	7.89	7.80	7.63	7.47	7.49	7.23	7.21	6.83	6.54	7.98	8.08	7.89
CO₂ content of dry waste gases: Per Cent	19.5	19.1	18.9	18.9	18.7	18.5	18.5	18.3	18.3	18.3	18.4	18.3	18.4	18.4	18.4	18.6	18.7	18.4	18.5
Composition of Moist Waste Gases: CO_2%	18.6	18.1	17.9	17.8	17.6	17.3	17.3	16.9	17.0	16.8	16.9	16.8	17.0	16.8	16.8	16.8	17.5	17.2	17.2
H_2O%	4.5	5.2	5.6	5.7	6.1	6.5	6.7	7.3	7.4	7.8	7.8	8.0	8.0	8.3	8.9	9.7	6.1	6.7	7.0
Oxides of N and S %	0.3	0.3	0.3	0.3	0.3	0.3	0.3	0.5	0.4	0.5	0.5	0.5	0.5	0.5	0.5	0.5	0.4	0.4	0.4
N_2%	76.6	76.4	76.2	76.2	76.0	75.9	75.7	75.3	75.2	74.9	74.8	74.7	74.5	74.4	73.8	73.2	76.0	75.7	75.4
Dewpoint of Waste Gases °C	31.0	34.0	35.0	35.5	37.0	38.0	38.5	40.0	40.0	41.0	41.0	42.0	42.0	42.5	44.0	44.0	37.0	38.5	39.0

17

BC Rank No	Remarks
400	On chain grates these coals are very unsatisfactory. The strong caking characteristic leads to the formation of large coke masses which will not burn out. This feature is particularly marked in the 401 group, and results in poor efficiency and boiler output. For PF, these coals are very satisfactory. The coals are soft and result in a lower carbon in dust.
500/600/700	The two latter groups are excellent on chain grate stokers. The caking property is sufficient to hold fines in the fuel bed. Less primary air is required and the coal can be fired at very high loadings per unit area of grate. The strong coking nature of the 500 rank could lead to a high carbon in ash at high grate loadings. These coals are all extremely satisfactory for PF and yield a low carbon-content PFA.
800/900	These low rank coals on chain grate stokers can give rise to grit 'carry-over', but secondary air can be used to retard this tendency. Grits carried over are very low in carbon, and these free-burning coals give a very low carbon in ash. Fired as PF, these coals are very good, the only drawback being the low inherent calorific value. Carbon in PFA is low.

The above comments concern combustion. In addition one needs to consider ash properties. The 'clinker' from chain grate combustion and the furnace bottom ash from pulverised fuel combustion are readily sold as 'hard core'. The finer pulverised fuel ash, which is carried through the boiler gas passes to the station's precipitators, also has many uses if it is of the right quality, as will be discussed later (Section 16 of this chapter). However, many of the uses require ash of a consistent colour, so that a low carbon in ash is desirable. For this reason the 'younger' coals are preferred to the anthracites (100) and steam coal (201) which give high carbon residues.

5 Preparation of coal

5.1 The purpose of coal cleaning for power station use

In the mid 1960s some 25% of the coal burnt at power stations was burnt on stoker-fired plant. This required purchase of significant amounts of washed coal, some 14% of the total tonnage then purchased. As noted previously in Section 4, such plant required better fuel than modern pulverised-fuel burning stations do and untreated coal was not always of adequate quality. In 1982–83 just under 10% of purchases were still clean coal (with an ash content below 12%). The majority of power stations were pulverised fuel burning, yet the reason for purchasing clean coal was still the same; that untreated coal was not always of adequate quality and some clean coal was necessary to improve it.

Raw coal as mined underground is a mixture varying in size from lumps perhaps 1 metre in length down to dust; it includes dirt or shale unavoidably removed from the roof and floor of the seam, or running as a dirt band in the coal seam itself. There is also a smaller amount of mineral matter distributed generally throughout the coal. This latter material largely arises from the mineral matter contained in the original vegetable detritus from which the coal is formed. The two mineral types differ in their behaviour during combustion. The shale minerals are alumino-silicates and have high melting points; the inherent mineral matter of the coal can be rich in alkali metals (wood ash is potassium carbonate) and their compounds have relatively low melting points. As a result, very clean coals may cause slagging in modern pulverised-fuel burning boilers, since molten ash particles adhere to the boiler tube surfaces. The average ash content of present fuel deliveries, 16.0%, is usually sufficient to ensure an adequately high ash fusion temperature.

The minerals in the shale differ from seam to seam. In some cases the roof or floor material may be sticky and if too much is present in the fines content of the coal it will be difficult to handle, and very likely to stick in the track hopper or bunkers of the power station. Coal preparation therefore aims to produce a power station coal which is free flowing, has a high ash fusion temperature, is easy to mill and is of good heat content.

5.2 Principles of coal cleaning

5.2.1 Screening

At collieries which have to provide a wide range of products, the run of mine (raw) coal is passed over a screen to separate large coal and then over further screens to separate the sizes required to make graded products, domestic coal for example. These potentially valuable sizes are then prepared for their markets. The remaining coal, which is the raw smalls, can either be sold direct for use in power stations or can be subjected to a cleaning process to improve it for industrial steam raising or carbonisation purposes, if it is of the correct rank and sulphur content. At some collieries or open cast sites the total preparation requirements may be as simple as provision of a screen to separate large from small coal (two products only being required), with the latter perhaps being mixed with a returned portion of the larger coal and crushed to a nominated top size to provide a suitable mixture for power station use. This approach is economical and suitable for seams which are relatively free from dirt. The large pieces of coal and the fines are of similar quality and it is only necessary to adjust the size distribution of the raw smalls to ensure a free flowing mixture.

Such a simple approach is not always possible, and at colliery preparation plants screening is carried out not only to remove large coal, but also fines, say coal less than 6 mm. The latter often require separate preparation from the 'middlings' coal. The subsequent plant then has three flows for the large, middlings and fines, and the cleaned outputs are brought together in a final blending unit to make the saleable fuel.

5.2.2 Density separation — wet processes

The next stage of coal preparation involves separating the components of the coal mixture by utilising the differences in their densities (specific gravities). Coal, as mined, has a specific gravity ranging from 1.2 to 1.7 depending upon its rank and inherent moisture content, and the nature and amount of the mineral matter which it contains. A 'pure' clean bituminous coal would typically have a density of 1.3. The shale associated with the coal might have a density of about 2.2 and the iron pyrites present would have a density of 4.9 (see Table 1.10).

TABLE 1.10

Specific gravities of the components which make up a coal seam

Component	Specific gravity
Coal macerals	1.15–1.5
Alumino silicates — shales, clay and sandstones	2.0 –2.6
Gypsum	2.3
Calcite	2.7
Pyrites	4.8 –5.0

Before coal cleaning can be planned, the preparation engineer first needs to know the density distribution of the material. Samples of screened coal are taken and put into liquids of known density, ranging from about 1.2 to 2.0, and the percentage of material which floats or sinks in each liquid mixture is noted. To take an example, the coal to be prepared might be of 37.5 mm top size. Three samples would be produced by further screening, 37.5 mm to 18 mm, 18 mm to 6 mm and 6 mm to 0.5 mm and each of these would be tested separately. The very finest material, below 0.5 mm, is difficult to deal with. It does not separate readily, even in the laboratory, and is normally considered to be a single material and is analysed as such.

The sized fractions are tested first in the liquid of lowest specific gravity, say 1.25. The material floating is collected, weighed, sized and analysed. The material which sinks is recovered and tested in the liquid of next density, say 1.3, and the process is repeated, with the 'floats' at each stage being characterised and the 'sinks' recovered for testing in the next highest density liquid. A graph can then be constructed to illustrate the distribution of material density within the dirty coal. A typical set of results is given in Table 1.11 and the derived graph illustrating the 'washability' of the coal is shown in Fig 1.8.

Curve (A), the increment curve, gives the ash content for the increment which would just float or sink at a specified density. The concept, though simple, is easier to understand from an example. Taking the density fraction 1.4 to 1.5 in Table 1.11, the ash content of the fraction which sank at density 1.4 but floated at 1.5 is 18.79 (column 4). As one-half of the material will have an ash content lower than this figure and one-half higher, it is assumed that the average ash of the fraction relates to the mid-range density, 1.45. The cumulative 'floats' to density 1.45 is thus 58.50% (from column 5) plus one-half of the fractional yield of 13.50% (column 3) making a total of 65.25%. The other points on curve A are calculated similarly.

The clean coal yield curve (B), is derived from weights from column 5 plotted against the cumulative 'float' ash content (column 6).

Curve (C) is a corresponding curve showing the cumulative weight of rejects (column 7) plotted against the cumulative 'sink' ash content.

Curve (D) shows the relationship between the yield of clean coal (column 5) and the specific gravity at the cut point (high figure in each case in column 2).

The engineer can use the graph to evaluate the options for coal cleaning. For example, if he wants to obtain an 80% yield of clean coal (curve B) then its ash content will be 9.0% and from curve D he will be able to achieve this with a liquid medium of 1.62 specific gravity. The rejects, curve C, will have an ash content of 60% and the increment of neutral buoyancy at the cut point, i.e., the dirtiest particle included in the clean coal, will have an ash content of 28% (curve A). The particular coal used in this example has a sulphur content approaching 5% in the raw coal and this can make it difficult to obtain good yields of clean coal if the pyrites is well distributed as fine particles throughout the coal seam rather than as distinct nodules which are readily separated.

A coal that is easy to clean is recognised by an L-shaped increment curve (A). This means that the ash content in the incremental particle increases rapidly for a small increase in yield in the middle portion of the curve. If the curve is more nearly straight, then the coal is hard to clean as the dirt is well distributed and a high proportion of 'middlings' is obtained (coal of medium to high ash content and with specific gravity midway between coal and shale).

TABLE 1.11

Typical results of a float/sink test of a bituminous coal [9]

Size of coal	Specific gravity fraction	Weight‡ %	Ash %	Cumulative float		Cumulative sink	
				Wt %	Ash %	Wt %	Ash %
Composite of	1.25	2.50	2.13	2.50	2.13	100.00	19.15
results i.e.,	1.25–1.30	29.20	2.98	31.70	2.91	97.5	19.58
37.5-0 mm	1.30–1.35	18.60	5.57	50.30	3.90	68.30	26.67
	1.35–1.40	8.20	9.58	58.50	4.69	49.70	34.57
	1.40–1.50	13.50	18.79	72.00	7.34	41.50	39.51
	1.50–1.60	6.50	21.49	78.50	8.51	28.00	49.50
	1.60–1.70	3.20	27.56	81.70	9.25	21.50	57.97
	1.70–1.80	2.90	32.46	84.60	10.05	18.30	63.29
	1.80–2.00	3.00	41.94	87.60	11.14	15.40	69.09
	2.00	12.40	75.66	100.00	19.15	12.40	75.66

‡Percent by weight which sank at the lower figure but floated at the higher specific gravity.

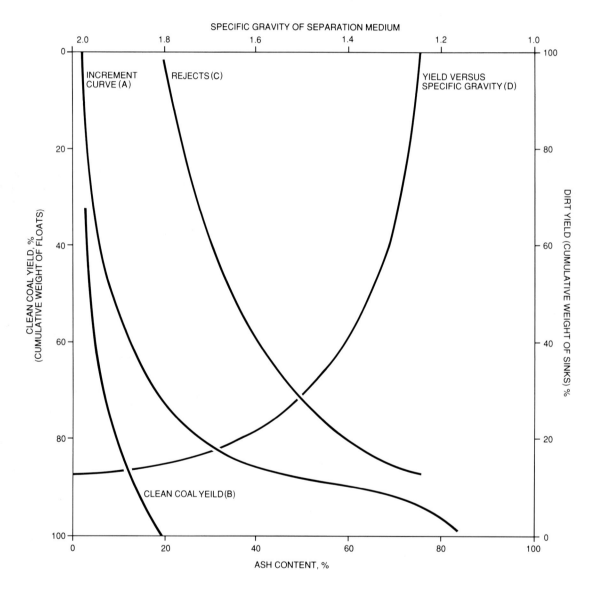

FIG. 1.8 Washability curves for coal data in Table 1.11

In any coal the limitation to improved cleaning is the variation of ash content between different sized fractions of coal. Only dirt that appears as separated particles, substantially free from coal, can be readily segregated. This is why a number of different particle size distributions are prepared by screening. Testing of these will establish an optimum strategy which may involve cleaning of only part of the coal by density separation in a wet process or it may be decided that the coal needs further crushing before separation in order to free the dirt from the coal.

5.2.3 Theory of settlement

The simplest means of separating the components of a coal in practice relies upon allowing particles to settle in a fluid. Small dense particles settle at a different rate from large light particles and a separation may be achieved.

A detailed approach to the theory of settlement requires assumptions to be made regarding the size, shape and density of particles. Since the assumptions cannot be confirmed there is little to gain from a complex approach. A semi-quantitative description is therefore given here, which is sufficient to illustrate the principles involved.

Three factors determine how a particle will sink in a liquid; gravity, countered by buoyancy gives a net downward force to accelerate the particle. Its velocity increases but is limited by the resistance to its motion arising from the viscous properties of the liquid. The faster the movement, the greater is the resistance and the particle soon reaches a terminal velocity. At this point the net downward force is balanced by the resistance of the liquid and no further acceleration is possible. The particle then falls at a steady rate.

The *force of gravity* (F) on a particle of mass 'm', density 'd' and volume 'v' is given below. The acceleration due to gravity is denoted by 'g' (9.81 ms^{-2})

$$F = mg = dvg \qquad (1.1)$$

Buoyancy is an upward force on the particle equal to the mass 'M' of liquid (density 'D') displaced by particle volume, i.e:

$$\text{Buoyancy} = Mg = Dvg \qquad (1.2)$$

The net downward force on the particle is thus

$$F_{net} = vg \, (d - D) \qquad (1.3)$$

The *resistance to motion* 'R' in the liquid is calculated from either Stokes's or Newton's formulae. Details can be found in a physics text book but suffice to say the formulae only readily apply to perfect liquids and spherical particles. As we have neither, we shall simply conclude that at the terminal velocity of v_t

$$R = F_{net} = vg \, (d - D) \qquad (1.4)$$

and as the resistance is proportional to the particle velocity

$$v_t \propto vg \, (d - D) \qquad (1.5)$$

The resistance is also proportional to particle dimensions. For spherical particles of radius 'r' Stokes's formula relates v_t to r^2 and Newton's has $v_t \propto r$.

The formulae apply within a limited range of conditions and give only an approximate indication of coal washing practices, but they do usefully illustrate certain points. Two cases may be considered, free settling and hindered settlement where the particles interfere with one another.

5.2.4 Free settling

If two spheres of specific gravity d_1 and d_2 and radii r_1 and r_2 fall separately in the same fluid and attain the same terminal velocity, then the ratio of their radii will be given according to Stokes's Law by:

$$r_1/r_2 = [(d_2 - D)/(d_1 - D)]^{\frac{1}{2}}$$

and from Newton's formula by:

$$r_1/r_2 = (d_2 - D)/(d_1 - D)$$

Generally behaviour between the two is found and

$$r_1/r_2 = [(d_2 - D)/(d_1 - D)]^m$$

where m varies between $\frac{1}{2}$ and 1. This is the free settling ratio.

If, instead of considering two spheres which attain the same terminal velocity, the case of two spheres of equal radius (i.e., $r_1 = r_2$) but different densities are considered, then the ratio of their terminal velocities v_t will be given from Equation (1.5) as

$$v_{t2}/v_{t1} = (d_2 - D)/(d_1 - D)$$

This equation can be used to compare the rate of settlement of the two spheres (see below).

5.2.5 Hindered settlement

In mentioning the free settling ratio above, it was noted that large light particles could fall at the same terminal velocity as small dense ones. This only occurs under free settling conditions. When the number of particles in the fluid is greater, so that they impede one another, the dense ones fall more quickly than their former lighter companions. This can be illustrated by two examples.

If a small number of equal-sized particles of shale, specific gravity 2.2, and coal, specific gravity 1.4, are allowed to settle freely in water, specific gravity 1.0, then

$$v_t(\text{shale})/v_t(\text{coal}) = \frac{2.2 - 1.0}{1.4 - 1.0} = 3.0$$

The shale sinks three times more quickly than the coal.

If the density of particles in the water is increased so that equal volumes of coal and shale together comprise 40% of the volume of the suspension, its effective density will be greater than one. Each unit of volume will contain 0.6×1 mass units of water, 0.2×2.2 of shale and 0.2×1.4 of coal so that the effective density of the suspension is 1.32 (i.e., $0.6 + 0.44 + 0.28$).

$$\text{Now } v_t(\text{shale})/v_t(\text{coal}) = \frac{2.2 - 1.32}{1.4 - 1.32} = 11$$

The separation in the hindered condition takes place over three times more quickly than in the free settling example. This result is exploited in dense medium washing plants where magnetite for example, specific gravity 5.2, is added to the suspension to increase the rate of separation of coal from its mineral contaminants.

5.2.6 Froth flotation

Finally the preparation engineer is left with the problem of cleaning fines, coal of particle size less than 0.5 mm. If, after screening out large coal and carrying out a wet density separation cleaning process on the smaller coal, say 37.5–0.5 mm, the ash content is sufficiently low, then it is usually satisfactory to add back the dry fines without further processing to the cleaned fraction, along with a proportion of crushed larger coal to ensure the mixture is free flowing. On the other hand, if the ash of the resulting mixture of dry fines and washed smalls is unsatisfactory, cleaning of the fines to improve ash levels and calorific value will be worthwhile. This is usually done by froth flotation.

Coal and shale particles have different surface properties, particularly in their behaviour on wetting. The surface of coal contains hydrocarbon molecules and their C–H bonds have a natural affinity for oils. The surfaces of the alumino-silicates in the shale particles on the other hand have many hydroxyl, –OH, groups which confer a strong affinity for water.

A second concept concerns how air interacts with these surfaces. It does not attach well to the wetted shale particles; bubbles of air adjacent to wetted shale have a small contact angle. In contrast, air bubbles attached to oily surfaces have a large contact angle and adhere well. Thus fine coal can be wetted by bubbles in the froth and the combined bubble and particle will float. Shale and dirt particles are not wetted and sink.

Oils such as light spindle or creosote are used as 'collectors' to coat and collect the coal particles. Frothing agents such as pine oil, eucalyptus oil or cresylic acid (from the creosote) are added, to improve the formation of a stable aerated foam. The process only works with fine particles of less than 0.5 mm since the combination of air bubble and coated coal particle has to have a density less than 1.0. Larger particles would not achieve this and would not be lifted to the surface. The froth containing the fine coal particles is skimmed off and collected by filtration. The filter cake containing about 25–30% of water, but perhaps as little as 5% of ash, is then mixed into the washed smalls for sale.

5.3 Screening and crushing methods

To separate coal into size fractions it is passed over bars, perforated plates, or wire mesh screens. Sizes smaller than the openings fall through and the larger coal is retained. 'Jiggers' are most commonly used, in which the screen is in the form of a rectangular tray, shaken with a reciprocating motion in its lengthwise direction. The coal is lightly handled by this device and breakage is minimised. Plate screens are of standard size (BS6620: 1985 or 150 7805/1 1984) and give good performance. Bar screens can be distorted in service and large coal may then enter the product. This causes difficulties in sampling the coal with mechanical bucket devices and should be corrected quickly.

Any screen, but particularly the smaller sizes, can block. A recent development, the BC Bretby rotating probability screen, avoids this problem. It consists of two bar screens (Fig 1.9) one of which rotates. It is the only variable-aperture device available and it gives considerable flexibility to the operations of the preparation plant.

Methods of reducing the size of coal are designed to minimise the amount of dust produced. Several types, the pick-breaker (Fig 1.10), the Bradford breaker (Fig 1.11) and the double roll crusher (Fig 1.12) are in common use. The first has strong pick blades mounted on a steel frame which moves slowly up and down. Coal passes under the frame on a horizontal plate conveyor. The amount of breakage is adjusted by the height to which the picks are raised before they fall onto the coal. Fines production is small with this device. The Bradford breaker is a large cylindrical screen fitted with longitudinal fins inside to cause the coal to be lifted and tumbled. As the pieces fall they break and fall through the screen. Large debris such as pit props pass through to waste. The double roller comprises two toothed or corrugated rollers. They are mounted horizontally and rotate towards one another. This produces a 'nip' at the top side which draws coal in to break it. Jaw crushers (Fig 1.12) are also used to reduce coal sizes. They consist of a vertical corrugated crushing plate and a moving 'jaw' which is attached at an angle. A lever system causes the jaw to open and close the gap between the plates with a reciprocating motion and coal falling through is 'chewed' to a smaller size.

5.4 Wet processes

Density separation processes use either the principle of free settlement in water (jig washers) or hindered settlement where addition of magnetite or other heavy solids to the water produces a stable suspension of fine particles (dense medium washers). In both cases the finest coal is screened out as this is not readily cleaned, and material of 37.5 to 6 mm is typically washed. The dry fines may be returned to the clean fraction after that is cleaned if the resultant ash is acceptable or,

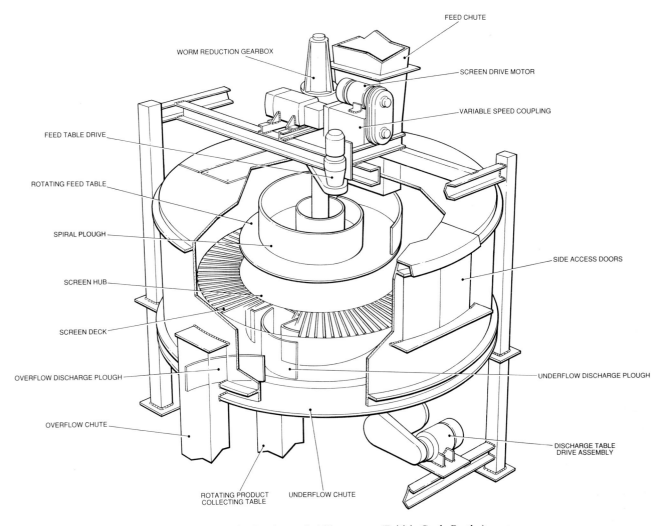

FIG. 1.9 The Bretby probability screen (British Coal, Bretby)

alternatively, may be cleaned by froth flotation. A coal preparation plant is shown in schematic form in Fig 1.13.

5.4.1 Dense medium washing

Dense medium plants are generally used for cleaning coal in the size range 200–25 mm by using suspensions of finely divided magnetite, barytes, sand, or even the shale itself, in water. The dense medium must be carefully maintained at the appropriate specific gravity (SG) levels. Separation into three products is achieved; coal, middlings and shale, usually in two steps. The raw (200–25 mm) coal is first fed to a dense medium of about 1.4 specific gravity where clean coal floats and the middlings and shale sink. The clean coal is removed by rakes or paddles and passed on to classifying screens for final grading. The 'sinks' are continuously removed by an elevator and passed to a second bath of specific gravity say 1.8, or whatever other figure has been chosen from the washability curves (Fig 1.8); the shale now sinks but the middlings float and are recovered. To achieve a further separation, the middlings must be crushed to 25 or 37.5 mm top size in order to further liberate the dirt particles. The smaller coal is then passed, along with the 'below 25 mm' coal from the original raw coal screening, to a jig washer. Well known types of dense medium plant are the Chance process which uses sand, SG 2.6, to form the suspension, Barvoys which uses barytes SG 4.2 along with 2% by volume of clay, and the Tromp process which uses magnetite, SG 5.2.

5.4.2 Jig washing

Jig washing uses water as the separating medium, by moving the fluid rapidly up and down through a bed of dirty coal on a perforated plate. Clean coal particles are lifted to the top of the moving bed and dirt settles to the bottom. Pulsation of the water may be produced by a mechanically driven piston or pneumatically as in the well known Baum washer (Fig 1.14).

The jig washer can handle coals of any close size range up to 150 mm, but most installations are designed for coals below 50 mm top size. Three products may again be produced (see Fig 1.14). Coal is

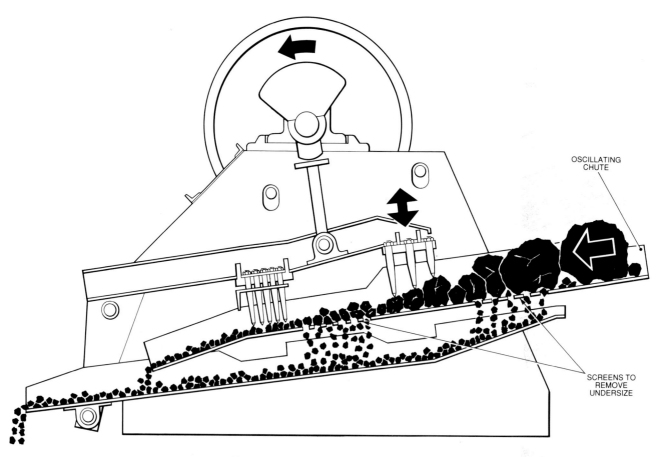

OSCILLATING
CHUTE

SCREENS TO
REMOVE
UNDERSIZE

FIG. 1.10 A pick-breaker (British Coal)

added at one end of the washing box and the heavy shale travels along the bottom of the bath close to the perforated plate, to be removed at the first weir. The middlings are more buoyant and clear this weir but settle later, to be collected at the second weir. The clean coal clears the last gate and is collected as a third fraction. The shale is disposed of, but the middlings are capable of further cleaning and are therefore screened at 6 mm or 12 mm and the larger fraction is crushed and recycled through the jig (Fig 1.15). The clean coal is dewatered either with a dewatering screen or better a centrifuge and then classified for sale.

5.4.3 Flocculation

The waste water from the dense medium or jig washing processes is taken to settling tanks. Fines (0.5 mm to 0) settle to form a thick slurry. This process is aided by the addition of starch or polyelectrolyte flocculating agents. The sediment is collected and dewatered in vacuum filters. It contains some residual coal and a certain amount is purchased by power stations each year (346 000 tonnes in 1982/83).

Generally, wet coal is considered highly undesirable for power station use. This is recognised and allowed for in the low price of slurries, but the moisture must be controlled. Wet coal in normally-priced coals is more expensive to transport, since useless weight is being moved, and it is difficult to unload and handle, particularly in cold weather. When burnt, extra mill drying capacity is required, and after combustion, latent heat is carried out of the stack with the water vapour. The average moisture, 11.6% in the coals purchased by the CEGB in 1982/83, is given in Table 1.5 which shows that the range of moisture within BC Areas is 9.9 to 15%. Those coals which have an inherently low calorific value, in addition to a high moisture content, such as the South Midlands Area of BC, require preparation of the coal to a lower ash content than most other coals in order for them to be acceptable.

5.4.4 Froth flotation

This relatively expensive process can clean coals that other washers cannot cope with, the 0.5–0 mm fines. The process has already been discussed in Section 5.3 of this chapter, and the main point to reiterate here is that the filtercake from a froth flotation plant still contains 25–30% of water. The energy losses which accompany its combustion should be taken into account when considering the economics of the process, in relation to power station coal.

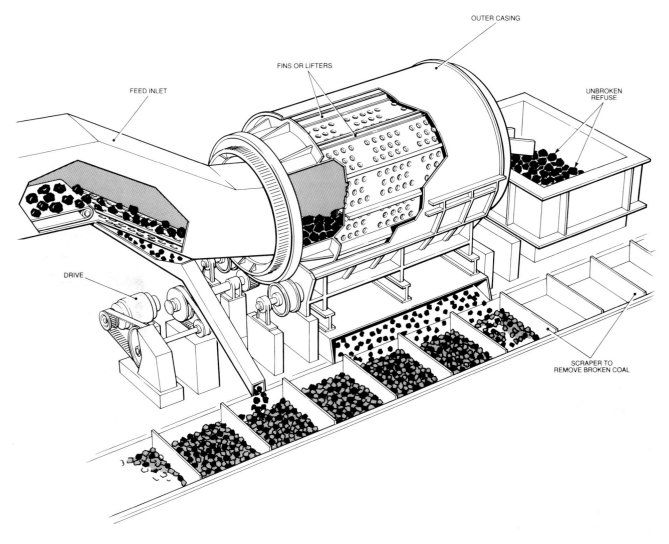

FIG. 1.11 The Bradford breaker (British Coal)

6 The price of coal

6.1 The industrial price structure

British Coal, formerly the National Coal Board, have several pricing formulae for their different qualities of coal. Power companies purchase the majority of their tonnage according to the industrial formula and this is discussed in detail below. Slurries are purchased in accordance with a schedule of slurry prices. The other formulae cover carbonisation coals and domestic coals.

The industrial price formula was first introduced in December 1951 and over the years it has been simplified to more nearly represent the value of the coal. At first, coals of similar quality from different BC Areas were priced differently to promote sales from the 'favoured' Area. This is no longer so and the process is now based on the gross calorific value of the coal on an 'as received' basis, with adjustments for the coal's ash and sulphur content.

For example, in 1985/86 the unadjusted process was evaluated on a net points basis where

$$\text{Net points} = \frac{\text{Gross calorific value (as received basis)}}{100}$$

Each point was charged at 20.32 pence, so a tonne of coal of 20 000 kJ/kg calorific value would cost £40.64 on an unadjusted basis. Allowance was next made for the ash content of the coal, again on an as received basis. As high ash coals lead to problems in power stations, increasing the burden on the precipitators and perhaps fouling boiler passes, and certainly increasing ash disposal costs, the ash adjustment is non-linear (Fig 1.16). There are breaks in the adjustment so that between 0 and 5% an ash allowance of £0.0625/percent of ash is applied to each tonne of coal, between 5 and 10% £0.1375/percent is deducted and above 10% £0.2112/percent.

A third adjustment is made for the grade of the coal:

	Washed	Blended	Untreated
	£	£	£
Graded coals	4.50	3.20	3.00
Smalls 25 mm & over	1.95	0.65	0.45
Smalls less than 25 mm	1.50	0.20	NIL

To complete the formula an allowance is made for the sulphur content of the coal:

Sulphur range %	£ per tonne
0.80 and below	+ 0.34
0.81 – 1.00	+ 0.17
1.01 – 1.80	DATUM
1.81 – 2.00	– 0.17
2.01 – 2.20	– 0.34
2.21 – 2.40	– 0.51
2.41 – 2.60	– 0.68
2.61 – 2.80	– 0.85
2.81 – 3.00	– 1.02
3.01 – 3.20	– 1.19
3.21 – 3.40	– 1.36
3.41 and above	– 1.53

The average quality of coal purchased in 1982/83 was moisture 11.6%, ash 15.9%, calorific value (gross) 24 508 kJ/kg and sulphur 1.6%. The major portion had a topsize of 37.5 mm. Thus the price would be:

			£
Heat	245.1 × 20.32 p	=	49.80
Ash adjustment‡		=	– 2.259
Grade adjustment (untreated)		=	– 0.45
Sulphur adjustment (datum)		=	0.00

$$\overline{\qquad\qquad}$$

£ 47.09/tonne

‡ from BC tables

6.2   The net heat value of coal purchased

Coal samples taken at the colliery or power station are analysed in the respective laboratories for their heat content by using a bomb calorimeter. This is a closed vessel and combustion takes place at constant volume. In a power station combustion is at atmospheric pressure and the gases are free to expand. Moisture present in the coal and arising from combustion condenses in the bomb calorimeter and the latent heat of the water vapour is liberated. In a power station the water vapour leaves the stack taking this heat with it. As a result the gross calorific value as determined in the laboratory does not represent the heat content of the coal that is available for conversion to electricity. An adjustment has to be applied to a 'net' basis, to allow for expansion of the gases at constant pressure and the latent heat losses.

The equation used is

CV net (p) =

$\qquad$ CV gross (v) – 212.1 H – 24.4(M + 0.1A) – 6

where CV net (p) = calorific value (kJ/kg) at constant pressure

$\qquad$ CV gross (v) = calorific value (kJ/kg) at constant volume as determined in the laboratory, i.e., after correction for the sulphur content of the coal

$\qquad$ H = percent of hydrogen in the fuel substance (i.e., excluding the hydrogen content of the moisture and mineral matter)

$\qquad$ M = moisture content of the coal

$\qquad$ A = ash content of the coal

The net calorific value is over 1000 kJ/kg less than the gross value and depends strongly upon the hydrogen content of the fuel and its moisture content. Neither of these parameters directly appear in the industrial coal price formula at present although in future moisture, at least, should appear.

The net calorific value may also be calculated from the proximate analysis of the coal, where V is volatile matter:

CV net (p) =

$\qquad$ CV gross (v) – 13.9V – 7.9A – 30.6 M + 6.12

Agreement between the net calorific values calculated by the two approaches is good.

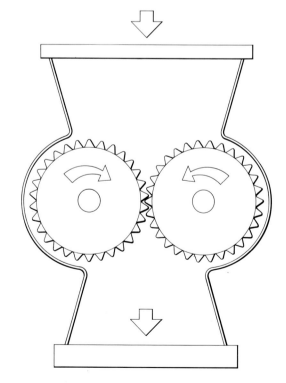

(a) Double roller crusher

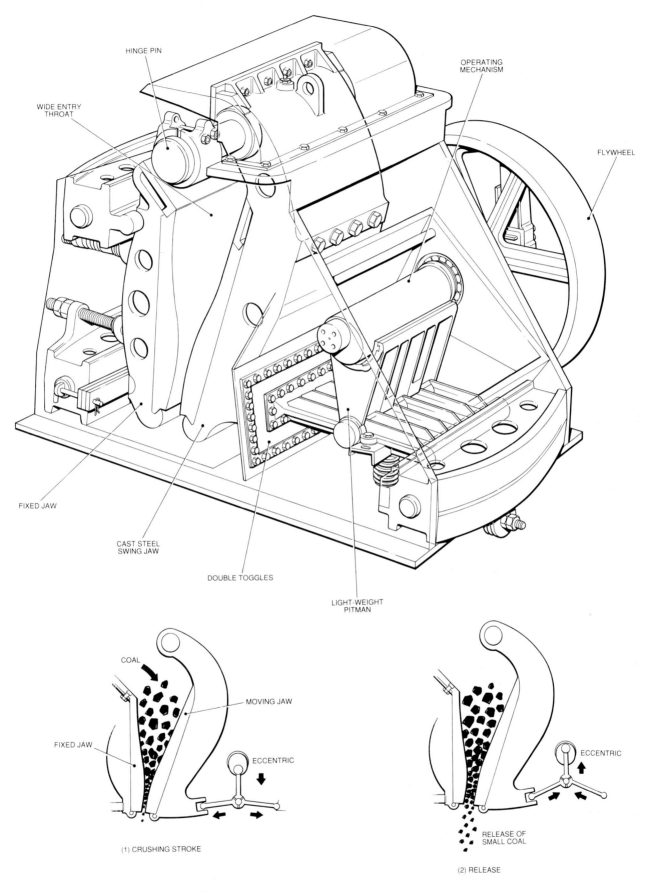

HINGE PIN

OPERATING
MECHANISM

WIDE ENTRY
THROAT

FLYWHEEL

FIXED JAW

CAST STEEL
SWING JAW

DOUBLE TOGGLES

LIGHT-WEIGHT
PITMAN

COAL

MOVING JAW

FIXED JAW

ECCENTRIC

(1) CRUSHING STROKE

ECCENTRIC

RELEASE OF
SMALL COAL

(2) RELEASE

(b) A jaw crusher (Brown Lenox & Co Ltd.)

FIG. 1.12 Coal crushers

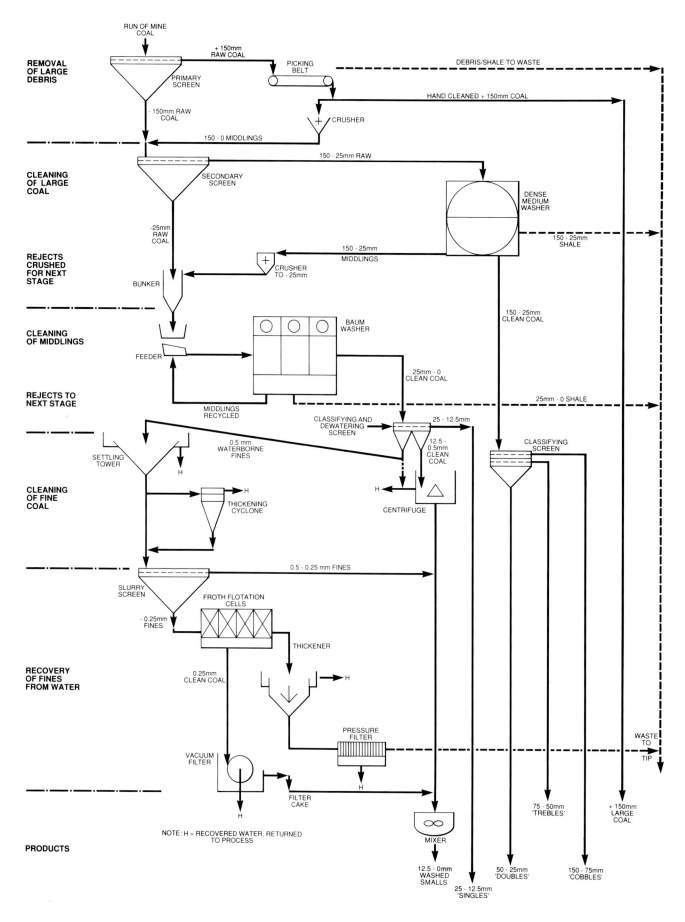

FIG. 1.13 Typical coal preparation plant

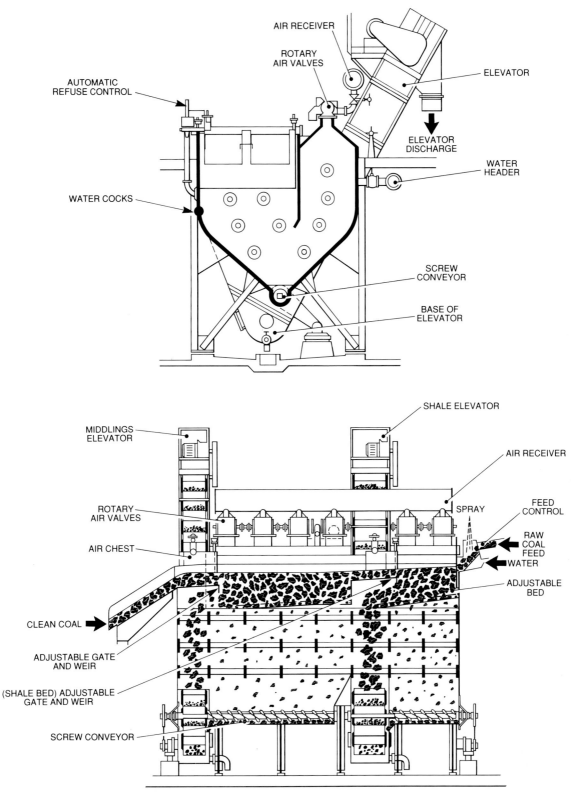

FIG. 1.14 Baum washer

7 Transport

7.1 Rail

The proportions of coal delivered by different means of transport is illustrated in Table 1.12.

Rail is by far the most important means of conveyance and the majority of tonnage is now successfully carried by 'merry-go-round' trains of 1000–1200 tonnes capacity. The wagons of these trains are automatically loaded at collieries from rapid loading bunkers and automatically discharged at the power station by means

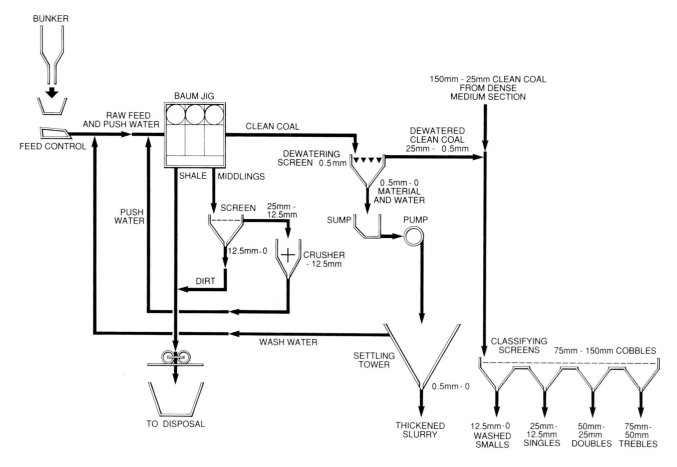

FIG. 1.15 Baum jig washer — flow diagram

TABLE 1.12

Proportions of coal carried to CEGB power stations according to method of transport (1982/83)

	Proportion	Average journey
Rail	77.5%	31 miles
Coastal shipping	9.2%	305 miles
Road	10.2%	26 miles
Canal	1.2%	7 miles
Conveyor	1.7%	1 mile

of trackside equipment. The coal falls directly into the station's track hopper, from which it is conveyed to bunkers or stock as required. The average journey in the UK is short and trains are unloaded in typically 40–60 minutes so that utilisation of the trains is good. By international standards the wagons are small (28 tonne capacity), see Fig 1.17, but they cope well with the small coal of high fines content burnt at British stations. The 'scissors' on the doors are designed to cut through the coal as the doors open and this breaks any bridges formed in wet or cold loads of coal which would otherwise prevent the coal from falling.

In very cold weather this design of wagon may cause problems if it is left standing, loaded. In UK winters coal freezes at 12.5–25 mm per hour and if a significant ice layer builds up the doors may not open at the track hopper, or they may open, but because of the weight of frozen coal on the door faces, the trackside equipment may not effect a reclosure. Fortunately, because the average UK journey is short the answer is simple, in cold weather the trains must be kept moving and not left to stand for any length of time. Successful deliveries are then ensured throughout the year.

7.2 Coastal shipping

The majority of large coal-fired power stations are sited on rivers close to the coalfields (Fig 1.2). There are however large stations, at the mouth of the Thames for example, that are typically supplied by sea from the BC NE Area, a journey of some 300 miles. Vessels of 6600 to 8000 tonnes capacity have been used until recently. The CEGB owned its own fleet, which it expanded. In 1984 it ordered three new colliers, each capable of carrying 19 000 tonnes of coal, from the Govan yard of British Shipbuilders.

Their final fleet numbered seven ships including the three new ones giving a total capacity of 85 000 tonnes. Other ships were chartered to increase capability when necessary.

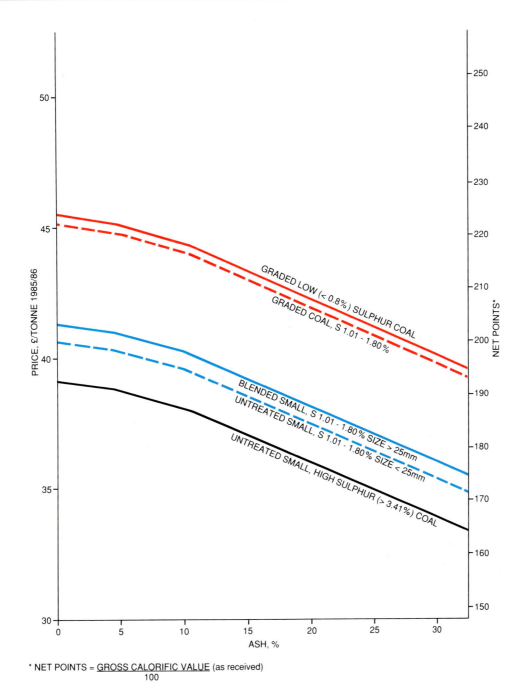

* NET POINTS = $\dfrac{\text{GROSS CALORIFIC VALUE (as received)}}{100}$

FIG. 1.16 The industrial coal 'ash adjustment' used for pricing in 1985/86

7.3 Road

In 1964 some 15% of the coal received at power stations came by road. In 1982/83 the figure had fallen to 10% due largely to the success of 'merry-go-round' trains. However, the roadborne proportion is unlikely to fall any lower, as some collieries are not connected to the rail system. Also, slurries or fines for example, do not discharge readily from rail wagons, so road delivery is preferred.

Conversely the roadborne proportion is not likely to increase significantly, as carriage of coal by road creates something of a nuisance and about 10–12% of roadborne coal appears to be a natural level.

7.4 Pipeline and conveyor

Where a colliery and power station are sited together, movement of coal by pipeline or conveyor may be considered. Slurry pipelines have been used successfully in the past but dewatering presented problems. Pneumatic movement is more promising and although not used within the electricity industry, has proven a successful means of increasing shaft capacity at BC collieries (Shirebrook for example). At Longannet in Scotland and at Rugeley A power station in England, substantial quantities of coal (2.6 m tonnes and 1.3 m tonnes per annum respectively) are moved by conveyor from the adjacent mines. The method is cheap and effective.

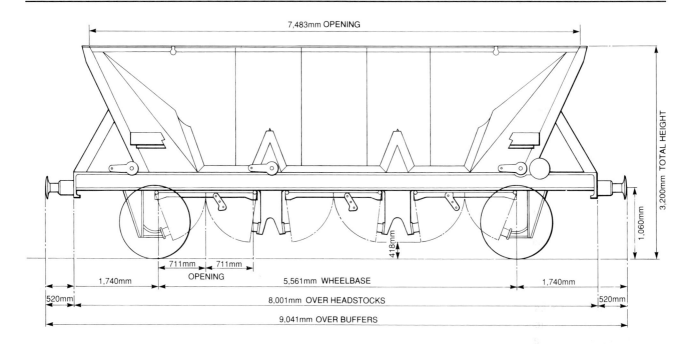

7,483mm OPENING

3,200mm TOTAL HEIGHT

1,060mm

418mm

711mm 711mm
OPENING

1,740mm

5,561mm WHEELBASE

1,740mm

520mm

8,001mm OVER HEADSTOCKS

520mm

9,041mm OVER BUFFERS

FIG. 1.17 British Rail HAA wagon

8 Coal storage and handling

There is a considerable variation between winter and summer in the demand for electricity (Fig 1.18). Collieries do not have large stockyards, so they produce coal steadily throughout the year. As they may have difficulty in delivering extra coal during cold weather, coal stocks at the power stations are built up during the summer months to cope with winter demand. In 1964, power stations generally kept stocks equivalent to four weeks' winter supply but this was increased in 1974 and again in 1984 so that stocks are now typically sufficient for four months. This allows power supplies to be maintained should there be temporary interruptions to coal production or supply.

Coal can deteriorate when stored incorrectly. Low rank, high volatile coals with a significant pyritic content are particularly prone to spontaneous combustion; the higher rank coals are less reactive. Nevertheless, any coal becomes reactive as its surface area increases and it follows, since the majority of tonnage purchased by power stations is small coal, that there is a potential problem when making this into a heap of over 1 million tonnes at a large power station. If heating were to exceed about 80°C a fire would certainly develop.

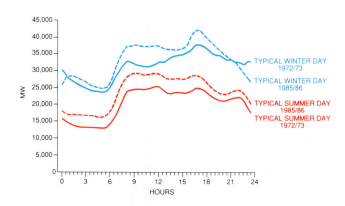

FIG. 1.18 Seasonal electricity demand curves

Twenty or thirty years ago, conventional wisdom held that stable stocks should consist of well ventilated, large coal, in heaps of less than 2.5 m. Large coal had a low surface area for its mass and free movement of air helped cool the stock. Many stations kept a reserve stock of large coal for seasonal use, but generally did not use it unnecessarily, as it also helped in stock accountancy to know that the quantity in reserve was constant. A separate stock was worked for day to day use and the reserve was only used when the normal stock proved difficult to handle. Some medium sized stations still operate this approach successfully.

Very large stations however have taken an alternative, equally successful, method of storing small coal. Spontaneous combustion cannot occur in the absence of air and by use of Terex machines (Fig 1.19) the coal stocks are built up by spreading thin layers and compacting these to high densities (SGs of 1.3 have been recorded). At these high densities air access to the coal is limited and heating does not occur. Stock heights over 20 m are achievable and stocks approaching 2 m tonnes can be managed. Continuous compaction is necessary to prevent air ingress and remedial work may be needed after very heavy rain, to eliminate channels, but generally such large stocks cause few problems (Fig 1.20).

Small coal in stocks of bulk density of about 1.0 prepared with bulldozers do oxidise to some extent. A loss of heat content of about 1% per annum has been estimated, but when densities in excess of 1.2 are achieved losses are about one-tenth of this value and the stocks can be considered to be stable.

FIG. 1.19 A Terex bowl scraper
(see also colour photograph between pp 208 and 209)

FIG. 1.20 Typical large CEGB coal stock
(see also colour photograph between pp 208 and 209)

9 Sampling of coal

9.1 The objective of sampling

Coal is sampled at power stations to ensure that the quality delivered correctly matches the quality which is paid for, and also to help account for the coal consumed by the station during any given period, i.e., for 'value for money' and heat accountancy purposes.

In addition, a knowledge of the quality of each variety enabled the national distribution of coal to be optimised by the CEGB. This was achieved with the aid of a computer program, SYMAC. Generally good coals, those with low ash and moisture, were distributed more widely than those of low heat content in order to minimise the unnecessary transport of inert material. Constraints could be built into the model to ensure that particular technical requirements of stations were also met. In this way the best possible use of the fuel was obtained.

Whatever the results of a coal analysis are to be used for, good sampling is essential. The principles which must be applied are given in British Standard 1017: 1977 Part 1. It is not possible to check every piece of coal in a consignment, or from a coal belt during a plant test, but nevertheless sampling must be

carried out so that each piece of coal has an equal chance of being included in the sample which is taken. For a train load of 1000 tonnes, a sample of almost 1 tonne may be required. This has to be crushed and reduced in size, according to rigorously defined rules (BS1017), to produce some 300 g of representative material for laboratory tests (BS1016). If this sampling and preparation is not carried out correctly, then the subsequent analysis of the material will be of little value.

9.2 Theory of sampling

9.2.1 Introduction

The correct procedures for sampling coal and coke are given in British Standard 1017: 1977 and its international counterpart ISO 1988. To obtain meaningful results from any sampling exercise the procedures should be followed precisely.

Coal is composed of two parts, combustible material and associated minerals; it follows that if sampling of the heterogeneous mixture can correctly determine the proportion of ash present, then other parameters such as calorific value or volatile matter must also be correctly determined. Standard procedures are based on ash analysis as the prime determinant. Moisture measurement has its own difficulties as samples can dry out if not handled correctly. Two samples are therefore prepared for subsequent analysis. The first, of small particle size (72 mesh) is suitable for ashing in a furnace, the second which has a larger particle size (6 mesh) provides a stable sample for moisture determination.

Coke is rarely purchased for use in modern power stations. It is an abrasive material and causes excessive mill wear, so its sampling is not considered in detail here. The reader will find a description of the principles in BS1017 Part 2 if required.

9.2.2 Representative sampling

The method of sampling coal at power stations is known as representative sampling. Consignments of fuel are divided into a number of roughly equal parts, grab loads, lorry loads, wagons or minutes of discharge from a conveyor belt, and an increment (part sample) is taken at random from each part; these are then combined to form a gross sample. To check that the procedure is giving a correct result, a second sample taken in the same way should give test results not significantly different from the first.

When sampling coal according to BS1017, the number of unbiased increments and the minimum weight of each which must be taken and combined to form a representative sample are prescribed. This ensures that the errors in the results obtained are not greater than a pre-selected level.

9.2.3 Unbiased increments

An unbiased increment is one in which all particles presented for sampling will have had an equal and random chance of inclusion. There should be no selective rejection of either large or small sizes, as this would lead to the results of the subsequent analyses being persistently higher or lower than the correct values. The best means of obtaining a representative sample is from a falling stream of coal, when the particles are falling freely.

Bias in sampling can only be detected by reference to a standard method of sampling which is known to be bias-free. There are no certain ways of eliminating bias, only the application of some general rules and commonsense. Periodic checking that sampling procedures are free from bias is therefore very important.

9.2.4 Weight of increments

Ideally, the mass of an increment should be sufficient to ensure that the size distribution of the collected particles is the same as in the unit of coal being sampled. In particular, large particles should not be excluded. To ensure that the largest particles may enter the sampling device (bucket or scoop) its aperture should be 2.5 times the top size of the coal. If two large pieces of coal are adjacent at the time of sampling then there is some free room for them to enter the device. Experimental work over the years has proved that this rule is entirely adequate to ensure a correct particle size distribution in the increment.

The minimum weight required in an increment has similarly been arrived at, through a commonsense approach confirmed by experiment. Taking a 120 mm coal size as the datum, the minimum weight of increment should be five times the weight of the largest particle. Assuming the density of coal to be 1.6 and the particle to be spherical then its mass is:

$$4/3 \, \pi \, (6)^3 \times 1.6 \text{ grams}$$
$$= 1448 \text{ g or } 1.448 \text{ kg}$$

The minimum mass of increment is then five times this amount namely 7.2 kg.

For coal sizes less than 120 mm

$$m = 0.06 \, I$$

where m is the minimum mass of increment in kg and I is the nominal upper size of the coal in millimetres. This relationship simply extends the first calculation downwards linearly to particles of smaller size.

For coal sizes greater than 120 mm, the increment size increases in proportion to the cube of the upper size relative to the 120 mm datum.

$$m = 7.2 \, (I/120)^3$$

It is not usual to calculate the required mass of increment on each sampling occasion. In BS1017, the nominal upper sizes of coal are divided into a series of discrete intervals and the minimum mass of increment appropriate to the upper end of each range is used for the interval.

It is important that increments in a set should be of approximately equal weight. Serious loss of accuracy arises if any increments are less than the stated minimum weight; on the other hand there is no appreciable gain in accuracy if the minimum weight is exceeded. Nevertheless, designers of automatic samplers generally arrange for increments to be overweight in an attempt to avoid biased sampling. This yields very large gross samples which require crushing, mixing and subdividing in order to yield a conveniently sized sub-sample for further preparation. The excess coal is returned to the coal handling system. Periodic checks are required to ensure that this dividing process is not introducing bias. This is done by sampling the reject coal to establish that it is not of significantly different quality to the sample stream.

9.2.5 Number of increments

Sampling a consignment of an 'unknown' coal

The number of increments which have to be taken to form the gross sample from a consignment depend upon the class of coal and how much is known about its variability. If the coal is 'unknown' and it is of a blended or untreated quality, 35 increments should be taken if it can be sampled from a falling stream at the end of a conveyor. Deliveries received in wagons and lorries, barges, or ships which can be sampled off the unloading conveyor, require 50 increments and if the ship has to be sampled from the hold, 65 increments are recommended.

To establish what precision has been achieved on the 'unknown' consignment at least *six* replicate samples are taken and the ash results are compared. The precision of the mean of the six ash results is given by:

$$P_m = \pm (ts/\sqrt{n})$$

where t is Student's t: numerical values for 't' at 95% confidence limits are given in standard statistical tables and also BS1017. The appropriate value is selected using

$$f \text{ (degrees of freedom)} = n - 1$$

where n is the number of results

s is the standard deviation of a single result calculated from the formula

$$s = \sqrt{[(1/n - 1) \times (\Sigma X^2 - (\Sigma X/n)^2)]}$$

where X is the sub-sample ash value.

For six samples t = 2.57 and

$$P_m = \pm 0.47 \sqrt{[\Sigma X^2 - (\Sigma X/6)^2]}$$

The ash results, X, for the six samples are added together to give ΣX and the square of each ash result is determined and the sum of the six squares gives ΣX^2. The processing of a set of results is described in BS1017 and is straightforward, but the taking of six samples, each containing 35 to 65 increments, is a considerable task.

Intermittent sampling of known coals*

Coal is delivered to power stations in the main by rail from a relatively small number of collieries. The station will consider some 12 to 15 collieries to be 'theirs' and they are familiar with each quality and its variability. Incremental rates can be set at lower levels than for unknown coals but periodic checks are required to establish that the required precision continues to be achieved.

In intermittent sampling only a proportion of the deliveries is sampled. As an example, a colliery may

produce one trainload per working day, sending 65 trains to the power station each quarter. If the quality is reasonably consistent only one sample per week need be taken, chosen at random, giving 13 samples per quarter. The precision obtained by an intermittent sampling regime depends upon the number of units sampled in a given period (usually a quarter) and the number of increments in the sample taken from each unit. Thus two checking procedures are required to be carried out periodically, but they use the same set of analytical results.

The precision checks are made by duplicate sampling. As a consignment is sampled, successive increments are alternately put into an A or B container until two full replicate samples are obtained. Ten successive pairs of duplicate samples are required in order to make the precision checks. These are analysed and the ash results are compared. For each pair the mean result is calculated and the difference between the duplicate values is determined. The mean difference between the ten duplicates is found 'd' and the range of the mean sample values 'c' is noted. An example calculation is given in BS1017 Appendix A5.

Testing results to see if number of units is correct

To check if the correct number of units has been sampled, in order to achieve the desired precision, the range that should theoretically occur 'C' is found from Table 17 of BS1017. A comparison between the theoretical and observed range is then made. If C/c is less than 0.6, then too few units have been sampled. If C/c is between 0.6 and 1.8 there is no evidence to suggest that the number of units sampled is incorrect. If C/c is greater than 1.8, more units will have been sampled than is necessary.

Testing results to see if number of increments is correct

Providing the above test is satisfactory (i.e., 0.6 < C/c < 1.8) a check may be made that sufficient increments are being taken. The ratio c/d is calculated from the set of results. If c/d is greater than 9.3, more increments have been taken from each unit than is necessary. If the ratio c/d is between 9.3 and 2.3, there is no evidence that the number of increments taken from each unit is incorrect. If the ratio c/d is less than 2.3, too few increments will have been taken from each unit.

The above tests of intermittent sampling are carried out on the station's most difficult coal; the one with the greatest variation in quality between the large size component and fines. The station's normal incremental rate is then set to be adequate for that coal's quality; this ensures that sampling is adequate for all other coals received.

9.2.6 Checking preparation

When a large automatic sampler produces increments of a weight greater than required, the sample must be crushed and subdivided in order to give manageable

---

*Coal is sampled at the collieries on a continuous basis; increments are taken at regular intervals from the coal produced throughout each shift. A procedure to prove that regular sampling is to the required standard is given in BS1017 (Appendix A4).

amounts of coal. The division procedure must be periodically checked to prove it is bias-free. This is also carried out by comparison of a two-sample set. A sample is taken from the normal stream and another at the same time from the coal which would normally be rejected. The two samples are worked up independently and a set of ten pairs of ash results is eventually obtained.

The mean difference 'h' of the ten pairs should be less than 0.37P, where P is the overall precision of sampling preparation and analysis. It is usual to carry out the procedure twice and if, with each set of ten results, a value of 'h' less than 0.37P is obtained, it may be assumed that the precision of the sample preparation and analysis is satisfactory.

9.3 Reference — stopped belt sampling

The tests described in the previous section ensure that the precision of procedures at a power station are satisfactory, but they do not of themselves prove that the result is accurate. To check accuracy at a station which takes its samples from a falling stream at the end of a belt is relatively straightforward. Two samples are prepared by stopping the belt at intervals. The reference set of increments is obtained by placing profiled boards into the coal on the stopped belt (Fig 1.21). The coal between the boards is carefully removed and constitutes the reference increments. When this has been removed, the belt is restarted and the automatic sampler is operated to take its increment from the coal stream, close to the position from which the reference was taken.

Two sets of ash results are obtained for the automatic and reference increments. These are compared pair by pair and the ash differences noted. In addition, the squares of the ash differences are calculated and the sum of the squares established so that the variance of the differences may be established. A sample calculation is given in BS1017: 1977 Appendix C5.

Four terms are calculated in order to determine if the normal automatic sampler is bias-free. Firstly the

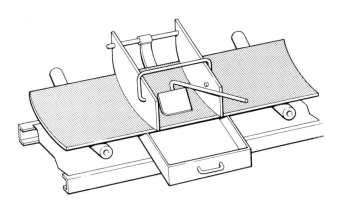

FIG. 1.21 Sampling frame for reference coal sampling

variance Vz of the difference between the ash results of 'n' pairs of samples is computed:

$$Vz = \frac{1}{n-1}\left[\sum_{r=1}^{r=n} Z_r^2 - \frac{1}{n}\left(\sum_{r=1}^{r=n} Z_r\right)^2\right]$$

where Z_r is the ash difference between the rth pair of increments.

The procedure is identical to that used in Section 9.2.5 of this chapter. Next the precision P of the two methods is calculated

$$P = \pm t \sqrt{(Vz/2)}$$

where t is student's t at 95% confidence level using f (degrees of freedom) = n − 1.

The value of Z, the mean difference in ash between the pairs of increments is noted and finally the precision of the estimate of bias 'η' is made

$$\eta = \pm t \sqrt{(Vz/n)}$$

The British Standard worked example shows how to carry out the calculation in detail.

The test is conclusive if the value of Z, the estimated bias obtained by taking the average of the ash differences, is greater than η and the true value of the bias will be in the range Z ± η. If Z is less than η the test is inconclusive because the precision of the estimate of the bias, η, is not good enough to establish a firm value for Z. More pairs of increments will then have to be taken to improve the precision. A satisfactory estimate of bias for an automatic sampler can usually be achieved with about 20 pairs of increments.

This test procedure is very valuable; it gives the precision of the sampling method P, information about the variability of the coal Vz, and of course puts limits on the level of bias in the sampling procedures. It should be carried out at least annually at all stations.

9.4 Primary sampling procedures

9.4.1 Hand sampling

From road or rail wagons

To ensure a representative sample is obtained the procedure used must allow each particle in the coal an equal and random chance of being included in an increment. This is most difficult to achieve by hand sampling from road or rail wagons. They cannot be sampled during filling or emptying and must be tackled either when they are tipped or, alternatively, by using an auger (Fig 1.22) to dig holes into the coal. Such samples are always suspect since there is not free access to the coal. Where two products have been loaded separately into one wagon, for example, untreated smalls and wet fines, it is virtually impossible

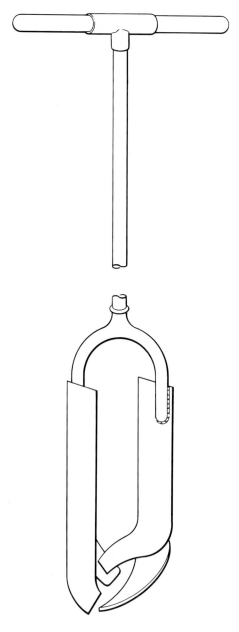

FIG. 1.22 Auger for sampling small coal up to 25 mm

to take a representative sample this way. Even for a fairly homogeneous coal, segregation during loading or by shunting of wagons can lead to a small bias.

The load can be sampled as it is tipped. A suitable device (a bucket on a pole, essentially) takes an increment from the 'falling stream'. Unfortunately the weight of coal falling is considerable, leading to safety problems, the bucket overfills and it is doubtful whether an accurate increment is ever obtained. A better approach is to automate the traditional hand sampling procedures by using a powered auger to sample the corners and centre of the wagon (Fig 1.23).

From conveyor or chutes

If the wagons referred to above are tipped onto a belt, hand sampling from beside the conveyor using a scoop (Fig 1.24) may be carried out successfully. The coal is now more readily accessible and provided the operative is instructed in the principles of sampling and can gain access to the full width of the belt good results may be obtained. The safety of the man carrying out the sampling should be carefully considered and hand sampling should not be attempted if the belt speed is greater than 1.2 m/s, the height of the coal is greater than 0.2 m or the flow is greater than 200 tonnes per hour.

9.4.2 Automatic sampling from a falling stream

The basic requirement of taking an unbiased increment using a sampling machine, is that the collecting bucket should traverse the falling stream of coal at a constant speed. It should be designed for its specific application, so that the whole width and depth of the coal stream is traversed and the increment taken should occupy no more than three-quarters of its capacity. Suitable machines are shown in Fig 1.25, while Fig 1.26 shows another design linked to crushers and dividers for reducing the mass of each increment before storage.

There are two older designs of sampler in existence, the Seaborne (Fig 1.27) and the Pollock (Fig 1.28). Each can develop bias and they are not favoured.

9.5 The moisture sample

The principles of sampling for moisture are the same as those of sampling for general analysis — the sample must be unbiased and made up of increments spaced evenly over the consignment. As moisture is more evenly spread through the coal than ash, fewer increments are usually needed for the same accuracy to be achieved. British Standard 1017 does not allow any relaxation in the number of increments in the moisture sample, however, for an 'unknown' coal.

The moisture sample may be taken as a separate sample to the analysis sample, or a gross common sample may be taken and the moisture sample abstracted from it. Most power stations using automatic samplers obtain a very large primary increment and it is convenient to take the moisture sample directly from this. The increment from the sampling bucket is discharged onto a smoothing device, a scroll feeder or smoothing belt, so that the coal enters the subsequent coal division system gradually. A flap valve diverts a proportion of the raw coal from each increment into a closed container, as it leaves the smoothing device.

The minimum mass of moisture increment which must be extracted from the common sample for total moisture is 1 kg for coals of 25–50 mm and is size related (see BS1017: Table 5). At least ten increments must be taken, but typically the flap valve on an automatic system takes a proportion from each increment of the gross sample so that more than ten increments of raw coal are obtained.

Subsequent handling of the moisture sample requires care to ensure there is no unmeasured moisture loss.

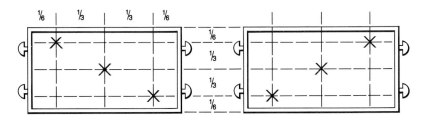

3 INCREMENTS PER WAGON (ALTERNATIVELY)

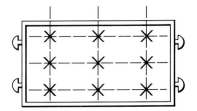

9 INCREMENTS PER WAGON

FIG. 1.23 Auger sampling from wagons
(see also colour photograph between pp 208 and 209)

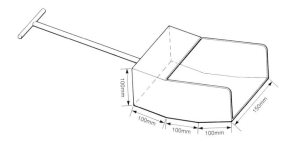

FIG. 1.24 Scoop for sampling coal up to 50 mm

Individual increments are accumulated in a closed air-tight container and stored in a cool place. In the case of the automatic system care is also necessary in the design of chutes following the flap diverter. These should be as short as possible so that no drying of the coal occurs as it falls. The moisture sample is therefore collected very close to the primary bucket. When complete, it is taken to a preparation room for further crushing and subdivision to produce about 300 g of coal of 3 mm top size for submission to the laboratory (see Section 10 of this chapter).

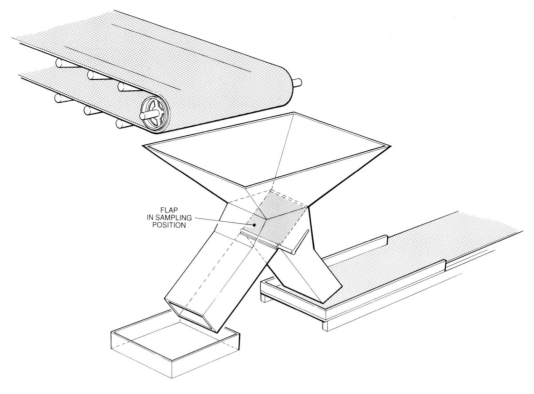

(a) Breeches chute

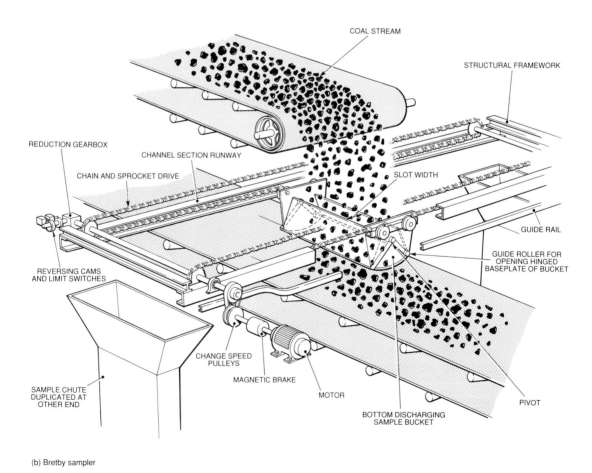

(b) Bretby sampler

FIG. 1.25 Equipment for automatically sampling falling streams of coal

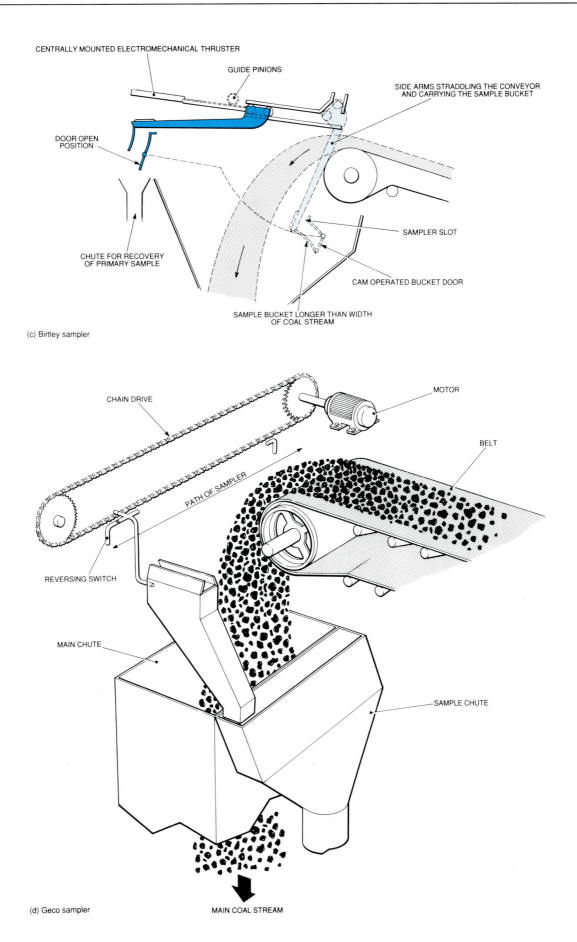

CENTRALLY MOUNTED ELECTROMECHANICAL THRUSTER

GUIDE PINIONS

SIDE ARMS STRADDLING THE CONVEYOR AND CARRYING THE SAMPLE BUCKET

DOOR OPEN POSITION

CHUTE FOR RECOVERY OF PRIMARY SAMPLE

SAMPLER SLOT

CAM OPERATED BUCKET DOOR

SAMPLE BUCKET LONGER THAN WIDTH OF COAL STREAM

(c) Birtley sampler

CHAIN DRIVE

MOTOR

BELT

PATH OF SAMPLER

REVERSING SWITCH

MAIN CHUTE

SAMPLE CHUTE

(d) Geco sampler

MAIN COAL STREAM

FIG. 1.25 (cont'd) Equipment for automatically sampling falling streams of coal

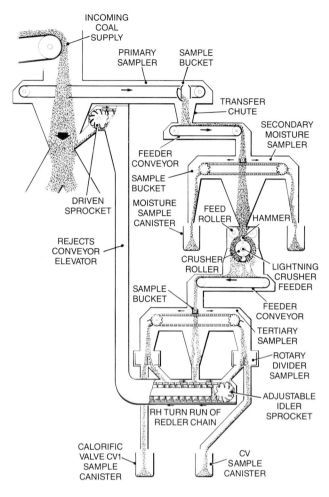

FIG. 1.26 Automatic sampler with sample reduction system

9.6   The analysis sample

Large automatic samplers produce a primary sample which needs reducing in size to make it convenient to handle. After abstraction of the moisture sample, subdivision to reduce the weight of coal must be accompanied by crushing and mixing. For each coal sizing, the British Standard prescribes a minimum amount which must be retained for the next stage of division. Thus for the raw coal sample of say 50 mm top size it will first be necessary to crush to 10 mm. The mass of the coal may then be reduced to 10 kg and it will require to be recrushed to 3 mm before further reduction to 2.0 kg (see BS1017: Table 6). The reduced sample is taken to a preparation room at one of the intermediate sizes (see next section) where further crushing and subdivision should produce 50–150 g of coal of 0.2 mm top size for laboratory analysis.

10   Sample preparation

10.1   The moisture sample

Whether obtained directly as a moisture sample or abstracted from the common sample, some 10 kg or

more of raw coal will be brought to the preparation room in a closed container for further treatment. The container and coal are weighed before opening. The coal is then transferred to a dry tray and spread evenly to a depth not exceeding 20 mm (except, of course, for lumps of greater than this size). The tray and coal are then weighed, and the container, its lid and the tray of coal are air dried to constant weight — this takes about 24 h at ambient temperature. The drying process requires control from two points of view. If it is carried out at ambient temperature in a ventilated cabinet, precautions must be taken to ensure that the air movement neither blows fine coal from the tray nor deposits dust from elsewhere in the building on to the coal. If the cabinet is heated its temperature should not exceed 30°C.

Air dried coal adhering to the container and lid is brushed into the tray and the dry empty container and lid are reweighed. The dry coal on the tray is then transferred to the container, again the lid is replaced and the full container is reweighed.

The percentage loss of moisture on air drying, M, is then:

$$M = (m_2 - m_3)/(m_2 - m_1) \times 100$$

where m_1 is the mass of the dry empty container plus lid, g

m_2 is the mass of coal and closed container before air drying, g

m_3 is the mass of coal and closed container after air drying, g

The coal is now in a most stable state, but should still be processed immediately. It is crushed in two stages. In each step it is necessary to avoid loss of fine coal and to ensure that the coal is not overheated. The coal is first crushed to 13 mm top size in a roll or jaw crusher. Air flow through this type of equipment is low and fines and moisture are not likely to be lost. The coal is then subdivided to 1 kg using a riffle, or better a rotary or other mechanical divider (Figs 1.29 to 1.32).

Mechanical dividers collect a large number of increments from a stream of coal and separate them into 'retained' and 'discard' portions. Adjustment to the settings provide a means of retaining any required fraction; in some machines this can be as low as 1/24 th. Riffles on the other hand can only halve a sample. If less than a half is required the sample must be repassed. The slot width of the riffle should be about four times the top size of the coal to ensure it is bias-free. Ratios very much greater than 5 or less than 3 may lead to bias.

The errors of division can be considerably reduced if the sample is thoroughly mixed beforehand. Hand mixing frequently leads to segregation and mechanical mixers, e.g., double-cone types (Fig 1.30), are preferred. Replicate riffling and reuniting of the halves is

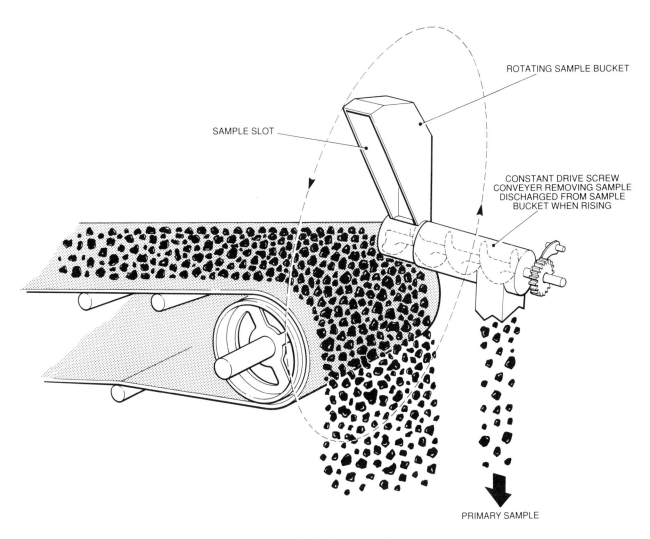

ROTATING SAMPLE BUCKET

SAMPLE SLOT

CONSTANT DRIVE SCREW
CONVEYER REMOVING SAMPLE
DISCHARGED FROM SAMPLE
BUCKET WHEN RISING

PRIMARY SAMPLE

FIG. 1.27 Seaborne sampler

an acceptable alternative mixing method provided the riffle apperture is appropriate to the coal top size.

The mixer, crusher and divider must be readily accessible so that they may be cleaned after use. This prevents contamination of the next sample. Any material removed must be added to the rest of its sample. Finally, the 1 kg of coal is crushed to 3 mm in a small laboratory mill and further subdivided to provide the 300 g sample for submission to the analysis laboratory. Details of the air dried moisture loss accompany the sample to the laboratory. The final stage of crushing may produce fines, and bag filters should be fitted to the mill to ensure that entrained material is recovered.

There are variations to the above procedures allowed in BS1017 and the standard should be referred to for further detail. Some locations, because their primary increment was very large, have in the past abstracted the moisture sample after crushing to 13 mm. This is not an approved procedure and frequent checks were necessary to show that moisture loss was not occurring due to windage in the large crushers. It is preferable to avoid this method, although BS1017 refers to a

closed mill which should be satisfactory. If this proves to be so, then a more convenient moisture sample reduction may be achieved in future.

10.2 The analysis sample

The object of preparation of the analysis sample is to produce 50–150 g of representative coal of 0.2 mm maximum particle size. This fine coal is not particularly stable with regard to its moisture content, but is intended to be suitable for ashing in a furnace. The small particle size facilitates complete removal of combustible or volatile matter during the laboratory analysis procedures.

Size reduction may be carried out in two or three steps and, as in the case of the moisture sample, crushing and mixing are required at each stage. With large automatic samplers the gross general analysis sample, after extraction of the moisture sample, weighs several hundred kilograms. It is not practical to air dry this amount of coal before crushing, although for smaller installations, if this can be done, it facilitates the subsequent crushing of the coal. Secondary samplers are

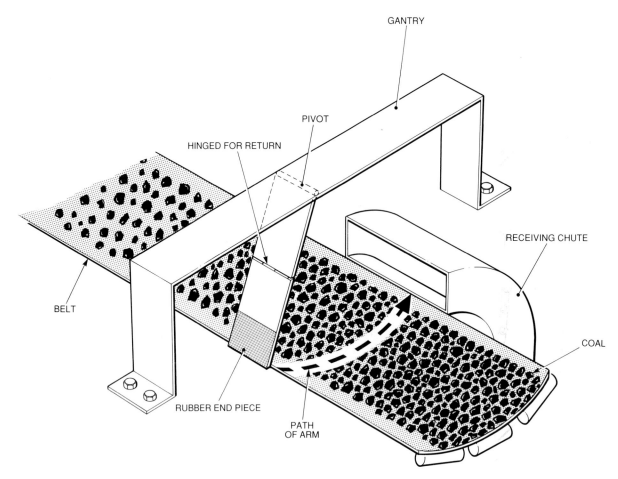

FIG. 1.28 Pollock sampler

available which can sub-sample the raw coal and reduce its mass by 20:1. The apertures of such devices have to meet the same rules as the primary sampler (i.e., 2.5 × the upper size of the coal) and a suitable design is shown in Fig 1.33. Coal preparation equipment following the secondary sampler can then be smaller. It is quite common as an alternative for the whole of the primary sample to be crushed to 10 mm before division. Roll crushers or jaw crushers are preferred as other mill types may generate excessive fines or heat. If the sample size after division is then small enough the coal will be taken to the preparation room, otherwise further crushing and division at 3 mm may be undertaken. The crushed coal is then air dried prior to final crushing in laboratory mills to 0.2 mm. This latter step may again be done in one step or two, with 1 mm being the intermediate size. A suitable laboratory mill is shown in Fig 1.34.

As with moisture sample preparation, precautions are necessary, particularly with the final mill to ensure no fines are lost. Bag filters should be fitted. These high speed mills also generate heat, particularly if the machinery is built to fine tolerances, and this can oxidise the coal. Time should be allowed between operations to allow mills to cool. Finally the preparation room and all equipment must be kept clean to avoid contamination of or losses from samples.

The minimum mass of coal which must be retained at each stage of division has been the subject of debate at least since 1909 when Bailey first suggested rules for minimum weights (E. G. Bailey J. Ind. Eng. Chem 1909 *1* 161). In 1951 alternative rules were suggested by Liddell (Fuel 1951 *30* 275) and the 1960 version of BS1017 reinforced the earlier work with much experimental evidence. The data presented then showed considerable 'scatter'. The current BS1017: 1977 recommends retention of less mass of coal than hitherto (see Table 6 of BS1017: 1977).

The figures given for mass retention for subsequent size analysis of the coal (Table 8 of BS1017: 1977) are identical with those given for coal of 0–5% ash content in the BS1017: 1960 recommendations, but the figures for mass retention when preparing the laboratory sample (Table 6 of BS1017: 1977) are less than the sets of figures given in 1960.

The smaller the mass retained at each stage of division, the more careful it is necessary to be, to obtain a representative sample. Mechanical dividers should be used and care must be taken to keep these free of contamination. The minimum masses presently recommended to be retained during sampling division are shown in Table 1.13.

It may be necessary to use larger amounts if the coal is particularly heterogeneous, i.e., if there is a big

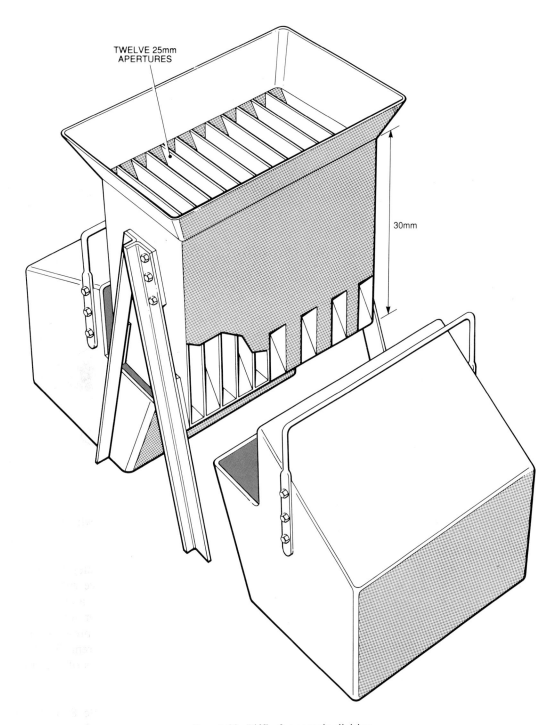

TWELVE 25mm
APERTURES

30mm

FIG. 1.29 Riffle for sample division

difference in ash content between the fines and larger size components of the coal. The checking of preparation procedures, as described in Section 9.2 of this chapter, should be undertaken frequently to ensure that the preparation precision is adequate.

11 Standards for the analysis and testing of coal

The analysis of coal is covered by British Standard 1016 (see Table 1.14). As many individual determina-

tions may be required this standard is extensive and has sixteen parts.

There are also international standards

ISO: R562, R579, R589, R687, R1171, R1994 and 154, 334, 351, 352, 587, 609, 925 and 1928.

and American standards

ASTM: D271, D720, D1756, D2015, D2234, D2361 and D2492.

Generally within the UK electricity industry tests are carried out according to the appropriate British Standard but in some cases alternative methods are

TABLE 1.13

Minimum mass of sample to be retained during sampling division

Maximum particle size after reduction	Minimum mass of sample to be retained			
	Two stage preparation		Three stage preparation	
	Clean coal ash < 10%	All other coal	Clean coal ash < 10%	All other coal
mm	kg	kg	kg	kg
10	1.5	10	2.5	15
5	0.5	3.5	0.75	5
3	0.3	2.0	0.45	3
1	0.15	0.6	0.25	1

TABLE 1.14

British standards for the analysis and testing of coal and coke

British Standard 1016	Subject covered
1973 Part 1	Total moisture of coal
1973 Part 2	Total moisture of coke
1973 Part 3[†]	Proximate analysis of coal
1973 Part 4[†]	Proximate analysis of coke
1977 Part 5	Gross calorific value of coal and coke
1977 Part 6	Ultimate analysis of coal
1977 Part 7	Ultimate analysis of coke
1977 Part 8	Chlorine in coal
1977 Part 9	Phosphorus in coal and coke
1977 Part 10	Arsenic in coal and coke
1977 Part 11	Forms of sulphur in coal
1980 Part 12	Caking and swelling properties of coal
1980 Part 13	Tests special to coke
1963 Part 14	Analysis of coal and coke ash
1970 Part 15	Fusibility of coal and coke ash
1981 Part 16	Reporting of results
1979 Part 17	Size analysis of coal
1981 Part 18	Size analysis of coke
1980 Part 19[*]	Determination of the index of abrasion of coal
1981 Part 20	Determination of Hardgrove grindability index of hard coals
1981 Part 21	Determination of moisture holding capacity of hard coals

[†] Parts 3 and 4 superseded 1991 1016 Sections 104.1
104.2
104.3
104.4

[*] Part 19 superseded 1990 by Section 111

used, for example, for chlorine analysis or fusibility measurements, where extra information is needed to cover power station needs. Of the tests referred to above, not all are carried out on each coal sample. For pricing purposes moisture, ash, calorific value and sulphur are determined routinely. Chlorine is determined frequently, particularly on coals of greater than 0.4% chlorine content, as chlorine contributes to furnace wall corrosion. Tests such as fusibility, grindability, abrasiveness, ash composition and trace element con-

tent are undertaken only occasionally, to keep general information about coal varieties up to date.

12 Proximate analysis of coal

Proximate analysis of coal is expressed in terms of moisture, volatile matter, ash and fixed carbon.

Ash is the residue remaining after the coal has been incinerated to constant weight under standard conditions.

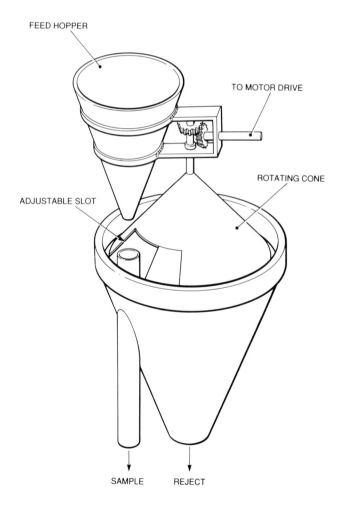

FIG. 1.30 Cone sample divider

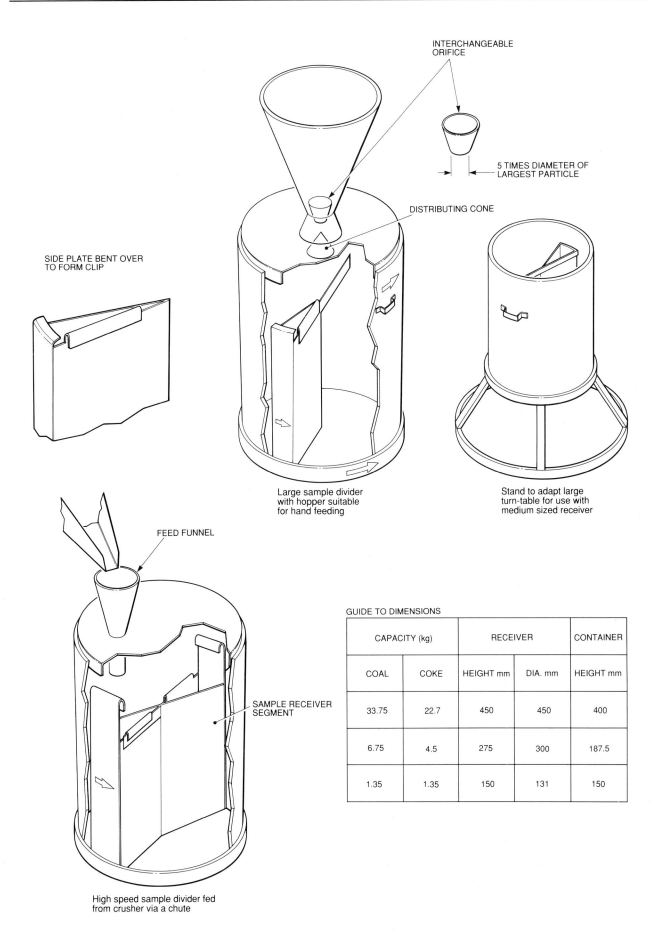

INTERCHANGEABLE ORIFICE

5 TIMES DIAMETER OF LARGEST PARTICLE

DISTRIBUTING CONE

SIDE PLATE BENT OVER TO FORM CLIP

Large sample divider with hopper suitable for hand feeding

Stand to adapt large turn-table for use with medium sized receiver

FEED FUNNEL

SAMPLE RECEIVER SEGMENT

High speed sample divider fed from crusher via a chute

GUIDE TO DIMENSIONS

CAPACITY (kg)		RECEIVER		CONTAINER
COAL	COKE	HEIGHT mm	DIA. mm	HEIGHT mm
33.75	22.7	450	450	400
6.75	4.5	275	300	187.5
1.35	1.35	150	131	150

FIG. 1.31 Rotary mechanical sample dividers

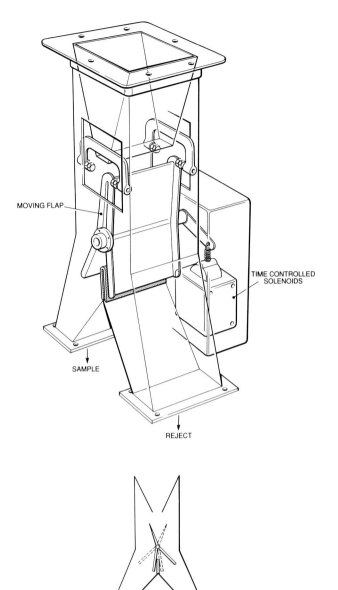

FIG. 1.32 Moving flap sample divider

Volatile matter is the loss in weight (other than that due to moisture) that occurs when coal is carbonised to a coke under standard conditions. The test is an arbitrary one; if conditions are changed the results will be different. Volatile matter is not, therefore, an entity contained in the coal which can be extracted and measured like water, and in consequence it is not strictly correct to speak of volatile matter 'content'.

Fixed carbon is a calculated figure:

Fixed carbon =

100 − (moisture + volatile matter + ash)

all expressed as per cent on the same basis. Its purpose is to measure the coke residue from the volatile matter determination; like volatile matter it is not a substance

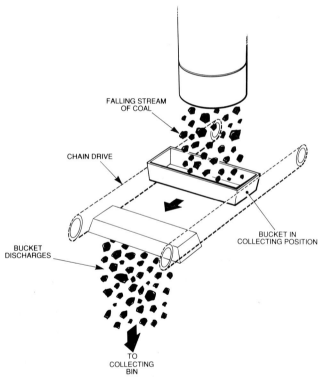

FIG. 1.33 Secondary coal samplers

contained in the coal. Since the two figures are inter-related, chemists tend to omit reporting fixed carbon. It is a relic of the days when the coke and gas quantities obtainable from a coal were important to the gas industry. With the advent of North Sea gas it has become of little use.

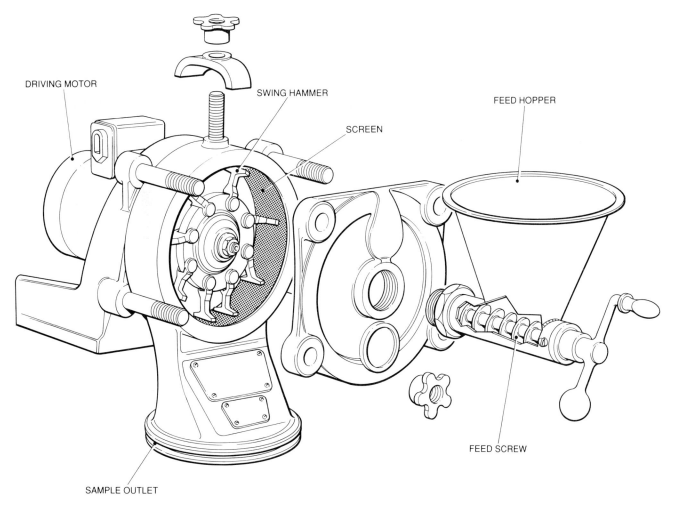

DRIVING MOTOR

SWING HAMMER

SCREEN

FEED HOPPER

FEED SCREW

SAMPLE OUTLET

FIG. 1.34 Laboratory coal grinder (Raymond Mill)

12.1 Moisture content

Moisture exists in all coals in the following forms:

- *Free moisture* 'surface' moisture, also present on apparently dry coals, which is dried off when coal is exposed to the air without heating.

- *Inherent moisture* the moisture retained in the pores of the coal substance when free moisture has evaporated.

- *Water of hydration* this is chemically bound to the shales in the coal. It is not driven off by heating to $105°–110°C$, as are free moisture and inherent moisture.

- *Air-dry moisture* is a term used to describe that part of the total moisture retained in the 72-mesh analysis sample after it has been exposed to the laboratory atmosphere and has attained approximate equilibrium with it. This is necessary so that there shall be no change in condition of the sample, either during the weighing of a portion for test or whilst the sample is standing between tests. It is not possible to handle a sample in a 'bone-dry' condition because finely-ground coal will pick up

moisture rapidly; on the other hand the presence of free moisture would lead to evaporation losses, not only during weighing but during preparation of the sample. In amount, the air-dry moisture is similar to the inherent moisture.

- *Total moisture* is the sum of free and inherent moisture. It is an inert constituent of coal and it reduces the calorific value; it costs as much as coal to transport and its latent heat of evaporation, when the coal is burned, contributes to flue gas losses.

12.1.1 Free moisture

All coals contain some free moisture. Its origin may be underground mine water, water infused into coal seams and sprays used for dust suppression, washing processes and exposure to rain or snow.

Free moisture is determined at the power stations as the first stage in arriving at the total moisture, and it is the proportion which is evaporated on bringing a sample into equilibrium with the ambient air. Free moisture is determined, as a percentage, from the loss in weight of a 10 kg sample after exposure to freely-circulating air at not more than $15°C$ above ambient temperature, for 16 to 24 hours. The sample is spread out into trays about 20 mm deep, and if very wet the drying time may be extended beyond 24 hours.

The difficulty in sampling coals with excessive free moisture, for example, coal from the drainage deck of a washer, has been noted. Drainage of free moisture from smalls with a high proportion of fine coal (passing 30-mesh BS sieve; 0.6 mm) is very slow and it is these coals which are difficult to handle (see Section 8 of this chapter).

12.1.2 Inherent moisture

This is related to the rank of the coal, being least in Ranks 200 to 300 (1–2%), rising to 3% for Rank 101, and to a maximum (10–12%) for Rank 900.

When moisture is determined using 12.5 mm coal, the water in a 1 kg sample is distilled off with toluene (boiling point 110°C) and the amount of condensed water collected in the receiver during 6–8 hours is noted. An alternative method suitable for high rank coals only (Rank 100–400), is to measure the loss in weight when a 1 kg sample is heated in an oven to 105°–110°C for 5–6 hours in a slow air current.

If the sample is crushed to 6-mesh size, as is the case with normal routine samples, two methods suitable for any hard coal may be used; distillation of 100 grams in toluene as above (time about $1-1\frac{1}{2}$ hours) or measuring the weight loss of *10 g* of sample when heated to 105°–110°C for $1\frac{1}{2}$ – 3 hours in an atmosphere of nitrogen. The toluene procedure would only be used very exceptionally as a reference, if it was suspected that oven drying was either causing some volatile loss or that the coal was gaining weight during preparation, by oxidation. Since water is collected in the toluene procedure the result is unambiguous.

12.1.3 Air-dry moisture

To determine the moisture content of the laboratory sample for general analysis (the purpose of this determination will be referred to in Section 13.6 of this chapter) three methods are available: drying *1 gram* in a nitrogen oven for $1\frac{1}{2}$ – 3 hours as above; drying in a vacuum oven in a similar manner; and lastly, a direct weighing of the moisture collected by an absorbent from dry nitrogen gas passed over the coal in a heated tube. For routine analysis the first provides the most convenient method.

It should be noted that when indirect methods based on loss of weight are used, the finely-ground coal in a bone-dry condition will pick up moisture from the laboratory atmosphere, and speedy cooling and weighing are essential.

When coal is heated in air to above 100°C but below ignition temperature, other changes occur besides the loss of moisture:

● A loss of weight due to the evolution of occluded gases and the slow decomposition of the coal.

● An increase in weight due to the formation of solid peroxides when the coal is heated in air. The use of nitrogen to exclude oxygen prevents this.

Both these changes are more marked the higher the temperature and the lower the rank of coal.

12.2 Ash content

Ash is of three types:

● *Inherent ash* — that ash content which cannot be reduced by any method of cleaning. It may be thought of as the mineral constituents of the vegetable matter from which the coal was derived plus the silt on which it grew. Inherent ash is usually constant for a given coal seam and can range from $\frac{1}{2}$% to nearly 20%.

● *Associated ash* — present in the coal seam as bands, lenticles and partings. One form of middlings consists of such mineral matter which has not been split off the coal lumps during mining (see Section 5 of this chapter on coal preparation).

● *Adventitious ash* — not present in the seam, but introduced from floor and roof during cutting, as a result of deliberate policy or because geological weaknesses dictate its removal to ensure a safe working roof. Adventitious ash may be fire-clay or carbonaceous shales from the clay deposited in shallow water when the coal was laid down.

The incineration of coal is accompanied by visible changes, burning of the volatile gases evolved and the slower combustion of the coke residue. There are, however, other less obvious changes taking place — water of hydration is driven off, carbonates (like limestone) break down to oxides and gaseous carbon dioxide, pyrites (iron sulphide) burns to iron oxide and oxides of sulphur, some sulphates may decompose to oxides and sulphur trioxide. The picture is further complicated by the possibility of the oxides from one sample picking up and 'fixing' the sulphur oxides emitted from an adjacent sample.

For these reasons the determination of ash has, to a lesser degree, the empirical nature of the determination of volatile matter, and has to be carried out under standard conditions if reproducible and comparable results are to be obtained.

Two methods are described. In the first a single muffle furnace is used; a known weight of coal (1–2 g) is heated in air to 500°C in 30 minutes; the temperature is then raised to 815°C in a further 60–90 minutes. The sample is heated to constant weight. In the second, two muffle furnaces are required. The sample is heated in one kept at 500°C for 30 minutes; the dish is then transferred to the other kept at 815°C and is heated to constant weight. The second method is preferable in laboratories which are required to test a large number of samples. In each case the furnace must be well ventilated to minimise fixation of oxides of sulphur.

12.3 Volatile matter

Volatile matter is used in most systems of coal classification. An early distinction was drawn between anthracites, gas/coking coals and free-burning steam-raising coals; which was based on their volatile matter contents. The present BC coal rank Code Number links volatile matter with coking behaviour.

It has been emphasised above that the test conditions must be standardised to yield reproducible and comparable results.

The determination is made by heating 1 g of coal in a crucible to 900°C for seven minutes, out of contact with air. The crucible has a lid to exclude air and stands in a wire frame. The volatile matter is calculated from the loss of weight and a deduction is made for the loss due to moisture.

12.4 Calorific value

12.4.1 Introduction

Calorific value is the basic standard of value of any fuel; it is a measure of its heating power and is the primary factor in coal pricing. It is the number of heat units liberated per unit weight of fuel when completely burned in oxygen. SI (Système International d'Unités) units are used in Britain and calorific value is expressed as kilo-joules per kilogram (kJ/kg). Elsewhere in the world thermometric units may be used, calories per gramme (cal/g) or Imperial units, British Thermal Units per pound (Btu/lb). The relationship between these units is:

$$1 \text{ cal/g} = 4.1868 \text{ kJ/kg}$$
$$1 \text{ Btu/lb} = 2.326 \text{ kJ/kg}$$
$$1 \text{ cal/g} = 1.8 \text{ Btu/lb}$$

The other unit of heat often encountered in describing gaseous fuels is the therm, equal to 100 000 Btu, or 105.5 MJ.

When a solid fuel is burned the energy released is taken up by the immediate environment. Some energy will be retained by the combustion gases. The amount will depend upon whether they are allowed to do work and expand, or are constrained to a constant volume. Similarly more energy will be released if the water formed is allowed to condense and give up its latent heat than if it remains as vapour. As a result it is necessary to distinguish between four calorific values:

• Gross calorific value at constant volume.

• Net calorific value at constant volume.

• Gross calorific value at constant pressure.

• Net calorific value at constant pressure.

The laboratory bomb calorimeter determination is the first of these — gross at constant volume; the others may be calculated from it if the composition of the fuel is known. The word 'gross' signifies that the latent heat of evaporation of the water present in the fuel plus that formed during combustion, is recovered by condensing the products; 'net' signifies that the latent heat is lost, the water being discharged as vapour. For a 'typical' Midlands coal (low rank, having 10% moisture and 4% hydrogen) with gross calorific value 24 420 kJ/kg, the net calorific value will be 23 375 kJ/kg.

The latent heat of evaporation is not recoverable under boiler operating conditions and this has led to Continental manufacturers reporting boiler efficiencies based on the lower or net calorific value. Their efficiencies are some 4% higher than the figures based on the higher, gross calorific value. In recent years the efficiencies reported for UK stations have also been on a net calorific value basis. When checking records or data from other countries the basis of the reported thermal efficiencies should be checked.

Combustion of fuel in a bomb differs from that in a boiler; the first is at constant volume, the latter at constant pressure. When the products of combustion are allowed to expand at constant pressure, work is done and the gross calorific value at constant pressure is higher than the bomb-determined calorific value by the heat equivalent of this work. The correction on this account is about 28 kJ/kg.

There are a number of formulae whereby the calorific value may be calculated to a fairly close approximation; it is necessary to know the ultimate analysis (q.v.) but as this is not routinely available, it is simpler and more reliable to carry out a direct determination of calorific value. Formulae also exist for predicting calorific value from the proximate analysis; the results may be in error by as much as ±6% and should be used only as a rough guide when a determination cannot be made.

It *is* necessary to calculate net calorific value from the gross calorific value determined in the laboratory so that the correct figures for heat accountancy may be produced. The formula used is

CV net(p) =

CVgross (v) − 212.1H − 24.4 (M + 0.1A) − 6

where H is the % hydrogen content of the fuel (i.e., excluding the hydrogen content of the moisture M and mineral matter) and

A is the percentage ash content.

Values of hydrogen content are available for all fuels and are periodically checked.

In the past calorific value determinations were time consuming and collieries were small. The calorific value of the pure coal substance (that is dry and free of all mineral matter) is virtually constant for the coal from any given seam in one locality. Once its value was established, this could be used with the total moisture and ash content to calculate the calorific value of any consignment from the same seam. This saved time and was an attractive approach. However times have changed, calorific values are now readily obtained and collieries are much larger. Several horizons may be worked in the same colliery. As production from each face changes, so the average intrinsic calorific value of the coal produced changes, since it is now a variable mixture of differing geological age. Laboratories must therefore routinely determine the calorific value of each sample, in order to obtain accurate data on the heat content of fuels delivered.

12.4.2 Determination of calorific value

There are two basic methods of determining calorific value, the 'isothermal' method and the 'adiabatic'. In each case 1 g of coal is burnt inside a metal 'bomb' in an atmosphere of oxygen. The temperature of the vessel rises as a result and if there were no heat losses from the bomb surface during the process, the energy released by the coal could be readily calculated from the measured increase in temperature. In practice, corrections have to be made for the heat loss from the calorimeter.

In the isothermal method a water jacket surrounds the vessel and is of sufficient mass so that its temperature remains essentially constant during the calorific value determination. The cooling correction is then simplified because the heat loss occurs to a constant temperature heat sink.

In the adiabatic method the surrounding jacket matches its temperature to that of the bomb during the determination. Sensors measure the temperature of the bomb and control heaters in the jacket to achieve this. As a result of there being no difference in temperature between the bomb and its environment the cooling correction is no longer required.

The isothermal calorimeter was generally used until the late 1950s. Thereafter the adiabatic calorimeter became the preferred method for laboratories processing large numbers of samples, since it obtained results more quickly. The advent of microprocessors in the present decade has produced an interesting situation. Rapid analyses may now be produced using the isothermal method, as a result of quick computer calculation of the cooling correction. This gives a mechanically simple system. Adiabatic calorimeter manufacturers have countered the development with equipment which can 'curve-fit', so that the operator does not have to wait for the rise in temperature of the calorimeter to reach its maximum. A result can be predicted accurately $3\frac{1}{2}$ minutes after firing of the bomb, but if necessary the calorimeter can be run normally in order to check results. Both systems are presently in use, with the adiabatic computer-controlled version proving to be particularly reliable.

The basic isothermal and adiabatic methods are described in detail in BS1016 Part 5, but since the heat content of its fuels is of such importance to the electricity supply industry, the methods will be described in some detail here.

12.4.3 The isothermal method

The essential parts of an isothermal bomb calorimeter are shown diagrammatically in Fig 1.35.

A small quantity of the sample is burnt in compressed oxygen inside a stainless steel cylinder, or 'bomb', which is immersed in water contained in a cylindrical calorimeter vessel. This vessel with its contents is situated inside a larger vessel which has hollow cylindrical walls filled with water or insulation to form a 'jacket'. The purpose of this jacket is to minimise heat transfer between the calorimeter vessel and the environment. The heat liberated on igniting the sample by an electrical fuse, is calculated from the temperature of the calorimeter water which is accurately measured before, during and after the rise in temperature resulting from combustion. The heat liberated is essentially equal to the temperature rise multiplied by the effective heat capacity, or 'water equivalent' of the instrument.

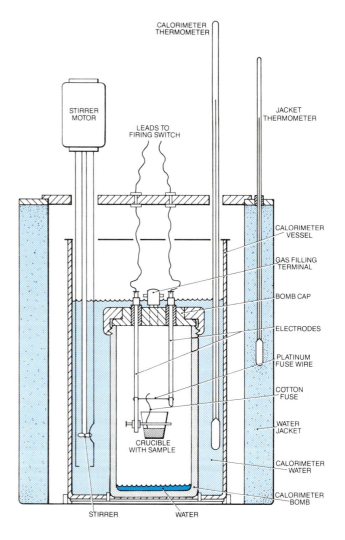

FIG. 1.35 Isothermal calorimeter

Procedure

One gram of the sample, ground to pass a 72-mesh sieve, is accurately weighed into the crucible which is then placed in the ring support attached to one of the electrodes, and a length of platinum wire is fastened to the electrodes, just above the top of the crucible, so that the wire is taut. A short piece of sewing cotton is tied to the centre of the wire and positioned so that a part of the cotton is in contact with the sample. 1 ml of distilled water is measured into the lower part of the bomb, which is then assembled with the cap carrying the electrodes and sample, and the whole is charged slowly with pure oxygen to a pressure of 25 atmospheres ($2.5 \times 10^6 \text{N/m}^2$); 30 atmospheres ($3 \times 10^6 \text{N/m}^2$) may be necessary for anthracite or coke.

Next, the calorimeter vessel is partially filled with a measured quantity of water (preferably distilled), sufficient to cover the upper surface of the bomb cap, and the vessel is then located centrally inside the water jacket. The bomb is placed inside the calorimeter vessel, electrical connections to the electrode terminals are made and the stirrer, calorimeter thermometer, and covers are placed in position. The temperature of the water in the calorimeter vessel is adjusted at the commencement of the procedure so that, when assembly is complete, it is about 1.5°C below that of the jacket.

The calorimeter thermometer, conforming to BS791, is of mercury-in-glass, graduated in 0.01°C and is read with the aid of a magnifying viewer, to 0.001°C.

The stirrer is operated throughout the determination and after 10 minutes of stirring, the calorimeter thermometer is read at intervals of 1 minute for 5 minutes, during which time there should be a uniform gentle rise in temperature. The charge is then ignited* by momentarily connecting the electrode leads to a 6–12 volts source of about 50 watts rating. Readings are continued without interruption at 1-minute intervals as the calorimeter water temperature rises to a maximum, which usually occurs 5–7 minutes after ignition, having risen by approximately 2.5°C. Readings are ended after a further five or six observations, during which a uniform fall in temperature should occur.

The bomb is then removed from the calorimeter, pressure is released slowly, and the bomb is dismantled. It can then be inspected for any sooty deposits on the internal parts and, in the absence of such deposits, combustion is regarded as having been satisfactory. The inner surfaces of the bomb and the internal fittings are carefully washed with distilled water into a beaker for the subsequent determination of the corrections for nitric and sulphuric acids, as mentioned below.

*As a safety precaution the firing key should be operated from a position some distance away from the calorimeter.

Corrections

A number of corrections must be applied in arriving at the calorific value, as follows:

- *Temperature corrections* A 'cooling correction' is calculated as shown in BS1016: Part 5, to allow for heat transfer between the calorimeter vessel and its environment during the main rise in temperature, and a further correction is made for errors in the thermometer. Calorimeter thermometers should be sent for test by the National Physical Laboratory (NPL) regularly — at 3 to 5 year intervals usually — and the thermometer correction is obtained from the test certificate which, at the points of calibration by the NPL, is accurate to $\pm 0.002^\circ$C.

- *Fixed corrections* These are required to correct for heat released by the fuse wire and combustion of the cotton. They are kept to a constant amount, about 80 J for every determination.

- *Nitric acid and sulphuric acid corrections* Some of the nitrogen in the sample and air enclosed in the bomb, is oxidised to nitric acid with the evolution of heat (about 40 J) whereas this reaction does not occur perceptibly in boiler furnaces.

  Additionally, sulphur in the sample is oxidised to sulphuric acid in the bomb, liberating more heat than it would in a boiler, where practically all of the sulphur is converted to sulphur dioxide. The correction here is usually between 60 and 250 Joules per gram of coal, depending on its sulphur content, and is either calculated (together with the nitric acid) from analysis of the bomb washings, or from the sulphur content of the sample, where this has been separately determined.

Mean effective heat capacity of the instrument

In order to complete the calculation of the calorific value, it is necessary to know what weight of water and metal (in terms of an equivalent weight of water), has absorbed the heat evolved during combustion in the bomb. This is obtained periodically as the mean of five determinations carried out as has been described, but substituting 1.2 grams of dry benzoic acid in place of the sample. The benzoic acid, which is a crystalline organic substance containing only carbon, hydrogen and oxygen, is specially purified for this purpose and is obtainable as a 'thermochemical standard' having a calorific value certified by the NPL, close to 26 467 J/g.

Calculation of the result

To summarise the calculation, we have:

$$\text{CV gross (v)} = \frac{(Tc \times H) - C_f - C_s - C_n}{g}$$

where,

CV gross (v) = Gross calorific value at constant volume, kJ/kg (numerically equal to J/g)

T_c = temperature rise in °C of calorimeter water, corrected for cooling and thermometer errors

H = mean effective heat capacity, in J/K

C_f = fixed correction, J

C_s = correction for sulphuric acid, J

C_n = correction for nitric acid, J

g = weight of sample, g

When determining the effective heat capacity, the certified calorific value of benzoic acid CV_{BA} is taken as a standard, the temperature rise and corrections are calculated and a measured weight of benzoic acid is used, g_{BA}; the effective heat capacity H, can thus be evaluated from the formula:

$$H = \frac{(CV_{BA} \times g_{BA}) + C_f + C_n}{T_c}$$

Having determined the mean effective heat capacity, the calorific value of coal or other fuel can be calculated. Sample calculations are given in BS1016: Part 5 1977. The difference between duplicates determined in one laboratory should not exceed 120 J/g and the difference between the means of duplicate determinations from different laboratories should not exceed 300 J/g; an experienced operator usually expects to obtain duplicate results within 50 J/g.

12.4.4 The adiabatic bomb calorimeter

In the determination of calorific value using the isothermal calorimeter, the magnitude and accuracy of the cooling correction depend on the temperature difference between the calorimeter water and the jacket, and it is not always convenient to control this so as to produce a small cooling correction, particularly when several determinations are to be made in one day. Further, the calculation of the cooling correction takes time, as does the need to read temperatures carefully at one-minute intervals during 15 to 20 minutes. For these reasons the 'adiabatic' bomb calorimeter, which eliminates the cooling correction, was developed (see Fig 1.36).

The innovation of the adiabatic calorimeter was in the provision of the means to change the jacket temperature rapidly, particularly to increase it. This is done by installing electric heaters in the jacket through which water is re-circulated. The uppermost cover of the calorimeter is also made in hollow form and the jacket water is passed through this in turn. Sensitive thermistors are placed, one in the calorimeter water and one in the jacket, and these control the jacket heaters

via a bridge circuit and relay, so that as the calorimeter water temperature rises, the jacket water is heated rapidly to the same temperature. In this way there is a negligible heat transfer between the calorimeter and the jacket, and it is only necessary to read the temperature accurately just before firing (when initial equilibrium has been reached), and again at a convenient time after the final equilibrium temperature has been attained. Calculation of the result is the same as has just been described, modified only by deleting the cooling correction.

As mentioned earlier, microchip technology has been applied to both the isothermal method, to allow the cooling corrections to be rapidly calculated, and to the adiabatic method, allowing cycle time to be reduced by predicting the outcome of the temperature rise and in each case the microprocessor also calculates the results. Examples of each modern version are given in Figs 1.37 and 1.38.

12.5 Sulphur

The determination of sulphur is part of the ultimate analysis of coal (q.v.) but is discussed separately because of its special importance in pricing, as well as technically. Sulphur in coal is found in three combinations:

● Sulphate sulphur (negligible).

● Organic sulphur (0.8% average and fairly constant).

● Pyritic sulphur (0.8% average but very variable).

Organic sulphur tends to be relatively constant at 1% of the coal substance but pyritic sulphur varies widely. Thus the minimum sulphur in coal as received of 0.8%, is all organic. The extra sulphur above this amount is almost all pyritic.

Sulphate sulphur exists as small quantities of ferrous sulphate ($FeSO_4 \cdot 7H_2O$) from the oxidation of iron pyrites (FeS) and as gypsum ($CaSO_4 \cdot 2H_2O$). They are found as thin plates in partings of the coal where the solution has evaporated.

Organic sulphur is combined with the carbon and nitrogen to form the coal substance. In consequence it is not removable by cleaning.

Pyrites (also marcasite) is ferrous sulphide (FeS). It occurs as massive nodules and lenticles, as 'brassy' bands, and as finely dispersed particles. The first form is easily removed by the various coal cleaning methods. Pyrites more closely associated with the coal is removable only with considerable loss of saleable carbonaceous material.

Methods are available for assessing the amount of each of the forms of sulphur but usually they are not important for power station purposes. Total sulphur may be determined in the bomb-washings from the determination of calorific value; this is a simple procedure and is often adopted since the result is also

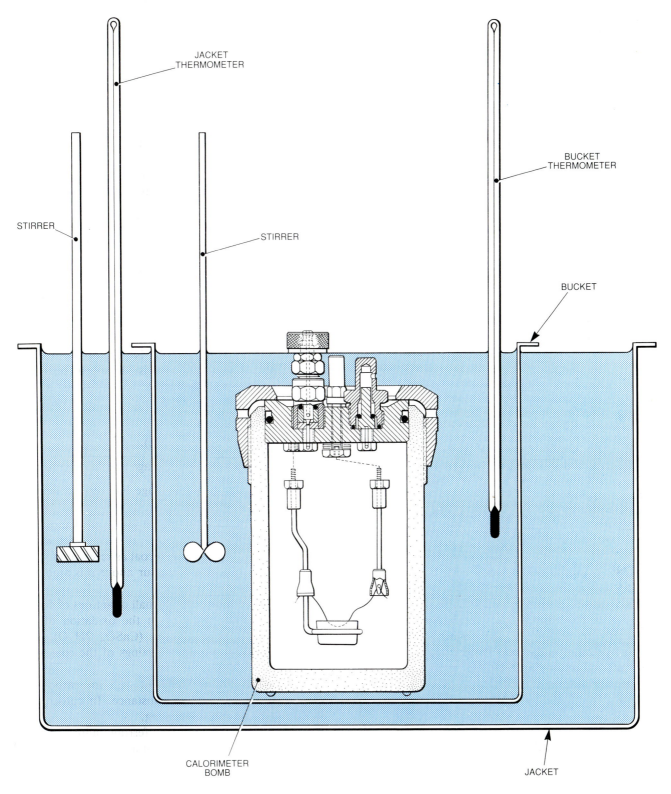

FIG. 1.36 Adiabatic calorimeter

required for the calorific value calculation. If total sulphur has to be determined, and the number of samples is not large, the 'Eschka' or bomb method is used; large numbers of determinations are more conveniently carried out by the High Temperature (Tube) method (see BS1016 Part 11).

In the Eschka method, 1 g of coal is burned by heating in a crucible with a mixture of magnesium oxide and sodium carbonate. This alkaline mixture retains all the sulphur as soluble sulphates which are determined by the standard chemical method of precipitation as barium sulphate, which is finally dried and weighed.

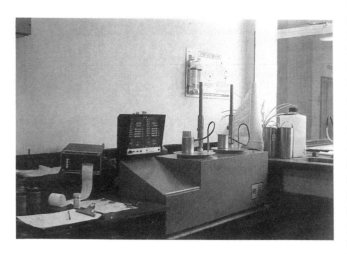

FIG. 1.37 Modern isothermal calorimeter

The High Temperature (Tube) method requires costly equipment which can be made largely automatic and needs little special training to operate. The justification for its use depends on the number of sulphur determinations to be carried out, or on results being required more rapidly than can be provided by the other methods. The coal sample (0.5 g mixed with alumina) in a small refractory 'boat', is pushed into a tube kept at 1350°C through which passes a rapid current of oxygen; the sulphur oxides formed are absorbed as the oxygen from the tube is bubbled through hydrogen peroxide, which converts them wholly to sulphuric acid. The acid is measured by titration with standard alkali solution, a correction being applied for chlorine carried forward by the oxygen. (This is also a method for the estimation of chlorine (q.v.).)

The high temperature method, using titration of sulphuric acid and hydrochloric acid, can be speeded up further for the measurement of sulphur. The combustion gases are passed through an infra-red detector cell, set for the absorption of the S-O band, and the concentration of sulphur is obtained instrumentally, i.e., no operator judgement is necessary. This is now the preferred routine procedure. If checks are required the Eschka method is used.

12.6 Chlorine

Chlorine has been regarded as a significant constituent of coal since about 1939 when a correlation between chlorine content of the coal and the degree of resulting gas side boiler fouling was observed. This aspect is discussed, with ash and sulphur contents, in a later section.

When BC proposed increasing the tonnage of high chlorine coal produced in its Western Area (1979/80), the behaviour of chlorine in coal was re-examined in the CEGB. Several established views of how chlorine existed in coal were shown to be incorrect. In the first instance the view that a significant amount of the chlorine was present as sodium chloride was found not to be so. If chlorine-containing coal is ground finely and heated to about 240°C, the chlorine volatilises. Clearly it is not present in inorganic form. When coal is examined under an electron microscope fitted with an energy dispersive X-ray analysis facility, the chlorine is found to be concentrated in different coal macerals to the sodium which is present. It is likely that chlorine is held in coal by its ion-exchange capacity. If the coal is ground finely, so that ready access to the solids is possible, then the chlorine can be completely removed by washing, along with balancing anions such as sodium. The resulting fine wet coal would not be a suitable fuel, however, and brine disposal would present problems, so large-scale washing to remove chlorine is not contemplated. Existing washeries at coal preparation plants deal with large coal and recycle the water so that little reduction in chlorine is achieved.

From an operational viewpoint the above observations have two consequences. Boiler fouling is due to low melting deposits which stick to metal surfaces. High concentrations of alkali metals in coal ash can lower its fusibility. It used to be thought that chlorine was a good indicator of the presence of alkali metals and hence fouling propensity, it being assumed that the chlorine arose from brines and therefore remained as sodium chloride. It is now realised that sodium/chlorine ratios can be very different from 1.0 and if boiler fouling is being studied actual sodium and potassium contents in the ash should be determined. Chlorine on the other hand contributes to furnace wall corrosion and is measured for this reason.

The methods available for estimation resemble those described for sulphur, namely the Eschka and High Temperature (Tube) methods (BS1016: Part 8). A rapid convenient procedure recently developed in the CEGB is to combust a sample in a calorimeter bomb and to then measure the chlorine content of the bomb washings using a specific ion electrode. This is now used routinely to monitor all coal varieties. The Eschka method is again used as a reference if required.

12.7 Instrumental methods

12.7.1 Instrumental methods for monitoring colliery output

If a colliery produces coal from a single horizon the dry-mineral-matter-free calorific value of the fuel will be fairly constant. Measurement of the inert matter in the coal, i.e., moisture and mineral content, should then allow the calorific value to be determined by difference.

Ash or mineral matter may be determined by X-ray back-scattering instruments or by X-ray fluorescence. In the first case a coal sample is obtained as described in Section 9 of this chapter. This is then crushed to about 6 mm and fed through the device (AERE, Gunson Sortex) on a continuously rotating table. The coal is irradiated by gamma rays, typically from an

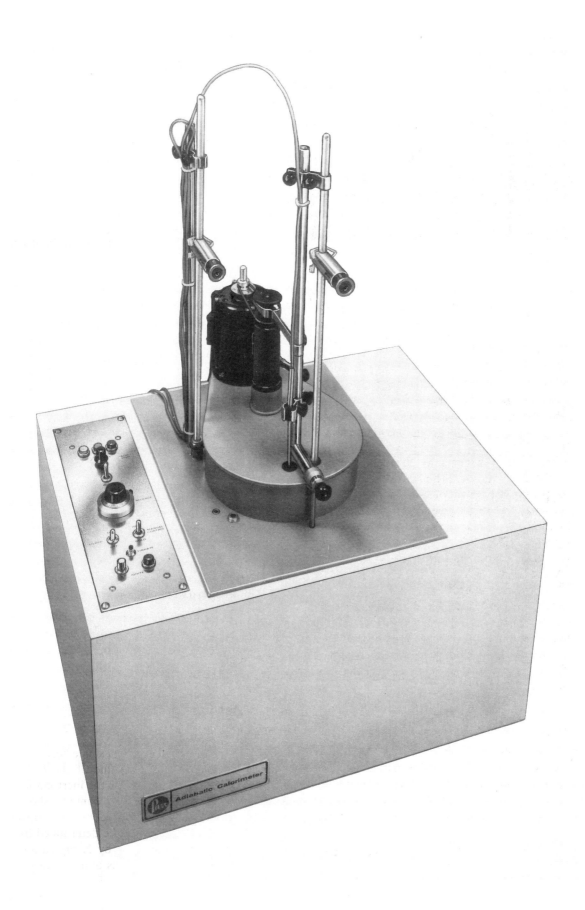

FIG. 1.38 Modern adiabatic calorimeter (Parr Instrument Company)

americium–241 source, and the amount of scattering is proportional to the number of heavier atoms in the material. The instrument requires calibration against ash figures obtained by British Standard laboratory analysis, but then gives adequate data to control the coal preparation plant. In an alternative approach (Telsec) the coal is quickly prepared to give a 72-mesh analysis sample, as described in Section 10 of this chapter, and a portion (about 20 g) is then irradiated to induce the elements comprising the mineral matter to emit characteristic X-ray fluorescent radiation. The emissions are measured and converted to an ash figure. Again checks against standard laboratory analysis are necessary but the procedure has proved of adequate accuracy and allows data to be sent to the colliery within a few hours (i.e., results will be obtained within the shift that the sample was obtained).

Moisture may be measured by devices which monitor the attenuation of a microwave signal passing through the coal stream. The attenuation is caused by the hydrogen to oxygen bonds predominantly and a good calibration can be obtained. Devices are available which measure the moisture of the bulk coal stream on the conveyor.

Moisture measurement may also be made on a laboratory sample (6-mesh) using nuclear magnetic resonance (Newport Instruments). This is a very accurate method except when the coal contains magnetite or is partially conducting (contains coke). As magnetite occasionally escapes from preparation plants erratic results can sometimes be obtained. The favoured methods at collieries are the on-line moisture method and the continuously sampling X-ray back-scatter method as these give more immediate response for the colliery preparation engineer.

12.7.2 On-line instrumentation for power stations

Power stations receive coals from many collieries and the equipment described above would need several calibrations to be applied to suit each quality delivered. This is not practical and devices which can cope with a variable mineral content are required. Ideally these should work on the full coal stream.

The industry is currently taking an active interest in several instruments although none have yet been purchased.

The simplest instrument on offer irradiates coal on the belt using an americium source and measures back-scattered gamma rays. It would be of use at a power station linked to a single colliery because only one calibration would be necessary. There is a further drawback, however, that only gamma rays emitted from the upper part of the coal on the belt are observed. If the coal is not adequately mixed an incorrect result will be obtained. A better device uses a stronger source above the belt and measures the forward scatter below the belt, so that the full coal stream is observed. As with any X-ray scattering method it requires calibration.

More sophisticated instruments use a californium–252 neutron source to excite prompt gamma emission, characteristic of each element in the mineral matter. These hold some promise as they may be able to cope with a variety of coals. They differ slightly in design. Some use sodium iodide (thallium-doped) detectors, others, high purity germanium. Most utilise a mechanical sampler to provide a reduced coal stream (6–30 tonne/h) to the instrument but there is one instrument capable of coping with 500 tonne/h and this should be suitable for power station use. Any of the above would require to be operated alongside a moisture measuring device, probably of the microwave type, to ensure that an 'as-received' analysis is obtained.

On-line instruments will detect anomalous deliveries and this should help power station operations. If excess air levels are set for 16% ash coal, for example, and a coal of high ash content arrives, then unnecessary stack losses and reduced steam output will occur. Similarly if a very low ash content coal arrives, sufficient air may not be provided and combustion will be incomplete.

However, the consistency of railborne supplies from large collieries is generally good and the stations know the average quality of each coal variety and its variability. As a result the amount of additional information gained from the on-line instrument may not be great. Furthermore, as mentioned in Section 17 of this chapter, boiler control is exercised by monitoring carbon monoxide and oxygen in the flue gases, to respond to changes in fuel quality. Nevertheless, early warning of unusual deliveries would be helpful and the use of instrumental coal analysis at the power stations will be carefully evaluated as the proposed methods are refined.

13 The ultimate analysis of coal

Strictly the ultimate analysis of coal should consist of its percentage elemental composition. In practice its fuel properties are adequately defined if the moisture, carbon, hydrogen, sulphur, nitrogen, oxygen and ash content are determined. Results may be reported on an ash-free or mineral-matter-free basis. For practical purposes the former is useful, for scientific use the latter is preferred.

The ultimate analysis is used in combustion calculations and heat balances. As already stated, calorific value can be calculated to a fairly close approximation from the ultimate analysis, but more importantly a knowledge of the hydrogen content of a coal is required in order to calculate its net calorific value.

13.1 Carbon and hydrogen

The determination of carbon and hydrogen depends upon combustion of the coal to give carbon dioxide and water, which are then estimated. The methods used are continually improving. Traditionally the water vapour and carbon dioxide are absorbed in weighing tubes and determined gravimetrically. This is slow but accurate. Modern high temperature procedures determine the two products by means of an infra-red detector. This has increased the speed of the method and allowed nitrogen to be determined also.

Two methods are described in BS1016 Part 6, combustion at 800°C (classical Liebig) and at 1350°C (High Temperature (Tube) method). In both methods the calculated result is 'total carbon', that is, it includes carbon from coal as well as from any carbonates present and a correction is needed at a later stage. In each the hydrogen is converted to water which is absorbed and weighed together with the water present, as moisture in the coal and as water of hydration, in the associated mineral matter and shale. In this case corrections have to be applied for these interfering factors.

In the low temperature method 0.3 g of coal is burned in a slow stream of oxygen in a heated tube 1250 mm long, capable of close temperature control in zones. The tube is packed so that the products of combustion pass through a plug of copper oxide (450 mm long) kept at 800°C and then successively over a lead chromate plug (100 mm long) and then a 25 mm roll of silver gauze, both kept at 600°C. The copper oxide ensures that all the carbon is converted to carbon dioxide and all the hydrogen to water. Oxides of sulphur are removed by lead chromate and chlorine by silver gauze.

The products of combustion carried by the stream of oxygen pass through an absorption train consisting of tubes which are packed with an absorbing substance. Water is first removed by magnesium perchlorate and measured by the increase in weight. Traces of nitrogen dioxide which would lead to erroneous results for carbon are then removed by manganese dioxide. Finally, carbon dioxide is absorbed by soda-asbestos; again the increase in weight of the absorption tube measures the carbon dioxide evolved.

The high temperature method differs from the Liebig method in having a shorter unpacked tube (only 650 mm long), using a rapid stream of oxygen, and operating at 1350°C. At this temperature both sulphur and chlorine are trapped by a roll of silver gauze; nitrogen dioxide is not formed under these conditions and no special precautions are therefore necessary. The assessments of water and carbon dioxide produced can be made as in the low temperature method, or can be carried out by automatic instruments if speed is a necessary factor.

13.1.1 Carbon present as carbonates

The carbon dioxide content of the mineral carbonates of a coal is required:

- To correct the determined % carbon when calculating to a dry, mineral-matter-free basis, and for calculating the combustible carbon.

- When calculating the volatile matter of low volatile coals and anthracites of high carbonated content, to a dry, ash-free basis.

- When calculating the volatile matter of a coal to the dry, mineral-matter-free basis.

The analysis is carried out in the laboratory by dissolving a known weight of the analysis sample in hydrochloric or phosphoric acids and absorbing the resultant carbon dioxide evolved. The increase in weight of the absorbent is directly due to the carbon dioxide liberated.

13.2 Nitrogen

During combustion of coal, nitrogen oxides are produced, which play a role in the chemical processes involved in air pollution. Depending upon the combustion temperature, about 40% of the nitrogen oxides are thought to come from the nitrogen contained in the coal (in the form of organic compounds) and 60% derives from nitrogen in the combustion air. A knowledge of the nitrogen content of the coals burnt is helpful in trials aimed at lowering the amount of nitrogen oxides produced in power stations.

The standard way of determining nitrogen in coal is the Kjeldahl method. This involves heating the coal with concentrated sulphuric acid to destroy the organic material; the presence of a catalyst (selenium, mercuric sulphate, vanadium pentoxide) ensures that all the nitrogen is rapidly converted to ammonium sulphate. When cool, excess sodium hydroxide is added to free the ammonia, which may then be steam-distilled. It is collected in the condensate for estimation by titration with standard acid solution.

The estimation may be carried out, with appropriate differences of detail, on 1 gram (macro-method) or 0.1 gram (semi-micro method) of coal. In practice because of the shorter times of digestion ($\frac{1}{2}$ hour as against 2 hours) and the smaller scale of the apparatus, nearly all laboratories use the semi-micro method.

A quicker procedure is associated with the carbon and hydrogen method already referred to. The high temperature combustion of the coal sample in a stream of helium and oxygen produces carbon dioxide, water and a mixture of nitrogen and nitrogen oxides. The first two gases are determined using an infra-red detector. The nitrogen compounds are passed over a bed of finely divided copper. This abstracts oxygen from the nitrogen oxides so that the only net product of the combustion is nitrogen. This is readily detected in the stream of helium carrier gas by use of a thermal conductivity detector. The latter is a hot wire carrying an electric current. When only helium is passing it loses heat at a steady high rate. Nitrogen conducts heat less well, so when passing the wire, its heat loss is less. The wire temperature rises and consequently its electrical resistance increases. The change in resistance is proportional to the nitrogen produced from the sample.

The latter method is quick and of adequate accuracy. This allows more data to be produced on the ultimate analysis of fuels than ever before. If this rapid convenient method needs to be checked, the BS1016 standard procedure is used, in this case the Kjeldahl method.

13.3 Phosphorus

Phosphorus occurs in British coals at levels between 0.005% and 0.15%. High phosphorus coals can be very troublesome in stoker-fired plants, especially in economisers where a slow but continuous build-up of hard deposits occurs. On gilled tubes a heavy draught loss occurs, which is sufficient to reduce the outputs of a boiler. The deposits are the result of chemical attack and digestion of fly ash particles. They are not removed by water washing, while chipping accessible parts is slow and laborious. The deposits, which always contain a high proportion of associated boron compounds, can be dissolved by alternately spraying with inhibited hydrochloric acid and hot caustic soda. This method has its difficulties and dangers and is only employed when fouling is exceptionally severe. A side-effect is that the cast iron gills suffer corrosion. As phosphorus compounds are only volatilised from deep fuel beds with high temperatures, the element causes no trouble with PF-firing.

For measurement, a known weight of coal ash is digested with a mixture of nitric and hydrofluoric acids to extract the phosphorus, which is precipitated under controlled conditions as ammonium phosphomolybdate. This is dissolved in an excess of standard sodium hydroxide solution and the amount of alkali remaining is titrated with standard nitric acid.

13.4 Arsenic and selenium

These two elements are present only in small amounts in British coal; arsenic, for example, is usually less than 10 PPM. It is necessary to determine their amount in coal only very occasionally. They become of interest in connection with ash disposal and it is usually the ash rather than the coal which is analysed. There are World Health Organisation limits on the allowable levels of arsenic which can drain from ash dumps into surface water and aquifers. Selenium on the whole is beneficial when it enters herbage growing on ash-recovered lands. In general, British soils have relatively little selenium and farmers may have to use animal feed containing added selenium. However, it can be troublesome to animals in too large amounts, so ash selenium levels are checked from time to time in connection with ash disposal schemes.

If coal is to be analysed it is subject to a wet digestion to remove combustible material as both arsenic

and selenium are relatively volatile and would be partly lost if the coal was ashed in a furnace. The resulting solution may then be reduced with a strong reducing agent to give arsine and hydrogen selenide. These gases are carried by a stream of nitrogen to an atomic absorption furnace where they can be determined with good accuracy. In an alternative method, the two elements are taken out of solution with a 'carrier' precipitate of tellurium. The precipitate collected is then placed in an X-ray spectrometer where the arsenic and selenium are conveniently determined.

A 'classical' procedure is described in BS1016 Part 10.

13.5 Other trace elements and minerals

The eight elements normally determined in the mineral matter associated with coal are silicon, aluminium, iron, titanium, calcium, magnesium, sodium and potassium. They certainly are not measured routinely, but if the ash is suspected of causing slagging or fouling in a boiler, then the ratio of the acidic oxides (for example, aluminium and silicon) to the basic oxides (calcium, magnesium, iron, sodium and potassium) can indicate its high temperature behaviour (see Sections 16.3 and 16.4 of this chapter). The coal is reduced to ash in a furnace and the elements are determined by placing the ash in an X-ray fluorescence spectrometer.

Amongst trace elements, boron is determined occasionally, but as with arsenic and selenium it is of interest in connection with ash disposal rather than for operational reasons although in the past, with stoker-fired plant, it gave hard economiser deposits.

A range of other elements is occasionally determined in connection with worries about environmental pollution. Most known elements may be found in coal but few are present in significant levels. In the past there was some interest in germanium recovery from 'fly-ash', i.e., the very finest ash particles, but modern semiconductor technology uses silicon and gallium arsenide devices; germanium devices are no longer fashionable.

13.6 Reporting of results

The results obtained from the tests discussed above will be on the basis of the air-dried condition of the laboratory sample. It is usual to convert them to the condition of the coal as it was sampled, i.e., with its full moisture content.

It is also possible to convert them to the 'dry' basis and to the 'dry ash-free' basis (DAF), in order to be able to compare results under standard conditions. For example, it may be required either to check whether a result for a particular coal is in line with the expected value, or to study the consistency of a single product (from one colliery) over a period of time. The 'dry' basis can be used to study variations in ash; the DAF basis to study variations in volatile matter and calorific value.

Unfortunately, the DAF calorific value can never be constant for any coal because the value changes with its ash content. However, the change can be quite regular and therefore serve as a guide to the usefulness of a result, or as to whether inconsistent results can be expected from a colliery. The following is an indication of how the DAF calorific value changes with ash content for a single colliery product. The results are illustrated in Fig 1.39.

Dry ash	Dry calorific value calculated from tests	DAF calorific value calculated
%	kJ/kg	kJ/kg
9.6	30 610	33 867
13.8	29 052	33 704
16.6	27 935	33 494
18.4	27 307	33 471
19.1	27 005	33 425
22.4	25 726	33 169
26.3	24 306	32 983

The reason for the progressive change in DAF calorific values is that ash in coal and the mineral matter originally present are not the same thing. The ash weighs less than the mineral matter from which it was derived, because of chemical reactions occurring during the combustion of the coal. The loss of water of hydration from the mineral matter is a predominant feature, accounting for a loss of almost 10%, so that as a broad approximation the weight of ash obtained from coal is about nine-tenths of the weight of the original mineral matter present.

Examination of Fig 1.39 shows in this case that the ash amounts to 90.7% of the mineral matter since this is the point where, by extrapolation, there is no heat content left in the material. Extrapolation to the point where there is no ash left in the substance gives a value of 34 234 kJ/kg for the dry mineral-matter-free calorific value of the coal. This parameter is a constant for a particular coal seam, providing no local oxidation has taken place.

If mineral matter could be used in the conversion calculations in place of ash, it would appear that the basis for comparisons would be improved. The best formulae for obtaining an approximation to the original mineral matter content of a coal sample require the determination of more constituents of the coal than are made in most routine checks; and unless the coal is low in sulphur, chlorine and carbonate content, the use of the more simple formulae give little or no advantage over the DAF method. Formulae for calculating mineral matter are included below:

Symbols

M	= % of total moisture as sampled
M_1	= % of moisture in the analysis sample
A	= % of ash in the analysis sample

Symbols

MM	= % mineral matter in the analysis sample
Cl	= % chlorine in the analysis sample
CO_2	= % carbon dioxide (from carbonate) in the analysis sample
Sp	= % pyritic sulphur in the analysis sample
S_a	= % ash sulphur in the analysis sample
S_s	= % sulphate sulphur in the analysis sample
S_t	= % total sulphur in the analysis sample
K	= a constant

'As sampled' basis

Multiply the % of the constituent estimated in the air dried analysis sample by:

$$\frac{100 - M}{100 - M_1}$$

'Dry' basis

Multiply the % of the constituent estimated in the analysis sample by:

$$\frac{100}{100 - M_1}$$

'Dry ash-free' basis

Multiply the % of the constituent estimated in the analysis sample by:

$$\frac{100}{100 - (M_1 + A)}$$

'Dry mineral-matter-free' basis

For calculation to the pure coal substance basis, the mineral matter content may be obtained from the following:

(a) $MM = 1.13A + 0.5Sp + 0.8CO_2 - 2.8S_a + 2.8S_s + 0.5Cl$

(b) $MM = 1.10A + 0.53S_t + 0.74CO_2 - 0.32$

(c) $MM = 1.1A + K$

Whenever possible, method (a) should be used. Method (c) is crude but effective for coals low in CO_2, Cl and S. For instance, for many South Wales coals it works well when K is taken as 0.2, 0.3 or 0.4.

The dry mineral-matter-free basis is calculated by multiplying the constituent estimated in the analysis sample by:

$$\frac{100}{100 - (M_1 + MM)}$$

Note that when the constituent being calculated is volatile matter of the coal substance, it is essential to make a deduction for the volatile constituents (such as CO_2 in carbonates) of the mineral matter, as shown in BS1016 Part 16.

Table 1.15 shows a coal analysis reported on various bases.

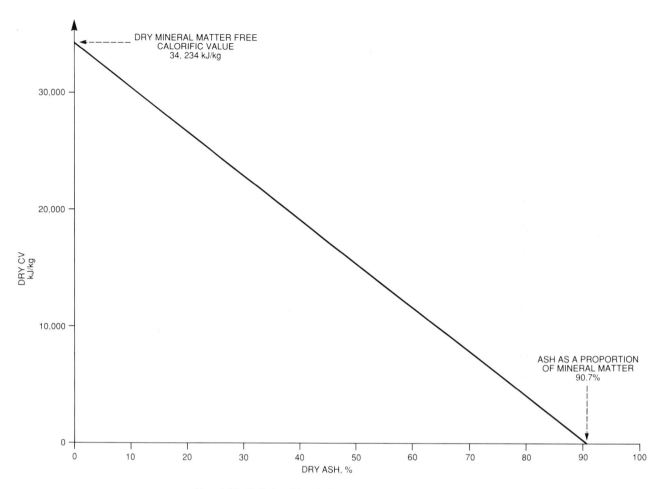

FIG. 1.39 Relationship of dry ash to dry calorific value

14 Testing of coal

In addition to determining the chemical composition of coal, its physical properties need to be determined. There are six main tests:

- Swelling Index.
- Gray-King coking test.
- Bulk density.
- Grindability.
- Abrasivity.
- Handleability.

14.1 Swelling Index

The swelling property of a coal indicates its behaviour as a fuel for stoker-fired boilers (see Section 4 of this chapter). For PF-firing, the index is not so important and it is no longer routinely determined for power stations needs. High-swelling coals lead to uneven and patchy fire beds on grates and water conditioning is essential for good combustion. This property also leads to difficulty in obtaining sufficient air flow through the fuel bed, while the burn-off of the massive cokes produced is slow and usually far from complete.

In the test, one gram of air-dried 72-mesh coal is heated under standard conditions in a closed crucible so that a temperature of 800°C is attained after $1\frac{1}{2}$ minutes, and 820°C within $2\frac{1}{2}$ minutes. After $2\frac{1}{2}$ minutes heating or when the flame of the burning volatile matter has died out, whichever is the longer period, the crucible is cooled and the coke button is compared with a series of standard profiles numbered 1 to 9 (Fig 1.40). If the residual coke is not adherent but is a powder, it is designated: Swelling Number, Zero.

14.2 Gray-King coking test

Devised originally for the gas industry, this test, together with the volatile matter, is the basis for the BC Coal Rank Code Number (Section 3.3 of this chapter). Both the Gray-King and Swelling Index tests involve caking phenomena — particle adherence, softening, pyrolytic swelling and shrinkage.

The Gray-King type is identified by the letters A to G (this last is subdivided into 11 sub-types). It is measured by heating 20 g of coal sample in a tube out of contact with air. Starting at 325°C the temperature is raised during 55 minutes to 600°C and maintained for a further 15 minutes. The coke formed is classified

TABLE 1.15

A coal analysis reported on various bases

Property	As received basis	Air-dried basis	Dry basis	Dry, ash-free basis	Dry, mineral-matter-free basis
Proximate analysis					
Moisture	13.0*	9.1	–	–	–
Volatile matter	32.8	34.3	37.7	41.6	40.3
Fixed carbon	46.1	48.1	52.9	58.4	59.7
Ash	8.1	8.5	9.4	–	–
Mineral matter	9.8	10.2	11.2	–	–
Calorific value (kJ/kg)	26 284	27 470	30 215	33 332	33 913
Sulphur (total)	1.69	1.77	1.95	–	–
Sulphate sulphur	0.08	0.08	0.09	–	–
Pyritic sulphur	0.77	0.80	0.88	–	–
Organic sulphur	0.84	0.89	0.98	–	1.10
Sulphur in ash (as sulphur in coal)	0.25	0.26	0.29	–	–
Carbon dioxide	0.47	0.49	0.54	–	–
Chlorine	0.58	0.61	0.67	–	0.19‡
Phosphorus	0.035	0.037	0.041	–	–
Ultimate analysis					
Moisture	13.0	9.1	–	–	–
Ash	8.1	8.5	9.4	–	–
Carbon	63.4	66.3	72.9	80.5	82.0
Hydrogen	4.3	4.5	5.0	5.5	5.5
Nitrogen	1.3	1.4	1.5	1.7	1.7
Sulphur (combustible)	1.4	1.5	1.7	1.8	1.1$^\phi$
Difference (oxygen, etc.)	8.5	8.7	9.5	10.5	9.7

Results are expressed as percentages unless otherwise indicated

* Free moisture 4.3, inherent moisture 8.7

‡ Assumed organic chlorine, i.e., in coal substance

ϕ Organic sulphur

by comparing it with photographs of a standard series of coke types. Type A coals do not form adherent coke, type C do and swell the most. If the coke is very swollen (higher than G2) the test is repeated with the coal mixed with a suitable quantity of inert electrode carbon. The minimum amount of electrode carbon needed to prevent swelling above the original volume of coal and electrode carbon mixture, is a further definition for these highly caking coals. The G sub-type figure indicates the number of parts of electrode carbon in 20 parts of coal/carbon mixture required to give a G-type (or standard) coke or carbonisation.

14.3 Bulk density

The bulk density of coal is straightforward to measure when it is in wagons or small heaps. In the former, if the surface is levelled, the density is the ratio of net weight to the volume calculated from the wagon dimensions and the levelled coal surface.

The same principle applies to small heaps of coal, but here the coal is removed and weighed and the volume of the hole is estimated. A square-sided hole is dug and then lined. It is filled with standard sand to the same level as the coal removed and the volume of sand required to fill the hole is noted.

Large stockpiles of perhaps 1 million tonnes present a more difficult task. They are formed by spreading the coal in layers and therefore can have an onion-like structure. Different layers have slightly differing properties. Density is measured using a gamma-ray backscattering device. A hole is drilled into the stock, usually with difficulty, and a tube is inserted. The gamma-ray source is lowered into the hole and the amount of scattering at different levels is noted. This can be related to the bulk density of the coal. If several holes are drilled and measurements made at more than one level, precisions of ±3% or better are obtainable.

The extent of scattering is proportional to the ratio of atomic number (Z) to atomic weight (A) of the scattering elements in the stockpile. For most elements Z/A = 0.5, for example, in the case of carbon Z = 6 A = 12. One element, hydrogen Z = 1 A = 1 does not fit this rule. As a result differences in moisture content between the layers of coal in the stockpile can affect the results significantly. This has to be taken into account when carrying out a stock survey.

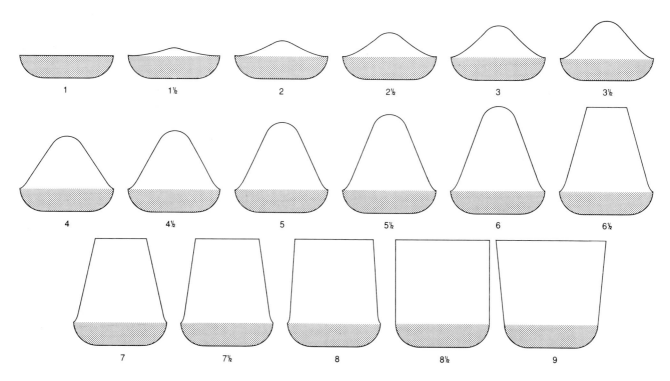

FIG. 1.40 Crucible swelling index profiles

14.4 Grindability

Most coal consumed in power station boilers is pulverised for firing. It is desirable to know how the properties of the coal and mineral matter affect the grinding process, how easily the fuel can be reduced to the required fineness and what power is needed to do this.

The best known tester for grindability is the Hardgrove machine which is a miniature ring-ball type pulveriser. The grindability index from this machine was originally based on the increase in surface area produced on the coal particles, but because its measurement was tedious a simplified procedure was introduced (ASTM Standard D409: 1971).

A 50 g air-dried sample, passing a No 16 US sieve and retained on a No 30 sieve (1.2 mm and 0.6 mm respectively) is ground for 60 revolutions at not more than 20 r/min with 29 kg loading on the ball-ring. The percentage amount of coal retained on a No 200 US sieve (75 micron) is measured (R) and the undersize material is weighed (W) to ensure that losses are acceptably low (< 0.02%). Although W could be used directly as a measure of the ease with which the coal was ground, in order to retain compatibility with the original index based on surface area it is adjusted by the formula:

$$\text{Hardgrove Index} = 13 + 3.465 \, (100 - R)$$
$$\text{or} = 13 + 6.93 \, W$$

where R is the % retained on a 75 micron sieve

W is the weight (in grams) passing a 75 micron sieve.

The apparatus used is illustrated in Fig 1.41. The index can range from 100 for easily ground coals (rank 200–300) down to less than 50 for more difficult coals. As BC rank is largely based on volatile matter there is also a link between the index and this parameter (Fig 1.42). An index of 54 is typical for most power station coals presently used and to a first approximation this figure means that 54% of the amount of coal can be milled per hour compared to a soft coal with an index of 100.

As a footnote; the BS1016 Part 20 version of the Hardgrove measurement does not rely on the ASTM formula given above. In 'Round-Robin' tests conducted in the UK considerable variability has been found between machines of nominally identical design. As a result there is now a UK reference machine and coals of known Hardgrove Index are prepared which are used to calibrate other individual machines. The test is not without its critics. The coal size used, 16/30 mesh, is artificial and not all of the coal is taken to final PF size. However, since the criticisms were voiced (Fitton, Hughes and Hurley, J. Inst. F. 1956 p 54 and Callcot, J. Inst. F. 1956 p 207) no better index has been proposed and it continues to be a useful general guide to a coal's likely milling performance.

14.5 Abrasivity

Hardgrove Index gives a guide to how coal will be reduced in size during milling but does not indicate to what extent the mill will wear. The Yancy-Gear and Price test is used for this purpose (BS1016; Part 19 1980). A sample of coal is placed in a bowl, containing

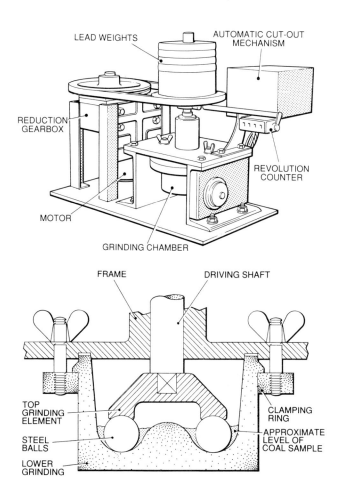

FIG. 1.41 Hardgrove grindability testing machine

a four-bladed paddle wheel (Fig 1.43). The paddles are demountable. For the test four paddle blades are prepared, $38 \times 38 \times 11$ mm made of carbon steel (type 060 A15) with a Vickers pyramid harness of 4160 ± 15. These are weighed before and after the test. The loss in weight of the four metal blades gives an index of grindability for the coal, in terms of metal loss in mg per kilogram of coal. For normal coals, say rank 900, an index of 20 is found but if coke, for example, is tested, figures in excess of 400 mg/kg are observed. For this reason coke is rarely knowingly purchased for use in pulverised fuel fired stations.

14.6 Handleability

14.6.1 In normal weather

The handling of coal has been referred in Sections 7.1 and 8 of this chapter. Occasionally handling problems may prove difficult to resolve. For example, a coal containing a large proportion of dry fines may prove easy to handle at the colliery; when tipped at the power station however, at the same time as a wet coal, the extra moisture acquired may cause it to become sticky and difficult to handle. In some cases a change of $\frac{1}{2}\%$ in moisture content can change a coal from easy to difficult to handle. A test procedure which can predict handleability is useful in solving such problems. There are several devices based upon flow through a conical hopper.

The Durham handleability testing machine developed by BC resembles most others. (A detailed description of its construction, use and test results is given by Hall and Cutress, J. Inst. F. 1960 p 63.) This type consists of a vibrating stainless steel conical hopper fitted with a slide on the 112 mm outlet. Twelve kilograms of coal (in the undried condition) is loaded, the slide door opened and the time to empty all the coal is noted. The mean of three tests is reported. Hall and Cutress report that the rate of flow for a given coal decreases as its free moisture increases, reaching a minimum between 12–17%. Thereafter flow rates increase as the coal tends towards liquid consistency. The time of flow is increased by a rise in the percentage of super fines (through 30-mesh sieve) present, but for a given size range the free moisture at the slowest flow rate varies with coal rank: 17% for rank 300–400 and 12% for rank 900. They also note that handleability is improved by oiling coal with small amounts (less than 1%) of gas-oil. The method is most effective with coals of high rank.

An alternative device, known as a shear cell, consists of a paddle which stirs a sample of coal in a cylindrical container. The power required to rotate the paddle in the cell is measured; when the required power rises, the coal is difficult to handle. Similar results to the Durham cone are obtained, i.e., that coal flow is generally poorest between 12 and 17% moisture.

Table 1.5 shows the average quality of coals received in 1982/3. South Midlands coals were within the prescribed range but are not always the most difficult coals to handle, as they have a high inherent moisture content. Handleability problems arise from free moisture in the coal. As a result, BC found that handleability tests were worth carrying out when deciding on preparation plant design at new collieries. The rule of thumb of 12–17% total moisture is only an average guideline.

14.6.2 In cold weather

Laboratory tests are helpful but are not a substitute for large trials. A refrigerator chamber at High Marnham power station has been used on several occasions during the last twenty years to check how coal wagons unload in cold weather (Fig 1.44). Freezing rates of coal in wagons at temperatures down to $-20°C$ were established using the equipment. The success (or lack of success) of measures designed to prevent frozen coal problems could be assessed. Spraying of coal with antifreeze (propylene glycol) for example was ineffective, because the coal fines soaked it up like a sponge and it was rapidly absorbed at economic treatment levels. Treatment of wagon sides and doors to prevent a metal-to-ice bond *was* effective. Oil or grease gave

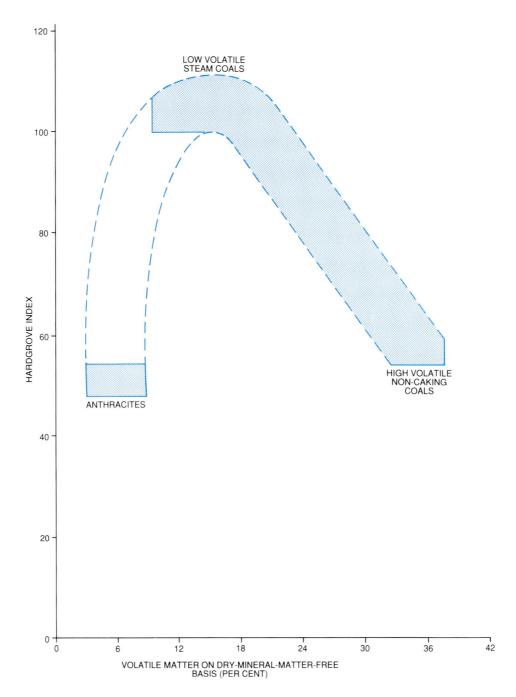

FIG. 1.42 Dependence of grindability on volatile matter content

benefit for at least three 'journeys' and a solid anti-freeze (urea) was effective at keeping door faces free of frozen coal. The most effective solution tried was a non-stick paint used on ice-breaker bows. This was painted on the wagon doors and sides (Fig 1.17) and allowed free movement of coal, provided the coal load was only frozen to a depth of 25–50 mm. The paint had a life of three years in trials.

The solutions described above involve extra work and prove to be uneconomic in Britain. The most practical solution, noted in Section 7.1 of this chapter, is to keep the wagons moving in cold weather, so that there is insufficient time for them to freeze. Trains which freeze in sidings are best left there until the temperature rises above freezing again.

15   Sizing analysis

15.1   Sizing of delivered coal

The size analysis of coal is important as a factor in its pricing and because it affects handleability. Over-size coal may also cause bias in coal sampling procedures. The following BS publications deal with the subject of sampling and sieving.

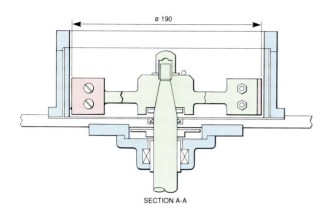

SECTION A-A

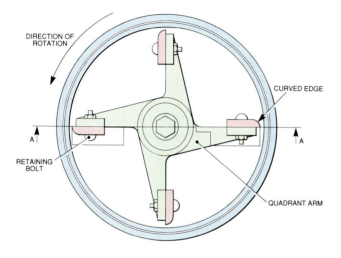

FIG. 1.43 Yancy-Gear and Price abrasiveness test
(BS1016 Part 19:1980)

BS410   : 1969 Test sieves

BS1796   :   Methods of use of test sieves

BS1293 and 2074 : 1965 Methods for the size analysis
                        of coal and coke

Sampling is covered in BS1017: 1977 Part 1. A minimum of 40 increments is recommended, in order to obtain a size analysis for a consignment of up to 1000 tonnes. The rules on bucket aperture and increment size are as for the general analysis sample. Preparation however requires more care, to ensure that the coal is not broken during handling. Air-drying is necessary as a first step to ensure that in subsequent sieving or division the finer particles do not adhere to the larger ones. If the nominal upper size of the coal is greater than 40 mm, sample division may only be carried out when the coal greater than 40 mm has been screened out. If the nominal upper size is between 12.5 and 40 mm the coal should be made into a heap or strip. Increments are removed in a prescribed manner using a shovel. If the coal is below 12.5 mm nominal top-size, then subdivision may be carried out using a

riffle or mechanical sample divider. Care must be taken to avoid breakage and loss of dust.

The minimum mass of sample to be retained for size analysis is somewhat less than for general analysis.

Minimum mass of sample to be retained for size analysis:

Nominal upper size of coal		Mass	
40	mm	180	kg
25	mm	70	kg
12.5	mm	16	kg
6.3	mm	2	kg
4.0	mm	1	kg
2.8	mm	0.25	kg

Finally, the precision of the size analysis should be checked in a similar manner to that described in Section 9.2.6 of this chapter for the preparation of the gross analysis sample. This checking procedure is the ultimate guarantee that a good result has been obtained. The preparation methods suggested in BS1017 *should* give a satisfactory result but this needs proving from time to time.

15.2 Sizing of pulverised fuel

The size analysis of powders may be carried out by a number of methods. For particle sizes greater than 45 μm, sieving may be used. Powders of smaller size require the use of instrumental techniques such as a sedimentation method, Coulter counter or Malvern counter. The methods involve dispersing the particles in a liquid and then utilising their sedimentation rate to separate the particles, measuring each particle's size by noting the variation in electrical properties caused as it passes through a cell, or noting the light scattering properties of a suspension. The instrumental techniques are specialised and are carried out in a laboratory.

For power station purposes the need is for a simple size indication to check mill performance and for combustion control, which depends greatly on particle size. Too fine a powder is wasteful of mill power, too coarse a powder does not burn completely in the combustion chamber and leads to an excessive proportion of unburned carbon in the precipitator dust. For most practical purposes in a PF station, a simple sieving test is adequate, in which the fractions (as percentage) retained on 100-mesh (150 μm) and passing 200-mesh (75 μm) are measured. About 75% of the pulverised fuel should be less than 75 μm.

The distribution of particle sizes in broken coal follows an exponential law (Rosin-Rammler) and the use of log-log versus logarithmic graph paper allows the full particle size distribution information to be presented in a straight line form if required (Fig 1.45). The Rosin-Rammler distribution is:

FIG. 1.44 Wagon freezing rig

$$R(\chi) = 100\ e^{-(\chi/\overline{\chi})^n}$$

where R (χ) = the percentage of particles with a size greater than χ

$\overline{\chi}$ = a constant known as the absolute size constant

n = a size distribution constant

when $\overline{\chi}$ = χ

$R(\overline{\chi})$ = $100\ e^{-1}$

= 36.79%

In practice $\overline{\chi}$ is the value of χ when R = 36.79%. Values of R are plotted on the log-log ordinate and values of χ on the logarithmic abscissa. A straight line should be obtained of gradient n, passing through $\overline{\chi}$ at R = 36.79%. The distribution is then fully described. Once n and $\overline{\chi}$ are known, estimates of R for any value of χ can be made.

16 Ash from coal

16.1 General chemical analysis

As noted in Section 13.5 of this chapter, ash is analysed for operational reasons and to help with ash sales and disposal.

16.1.1 Operational requirements

The object of combustion of coal is to liberate its heat content. Consequently the most important determination carried out on ash is carbon or more usually loss on ignition. This should be as low as possible. Any residual carbon in ash represents a loss of heat.

The elemental composition of the ash is determined next, in order to assess how the ash will behave in the furnace. Results on nine elements are required for pulverised fuel firing and, in addition, information on phosphorus is obtained for stoker-fired stations. Typical results are given in Table 1.16. They will be discussed in some detail in the following sections.

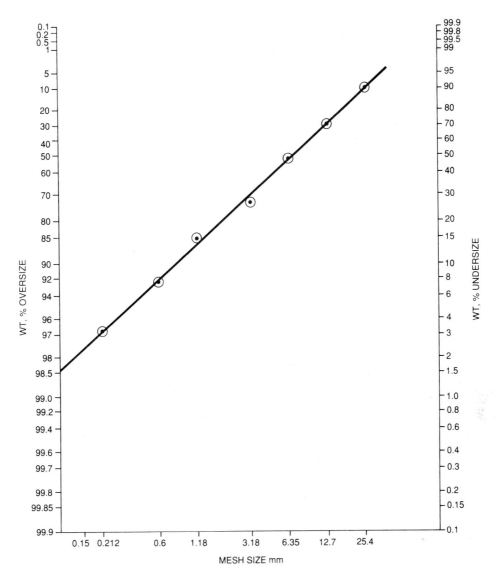

Fig. 1.45 Plot of coal particle size distribution obtained during a mill test at a power station (Rosin-Rammler plot)

16.1.2 Ash sales

Carbon in ash may give it a dark colour. It is then unsuitable for some uses, in paving slabs or building blocks for example, where consistency of colour is important. The colour may also vary from almost pure white to a dark chocolate colour as a result of its increasing iron content. Carbon and iron are therefore determined from time to time when a station begins to produce dark ash.

The normal tests of ash are for characterising its usefulness as an additive to cement, according to British Standard 3892.

For structural concrete the standard requirements are:

Moisture content	0.5% max.
Loss on ignition	7.0% max.
Magnesium (MgO)	4.0% max.
Sulphate (as SO₃)	2.5% max.
Sieve retention	12.5% max. retained on a 45 μm sieve

16.1.3 Ash disposal

When ash is disposed of it will sooner or later become part of the natural environment. Rain water will slowly percolate through it before entering aquifers and surface waters, and plant roots will penetrate it in their search for nutrients. A range of trace element analyses are carried out to ensure that no dangers arise from disposal schemes. Boron, arsenic and selenium have already been referred to in Section 13.5 of this chapter; in addition metals such as calcium, zinc, chromium, lead, molybdenum and copper are determined on occasion.

16.2 Loss on ignition

To determine loss on ignition, dry ash samples are heated at 800°C for $1\frac{1}{2}$ h in a furnace. Any residual carbonaceous matter burns off. Typical results depend upon station design, particularly the mill type in use and its state of wear (see Table 1.17).

TABLE 1.16

Analysis of coal ash and related fusion characteristics

Sample No		1	2	3	4	5	6
		%	%	%	%	%	%
Silica,	SiO_2	30.0	33.7	38.6	45.3	45.0	49.4
Alumina,	Al_2O_3	20.2	22.4	21.8	25.5	29.7	29.2
Ferric oxide,	Fe_2O	20.7	16.7	9.1	7.5	6.7	3.7
Calcium oxide,	CaO	16.3	12.1	16.2	11.6	3.4	2.8
Magnesium oxide,	MgO	5.6	6.4	6.9	3.1	6.4	6.5
Sodium oxide,	Na_2O	0.9	1.5	0.8	0.9	2.3	2.6
Potassium oxide,	K_2O	0.8	1.4	2.1	3.9	2.6	1.5
Titanium oxide,	TiO_2	0.7	0.7	1.4	0.8	1.0	1.2
Manganese oxide,	Mn_3O_4	0.4	0.4	0.2	0.1	0.1	0.0
Sulphate,	SO_3	4.1	4.0	2.6	1.1	2.1	2.1
Phosphate,	P_2O_5	0.3	0.8	0.3	0.2	0.7	0.9
Silica Ratio,		41.1	48.9	54.4	66.9	72.9	79.1
Fusion characteristics:							
		°C	°C	°C	°C	°C	°C
Deformation		1110	1140	1210	1200	1200	1150
Hemisphere		1140	1190	1250	1280	1360	1400
Flow		>1550	>1550	>1550	>1550	>1550	>1550

TABLE 1.17

Loss on ignition results 1983/84

Station	% loss on ignition			Mill type
	Max	Min	Avg	
L	5.0	1.5	3.0	Verticle spindle-roller
M	4.8	1.4	2.5	Tube ball
N	7.0	0.8	2.5	Tube ball
O[‡]	12.7	2.3	4.5	Rotating table ball

[‡]Average of seven years' data

Tube ball mills generally give the lowest carbon in ash values, whereas rotating table ball mills give the highest. However, all modern mills are a considerable improvement on figures obtained from plant 35 years ago. Then an average loss on ignition figure was 8.3%.

16.3 Ash fusion test

The fusion characteristics of ash are determined as a guide to possible deterioration of combustion chamber performance by slagging, and to the propensity of a coal to produce clinker. These characteristics are considered when selecting fuels for the various designs of boiler. For most boilers comparatively high fusion temperatures are obviously desirable, but for slag-tap and cyclone types of furnace, a readily fusible ash is necessary, and it is desirable that the molten slag should flow readily. Reid and Cohen in the USA found (1940) that slag viscosity was related to the 'Silica Ratio' defined as:

$$\frac{100 \times SiO_2}{SiO_2 + Fe_2O_3 + CaO + MgO}$$

where the components are expressed as per cent of each oxide (see Table 1.16). BCURA also carried out extensive work on this subject for the CEGB. (The Chemical Composition and the Viscometric Properties of Slags — 1963.)

The determination of ash fusion temperature is carried out in the laboratory according to BS1016, Part 15, 1960, using ash obtained under standard conditions, ensuring complete combustion of the prepared coal sample. In brief, the method consists of the controlled raising of temperature of a specimen of the ash (moulded to a standard pyramid shape), first to the point of deformation of the apex, then through the temperature at which the sample has formed a hemisphere, to the final flow temperature (see Fig 1.46). This is carried out in a small furnace containing a reducing atmosphere, consisting of carbon dioxide, 50%, and hydrogen, 50%, and the temperature is measured by means of a suitable thermocouple or an optical pyrometer.

In an alternative method, the ash is moulded into a 3 mm cube or cylinder and melted in a Leitz 'heating' microscope with a controlled atmosphere. The profile of the ash specimen can be examined visually or photographed as a permanent record of its flow characteristics.

Owing to the very different conditions in which ash fusion temperature is measured, compared with combustion conditions in the boiler, a close correlation between measured fusion temperature and experience in operation is not usually obtained. Nevertheless, this measurement provides an initial guide to the likely slagging propensity of many coals, particularly in PF boilers.

FUEL AND OIL

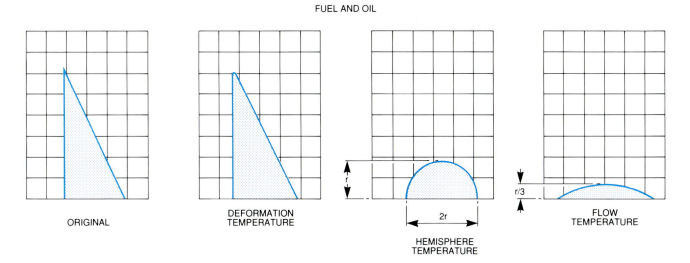

ORIGINAL

DEFORMATION
TEMPERATURE

HEMISPHERE
TEMPERATURE

FLOW
TEMPERATURE

FIG. 1.46 Ash fusion test cone profiles

16.4 Slagging and fouling tests

To provide better guidance as to the likely slagging and fouling behaviour of fuels than is provided by the ash fusion test, more detailed studies have been carried out (for example, Gibb 1981[10]). Slagging is a high temperature (1500–1600°C) phenomenon occurring in the furnace of a PF-fired boiler, whereas fouling occurs at a lower gas temperature (1000–1400°C) in the super-heaters (Fig 1.47).

16.4.1 Slagging

During pulverised fuel combustion the mineral matter within the coal particles is heated to 1500–1600°C. This temperature is sufficient to melt all of the mineral components, apart from a few large quartz particles. Droplets of low viscosity molten glass are formed and a proportion adhere to the boiler tube surfaces. If their temperature is high enough adjacent particles can fuse together, thus strengthening the mechanical bonding in the deposits and ultimately thick layers may be built up, reducing the heat transfer efficiency of the boiler surfaces.

The rate at which glass droplets fuse together is directly dependent upon the surface tension of the liquid glass and the time available for bonding to develop, and is inversely proportional to the viscosity of the melt, i.e., the more fluid the droplets are the more readily they will run together. Since surface tension does not vary greatly with temperature or glass type, the variation in the viscosity of the molten glass becomes the major factor determining the rate of slagging. The BCURA viscosity studies already referred to in the previous section have therefore been extended over the years.

The viscosity of molten ash samples is measured using a rod penetration viscometer (see Fig 1.48). This tech-nique enables viscosities in the range 10^3–10^8 Nsm^{-2}

to be measured, on molten ash samples cooling from 1550°C. Some ashes exhibit supercooling and remain fluid at a lower temperature than would be expected from an ash fusion test. The rod penetration visco-meter allows the cooling phenomenon to be studied in some detail. Control of the atmosphere in the equip-ment is possible and measurements may be made under reducing (50% CO_2/H_2 as for the ash fusion test) or oxidising conditions (pure carbon dioxide).

Slag deposits on boiler tube surfaces typically sinter together at viscosities of 10^6 Nsm^{-2} and the results of viscosity measurements on coal ash melts are con-veniently reported in terms of the temperature at which the cooling melt reaches a viscosity of this value. Typical results are given in Table 1.18. This shows that there is a range of several hundred degrees between the temperatures at which individual coals exhibit similar viscosity values.

The chemical composition of the ashes and the oxidation potential of the gaseous environment have a major influence on viscosity behaviour. The presence of iron oxide in a melt, under reducing conditions for example, can lower the viscosity considerably. Pye Hill coal ash in Table 1.18 has a viscosity of 10^6 Nsm^{-2} at 1200°C under oxidising conditions, whereas under reducing conditions, this value is not attained until the molten ash has cooled to 860°C.

The silica ratio (below) has also been used by Reid and Cohen

$$\frac{\% \ SiO_2}{\% \ SiO_2 + Fe_2O_3 + CaO + MgO}$$

(1944) to predict the viscosity behaviour of ash melts in the USA. Watt and Fereday (1969) extended the earlier work and published nomograms for predicting the viscosity behaviour of British coal ashes. This latter work includes Al_2O_3 in the ratio. Both studies are

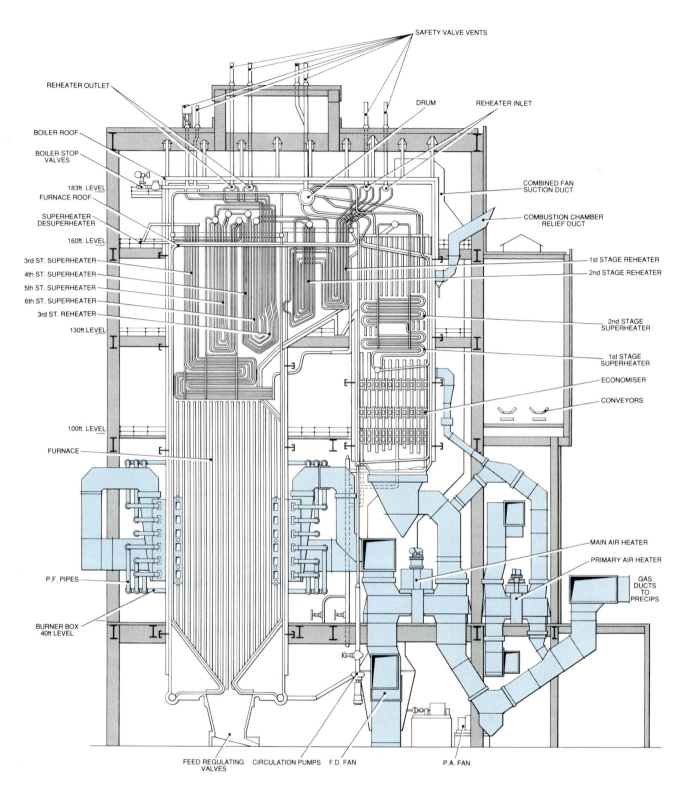

FIG. 1.47 Section through a power station boiler

aimed at selecting fuels for slag-tap boilers, i.e., finding coals with very fluid ashes (viscosity < 1000 Nsm^{-2}).

If the chemical composition of an ash is determined, then as a guide to viscosity behaviour, those coals with a high content of fluxing oxides CaO, MgO, and FeO (under reducing conditions) will give low viscosity melts leading to slagging. For a given level of fluxing oxides in a coal, as the Al_2O_3/SiO_2 ratio increases, the vis-

cosity will increase and the slagging will be less likely. (Al_2O_3, melting point 2050°C, is more refractory than SiO_2.)

If quantitative data is required, to aid in resolving a slagging problem in a pulverised fuel-fired power station boiler, then coal samples should be submitted to a laboratory for direct viscosity measurement, using a penetrating rod viscometer. This determines viscosity

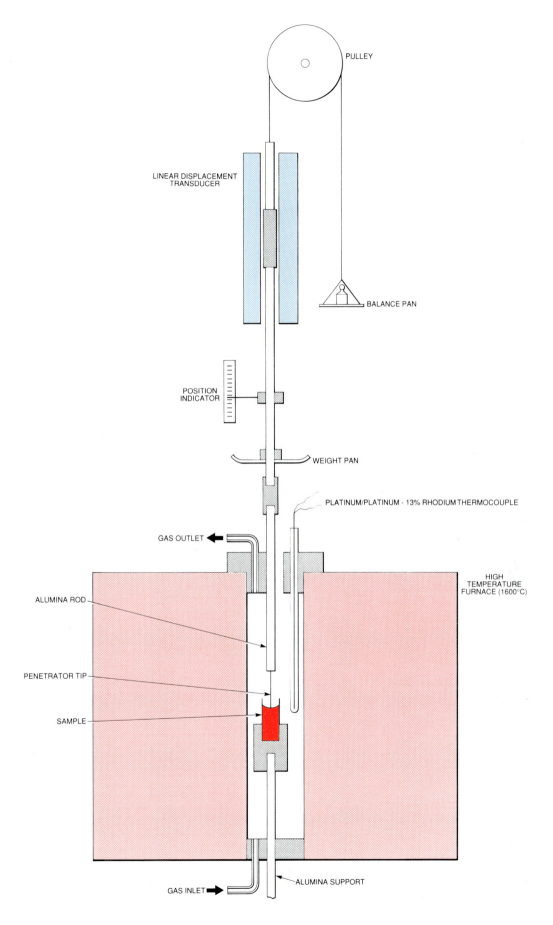

PULLEY

LINEAR DISPLACEMENT
TRANSDUCER

BALANCE PAN

POSITION
INDICATOR

WEIGHT PAN

PLATINUM/PLATINUM - 13% RHODIUM THERMOCOUPLE

GAS OUTLET

HIGH
TEMPERATURE
FURNACE (1600°C)

ALUMINA ROD

PENETRATOR TIP

SAMPLE

GAS INLET

ALUMINA SUPPORT

FIG. 1.48 Rod penetration viscometer

TABLE 1.18

Slagging indices based on the temperature at which viscosity
of the molten ash is 10^6 Nsm^{-2}

Coal	Area	Temperature (10^6) °C		Fe_2O_3 %
		Oxidising	Reducing	
Daw Mill	S Midlands	950		
Lea Hall	Cannock	955	930	7.3
Bagworth	S Midlands	980		
Shireoaks	N Notts	1050	1050	5.6
Hem Heath	N Staffs	1110	825	14.7
Pye Hill	S Notts	1200	860	13.7
Silverdale	N Staffs	1175		
Granville	Shropshire	1175		
Newstead	S Notts	1300		

behaviour in the relevant range for slagging behaviour, $10^3 - 10^8$ Nsm^{-2}. A much better insight into the coal/ash behaviour can then be obtained than would be given by the ash fusion test.

16.4.2 Fouling

Superheater fouling occurs at a lower gas temperature (1000–1400°C) than furnace slagging. It is initiated by a sticky layer of low melting alkali sulphates, formed during combustion in the furnace, being deposited on the superheater surfaces. The layer acts as a 'flypaper' for ash particles which, once trapped, slowly sinter together. The sintering process becomes more effective as the deposits increase in thickness. The inner part is cooled by the boiler tube surfaces but the outer layers insulated by the trapped ash particles approach the gas temperature. If the outer temperature becomes sufficiently high then bulk sintering occurs, as in furnace slagging.

A sintering test has been considered (Gibb 1981) in which pellets are formed from fine ash (< 212 micron; 72-mesh) in a die, by using a lever press. They are then heated to 500°C for one hour to sinter. After cooling the pellets are cut to a standard length and their strengths are measured. The stronger the pellet the more likely it is that the ash will cause boiler fouling.

This test was simple but the chemical composition of the ashes studied showed that an even simpler approach was possible. The strength of the test pellets correlated well with the sodium content of the ashes. Sodium is a mobile ion and its salts are low melting. These low melting salts make a major contribution to the bonding together of ash particles. Measurement of the sodium content of coal ash is therefore a very good measure of its fouling propensity. High sulphur and iron content in the coal will also give rise, during combustion, to low-melting sulphates. These form the sticky layers that initiate the fouling process.

When boiler fouling is observed, coal samples should be submitted for ash analysis in order to identify which of the several coal varieties delivered to the power station are likely to be causing the problem.

16.5 Commercial uses of ash

16.5.1 Sources of ash

A modern 2000 MW pulverised fuel fired power station will burn 20 000 tonnes of coal per day at full load. An average ash content for the coal would be 16% so that 3200 tonnes of ash would typically be produced. The ash arises at several points in the combustion process (Fig 1.49) and the nature of the ash is different at each point.

As the coal particles burn, glassy droplets of ash are produced. Those that impinge on the furnace wall begin to form deposits of slag, as discussed in the previous section. These either fall off or begin to accumulate on the walls. To control the build-up, the deposits are blown off periodically with steam or air by the sootblowers and they fall to the ash hoppers as furnace bottom ash (FBA). About 20% of the total ash made is collected in this way. The material is coarse and some pieces may be very large, 0.25 to 0.5 m in length. The ash hopper is shielded from the direct radiation of the furnace to allow the hot ash to be sprayed with water. This induces thermal shock in the glassy particles which makes them easier to crush. The FBA is sluiced out of the ash hoppers through crushers which reduce it to less than 50 mm top-size, and it is then swept into tanks to complete its cooling. It is removed by a grab and taken away for size grading and sale.

The remaining 80% of fine ash, which has an average particle diameter of 30–100 μm or so, is swept through to the back passes of the boiler. The larger sized particles and deposits falling from the economisers are collected in the economiser hopper. This ash is not so massive as the deposits which fall into the furnace ash hopper, but it is coarser than the fine material entrained in the gas stream which can only be collected by electrostatic precipitators and is known as pulverised fuel ash (see Fig 1.49).

Electrostatic precipitators vary in design and the first zone of some designs is a purely mechanical collector. The most modern designs have four electrostatic

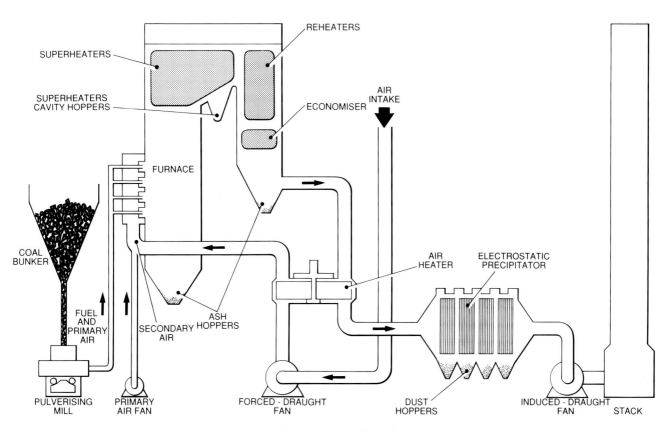

FIG. 1.49 Boiler ash collection points

zones. These charge the dust particles and use electrostatic attraction to remove 99.5% from the fuel gas. Most of the dust will be collected in the first zone of the precipitator and lesser amounts in later zones (Table 1.19).

TABLE 1.19

Typical electrostatic precipitator dust removal rates for a 500 MW boiler

Zone	PFA tonnes/day	% of total
1	560	82
2	95	14
3	21	3
4	5	1

The size distribution of the material from each zone differs. The coarsest material is collected in the first zone hoppers and the finest in the outlet zone. Typical distributions are illustrated in Fig 1.50.

16.5.2 Marketing of ash

The differing nature of the ashes produced determines their marketability. About 12 million tonnes of ash is produced each year from power stations in England and Wales, and in recent years over 40% has been used in commercial applications. The remaining 60% is disposed of as landfill (see Table 1.20).

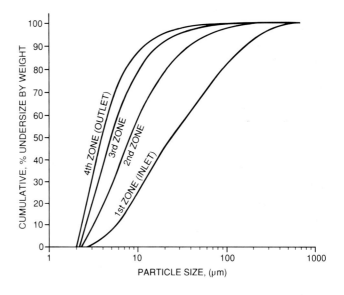

FIG. 1.50 Ash particle size range from a power station precipitator

Furnace bottom ash and clinker The coarsest materials, furnace bottom ash and clinker from stoker-fired stations, find a ready sale as aggregate. They are usually screened to provide the customer with grades of suitable size. Because of the thermal shock induced during quenching in the ash hoppers FBA is a sharp material, abrasive, and requires careful handling but in every other respect is a good aggregate used either as hard core or as an addition to cement to make concrete.

Pulverised fuel ash (PFA) has many uses. A flexible sales policy is necessary, however, as markets may fluctuate widely. When large PF-fired power stations began to produce PFA in the 1960s, for example, motorway construction was at a peak. PFA found a ready use in forming embankments and bridge abutments and large tonnages were sold. PFA possesses outstanding properties as a *load bearing fill* as it is light in weight and in most cases exhibits self-hardening characteristics. When used in bridge abutments, these properties allow lighter piling to be used as there will be little thrust on the containing structures. In addition, its low density is valuable for construction on land of low load-bearing capacity, as subsequent settlement is limited.

Although motorway construction has passed its peak it is still a large user of PFA and other load bearing fill markets are continually being found, the Thames barrage and its approaches providing an excellent outlet in recent years.

Before it can be laid and compacted PFA must possess the correct moisture content. This varies for ashes from different sources but the optimum value is generally in the range 10–20%. Two main types of PFA are supplied for load bearing fill; conditioned PFA produced by the addition of the optimum amount of water at the loading point (ash bunker or silo) and PFA reclaimed from stockpiles. The latter often contains a proportion of the coarser furnace bottom ash.

By adding some 10% of cement to PFA, a material known as *cement stabilised PFA* is obtained. This is used for road base, sub-base hard shoulders, site roads, footpaths, bridge abutments and many similar constructions. Success is achieved by again controlling moisture to the optimum level for the ash.

Lightweight aggregates are an important market. An aggregate of sintered PFA is half the weight of gravel and finds many uses in cladding panels and the frameworks of multistorey buildings. The production process is relatively simple, PFA mixed with a small amount of water is pelletised on a rotating inclined dish. The pellets are then fired on a horizontal moving grate or 'sinterstrand' at a temperature adequate to cause the pellets to develop a sintered strength, but not sufficient to cause melting. The resulting rough, slightly porous, material provides a good key for cement. As in the case for FBA, the material is size graded before packaging and sale.

Lightweight building blocks are prepared by mixing PFA with cement and a small amount of aluminium powder. The blocks are then heated in an autoclave and the aluminium reacts with free lime to liberate hydrogen which aerates the finished block to give it a light open structure. Such blocks are readily cut and drilled and find many uses in the building industry. Non-aerated blocks incorporating PFA are also produced. Consistent colour in the PFA supplies for this market is an important factor.

The use of PFA in cement and concrete manufacture has a long history and a difficult one from the point of view of gaining acceptance of the value of PFA. However, the issue of British Standard 3892 (see Section 16.1) has provided specifications for ash which assure that the customer will obtain a good product by using PFA as a partial cement replacement in making concrete products. PFA has three beneficial effects in a cement mix, it is able to combine with free lime (i.e., it exhibits pozzolanic activity), its rounded particle shape improves the flow properties of the mix and it reduces the water demand of the mix. This latter point allows the mix to develop good strength and also reduces the heat of hydration. This is very helpful to the engineering of massive concrete structures. Amongst the uses of PFA in the cement products market are the production of kerb and flagstones, partial replacement of cement in the ready-mix industry, and substantial replacement of cement in high fly-ash content cement.

PFA has long been used in *grouting*. Its particle shape and small size aid the flow and penetrating properties of the grout. PFA-containing grout is lighter than alternatives and its ability to react with free lime, generated during cement setting, conveys sulphate resistance to the product.

Finally there are 'clever' uses for PFA. A proportion of the ash particles are hollow and known as cenospheres. These are used in specialist insulation, as fillers for plastics (paintbrush handles for example) and in paints where the 'bubble' light-refracting properties give the paint opacity. Certain fractions of ash may find a use in brick making, partially replacing clay and thus extending the life of clay-pits. Other portions are magnetic and their high iron content can be used to produce dense cements. New uses are continually being found for ash and it is becoming much less thought of as a waste product and is increasingly viewed as a useful industrial material.

The proportion of ash sold in various markets over a ten year period is shown in Table 1.20.

16.6 Disposal of ash

Ash which cannot be sold must be disposed of. In the UK there are several major schemes of land reclamation using ash, producing valuable industrial building, recreational and agricultural land. As an example of the former, Longannet power station (see Fig 1.3) was constructed on the site of the former ash lagoons of Kincardine power station. No structural problems attributable to foundation difficulties have been experienced.

Using ash to fill old gravel or clay pits allows land to be reclaimed and provision of recreational areas (playing fields) is a satisfactory end use which is often achieved. On other occasions, such as at the CEGB ash disposal scheme at Peterborough, filled clay pits

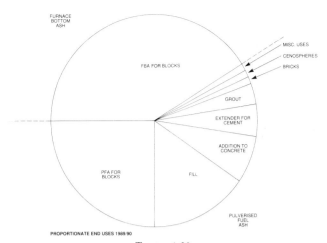

FURNACE
BOTTOM
ASH

FBA FOR BLOCKS

MISC. USES

CENOSPHERES

BRICKS

GROUT

EXTENDER FOR
CEMENT

ADDITION TO
CONCRETE

PFA FOR
BLOCKS

FILL

PULVERISED
FUEL
ASH

PROPORTIONATE END USES 1989/90

TABLE 1.20

Ash marketing 1976–1990 (quantities in tonnes × million)

	1976/77	1977/78	1978/79	1979/80	1980/81	1981/82	1982/83	1983/84	1984/85	1985/86	1986/87
Production (dry)	12.4	12.34	13.51	14.80	14.07	13.16	12.87	13.04	6.68	12.64	12.55
Ash used											
external sale	3.91	3.96	4.80	4.52	3.76	4.70	4.38	4.70	3.55	4.68	5.24
CEGB works*	1.10	1.12	1.37	1.61	1.59	1.52	1.49	1.33	0.29	1.05	1.03
Ash disposal	7.39	7.26	7.34	8.67	8.72	6.93	7.01	7.01	3.69	6.91	6.28
% used	40.4%	41.2%	45.7%	41.4%	38.0%	47.3%	45.5%	46.2%	‡	45.3%	50.0%

* Most ash used at Gale Common, Brotherton Ings, Devils Dingle and Drax completion

‡ Figures do not balance due to sales from stockpiles

have a top layer of a few inches of soil added and are then returned to agricultural use.

Ash is transported to the disposal point by a variety of means. It is carried dry in 'Presflo' rail wagons from large power stations in the Midlands to the Peterborough disposal scheme. On arrival it is removed by fluidising, using compressed air, and it is then mixed with water and the slurry is sent by pipeline to the lagoons. Slurry pipelines are also used at several large stations to carry ash to nearby lagoons. An alternative is to condition the fresh ash by adding water and to then transport it by conveyor. This approach is being used to take ash from Drax power station to a disposal site at Barlow where an artificial hill is being constructed from the ash. The ash is conveyed to its final position by a boomstacker. The water prevents dust blowing around in windy conditions.

A number of problems have to be attended to before a proposed disposal site can be completed, restored and put to new use. In the first instance a planning consent is required from the regulating authorities. There will be two constraints to consider, first the allowable effect of ash disposal on groundwater flows and secondly the control of any dust nuisance during the filling operation.

About 2% by weight of ash consists of water soluble material. New ash consists of an absorbed acidic layer surrounding an alkaline core. The acidity arises from condensation of acid sulphates formed during the combustion process. In water, over a period of about one hour or less, neutralisation occurs and ash lagoons are normally strongly alkaline. Many trace metals are immobilised as hydroxides, but small amounts of trace elements do enter the lagoon water. If water from the ash lagoon enters the groundwater or nearby streams or rivers, then the amounts of dissolved metals and suspended solids must be controlled to low levels to satisfy the water authority consent conditions. Typical consent conditions are given in Table 1.21.

During the planning stages of a new disposal scheme tests may be carried out to establish the hydrogeology of the site, i.e., the water table level and directions and volumes of ground and surface water flows. The ash quality will be known and calculations can then be made to ensure that sufficient dilution of any lagoon water discharges occurs to ensure that the limits set in Table 1.21 are not exceeded.

Once planning consent is obtained and disposal is under way, monitoring of the site is necessary to ensure that control limits are being maintained. Towards

TABLE 1.21

Typical consent conditions for discharges of trade waste into rivers

Parameter	Limit
Biochemical oxygen demand	20 mg/l
Arsenic, cadmium, chromium, copper, lead, nickel, and zinc	1.0 mg/l (as total or individually)
Cyanide	0.1 mg/l
Non-volatile matter extractable by carbon tetrachloride (e.g., oil)	5 mg/l
pH	5 to 9
Temperature	30°C
Suspended solids	30–100 mg/l (depending on type of discharge and flow rate)

the end of the filling of a lagoon, as it begins to dry out, windborne dust may become troublesome. The hollow ash particles (cenospheres) float on the lagoon water and although they are frequently 'harvested' for sale, sufficient light particles remain at the surface of a dry lagoon to cause some problems. It may then be necessary to spray the surface with a binding agent to control the dust. Typical materials used contain water soluble polymers. As soon as practical, however, a thin layer of soil is spread and then grass is sown. This completes the filling stage.

As ash is formed at high temperature it contains no nitrogenous material. Plant growth on bare ash is therefore poor apart from leguminous plants which provide their own nitrogen. Other plant foods are generally present in adequate amount. Spreading soil on the ash surface introduces soil bacteria to the area and when the grass is sown, clover (a legume) is included in the seed mix. A nitrogenous fertiliser is then added so that good cover can be quickly obtained.

Boron is one element mobilised during combustion which can damage plants. As it is very water soluble its amount in the ash decreases rapidly during the first year, but if boron levels are high it may be necessary to delay seeding for some months. Ash analysis is carried out before a seeding programme is planned, to check this point.

Once grass is established the site is monitored to ensure that water leaching, dust nuisance and plant growth are all proceeding as planned. To date some 15 km² of ash reclaimed land have been successfully restored by the industry and this can be considered a useful benefit to the country. Generally, however, ash sales are preferable to disposal as the material has such useful properties.

16.7 Cleaning ash from the sides of boilers

16.7.1 On-load cleaning

Accumulations of ash reduce boiler efficiency and must be removed. On-load gas side cleaning is normally carried out by means of a sequence of sootblowing or water lancing. The principle is that mechanical and thermal shock weakens deposits, so that by directing high pressure steam or water on to the metal surfaces the deposits may be removed. This is usually quite effective in controlling the secondary deposits, but once fused fly-ash deposits are formed, on-load cleaning becomes less effective and boiler performance falls.

16.7.2 Off-load cleaning

Boiler furnaces, superheaters, airheaters and economisers can be cleaned more effectively during the annual overhaul period when the plant is accessible and cool. Methods vary in detail between stations but in principle consist mainly of the following techniques:

- Manual cleaning.
- Manual cleaning followed by water washing.
- Water soaking.
- Steam soaking.

In water washing techniques, a convenient source of fresh water may be used and the process is continued until the pH value of the water leaving the plant is comparable with the clean water entering, or until the areas treated are seen to be clean. The final effluent should be clean and free of visible solids. If the deposit is not removed by direct water washing, water soaking may be adopted. This consists of soaking the deposits with slowly running jets of water for a period of about three days. Manual cleaning, however, may be necessary following water soaking, if the deposit is particularly troublesome. The addition of a wetting agent in the washing process is often useful and ensures deeper penetration of the water through the deposit to the metal surface where the sticky acid deposits are located. Soda ash is sometimes used in the initial stages of water washing to assist in neutralising acidic deposits, and to liberate carbon dioxide in order to break up the deposit. All water washing processes should be followed immediately by a brief 'drying out' period to dry out the boiler and minimise superficial rusting.

When water washing fails, steam soaking may be attempted. The principle of the method is to fill the boiler with cold water and inject low-pressure steam into the furnace so that it penetrates the deposits and condenses on the metal surfaces of the tubes, thus causing loss of adhesion of the deposit. This process may take two or three days and in difficult cases may have to be followed by further manual cleaning.

In all of these techniques, care must be taken to protect other metal surfaces, ducting and refractories from the deleterious effects of the wash water and effluents.

17 Coal firing — combustion calculations

17.1 Combustion calculations — general data

In previous sections, methods of determining the composition of coal and of predicting how its mineral components will behave during combustion have been given. Armed with all this data one further piece of information is still required before operations can begin, and that is the quantity of air which must be supplied to ensure complete combustion of the fuel.

Coal consists of carbon, hydrogen, oxygen, nitrogen, sulphur and mineral components. Thermochemical data is therefore needed about the main reactions of these combustible elements:

$$C + O_2 \rightarrow CO_2 \quad + 372.7 \text{ MJ} \qquad (1.6)$$

$$2H_2 + O_2 \rightarrow 2H_2O + 525.4 \text{ MJ} \qquad (1.7)$$

$$S + O_2 \rightarrow SO_2 \quad + 26.9 \text{ MJ} \qquad (1.8)$$

and also about the reaction leading to partial combustion:

$$C + \tfrac{1}{2}O_2 \rightarrow CO + 112.0 \text{ MJ} \qquad (1.9)$$

These equations relate the heat produced by the reactions to the masses of each element involved. Thus for Equation (1.6), 12 kg of carbon combines with 32 kg of oxygen to give 44 kg of carbon dioxide and liberate 372.7 MJ of heat energy. Similarly for Equation (1.7) 4 kg of hydrogen combines with 32 kg of oxygen to give 36 kg of water and 525.4 MJ.

If the amount of air provided for combustion is less than required by Equations (1.6–1.8), some of the carbon will burn incompletely to carbon monoxide, Equation (1.9). Taking the latter for comparison with Equation (1.6) it will be seen that 12 kg of carbon completely combusted gives 372.2 MJ but on partial conversion to carbon monoxide only 112.0 MJ is evolved, a difference of 260.7 MJ. This can be retrieved by completing the combustion:

$$CO + \tfrac{1}{2} O_2 \rightarrow CO_2 + 260.7 \text{ MJ} \qquad (1.10)$$

Incomplete combustion can lead to considerable heat loss and this is why calculation of theoretical air requirement is of importance.

To make the necessary calculations, the amount of air required for combustion, and the quantities of oxygen and nitrogen in the fuel must be considered.

The composition of standard dry air is shown in Table 1.22.

TABLE 1.22

Composition of standard dry air
(density 1.293 kg m^{-3} at STP)

Constituent	Mole fraction#	Mass fraction
Nitrogen	0.7809	0.7552
Oxygen	0.2095	0.2315
Argon	0.0093	0.0128
Carbon dioxide	0.0003	0.0005

#Assuming ideal gas behaviour, this is the same as the proportionate volume analysis

For simplicity nitrogen and argon may be taken together as 'nitrogen' so that, from Table 1.22, air can be said to consist of 79% by volume of nitrogen. The volume ratio of nitrogen to oxygen is then 3.77:1 and

the total volumes of gases involved when fuel burns in air as opposed to oxygen can then be expressed. For example, Equation (1.6) becomes:

$$C + O_2 + 3.77 \text{ N}_2 \rightarrow CO_2 + 3.77 \text{ N}_2 + 372.7 \text{ MJ}$$

This means that 12 kg of carbon burns in 32 kg of oxygen plus 105.6 kg of nitrogen (i.e., 3.77 × 28 kg) to form 44 kg of carbon dioxide and 105.6 kg of nitrogen whilst liberating 372.7 MJ of heat. There is no change in the energy liberated compared with Equation (1.6), but as this heat is now shared with a greater mass of gas, combustion temperatures will be lowered.

Generally volumes will be calculated at standard temperature and pressure (0°C and one atmosphere, 101 325 N/m^2). Adjustments are made using the gas laws, so that if the calculated standard volume is V_0 at T_0 (0°C) and P_0 (1 atm) then at any other temperature T_1 or pressure P_1 the volume V_1 will become:

$$V_1 = V_0 (T_1/T_0) \times (P_0/P_1)$$

17.2 Detailed calculation of air requirements

As an example of the calculation of the amount of air required to burn 1 kg of fuel, take the case of 902 non-caking coal in Table 1.9. Its composition is reported as:

	Analysis %	kg constituent/kg fuel
Moisture	18	0.18
Ash	8	0.08
Carbon	59	0.59
Hydrogen	3.7	0.037
Nitrogen	1.2	0.012
Sulphur	1.7	0.017
Oxygen (balance)	8.4	0.084

The first step in the calculation is to determine the amounts of gases which react with, or are produced from, reaction with 1 kg of carbon, hydrogen or sulphur in the fuel using Equations (1.6) to (1.10). The results are given in Table 1.23.

TABLE 1.23

Mass balances for combustion reactions

Combustible element	Equivalent amount (kg) of:						
	O$_2$	N$_2$	Air	CO$_2$	CO	H$_2$O	SO$_2$
Carbon (1)	2.66	8.85	11.51	3.66			
Carbon (2)	1.33	4.43	5.76		2.33		
Hydrogen	8	26.38	34.38			9	
Sulphur	1	3.32	4.32				2

(1) when burned to CO$_2$
(2) when burned to CO

The amount of air required can now be calculated from the analysis using Table 1.23 as follows:

0.59 kg carbon requires	0.59×2.66 kg O_2	$= 1.573$ kg O_2
0.037 kg hydrogen requires	0.037×8 kg O_2	$= 0.296$ kg O_2
0.017 kg sulphur requires	0.017×1 kg O_2	$= 0.017$ kg O_2
	Total weight O_2	$= 1.8863$ kg O_2

The amount of oxygen in the fuel must be deducted, as this gives rise to carbon dioxide directly, thereby reducing the amount of gaseous oxygen that needs to be supplied. In normal routine calculations the oxygen reported, 8.4%, will be deducted, so that the total weight of oxygen required from the air will be judged to be $1.8863 - 0.084 = 1.8023$ kg O_2. To be more precise, however, it is necessary to correct the ash to mineral matter.

$$MM = 1.10\ A + 0.53\ S_t + 0.74\ CO_2 - 0.32$$

The CO_2 correction may be omitted, since it will then be rightly assigned to the oxygen in the fuel when this is calculated by difference. Using the figures for the 902 coal:

Mineral matter =

$$1.10 \times 8 + 0.53 \times 1.7 - 0.32 = 9.38\%$$

The amended composition of 902 coal is then,

	Analysis %	kg constituent/kg fuel
Moisture	18	0.18
Ash	9.38	0.0938
Carbon	59	0.59
Hydrogen	3.7	0.037
Nitrogen	1.2	0.012
Sulphur	1.7	0.017
Oxygen (balance)	7.02	0.0702
Total	100.00	1.0000

and the correction for oxygen in the fuel is:

Total weight O_2 required/kg fuel	$= 1.8863$ kg O_2
less O_2 content of fuel	$= 0.0702$ kg O_2

Amount of O_2 to be supplied by air $= 1.8161$ kg O_2

Weight of air required to burn 1 kg of fuel, using data from Table 1.22 is

$$1.8161 \times \frac{100}{23.15} = 7.85 \text{ kg}$$

The volume of air at STP will be

$$\frac{7.85}{1.293} = 6.07 \text{ m}^3$$

The reader will see that these are the values given in Table 1.9.

17.3 Calculation of the composition of the waste gases

Continuing with the above example, the waste gases from burning 1 kg of the 902 coal will consist of the nitrogen remaining from the combustion air, a small amount of nitrogen originating from the fuel and the products of oxidation; carbon dioxide, water and sulphur dioxide. Referring to the analysis above:

	Weight of gases, kg	Volume of gases m³ (Note 2)
Nitrogen in coal	0.012	
Nitrogen in air = $7.85 - 1.816$ (air $- O_2$)	= 6.034	
Total nitrogen in flue gases	= 6.046	4.789
Wt[1] of CO_2 produced $= 0.59 \times 3.66$	= 2.16	1.099
Wt[1] of H_2O produced $= 0.037 \times 9$	= 0.333 ⎞	
Wt of H_2O in coal	= 0.18 ⎠	0.636
Wt[1] of SO_2 produced $= 0.017 \times 2$	= 0.034	0.012
Total products of combustion (wet)	= 8.753 kg	6.54 m³
Total products on a dry basis	= 8.24 kg	5.9 m³

Notes

1 Factors from Table 1.23

2 Avogadro's Hypothesis states that the molecular weight in kg of any gas will occupy 22.41 m³ at STP. For the above computation the use of the specific volume of each gas is helpful. This is the volume occupied by 1 kg of the gas (equal to 22.41 m³/molecular weight in kg), see Table 1.24.

TABLE 1.24
Specific volume of gases

Gas	Molecular weight (approx)	Specific volume m³ kg⁻¹ at STP
H_2	2	11.11
O_2	32	0.70
N_2*	28.3	0.792
CO_2	44	0.509
SO_2	64	0.343
H_2O	18	1.24
CO	28	0.80
CH_4	16	1.40
Air (dry)	28 964	0.77

*including the rare gases

The total volume of waste gases, 6.54 m³, is the figure which appears in Table 1.9.

Carbon dioxide in flue gas is measured on a *dry* basis and from the above calculation is present to the extent of $1.099/5.9 \times 100\%$, i.e., 18.6% by weight, the figure in Table 1.9. The composition of the wet gases is calculated from the above results with the exception of the concentrations of NO_x and SO_x. The calculation of the concentrations of these gases is difficult and a combined figure of 0.3% is given. This is an arbitrary figure (to allow the analysis to equal 100%); it is of the right order but a figure of 0.16 to 0.2% is perhaps more typical of actual waste gases as they, of course, contain excess air.

17.4 The significance of carbon dioxide levels in flue gases

When a fuel is burnt with the exact amount of air to ensure complete combustion the flue gases will contain a fixed and characteristic level of carbon dioxide, known as the 'theoretical percentage of carbon dioxide'. As one volume of oxygen combines with carbon to give one volume of carbon dioxide (Equation (1.6)), when pure dry carbon is burnt the oxygen of the air will be replaced by an equivalent volume of carbon dioxide, i.e., 20.95% (Table 1.22). Any excess air will dilute the carbon dioxide to a lower level and if there is sufficient air for complete combustion the carbon dioxide level will also fall as carbon monoxide is produced instead. The theoretical percentage of carbon dioxide represents a datum for good combustion and the plant performance may be assessed by use of this parameter.

With fuels containing hydrogen in addition to carbon, the theoretical CO_2 is less than 20.95% and the calculation in the previous section gave a value of 18.6% for the 902 coal from Table 1.9. Inspection of Table 1.9 shows that for British bituminous coals the range of theoretical CO_2 is $18.6 \pm 0.3\%$ and for anthracite is between 19 and 19.5%.

17.5 Calculation of excess air

In practice coal is never burnt with an exactly stoichiometric quantity of air. Mixing of fuel and air in the burners is not perfect and an excess of air is used to encourage complete combustion. Nevertheless some carbon monoxide is always produced in the process. The amount of excess air may vary from 5 to 50% depending upon plant type and the fuel burned. In a large PF-fired boiler, about 10 to 20% excess air is typical. Excess air is defined as:

$$\text{Excess air (\%)} = \frac{(\text{air supplied} - \text{theoretical air})}{\text{theoretical air}} \times 100$$

17.5.1 Inefficient combustion

In small boilers, with inefficient gas mixing, the fuel may not be completely burned, despite there being an excess of air. To determine the amount of excess air supplied to such a boiler an analysis of both the fuel and flue gases is required. The precise calculations made above can be compared to these analyses to determine the combustion efficiency. The excess air is then calculated by determining the amount of carbon in 1 kg of the coal and then finding the volume of flue gases which also contains this amount of carbon. By comparing the volume with that calculated in Section 17.3 of this chapter, an estimate of the excess air may be made.

Taking coal 902 from Table 1.9 as an example again, when burned in a small boiler it might give flue gases with the following composition:

$$
\begin{array}{ll}
CO_2 & 12.5\% \\
CO & 1.0\% \\
O_2 & 6.5\% \\
N_2 & 80.0\%
\end{array}
$$

The dry flue gases produced from burning 1 kg of the coal must contain the same weight of carbon as was originally present in the coal.

The amount of carbon in 1 kg of the 902 coal is 0.59 kg. The weight composition of the dry flue gas is computed as follows (Table 1.25).

The weight of carbon in 1 kg of the flue gas is deduced from the amounts of CO and CO_2 present.

Carbon as CO_2 and CO in 1 kg

$$= 0.1803 \times \frac{12}{44} + 0.0092 \times \frac{12}{28}$$

$$= 0.0531 \text{ kg}$$

The amount of dry flue gas which contains the same amount of carbon (0.59 kg) as 1 kg of coal is therefore:

$$\frac{0.59}{0.0531} = 11.11 \text{ kg}$$

From Table 1.25 the mean molecular weight (mw) of the gas is 30.50, i.e., $(\Sigma \text{ Vol} \times \text{mw})/100$. This weight will occupy 22.41 m^3 (according to Avogadro's Hypothesis). The volume of dry flue gas is therefore:

$$\frac{22.41}{30.50} \times 11.11 = 8.164 \text{ m}^3$$

From Section 17.3 of this chapter we know the coal also gives rise to 0.636 m^3 of water on combustion, so the total wet combustion gas volume must have been 8.8 m^3 at STP (8.164 + 0.636 m^3).

From Section 17.3 the *theoretical* wet combustion gas volume was calculated as being 6.54 m^3. Therefore,

$$\text{Amount of excess air} = \frac{8.8 - 6.54}{6.54} = 34.5\%$$

17.5.2 Efficient combustion

In a modern 500 or 660 MW boiler at least 90% of the heat of the coal is extracted and little carbon monoxide is produced. Since carbon dioxide is predominantly formed, some shortcuts may be taken with the combustion calculations.

TABLE 1.25

Computation of composition of dry flue gas

Gas	% by volume	Molecular weight, mw	Weight product (vol × mw)	% by weight
CO_2	12.5	44	550	18.03
CO	1.0	28	28	0.92
O_2	6.5	32	208	6.82
N_2	80.0	28.3[‡]	2264	74.23
		Total	3050	100.00

[‡] including the rare gases

In general operations the fuel analysis is not available and calculations of excess air are made from the dry flue gas analysis. The amounts of oxygen, carbon dioxide and carbon monoxide in the gases are measured and nitrogen is determined by difference. The excess air on a dry basis may then be estimated by assuming that all the carbon has been converted to carbon oxides. The volume of CO_2 produced will be equal to the volume of oxygen consumed (Equation (1.6)), but if carbon monoxide is formed then only half the volume of oxygen will have been consumed (Equation (1.9)). The apparent excess air will be enhanced in the latter case and an adjustment for the volume of carbon monoxide formed must be made, i.e.,

Corrected excess oxygen =

(% by vol O_2 in flue gas) − ($\frac{1}{2}$ measured CO%)

The theoretical oxygen is the amount present in the air supplied, less the corrected excess of oxygen just calculated. The amount in the dry air supplied can be deduced from the nitrogen content of the flue gases since (from Table 1.22) the ratio of oxygen to nitrogen in the original air must have been 0.264. Thus:

$$\text{Excess air} = \frac{(O_2 - CO/2)}{0.264\, N_2 - (O_2 - CO/2)} \times 100$$

The equation ignores the small amount of nitrogen present in the fuel but it turns out to be sufficiently accurate for day to day purposes.

17.6   Measurement of boiler efficiency

If the boiler efficiency is to be measured then a special trial is undertaken. The fuel supplied to the burners is analysed and the flue gases are monitored carefully. Details of the procedures are described elsewhere, but the general chemical principles are worth noting. If the fuel is analysed as described earlier then the heat input to the boiler is known. The efficiency may then be estimated by determining the percentage loss of heat due to each of the following seven possible causes, and deducting their sum from 100:

L_1 — loss due to the latent heat of moisture in the fuel

L_2 — loss due to the latent heat of water in the flue gas arising from hydrogen in the fuel

L_3 — loss due to the elevated temperature of the dry flue gases

L_4 — loss due to incomplete combustion of the fuel resulting in carbon monoxide in the flue gas

L_5 — loss due to incomplete combustion of the fuel resulting in carbon in the refuse

L_6 — loss due to elevated temperature of the refuse

L_7 — loss due to radiation

Losses L_1 and L_2 require measurement of the gas temperature of the flue gases leaving the boiler, determination of the dewpoint of the gases, and measurement of their water content. The gas temperature is measured at the airheater, but the gas composition and dewpoint are measured before this point as air often leaks into the system after the boiler, causing a dilution. The acid dewpoint is typically 120°C and flue gases are maintained above this temperature to prevent corrosion of the ductwork. The true dewpoint is typically 50°C, so all of the water present in the flue gases leaves the stack, taking latent heat with it. As it takes about 1 tonne of coal to evaporate 10 tonnes of water this represents one of the larger causes of heat loss from a boiler.

Loss L_3 is estimated from gas temperature and the specific heat of the flue gases.

The carbon monoxide concentrations in gases leaving a modern boiler are very low. On-line measurement using an infra-red detector is used to control the boiler. Levels of carbon monoxide above 50–100 PPM by volume are taken as showing the onset of incomplete combustion and the air supplies to the burners are adjusted to reduce the concentrations to below these figures. Oxygen is also measured in order to estimate the excess air level. Figures of 3% O_2 and 10–20% excess air are typical. The oxygen is measured on-line using either a solid state electrolyte (yttrium-doped garnet) cell or a paramagnetic instrument. Carbon dioxide may also be measured using an infra-red detector to complete the flue gas characterisation.

The measurement of the carbon in ash and grits requires care in sampling in order to determine the fifth loss. Flue gas samples must be withdrawn at a velocity equal to that of the flow in the gas ducts. This is call isokinetic sampling. If a sample gas flow is withdrawn from the duct at a velocity greatly different from that of the bulk gas, then the entrained ash's particle distribution will not be reflected in the sample and incorrect analyses will be obtained. The gas flow extracted is passed through a small cyclone to collect the solid particles for analysis. The samples of ash, dust and grit are collected at the appropriate points in the boiler (see Fig 1.49). Carbon levels in PFA from modern power stations are typically 2.5 to 4.5%. As ash levels are 16% on average in contemporary coal deliveries, this represents a loss of about 0.5% of the heat.

The loss of heat in hot ash is measurable, but one of the smaller losses and the final loss due to radiation is about 0.2% for a large 500 or 660 MW boiler.

18 Air pollution

18.1 Introduction

The behaviours of the main air pollutants emitted from coal and oil fired power stations are considered in this section. There have been major changes in the last 25 years in the measured level of air pollutants, in the methods available for monitoring such pollutants, and attitudes and legislation concerning them. At the start of this period, the Clean Air Act of 1956 was introduced in the UK to control 'traditional' pollution from combustion of coal by industry and domestic users. In 1991 the scene is more complex; for example, the motor car is now a major polluter and pollution issues have to be tackled on a global scale. Such issues are now covered by the Environmental Protection Act (1990).

18.1.1 Developments following the Clean Air Act of 1956

In the mid-1960s, much air pollution monitoring in Britain was at fixed sites set up to measure the contributions from individual works in their immediate surroundings. Dustfall, smoke and sulphur dioxide were seen as the main items of concern. The Clean Air Act of 1956 had already led many householders to cease burning coal inefficiently in open grates. As a result, the black smoke in the towns was seen to be decreasing. Intense 'smogs', lasting a few days at a time, and having detectable adverse health effects, still occurred in London in the winter of 1962/63 and in Sheffield in the winter of 1963/64, but as it turned out, they were the last of such intense smoke pollution episodes.

18.1.2 The present situation

Today, *local* pollution from individual works, and in towns, may still be monitored at a limited number of sites. At some sites this is to follow trends, at others to demonstrate compliance with European Community Directives and some are used for studies into any residual air pollution related health effects. The most intensive studies currently being made are to assess the extent of the long-range drift or transport of mixtures of air pollutants from one region or country to another. Claims have been made that such long-range drift could be responsible for effects seen in certain distant sensitive rural areas, such as the loss of fish from lakes in some parts of Scandinavia or damage to trees in some parts of Germany. Pollutants also travel in the reverse direction and effects in the UK can be caused by air movement from Europe. The main pollutants of concern today which arise directly or indirectly from combustion are 'acid rain', ozone, oxides of nitrogen, hydrocarbons and sulphur dioxide. In addition, the role of carbon dioxide which contributes to the 'greenhouse effect' of the atmosphere is subject to much debate. Fixed sites for monitoring are still used, but intermittent sampling from specially instrumented vehicles and aircraft is also employed. The mathematical modelling of air pollution pathways is increasingly being used in conjunction with actual measurements. Modelling requires a knowledge of emissions, weather factors and the fate of pollutants (discussed later).

18.2 Emissions

The emissions of individual pollutants from power station chimneys are not monitored continuously, but intermittent analyses are made.

18.2.1 Flue gas constituents

An approximate composition of the stack gases from a modern British coal-fired power station is shown in Table 1.26.

The precise composition at any time will depend on the composition of the coal being burned and the furnace conditions. The hydrogen chloride will be an order of magnitude less with some coals. The gas temperature at the stack top will be around 105°C.

18.2.2 Rates of emission and dispersal of pollutants

The effects at ground level depend on how the plume from the stack disperses and not on the gas composition as such. To a first approximation, ground level concentrations are proportional to the rate of emission divided by the square of the effective height of discharge. This height includes the buoyant rise of the waste gases due to their excess temperature over that of the ambient air. Examples for a modern power station

TABLE 1.26

Composition of stack gases from a typical coal-fired power station

Flue gas constituents	Quantities
Nitrogen	Around 76% by volume
Carbon dioxide	13%
Water vapour	7%
Oxygen	4%
Sulphur dioxide	Around 0.12% or 1200 PPM* or 3400 mg m^{-3} at NTP
Nitric oxide	400 PPM* or 570 mg m^{-3} at NTP
Hydrogen chloride	200 PPM*
Carbon monoxide	50 PPM*
Hydrogen	50 PPM*
Nitrogen dioxide	20 PPM*
Together with ash particles: around 100 mg m^{-3}	

*Parts per million, by volume

and a coal-burning house in the same weather conditions are given in Table 1.27.

The coal burnt (column 2) in the power station is equivalent to that burnt in nearly half a million coal-burning houses. If the houses were concentrated at one site, the total sulphur dioxide depositing per square metre at ground level would reach 2800 times that produced by the power station (comparisons in column 5 of Table 1.27). This illustrates why, prior to 1956, the burning of coal in domestic grates made our cities so dirty.

Today, many houses burn natural gas and these have an emission much weaker in sulphur dioxide, which is good news. However, these emissions are just as strong in oxides of nitrogen, compared with coal burning. The emission rates for oxides of nitrogen for the power station and a gas-burning house would be in similar proportion to, but smaller than, the emission rates for sulphur dioxide in Table 1.27. Calculation shows as before, that the low-level emission from houses has a much greater local effect per square metre at ground level than does the power station.

18.2.3 Legal controls applied to emissions

The legal control of emissions from large (>50 MW thermal) power stations is administered by HM Inspectorate of Pollution. The current laws deriving from the Environmental Protection Act (1990) still require that 'best available techniques' be used to prevent the emissions but economic considerations are taken into account by the BATNEEC concept; that 'Best Available Techniques Not Entailing Excessive Cost' may be employed. Numerical 'presumptive emission limits' for particulate matter emissions from power stations have been set by the Inspectorate for some time and will continue unless lower limits are negotiated. Operating within these limits 'would normally be accepted by the Inspectorate as presumptive indication that best practicable means are being used' (Health and Safety Executive, 1985). The current particulate matter emission limits are 115 mg m^{-3} for post-1974 constructed power stations, 460 mg m^{-3} for pre-1958 and individually set values for those constructed in between. The flue dust burden is monitored continuously but indirectly, using the obscuration of light across a duct

TABLE 1.27

A comparison of power station and domestic contributions to sulphur dioxide deposition rates

Source	Coal burnt	Sulphur dioxide emitted Q	Effective height H	Q/H^2
1	2	3	4	5
2000 MW power station, full load, single stack	240 kg s^{-1} (Sulphur content 1.5%, 10% sulphur retained in ash)	6.6 kg s^{-1} (Total gases emitted around 2000 m^3 s^{-1})	400 m (200 m actual height plus 200 m buoyant rise)	41 mg m^{-2} s^{-1}
Domestic chimney	0.5 g s^{-1} (winter rate)	15 mg s^{-1}	8 m (little or no buoyant rise)	0.23 mg m^{-2} s^{-1}

or on a prepared surface presented to the flue gases. Direct gravimetric checking is carried out regularly by a special testing team using British Standard 893 (1978).

Legislation is being tightened further and, for example, the Commission of the European Communities 'Large Combustion Plant Directive' (88/609EEC) proposes new emission limit values. These are, for coal-fired power stations over 300 MW, dust 50 mg m^{-3} and sulphur dioxide 400 mg m^{-3} (200 after 1995). Larger emissions are permitted from smaller power stations. The purpose is to reduce the total annual emissions from large combustion plant; for SO_2 by 63% by 2003, and for NO_x and dust by ~40% by 1988, taking 1980 levels as a base line.

18.2.4 Emission trends

The annual rate of emission of sulphur dioxide from the UK as a whole has been estimated for recent years. Some of these annual values are quoted in Table 1.28.

TABLE 1.28

Trends in annual sulphur dioxide production in the UK

Year	Annual emissions of SO_2, million tonnes
1950	4.7
1955	5.1
1960	5.6
1965	5.9
1970	6.1 (peak year)
1975	5.2
1980	4.7
1985	3.7
1989	3.7

The latest value is some 40% less than the peak (1970) value. The reasons for the decrease given by the Department of the Environment are:

- Industrial modernisation.

- Changes in fuel use patterns.

- Energy economies and reductions in the sulphur content of fuels.

Most of the recent decreases have occurred in the domestic, commercial and smaller industry sectors. This has resulted in the proportion of the total UK SO_2 emissions due to power stations changing from around 45% in 1970 to almost 70% in 1990, despite the amounts of sulphur emitted by power stations having declined. Further reductions will be achieved as gas-fired power stations are commissioned progressively from 1992, and when flue gas desulphurisation plants on coal-fired stations (e.g., Drax and Ratcliffe) start operating.

Recent annual emissions of oxides of nitrogen from the UK as a whole have also been estimated, but with much less certainty since those due to motor vehicles vary considerably with road speed and the state of tune of the engine, and these factors are not well known throughout the UK. In 1983 some 1.9 million tonnes per year, expressed as NO_2, was estimated to have been emitted. The proportion of this total due to power stations was 40%. Emissions are still high in 1991 and the total NO_2 emission may not have reached its peak. Improvements in burner design (low NO_x burners) are reducing the proportion due to power station emissions and the gradual introduction of pollution control measures on motor cars should be accelerated shortly, to ensure that overall NO_2 emissions start to decline.

18.3 Measurement of pollutants

The methods in regular routine use around power stations are briefly described in the following subsections.

18.3.1 Total dustfall

Daily, over each month, the insoluble matter which settles under its own weight, or is washed down by rain into the collecting bowl of a British Standard deposit gauge is collected. The bowl is mounted 1.2 m from the ground to minimise the collection of wind-lifted particles (BS 1747: Part 1: 1969). Filtering and weighing the collected deposit gives values which are expressed as mg m^{-2} day^{-1}, monthly average.

18.3.2 Daily average sulphur dioxide

Air is sampled at 2 m above the ground and drawn by pump continuously through a reagent in the British Standard volumetric apparatus (BS 1747: Part 3: 1969). Sulphur dioxide is trapped in the acidifying reagent and determined afterwards in a laboratory, usually by titration with standard alkali solution, to give μg SO_2 m^{-3}, daily average.

18.3.3 Daily average smoke

The insertion of a filter paper in a clamp ahead of the SO_2 absorbing reagent, traps smoke particles. The stains on the daily exposed papers are afterwards analysed with an optical reflectometer (BS1747: Part 2: 1969) to give μg smoke m^{-3}, daily average.

18.3.4 Hourly average sulphur dioxide

At a number of selected sites, distributed nationally, commercial continuous sampler-analysers are used. Sulphur dioxide-containing air is introduced into a hydrogen flame (or in another instrument, into a zone of pulsed ultra-violet light energy). This immediately emits specific radiations in proportion to the concentration, and these are monitored by photomultiplier tubes. The instruments require calibrating with standard gas and zeroing with absorbants which provide a cleaned 'zero' gas sample. They give a considerable amount of data, consequently data logging and computing facilities are routinely used. Although minute by minute values are obtained, hourly averages are conveniently used to compare with meteorological factors like wind directions, which are relatively steady over periods of an hour. The results are usually recorded in PPB (parts per billion or 10^9, by volume). 1 PPB SO_2 = 2.9 μg SO_2 m^{-3} at NTP.

18.3.5 Hourly average oxides of nitrogen

The instruments used are broadly similar to those described for sulphur dioxide. Nitric oxide (NO) in air reacts with excess ozone from a generator in the instrument to give immediate chemiluminescence, monitored by photomultiplier tubes. In the same instrument, nitrogen dioxide (NO_2) in air is catalytically converted to nitric oxide and determined as above in another channel. 1 PPB NO = 1.4 μg NO m^{-3} at NTP; 1 PPB NO_2 = 2.1 μg NO_2 m^{-3} at NTP.

18.3.6 Hourly average ozone

Broadly similar instruments are used to those described above. Ozone (O_3) in air reacts with excess ethylene gas (from cylinders) in the instrument to give immediate chemiluminescence, monitored by photomultiplier tubes. In another type of instrument, ozone in air is determined by monitoring its absorption of ultra-violet light. 1 PPB O_3 = 2.1 μg O_3 m^{-3} at NTP.

18.3.7 Daily average rainwater constituents

Rainwater falling in British Standard deposit gauge bowls (see Total dustfall above) is collected daily, or diverted automatically into different containers for each day. Determinations of acidity by pH meter and some other chemical constituents are made later in a laboratory. At some sites, lids over the bowls are used, which open automatically when rain falls on to a sensor. This minimises the collection of dust and other materials which deposit into the bowl when it is not raining. The pH units are logarithmic and are converted into linear units of micro equivalents of hydrogen ion per litre, μeq H l^{-1} (pH 4.0 = 100 μeq H l^{-1}; pH 5.0 = 10 μeq H l^{-1}). Rain volume is also recorded in mm rain (litres per m^2 of horizontal collector area). Multiplication of rain volume and acidity gives deposited acidity, expressed as meq H m^{-2} of 'ground'.

18.3.8 The avoidance of local effects

Since results can be affected by local shielding or exposures, the advice of the British Standard on siting deposit gauges (BS1747: Part 1: 1969) is followed:

- 'The distance of the deposit gauge' (or collector or probe or wind recorder) 'from any object shall be more than twice the height of the object above the rim of the collecting bowl' (or probe or sensing head)

- 'The rim of the collecting bowl should not be more than 5 m above the ground'. (The airflow around the buildings may be disturbed, with gusts of wind from directions not experienced by a properly exposed wind recorder.)

In addition, local source effects may be minimised (unless attempts are being made to monitor them) by having sites at least:

- 100 m from emissions from any single house, silo, non-electric engine or pump, sewerage farm, wood, stock-pile, quarry, busy road or railway.

- 1 km from any village.

- 10 km from any town, large works or the sea.

18.4 Pollution at ground level

18.4.1 Measurements at typical sites

These vary widely, with smaller values generally in rural districts and larger values generally in and near towns or busy roads. The reverse is true of ozone since it reacts with pollutants and is therefore more prevalent in rural areas. Annual average values typical of the mid 1980s for a range of sites are given in Table 1.29. Shorter term values cover a wider range.

Power station contributions to ground level pollution for most constituents are largest at some point near the station, depending on wind speed and other factors.

For coal-fired power stations, typical annual contributions to total dustfall in the 'worst' place (2 to 3 km from a stack) have been around 10 mg m^{-2} day^{-1} (up to 20% of the total), with less at all other places. These power station contributions are recognised under the microscope as they contain distinctive fused ash spheres (cenospheres), commonly around 10 μm diameter. Most of the dustfall consists of angular soil particles. Oil-firing in power stations has been comparatively costly when compared with coal-firing and is relatively little used. It produces little actual ashfall, but a problem in the past from oil-firing was 'acid smut' emission, i.e., thin flakes of carbon with sulphuric acid condensed on them. This has been largely overcome by careful attention to combustion conditions and by the use of an alkaline combustion additive, dolomite. However, 'acid smuts' can also occur from domestic oil-fired stacks and can be detected occasionally from oil-fired power stations, especially if they are operating intermittently.

For sulphur dioxide, typical annual contributions in the 'worst' place (5 to 9 km from a stack) have been around 7 μg m^{-3} (15% of the total) with less at all other places. The power station contributions are recognised from the larger peak to mean patterns during an hour, found from elevated sources along the 'right' wind direction. Smaller peak to mean ratios are found from other and distant sources. For nitrogen oxides, the 'worst' annual power station contributions (at similar

TABLE 1.29

Site measurement of air pollutants

Constituent	Units	Annual average values		
		Smaller	Medium	Larger
Total dustfall	mg m^{-2} day^{-1}	20	50	100
Smoke	μg m^{-3}	10	20	50
Sulphur dioxide	μg SO$_2$ m^{-3}	10	40	80
	PPB SO$_2$	3	15	30
Nitric oxide	PPB NO	3	10	40
Nitrogen dioxide	PPB NO$_2$	3	15	50
Ozone	PPB O$_3$	10	20	30
Rainwater acidity	μeq H l^{-1}	20	40	80
	Equivalent pH	4.7	4.4	4.1
Deposited acidity	meq H m^{-2}	20	40	80

distances) have been 0.4 PPB NO (4% of the total) and 0.6 PPB NO$_2$ (5% of the total). No annual contributions to smoke values by power stations have been detected by measurements. Ozone is partly formed (at some tens of kilometres from sources) from complex reactions of certain constituents in the air, including hydrocarbons and oxides of nitrogen. The contributions of different sources to ozone and to rainwater acidity are still the subject of research.

18.4.2 Legal controls of air quality: smoke and sulphur dioxide

Legal air quality standards arise from the Council of the European Communities' Large Combustion Plant Directive of 1988 which will be complied with in the UK through regulations promulgated in the Environmental Protection Act of 1990. Each station in the UK will initially be given an allocation of emissions specified by the Chief Inspector on the basis of its expected use during the year. Whilst these allocations are being agreed, earlier quality standards laid down by an EC Directive in 1980 continue to apply. This Directive gives limit values for smoke and sulphur dioxide in the air (Directive 80/779), interpreted by the

Department of Environment Circular 11/81 for the smoke method in use in the UK, as shown in Table 1.30.

Measurements have to be made in places likely to exceed the values and all breaches reported to the European Commission, together with the measures taken to avoid their recurrence. The Department of Trade and Industry publish values each year and report on exceedances. Improvement, even in the first year 1984, was evident when results were compared with the 1960s. For 1984, the 98 percentile of smoke limit value was exceeded at 10 rural sites (6 in South Yorkshire, 2 in Northumberland, 1 in West Yorkshire and 1 in Cumbria). All other limit values were met, including the winter mean smoke limit value. For comparison, for the year ending March 1963 (a peak pollution year), the new winter mean smoke limit value was exceeded at 458 sites out of 509 available. The 98 percentile smoke values are not available for this earlier year.

18.4.3 Legal controls of air quality: nitrogen dioxide

Another current Council Directive (85/203), introduced in July 1987, gives limit values for nitrogen dioxide. The reference period is a year, and the 98 percentile

TABLE 1.30

EC control limits for air quality

Reference period	Limit values in μg m^{-3}, not to be exceeded for:	
	Smoke	Sulphur dioxide
Year (median of daily values)	68	If smoke less than 34, then 120 If smoke more than 34, then 80
Winter (median of daily values, October to March)	111	If smoke less than 51, then 180 If smoke more than 51, then 130
Year (Peak) (98 percentile of daily values, i.e., not to be exceeded more than 7 days in a year)	213	If smoke less than 128, then 350 If smoke more than 128, then 250

of *hourly* values must not exceed 200 μg m^{-3}. Nitric oxide, its intermediary in formation, must also be monitored, although it is not at present included in the limits. Monitoring to meet Directive 85/203 indicates that road traffic makes the greatest contribution to ambient levels. This is to be expected in towns; at a kerbside site in London, the 98% level of NO$_2$ alone can be 440 μg m^{-3} (a breach). Even at some distance from towns, levels of NO$_2$ may be high and during anticyclone conditions in 1990, high values were observed in the Thames Valley due to accumulation of pollution in the relatively stagnant air. Power stations have been found to make little contribution to high NO$_2$ levels.

18.4.4 Control of ozone

There is currently no legal limit in the UK for ozone. However, the World Health Organisation have published environmental health criteria, suggesting that ozone levels over one hour should not exceed 100 PPB more than once per month. Such levels have been exceeded at UK rural sites in several summer months during most of the last 20 years, since monitoring started. Many of the large levels occurred in 'episodes' of simultaneously large levels at many sites in Europe as well as in the UK. The present understanding is that ozone may be limited by controlling the exhausts from petrol and diesel engines (vehicles, aircraft, boats). The Governments of the European Communities have already agreed to the principle of controlling vehicle exhausts, although the precise ways and timings are still under discussion.

Acidity in rain is also 'episodal', but in a slightly different sense from the ozone. At several rural sites monitoring rain acidity in the UK, 30% of the total acidity deposited there each year arrives in six or less individual days of rain.

18.5   The fate of pollutants

Sooner or later, all emissions should be deposited at the ground in some form or other. An indication of how soon, is obtained from calculations based on measurements of sulphur dioxide. Taking the sulphur dioxide emissions from a 2000 MW power station, then some 0.1% of these are estimated to be deposited in a circle of radius 10 km around the station. This is the area of maximum ground-level concentrations from the station, with comparatively less further away. Taking the total sulphur dioxide emissions from all sources in Britain, then around 25% of these have been estimated to be deposited in Britain itself, with the rest depositing on the seas and on countries further away. Some 2% of the total British sulphur emission has been estimated to be deposited on Norway as a whole. This contribution was then some 17% of the sulphur from all sources depositing onto Norway. The British contributions are deposited on the nearer parts of Southern

Norway rather than on the more distant parts of Northern Norway.

Some of the sulphur compounds are brought down by rain and some are deposited 'dry', i.e., in the absence of rain. Of the sulphur compounds depositing on Britain, one-third is estimated to be brought down by rain and two-thirds to be deposited 'dry', on average. The 'dry' contribution is greater nearer sources and the 'wet' contribution is greater in large rainfall areas.

Fewer measurements of nitrogen compounds have been made, but one model indicates the total deposition of nitrogen compounds over Britain to be equivalent to around one-third of the total deposition of sulphur compounds. The estimated average proportions of 'dry' and 'wet' deposited nitrogen compounds seem to be similar to those for sulphur compounds, i.e., one-third wet, two-thirds dry.

Ozone gas is removed from the air by being deposited 'dry'. After atmospheric reactions, its products are deposited both 'dry' and 'wet'. The relative amounts deposited by each mechanism have not been estimated.

Much research is being undertaken throughout the world to evaluate whether or not the combinations and amounts of pollutants which are deposited in different areas could be responsible for adverse effects which have been seen and, if so, which items are critical and should be further controlled.

18.6   'CEGB' research and development on pollutants

This includes monitoring and research by power industry staff, the funding of others to carry out research, the development of new technology, and the evaluation and dissemination of information.

18.6.1 Research projects

Some examples of monitoring have already been given. Other major research includes studies of long range dispersion of pollutants by tracking tracers released from a 2000 MW coal-burning power station stack, using specially instrumented aircraft. There are also studies of the direct effects of mixed pollutants on vegetation and on metal and building materials by laboratory and field exposures of specimens.

Major funding of other projects extends co-operation to include BC. For example, funds have been jointly provided to be allocated by the British Royal Society for research into the causes of surface water acidification in Scandinavia. The Loch Fleet Project funded jointly by PowerGen, National Power, British Coal and the Scottish electricity companies, provides funds for external contract work aimed at finding out how best to counter water acidification locally. Other studies aim to measure the concentrations of ozone-forming chemicals, including nitrogen oxides and hydrocarbons, and to follow the reactions actually in the air, using a specially instrumented aircraft. These projects are additional to an underlying joint monitoring of ground level air pollution, so that trends can be steadily recorded.

18.6.2 New technology

Privatisation of the electricity industry has caused a reappraisal of how best to control air pollution. New gas-fired power stations using combined cycle gas-turbines are more efficient. Little or no sulphur dioxide will be emitted (this depends on the quality of gas burned) and carbon dioxide and nitrogen oxide emissions per MW of electricity generated (see Table 1.41) will be reduced. For existing coal-fired power stations, low NO_x burners have been designed and fitted which reduce emissions by reducing the flame temperatures in the boiler furnace. Sulphur emissions from Drax and Ratcliffe power stations are to be controlled by limestone-gypsum flue gas desulphurising plant. The gypsum produced will largely be used to make building products. The plants will also remove hydrochloric acid from the flue gases as an added benefit.

Other developments, such as pressurised fluidised bed combustion of coal have been evaluated. The latter, studied jointly by the CEGB and British Coal at Grimethorpe in Yorkshire, showed that nitrogen oxide emissions were lowered and sulphur emissions could be reduced by limestone injection. There are no plans at present to build large scale plants based on this technology.

Gasifiers of various types, which can remove sulphur prior to combustion of the fuel, are also being evaluated. Whether such plant is built to complement existing power stations, will depend upon the relative economics of the technology compared to post-combustion flue-gas cleaning, or a conversion to clean fuels such as gas.

18.6.3 Public awareness

During the time that the electricity supply industry was nationalised, close liaison with government laboratories allowed working groups such as the Review Group on Acid Rain, the Photochemical Oxidant Review Group, the Building Effects Review Group, the Acid Waters Review Group and the Atmospheric Effects Review Group to prepare reports jointly. Some were written to be suitable for non-specialists in order to increase public awareness of the complexities of issues such as air pollution control or acid rain.

The review groups, if they survive at all, no longer have joint industry/government participation. Public information will in future come either from government in the form of annual reports or from industry in the form of press releases at appropriate times.

As a footnote to this section, the CEGB environmental studies are continuing following privatisation of the industry in a co-operatively funded Joint Environmental Programme between PowerGen and National Power.

19 Oil fuels

19.1 Introduction

The mechanism of formation of crude oil under the earth's crust is not unanimously agreed by geologists. Alternative hypotheses are still put forward from time to time, but a consensus view would be that oil is formed predominantly from marine animal and vegetable debris accumulating quietly in sea basins or estuaries. The debris, intermingled with fine silt, would then decay anaerobically. The hydrogen to carbon ratio is higher if bacteria are able to act under reducing conditions, since the oxygen present in the original material is removed. Coal in contrast is formed from land animal or vegetable remains, in fresh or brackish water under oxidising or only mildly reducing conditions. The age span of crude oils is somewhat wider than that of coals with the oldest crudes being some 440 million years old and the youngest some 10 million years (compare Table 1.1). Because it is fluid, oil may not be found at its point of origin, as it can migrate through porous strata to a collecting point. It is probably formed under conditions of temperature and pressure which only occur at more than 1500 metres below the surface, but it is found at depths ranging from 600 to 3000 metres.

Crude oils from differing sources vary in their properties, with Libyan and North Sea crudes being fluid and light and South American deposits being particularly heavy. A typical composition is within the range:

Carbon	80	– 89%
Hydrogen	12	– 14%
Nitrogen	0.3	– 1%
Sulphur	0.3	– 3%
Oxygen	2	– 3%
Density	0.8 to 1.0 kg/l	

19.2 Oil resources

World reserves of oil amount to some 8×10^{10} tonnes and daily usage is one ten-thousandth of this amount, i.e., 8×10^6 tonnes. It follows that current reserves will last for 27 years. Coal is more plentiful and known reserves are sufficient for 250 years. As oil is a valuable chemical feedstock, its use as an energy source ought to decline, with coal and nuclear power taking its place. Conservation of energy will also become of increasing importance as oil becomes increasingly expensive. The present generation of oil-burning power stations are unlikely to be replaced, but whilst they are available they provide a flexibility of operation that is not easy to obtain from alternative plant.

The present pattern of world oil production is indicated in Table 1.31.

TABLE 1.31

World oil production

	Proven reserves*		Production		Ratio of proven reserves to production
	billion barrels	%	million barrels daily	%	
USA	26	4.5	8.6	14.6	8:1
Canada	6	1.0	1.3	2.2	13:1
Latin America	71	12.0	5.6	9.4	35:1
of which: Mexico	44		1.9		62:1
Venezuela	20		2.2		25:1
Western Europe	17	2.8	2.4	4.0	19:1
of which: UK	8		1.6		14:1
Norway	7		0.5		37:1
Africa	58	9.9	6.0	10.2	26:1
of which: Libya	26		1.8		39:1
Algeria	12		1.0		32:1
Nigeria	11		2.1		15:1
Middle East	307	52.0	18.4	31.1	46:1
of which: Saudi Arabia	113		9.6		32:1
Kuwait	68		1.4		134:1
Iran	40		1.5		74:1
Abu Dhabi	35		1.3		73:1
Iraq	34		2.6		35:1
Asia and Australasia	20	3.3	2.6	4.4	20:1
of which: Indonesia	11		1.5		19:1
Communist areas	86	14.5	14.3	24.1	16:1
of which: USSR	65		11.8		15:1
China	19		2.1		25:1
World total	591	100.0	59.2	100.0	27:1
of which: OPEC	383	64.8	26.7	45.2	39:1
*at year end (1980)					

Note: 1 barrel = 35 Imperial gallons or 159 litres. Assuming a density of 0.87 kg/l, typical of a Middle Eastern crude, 1 barrel = 138 kg.

19.3 Utilisation of oil in British power stations

The amount of oil burned by power stations varies greatly from year to year (Table 1.32) and to a greater extent than the amount of coal which is burned. This occurs because of economic and political factors; oil is burned when its effective cost is competitive with coal or whenever sufficient coal is not available, as in 1984/85.

When oil is expensive, coal is burned preferentially and oil-fired plant is used solely to ensure flexibility of operation. Typical deliveries to the CEGB in a recent year (1983/84) are shown in Table 1.33.

Gas-oil is used as a fuel for gas turbines primarily, but also as a light-up oil at a few older stations. It is expensive to use for this latter purpose. Light fuel oil is used as a light-up oil at a few modern coal-fired stations. It is easy to store compared to heavier oils but again is relatively expensive and its use is likely to decline. Medium and particularly heavy fuel oils are the most economic fuels for lighting coal-fired burners and for flame stabilisation during start-up and shut-down. They need to be stored and handled at elevated temperature because of their relatively high viscosity but they are a cheaper source of energy than their lighter counterparts. Finally residual fuel oil, which is very

TABLE 1.32

Comparison of coal and oil usage in power stations

Year	Coal thousand tonnes	Oil thousand tonnes
1964/5	65 014	5 521
1965/6	64 350	5 912
1966/7	63 140	6 234
1967/8	66 005	5 674
1968/9	70 066	5 253
1969/70	69 994	7 663
1970/1	68 331	11 228
1971/2	62 218	13 244
1972/3	62 696	14 822
1973/4	63 450	13 755
1974/5	64 586	13 306
1975/6	67 280	9 572
1976/7	70 330	7 973
1977/8	70 005	8 889
1978/9	75 240	9 607
1979/80	80 367	6 755
1980/1	79 460	4 500
1981/2	76 782	4 412
1982/3	75 252	2 983
1983/4	77 183	2 674
1984/5	40 490	22 037

viscous, is the main material burned in oil-fired power stations. It has to be delivered and handled at elevated temperature, but as it is the cheapest petroleum-

TABLE 1.33

Proportions of oil burned in power stations in England and Wales

Type	Gas oil	Light fuel oil	Medium fuel oil	Heavy fuel oil	Residual fuel oil
Class*	D	E	F	G	H
Thousand tonnes	80.3	83.6	154.1	427.8	2015

*see Section 20 of this chapter

derived fuel, the cost savings more than counteract its handling difficulties.

20 Classification of oil

20.1 British Standard classification of fuel oil

Fuel oils in the UK are specified by reference to BS2869:1983. This is a 'master document' and may be consulted to obtain references to other standards for sampling procedures and test methods. Fuel oils are classified into classes A to H in the standard, with A being the lightest, most highly refined, distillate fuel and H being the heaviest, residual, oil. Gasolines (petrol) the lightest liquid fuel are not included in this standard.

20.1.1 Engine fuels

The two Class A fuels are marketed specifically as oil-engine fuels (Diesel). Class A1 is intended primarily as an automotive diesel fuel and is of higher quality, whilst Class A2 is intended as a general purpose diesel fuel. The latter, for example, is used in static diesel engines at sub-stations to provide power for air compressors. There are two grades of Class A fuels to cover winter and summer use. They differ in their propensity to deposit wax when cold. The summer grades, marketed from March until September, are satisfactory down to 0°C and the winter grades available from October to February are suitable to −9°C. This provides adequate protection for most sites, but particularly exposed locations will require storage tanks to be lagged or otherwise protected from extreme cold to ensure that standby engines start as required. If grades have been mixed in storage tanks then it will be necessary to test the fuel at the onset of winter to ensure it has adequate cold weather performance (see Section 24.7 of this chapter).

20.1.2 Burner fuels

Classes B1 and B2 are omitted from the present fuel oil standard and are instead regraded Classes M2 and M3 in BSMA100; they are intended for marine diesel use. The first burner fuels in the fuel oil standard now are the Class C grades, which are divided into

two sub-classes in BS2869: 1983. They are both distillate fuels of the kerosine type, intended for flueless domestic heating in the case of Class C1 and for vaporising and atomising burners in domestic appliances with flues in the case of Class C2.

Class D is a burner fuel (gas-oil) which may be used in gas turbines but it is to all intents and purposes identical to Class A2 and may also be used in diesel engines.

Classes E, F and G are fuel oils which are traditionally differentiated by their Redwood No. 1 viscosity ratings (see Section 24.4 of this chapter) determined at 40°C, of 220 s, 950 s and 3500 s respectively into light, medium and heavy fuel oils. Finally Class H is a residual fuel oil which is particularly heavy and can only be atomised when hot. It may have a viscosity in the region of 6000 s at 40°C. (Note that BS2869:1983 specifies viscosities of Classes E, F, G and H as 13.5, 35, 85 and 130 cSt respectively at 80°C. This is the accepted modern definition of kinematic viscosity and the use of Redwood seconds should be discouraged.) The classes of fuel oil which find uses in power station boilers range from light-up oil and flame stabilisation in coal-fired stations (Classes E and F, formerly '220 s' and '950 s'), through to 'oil overburn' in coal stations and as the main fuel in oil-burning power stations (Classes G and H respectively).

The British Standard gives graphs to allow the viscosity of burner oils to be estimated at various temperatures other than 80°C. This allows the amount of pre-heat to be determined to adjust the burner oil viscosity to the makers recommended value for good atomisation at the burner. The viscosity of an oil may be specified in several ways in addition to the preferred units of centistokes, depending upon the viscometer used for the measurement (Redwood, Saybolt, etc.) and BS2869Z gives conversion methods that can be used to correlate the various results (Fig 1.51).

20.1.3 Compatibility with overseas oils

Any fuel oil produced to the BS2869 specification will be miscible with similarly specified fuel from alternative suppliers in the UK. Occasionally the world price of oil fluctuates widely and it is possible to buy tanker loads on the 'spot market' at advantageous prices. These oils may not be produced to UK-compatible specifications and on mixing with local supplies, in

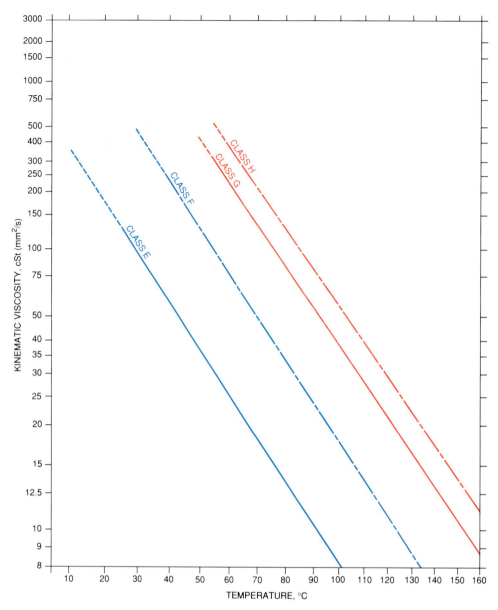

FIG. 1.51 Viscosity conversion chart for fuel oils

storage tanks, a sediment of wax or bitumen could form. To avoid this the CEGB had its own 'spot market' specification (Table 1.34), which was used as a basis for contract whenever such purchases were made.

20.2 Classification by chemical type

As already mentioned, crude oils differ in type; they are classified into three main groups paraffinic, naphthenic and asphaltic on the basis of their predominant chemical constitution. Most crudes contain a variety of hydrocarbons ranging from the simplest gas (methane CH_4) to the most complex solid, paraffin wax or bitumen. Many crudes exhibit just one of the three characteristics but some have a dual character. Their lighter boiling constituents may be paraffinic in

nature, whereas the heavier fractions may tend to be naphthenic. The three oil types differ in the mode of combination of their carbon and hydrogen atoms.

Paraffinic crudes consist largely (>75%) of aliphatic chains of carbon atoms which may be straight or branched and have the general formula C_nH_{2n+2}, for example:

$$H_3C - \underset{\underset{CH_3}{|}}{CH} - (CH_2)_m - \underset{\underset{C_2H_5}{\overset{CH_3}{|}}}{C} - (CH_2)_l - CH_3$$

The ratio of hydrogen to carbon atoms approaches 2.0 for this type of crude.

92

TABLE 1.34

Purchase specification for CEGB fuel oils

			Test method
Density	kg/l @ 15°C	min 0.93 max 0.998	IP160 (ASTM D1298)
Viscosity	cSt @ 80°C	min 38 max 120	IP71 (ASTM D445)
Flash point closed Pensky-Martens	°C	min 66	IP34 (ASTM D93)
Sulphur content	% mass	max 3.0*	IP61 (ASTM D129)
Sediment and water content	% volume	max 1.0	IP75 (ASTM D1796)
Ash content	% mass	max 0.15	IP4 (ASTM D482)
Sodium content	% mass	max 0.015	IP288 or ASTM D1318
Vanadium content	% mass	max 0.030	IP288 or ASTM D1548
Pour-point	°C	max 30	IP15 (ASTM D97)
Asphaltenes	% mass	max 10	IP143
Acid number, strong	mgm KOH/g	max nil	IP177 (ASTM D664)
Calorific value, gross	kJ/kg	min 42 000	IP12 (ASTM D240)

*The CEGB at its sole discretion was prepared to consider offers of oil up to 3.5% sulphur

Naphthenic crudes are largely (>70%) composed of rings of carbon atoms with a full complement of hydrogen (alicyclic rings). They have the general formula $C_nH_{2n-2(r-1)}$ where n is an integer and r is the number of rings, for example:

cyclohexane
C_6H_{12} (r = 1)

perhydronaphthylene
$C_{10}H_{18}$ (r = 2)

The ratio of hydrogen to carbon atoms is about 1.7 to 1.8 for this type of crude.

Aromatic crudes are >60% comprised of rings of carbon atoms held together by double bonds and as a result they contain even less hydrogen. They have the general formula $C_nH_{n-2(r-1)}$, where n is an integer and r is the number of rings. The atomic ratio of hydrogen to carbon atoms is low for such crudes, for example:

benzene
C_6H_6 (r = 1)

naphthylene
$C_{10}H_8$ (r = 2)

The properties of the crude oils derive from their characteristic chemical structures and are summarised for paraffinic and asphaltic types in Table 1.35. Naphthenic crudes are intermediate in their properties.

TABLE 1.35

Characteristics of paraffinic and asphaltic crudes

Characteristic	Paraffinic	Asphaltic
Density	low	high
Yield of gasoline	high	low
Octane number of gasoline	high	low
Sulphur content	low	high
Ratio of hydrogen to carbon	high	low
Centane value of gas-oil	high	low
Yield of lubricants	high	low
Solid product	wax	Bitumen

Typical fuel properties which might be exhibited by burner oils produced from the differing types of crude are given in Table 1.36.

21 Refining of oil

A typical refinery flow diagram is illustrated in Fig 1.52. The purpose of the refinery is to take crude oil, from perhaps several selected sources, in order to convert it into the products at the right of the flow sheet, in the correct proportions for the current market. Light products such as natural gas, gasoline, diesel fuel and aviation fuels have a ready sale; and naphtha is a valuable feedstock for the chemical industry. The oil companies endeavour to promote their sales into these

TABLE 1.36
Some typical analyses of light fuels derived from specific crudes

Crude oil	Carbon %	Hydrogen %	Sulphur %	Specific gravity 15°C	CV kJ/kg	H/C molar ratio
Paraffinic	84	13.8	1.0	0.82	45 100	1.97
Naphthenic	86.3	12.2	1.3	0.86	44 700	1.70
Asphaltic	86.7	10.7	2.1	0.96	44 000	1.48

Note that the fuels with the highest hydrogen to carbon ratio have the greatest calorific value whilst the high aromatic-content fuels have the greatest density

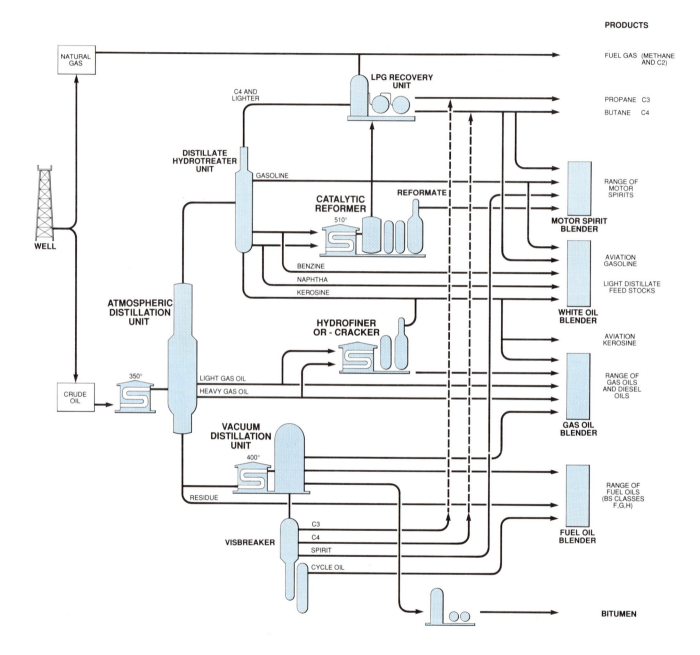

FIG. 1.52 Refinery flow diagram illustrating how hydrocarbon fuels are produced

profitable sectors and the heavier residues from the initial distillation of crude oil are processed to break them down to the favoured lighter products. The ultimate heavy fuel oils left for use in power stations have a character determined by the refinery processes. It will help the engineer to understand the storage and handling of such fuels if he has a knowledge of these refinery processes.

21.1 Preliminary treatment

Crude oil, of whatever type, when it issues from the well contains dissolved gases, water and various solids. Before refining proper may be carried out, a preliminary treatment is necessary to separate these components. The gases and solids will separate from the liquid petroleum in storage tanks, but it is more effective to pass the crude oil through a centrifuge before storing it. The water and solids (sand, wax or bitumen) are readily removed along with much of the dissolved wet gases. The remaining stabilised crude oil may then either be distilled on site, or sent to refineries for distillation and further processing.

The wet gases consist of the lighter hydrocarbons methane to pentane, along with nitrogen, helium, hydrogen, hydrogen sulphide and carbon dioxide. These are separated by standard procedures of compression and adiabatic cooling into dry gas (methane and ethane), water and, after further fractionation, 'natural gasoline' (butane and pentane) and 'bottled gas' (propane and butane). If the well contains a significant amount of helium this may also be recovered from the waste gases.

The crude oil on distillation will give differing proportions of products depending upon its chemical type (see Section 20.2 of this chapter). Table 1.37 illustrates this. The residue may be further distilled under reduced pressure to give light, medium and viscous lubricating oils and a residuum which can be subjected to cracking processes to produce the range of fuel oils (Classes E, F, G, and H) used in power generation.

Some variation in outputs from the refinery may be achieved by varying distillation conditions, for example, by using steam distillation, fractionation under reduced pressure or distillation at pressures above atmospheric. In the latter case higher temperatures are achieved and a measure of 'cracking' can be induced. However, whatever the procedures used, the yields will broadly be as shown in Table 1.37; with light paraffinic crudes being a prime source of petrol naphtha and kerosine, and the heavy asphaltics a prime source of gas-oil, lubricants and burner oils. Naphthenic crudes are intermediate in properties; they give a large residuum on distillation and are notable from the point of view of the electricity industry in being the source of most of the insulating oils used, as they have a good capacity for dissolving gases generated in equipment such as transformers and cables.

21.2 Refinery operations

The motor car is the major consumer of petroleum products and oil companies must balance their output to meet the demand for petrol without generating a glut of unwanted heavy fuel oil simultaneously. Distillation is initially used to produce 'straight-run' products and then some of the heavier materials (boiling point $> 350°C$) must be converted to lighter ones in order to maximise petrol, diesel and aviation fuel production.

The first conversion process used consisted of 'thermal cracking'. Heavy petroleum oils are heated to $560°C$ under pressure (up to 7 MPa) and the heavy molecules are split.

$$R \cdot CH_2 \cdot CH_2 \cdot CH_2 R^1 \rightarrow R \cdot CH = CH_2 + CH_3 R^1$$

The octane number of such gasoline products is low (65–70) but may be improved by adding tetra-ethyl lead to prevent knocking.

A higher octane product (~ 80) is obtained by taking a lighter oil feedstock (naphtha) and using slightly higher temperatures and pressures. This is known as 'thermal reforming'.

By using catalysts in the cracking process such as artificial clays, bentonite or montmorillonite activated by sulphuric acid, even higher octane fuel may be produced. This is known as catalytic cracking.

Finally, in order to obtain 100 octane fuel without the need for tetra-ethyl lead addition, more sophisticated reforming processes may be undertaken. There are several; the feedstock may be broken down to C_3 and C_4 alkenes and these may then be polymerised to give a branched chain product. This is known as 'polymerisation' and is achieved at modest temperature and pressure (1.5 MPa) using a Kieselguhr/acid catalyst.

Alternatively C_3 and C_4 alkanes can be added to their corresponding alkenes in an acid-catalysed 'alkylation' reaction.

Straight chain (low octane) hydrocarbons may be 'isomerised' to branched chain products of excellent quality (ON 95–107) in one step using Friedel Crafts type catalysts (aluminium chloride activated by anhydrous hydrogen chloride). Finally low octane products may be converted in greater yield to high octane products in a 'hydroforming' process. The use of hydrogen and a catalyst (8–10% molybdena or alumina at $\sim 500°C$ and 0.2 MPa) allows good yields and adequate quality products.

TABLE 1.37

Types of crude oils and the yields of products obtained by distillation

Density type		Low	Intermediate			High
Origin		North Africa	North Sea	Middle East	North America	South America
Crude oil						
Density at 15°C	kg/l	0.801	0.842	0.869	0.890	1.000
Total sulphur	% wt	0.1	0.3	2.5	1.0	5.5
Viscosity at 38°C	cSt	1.4	4.5	9.6	13.4	19 400
Pour-point	°C	− 51	0	− 24	0	15
Wax	% wt	3	9	6	7	2
Light gasoline **(0 to 70°C TBP)**						
Yield	% wt	8.8	5.8	4.7	2.4	0.1
Total sulphur	% wt	0.001	0.001	0.020	0.002	0.100
Octane No (Res clear)		73	76	72	75	−
Naphtha **(70–140°C TBP)**						
Yield	% wt	16.0	11.0	7.9	6.5	1.1
Total sulphur	% wt	0.002	0.001	0.020	0.005	0.450
Hydrocarbon type:						
Alkanes (paraffins)	% wt	56	46	72	45	−
Cyclo-alkanes (naphthenes)	% wt	35	42	19	36	−
Aromatics	% wt	9	12	9	19	−
Kerosine **(140–250°C TBP)**						
Yield	% wt	26.3	18.6	16.4	15.6	4.4
Total sulphur	% wt	0.01	0.02	0.20	0.06	2.50
Aromatics	% vol	17	20	17	23	−
Smoke point	mm	24	23	28	21	−
Diesel fuel or gas oil **(250–350°C TBP)**						
Yield	% wt	18.2	19.1	15.3	19.6	9.6
Total sulphur	% wt	0.10	0.18	1.40	0.49	4.40
Cloud point	°C	− 13	− 10	− 7	− 12	− 24
Diesel index		55	53	58	45	30
Residue **(above 350°C TBP)**						
Yield	% wt	27.5	43.5	54.4	55.5	84.8
Total sulphur	% wt	0.3	0.6	4.1	1.5	6.0
Viscosity at 50°C	cSt	74	103	545	370	120 000
Pour-point	°C	18	24	15	21	42
Metals: Vanadium	PPM	< 2	8	50	25	1415
Nickel	PPM	< 2	4	13	13	177

TBP = True Boiling Point at NTP

As refineries have developed the above processes to increase the amount of lighter products, the heavy residues have become increasingly more viscous. It is now necessary to reduce the viscosity of the residual oil by means of a 'Visbreaker'. This is a thermal cracking process which breaks down the larger complex molecules to smaller ones, thus reducing the viscosity of the oil. The product may then be fed for further cracking to produce burner oils, without the need to blend it first with kerosine or gas oil to improve its flow. The residues of Visbreaking and catalytic cracking can be of very high density (greater than 1.3) and need blending before use as heavy fuel oil (Class H).

The schematic refinery layout illustrated in Fig 1.52 includes each of the processes described so far. Possible problems which may arise with fuel performance as a result of refinery operations are referred to in the next section.

21.3 United Kingdom consumption of petroleum products

The extent to which refining of crude oil is necessary may be judged by comparing the yields of various petroleum products obtained by initial distillation (Table 1.37) with the actual consumption in the United Kingdom in a recent year (Table 1.38).

Even North African crude will only give 8.8% of light gasoline and 16.0% of napththa by distillation,

TABLE 1.38

Consumption of petroleum products in the United Kingdom

Product	1983 tonnes $\times 10^6$	%	1984 tonnes $\times 10^6$	%
Gases (refinery gases, ethane, propane and butane)	2.217	3.2	2.560	2.9
Aviation fuels	4.596	6.6	4.856	5.6
Motor spirit	19.566	28.0	20.226	23.3
Naphtha and industrial spirits	3.765	5.4	3.557	4.1
Kerosine	1.662	2.4	1.710	2.0
DERV	6.183	8.9	6.755	7.8
Gas/diesel oil	10.432	14.9	10.499	12.1
Fuel oils (Classes E to H)	12.524	17.9	27.864	32.1
Lubricants	0.818	1.2	0.818	0.9
Bitumen	1.986	2.8	1.900	2.2
Petroleum coke	0.142	0.2	0.104	0.1
Miscellaneous	0.573	0.8	0.586	0.7
Refinery consumption	5.343	7.7	5.350	6.2
Totals	69.807	100	86.785	100

yet the UK market for motor spirit in 1983 was 28% of total production. This proportion can only be achieved by means of the many refinery processes discussed earlier in this section, thus illustrating how the refinery is essential to flexible production. The distribution of products in 1983 was typical, but in 1984 was distorted by a high demand for fuel oils as a result of limited coal availability.

The refineries which at present supply fuel oil in bulk to the electricity supply industry are:

Fawley (Southampton Water)
Milford Haven (Pembrokeshire)
Stanlow (Merseyside)
Shellhaven (Thames Estuary).

22 The uses of fuel oil by the CEGB

22.1 Lighting-up uses in power stations

Burner oils of Classes E to G may be used for lighting pulverised coal burners. The lightest Class E, of 13.50 cSt viscosity at 80°C, is used for lighting-up at, for example, Drakelow power station and for flame stabilisation at Ratcliffe and Willington. At one time, Class E fuel was produced directly but it is now a blended product. As a result stocks should be checked periodically for evidence of waxing. The asphaltene content may also be higher than previously, but as tonnages used are small, this should not cause problems. The oil may be used without heating, a temperature of 10°C is satisfactory for handling and it is perhaps the most convenient light-up fuel, but also the most expensive.

Class F (35 cSt/80°C) is used for lighting-up at Castle Donington, High Marnham and Fiddler's Ferry power stations, for example. It is directly produced so should not suffer from waxing but it requires heating above 30°C to enable easy flow.

Class G (85 cSt/80°C) is used for lighting-up at Cottam, Drax, Ferrybridge and Thorpe Marsh power stations and for flame stabilisation. It requires heating above 50°C before it flows satisfactorily, but is cheap to use.

The minimum flow temperatures given above are specification values for the classes but favourable lower temperatures may often be achieved. Deliveries can be checked using a pumpability test (BS2000 Part 230) whenever possible if it is desired to reduce heating costs.

Although the prime use of fuel oil in coal-fired stations is for lighting the coal burners and stabilising combustion on start-up or shutdown, many large stations can now burn significant amounts (~20%) of oil as a primary fuel. This oil 'overburn' allows a more flexible fuel policy. There are many advantages and few disadvantages of this dual-firing but one point should be mentioned. If oil is being burned to stabilise combustion it is possible that incompletely-burned oil coke spheres may collect in the precipitator ash hoppers along with the coal ash. Great care must then be exercised during 'ashing-out', as such a mixture, when hot, can ignite spontaneously. The ash hoppers should only be emptied via a water cyclone.

22.2 Heavy fuel oil use in oil-fired boilers

The major oil-fired stations at Fawley, Grain, Ince, Littlebrook and Pembroke burn Class H residual fuel oil (130 cSt/80°C). This viscous material usually requires heating to 55°C before flowing readily. Since the stations were commissioned (from 1969 onwards), changes in refinery practice have changed the character of their fuel.

The introduction of visbreaking has tended to increase the asphaltene content of residual oils. Combustion is less easy when asphaltenes are present and grits may be emitted. Marchwood Engineering Laboratories have designed new burners for Pembroke and Fawley which reduce the oil droplet size and aid combustion, thus avoiding grit emission.

Catalytic crackers are able to produce high density fluids of substantial aromatic content now and densities approaching 1.6 have been observed, although this is extreme. The aromatic nature of the fluids gives them a high 'cutting power' and relatively low viscosity and they are useful for cleaning pipelines. The mixture may then be blended into Class H oil for sale. Heavy fuel oil purchased by the power stations should not have a density greater than 1.005 at 15°C otherwise it may not float on water. (The large oil-fired stations are on the coast or estuaries.) Should there be an oil spill with over-dense oil it is more difficult to clean up, and density checks of deliveries have become important.

A final point to note with heavy oils is that they are residues of refinery processes. As a result they can contain catalyst residues, artificial clays, montmorillonite or bentonite (see Section 21 of this chapter). When such contaminated oils are burned, slag deposits can form in the boiler. Slagging is an expected phenomenon in coal-fired boilers, but oil-fired boilers are not designed to cope with such difficulties and catalyst residues are therefore a problem. Deliveries should be checked for solids content to protect against possible slagging.

22.3 Light oil used in gas turbines

Gas turbines meet three requirements. Installations such as those at Ocker Hill, Leicester or Bulls Bridge are intended to meet peak loads. Since the Dinorwig pumped-storage scheme was commissioned they are less necessary than in the past, but they still perform a useful function at times of maximum demand although they are expensive to run. Gas turbines on fossil-fired power station sites are intended to bring the plant up from a 'black-start', i.e., when no assistance is available from the grid. Those on nuclear sites are there to provide auxiliary power in the case of a reactor shutdown, although prior to commissioning they will be available for meeting peak-load demands.

In each case, fuel quality is of great importance to ensure an instant start. The fuel must be free from sediment and should not precipitate wax when cold. As noted earlier, Class D fuel is supplied in winter or summer grades which have cold filter plug points of −9°C or 0°C respectively. As temperatures below −9°C have occurred more frequently in recent winters, tanks are now often lagged and trace heating is provided for pipe runs.

The fuel must also be free of sodium, as sodium salts contribute to corrosion of turbine blades. A limit of 0.2 PPM is set for deliveries and checking must be carried out as a routine.

22.4 Diesel fuel

The CEGB uses diesel fuel in Terex bowl scrapers (Fig 1.19), for transmission district air compressors used on air-blast circuit-breakers and at some nuclear installations where diesels rather than gas turbines are used for standby power. As noted in Section 22.3 of this chapter, fuel must be free of sediment and of the appropriate seasonal grade. Tanks require lagging in cold weather and on occasion the use of additives has been considered to lower the cold filter plug point. However, the additive effectiveness is difficult to predict and this is not a recommended policy.

A specific point to note concerns long term storage for standby diesel generators. Twenty-five years ago, refineries were producing 'straight cut' diesel fuel by distillation of the crude. Nowadays extra diesel output is obtained by cracking and reforming. This introduces olefines (unsaturated hydrocarbons) into the product which are more susceptible to atmospheric oxidation leading to the production of gums. Modern standby stores will have a shorter life than 'vintage' material. As a result all long-term stores should be checked for gum formation every two years.

23 Delivery and sampling of oil

23.1 Delivery

Petroleum products are delivered to power stations by ship, pipeline, rail or road. Large oil-fired stations, such as Fawley, are linked to nearby refineries and receive their heavy fuel oil by pipeline. Such stations are also supplied by ships of 500 to 40 000 tonne capacity. The smaller ships are used for estuarial transport of heavy fuel oil. Gas-oil for use in the large gas turbines is delivered direct from the refinery in trainloads of 500 tonnes or more.

Practically all of the other fuel oils used by the industry are carried either in road tankers of up to 22 tonnes capacity or by rail in trains of up to 1000 tonnes capacity.

Oil companies must ensure that the heavier fuel oils arrive for off-loading at a sufficiently high temperature to enable the fuel to be effectively and swiftly discharged. Whilst most ships' tanks are provided with steam heating coils which can be used (if required) to maintain the correct temperature, neither rail tank cars nor road tankers can be heated *en route*. Because there can be delays on the railways, steam coils are fitted in the cars and the customer has to provide his own steam supply for emergency use. Road tanker vehicles are not usually subject to serious delays and the customer is able to reject any consignment which has overcooled.

23.2 Measurement for pricing purposes

The oil companies normally sell oil fuels by 'volume', but the Government apply a tax of £7.75 per tonne. This means that the user should check not only the volume delivered but also the oil temperature, in order to correct the measured volume to the volume at 15°C which is used in the method of billing.

Pricing by volume (litres) is satisfactory for small quantities of light-up fuel but large quantities of fuels are priced on a 'per tonne' basis. The method of checking deliveries is the same so far as ascertaining the volume supplied at 15°C, but in addition a specific gravity measurement is required so that the number of litres per tonne can be derived. Specific gravity can vary considerably from consignment to consignment, even of the same product, and the temperature of a consignment may not be even throughout. Consequently, attention must be given to ensuring that the values used for specific gravity and temperature are properly representative of the whole consignment. Several measurements of each parameter will be needed from each train, for example.

23.3 Sampling — general

In addition to sampling for specific gravity and temperature for the purpose of measurement checks, samples are taken to obtain small representative quantities, from a part or whole of a quantity of oil, for visual or laboratory examination. It cannot be over-emphasised that an incorrect sample invalidates all of the work ultimately done in the laboratory on the oil. The sample therefore must not contain any adventitious material or be altered in the sampling procedure by evaporation or by oxidation.

The Institute of Petroleum's handbook 'IP Standards for Petroleum and its Products: Part IV: Methods of Sampling' (revised annually) gives detailed procedures for sampling oils from cans, drums, rail and road tankers, storage tanks, ships, barges and pipelines, and describes appropriate sampling equipment. Cleanliness is essential during sampling, sampling equipment and containers therefore must be chemically clean. Responsible personnel should be trained in the techniques of sampling and compositing, and be skilled in the preparation of final samples for the laboratory. Because of the fire hazard, necessary safety precautions should be implemented at each stage of operation.

23.4 Sampling from pipelines

Samples may be drawn continuously from pipelines by means of a drip feed system, or intermittently at timed intervals, by opening a sampling valve to give individual weighted increments. The pipeline sampling equipment consists of a 12.5 mm pipe which projects into the pipeline internally to one-third of the pipeline diameter. The inner end is bent at 90° to face the flow of the oil stream and a sampling cock is fitted

outside. Figure 1.53 shows a typical pipeline sampling arrangement for continuous or intermittent sampling.

If maximum accuracy is required, a micropump may be adapted for continuous sampling to match the flow rate in the sampling line with that in the pipe (isokinetic sampling).

23.5 Sampling from ships

An oil cargo may be shipped in several separate tanks on the vessel and is sampled at the suppliers' end before despatch. On arrival at a power station wharf, each hold is sampled separately before discharge and the oil is then discharged to the station where it may again be check-sampled by a continuous sampling technique. For sampling from holds, the weighted sampling-can method is conventionally used. The can, of 1 litre capacity and suitably weighted (Fig 1.54), is fitted with a long cord attached to a stopper in its neck in such

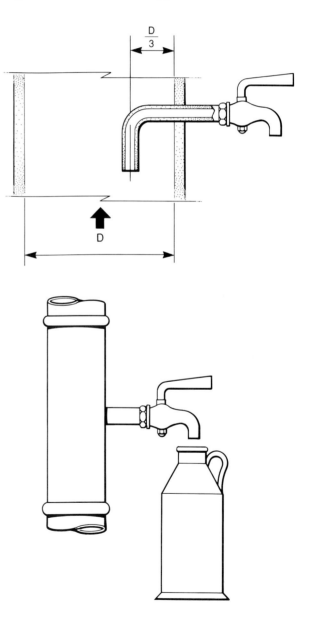

Fig. 1.53 Pipeline sampling

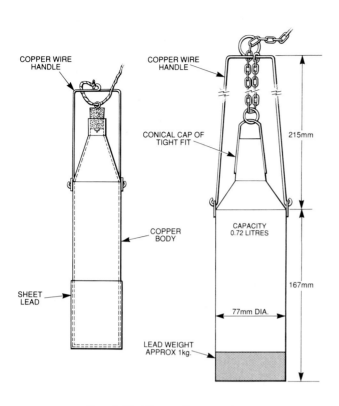

FIG. 1.54 Oil sampling containers

a way that the stopper can be jerked out at any desired level. Upper, middle and lower samples are taken from each hold at one-sixth, one-half and five-sixths depths, respectively, after first rinsing out the can at each depth before drawing the sample. The samples may then be composited on a weighted basis and mixed to give three final samples; one for the supplier, one for the consumer and the third retained by the consumer as a referee sample in case of dispute.

When sampling from holds, tankers and storage tanks, it is essential to sample systematically, from top to bottom in order, so that each sample is obtained before the oil at that level is disturbed.

Where tanks are horizontal, cylindrical or irregular in shape, the tank dip should be translated into volume from the tank calibration table and the samples taken at levels equivalent to one-sixth, one-half and five-sixths of that volume.

23.6 Sampling from road and rail tankers

The general principle of the one-sixth, one-half and five-sixths sampling depth procedure may be applied but it is sometimes more convenient to discharge the tankers into a site storage tank before sampling. However, the latter method has the disadvantage that the storage tank is likely to contain some oil already. In addition to undertaking sampling, one should also check that *discharge* is complete. The old adage, 'Never mind the quality, feel the width' contains more than one message.

23.7 Sampling from storage tanks

A number of samples are taken from different parts of the tank on the one-sixth, etc., basis and combined in suitable proportions to provide a single tank composite oil sample.

23.8 Sample containers

Containers should be of glass or metal and fitted with a suitable cap or lid. If of glass, the bottles must be scrupulously clean and treatment with chromic acid followed by copious water rinsing is recommended. After washing and draining, the bottle is dried in an air oven at 105°C. If cleaning used oil bottles, pre-cleaning with an oil solvent before the above treatment is essential. Cans are often used as sample containers and these should have external seams. Particular care is necessary in cleaning, particularly new cans, for it is essential that all traces of rust, soldering flux, etc., be removed.

24 Analysis and testing of oil

24.1 Standards applicable

The Institute of Petroleum produces IP standards for petroleum and its products in several parts. Part 1 covers Methods for Analysis and Testing and is revised annually. The introduction to the methods lists those British Standards which are identical or technically identical to IP or joint IP-ASTM methods.

24.2 Safety tests

It is seldom necessary for regular routine tests to be made for all the main characteristics and properties of the oil fuels. However, an inherent danger of oil fuels is their fire risk. For this reason it is wise either to check every consignment before accepting delivery or, because of the delays this would involve in many cases, to take deliveries into holding tanks at first. The oil can then be tested and will not be released for use unless satisfactory. The check most desirable is the 'closed flashpoint' test, and fuel oils and gas-oil should have a closed flashpoint (Pensky-Martens) of not less than 66°C for fuels of Classes E to H or 56°C for diesel fuel. This test is not simple and a rapid test has been devised (IP303) which can be used to determine whether the oil will or will not flash at the specified temperatures of 66° or 56°C respectively.

Other quick tests may be appropriate to give an initial indication of a wrong delivery; for example, smell and sight tests, and a specific gravity test by hydrometer will, on occasion, show whether or not the correct grade of oil has been delivered. In addition, the details of the delivery note should be carefully examined. However, none of these simple precautionary checks would provide a safeguard against receipt of the correct grade of fuel contaminated with a low flashpoint product; therefore, the flashpoint is the only true indication. Oil companies are well aware of the danger and do all that is possible to ensure that correct and uncontaminated deliveries of fuel oil are made.

24.3 Density

This is primarily required when a fuel oil is sold on the 'per tonne' basis but is also needed for 'per litre' sales in order to correct weight to volume at 15°C. A hydrometer or pyknometer procedure may be used as appropriate (IP160 or IP190). A check should have been made at the time of sampling that the tank contained only oil and that there is no water present, but a further check should also be made on the sample in the laboratory to see if any water has separated with standing.

24.4 Viscosity

Viscosity is the main characteristic by which fuel oils are classified. Its measurement is required routinely to show that the correct grade of oil is in use and that the temperature of preheat is satisfactory to give the correct viscosity of the fuel at the burners. The viscosity of oil is strongly affected by temperature and diminishes as the temperature rises (Fig 1.51). The preferred unit of measurement is the centistoke and Fig 1.51 gives conversions to earlier units if required. Burners in Britain are designed to give good atomisation with oil of 12 to 24 cSt kinematic viscosity (60–100 Redwood) and the fuel supplied to the burner must be heated to a sufficient temperature to reduce its viscosity until it is within this range. The measurement of viscosity is described in the Institute of Petroleum test IP71.

24.5 Flash point

The flash point is the minimum temperature at which an oil gives off sufficient vapour to ignite momentarily (i.e., flash) on the application of a 4 mm bead-sized flame, when the oil is heated at a defined constant rate in a standard apparatus. The sample is placed in a cup which is heated by an air bath and the flame is introduced at regular intervals as the oil warms up.

The flash point observed depends upon the type of instrument used for the test. In particular it may have a 'closed' or 'open' cup to hold the sample. The 'closed' flash point is most commonly used for fuel oils as it always gives a lower reading than the 'open' and therefore gives a better indication of the worst hazard in handling the fuel.

There are two commonly used standard forms of apparatus, the Abel used for petroleum products with low flash point (−18 to 70°C) described in IP33 and IP170 and the Pensky-Martens which is used for higher flash points (to 371°C). The Pensky-Martens

equipment is illustrated in Fig 1.55. Its use as a closed tester is described in Institute of Petroleum Standard IP34, and as an open tester in IP35. The standard methods, particularly the use of the Pensky-Martens closed-cup (IP34) would be used as the referee procedure if there were a dispute about the quality of supplies. In routine use, however, miniflash testers are preferred (IP304). These are automated scaled down versions of the Pensky-Martens apparatus which use

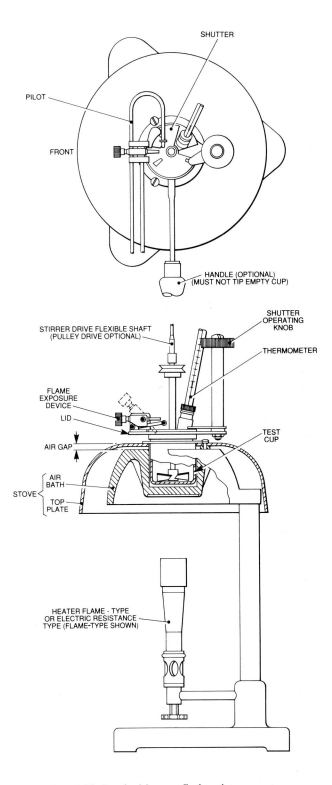

FIG. 1.55 Pensky-Martens flash point apparatus

only 2 ml of sample. They can give a rapid 'go-no-go' result as to whether a delivery is to specification, whilst the consignment of oil is awaiting discharge.

24.6 Water content

The water content of fuel oil is determined by the Dean and Stark method (IP74), where a standard amount of sample, typically 100 ml, is co-distilled with 100 ml of a standard boiling range petroleum spirit. The condensate falls into a graduated tube; the water sinks to the bottom enabling the amount to be read off (Fig 1.56).

Note that whilst petroleum spirit is used to distil water from fuel oil in this method, xylene is used for more aromatic materials. Supplies of emulsified fuels consisting of water-bitumen mixtures for example, would be determined using xylene and not petroleum spirit.

24.7 Cloud point, pour point and cold-filter plug point

In cold weather the waxes normally dissolved in engine oils begin to solidify into small crystals, causing the oil to become cloudy or hazy. The crystals continue to grow as the temperature falls further and will block

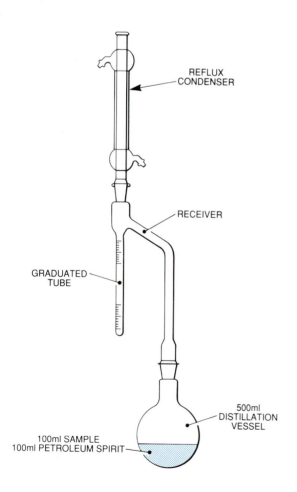

FIG. 1.56 Dean and Stark apparatus

REFLUX CONDENSER

RECEIVER

GRADUATED TUBE

500ml DISTILLATION VESSEL

100ml SAMPLE 100ml PETROLEUM SPIRIT

the fine strainers used in oil engines. Such strainers may have apertures as small as 5 microns, in gas turbines for example.

Additives are added to engine oils to inhibit the growth of the wax crystals so that the oil can be used at lower temperatures. It is necessary to check supplies to ensure they are fit for use in cold weather. The cloud point test (IP219) is simple to carry out but is not so useful as the cold-filter plug point test (IP309). The latter is now routinely used by the electricity supply industry to confirm that fuel stocks are appropriate for winter use.

The pour point test (IP15) is a simple test of the 'pumpability' of an oil at lower temperatures. Samples of oil are cooled and the 'pour point' is the lowest temperature at which movement of the oil can be observed. It is not an ideal test since it has been found that most oils can be used at temperatures considerably below their pour point, but the pumping rate of others (notably those from the North and West African crudes) becomes abnormal because of thickening at temperatures above the pour point. The reasons for the anomalies are connected with the oils' responses to shear during pumping. However, if the limitations of the pour point test are allowed for, it is of value in indicating minimum temperatures for storage and handling of supplies.

24.8 Calorific value

The measurement of the calorific value of a sample of fuel oil is carried out in a similar manner to that used for coal (see Section 12.4 of this chapter). A weighed quantity of the sample is burned in oxygen in a bomb calorimeter under controlled conditions. The heat of combustion is calculated from the weight of the sample and the rise in temperature, after allowances for heat transfer and the formation of nitric and sulphuric acids in the bomb. As with coal, the value obtained is the gross calorific value at constant volume. A different standard applies to that used for coal determination, in this case BS4379 or IP12. As for coal, the bomb calorimeter is calibrated using a benzoic acid standard.

The main points of difference compared to analysing coal are associated with handling the liquid fuel. The lighter fuel oils of low flash point may evaporate during weighing. To avoid this a 'lid' is used on the combustion cup, made from cellulose. Typical gas-oils can sputter during combustion so that burning is incomplete and it is sometimes an advantage to use alumina powder in the cup to prevent this.

If a 'quick' analysis of a fuel is required it is possible to determine the calorific value indirectly. There is a correlation between the density of a fuel oil and its calorific value, provided its sulphur content is known. The latter can readily be measured by X-ray fluorescence methods and the density is measured by pyknometry. However, in practice the procedure has not been found to be quick for small numbers of samples, as cleaning the pyknometers after use takes time. If the cleaning were automated in a busy laboratory this approach might have advantages. The correlation graphs are shown in Fig 1.57.

24.9 Sulphur

The sulphur content of fuel oil is determined to correct the measurement of its calorific value, as mentioned above, to provide data on sulphur emissions from the station with regard to air pollution control and to give information on likely corrosion in the boiler and back-end ductwork. Molten sulphates cause corrosion in the boiler and sulphuric acid causes corrosion in the ductwork, when the combustion gases cool to temperatures below the acid dewpoint.

There are several ways of measuring sulphur in fuel oil. As already noted, X-ray fluorescence determination of an oil sample is a quick procedure. Alternatively, after the calorific value measurement, sulphur can be determined in the bomb washings using a barium sulphate precipitation (IP61).

There are also quick modern methods, using infra red absorption evaluation of the SO_2 produced on combustion.

24.10 Sodium

Gas-oils used in gas turbines should be free of sodium, otherwise molten sodium salts depositing on the turbine blades will cause accelerated corrosion of the Nimonic alloys. The sodium content of deliveries is limited to 0.2 PPM. Samples of deliveries are analysed using atomic absorption spectroscopy. Deliveries are initially held in a receiving tank. If they are satisfactory they are transferred to storage. If they contain high sodium levels it may be necessary to pass them through a treatment plant before they can be accepted for storage.

Heavy fuel oil deliveries to power stations can contain sodium in the range 40 to 150 PPM. On combustion, molten sodium salts may cause boiler fouling. For this reason the sodium content is monitored and efforts are made to prevent the levels exceeding 100 PPM. Above this figure boiler fouling becomes an increasing problem. The sodium tends to arise predominantly from sea water contamination in the tankers during transportation.

Crude oil does not contain large amounts of sodium. However some refinery neutralisation processes may introduce a certain amount of sodium if they are not properly controlled.

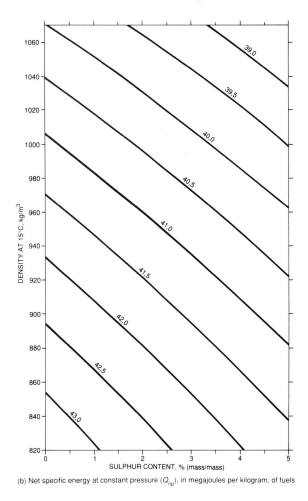

(a) Gross specific energy at constant volume (Q_{gv}), in megajoules per kilogram, of fuels

(b) Net specific energy at constant pressure (Q_{np}), in megajoules per kilogram, of fuels

FIG. 1.57 Nomograph for estimating the calorific value of fuel oil from its density and sulphur content

24.11 Vanadium

Whilst modern mammals contain iron in the hae-moglobin in their blood, the molluscs present in the marine sediments associated with oil formation, appear to have contained vanadium-porphyrin compounds. Crude oils therefore contain vanadium to varying degrees. Venezuelan crudes may contain 800 PPM by weight (although 500 PPM is a more normal maximum). More typical figures for the heavy BS Class H oils used in power stations are 50 to 250 PPM.

On combustion, sodium vanadyl sulphates form in the boiler and contribute to molten salt deposits at temperatures above about 566°C. The salts in combination with 'catalytic fines' left in the oil may also give black glassy deposits which affect heat transfer and hence boiler efficiency. The molten salt deposits cause boiler tube corrosion.

Vanadium may be determined at concentrations above 10 mg/kg in fuel oils by igniting the sample in the presence of sulphur. The ash is then dissolved in sulphuric acid and the vanadium is determined by the photometric phosphotungstate method. Nickel can be determined using the same solution, by the photometric dimethylglyoxime method (IP285). Where they are available, modern instrumental methods such as atomic absorption may be preferred for the determination of the metals in solution.

24.12 Ash content

The ash content of fuel oil is determined in an analogous manner to coal (Section 12.2 of this chapter) but as there is much less ash in oil, larger samples are required. The sample crucible may be up to 120 ml capacity and more than one filling may be required in order to obtain 20 mg ash. Before final ashing the weighed sample in the crucible is warmed and allowed to ignite. A carbonaceous residue is left which is then heated to 775°C ±25°C to complete combustion. The remaining ash in the crucible is then weighed to complete the analysis (IP4). Analysis is usually straightforward but if there is water in the sample it may foam during the initial combustion. Addition of isopropyl alcohol to such a sample is helpful. Several strips of ashless filter paper are then added to the mixture and it is heated; when the paper begins to burn the greater part of the water will have been removed.

24.13 Sediment

Fuel oils may contain 0 to 0.4% of components which are of relatively high molecular weight and which are insoluble in aromatic solvents. They can cause sediment formation during storage. They are determined by taking a 10 ml sample of oil and extracting it with refluxing toluene until the residue of sediment reaches constant weight. The test is described in the Institute of Petroleum Standard IP53.

24.14 Asphaltene content

Fuel oils also contain rather larger quantities of high molecular weight components which are soluble in aromatic solvents but insoluble in alphatic ones. These are called asphaltenes. During combustion asphaltenes may not be completely oxidised, so that acid smuts may escape from the stack. A knowledge of the asphaltene content of fuel oils is helpful to the combustion chemist, who must ensure such problems are minimised.

To determine the asphaltene content of an oil sample it is heated under reflux with thirty times its weight of n-heptane. This precipitates the asphaltenes plus some waxy material both of which are collected on a filter paper. The solids on the filter paper are extracted first with refluxing n-heptane to remove the wax and then with benzene or another aromatic solvent, to dissolve the asphaltene. The aromatic solvent is then evaporated and the residue of asphaltene is weighed (IP143).

25 Storage and handling of oil

General specifications for oil fuel used in power stations will be found in Sections 20.1 and 22 of this chapter. Specific aspects relevant to storage and handling are summarised in this section.

25.1 Heavy oil

Heavy fuel oils are costly to store, not only because of their capital cost, which includes tax on the fuel delivered, but also because the tanks necessary to hold the oil are expensive. In addition, because heavy oils are of high viscosity, they must be received and handled hot with all lines external to the tanks trace heated to prevent slow flows or blockages in cold weather. Suitable storage temperatures are noted in Table 1.39.

It is possible to purchase winter and summer grades of fuel but these are only available on a seasonal basis. To meet EC rules it is necessary to keep a 60 day stock of strategic materials, which includes fuel oils, and during restocking some mixing of grades may occur, consequently a knowledge of the cold-filter plug point of the materials in stock is required by stations prior to winter. They are then able to plan their oil-handling procedures more effectively.

Modern refinery practice has changed the nature of many oil fuels. Straight cut fuels, produced by distillation essentially, were easier to store than cracked fuels and the heavy products of visbreaking. The tendency to form gums from unsaturated compounds in the cracked material is probably greater now and sediment testing may be necessary at some locations, depending upon how quickly stocks are used and replaced and whether any mixing of supplies takes place. The heavy product from visbreaking can have high densities, greater than that of water, so that surface water drains designed to prevent fuel oil escapes, by skimming the water surface before the station outfall,

TABLE 1.39

Storage, handling and atomisation temperatures required for various classes of fuel oil

BS class	Minimum temperature for storage °C	Minimum temperature for outflow from storage and for handling, °C	Atomisation temperature, °C
D	Maintain above cold filter plug point temperature		
E	10	10	60–65
F	25	30	90–95
G	40	50	115–120
H	45	55	125–140

will not be effective. The heavy oil will roll underneath the barriers. Density measurements in addition to being required commercially, will alert a station to this potential problem.

The 'spot market' specification used for purchases of single consignments (Table 1.34) is intended to ensure that the above dangers are avoided. Regular deliveries are to the relevant British Standards noted in Section 20.1 of this chapter.

25.2 Light oil

Light oils are used for standby equipment, such as diesel generators. Such equipment may rarely be used and stocks could be many years old. One diesel tank, containing fuel over twenty years old, when tested for sediment and waxing potential was found to be in excellent condition. More recently produced fuel, the product of cracking rather than distillation, has a greater tendency to form gums. Five yearly checks of long term stocks for gum formation is necessary. Prior to winter, cold-filter plug points should be checked and trace heating of fuel lines must be installed at exposed locations.

Light oil used for lighting-up purposes is often blended. Checks for potential sediment formation are required before oil deliveries are mixed.

Light oils used as gas turbine fuels present particular problems. Many of the gas turbines in use are designed to aviation standards. Aircraft use kerosene, which is less viscous and of lower specific gravity than the light gas-oils used for power generation. Any solid particles contained in kerosine settle better and are easier to remove than similar suspensions in gas-oil. Aviation filters are typically capable of removing 5 micron particles. Such filters are readily blocked when used with gas-oils, which may contain debris from rusting of the fuel tanks and pipework. As a result, prefilters are fitted to remove +10 micron material prior to the finer 5 micron filter, and precautions must be taken to prevent rusting.

All petroleum products exposed to moist air will take up some water; the heavier the product the greater the amount. Gas-oil contains sufficient to rust tanks and pipework and to form ice particles in cold weather. The latter can also block fine filters. Trace heating of exposed pipework must be installed to keep the fuel free-flowing in cold weather. Most oil fuels used on

power stations are stored at elevated temperatures to prevent flow problems. The conditions are summarised in Table 1.39.

A final problem caused by water contamination in gas turbine fuel is the possible introduction of sodium into the engine. This causes enhanced corrosion of the turbine blades. A rigorous control is applied to deliveries. They are first put into a holding tank. If dissolved sodium levels above 5 PPM are detected, the oil is washed with deionised water and centrifuged to dry it before it is passed to storage tanks ready for use.

25.3 Safety

When storage tanks are emptied for cleaning or maintenance, personnel must not be allowed in until checks have established that it is safe to enter. There are two principal hazards, explosion and asphyxiation. In addition protective clothing must be worn as skin contact with residual oily materials must be avoided.

The tank itself is recognisably a confined space within the safety rules. When the tank is empty, man-access covers are removed and the vessel is allowed to vent for 24 hours or more to ensure that vapours escape. When it is safe to enter, further checking may be necessary to ensure that cleaning operations do not disturb sediments and re-create high vapour levels.

A point that should be noted is that large storage tanks have a bund wall surrounding them. In case of a tank leak, the bund will retain the spillage. If the wall is tall compared with the gap between the tank and bund walls, then the space within the bund, in still air conditions, can retain high vapour concentrations. It should be checked in addition to the tank.

Finally, as heavy fuels are handled hot to lower their viscosity, records must be kept of the closed-cup flash point of deliveries. The oil fuel in store should never be heated to near to its flash point because of the fire hazard.

26 Oil-firing — combustion and associated problems

26.1 Combustion

The fuel oil burned in power stations is very heavy and will only flow at elevated temperature. The hot oil is delivered to burners which atomise it, so that the small droplets burn completely. The aim, as with any other fuel, is to ensure complete combustion with as little excess air as possible.

Environmental issues, however, are also important and it is worth considering first how oil compares with other fuels in this respect.

The differences may be illustrated by considering three typical examples of fuels currently being burned; a non-caking high volatile coal (902), a Class H heavy fuel oil with a high density, typical of modern refinery practice, and natural gas; see Table 1.40.

TABLE 1.40

Three typical power station fuels

Content	Coal	Oil	Gas
	%	%	%
Moisture	12	0.2	–
Ash	14	0.05	–
Carbon	59	86.0	75
Hydrogen	3.7	10.0	25
Nitrogen	1.2	0.4	–
Sulphur	1.7	3.0	–
Oxygen	8.4	0.35	–
Calorific value*	23 840 kJ/kg	42 000 kJ/kg	54 040 kJ/k
Specific gravity		0.995	

* gross value, constant volume

To obtain the same heat input to the boiler as provided by one tonne of oil, 1.76 tonnes of coal or 0.833 tonnes of gas must be burned. The products are compared on this equal (gross) heat input basis in Table 1.41.

TABLE 1.41

Products of combustion from examples of fuel given in Table 1.40

	Coal	Oil	Gas
Amount burned (tonnes)	1.76	1.0	0.833
Heat released (GJ)	42.0	42.0	42.0
Products:	kg	kg	kg
Total moisture	797.2	902.0	1874
Ash	246.4	0.5	nil
Carbon dioxide	3484.8	3153.3	2290
Sulphur dioxide	59.8	60.0	nil

Solid waste is a feature of coal burning, but little or no solid material is generated by oil or gas burning. Consequently electrostatic precipitators are only necessary with coal burning, to protect the environment from dust deposition.

Carbon dioxide is produced in greatest amount by coal burning, but note that the use of visbreakers in refineries, to produce a heavy oil with less hydrogen content, has led to similar amounts of carbon dioxide arising from both fuels. Class H oils produce about 90% of the carbon dioxide generated during equivalent coal burning. Lighter oils (gas-oils to Class G) contain hydrogen in reducing proportion to their specific gravity (d) such that % H = 26 – 15d.

They also have a greater calorific value. Making allowance for this and the lower C/H ratio leads the lightest of them to produce about 80% of the carbon dioxide generated from coal. Finally natural gas, allowing for stack losses, produces only 70% of the carbon dioxide that would be produced by coal. It is for this reason that gas firing is promoted as a means of alleviating the greenhouse effect.

The *sulphur dioxide* produced from coal and oil burning is similar (Table 1.41) but again natural gas has minimal environmental impact. Heavy oil contains more sulphur than coal per tonne (Table 1.40), but the need to burn more coal than oil to liberate the same amount of heat leads to almost the same sulphur emissions.

Nitrogen oxide emissions are difficult to predict as they depend upon combustion conditions. Part of the nitrogen oxide emissions arise from nitrogen in the fuel and the remainder from oxidation of nitrogen in the air. Coal contains much more combined nitrogen than oil and gas contains none, so coal burning tends to produce more oxides of nitrogen than combustion of the other fuels.

Next we should look at operational factors.

Stack losses, that is loss of heat carried away by moisture as latent heat, are least with coal and greatest with natural gas burning (Table 1.41), if the fuel is burned in plant with a back-end temperature above the dewpoint for water vapour. However, natural gas is such a clean fuel that condensation of water in the airheaters would cause few problems, in which case the latent heat might be recovered. Burnt in this mode gas is very much a premium fuel. Oil fuels are intermediate between coal and gas. Stack losses with the very heavy grades (Class H) are almost as low as coal (Table 1.41). Lighter grades have greater losses. Oil is an easier fuel to control than coal. It is less variable in its composition and although the heavy grades have to be heated before they will flow, they atomise well and combustion can be optimised with a smaller margin of excess air.

Oil burners atomise the fuel either by direct pressure, injection of steam or with a rotating cup (Fig 1.58). The aim in each case is to produce droplets 30 to 150 microns in diameter which readily vaporise to facilitate combustion. The burner also directs the fuel evenly into the preferred combustion zone.

On starting a new boiler from cold, atomisation may not at first be easy to achieve and carryover of oil droplets into colder parts of the boiler can occur.

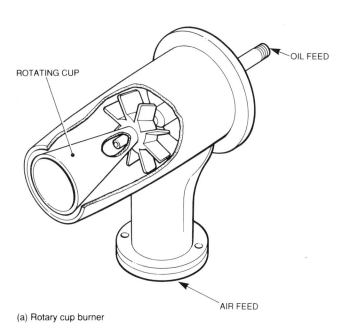

(a) Rotary cup burner

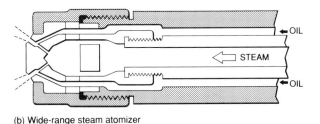

(b) Wide-range steam atomizer

FIG. 1.58 Types of oil burners

These may accumulate in airheaters and lead to serious fires. The risk can be minimised by the use of sootblowers during this period, to heat the depositing fuel.

Poor combustion can easily occur during the commissioning of new plant when cold starts are frequent. Use of a lighter, more easily atomised fuel may make control easier whilst experience of operating the new plant is gained.

Finally, because of the high luminosity of oil flames, radiant heat transfer rates are high. As a result oil-fired boilers are smaller than coal-fired and, because oil contains little ash, they remain cleaner during operation.

26.2   High temperature fouling and corrosion

Although little ash arises from burning heavy fuel oil, it is chemically reactive and may cause operational problems. It contains trace amounts of transition elements especially V, Ni and Fe along with the alkali metals, Na and K. During combustion the latter oxidise and react with sulphur oxides in the flue gases to give, e.g., Na_2SO_4. This has a low melting point and it deposits as a sticky film on heat transfer surfaces in the boiler furnace. Other oxide particles stick to this film on impact and an accumulation of deposit begins.

Iron and vanadium in the trapped material catalytically oxidise SO_2 to SO_3 which is then incorporated into more complex molten sulphates. $Na_3Fe(SO_4)_3$ in particular, which melts at about 860 K, is highly corrosive to boiler tube steels. Vanadium salt formed from V_2O_5 entering the melt as sodium vanadates also have low melting points, dependent upon their Na/V ratio and further contribute to the corrosiveness of the deposits.

As fouling increases, heat transfer to the metal boiler tube surface is impeded. The outside temperature of the deposit eventually approaches that of the bulk gas, and the catalytic behaviour changes.

It is difficult to predict exactly how an oil-fired boiler will behave but, in general, fouling will be more likely if the vanadium and sodium content of the oil supplied is high. Rates of corrosion can be reduced by lowering the metal temperatures of the superheater and reheater tubes. If this is not successful then the use of additives may be considered. Magnesium hydroxide or oxide (magnesia) have been found effective in reducing corrosion and making the deposits easier to remove.

26.3   Low temperature fouling and corrosion

Sulphur on combustion produces sulphur dioxide and a small proportion of sulphur trioxide. The latter condenses, along with moisture, as sulphuric acid in the cooler parts of the back-end of plant when the 'acid dewpoint' is reached. In coal-fired plant this causes no problems, as alkalinity in the ash neutralises the acid. In oil-fired plant however, the few parts per million of sulphur trioxide formed at the high temperatures of the furnace are a serious problem.

The acid forms a sticky film on airheater and cool duct surfaces. It corrodes the metal, to form iron sulphates, and promotes the adhesion of fine unburnt carbonaceous particles from the flue gases, to foul the surfaces further. The absorbed particles may be held for several weeks, accumulating acid up to 30% of their weight. Eventually, flakes of deposit break loose and are carried out of the stack by the flue gases, to fall as 'acid smuts' on surrounding areas.

The acid dewpoint depends on several factors. When it is high, the station has to operate with a high back-end temperature to avoid condensation. This results in a costly loss of efficiency. Alternatively, additives may be used to control the acidity, but this also adds to station costs.

26.4   Factors affecting the acid dewpoint

The acid dewpoint is the maximum temperature at which condensed acid and its vapour can exist in equilibrium. Above the dewpoint acid will tend to re-evaporate; below the dewpoint, vapour will tend to condense.

Its precise value depends on several factors; the water and sulphur trioxide concentration of the flue gas and to some extent, also, the strength of acid solution already condensed. During two-shifting the acid film may not be in equilibrium with the flue gases at certain times, thus making exact predictions of the dewpoint difficult.

Laboratory experiments show that for typical flue gases containing between 5 and 20 parts per million of sulphuric acid and about 11 percent of water vapour, the dewpoint will range between 125°C and 138°C, with an equilibrium concentration of condensed acid of just less than 80 percent (Fig 1.59).

In practice, the presence of corrosion products and carbonaceous deposits in the condensed acid complicates the situation. Nevertheless, Fig 1.59 will normally predict the dewpoint to within 5°C, i.e., operating with a back-end temperature 5°C greater than predicted should ensure condensation does not occur.

The main means of controlling the dewpoint is to reduce the sulphur trioxide concentration in the flue gases by tight combustion control. Excess oxygen levels should be minimised to reduce its formation by the reaction $SO_2 + \frac{1}{2}O_2 \rightarrow SO_3$. Thereafter not a lot can be done as some sulphur trioxide must be formed due to the high furnace temperatures (Fig 1.60).

Table 1.42 shows typical levels of oxygen, sulphur dioxide and sulphur trioxide from a set of residual oil-fired boilers.

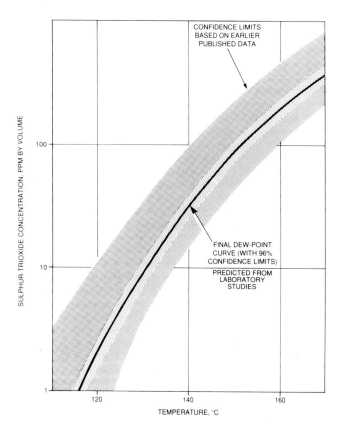

FIG. 1.59 Variation of acid dewpoint with sulphur trioxide concentration in flue gas

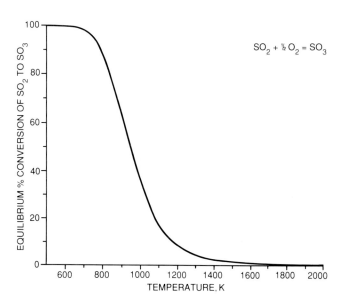

FIG. 1.60 Equilibrium conversion of SO_2 to SO_3 in 10% oxygen as a function of temperature

TABLE 1.42

Oxygen, sulphur dioxide and sulphur trioxide levels sampled at the economiser inlet

Boiler	% O_2	PPM SO_3	PPM SO_2
1	1.0	3	2560
2	0.7	4	2750
3	0.5	3	2540
4	1.3	4	2630
5	0.7	7	2780
6	2.8	10	2780
7	1.0	6	2620
8	1.0	5	2690

26.5 Combating back-end corrosion

There are four approaches to reducing low temperature corrosion in oil-fired plant, these are:

(a) Selection of low sulphur fuels.

(b) Operational means.

(c) Use of additives.

(d) Use of corrosion-resistant materials for construction.

26.5.1 Low sulphur fuels

This option is rarely viable. Fuel is purchased to optimise costs and the cheaper heavy oils tend to contain sulphur at about 3.0%. There is limited scope for improving on this.

26.5.2 Operational means

If the plant can be operated so that the flue gas temperature does not fall below the dewpoint until after it has passed the more vulnerable parts of the system (i.e., airheaters and economisers), then this will considerably reduce corrosion. Good maintenance helps in this respect. Lagging on ductwork should be kept sound to prevent cold spots which would cause condensation; and air inleakage should be prevented, since this also reduces temperatures.

As already mentioned, excess combustion air should be kept to a minimum to reduce the sulphur trioxide production in the boiler; and back-end temperatures in the plant may be increased. The latter is a high cost option, because it reduces station efficiency.

Finally, vulnerable areas can be hosed down periodically to remove the acidic deposits. Airheaters and economisers can be washed off-load. The water must be well distributed and large amounts should be used in a short time to remove the acid quickly and thoroughly. This can be a very effective approach.

26.5.3 Use of additives

Additives are intended to neutralise the acid in the vapour phase, thus preventing the later formation of acidic deposits on ductwork and airheaters. They are usually injected into the gas stream at the airheater inlet, although additions to the fuel itself have also been found effective.

Over the years magnesium carbonate, magnesium oxide and hydroxide, dolomite (magnesian limestone), zinc metal powder, tertiary amines, pyridine and ammonia have all been tested in extensive power station trials. Experience has shown magnesium oxide or hydroxide to be the most effective.

The task of neutralising the sulphuric acid in the vapour phase requires the magnesia to be finely ground, to increase its available surface area. A mean diameter of 5 μm is the normal aim and to ensure effective neutralisation a molar ratio of Mg/SO$_3$ in the flue gas of 2 is required.

Equipment (Fig 1.61) has been developed which will inject up to 180 kg h^{-1} of magnesia powder into the airheater inlet. It consists of a hopper fitted with a paddle and screw feeder, to deliver the powder to a venturi entrainment pipe. Here the powder is mixed with a jet of air and propelled into the flue gases. This rate of injection is sufficient to control acidity in boilers of up to 660 MW size.

The above treatment will prevent corrosion in plant which is clean when it starts operating. A plant which already contains acidic back-end deposits will not respond so well. This is because 5 μm magnesia particles do not readily deposit on to surfaces. This particle size was deliberately chosen for ensuring acid neutralisation in the flue gases. If it is judged necessary to promote settlement of magnesia on to surfaces, to neutralise established deposits, then coarser particles ($\sim$ 50 μm) should be utilised.

Finally a comment on the use of ammonia injection. This has been successful in some trials, but in others

it has perversely, promoted fouling. The problem is in ensuring that ammonium bisulphate does not form. This material is corrosive and promotes further fouling if it deposits on to cool surfaces. In successful trials, recommended doses of 0.06–0.08% by weight of fuel burned, depending upon its sulphur content, were effective.

26.5.4 Use of corrosion-resistant materials for construction

Corrosion-resistant materials of many kinds have been considered for corrosion in rotary airheaters. This item of plant is prone to attack because of the temperature cycle undergone once or twice per minute as the elements turn (Fig. 1.62). Design of the elements is important. For good heat transfer, convoluted designs perform well, but may encourage deposits to form. Designs which are readily washed, to remove deposits, are probably more cost effective, although the balance will depend upon how the station operates and the fuel used.

Experience has led to a choice of low alloy steels as the preferred material for construction. These give a good balance between service life and low cost. Steels of the 3% Cr type with or without $\frac{1}{2}$% Mo have been used, but the favoured material is Corten (C, 0.12 max; Si, 0.25/0.75; Mn, 0.2/0.5; P, 0.07/0.15; S, 0.05 max; Cr, 0.30/1.25; Ni, 0.65 max; Cu, 0.25/0.55) a formable high-strength low-alloy weldable steel possessing a resistance to corrosion estimated at three times that of mild steel.

26.6 Smut emissions

The mechanism by which smuts form is described in Section 26.3 of this chapter. An acid smut has two components, absorbed acidity and a carbonaceous matrix. The amount of acidity depends upon how much sulphur is present in the fuel and its sodium and vanadium content; the carbonaceous particle formation is related to the fuel's asphaltene content. Analysis for these parameters (see Sections 24.7, 24.8, 24.9 and 24.12 of this chapter respectively) is required, to assess whether smut emissions are likely to pose a problem.

Counter measures are essentially the same as for reducing back-end corrosion. A combination of low excess air, control of back-end temperatures, washing and the use of additives (magnesia) should be considered.

26.7 Gas-side cleaning of oil-fired boilers

The fouling of oil-fired boilers derives from the ash content of the oil, from carbon and from any solid additives such as magnesia which may be injected into the combustion chamber. On-load cleaning resembles that for coal-fired boilers, sootblowing perhaps supplemented by water lancing. Off-load cleaning differs only in the need to take more stringent precautions

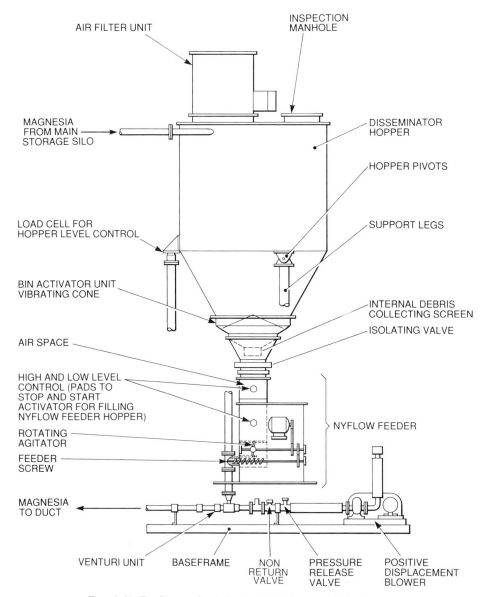

FIG. 1.61 Equipment for injecting MgO into an oil-fired boiler

against exposing cleaners to the dust hazard. The vanadium content of the deposits is the main health hazard. This is almost always over 3%, the level at which control procedures should be implemented. Workers should wear goggles, breathing masks and protective clothing. Over the years, this working procedure has given adequate protection to all involved.

26.8 Ash disposal

The amounts of ash in heavy fuel oils are low compared with the ash contents of coal. Consequently much smaller amounts of ash need to be disposed of from oil-fired stations; in the range of 20–80 tonnes per week for a 2000 MW (e) station. Disposal is nevertheless a problem because of the toxicity of the vanadium in the ash. This must not be allowed to leach from the disposal site into underground water, as this could subsequently be used for drinking purposes.

Vanadium is an expensive metal and ashes containing greater than 24% V_2O_5 are saleable to the metal producers. Sales depend upon supplies of ore to the world markets, so that ash sales from power stations are not always possible. Nevertheless, when they are, selling is the most economic, environmentally sound solution.

27 Testing of flue gases

27.1 Reasons for testing

There are three reasons for testing; to optimise the combustion process, to control fouling and corrosion of the boiler and to ensure that environmental standards are met.

Taking control of combustion first; the boiler operator needs to know exactly how much air to supply to the burners and to confirm that this has been achieved

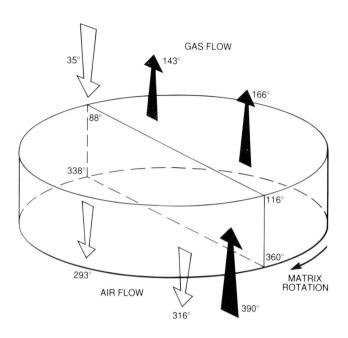

GAS FLOW

35°
143°
166°
88°
338°
116°
293°
360°
MATRIX ROTATION
AIR FLOW
316°
390°

FIG. 1.62 Temperature variation in a rotary airheater

he must, in theory, measure the CO_2 in the flue gases leaving the boiler to check that the expected amount is present.

To calculate the required amount of air, we start from considering the three combustion reactions:

$$C + O_2 \rightarrow CO_2 \qquad (1.11)$$
$$4H + O_2 \rightarrow 2H_2O \qquad (1.12)$$
$$S + O_2 \rightarrow SO_2 \qquad (1.13)$$

The atomic weights of carbon, hydrogen, sulphur and oxygen are respectively 12, 1, 32 and 16. Thus for complete combustion:

from Equation (1.11),
1 kg of carbon requires 8/3 kg oxygen

from Equation (1.12),
1 kg of hydrogen requires 8 kg oxygen

from Equation (1.13),
1 kg of sulphur requires 1 kg oxygen

When fuel is burned the amount of oxygen required to be supplied, making allowance for the oxygen content of the fuel itself, is therefore:

$$(8/3)C + 8H + S - O \text{ kg/100 kg of fuel burned}$$

where C, H, S and O are respectively the percentage amounts of each element in the fuel.

As 100 kg of air contains 23.2 kg of oxygen, i.e., it is $100/23.2 = 4.31$ times heavier:

Theoretical air required =

$$4.31 [(8/3)C + 8H + S - O] \text{ kg/100 kg of oil}$$

Applying this formula to the heavy fuel oil described in Table 1.40

Theoretical air =

$$4.31 [(8/3)86 + (8 \times 10) + 3 - 0.35]$$

i.e., 1344.6 kg of air should be required for complete combustion of each 100 kg of fuel. By dividing by 4.31, this contains 311.98 kg O_2.

Estimation of the theoretical carbon dioxide level in the flue gas is also straightforward. On a dry basis the flue gas produced would consist essentially of nitrogen $[(1344.6 \text{ air}) - (311.98 \text{ kg } O_2)] = 1032.6$ kg N_2, carbon dioxide and sulphur dioxide. The volume composition per 100 kg of fuel burned can be calculated from their molar bases, i.e.,

				% by volume
moles of nitrogen	=	1032.6/28	= 36.88	83.6
moles of carbon dioxide (86%C in fuel)	=	86/12	= 7.16	16.2
moles of sulphur dioxide	=	3/32	= 0.09	0.2
		Total	44.14	100

In principle, knowledge of the mass of air required, 1344.6 kg/100 kg fuel, and the theoretical carbon dioxide level to be expected in the flue gases leaving the boiler, 16.2%, should allow operations to be carefully controlled.

In practice, a proportion of excess air is necessary, to ensure complete combustion, as perfect mixing of fuel and air is difficult to achieve in a large boiler. The amount of excess air can be monitored by measuring oxygen in the flue gas and, in theory at least, the carbon dioxide level. Since the excess air dilutes the flue gases, the carbon dioxide levels measured will be less than the theoretical figure of 16.2%. The excess air, allowing for dilution, will be:

Excess air = [(Theoretical CO_2%)/(Actual CO_2%) − 1]

So far the calculation is tidy and if the boiler were airtight, control would require only carbon dioxide and oxygen levels to be measured in the exit flue gases. Unfortunately air leaks do occur, due to the action of the induced draught fans, which lower the pressure in the boiler. As a result oxygen is always found at the boiler exit. An extra measurement is therefore required to enable complete control; the carbon monoxide level.

There is always a background level of carbon monoxide in boiler flue gases. In coal-fired plant the level is less than 100 PPM by volume. For oil, the value is about 400 PPM. As the operator reduces the apparent amount of excess air there may be a slight increase in carbon monoxide, but eventually, at about 3.0% measured oxygen content at the airheater inlet, the carbon monoxide level will suddenly rise to a few thousand PPM. The exact oxygen level at which this occurs is the CO breakpoint (see Fig 1.63). It is an indication that, although there is a measurable oxygen

111

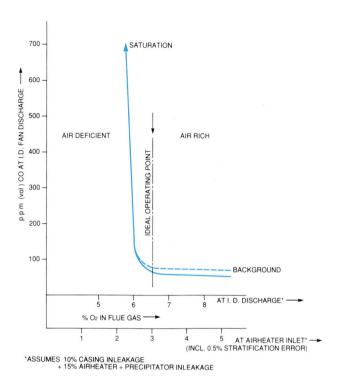

FIG. 1.63 Variation of CO with O_2 in flue gas

level at the airheater, it is only as a consequence of air ingress. At the burner, the air supply must have fallen below the theoretical amount required (1344.6 kg/100 kg of fuel in our example).

The second reason for testing flue gases is *to control fouling and corrosion* in the boiler. Measurement of sulphur dioxide and trioxide levels will indicate if production of the latter is being adequately controlled. Measurement of back-end temperatures and the acidity of deposits will indicate the extent to which low temperature fouling is a problem. Evaluation of high temperature corrosion is achieved using a corrosion probe.

The third reason for testing is *to comply with environmental standards*. For oil firing this means measuring the dust burden in the gases, the amount of acid smuts formed and the levels of nitrogen oxides in the stack gases.

27.2 Dewpoint measurement and rate of acid build up

Measurement of the dewpoint of flue gases is achieved by a conductivity measurement. The device used was developed in 1952 by the British Coal Utilisation Research Association (BCURA) and is marketed commercially by Land Pyrometers Limited. It consists of a glass or ceramic thimble into which two electrodes are embedded. One electrode wire is at the centre of the thimble and the other is in the form of a ring, surrounding the first. Two platinum/platinum-rhodium thermocouples are attached to the electrodes to measure

their temperature. The general appearance of such equipment and a wiring diagram are shown in Fig 1.64.

The thimble fits to the end of a metal probe and is inserted into the flue gases at a suitable sampling position. Cooling air is fed to the inside of the thimble and a potential of ten volts is applied between the electrodes. As the thimble cools, a point is reached when current starts to flow through the film of condensed acidity which forms on the outside of the thimble. The temperature at which a current is first registered on the μA meter is noted. This is the 'acid dewpoint' for the gases.

Corrosion is not a maximum at the 'acid dewpoint' temperature. If the temperature of the thimble is lowered in steps, by adjusting the flow of cooling air, the rate at which acid builds up at each temperature may be monitored. At each of the set lower temperatures, readings of current are noted over a five minute period. The current increases over the period, as acid condensation builds up and the rate of increase (in $\mu A/min$) is recorded. This is repeated for several temperatures in order to construct a graph (see Fig 1.65).

As a general rule the results will indicate possible corrosion rates as follows:

$$< 100 \ \mu A/min \ - \ minor$$
$$100-500 \ \mu A/min \ - \ moderate$$
$$> 500 \ \mu A/min \ - \ serious$$

An alternative to the dewpoint/rate of build up measurements using the BCURA probe, was introduced in 1959 by the Central Electricity Research Laboratories (CERL). This is an austenitic stainless steel air cooled lance of 25 mm diameter, about 2.4 m long, which is inserted into the flue duct (Fig 1.66). Cooling air is supplied via the central tube and withdrawn via the inside of the outer tube, thus cooling its external surface. Thermocouples measure the temperatures of a number of defined sections. After a timed period of thirty minutes in the gas stream, the lance is removed and condensed acid is washed from each of the tube sections. The area of each section is known, so that when washings are titrated for total acidity a graph of acid deposition (expressed as micrograms of $SO_3/cm^2/h$) may be plotted against temperature. From this graph, the maximum rate of build up and dewpoint temperature may be estimated in an analogous manner to Fig 1.65.

The BCURA dewpoint and CERL lance measurements give a good indication of the corrosive potential of the flue gases and an early warning perhaps of acid smut formation. However, they need supplementing by other tests.

27.3 Corrosion probe tests

Rate of acid build up (RBU) tests cover only short time periods. To determine how corrosion is developing over somewhat longer periods a corrosion probe is used

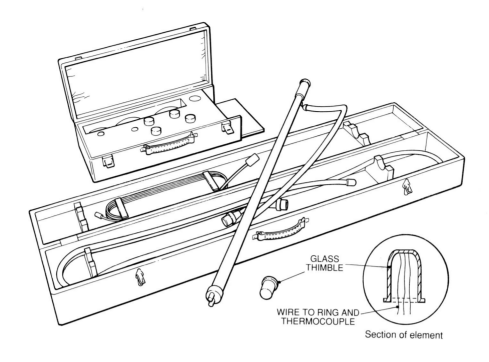

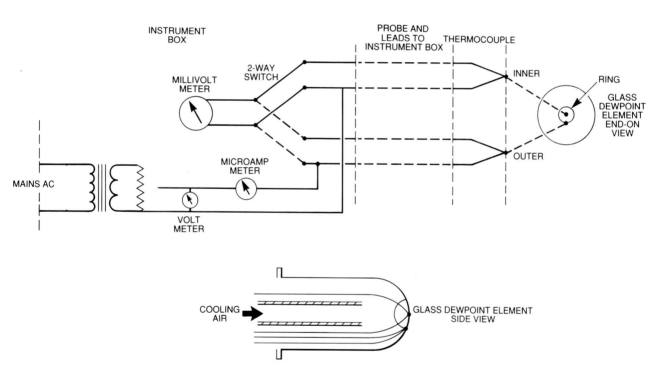

FIG. 1.64 Dewpoint meter for flue gas testing

(Fig 1.67). This also originated from BCURA (1952). The cooled metal thimble is the test piece. It is cooled by a flow of air and is thermally insulated from the probe by a refractory collar. Its temperature on insertion into the flue gas duct is measured by an iron/constantan thermocouple. At the end of the test period, of typically 1–2 hours, the probe is removed from the duct and the corrosion rate determined, either by weight loss of the thimble after washing, or direct analysis of the corrosion products. Some typical results are illustrated in Fig 1.68. They correlate with acid deposition rate but again do not predict longer term plant corrosion experience well.

27.4 Acid and solid deposition sampler

The measurements described so far take place in the bulk of the flue gas flow. Corrosion and deposition, in practice, occur at surfaces, generally in regions of laminar flow. A sampler has been designed at CERL

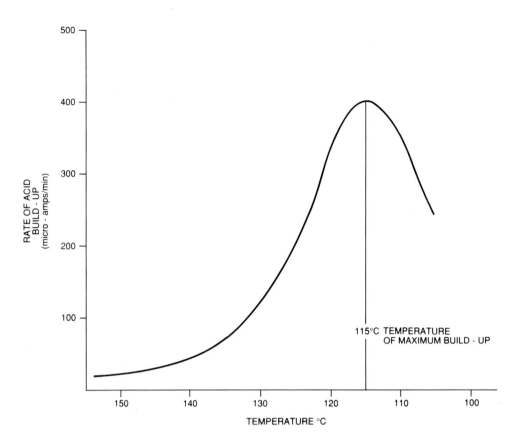

FIG. 1.65 Rate of acid build-up with temperature

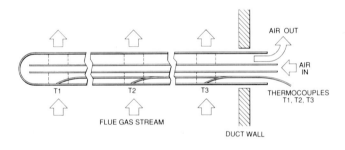

FIG. 1.66 CERL low temperature deposition probe

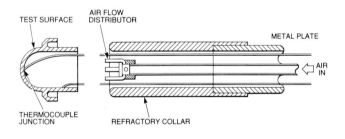

FIG. 1.67 A typical low temperature corrosion probe

(1978) to obtain data from samples exposed integrally with the duct wall, under laminar flow conditions (Fig 1.69).

The device slides in and out of the duct in order to expose the deposition plate or corrosion test disc. The deposition disc is gold-plated and is moved in to be flush-mounted in the duct wall. Its temperature is controlled by a hot air blower.

The unit was designed primarily to study how acid smuts form on internal surfaces wetted by condensed sulphuric acid and to assess the effectiveness of additives. The method has proved helpful in both respects.

For corrosion studies the deposition disc is replaced by a corrosion test disc. Good agreement has been found between measured dewpoint corrosion rates and rates of acid deposition but again the data were for relatively short time periods. The data complement those obtained by the BCURA probe.

27.5 High temperature deposition measurement

Corrosion of the convective pass tube surfaces in oil-fired boilers is caused by deposition of molten salts (Section 26.2 of this chapter). A similar design of probe to that used for measuring acid deposition rates in the back-end of plant (Fig 1.66) has been developed by CERL (1961) to study the deposition of molten salts. It is inserted into the gas stream for a period of time, possibly as low as 15 minutes, dependent upon the rate of deposition expected. After withdrawal, the deposit adhering to selected sections is removed by brushing or washing with water. The rate of deposition is determined by weighing and/or chemical analysis of the collected material.

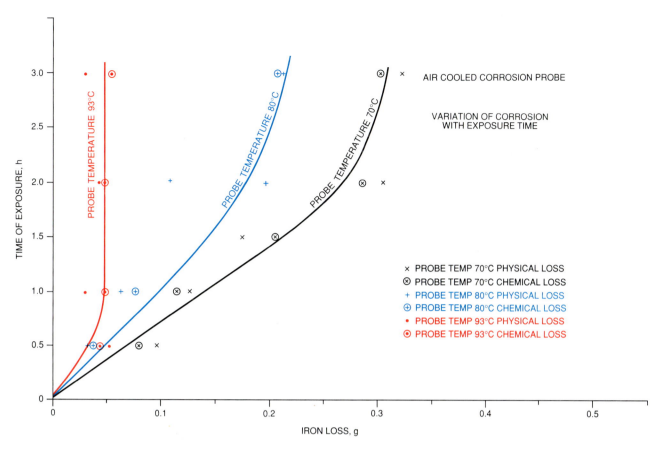

FIG. 1.68 A low temperature corrosion measurement

27.6   Dust burden measurement

Oil-fired boilers are not fitted with electrostatic pre-cipitators, as the amount of ash in the fuel is low. Mechanical dust-arrestors provide adequate removal efficiency which must be monitored by measuring the dust burden of the flue-gases before and after the arrestors.

The 'classical' method (BS893) of measuring dust-burden is to draw flue gas isokinetically from the centre of the duct on to a filter. The filter must be kept hot to prevent acid condensation. The gas flow drawn through the filter is measured using an orifice plate flow meter. After a measured time the filter is removed and the weight of dust collected from the measured volume of gas is determined.

Dust from oil-fired plant is fine and acidic and is more likely to block filters than pulverised fuel ash from coal-fired plant. The filter material used is usually coarser than for coal-firing but is of greater depth.

Continuous dust monitoring is desirable to show that a station is complying with emission standards. In the UK two monitors are in general use, the CERL monitor and the SEROP. They are complementary, in that each responds best to a different particle size spectrum.

The CERL monitor (Fig 1.70) responds better to larger particles (10–100 μm diameter). Dust from the gas passing through the device deposits on to a glass plate. The resultant obscuration is integrated over 15 min, after which the glass is cleaned by an air purge and the cycle is repeated.

The SEROP (South East Region Optical Probe) monitor (Fig 1.71) responds preferentially to smoke and very small particles (0–20 μm diameter). It consists of a tube 1.5 metres long inserted into the flue duct. Gases pass through a slot in the tube. Light from a lamp at the far end of the probe is detected by a photoelectric cell mounted on the outside of the duct. The amount of light detected is reduced in proportion to the dust burden of the gases.

Finally it should be mentioned that there are modern cross-duct monitors. These send a beam across a duct to a receiver at the opposite side. Dust in the gases obscures the beam in proportion to the amount entrained. Power station ducts are not the best of environments for scientific equipment, as vibration and thermal movements make alignment difficult. The equipment manufactured by Erwin Sick has provision for continual realignment and is less prone to such problems. It is in use at some stations.

27.7   Gas analysis

For efficient operation of the boilers, the oxygen and carbon monoxide content of flue gases at the airheater

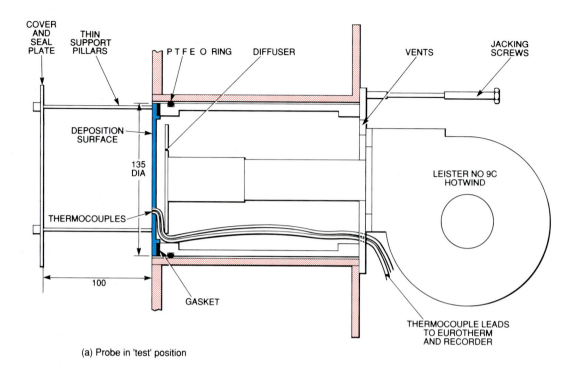

(a) Probe in 'test' position

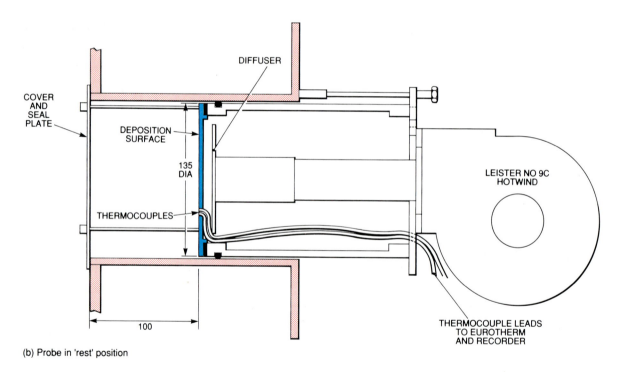

(b) Probe in 'rest' position

FIG. 1.69 Acid deposition sampler

inlet or ID fan discharge must be measured. On occasion carbon dioxide may also be measured. For environmental reasons sulphur dioxide and nitrogen oxides must be monitored and to combat back-end dewpoint corrosion, measurements of sulphur trioxide are required.

Oxygen This is measured by one of two methods. Zirconia cells provide a simple method. At high tem-

perature this material behaves as a solid state electrolyte. The charge carriers are oxygen anions, O^{2-}. The measuring device is a concentration cell, with an air flow at one side of the zirconia and flue gas at the other. The potential of the cell, E, provides a measure of the partial pressure of the oxygen, p, through the Nernst equation:

$$E = E^\circ - [RT/nF] \log_n [p(\text{flue gas})/p(\text{atm})]$$

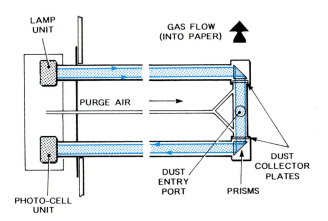

FIG. 1.70 CERL dust monitor

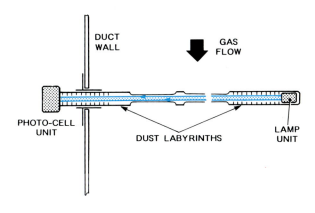

FIG. 1.71 SEROP monitor

The alternative method measures oxygen through its paramagnetism. Both types of instrument are in use for on-line monitoring in power stations.

Carbon dioxide This is not normally measured, since boilers are not airtight and cannot be controlled to give theoretical amounts of carbon dioxide at the exit. However, spot measurements are sometimes made to check the mass balance of gas flows. Samples are withdrawn from the plant and carbon dioxide is determined spectrophotometrically using infra-red detectors.

Carbon monoxide This is also measured spectrophotometrically using infra-red detection. It is present in small amount, 400 PPM, and instruments must be capable of measuring with a precision of better than 10–20 PPM. There are several commercial instruments which will do this. Some devices measure carbon monoxide by sending a beam across the duct, others require gas to be withdrawn to an external instrument. Carbon monoxide is usually measured after the ID fan. Before the airheaters the flue gases are stratified, i.e., they behave as though they were a series of parallel gas flows. Measurement of carbon monoxide at points across the duct would show different results at each sampling position. The ID fan mixes the flue

gases, and gas monitoring after the fan gives consistent readings.

Sulphur dioxide This is present to the extent of 0.2% in flue gas. It is a very soluble gas and if an external infra-red instrument is used for measurement, avoidance of condensation in sampling lines is important. Cross-duct analysers using ultra-violet (or infra-red) spectrophotometry are used satisfactorily and avoid the condensation problems.

Sulphur trioxide This is present in oil-fired station ducts at ~10 PPM by volume. It is measured using a colorimetric procedure. The sulphur trioxide is absorbed into an iso-propylalcohol solution of barium chloranilate, containing a little moisture. Sulphuric acid is formed, which reacts with the barium ions to give a precipitate of barium sulphate. The chloranilic acid liberated is purple and is measured photometrically in order to estimate the sulphur trioxide absorbed.

$$SO_3 + H_2O \rightarrow H_2SO_4$$

$$H^+ + SO_4{}^{2-} + BaC_6O_4Cl_2 \rightarrow BaSO_4 + HC_6O_4Cl_2{}^-$$
$$\text{(purple)}$$

The method has a limit of detection of better than 2 PPM. As with other gases, condensation in gas lines from duct to instrument could cause problems. It is usually adequate, however, to use PTFE polymeric tubing. A layer of condensed acid forms on the tubing but it 'conditions' the surface and an equilibrium is established which allows the sulphur trioxide to reach the instrument without loss.

Nitrogen oxides are measured from time to time to give a measure of emissions but are not usually measured continuously. A typical level in oil-fired plant flue gases would be 400 PPM. About 5% maximum is nitrogen dioxide; the bulk is NO.

For most purposes it is sufficient to measure only the nitric oxide, using an ultra-violet instrument. When complete measurement of oxides is required, a chemiluminescent procedure is used. At low pressure when NO reacts with ozone, the NO_2 produced is formed in an excited state. This loses energy by emitting light which is measured photometrically, to complete the determination:

$$NO + O_3 \rightarrow NO_2{}^* + O_2$$
$$NO_2{}^* \rightarrow NO_2 + h\nu$$

One sample of gas is introduced to the test cell directly to measure the nitric oxide present. A second sample is passed over a reducing agent to reduce any nitrogen dioxide to nitric oxide. The treated sample is then passed to the test cell to allow measurement of the total oxides present. The difference between the two measurements provides a measure of the amount of NO_2 present.

28 Lubricating oils and fire-resistant fluids

28.1 Introduction

When a new power station is built the manufacturers specify the grades of lubricants to be used in each item of plant they install. A lubricants supplier is nominated for the site and in consultation with the plant contractors, schedules are prepared showing all the lubrication points at the station and the appropriate lubricants for each application.

Public Authority Standards (PAS) and International Standards Organisation (ISO) viscosity classifications (ISO 3448) are primarily used to specify the lubricants. Products are identified by their kinetic viscosity in centistokes (cSt) at 40°C. Prefix letters denote the type of oil. The recognised grades are:

2	3	5	7	10	15
22	32	46	68	100	150
220	320	460	680	1000	1500

Additional standards apply to Lubricating Grease (Lithium base), Fire Resistant Fluid (phosphate ester type), Nuclear Gas Circulator Oils and Gas Generator (gas turbine) Synthetic Lubricants, as no PA Standards are available for these.

The simplest lubricants are pure mineral hydrocarbon oils. For more demanding uses additives and inhibitors are employed to 'tailor' the properties of the lubricant to meet the higher specifications.

Examples of the range of lubricants used in power stations are given below and then the tests required and care which must be given to maintain the oils within specification are described. The descriptions are not intended to be comprehensive, but sufficient information is given to cover typical operational requirements.

28.2 Straight mineral lubricating oils

All straight mineral lubricating oils are purchased to Public Authority Standard 2 (PAS 2). The eight grades are used in almost every type of equipment. They are of medium quality and have to pass an oxidation test (IP306) to ensure satisfactory service performance between plant overhauls. Their characterisitics are set out in Table 1.43. They are pure mineral hydrocarbon oils, but may contain an additive to serve as a pour point depressant.

Typical uses are as follows:

- *CSB 10* Hydraulic couplings, lubrication of lightly loaded, high speed bearings; refrigerating machinery requiring a low viscosity, low pour-point oil; flushing of gear boxes, engines and the like.

- *CSB 32* Similar to CSB 10 but where a more viscous oil is required.

- *CSB 68X* Ring-oiled bearings; oil-lubricated ball and roller bearings; certain small compressors; fluid couplings requiring a more viscous grade than CSB 10 or CSB 32.

- *CSB 100X* Crankcases of most air compressors (see section for cylinder lubrication); low rated rotary air compressors; ring-oiled bearings; medium speed, light duty enclosed gears.

- *CSB 220* Similar to CSB 100 where a more viscous grade of oil is required.

- *CSB 320* Medium speed gearboxes.

TABLE 1.43

Characteristics of straight mineral lubricating oils

Type prefix ISO 3448 viscosity grade designation		CSB 10	CSB 32	CSB 68 X*	CSB 100 X*	CSB 220	CSB 320	CSB 460	CSB 1000	Test Method British Standard	IP Method
Viscosity, kinematic	min	9.0	28.8	61.2	90	198	288	414	900	4708	IP71
at 40°C (cSt)	max	11.0	35.2	74.8	110	242	352	506	1100		
Viscosity Index	min	70	70	70	70	70	70	70	70	4459	IP226
Flash point, closed, °C, Pensky-Martens	min	120	165	192	204	228	237	237	260	2839	IP34
Pour-point °C	max	−18	−15	−12	−9	−6	−6	−6	+6	4452	IP15
Total acidity, mg KOH/g	max	0.1	0.1	0.1	0.1	0.1	0.1	0.1	0.1	4705/A	IP1/A
Demulsification number (seconds)	max			360	360					4381	IP19
Oxidation characteristics: % Total oxidation products (TOP) in 48 h test with										5301	IP306
no catalyst	max	0.10	0.10	0.10	0.10	0.10	0.10	0.10	0.20		
Solid copper catalyst	max	1.00	1.00	1.00	1.00	1.00	1.00	1.00	2.00		

- *CSB 460* Gear boxes requiring a plain mineral oil of this grade in order to give the correct viscosity at higher operating temperature; slow speed, heavily loaded, reduction gear boxes; worn gears and final drives of vehicles other than those having hypoid gears; cylinders of compressors of high compression ratio.

- *CSB 1000* Gearboxes requiring a plain mineral oil of this viscosity.

28.3 Mild extreme pressure gear oils

These oils are purchased to PAS4. They are prepared from base oils which, in regard to their viscosity behaviour and oxidation characterisitics, comply with CSB 220, CSB 320 and CSB 1000; but in addition a sulphur-phosphorus type of extreme pressure additive is included to improve their load-carrying ability. Anti-foam and pour-point depressant additives may also be present. Their characteristics are set out in Table 1.44.

Typical uses are as follows:

- *GSP 220* Gear boxes of certain pulverised fuel mills.

- *GSP 320* Similar to GSP 320 but where a more viscous oil is required.

- *GSP 1000* Similar to GSP 320 but where an even more viscous oil is required.

28.4 Heavy duty engine oils

These oils are purchased to PAS1 Part 4, Part 2 or Part 5, which in turn are based on military specifica-tions. They are mineral or synthetic hydrocarbon oils containing detergents, dispersants, oxidation and cor-rosion inhibitors, as necessary, to pass the specified qualifying requirements. Each formulation is approved by the relevant military qualification procedure. As a result purchasers are able to be supplied with an ap-proved dated certificate number. The characteristics of each one are given in Table 1.45.

PAS 1, Part 4 oils are for use in supercharged diesel engines. The base hydrocarbon oil may be mineral or synthetic. The lighter SAE 20W grade (Table 1.45) is suitable where an oil of SAE 20, 20W or 20/20W has been recommended by the engine manufacturer. The heavier SAE 30 grade is also for use in supercharged diesels, as recommended by the manufacturer.

PAS 1, Part 2 mineral hydrocarbon oils are for use in the two-stroke Detroit Diesel engines of coal moving equipment used at power stations. They may also be recommended by manufacturers of naturally aspirated diesel engines. The SAE 30 grade is preferred for Detroit Diesels where cold starting is a problem. The SAE 40 grade is preferred for Detroit Diesels where valve wear is a problem.

PAS 1, Part 5 mineral hydrocarbon oils are used in petrol and naturally aspirated diesel engines. The SAE 20W/20 grade is suitable where an oil of SAE 20, 20W or 20W/20 is recommended. The heavier SAE 20W/40 oil is 'multigrade'. It contains a viscosity index improver. It should only be used if recommended by the engine manufacturer.

TABLE 1.44

Characteristics of mild extreme pressure gear oils

Type prefix ISO 3448 viscosity grade designation		GSP 220	GSP 320	GSP 1000	Test Method British Standard	IP Method
Viscosity, kinematic	min	198	288	900	4708	IP71
at 40°C (cSt)	max	242	352	1100		
Viscosity Index	min	70	70	70	4459	IP226
Flash point, closed, °C Pensky-Martens	min	183	183	204	2839	IP34
Pour-point, °C	max	−3	−3	+3	4452	IP15
Oxidation characteristics		see PAS2 CSB220, 320, 1000 (Table 1.43)			5301	IP306
Storage stability solid separated % w/w of additive in oil	max	0.2	0.2	0.2	−	DEF 2000 Method 5
Copper corrosion, classification (3 h at 100°C)	max	2	2	2	4351	IP154
Steel corrosion, classification	max	2	2	2	−	MOD Chem Spec. CS3000B Annex B
Foaming tendency Foam, ml at 24°C	max	40	40	40	5092	IP146
ml at 93.5°C	max	5	5	5		
ml at 24°C after 93.5°C test	max	40	40	40		
Foam stability, at all test temps.	max	nil	nil	nil	5092	IP 146
Timken OK Load	min	40	40	40	−	IP 240
FZG test A/8.3/90 (stages passed)	min	12	12	12	−	DIN 5135

TABLE 1.45

Characteristics of heavy duty engine oils

PAS 1, Part		4		2		5		Test Method	
Type		MIL-L-2104C		MIL-L-2104B		MIL-L-46152			
SAE viscosity grade*		20W	30	30	40	20W/20	20W/40	British Standard	IP Method
Viscosity, kinematic at 100°C	min	7.4	9.3	9.3	12.5	7.4	12.5	4708	IP71
mm²/s (cSt)	max	9.3	12.5	12.5	16.3	9.3	16.3		
at −18°C, mN s/m²(CP)	min		−	−	−	2500	2500		
	max	10 000	−	−	−	10 000	10 000		
Viscosity Index	min	75	75	−	−	−	−	4459	IP226
Pour-point, °C	max	−21	−18	−18	−9	−23	−23	4452	IP15
Flash point, °C open	min	182	220	199	199	215	215	2839	IP36
approx. Pensky-Martens closed	min	172	210	189	189	205	205	2839	IP34
Foaming tendency:									
Foam, ml at 24°C	max	25	25	−	−	−	−	5092	IP146
ml at 93.5°C	max	150	150	−	−	−	−	5092	IP146
ml at 24°C after 93.5°C test	max	25	25	−	−	−	−	5092	IP146
Foam stability, at all test temps.	max	Nil	Nil	−	−	−	−		
Foam stability after 600									
Foam, ml at 24°C	max	−	−	300	300	25	25		IP146
ml at 93.5°C	max	−	−	25	25	150	150		IP146
ml at 24°C after 93.5°C test	max	−	−	300	300	25	25		IP146
Zinc % by weight	min	−	−	0.07	0.07	−	−		IP117
Sulphated ash % by weight	max	−	−	1.0	1.0	−	−	4716	IP163

*Automotive lubricants are identified by an SAE number which does not directly refer to a viscosity

28.5 Extreme pressure gear oil

This oil is composed of refined petroleum and/or synthetic products, containing appropriate additives. It is suitable for use in the hypoid and spiral level gears of rear axles of motor vehicles where SAE 90 viscosity grade oil is recommended by the manufacturer. The oil meets the requirements of Chemical Inspectorate Specification CS3000B. Its characteristics are given in Table 1.46.

28.6 Mineral turbine oils

These oils are purchased to PAS 5. They are high quality oxidation- and rust-inhibited materials. The four grade numbers indicate their kinematic viscosity at 40°C (ISO 3448).

The turbine manufacturers specify which of the four grades of oil are to be used in their machines. Very large quantities are employed and unlike automobile engines the oil is not changed at service intervals. Instead the oil itself is serviced. Rust inhibition and oxidation tests are carried out periodically and if necessary inhibitors are added to restore the oil properties. In addition the oil circuit contains a centrifuge to remove water continuously, so that the oil stays in good order.

The characterisitics of mineral turbine oils are given in Table 1.47.

TABLE 1.46

Characteristics of extreme pressure gear oil

SAE viscosity grade		90	Test Method	
			British Standard	IP Method
Viscosity, kinematic, at 100°C, mm²/s (cSt)	min	16.8	4708	IP71
	max	19.2		
Viscosity Index	min	85	4459	IP226
Flash point, open, °C	min	150	2839	IP36
Pensky-Martens approximate closed cup value °C	min	140		IP34
Foam stability:				
Foam, ml at 24°C	max	300	5092	IP146
ml at 93.5°C	max	50	5092	IP146
ml at 24°C after test at 93.5°C	max	300	5092	IP146
Copper corrosion, (3 h at 121°C), classification	max	2	4351	IP154

TABLE 1.47

Characteristics of mineral turbine oils

Type prefix ISO viscosity grade designation (ISO 3448)		TO 32	TO 46	TO 68	TO 100	Test Method British Standard	IP Method
Viscosity, kinematic at 40°C, mm²/s (cSt)	min	28.8	41.4	61.2	90	4708	IP71
	max	35.2	50.6	74.8	110		
Viscosity Index	min	70	70	70	70	4459	IP226
Flash point, closed °C, Pensky-Martens	min	168	168	168	168	2839	IP34
Pour-point, °C	max	−6	−6	−6	−6	4452	IP15
Total acidity, mg KOH/g	max	0.20	0.20	0.20	0.20	4705, Method A	IP1, Method A
Copper corrosion, classification (3 hours at 100°C)	max	2	2	2	2	4351	IP154
Rust-preventing characteristics		rusting absent	rusting absent	rusting absent	rusting absent	4387	IP135 Procedure B
Demulsification number, s	max	300	300	360	360	4381	IP19
Foaming tendency: Foam, ml at 24°C	max	450	450	600	600	5092	IP146
ml at 93.5°C	max	50	50	100	100	5092	IP146
ml at 24°C after test at 93.5°C	max	450	450	600	600	5092	IP146
Foam stability after 300 s Foam, ml at 24°C	max	nil	nil	100	100	5092	IP146
ml at 93.5°C	max	nil	nil	25	25	5092	IP146
ml at 24°C, after test at 93.5°C	max	nil	nil	100	100	5092	IP146
*Air release properties, minutes to 0.2% air content at 50°C		5	6	7	10	5303	IP313
Oxidation characteristics: Total oxidation products (TOP) with sludge limited to 40% of the determined TOP No catalyst, % ⎫ duration of tests	max	0.1	0.1	0.1	0.1	5301	IP306
Solid copper catalyst % ⎬ 164 hours	max	1.00	1.00	1.00	1.00	5301	IP306
‡Soluble metal catalysts, % ⎭	max	1.00	1.00	1.00	1.00	5095	IP280

* Where difficulties are encountered with starting up at low temperatures, the air release value at 25°C should be agreed between the purchaser and the vendor.

‡ 'Total oxidation products' to be calculated from the 'total sludge' and 'total acidity' results given by BS5095 as described in BS5301.

28.7 Nuclear gas circulator oils

Nuclear gas circulator oils are purchased to a specification based on PAS5, with additional requirements to ensure they are suitable for use in gas cooled reactors. The two viscosity grades typically used are NGC046 and NGC068. Their specification, in the main is identical with that for T046 and T068 (Table 1.47).

In a reactor they have to withstand a degree of contamination by carbon dioxide at pressures up to 50 bar, and must have the ability to release the dissolved gas when depressurised, without excessive foaming. They must exhibit special properties in respect of radiation resistance, and freedom from the development of corrosive substances when degraded in the reactor. They must have a low vapour pressure, as it is undesirable to have volatile organic compounds in the reactor gas and finally they should contain low levels of sulphur and nitrogen. The latter elements, in the reactor, give rise to an increased activity in the coolant gas.

The extra qualities required of grades NGC046 and NG068 are given in Table 1.48. Both grades are used in motor drives and blower units of nuclear gas circulator units. The equipment manufacturer specifies which viscosity grade is required.

28.8 Synthetic gas turbine gas-generator lubricants

These lubricants are synthetic organic esters with additives, to meet their specification and performance requirements. They were specifically developed for aviation gas turbines and are approved by the engine builder for individual engines.

They are purchased against a military specification. The viscosity grades refer to a temperature of 100°C rather than 40°C as for other oils mentioned earlier.

SGTL3 is used for lubrication of Avon engines.

SGTL5 is the most common lubricant of this type and is used in Avon and Olympus gas generators.

SGTL7.5 is used in Olympus and Proteus engines.

The characteristics of the oils are given in Table 1.49.

28.9 Fire-resistant fluids

These are aromatic esters of phosphoric acid. The 'cresol' esters are typical. The fluids are used in hydraulic control systems where leakage to a hot surface

TABLE 1.48

Characteristics of nuclear gas circulator oils

Type prefix		NGCO	NGCO	Test Method	
ISO viscosity grade designation		46	68	British Standard	IP Method (or other)
Viscosity				4708	IP71
Viscosity Index				4459	IP226
Flash point				2839	IP34
Pour-point				4452	IP15
Total acidity		As for TO46, TO68 grades in Table 1.47		4705 Method A	IP1 Method A
Copper corrosion				4351	IP154
Rust prevention				4387	IP135
Demulsification number				4381	IP19
Foaming tendency				5092	IP146
Foam stability after 300				5092	IP146
Oxidation characteristics				5301	IP306
Air release properties min to 0.2% air content at 50°C	max	6	10	5303	IP313
Vapour pressure, mbar 100°C	max	1.33×10^{-2}	1.33×10^{-2}		State Method
Radiation resistance (after 2×10^8 rads at 40°C):					
Viscosity increase %	max	50	50	—	—
Oxidation, TOP % (24 h test)	max	0.50	0.50	5301	IP306 non-catalysed
Sulphur %					IP336 or 61
Nitrogen %		To be reported			State method
Trace metals PPM					State method

TABLE 1.49

Characteristics of synthetic gas turbine gas-generator lubricants

Type prefix		SGTL		SGTL		SGTL		Test Method	
Grade designation		3		5		7.5		British Standard	Technically identical to
Viscosity, mm²/s (cSt) at:									
40°C		15	approx.	25	approx.	35	approx.	4708	IP71
100°C		3.0	approx.	5.0	approx.	7.5	approx.	4708	IP71
Pour-point, °C	max	−85	approx.	−50	approx.	−50	approx.	4452	IP15
Flash point, °C									
open	min	230	approx.	—		220	approx.	—	IP36
closed	min	184	approx.	260	approx.	170	approx.	2839	IP34
Relevant Standard		—		MIL-L-23699B and D. Eng RD2497		D. Eng. RD2487			

could occur. As they have a high auto-ignition temperature the danger of fire is greatly reduced.

They need careful handling and must be kept dry to avoid hydrolysis, which would produce phosphoric acid and lead to corrosion of components in the hydraulic system. They are toxic and skin contact should be avoided. Swallowing constitutes the greatest risk and smoking eating or drinking is prohibited when handling the fluids.

Their physical and chemical properties are given in Table 1.50.

28.10 Additives and inhibitors

The properties of power station lubricating fluids are largely achieved by the use of additives. Although the base oil may be highly refined and a good lubricant in its own right, it can only meet the full specification required for it to perform well in specialised plant by the addition of several inorganic and organic materials. These additives are 'consumable' and may have to be replaced within the lifetime of the parent oil. Many of the tests referred to in previous sections are undertaken to establish if the additives need replenishing. If a lubricant is found to be 'out of specification', the oil company will usually advise on the quantity and type of additive needed to restore its properties.

In large plant, steam turbines for example, additives and inhibitors are replenished several times in the lifetime of the equipment. However, it would be unusual for the oil itself ever to be replaced. Very large quantities of lubricant are employed and an oil-change would be expensive.

Briefly, additives are used for the following purposes.

28.10.1 Oxidation and rust inhibition

Straight mineral oils lack oxidation resistance and do not inhibit rust in the presence of water under modern

TABLE 1.50

Characteristics of fire-resistant fluids (FRF)

Type prefix		PE	PE	PE	Test Method British Standard	Technically
ISO viscosity grade designation (BS4231)		22	32	46		identical to
Flash point, open, °C	min	180	180	180	4688	IP35
Fire-point, °C	min	290	290	290	4688	IP35
Auto-ignition temperature, °C	min	580	530	470	—	ASTM D2155
Density at 15°C, g cm^{-3}	max	1.25	1.25	1.25	—	IP59
Viscosity, kinematic at 0°C, mm^2/s (cSt)	max	300	850	2000	4708	IP71
Viscosity, kinematic at 40°C, mm^2/s (cSt)	min	19.8	28.8	41.4	4708	IP71
	max	24.2	35.2	50.6		
Viscosity, kinematic at 100°C, mm^2/s (cSt)	min	2.5	3.0	4.0	4708	IP71
Minimum pumping temperature, °C viscosity of 850 mm^{2}s (cSt)	max	−5	0	7	—	
Pour-point, °C	max	−20	−12	−7	4452	IP15
Boiling point, °C	min	325	325	325	—	IP123
Shear stability, % at 40°C	max	5	5	5	—	IP294
(100 cycles) % at 100°C	max	5	5	5	—	IP294
Foaming tendency:						
Foam, ml at 24°C	max	150	150	300	5092	IP146
ml at 93.5°C	max	25	25	50		
ml at 24°C after test at 93.5°C	max	150	150	300		
Foam stability:						
Foam, ml at 24°C	max	Nil	Nil	Nil	5092	IP146
ml at 93.5°C	max	Nil	Nil	Nil		
ml at 24°C after test at 93.5°C	max	Nil	Nil	Nil		
Demulsification number, s	max	300	300	600	4381	IP19
Bulk modulus at 15°C and 1.013 bar, GN/m^2	min	2.41	2.41	2.41	—	
Air release value, minutes to 0.2% air content at 50°C	max	5	10	10	5303	IP313
Total acidity, mg KOH/g	max	0.20	0.20	0.20	4705 Method A	IP1 Method A
Stability-corrosion						
Procedure *A*, oxidation and thermal					—	IP331
Total acidity, mg KOH/g	max	1.00	0.50	0.50		
Deposit, %	max	0.10	0.05	0.05		
Corrosivity, mg, CC	max	±5	±5	±5		
mg, CD	max	5	5	5		
Change in viscosity, %	max	±10	±10	±10		
Resistivity at 20°C, MΩ m*	min	—	—	50		
Procedure *B*, hydrolysis					—	IP331
Total acidity, mg KOH/g	max	1.50	1.00	1.00		
Deposit, %	max	0.30	0.20	0.20		
Corrosivity, mg, CC	max	±5	±5	±5		
mg, CD	max	5	5	5		
Rust preventing characteristics, rusting		absent	absent	absent	4387	IP135 Procedure A
Water content, % ξ	max	0.10	0.10	0.10	2511	
Resistivity at 20°C, MΩ m*	min	—	—	50		Appendix D

* Applicable to high resistivity grades only which are used in fine clearance (10 μm) servo valve systems operating at high fluid pressures when the suffix HR should be added to the grade designation

ξ Due to the water absorption characteristics of these fluids this limit can only be maintained by storage under suitable dry conditions

operating conditions. An oxidation inhibitor stabilises the oil against oxidation and also passivates the metal surfaces with which it is in contact. Otherwise these surfaces would act catalytically to increase oxidation further. The inhibitors also maintain a low neutralisation value for the oil (acidity) over many years in service.

A rust inhibitor, as its name indicates, protects steel surfaces from rusting when in contact with wet oil. A typical oxidation inhibitor is 2–6 ditertiary-butyl para-cresol, used in amounts of 0.2 to 1% by weight. Rust inhibitors include sulphur-organic compounds and certain metal salts. Both types of inhibitor may be found in turbine, hydraulic, transformer and other highly refined oils.

28.10.2 Extreme pressure

Additives, which are polar in nature and contain compounds of sulphur, chlorine, phosphorous and zinc, are added to gear oils to increase their load bearing characteristics. They ensure non-corrosivity, good oxidation resistance, high temperature stability, good water tolerance, and constant viscosity characteristics. This latter property ensures that the lubricating film on the gears is maintained even when the oil is worked hard and becomes hot.

28.10.3 'Cling' promotion

Industrial oil improvers confer better adhesion to metal surfaces. They are added when straight oils are likely to be lost, by dripping or being flung off moving surfaces. They are high molecular weight polymers, of isobutylene, for example, or heavy synthetic esters added in amounts up to 5%.

28.10.4 Detergents

These additives are of particular use in diesel engine lubrication to inhibit high temperature oxidation, formation of sludge and deposition of contaminants. A typical example is zinc dialkyl phosphate but other proprietary compounds containing zinc, barium, calcium, phosphorus or sulphur organic compounds may be used.

28.10.5 Viscosity index improvement

Oils thin out as the temperature rises. Those which thin considerably have a low viscosity index and those which show little change have a high viscosity index. By adding linear isobutylene polymers of controlled molecular weight the viscosity index can be controlled. They have an added advantage of promoting engine cleanliness in intermittent service.

28.10.6 Pour-point depression

Oils contain waxes and high molecular weight aromatic substances. On cooling they do not have a freezing point, but instead gradually thicken until they become immobile. This is termed their 'pour-point'. By adding additives, typically composed of a hydrocarbon wax naphthalene condensate, in amounts of about 1%, a considerable lowering of the pour-point can be achieved to improve the cold weather behaviour of the lubricant.

28.10.7 Emulsification and demulsification

To prevent metal corrosion by wet oil, emulsification agents, typically metallic soaps, tallow, sulfinates or rape seed oil, are added. These substances surround the water droplets and keep them in suspension in the oil as an emulsion. The ability of the water to contact the metal and then to cause corrosion is much reduced.

Ideally water is separated from lubricating oil to prevent it causing problems. In the case of steam turbine lubricants, this is carried out in a centrifuge. However, if the oil-water mixture forms an emulsion, separation becomes very difficult. Warming the oil to 80-100°C often breaks the emulsion, allowing the water to separate. If this does not work then a small amount (less than 0.1%) of a demulsifying additive may be required. Investigations of the cause of the problem will be necessary as there is a no uniform cure. Several types of additive are available and the most appropriate one must be identified.

28.10.8 Prevention of foaming

In a large oil-circulating system, air may be entrained and a foam can form. A very small amount of a sili-

cone oil (about 0.003% by volume) is added to lubricants to prevent foaming from occurring. Interestingly larger additions are not effective.

28.11 Testing of oils

The sampling and testing of lubricants is carried out much as described in Section 23 of this chapter, for fuel oils. For new oil, delivered in drums, a proportion of the drums in a consignment should be sampled as below:

No of drums	No to be sampled
1	1
2 – 5	2
6 – 20	3
21 – 50	4
51 – 100	7
101 – 200	10
201 – 400	15
over 400	20

If the oil is delivered by tanker, each compartment must be sampled and in addition a sample should be taken from the outlet of the flexible pipework or the tanker bottom valve manifold. This latter sampling checks for any contamination due to residual material. This can occur when different products are being carried in separate compartments or deliveries of a different product have been made to other locations without adequate cleaning and flushing.

On delivery The samples are immediately checked for appearance. They should not contain particulate material. A value of 0.0007% by weight would be used as a criterion for rejection. The main laboratory tests undertaken are then for viscosity (IP71) flash point (IP34, IP35 or IP36) acidity (IP1) and where applicable, demulsification number (IP19). This set of tests is sufficient to characterise the delivery to ensure that it is the grade expected.

In service The quality of the oil should be monitored by obtaining oil samples from the oil circulating system at suitable sampling points. Viscosity, water content, acidity (neutralisation value) flash point, demulsification number, pour-point, oxidation stability and rust preventing characteristics may be determined. Stations set their own routines for testing, appropriate to their plant operating conditions. Usually only the first five tests are carried out routinely. Other testing referred to in the specifications above are only carried out in the event of an investigation, should the lubricant behave abnormally.

It is not necessary to describe the test methods. Reference should be made to the appropriate Institute of Petroleum Standards for details. The results obtained should be compared to the purchasing specifications (Tables 1.43 to 1.50) and should the lubricant be 'out of specification' help should be sought either from the supplier or the 'in house' laboratory which supports the power station.

28.12 Purification of lubricating oils and other fluids

In service, lubricating oils become wet, oxidised and contaminated by fine particles of metal and deposits of sludge. Their acidity will also increase due to oxidation. The normal method of purification for steam turbine lubricants, is the use of a centrifuge. These operate on a by-pass loop from the main lubricating oil flow, to provide continuous cleaning.

A De-Laval centrifuge is typically used. The oil is spun at high speed and water and deposits are separated. In earlier days centrifuges were deliberately run wet as this helped the separation process. In addition, acidity in the oil was removed into the separated water. This treatment can be overdone, leading to 'washing-out' of inhibitors. When excessive water is collected from a centrifuge on a steam turbine lubricating system, due perhaps to a steam leak into the oil, it is good practice to check that rust and oxidation inhibitor levels are still adequate.

Off-line treatment of lubricants and fire-resistant fluids is occasionally required. If contaminating water is not separating adequately from an oil, a coalescer may prove effective. The oil-water emulsion is forced through a fibrous membrane and the droplets of water coalesce.

The problem with fire-resistant fluids is a rise in acidity due to water contamination. They can be improved by draining the system and filtering, using a bed of fuller's earth. The purified fluid can then be re-used.

29 Greases and solid lubricants

29.1 Lithium greases and a few alternatives

Greases are semi-solid mixtures of oily or fatty materials, used as lubricants. They do not consist of a single substance. Most greases used in power stations are a mixture of mineral oil thickened with soaps. Some heavy lubricants consisting of very viscous oils or mixtures of oil and bitumen may also be referred to as greases. Lime soaps and soda soaps are commonly used in greases but lithium soaps have superior properties and despite a higher cost are the preferred soap base in most greases used in power stations. Thickeners such as bentonite may also be added to a grease to achieve an appropriate consistency.

Grease is normally used in applications where oil would be unsuitable, for example, where the oil cannot be retained and would run away. Roller bearings are perhaps the main item of plant lubricated by greases.

A method of specification of a grease has been established by the National Lubricating Grease Institute

for all greases, irrespective of type. The softest consistency is Grade 0 and the stiffest is Grade 6. Grades 2 and 3 are most commonly used (Table 1.51).

TABLE 1.51
National Lubricating Grease Institute consistency classification

Consistency Number	ASTM D217 and IP50 worked penetration (at 25°C)	Remarks
0	355 – 385	Semi – fluid greases
1	310 – 340	
2	265 – 295	
3	220 – 250	
4	175 – 205	
5	130 – 160	
6	85 – 115	Block-greases

Because greases are mixtures, their properties are readily changed by contamination with other lubricants. For this reason it is desirable to stick to one type of grease for a particular application. If it is necessary to change a grease then the item of plant must be thoroughly cleaned before the replacement lubricant is introduced.

The specifications of lithium based lubricating greases suitable for use in ball and roller bearings are given in Table 1.52.

They are soft buttery type greases with a melting point of ~175°C and are very resistant to water washing. They can be used in high speed bearings (say up to 3000 r/min with a 50 mm shaft) within a wide temperature range of −20° to +120°C. They are also suitable for general high temperature applications, say up to 150°C.

Alternative greases There are several formulations which may be used for other applications. *Aluminium soaps* provide greases with good adhesion to metal surfaces. They are particularly useful for lubricating slow moving gears, but are unsuitable for high speed applications.

Complex greases containing more than one type of metal soap, along with metal salts of low molecular weight fatty acids, have a good high temperature behaviour and resistance to oxidation. They have a broad range of applications.

Block greases can be handled as solids. They are viscous mineral oils heavily thickened with metallic soap. They are suitable for use in slow speed bearings.

Bentonite greases are formed from bentonite clay treated with an organic surfactant which allows the clay to be dispersed in a mineral oil base to form a stable gel. Such materials combine good lubricating properties with outstanding heat resistance.

TABLE 1.52
Characteristics of lithium-based lubricating greases

Test		Test limits		IP Test Method
		Grade No 2	Grade No 3	
Consistency		265–295	220–250	IP50 (worked penetration at 25°C)
Drop point °C	min	175	175	IP132
Water content, % m/m	max	0.50	0.50	IP75
Base oil: Content, % m/m Viscosity at 40°C, mm^2/s (cSt)	min	to be advised by Contractor		
Corrosion test		Negative	Negative	IP112 (24 h at 100 ±5°C with BS970 En 33 steel and BS3919 copper)
Oil separation, % m/m	max	6	6	IP 121 (after 7 days at 40° ±1°C)

29.2 Solid lubricants

Graphite and molybdenum disulphide are solid lubricants suitable for use at high temperatures or at high loads to prevent components from seizing.

They should only be used after taking advice from the supplier of the lubricant and the manufacturer of the plant in which they are to be used. Applications which have been identified are:

● Pedestal bearing — sliding feet.

● Mechanical parts of governor gear, not lubricated by turbine oil.

● Studbolts and other fasteners liable to seizure (up to 400°C).

● Valve spindles (up to 400°C).

● Valve spindle drives.

29.3 Testing of greases

Because greases are mixtures, separation of their components is the most important thing to check for on delivery. Samples should be taken and examined to ensure that the grease is smooth, homogenous and free from lumps. Afterwards a check may be made in the laboratory that the consistency is of the required grade (Table 1.52) using the worked penetration test of IP50, or the drop point may be checked (IP132) using the apparatus shown in Fig 1.72. The samples should also be free from mineral acidity, mineral filling matter, abrasive impurities and objectionable odour. These simple checks should be sufficient to confirm that the supply is acceptable.

In service the main cause of problems is again separation. This can usually be detected visually; consistency and drop point checks would confirm if the grease were unacceptable.

30 Insulating oils

30.1 Introduction

Insulating oil fulfils several functions within electrical equipment. Primarily, of course, it meets the electrical insulation needs but it is also the heat transfer fluid in a transformer, taking heat from the core and conductors to the outside cooling surfaces, where it is dissipated to the atmosphere. It must also have a good capacity to absorb gases, as the formation of bubbles within equipment would seriously reduce the insulation resistance. Insulating oil has a satisfactory specific heat coupled with excellent electric strength.

In switchgear and circuit-breakers, the main function of the oil is to extinguish the arc formed between contacts when an electric circuit is opened. Large switchgear uses an airblast for this purpose, but small equipment and even some large circuit-breakers are oil-filled.

Special viscous oils are used as insulants in metal-clad busbar chambers where exposure to high temperatures and stresses is absent and also in some cable-end joints where their purpose is to restrict movement of oil through the seals.

Specially compounded insulating oils are also used as dielectrics in capacitors and in cables, for resistance to prolonged high electric stress.

For each application the insulating oil must be of high purity, free of suspended matter and dry. New transformer oil, for example, is required to meet a strict standard (e.g., BS148: 1984) and other standards give guidance for maintaining oil in good condition for further use (e.g., BS5739: 1979). Further periodic testing is also carried out to determine if gases have been produced within the equipment. This can occur when oil or insulation is overheated; the gas measurements aid diagnosis of equipment faults (see BS5574: 1978).

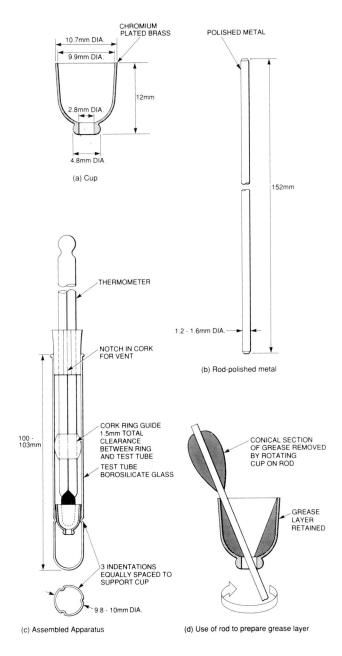

CHROMIUM PLATED BRASS

POLISHED METAL

10.7mm DIA.

9.9mm DIA.

12mm

2.8mm DIA.

4.8mm DIA.

(a) Cup

THERMOMETER

152mm

NOTCH IN CORK FOR VENT

1.2 - 1.6mm DIA.

(b) Rod-polished metal

100 - 103mm

CORK RING GUIDE 1.5mm TOTAL CLEARANCE BETWEEN RING AND TEST TUBE

TEST TUBE BOROSILICATE GLASS

CONICAL SECTION OF GREASE REMOVED BY ROTATING CUP ON ROD

GREASE LAYER RETAINED

3 INDENTATIONS EQUALLY SPACED TO SUPPORT CUP

9.8 - 10mm DIA.

(c) Assembled Apparatus

(d) Use of rod to prepare grease layer

FIG. 1.72 Grease drop point test (IP132)

30.2 British Standard insulating oil specifications

British Standard insulating oil specifications provide a basis for understanding which factors are important in selecting an oil. Within the electricity supply industry equipment is frequently highly stressed and more appropriate, stringent standards may apply to the various classes of equipment where necessary. International standards may also apply.

30.2.1 New oil

Insulating oil for transformers and switchgear is naphthenic in character, usually pure and without additives. It may be specified by reference to British Standard

148: 1984. The limiting characteristics are summarised in Table 1.53.

Three classes of oil are included in the table. They differ in flash point, pour-point and viscosity behaviour. Generally Class 1 oils are used in Britain. Inhibited oils would be indicated by an A (for anti-oxidant) following the class number. British Standard 148: 1984 gives a specification for inhibited oils, although their use in the UK is small. The anti-oxidant stabilises an oil against oxidation and deactivates the catalytic action of metals on the oxidation process.

New insulating oil for cables may be of mineral or synthetic (e.g., dodecylbenzene) origin. They are specified for example in IEC 465: 1986. National Standards are similar to IEC standards but again the industry may apply more stringent standards for highly stressed equipment. The limiting characteristics are summarised in Table 1.54.

Three grades of product are considered, classified according to their viscosity.

30.2.2 Maintenance of oil in use

The condition of oil and paper insulation can deteriorate with time and as a consequence of the operating stresses within the oil-insulated equipment. British Standard 5730 provides limiting specifications against which to judge whether oil is fit for further use in transformers. In general, the industry applies more stringent standards than BS5730 but the principles of the document are followed. Although breakdown voltage is a prime specification, electrical resistivity (the CERL test: Forrest 1960) is a more useful test with oils being categorised as:

Good oil — better than 200 giga-ohm m at 20°C.

Fair oil — between 20 and 200 giga-ohm m at 20°C.

Bad oil — worse than 20 giga-ohm m at 20°C.

High acidity, water, fibre and metallic content can all result in poor oil resistivity. Subsidiary testing is therefore carried out before deciding upon a course of action, which may also depend upon the transformer type, i.e., 'fair' oil may be usable in a small low-rated transformer but certainly not in a highly rated one.

In Table 1.55, examples of specifications for differently rated transformers are given, where the transformer classes are:

A1 All transformers fitted with refrigerated breathers.

A2 All transformers above 170 kV fitted with silica gel breathers.

B Transformers with primary voltages between 80 and 170 kV.

C Transformers with primary voltages below 80 kV.

30.3 Sampling from storage

This is described in detail in British Standard 5263 and the principles follow closely those previously described

TABLE 1.53
Uninhibited mineral insulating oils

Property	Test method	Limiting values for measured characteristics		
		Class I	Class II	Class III
Kinematic viscosity, mm^2/s at 40°C*	BS2000 Pt 71	≤ 16.5	≤ 11.0	≤ 3.5
− 15°C		≤ 800		
− 30°C			≤ 1800	
− 40°C				≤ 150
Flash point °C	BS2000 Pt 34	≥ 140	≥ 130	≥ 95
Pour-point, °C	BS2000 Pt 15	≤ − 30	≤ − 45	≤ − 60
Appearance	View through 100 mm thickness	Clear, free from sediment and suspended matter		
Density, kg/dm^3 at 20°C	BS4714	≤ 0.895		
Neutralisation value, mg KOH/g oil	BS2000 Pt 1 BS148	≤ 0.03		
Corrosive sulphur	BS5680	Non-corrosive		
Water content, mg/kg bulk delivery drum delivery	BS6470	≤ 30 ≤ 40		
Anti-oxidant additives	BS5984	Not detectable		
Oxidation stability, 164 h total acidity, mg KOH/g oil sludge % by mass	BS148	≤ 1.5 ≤ 1.0		
Breakdown voltage, as delivered, kV	BS5874	≥ 30		
Dissipation factor at 90°C and 40 Hz to 62 Hz	BS5737	≤ 0.005		
Gassing tendency at 50 Hz after 120 min, mm^3/min	BS5797	≤ + 5		

*Corresponding viscosity values at 20°C are approximate. 40 mm^2/s for Class I oil 25 mm^2/s for Class II oil and 6 mm^2/s for Class III oil.

for fuel oil and turbine oil. Much more care must be exercised for insulating oil, however, as even slight contamination by moisture or atmospheric pollution may cause the sample to be out of specification. Various sampling thiefs are available for sampling from bulk tanks or drums. Figure 1.73 shows a typical design. When the valve rod strikes the bottom of the tank the top and bottom valves are opened allowing oil to enter and air to escape. Raising the device causes the valves to close again, thus protecting the sample. Bulk tanks may be sampled using the arrangement shown in Fig 1.74. The apparatus, of stainless steel or glass tubing, is made to withdraw oil from close (~3 mm) to the bottom of the tank so as to detect any contamination. Barrels may be sampled similarly, in this case using a sampling tube fitted with taps as shown in Fig 1.75 instead of a bottle. About 2.5 l of oil is required to carry out all tests.

30.4 Testing

The tests required for insulating oils have been given in Tables 1.53 and 1.54. Of these, flash point, elec-

tric strength, neutralisation value (or acidity) and water content are those most commonly made as routine.

30.4.1 Electric strength

The electric strength test provides a measure of the electrical insulating properties of an oil. Two tests are described in BS5874: the breakdown voltage test, in which an alternating voltage is applied to two metal spheres submerged in a sample of the oil using equipment of standardised design and characteristics and the withstand voltage test. In the former the voltage is raised from 15 kV at a rate of 1 kV per second until the sample breaks down. The test is carried out six times on one sample. In the latter, using the same equipment, the voltage is raised from 15 kV to the test voltage in 10–15 seconds and maintained at this value for 1 minute (see Table 1.55).

30.4.2 Resistivity

Resistivity is measured routinely on oils in service (BS5737) and provides the primary indication that oils

TABLE 1.54

New mineral oils for cables with oil ducts

Characteristic	Test method	Class I	Class II	Class III
Kinematic viscosity	IEC 296 (1969)	> 30 (*1*)	13–30 (*1*)	< 13 (*1*)
Flash point, °C (*2*)				
open cup	ASTM 92–72	⩾ 150	⩾ 135	⩾ 110
closed cup	IEC 296A	⩾ 140	⩾ 125	⩾ 100
Pour-point, °C	IEC 296A	⩽ − 30	⩽ − 30	⩽ − 40
Density, kg/dm^3 at 20°C	ISO R649	0.87–0.90	0.86–0.90	0.85–0.88
Appearance	IEC 296	Clear no suspended matter		
Cloud point, °C	ASTM D 2500–66	⩽ − 20 (*3*)		
Auto-ignition temperature, °C	ASTM D 2155–66	⩾ 250		
Neutralisation value, mg KOH/g oil	IEC 296	⩽ 0.03		
Corrosive sulphur	IEC 296A	Non-corrosive		
Aromatic hydrocarbons content	Method under consideration			
Oxidation stability	Method under consideration			
Breakdown voltage, kV	IEC 156	⩾ 30 (*4*)		
Dissipation factor (90°C, 50 Hz)	IEC 250, 247	⩽ 0.002 (*4*)		
Resistivity, GΩm (90°C)	IEC 93, 247	⩾ 50 (*4*)		
Stability under electric stress	Method under consideration			

Notes

1   For oils covered by this table, the following values are the most usual and generally recommended:

        Class   I   40 ±6 cSt
        Class   II 16 ±2.5 cSt and 25 ±4 cSt
        Class   III   6 ±1 cSt and 10.5 ±1.5 cSt

2   Only one of the methods need be used.
3   Only a slight opalescence, no separation for solid products.
4   For new, untreated oil, as received.

TABLE 1.55

Further use of transformer oils

Property	Test method	Limiting values for measured characteristics to allow further use in transformers of Type:			
		A1	A2	B	C
Resistivity	BS5730 and	> 200	> 60	> 20	> 20
giga-ohm m at 20°C	BS5737				
Neutralisation value	BS2000 Pt 1	< 0.15	< 0.20	< 0.40	< 0.40
mg KOH per g oil					
Water content mg/kg ‡	BS6470	< 15	< 23	< 23	< 23
at 20°C					
Breakdown voltage *, kV	BS5874	⩾ 65	⩾ 55	⩾ 45	⩾ 40
or					
Withstand voltage, kV					
for 1 min (2.5 mm gap)		> 40	> 37	> 30	> 22

‡ Limits apply where acidity < 0.1 mg KOH per g oil

* Average of six breakdowns on one cell filling with 2.5mm gap spacing.

may not be fit for further use. Chemical investigations are then undertaken to determine why the insulation is deteriorating.

30.4.3 Neutralisation value (acidity)

The development of acidity in an insulating oil is an indication of deterioration due to oxidation. The test described in BS148 is a slight modification of that in BS2000 Part 1 used for lubricating oil. The oil sample is mixed with toluene and ethanol and neutralised with a standard solution of alcoholic potassium hydroxide

using alkali blue 6B as indicator. The result is expressed as mg KOH per gram of oil.

High acidity will reduce the insulating properties of an oil and increase its ability to absorb moisture. This latter behaviour is taken into account when setting moisture limits for oil (see Table 1.54 above).

30.4.4 Water content

The moisture content of samples is measured at supporting laboratories, by use of an automated Karl Fischer method (BS6470). Results are reported by weight, i.e.,

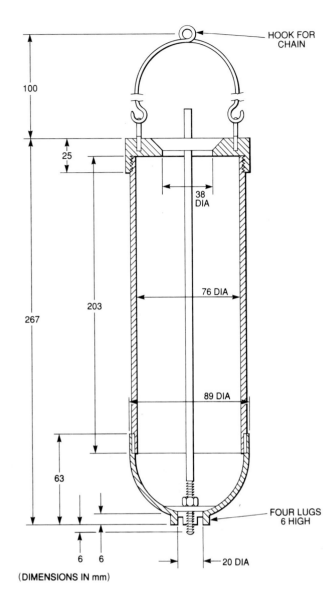

100

25

38
DIA

203

267

76 DIA

89 DIA

63

FOUR LUGS
6 HIGH

6   6

20 DIA

HOOK FOR
CHAIN

(DIMENSIONS IN mm)

FIG. 1.73 'Thief' dipper for sampling oil in bulk

as mg water per kg of oil. Limits are set in these terms also (see Tables 1.53 and 1.55), but care may be required in interpreting results. The objective of setting moisture limits is to ensure that free water cannot form within the oil-insulated equipment. Such water as is present will distribute itself between the paper insulation and the oil phase, so that the water vapour pressure over each phase is in equilibrium *at the working temperature of the equipment* at the time of sampling. Then:

$$\frac{\text{Actual moisture content in paper}}{\text{Saturation moisture of paper}} = \frac{\text{Actual moisture content in oil}}{\text{Saturation moisture in oil}}$$

If the equipment was switched off and cooled to ambient temperature, a different equilibrium would be

established and it might be necessary to consider if free water could then form.

The *saturation* level of paper does not vary greatly with temperature and is typically 16% by weight. The actual moisture level in the paper insulation of a transformer is therefore determined by the oil properties, since from the above equation:

Actual moisture content in paper =

$$16\% \times \frac{\text{Actual moisture in oil}}{\text{Saturation moisture in oil}}$$

The electric strength of paper decreases as the moisture content increases. A content of 2% by weight is desirable, equivalent to a saturation of 0.125, i.e., 2/16. If the moisture content rises to 6% the electric strength will fall by a quarter and decrease even more rapidly with further rises.

Interpretation of the moisture in oil measurements to allow for temperature is straightforward. If D_n represents the degree of saturation at a temperature T_n (K) and 'C' or '$C_{sat(Tn)}$' are respectively the actual and saturated oil concentrations at the appropriate temperatures, then we know:

$$D_1 = C/C_{sat(T1)} \quad \text{and} \quad D_2 = C/C_{sat(T2)}$$

$$\text{so} \quad D_2 = D_1 (C_{sat(T1)})/(C_{sat(T2)}) \quad (1.14)$$

As the saturation level of moisture in oil obeys the relationship:

$$\log C_{sat} = B - A/T \quad (1.15)$$

where A and B are constants, it follows from Equations (1.14) and (1.15) that:

$$\text{Log } D_2 = \log D_1 - A [(1/T1) - (1/T2)] \quad (1.16)$$

The value of A is about 800. Figure 1.76 illustrates Equation (1.15) and gives values of A of 868 for a new oil within specification and 732 for a used oil at the limit of acceptability.

To take an example, suppose a small transformer was found to contain 23 mg kg^{-1} of moisture in its insulating oil, which is the limit from Table 1.55. The measuring laboratory would shake the oil sample with more water and measure how much was taken up by the oil, the saturation level and the temperature would be noted. If the temperature was 20°C, then from Fig 1.76 if the oil was new (acidity $\leqslant 0.03$ mg KOH/g oil Table 1.53) the saturation moisture content would be found to be 55.9 mg kg^{-1}. The degree of saturation is therefore 0.41 (23/55.9) at 20°C. If the oil was very old (with an acidity of ~0.4 mg KOH/g oil) then, from line (a) in Fig. 1.76, a saturation of 132 mg kg^{-1} would be found and the degree of saturation would be 0.17 (i.e., 23/132).

The actual moisture content of the paper is 16% × the degree of saturation, in the new oil case the paper would contain 6.5% of moisture at 20°C (i.e., 16 ×

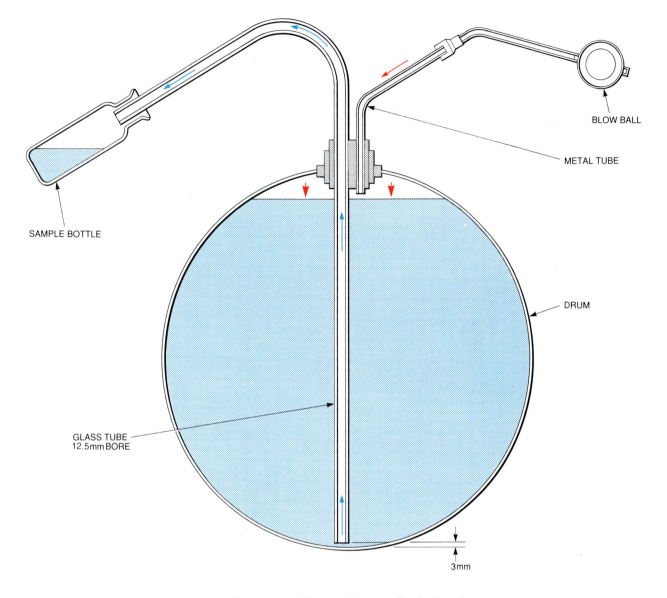

SAMPLE BOTTLE

BLOW BALL

METAL TUBE

DRUM

GLASS TUBE
12.5mm BORE

3mm

FIG. 1.74 Pressure 'thief' used for sampling bulk tanks

0.41) which is not good, whereas in the used oil example it would contain 2.7% (i.e., 16×0.17) which is acceptable. Degrees of saturation at other temperatures would be calculated using Equation (1.16).

In practice the laboratory would advise action for both examples. For the new oil case the engineers would be advised to reduce the moisture level in the oil, in the latter case the acidity level is at the limit (Table 1.55) and the oil would need purification.

30.5 Purification

All insulating oil must be pure and for transformers operating at over 250 kV particular care is needed in its purification and handling. Moisture levels in new oil which is to be used to fill large transformers may be specified to be below 5 mg kg^{-1} and almost complete removal of dissolved air may be necessary; in-service, moisture levels below 11 mg kg^{-1} may be needed to maintain trouble-free operation. Specialist services are required to meet these demanding standards.

On delivery, transformer oil will be saturated with air and contain about 25 PPM of water. During handling it may pick up small amounts of dirt or fibre. Thus, before it is added to a new transformer or used to refill an existing one, the oil must be filtered, degassed and dried. The operations are carried out in this order and a schematic layout of the equipment used is shown in Fig 1.77.

30.5.1 Filtration

This has two purposes, removal of dirt and removal of free water. The oil is passed between two thicknesses of paper in a suitably designed filter pack. As the oil percolates out, through the paper, solid particles, down to 1 micron diameter, are removed. The paper is also water absorbent, and any free water droplets are taken up by the paper. The process is

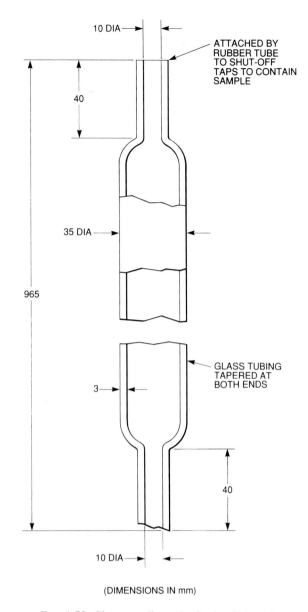

(DIMENSIONS IN mm)

FIG. 1.75 Glass sampling tube for insulating oils

readily controlled, since when the paper is saturated with water it becomes oil repellent and the pressure drop across the filter increases. By monitoring this pressure drop, the filter condition is immediately apparent and action can be taken to renew the paper sheets.

30.5.2 Degassing

The oil leaving the filter is passed to a vacuum chamber. Here, the vapour pressure exerted by dissolved water, air and the lighter fractions of the oil come to equilibrium. The aim is to remove as much water and air as possible without unduly affecting the oil composition. The process is one of fractional distillation and consideration of its thermodynamics suggests that separation should be most easily achieved under high vacuum and low temperatures. In practice the oil foams excessively under the best theoretical conditions and it is found desirable to raise the temperature to 50°C and to use a vacuum of 7.6 Torr (one hundreth

of atmospheric pressure). Under these conditions dissolved gases are reduced from typically 10% (i.e., 100 ml/l of oil) to 0.1% and dissolved moisture is reduced to typically 1.5 mg kg^{-1}. The process relies on the degassing of a thin film of oil, so that gas diffusion is aided and vapour pressures are consistent throughout the process. Droplet formation is minimised, since the vapour pressure of a dissolved gas over a droplet varies with its size and the process would not then proceed as uniformly as it does over a thin film of oil.

30.5.3 Heating

The heating of the oil to 50°C prior to its degassing must be carried out with care. Transformer oil begins to degrade above 140°C and therefore a heat transfer medium is usually employed to transfer energy to the heat exchange surfaces. The direct use of electric elements, for example, could lead to hot spots and consequent oil damage.

When each of the above steps is completed the oil is ready to be introduced to the equipment.

30.6 PCB synthetic insulating oils

Some smaller items of electrical equipment, capacitors for example, are filled with non-flammable insulating oil to reduce fire hazards. These are not mineral oils; they are usually mixtures of chlorinated aromatic compounds with pentachlorobiphenyl and trichlorobenzene being prominent components. They are referred to generically as polychlorinated biphenyls (PCBs). The materials are damaging to the environment and no new insulants of this type are to be purchased.

Their electrical properties are comparable to mineral oil insulants. In service they have some disadvantages. Arcing can lead to their thermal decomposition, producing hydrochloric acid which in the presence of moisture causes serious corrosion. The amount retained in use is declining and disposal of unwanted PCBs requires care and advice. They are not degraded naturally and if ingested by animals, accumulate in their body fat, ultimately causing problems with the top carnivores of a food chain. Currently, waste PCBs must be sent to an approved incinerator where they can be completely decomposed.

Chemical treatment to remove PCBs from contaminated hydrocarbon oils is possible and desirable, as the oil is recovered for re-use. Suitable plant is not available in the UK at the moment, but the situation could change if the balance of the economics of incineration versus treatment and recovery tipped in favour of treatment.

31 In-service monitoring of insulator oils

31.1 Buchholz gas analysis

When a fault occurs inside a transformer, the energy released will decompose oil molecules or degrade paper,

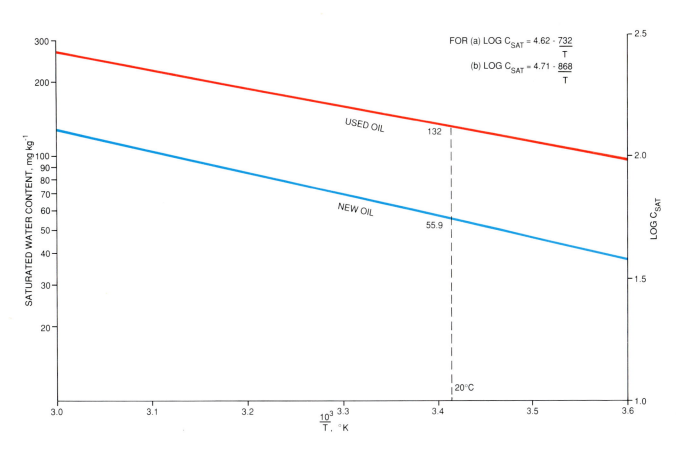

FOR (a) $\text{LOG } C_{SAT} = 4.62 - \dfrac{732}{T}$

(b) $\text{LOG } C_{SAT} = 4.71 - \dfrac{868}{T}$

FIG. 1.76 Solubility of water in insulating oils

in both cases producing free gases. Detection of this gas at an early stage, by a device which can also automatically trip the transformer, may avoid serious damage to the equipment. The Buchholz relay shown in Fig 1.78 is the device used for this purpose. Under normal conditions it is filled with oil, but when gas is produced it collects in the upper part of the chamber and displaces the oil. As the oil level falls, the hollow float swings down and when contact is made by the mercury switch an alarm is activated. If gas is generated by a more serious fault in a sudden surge, then the flap valve is raised as oil and gas push past; the mercury switch attached to the flap trips the power when contact is made.

Following either a gas alarm or a power trip, a gas sample is taken from the sampling cock at the top of the Buchholz relay and submitted to a supporting laboratory for analysis. At the same time oil samples are taken and the gases dissolved in the oil are determined. If gases arising from the decomposition of oil or paper insulation are found the transformer will be taken out of service for inspection.

On occasion the gas alarm will be as a result of air in the system becoming displaced and collecting in the Buchholz relay. Gas analysis will quickly show that only air is present and the oil will be checked in case there is a small fault beginning to develop. If all is well the transformer can be returned to service. It is often

worth checking the nitrogen to oxygen ratio when air is the cause of an alarm. If an air pocket has been present in a transformer for some time before being displaced, oxygen will have been consumed by the oil, and the nitrogen to oxygen ratio will be less than the 4:1 value of normal air. If new air has been drawn into the system, then the ratio will be normal and although in all other ways the transformer may appear healthy, it will be necessary to consider where the air inleakage is occurring.

31.2 Routine gas in oil analysis for faults

31.2.1 Introduction

Before a damaging fault occurs in a transformer, there may be a period of months or even years where a minor fault condition produces enough energy to break down insulation, oil or paper. During this period, as the gases are produced slowly, they have time to dissolve in the oil and no free gas bubbles are formed. By monitoring oil samples from key transformers quarterly, or yearly in the case of less important transformers, trends in the quantities and types of dissolved gases may be followed and the nature of the developing fault can be deduced. Guidelines are laid down for the sampling frequency for each class of transformer based upon their importance to the transmission system.

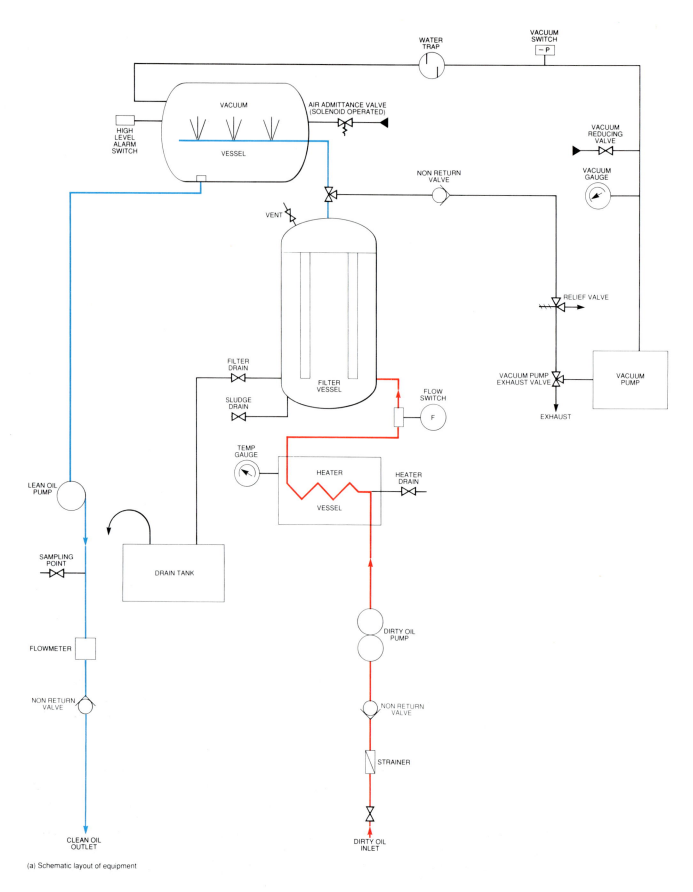

(a) Schematic layout of equipment

FIG. 1.77 Transformer oil degassing and purifying equipment

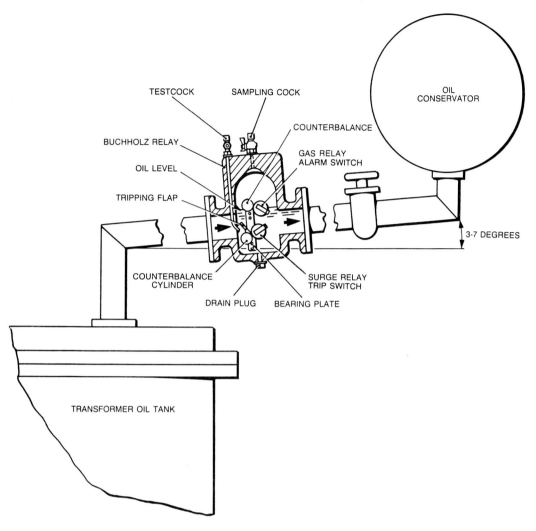

FIG. 1.78 Buchholz protection relay

The technique is particularly sensitive. To break down an oil molecule to produce hydrogen, for example:

$$RCH_2 - CH_3 \rightarrow RCH_2 - CH_2{}^{\cdot} + H^{\cdot}$$

requires about 420 kJ of energy per mole (1 g) of hydrogen formed. Subsequent reactions produce 11 200 ml of hydrogen gas and about 3×10^{-2} ml of gas would thus be expected to form, per joule dissipated from an arc immersed in oil. This has been confirmed by experiment (Waddington and Allan 1969). Since a typical large transformer might contain 50 000 l of oil and present analytical techniques can easily detect 1 PPM (1 μl of gas/litre of oil) of dissolved gas, this means that energy releases as low as 2 kJ can readily be detected.

The above comments apply to transformers in service. The operation of their coolers ensures that any dissolved gases are well mixed into the bulk of the oil and obtaining a representative sample of oil does not present a problem. In an oil-filled cable, or other equipment where there is no imposed oil circulation, gases can only move from their point of origin by diffusion.

This is an extremely slow process; movement is of the order of 1 cm per day. If a fault was suspected to be present in a cable stop-joint, for example, then a calculation of the amount of oil in the sampling line would have to be made. This amount of oil would have to be withdrawn and discarded before a volume representative of the suspected faulty location could be reached. Consideration of dissolved gas movement is necessary whenever equipment containing static oil is sampled.

31.2.2 Sampling requirements

If it is necessary to determine the quantity of air dissolved in an oil sample from a transformer, then oil is taken from the main tank and allowed to fill a sampling vessel (Fig 1.75) from the bottom upwards thus displacing any air. At least 500 ml of oil is allowed to overflow as this will have been in contact with air and therefore be untypical. The vessel is then sent to a supporting laboratory for analysis.

For routine monitoring, dissolved air is not a problem, but avoiding contamination by moisture is important. Normally, therefore, amber glass bottles of

135

500 ml capacity are used for containing the sample. A flange (Fig 1.79) is fitted to the main filling valve of the transformer and silicon rubber tubing is used to connect the bottle to the valve. The bottles are of amber glass to prevent the ultra-violet component of sunlight generating gas in the oil sample. This has been found to occur when clear glass bottles are used, even with British sunlight! The bottles are used solely for oil sampling and are dried and conditioned with dry oil to flow through the system (Fig 1.79) again rejecting about 1 litre as before, since this may be untypical if it has been in the static region near to the valve. When the bottle is full a black plastic cap is quickly fitted and about 0.5–1 cm space may be left at the top, so that oil expansion in the event of a temperature change does not damage the bottle by hydrostatic pressure. Alternatively, the bottle may be completely filled but the top will not be tightly secured. The final step is to label the bottle clearly. The date of sampling, the transformer identification, sampling point, the operating temperature and any other significant observations must be clearly noted so that the analyst can interpret the results to best advantage.

An alternative sample containment, used in countries with large diurnal temperature variation or when the sample has to be transported long distances, is the syringe (Fig 1.80). This is filled taking similar precautions to bottle filling. During transport the syringe piston moves to accommodate oil volume changes as temperatures or pressures vary.

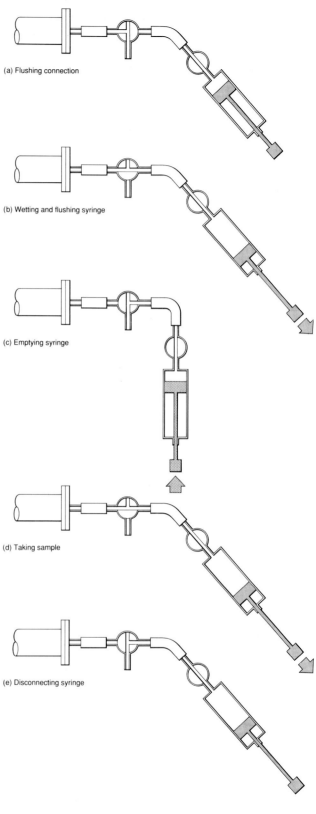

(a) Flushing connection

(b) Wetting and flushing syringe

(c) Emptying syringe

(d) Taking sample

(e) Disconnecting syringe

FIG. 1.80 Sampling by syringe

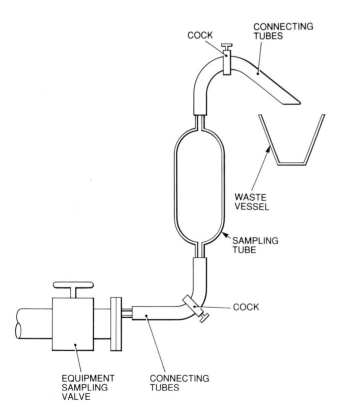

FIG. 1.79 Sampling flange arrangements

31.2.3 Analysis of dissolved gases in the oil

There are at least three sampling and analysis situations. The first is the monitoring of new transformers under test. No more than 2 PPM of gases (2 μl of gas at STP/l of oil) may be generated during the temperature rise tests and more care than usual is required to obtain good results. The second case is routine monitoring. Typical gas levels are in the range 10 to 100 PPM and adequate accuracy is obtained with smaller samples than are needed for factory tests. The third case is routine testing by rapid methods. This is often necessary when a laboratory has a heavy workload and needs to survey many samples quickly.

31.2.4 The Toepler method

In the first two cases the apparatus used is shown schematically in Fig 1.81.

For the factory test situation, an oil sample of 1500 ml is required usually taken using a vessel such as shown in Fig 1.75 and in the second case 250 ml of oil is used. The sample is transferred into the measuring funnel and the apparatus is evacuated. Oil is then introduced into the degassing flask and agitated by means of a magnetic stirrer. The dissolved gases equilibrate between the free space in the apparatus and solution in the oil. After 3 minutes the free gases are removed by means of the Toepler pump. This operates as a piston pump. When the mercury level is lowered the gases expand into the Toepler barrel and then when the mercury level is raised the gases are trapped, compressed and pushed into the gas collecting burette.

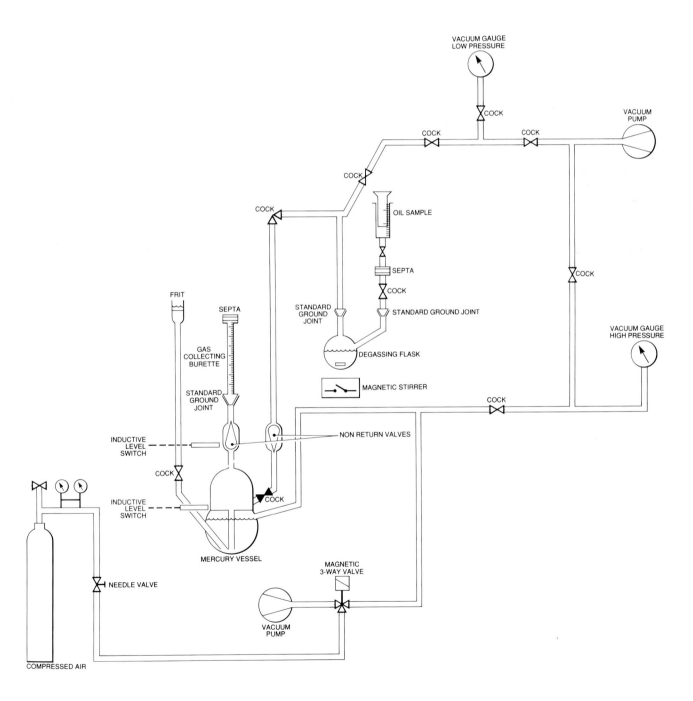

FIG. 1.81 Toepler pump apparatus used for gas-in-oil analyses

Four further extractions may be carried out at 5 minute intervals before the majority of the gases are removed. Different laboratories use different numbers of extractions and the dimensions of their Toepler pumps and extraction vessels may vary. As a result, although within a laboratory reproducible results may be obtained on gas extraction, comparison between laboratories may show considerable variation. Within an organisation a 'round-robin' exchange of samples is desirable to establish the compatibility of results.

The volume of gas extracted at a pressure 'p' and at laboratory temperature is noted and the volume is reported at STP (standard temperature and pressure). The gases are then transferred from the gas burette and put into the sample loop of a chromatograph to determine the type of gases present; hydrogen, methane, ethane, ethylene, acetylene and carbon oxides. To determine the proportion of each type of gas in the mixture an inert carrier gas (argon or helium) is used to sweep the gas sample through one of two columns, each about 2 m long and 4.5 mm diameter (Fig 1.82). As gases are carried through the first column, which is packed with Porapak N, the carbon oxides and hydrocarbon gases are partially absorbed. The result is a 'braking' effect, such that some gases in the mixture move faster than others and a separation is achieved. At the outlet of the column the gas flow is passed through a flame. When the separated hydrocarbons reach the flame (in the order methane, ethylene, ethane and acetylene) they burn and produce ionised gases. Two electrodes placed in the flame, with a voltage applied, cause a current to flow when ionisation occurs and the magnitude of the current is in proportion to the amount of hydrocarbon present in the 'peak' of separated gas leaving the column.

This type of detector is called a flame ionisation detector (FID). Carbon dioxide leaves the column after methane and before ethylene. It does not cause ionisation in a flame so a timer switch diverts the gas flow temporarily to a thermal conductivity detector (see below) in order to measure the amount of carbon dioxide present; the flow is returned to the FID in time to measure the ethylene and subsequent gases.

A second sample of gas passes through the other column (Fig 1.82) which is packed with molecular sieve, grade 5A, or perhaps better, grade 13X. This separates the gases hydrogen, oxygen, nitrogen and carbon monoxide in that order. To detect when they emerge from the column the gas stream is passed over a heated wire. Its resistance is steady when the carrier gas is passing, but when another gas 'peak' passes the rate of heat loss changes, due to the different thermal conductivity of the gas. The wire's resistance consequently changes as its temperature alters and this is detected and acts as a measure of the quantity of gas in the peak.

When the proportion of methane, carbon dioxide, ethane, ethylene and acetylene and hydrogen, oxygen, nitrogen and carbon monoxide have been determined in the gas, the amounts present in the original oil sample may be calculated. Results are reported as parts per million by volume and are in units of microlitres of dissolved gas, at standard temperature and pressure, per litre of oil.

Rapid methods

Instead of using the Toepler pump to extract gas from the sample, the carrier gas to the chromatograph can be routed through the oil sample (Fig 1.83) and used

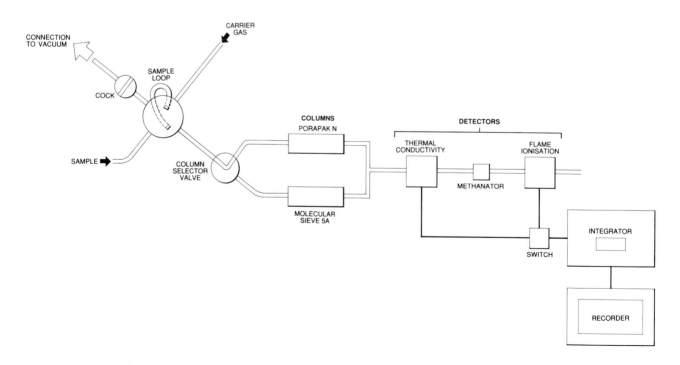

FIG. 1.82 Arrangement of chromatograph columns for gas-in-oil analysis

to sweep the dissolved gases out of solution. Only 0.25 ml of oil is necessary for this method. The sample is introduced to the stripper by means of a syringe. The needle is inserted through the septum and the sample is quickly injected. Removal of gases is rapid and adequate results can be obtained. The method is not absolute however. Standard oil mixtures of known gas content have to be prepared to calibrate the procedure. Furthermore, because the sample size is so small, carbon oxide detection is difficult. A 'methanator' is employed to convert the carbon monoxide or dioxide in the gas stream leaving the columns into methane. This gas is then detected with greater sensitivity by a flame ionisation detector than are the carbon oxides by a thermal conductivity detector. Sufficient sensitivity can then be obtained. The method has advantages of speed but for reference the Toepler procedures are presently preferred in the UK.

Interpretation of results

If a system containing carbon, hydrogen and the gases CH_4, C_2H_4, C_2H_2 and C_2H_6 is considered from a thermodynamic viewpoint, predictions can be made as to what the relative proportions of each present at equilibrium will be at various temperatures (Halstead 1973: see Fig 1.84). This mirrors the decomposition of insulating oil in a transformer during fault conditions. Examination of Fig 1.84 shows that for low temperature faults, oil will break down predominantly to hydrogen and methane. Typical conditions might be circulating currents in a transformer core. At high temperatures, such as those produced by an electric arc, acetylene C_2H_2, becomes a prominent product.

Starting from the suggestions arising from this thermodynamic approach, CEGB staff accumulated data on the gas proportions observed for dissolved gases arising from many different types of transformer fault. This work commenced about 1969. The results were collated on a statistical basis (Rogers 1975) and it was found that faults could be classified by the ratios of the gases produced. The gases used are methane (CH_4), ethane (C_2H_6) ethylene (C_2H_4) and acetylene (C_2H_2). Table 1.56 illustrates the method.

TABLE 1.56

'Rogers' ratios and codes
(a) Refined code

Gas ratio	Range	Code
CH_4/H_2	Not greater than 0.1	5
	Between 0.1 and 1.0	0
	Between 1.0 and 3.0	1
	Not less than 3.0	2
C_2H_6/CH_4	Less than 1.0	0
	Not less than 1.0	1
C_2H_4/C_2H_6	Less than 1.0	0
	Between 1.0 and 3.0	1
	Not less than 3.0	2
C_2H_2/C_2H_4	Less than 0.5	0
	Between 0.5 and 3.0	1
	Not less than 3.0	2

(b) Fault diagnosis table based on Code given in (a)

CH_4/H_2	C_2H_6/CH_4	C_2H_4/C_2H_6	C_2H_2/C_2H_4	Diagnosis
0	0	0	0	Normal deterioration
5	0	0	0	Partial discharge
$\frac{1}{2}$	0	0	0	Slight overheating — below 150°C (?)
$\frac{1}{2}$	1	0	0	150°C–200°C (?)
0	1	0	0	200°C–300°C (?)
0	0	1	0	General conductor overheating
1	0	1	0	Winding circulating currents, overheated joints
1	0	2	0	Core and tank circulating currents
0	0	0	1	Flashover without power follow-through
0	0	$\frac{1}{2}$	$\frac{1}{2}$	Arc with power follow through
0	0	2	2	Continuous sparking to floating potential
5	0	0	$\frac{1}{2}$	Partial discharge with tracking

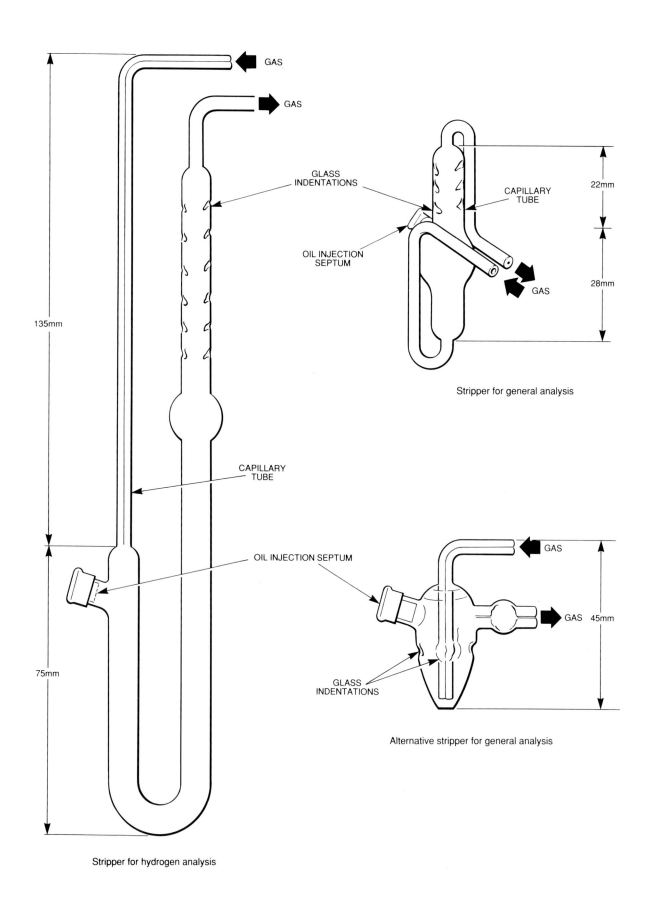

FIG. 1.83 Types of gas strippers; to remove dissolved gases from transformer oil prior to analysis

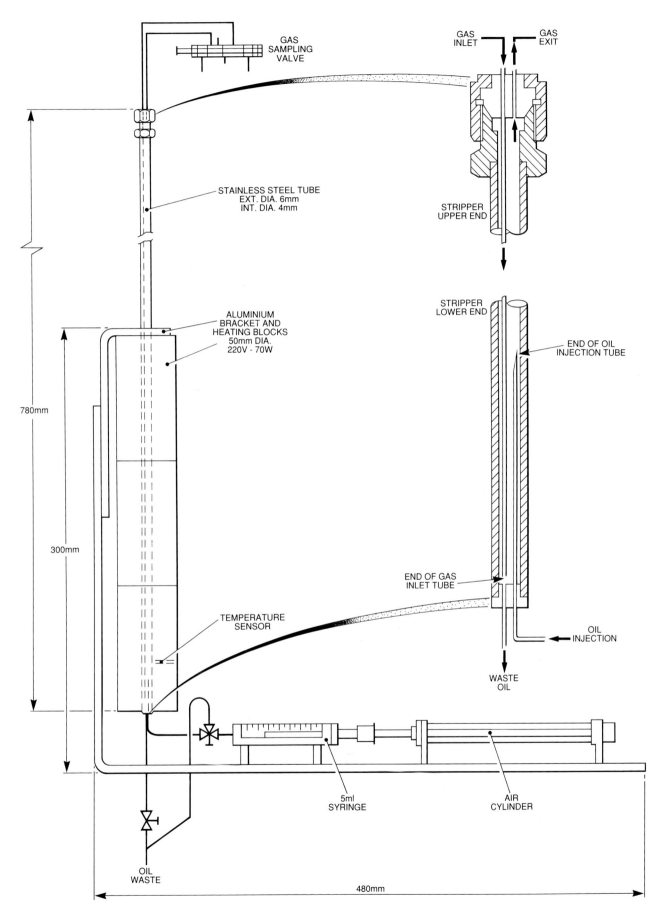

FIG. 1.83 (*cont'd*) Types of gas strippers; to remove dissolved gases from transformer oil prior to analysis

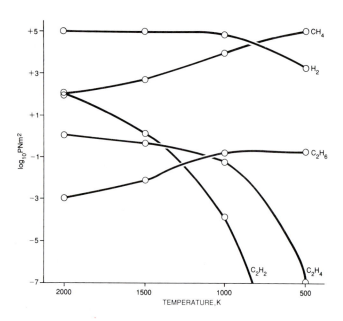

FIG. 1.84 Thermodynamic equilibria of gases produced from hot oils — pressure 1×10^5 N/m^2

The ratio of the reported laboratory results are calculated and compared with the ranges given in Table 1.56 (a). Four code numbers are thus obtained, to check with Table 1.56 (b) and from experience a diagnosis of a fault in the transformer may be deduced. Some care is necessary in applying the method however.

If a transformer contains very little dissolved gas then it follows that even if the gas ratios indicate a fault condition, the energy being dissipated must be very small. The laboratory would advise either ignoring the results or perhaps resampling three months later to establish the rate of gas generation. If only 1 or 2 PPM of gas is generated per three months then the 'fault' is clearly insignificant.

A more difficult situation arises when, as a result of past incidents, a transformer may contain a moderately high gas level. If a new fault develops it will be necessary to inspect past records of the transformer's dissolved gas results. Background levels of gas are then deducted and ratios calculated on the basis of the currently measured rates of rise of each gas.

Finally the levels of carbon monoxide and dioxide are noted. These gases can be the end product of paper deterioration (carbon monoxide) or of high temperature oxidation (carbon dioxide). In a healthy transformer some carbon dioxide will be present due to dissolution from the atmosphere and typically the ratio of carbon dioxide to carbon monoxide is between 4 and 11. If the carbon monoxide level starts to rise sharply, paper deterioration would be suspected; if the carbon dioxide level increases, then more severe thermal oxidation of the cellulose insulation must be suspected. Other indicators of paper breakdown will then be looked for to confirm the diagnosis.

31.3   On-line monitoring

If routine monitoring uncovers an incipient fault in a transformer it may be necessary to keep the equipment in service until a suitable time is available for inspection. During the waiting period it is now possible to monitor the dissolved gases continuously. This is achieved (Fig. 1.85) by passing oil from the transformer through a semi-permeable tube. The dissolved gases are able to diffuse through the tube wall and give rise to their equilibrium pressure in the space between the membrane and outer tube wall. A slow flow of carrier gas sweeps dissolved gases to the sampling loop of a chromatograph and by use of a Porapak column and flame ionisation detector the methane,

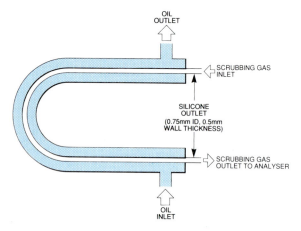

(a) Membrane diffusion cell for on-line gas-in-oil analysis

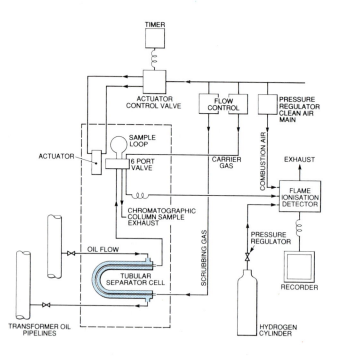

(b) Diagrammatic arrangement of system

FIG. 1.85   Membrane diffusion cell for on-line gas-in-oil analysis

ethylene, ethane and acetylene are determined every twenty minutes. Hydrogen is determined by means of a fuel cell, continuously. 'Rogers' ratios are thus obtained and displayed or recorded on tape by means of a microprocessor. Alarm levels may be set to protect the transformer or alternatively to alert engineers to any increase in the rate of gassing, which would indicate that the incipient fault has passed its early phase and was progressing to a point where damage might occur. The fully weatherproofed unit is illustrated in Fig 1.86.

31.4 Analysis of other insulation breakdown products

When paper is heated to $100^\circ - 300^\circ C$ in oil containing some moisture and acidity, the cellulose is degraded (Shroff 1985). Pentose sugars give furfuraldehyde (I) as an oil-soluble end product of the degradation. The cellulose structure is weakened as it progressively de-polymerises. By monitoring the furfuraldehyde (FFA) level in a transformer's oil samples, valuable information may be obtained about the state of its paper insulation [11]. The technique for determining furfuraldehyde, High Performance Liquid Chromatography (HPLC), is a specialised one, provided by a

FIG. 1.86 A complete on-line monitoring unit

supporting laboratory. If significant amounts of FFA are present, however, a simple colour test is sensitive to the presence of 1 PPM. The oil sample is shaken with glacial acetic acid and aniline and a pink/red colour indicates the presence of FFA. Interferences can occur, so the indicative colour test is not an alternative to HPLC, but it may prove useful when a quick judgement is required on-site.

The laboratory method of HPLC involves taking an oil sample, the same one used for the gas analysis suffices, and shaking it with a small quantity of acetonitrile to extract the furfuraldehyde. The concentrated solution is then passed down a fine chromatographic column under pressure. A variable mixture of water and methanol is used to carry the FFA through the column, and the 'peak' of concentration when it emerges from the column is measured using an ultra-violet detector.

The method is very sensitive, 0.1 PPM may be determined in an oil sample, and the technique should prove increasingly valuable in monitoring transformer insulation. The same method also detects phenols and cresols produced when synthetic resin bonded paper board is degraded. The amount of insulation-condition information which is obtained is therefore considerable. Typical 'high' FFA levels in a transformer might be 1 PPM at 10 years and 4 PPM at 40 years. Actual levels vary with transformer design and service, and some 20 year old transformers may have very little or no furfuraldehyde if they have operated with good cooling and within design specification.

32 Monitoring the occupational environment

32.1 Introduction

Although at times there appear to be too many safety regulations, the body of legislation consists essentially of statements of common sense. These are put together in much the same ways for each regulation, so that the principles of safe working emerge as a common unifying pattern.

Regulations first require that hazards in the work place should be identified and associated risks assessed. Usually a control limit is specified, for example, a concentration of gas fibres or dust in the air near to where people are working, which must not be exceeded.

Controls have then to be established. These could involve restricting entry into certain areas; confining the hazardous material in some way; providing local exhaust ventilation to remove vapours or fume, or perhaps providing personal protective equipment to staff to ensure their safety.

Responsibilities must be defined. Someone should be nominated in writing to ensure the controls are applied and another person should have the job of ensuring that any control equipment is maintained in effective order.

Monitoring will be required to demonstrate that controls are effective and records must be kept. If substances are in use which could affect the health of employees, the regulations will also require that in addition to monitoring the work area, health surveillance of staff should be undertaken. Again records must be kept.

At prescribed periods the records and control procedures should be reviewed and, if necessary, working practices revised to improve the safety and health of employees.

Finally there is a need for good communication. Staff must be made aware of their individual responsibilities and trained in safe working practices. To ensure this is done, training records must be kept and updated as part of the review already mentioned.

The power station chemist is usually given the tasks of controlling the exposure of the workforce to dusts, fumes, chemical vapours, asbestos fibres and toxic liquids, and the implementation of procedures to minimise microbiological hazards and risks associated with entry into confined spaces.

The following sections illustrate how this is done; with the section on asbestos being the most detailed to show the full extent of the efforts required to set up safe systems of work.

32.2 Asbestos and man-made mineral fibre

The most recent control of asbestos at work regulations CAWR came into force in 1988. Two Codes of Practice support the regulations. Guidance on exposure limits and measurement is given in the UK Health and Safety Executive guidance note EH10. In its (1990) edition, asbestos fibre concentrations in the workplace air are subject to the following control limits.

For asbestos consisting of or containing any crocidolite or amosite:

- 0.2 fibres per millilitre of air averaged over any continuous period of 4 hours.

- 0.6 fibres per millilitre of air averaged over any continuous period of 10 minutes.

For asbestos consisting of or containing other types of asbestos but not crocidolite or amosite (e.g., chrysotile)

- 0.5 fibres per millilitre of air averaged over any continuous period of 4 hours.

- 1.5 fibres per millilitre of air averaged over any continuous period of 10 minutes.

When work is to be undertaken in a power station which may involve asbestos, the person nominated to control the situation, typically the chemist, must first develop a strategy. The steps required were outlined previously in the introduction to this section.

This first step, *assessment*, is to find out if asbestos is present, and if so what kind it is. Let us say work

is to be undertaken on a lagged steam pipe. Samples must be taken with a 'corer' to obtain a full cross-section of material right down to the pipe surface. The sample is then identified in a laboratory either by microscopy or X-ray diffraction techniques.

Microscopy involves putting fibres from the sample onto a slide, covering this with a cover plate and then introducing a small drop of Cargille liquid of known refractive index to immerse the fibre by capillary action. The covered slide is placed under a binocular microscope, between crossed-polars. The lower polarising filter polarises light in the east–west direction and the upper analysing filter in the north–south. In the absence of the slide, the field of view appears dark to the observer because no light can pass the analyser. Asbestos fibres rotate polarised light, allowing it to pass the analyser, whereas man-made mineral fibres (MMMF) do not; they are polycrystalline. If bundles of fibres appear white when viewed in the microscope they are asbestos. The MMMF are barely visible and distinction is straightforward. Wool and cotton fibres may be visible but their morphology is very different from bundles of asbestos fibres.

To identify which kind of asbestos fibres are present, a tint plate (first-order red) is added to distinguish between crocidolite (blue) asbestos and other types. Examination of the slide under these conditions enables the 'sign of elongation' to be determined, i.e., whether light travels faster down the fibre (length fast) or across (length slow). Crocidolite is length fast, amosite and chrysotile are slow. The tint plate is inserted into the objective turret of the microscope with its fast axis NE/SW. A fibre is aligned with its long axis parallel to the fast direction of the tint plate and its colour is noted. It is then rotated 90° to be at right angles to the fast axis of the plate and the colour is again noted. Fast fibres show first-order oranges and yellows when aligned parallel with the plate and second-order blues at right angles. The opposite is true for slow fibres (see Table 1.57 and Fig 1.87).

TABLE 1.57

Colours observed from asbestos fibres using crossed-polars and a sensitive tint plate

	Colour: parallel NE/SW	Colour: NW/S
Amosite	Blue/green	Yellow
Chrysotile	Blue	Yellow/orange
Crocidolite	Yellow	Blue/green

The above colours are enhanced by using appropriate Cargille refractive index liquid of N_D 1.67, 1.55 and 1.70 respectively for amosite, chrysotile and crocidolite.

Use of a dispersion staining objective on the microscope, along with the polariser and appropriate liquids, allows final identifications by observing fibre colour changes in the E/W and N/S directions (Fig 1.88).

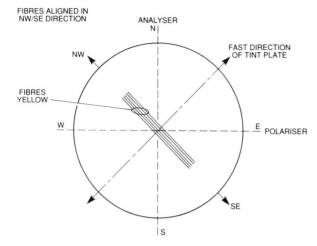

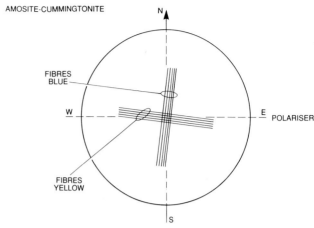

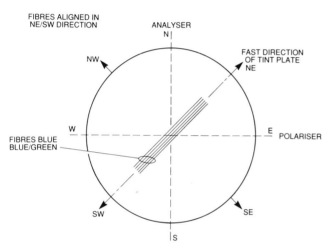

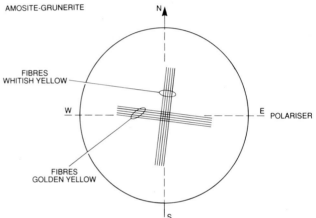

FIG. 1.87 Identification of asbestos fibres: polarisation colours produced by amosite using a red-tint plate

FIG. 1.88 Identification of asbestos fibres: dispersion staining colours produced by types of amosite — mounting liquid RI 1.67

The alternative identification technique of X-ray diffraction involves putting fibres in a beam of X-rays and noting the angles of diffraction of the beam. These are quite characteristic for each asbestos type and the technique is reliable.

Visual inspection of a sample of fibres can give a guide to which type is present but only careful microscopy or X-ray methods can give a sure identification (see Table 1.58 and Fig 1.89).

The next step in managing safe working is to establish *controls*. All of the measures mentioned in the introduction are used. Asbestos-containing material is wetted, to contain the fibres as far as possible. A plastic-sheet tent is placed around the work area to contain fibres further and it is ventilated by suction to ensure that reduced pressure inside prevents escape of fibres. Workers inside the tent are provided with personal protective equipment; face masks and overalls. Finally, to protect other power station staff, barriers are erected to delineate the working area. Outside the barriers, fibre counts must be well below the control limits (typically lower than 0.01 fibres/ml).

TABLE 1.58

Colour and appearance of fibres under a low power microscope

Fibre	Colour	Appearance
Chrysotile	Colourless through white to greenish	Silky, flexible fibre often split. Fibrils easily visible
Amosite	Colourless to grey-brown	Stiff and spiky. Possible to see fibrils. Ends have distinctive split, paint-brush appearance
Crocidolite	Lavender blue	Spiky fibre varying from flexible to stiff. Possible to see fibrils. Ends have distinctive split paint-brush appearance

Responsibilities are defined. Workers must be classified asbestos workers, to undertake the job. The people who carry out asbestos monitoring must be part of a quality control scheme. This involves circulating typical microscope slides used for counting (see below)

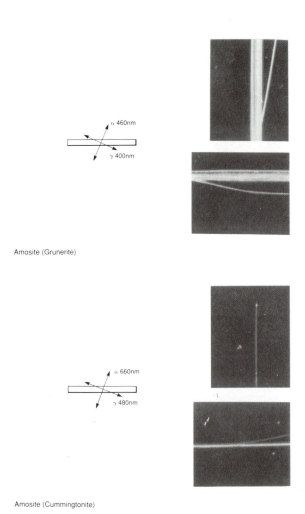

Amosite (Grunerite)

α 460nm
γ 400nm

α 660nm
γ 480nm

Amosite (Cummingtonite)

Fig. 1.89 Appearance of asbestos fibres in a
microscope (as per Fig 1.88)
(see also colour photograph between pp 208 and 209)

between laboratories to ensure the 'correct' results are obtained by all workers. If an individual's results fall outside of agreed statistical limits he/she is retrained. The pumps used in the monitoring process must also be calibrated as part of the quality control scheme and a staff member is responsible for this activity.

Monitoring of asbestos work is undertaken by trained staff using standard procedures which derive from recommendations issued by the Asbestosis Research Council in 1971. It is important that this method is preserved, since the development of illness in workers is related back to their exposure earlier in their working lives. If counting methods were to change, to give different fibre counts which could not be related to earlier methods, the whole basis of the long term monitoring of workers would be compromised.

Briefly the method involves drawing a measured sample of air through a membrane filter using a battery-operated pump in order to trap any entrained fibres. The filter is later rendered transparent by using

a suitable clearing agent. Fibres are then counted using a phase contrast microscope. The result is expressed as fibres per millilitre of air, calculated from the number counted, the ratio of the counted area to the total filter area and the measured volume of air sampled.

Success in counting depends upon good quality control and complying strictly with counting rules. Sampling pumps capable of drawing 0.5 to 2.0 1/minute of air through a membrane filter are employed. They are calibrated using a simple soap-film flowmeter (Fig 1.90). A film is drawn across the mouth of the sampling tube and the pump switched on. The time taken for the film to be drawn past the calibration marks is noted, so that the volume of air drawn in by the pump per minute can be calculated.

Membrane filters printed with a green grid (25 mm in diameter, 0.8 μm pore size) are mounted in the sample head shown in Fig 1.91. The head is attached to the calibrated sampling pump and sampling takes place at a set flow for a timed period. The total volume of air drawn through the membrane is noted.

Asbestos fibres stick firmly to the filter and in normal circumstances are not dislodged during their journey to the laboratory. Here they are rendered transparent by placing the membrane on to a microscope slide and exposing it to acetone vapour for three to five seconds. Then using a hypodermic syringe one to three drops of triacetin are placed on the cleared

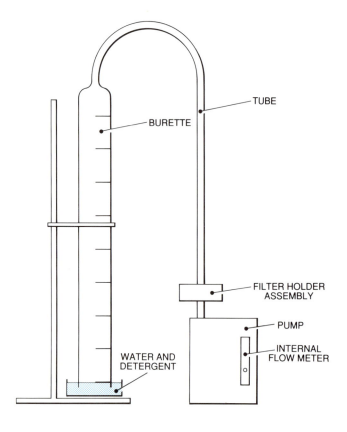

TUBE
BURETTE
FILTER HOLDER
ASSEMBLY
PUMP
INTERNAL
FLOW METER
WATER AND
DETERGENT

Fig. 1.90 Film flow meter used for calibrating asbestos
monitoring pumps (from AIA RTM No 1)

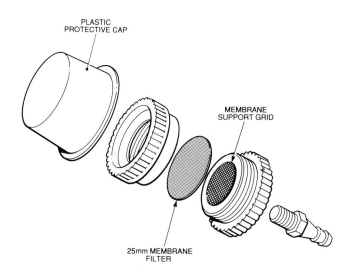

FIG. 1.91 Filter holder — exploded view

slide to fix it. A clean cover slip is placed over the slide and it is warmed at 50°C for fifteen minutes to complete the preparation.

Counting takes place using an approved microscope. Essentially this means one which can resolve block 6 of the standard (MK II) phase contrast slide.

Counting is aided by a Walton and Beckett eyepiece graticule (Fig 1.92) which is engraved and calibrated to match the microscope used, at a nominal 40 × objective magnification. A distance of 100 μm on the stage then becomes 4 mm on the eyepiece graticule. The exact calibration of the microscope needs to be checked against a stage graticule and the result noted, however before the Walton and Beckett graticule is ordered, to ensure a correct match.

Subject to this exact calibration, the graticule (Fig 1.92) shown as viewed through a 10 × or 12.5 × eyepiece, represents a circle 100 μm in diameter and the shapes around the outside of the counting area

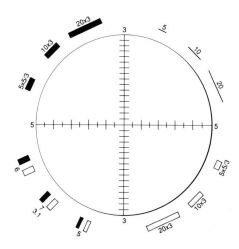

FIG. 1.92 Walton and Beckett graticule for evaluating fibrous dusts

represent examples of fibres of critical dimensions. This aids decisions as to whether an observed fibre should be counted or not. The NS diameter scale divisions represent 3 μm and the EW divisions 5 μm.

The slide is mounted on the microscope and viewed through the Walton and Beckett graticule. The number of fibres in the circle are counted. This is not as simple as it sounds, as the slide will contain dust as well as fibres. A fibre is defined as having a length greater than 5 μm with the length/breadth ratio of at least 3:1. There is no fixed upper limit on length but a maximum diameter of 3 μm is defined.

A guide to what constitutes a countable fibre is given in Fig 1.93. This derives from the Asbestos International Association Health and Safety Publication RTM1 (1979). The same document gives rules on how to count fibres straddling the edge of the counting field. Counting of fields is carried out until 100 fibres have been counted or is stopped at 100 fields if a count of 100 fibres is not reached. Counting of fields is carried out along four diameters as shown in Fig 1.94. Twenty-five fields are examined along each line. The reason for this is to cover the membrane area as well as possible. Bearing in mind that one hundred Walton and Beckett areas amount to less than 0.2% of the membrane area, the counting must be disciplined in this way to get an acceptable result.

The fibre count can now be calculated as

$$C = (D^2/d^2) \times (N/n) \times (1/V)$$

where C = fibre concentration in fibres/ml

    D = effective diameter of filter in mm allowing for filter holder masking (Fig 1.94)

    d = graticule diameter in mm

    N = total number of fibres counted

    n = number of graticule areas counted

    V = volume of air sampled in ml

The final steps in the exercise are now to keep a record of the results and a note of staff involved. The classified workers are subject to health surveillance and a full record of monitoring is kept for many years. If results at the barrier, i.e., the edge of the working area, exceed limits, work must stop and the controls must be re-established. This is a most unlikely occurrence but occasionally, as an example, tents are observed to have tears, and review and improvement of procedures must be practised continuously. If necessary the chemist will arrange for training of any staff involved in the work, control or monitoring procedures whose performance needs restoring to 'acceptable' levels indicated by the quality control procedures.

The control of man-made mineral fibre involves similar sampling procedures to those for asbestos. The control limit, however, is based on weight; 5 mg m^{-3} for 8 hours. In this respect control is closer to that practised for dusts (Section 32.3.4 of this chapter).

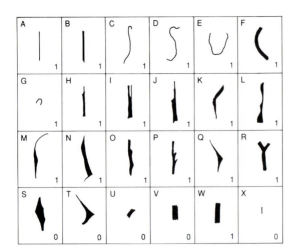

SCALE: ⌐5μm
(a) Single fibres

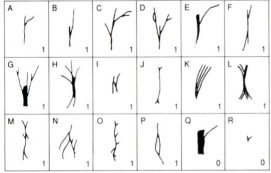

SCALE: ⌐5μm
(b) Split fibres

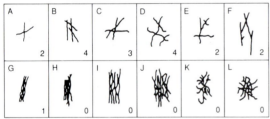

SCALE: ⌐5μm
(c) Grouped fibres

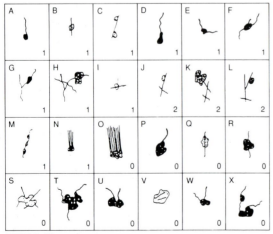

SCALE: ⌐5μm
(d) Fibres with other particles

NOTE:
THE NUMBER IN THE RIGHT BOTTOM CORNER OF EACH DRAWING
INDICATES THE NUMBER OF FIBRES (AS DEFINED) COUNTED.

FIG. 1.93 A guide to the assessment of countable
fibres on a cleared membrane filter

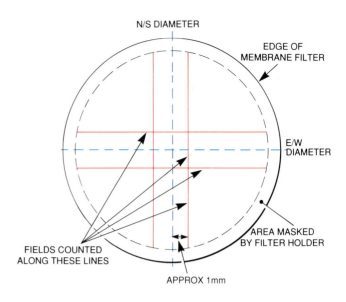

FIG. 1.94 Membrane filter counting

32.3 Control of substances hazardous to health

32.3.1 General

The Control of Substances Hazardous to Health (COSHH) Regulations published in 1988 came into force in 1989. They formalise how an employer should assess the risk arising from any hazardous substances associated with work undertaken by his staff. The steps which should be taken to minimise these risks, both to employees and anyone else who might be affected, are indicated in Fig 1.95. The regulations do not cover the control of asbestos, radioactive substances, the use of lead-containing materials, explosive or flammable materials or the effects of high or low temperature or high pressure in equipment. However, this still leaves the major part of the duties to care for staff exposed to chemical and microbiological hazards at a power station, covered by the COSHH Regulations.

The COSHH ladder (Fig 1.95) shows how safe working should be achieved and maintained. The principles are those outlined in the introduction to this section. Like so many tasks, the first step on the ladder is the most difficult one. Assessment is to do with how *work* is undertaken. It is not directly about assessing the hazardous substances. Indeed, one substance may require several assessments, depending upon how it is to be used. A simple example will illustrate this.

A 250 g bottle of potassium cyanide on a shelf presents a considerable *hazard* since it could poison a large number of people. If, however, the material is only to be used by professional staff in small laboratory experiments with the bulk stored in a secure poisons cupboard, and if the key to the cupboard and the issue of small amounts of reagent are controlled by a responsible person, then there is very little *risk*.

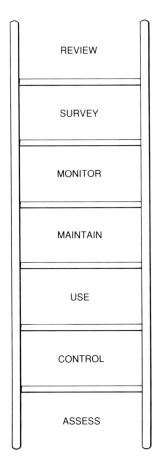

REVIEW

SURVEY

MONITOR

MAINTAIN

USE

CONTROL

ASSESS

FIG. 1.95 The COSHH ladder

If, however, all of the cyanide were to be used to prepare an electroplating solution in a production process, a second very different assessment would be required. Issue of the cyanide would again be controlled. In addition, work instructions would be prepared and staff would be instructed on the safe handling and use of the chemical. A copy of the work instructions and dates and details of training of each person would be kept on record. The plating process would be carefully controlled, with local exhaust ventilation to remove any fumes and workers would be provided with protective clothing. Finally, disposal of spent solutions would be controlled and details recorded.

The examples given above illustrate two ways in which hazardous substances may be controlled. In a laboratory there may be thousands of chemical reagents. Once an inventory has been prepared these can be controlled *generically*. Using the symbols on the bottles (Fig 1.96) one set of rules can be set, for example, for the handling of all poisons. The precautions required for small scale use by professional staff are generally the same for each poison, so it is economic to have the one set of rules. Similarly all flammable materials, although they are not covered by the COSHH regulations, can be controlled by another set of rules. The poisons will be kept in the poisons cupboard and the rules will nominate who is responsible for purchase, control and disposal. The flammable materials will be

kept in a flame-proof cupboard and the rules will state how much material may be kept in any one location and will record the total amount that the site is allowed to hold. The other hazards (Fig 1.96) are handled similarly, resulting in a compact set of safety documents from which staff can readily work.

The second way of control requires *specific* assessment of the process being undertaken, particularly if large amounts of material are being handled by industrial staff. The chemist on a power station will be closely involved in these specific assessments. The techniques which might be used and the control procedures are outlined in the next sections.

32.3.2 Control of gases

Air consists of 75.5% nitrogen, 23% oxygen and 1.4% argon along with a residue of other rare gases and carbon dioxide. If anything is added or removed it may prove hazardous to breathe. A 2000 MW coal-fired power station burning 20 000 tonnes of fuel per day, uses 250 000 tonnes of air in the process. Consequently there are many possibilities for the workplace atmosphere to be modified. The chemist is usually nominated to make measurements to ensure that the atmosphere, particularly in any confined space, is safe.

Before workers are allowed to enter a confined work area, tests are carried out to determine that the correct amount of *oxygen* is present. Too little would lead to asphyxiation; too much, from a leaking welding line for example, could lead to a serious fire hazard. There are specific portable oxygen analysers available which either utilise the paramagnetic property of O_2, or electrochemical instruments which determine it by means of its oxidation potential. Calibration is simple. Clean dry air drawn through the instrument contains 23% O_2.

Tests for toxic containers are carried out next. *Carbon monoxide* and *sulphur dioxide* are likely to be present in leaking flue gases from a boiler or gas ducts. Usually only the former is measured. There are several portable instruments available for making specific measurements of carbon monoxide and pumped samples using detector tubes may also be used (Fig 1.97).

The third type of measurement which is frequently required is for *flammable gases*. Again portable instruments are available. Ideally they should be calibrated using the appropriate flammable gas; hydrogen if a generator leak is being evaluated, methane if a leak from a gas line or groundwater is suspected. In all cases the instrument must be electrically safe, so as not to ignite a potentially explosive gas mixture.

Water conditioning chemicals used in power stations require monitoring. *Chlorine* levels in the air in a water treatment plant may be measured by fixed dedicated instruments, or again a pump sampler may be used for spot checks (Fig 1.97). *Hydrazine* levels in the atmosphere may be measured using a fixed instrument. This measures colour changes induced on a special paper tape. The measurement is very important because of

Corrosive

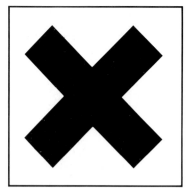

Harmful

Oxidising

Explosive

Toxic

Irritant

Flammable

Highly flammable

Biohazard

Fig. 1.96 Hazard symbols

the potential long-term health risks. The levels of hydrazine must be kept very low indeed (below 0.1 PPM in air).

Finally the chemist must be able to cope with the unexpected. Workers in an enclosed area may complain of feeling unwell. They may also report an unusual smell, but there could be little further information. The instrument shown in Fig 1.98 is most helpful in investigating such incidents. It is a long pathlength (10 m) infra-red gas analyser. A computer is included which has in its memory the spectra of most commonly encountered gases. The substance causing the problem

FIG. 1.97 Operator with personal dust sampler and portable gas detector
(see also colour photograph between pp 208 and 209)

can usually be identified quickly and appropriate action taken. Most gaseous molecules absorb infra-red energy and can be measured, but a small number, for example, oxygen and chlorine, would not be detected.

The control limits for gaseous atmospheric contaminants are issued each year in the United Kingdom's Health and Safety Executive Guidance Note EH40.

For 1990 the control limits for the gases mentioned above are listed in Table 1.59.

32.3.3 Control of liquids

There are only a handful of hazardous liquids which can cause risks to health on a power station. Some electrical equipment such as condenser banks still con-

FIG. 1.98 The MIRAN long pathlength infra-red analyser in use
(see also colour photograph between pp 208 and 209)

tain small amounts of *polychlorinated biphenyls* (PCBs). These can be identified by infra-red spectroscopy and determined using chromatographic methods. Skin contact should be avoided and disposal must be via an approved incinerator.

Some control equipment contains *fire-resistant fluid* (FRF) which is a phosphate ester. Skin contact should be avoided. Identification by infra-red and analysis for phosphorous is simple. It is not necessary to make any further measurement.

Hydrazine can be absorbed through the skin in addition to giving hazardous vapour. Skin contact must be avoided.

Oils can cause skin cancer. Casual exposure would have no consequence, but workers using oils everyday should maintain good hygiene. Skin contact should be minimised and oily overalls should be exchanged for clean ones.

Ortho-tolidine; this substance was formerly used in some analytical procedures. It causes bladder cancer and as a result of having used the material in the past, some staff are subject to medical surveillance. Ortho-tolidine is no longer used and should not be encountered in future, except that it is present in some detector tubes used for estimating trichloroethylene and trichloroethane. The tubes require careful disposal, supervised by a competent chemist.

TABLE 1.59

*Control limits for some gases encountered
on power stations*

Gas	Formula	Long-term exposure limit (8 hour TWA† reference period)		Short-term exposure-limit (10 minute reference period)	
		PPM	$mg\ m^{-3}$	PPM	$m\ gm^{-3}$
Carbon monoxide	CO	50	55	300	330
Sulphur dioxide	SO_2	2	5	5	13
Chlorine	Cl_2	1	3	3	9
Hydrazine	$H_2N.NH_2$	0.1	0.1	Sk‡	

† Time weighted average ‡ Can be absorbed through skin

32.3.4 Control of dusts and fume

Health and Safety Executive Guidance Note EH40 gives control limits for dusts and fume in addition to those for gases. Probably the most significant risk on a power station is associated with welding. Metal oxide fume is produced which can endanger health. There are two control strategies; local exhaust ventilation to remove fume from the weld area or the wearing of a mask. Both may be necessary to protect the welder and workers nearby.

Measurement is by withdrawing samples through a filter, rather as for asbestos monitoring (Fig 1.91), although some modifications are usually made for personal sampling (Fig 1.99). There is a choice of membrane filter. If the fume collected is to be weighed, a glass fibre membrane filter is sufficiently stable in weight for the fume collected to be estimated using a 'six figure' balance. If identification is the prime concern the glass will interfere. A carbon based filter (cellulose nitrate or similar) is therefore used and the metal oxides collected can be identified by X-ray fluorescence or other standard analytical procedures. Health and Safety Executive Guidance Note EH54 gives specific advice.

Chromium from welding and cadmium fume from brazing present particular hazards. There are specific guidance documents (MDHS — Methods for Determining Hazardous Substances No 10, 11, 12 and 13) covering estimation by either atomic absorption or X-ray fluorescence.

Dust on power stations is usually a nuisance rather than a hazard and is controlled to below 10 mg m^{-3} (total) or 5 mg m^{-3} (respirable). Measurement is by means of a portable sampling pump drawing air through a filter held in a personal sampling head (Fig 1.99). The filter is then weighed to give a result for total dust. Respirable dust is collected similarly but a small cyclone sampler is used instead of the sampling head shown in Fig 1.99. Typical dusts which are measured are coal dust and pulverised fuel ash.

32.4 Control of noise

Although the logic is perhaps doubtful, the task of measuring noise exposure on plant is often given to the station chemist, since it fits in with his other occupational monitoring tasks. Noise is measured in decibels which is a measure of the intensity of sound. Typical levels are illustrated in Table 1.60 to give the reader some context from which to judge measurements they may encounter in their work.

The relevant regulations are the Noise at Work Regulations 1989. The aim of control is to identify areas where noise in excess of 90 dBA can be expected when plant is operating. In these areas staff *must* wear hearing protection. In a second zone in which noise levels greater than 85 dBA but less than 90 dBA exist, ear protection is advised but not mandatory. The resulting noise zones around a typical turbine are illustrated in Fig 1.100. A HSE Guidance Note EH14 gives advice on making the necessary measurements.

32.5 Microbiological hazards

Cooling water in power stations is often taken directly from a river or the sea. Bacteria, algae, parasites and other microorganisms may be present. When work is undertaken on plant this has to to taken into account. Condenser cleaning presents a particular risk, of Weil's disease (leptospirosis). This is caused by a parasite present in rats which is passed on from their urine. Infection occurs through cuts or abrasions. Control is effected by chlorinating the water in the condenser for some hours prior to draining, to allow work to begin. Sodium hypochlorite is the preferred biocide. In addition workers are given instruction on the risk and provided with protective gloves and boots. Infection is now very rare.

The other risk on a power station relates to hot water and heating systems. As in any modern centrally heated building where water may be kept warm continuously, there is a risk of legionnaires' disease. Chlorination is the solution to the problem and the Health and Safety Executive issue guidance (EH48). Particular attention has to be paid to shower heads in frequently used facilities. The heads are removed and sterilised in hypochlorite (50 PPM) overnight. A spare shower head set enables this to be done without interrupting the availability of the shower.

153

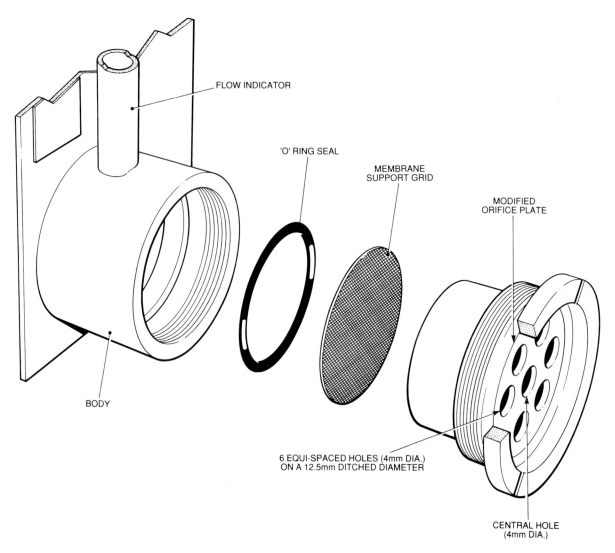

FLOW INDICATOR

'O' RING SEAL

MEMBRANE
SUPPORT GRID

MODIFIED
ORIFICE PLATE

BODY

6 EQUI-SPACED HOLES (4mm DIA.)
ON A 12.5mm DITCHED DIAMETER

CENTRAL HOLE
(4mm DIA.)

FIG. 1.99 Modified UKAEA personal sampling head used for gravimetric dust sampling

TABLE 1.60
Decibel rating of common sounds

Sound pressure level dB	Condition	Sound pressure N/m	General class
120	Threshold of pain	20.0	Deafening
110	Thunder, Artillery		
100	Steel riveter at 5 m	2.0	
90	Noisy factory		
80	Ringing alarm clock at 60 cm	0.2	Distracting
70	Inside small saloon car at 50 kph		
60	Loud	0.02	Range of conversation
50	Average office		
40	Average living room	0.002	
30	Private office		Extreme quiet
20	Whisper	0.0002	
10	Soundproof room		Soundproof chambers
0	Threshold of audibility	0.00002	

FIG 1.100 Noise zones around a steam turbine

Hot water systems and air conditioning humidifiers may have to be monitored depending upon the design of the systems. The chemist has a key role in watching the system but it is essential that professional microbiological advice is available also. The chemist can be expected to treat the water systems but the microbiologist is required to establish that the control procedures are working.

The control of microbiological hazards is part of the COSHH requirements but the chemist cannot be expected to control this risk unaided and this is perhaps a fitting point to end the chapter. The station chemist carries out many duties which are clearly chemical in nature but it will have been seen in the course of this chapter that station chemistry is a bridge between many disciplines. As with all workers on a power station, the chemist is very much one of the team.

33 References

[1] Stopes, M. C. and Wheeler, R. V.: Monograph on the constitution of coal: HMSO: 1918

[2] Wheeler, R. V. *et al* as reported in Francis, W. and Peters, M. C.: Fuels and fuel technology, 2nd Edition: Pergamon: 1980

[3] Francis, W.: Coal — its formulation and composition: Edward Arnold: 1961

[4] Regnault, V.: Ann. Chim. Phys., *66*(2) pp 337–365: 1837
     Regnault, V.: Ann. Mines, *12*(3) pp 161–240: 1837

[5] Gruner, L.: Ann. Mines, *4*(7) pp 169–207: 1874

[6] Seyler, C. A.: Proc. South Wales Institute of Engineering;

     21 pp 483: 1900
     22 pp 112: 1901
     Colliery Guardian; *80* pp 17–19, 80–82, 134–136: 1900
     Fuel; *3* pp 15, 41, 79: 1924
     Proc. South Wales Institute of Engineering; *47*, pp 547: 1931
     53 (4): 1938
     Fuel; *17* pp 177, 200, 235: 1938

[7] Mott, R. A.: Coal assessment: Institute of Fuel: 1948

[8] Mott, R. A. and Spooner, C. E.: Fuel; *23* pp 9: 1944

[9] Meyers, R. A.: Coal handbook: Marcel Dekker: 1981

[10] Gibb, W. H.: Power Industry Research; *1* pp 29–42: 1981

[11] Burton, *et al*: CIGRE International Conference on Large High Voltage Electric Systems, 15–08: 1988

CHAPTER 2

Corrosion: feed and boiler water chemistry

Introduction

In this chapter the field of water chemistry is reviewed with particular emphasis on operation of the main water/steam circuits. The theoretical principles of aqueous solutions of ionic compounds are discussed with reference to the concept of acidity and alkalinity, and the effects of temperature on the solubility of ionic compounds. The mechanisms of metallic corrosion in aqueous solutions are also reviewed. This is followed by a discussion of the operation of both drum boiler and once-through boiler plant, highlighting specific corrosion problems that can arise and detailing boiler and feedwater chemical treatment used to give safe, corrosion-free operation. Information is also presented on such subjects as off-load plant storage, sampling and analysis of water and steam, and chemical instrumentation.

In the early history of steam boilers frequent plant failures resulted from two main causes. The first, a limited knowledge of design, operation and maintenance requirements, was gradually overcome. Increasing experience in the operation of high pressure plant, a fuller understanding of the mechanical properties of metals and the development of stronger alloys led to improved plant designs.

The second cause of failures, inadequate control of water chemistry and insufficient knowledge of corrosion processes, took longer to correct. The mechanisms of scale formation and corrosion occurring in steam raising plant began to be understood in the 1920s. Subsequent research led to the development of boiler and feedwater chemical treatments that allow safe and efficient operation of plant over a wide range of operating conditions. However, corrosion can still occur, particularly in new plant designs, and this development work is continuing.

Although a range of feed and boiler water chemical treatments are considered the chapter is most concerned with plant operating at 6 MPa (60 bar) or above.

1 Chemistry of reactions in solutions [1,2]

1.1 A simple model of the atom

From the chemical point of view an atom is essentially made up of three types of particles; the positively charged protons, negatively charged electrons (the charges being equal and opposite) and uncharged neutrons. An atom of a particular element contains a fixed, characteristic number of protons associated with a number of neutrons in the nucleus of the atom. Around the nucleus a number of electrons travel in various orbits, the number of electrons being equal to the number of protons in the nucleus, i.e., an atom has no resultant electric charge.

Protons and neutrons are of approximately equal mass and very much heavier than electrons. Consequently, the mass of the atom is almost entirely centred in the nucleus.

The unique number of protons in the nucleus is referred to as the *atomic number* with no two elements having the same number (see Table 2.1). The total number of protons plus neutrons very largely determines the atomic weight of the element. The orbiting electrons determine the chemical properties of each element.

The simplest atom is the hydrogen atom which contains one proton in the nucleus and a single electron. The oxygen atom has eight protons and eight electrons and the iron atom has 26 protons and 26 electrons.

TABLE 2.1
Atomic number/atomic weight/valency

Element	Atomic No	Atomic weight	Valency
Hydrogen	1	1.008	1
Carbon	6	12.011	2,3 or 4
Nitrogen	7	14.007	3 or 5
Oxygen	8	15.999	2
Sodium	11	22.989	1
Magnesium	12	24.312	2
Silicon	14	28.086	4
Phosphorus	15	30.974	3 or 5
Sulphur	16	32.064	2,4 or 6
Chlorine	17	35.453	1,3,5 or 7
Calcium	20	40.08	2
Iron	26	55.85	2 or 3

Although carrying the same negative charge, the electrons in an atom can have different orbiting energies. With increasing numbers in the atom, the electrons are divided up into groups and within each particular group the electrons all have the same orbiting energy.

All chemical reactions involve interaction of the electrons in the reacting atoms, leading to partial or total transfer of electrons between the atoms. The capacity of an element to combine with other elements is known as its *valancy* and is controlled by the extent to which its atoms can lose or gain electrons, in specific chemical reactions for example:

● In water (H_2O) one atom of oxygen combines with two atoms of hydrogen.

● In methane (CH_4) one atom of carbon combines with four atoms of hydrogen.

If hydrogen is given a valency of one (i.e., univalent) then oxygen has a valency of two (bivalent) and carbon has a valency of four (tetravalent) (see Table 2.1).

1.2 Ions and compounds

When elements combine they produce compounds (e.g., H_2O, CH_4) and the extent of transfer of electrons between the atoms determines the type of compound formed. If the reacting electrons are only partially transferred and are shared between the reacting atoms, then a *covalent* neutral compound is formed such as methane (CH_4). However, if complete transfer of electrons occurs, as in sodium chloride (NaCl), the product is neutral, but is made up of equal numbers of positively charged Na^+ atoms and negatively charged Cl^- atoms. This is the result of the complete transfer of an electron from each neutral sodium atom to each neutral chlorine atom. Sodium chloride is known as an *ionic* compound and the charged atoms are known as *ions*. Na^+ the sodium ion is a *cation*, and Cl^- the chloride ion is an *anion*.

Most elements will form ions. Metals lose electrons to form cations, e.g., the calcium atom can lose two electrons to form the calcium ion (Ca^{2+}). Non-metals such as chlorine and oxygen gain electrons and form anions, i.e., Cl^- and O^{2-}. Hydrogen forms a cation, (H^+).

Combined groups of atoms can also form ions of differing valencies. Of particular importance in the present chapter are the oxy-anions sulphate (SO_4^{2-}), carbonate (CO_3^{2-}), phosphate (PO_4^{3-}) and the ammonium cation (NH_4^+).

1.3 Ionic compounds in solution

Sodium chloride is a simple example of an ionic compound. In solid sodium chloride the Na^+ cations and Cl^- anions are arranged in a regular 3-dimensional lattice (Fig 2.1). However, when this compound is dissolved in water, the lattice is destroyed and the sodium and chloride ions are free to move in the solution as separate electrically-charged entities. It can be shown that a solution of an ionic compound in water will conduct electricity, the current being carried by the individual ions present. The conductivity of a solution depends on the mobility of the ions present and is defined as the reciprocal of the resistance, in ohms, of 1 m^3 of solution retained between two 1 m^2 electrodes, the units being Siemens/m (i.e., S/m). For the convenience of using suitably-sized numbers, conductivities are frequently quoted in μS/cm. In general, the greater the concentration of ions in a solution, the greater its conductivity, i.e., seawater contains some 2.5 g/dm^3 of dissolved salts and has a con-

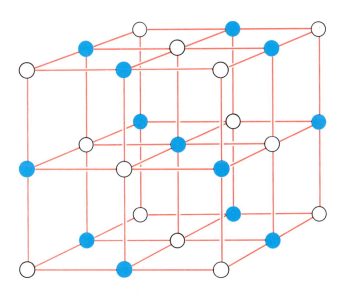

KEY

 SODIUM, Na$^+$

⬤ CHLORIDE, Cl$^-$

FIG. 2.1 The ionic lattice in solid sodium chloride

ductivity of 45 000 μS/cm. Typical towns water on the other hand contains only 100 mg/dm^3 of dissolved salts, with a conductivity of 150 μS/cm.

If, when dissolved in water, a compound is very largely or entirely converted into freely moving ions it is called a *strong electrolyte*. If only a proportion of the compound is converted into ions, it is termed a *weak electrolyte*, and if no ions are formed, a *non-electrolyte*. Solutions of non-electrolytes are therefore essentially non-conducting.

The formation of ions by a weak electrolyte is an equilibrium reversible reaction, i.e.,

$$BA \leftrightarrow B^+ \rightarrow + A^-$$

The 'dissociation constant' for this reaction is defined as

$$K = \frac{(\text{conc } B^+) \times (\text{conc } A^-)}{(\text{conc undissociated BA})}$$

and is essentially independent of the total concentration of the dissolved salt. Consequently the proportion of the weak electrolyte converted into ions increases with decreasing concentration.

The *molecular weight* of a compound is defined as *being equal to the sum of the atomic weights of the constituent atoms*. For example, from Table 2.1 sodium chloride (NaCl) has a molecular weight of (22.99 + 35.45) = 58.44; methane (CH_4) has a molecular weight of [12.01 + (4 × 1.008)] = 16.04. If an aqueous solution contains a weight of a compound in g/dm^3 equal to the molecular weight, the solution is said to contain 'lg.mol/dm^3' of that compound. Therefore a lg.mol/dm^3 solution of sodium chloride would contain 58.44 g NaCl as 22.99 g/dm^3 Na^+ cations and 35.45 g/dm^3 Cl^- anions. Such a solution is said to contain l g.ion/dm^3 Na^+ and 1 g.ion/dm^3 Cl^-. When a concentration of a solution is quoted in g.mol/dm^3 or g.ion/dm^3 it can be converted into g/dm^3 by multiplying by the appropriate molecular or atomic weight. This chapter is particularly concerned with the properties of aqueous solutions of electrolytes and the term 'g.ion/dm^3' is a frequently used method of presenting ionic concentrations.

Water itself is a very weak electrolyte forming very low concentrations of hydrogen ions (H^+) and hydroxyl ions (OH^-):

$$H_2O \leftrightarrow H^+ + OH^-$$

water   hydrogen   hydroxyl
molecule   cation   anion

At 25°C the concentration of H^+ and OH^- ions is 1×10^{-7} g.ion/dm^3 and the dissociation constant, Kw, is defined simply as the mathematical product of the concentrations of the two ions, i.e.,

$$(1 \times 10^{-7}) \times (1 \times 10^{-7}) = 1 \times 10^{-14}$$

This product is referred to as the *ionic product* of water.

1.4 Acids and alkalis

If an electrolyte when dissolved in water produces hydrogen ions, e.g.,

$$HCl \text{ (hydrogen chloride)} \rightarrow H^+ + Cl^-$$

then the concentration of hydroxyl ions in that solution will be reduced to maintain the dissociation equilibrium of water. For example, if the resulting solution contains 1×10^{-4} g.ion/dm^3 H$^+$, then the hydroxyl ion concentration will be reduced to 1×10^{-10} g.ion/dm^3 OH$^-$. Such a solution is said to be an *acid solution* and hydrogen chloride and compounds with similar properties are called *acids*.

Conversely, if the electrolyte produced hydroxyl ions, e.g.,

$$NaOH \text{ (sodium hydroxide)} \rightarrow Na^+ + OH^-$$

the concentration of hydrogen ions would be reduced. When the resulting solution contains 1×10^{-3} g.ion/dm^3 OH$^-$ the hydrogen ion concentration would be 1×10^{-11} g.ion/dm^3 H$^+$. This solution is termed *alkaline* and sodium hydroxide is an *alkali*.

In considering the corrosion of metals in aqueous solutions the acidity or alkalinity of the solution is of prime importance. To avoid the inconvenience of dealing with the negative powers of 10 in concentration terms, Sorenson, in 1909 introduced a shorthand notation of hydrogen ion concentration to define the acidity or alkalinity of a solution. He proposed the use of the term pH, which is defined as the logarithm to base 10 of the hydrogen ion concentration, with the sign reversed, to express acidity or alkalinity. Thus a hydrogen ion concentration of 1×10^{-7} g.ion/dm^3 H$^+$ as in pure water is expressed as pH = 7.0, a neutral solution containing equal numbers of hydrogen and hydroxyl ions. A hydrogen ion concentration of 1×10^{-6} g.ion/dm^3 H$^+$ has a pH = 6.0 and is a slightly acid solution. A solution having a hydrogen ion concentration of 1×10^{-9} g.ion/dm^3 H$^+$ has a pH = 9.0 and is an alkaline solution. The range of most commonly occurring pH values is given in Table 2.2 along with the corresponding hydrogen electrode potentials.

It should be remembered that the scale is logarithmic so that a decrease in the pH value by one unit corresponds to a tenfold increase in hydrogen ion concentration.

When quoting pH values, the temperature of the solution should also be stated. Where this is not done it is usually assumed to be 25°C. As the temperature of water increases, the extent of formation of H$^+$ and OH$^-$ ions also increases. Consequently, the hydrogen ion concentration (and pH) of a neutral solution changes together with the ionic product of water, Kw as shown in Table 2.3.

To avoid confusion arising from the changes in the pH scale with changing temperature it is common practice to refer all pH values to their equivalent values at 25°C.

TABLE 2.2

Relationship between concentration of hydrogen ions, pH and hydrogen electrode potential in aqueous solutions at 25°C

Concentration of hydrogen ions, g.ion/dm^3	pH value	Hydrogen electrode potential, V
1×10^0	0	0.000
1×10^{-1}	1	-0.059
1×10^{-2}	2	-0.118
1×10^{-3}	3	-0.177
1×10^{-4}	4	-0.236
1×10^{-5}	5	-0.295
1×10^{-6}	6	-0.354
1×10^{-7}	7 Neutral	-0.413
1×10^{-8}	8	-0.422
1×10^{-9}	9	-0.531
1×10^{-10}	10	-0.590
1×10^{-11}	11	-0.649
1×10^{-12}	12	-0.708
1×10^{-13}	13	-0.767
1×10^{-14}	14	-0.826

TABLE 2.3

Ionic product of water

Temperature (°C)	pH of neutral solution	Kw
25	7.0	1×10^{-14}
50	6.63	5.5×10^{-14}
100	6.10	6.3×10^{-13}
200	5.65	5.0×10^{-12}
250	5.60	6.3×10^{-12}
300	5.65	5.0×10^{-12}
350	5.92	1.4×10^{-12}

1.5 pH measurement

There are two main methods of determining the pH of a solution. The first, historically, was the use of chemical dyes (such as litmus) which show characteristic colours in solutions of a particular pH range. This is only an approximate method but gives a very rapid result. The most widely used method today for accurate determinations is a potentiometric technique and is described in Section 12 of this chapter.

Corrosion of a metal involves the metal being attacked by the surrounding fluid to produce metal oxides or salts. In aqueous solution these reactions take place more readily in particular pH ranges with the production of hydrogen. In terms of corrosion processes the importance of the pH value lies in the electrode potential associated with it (Table 2.2). The concept of electrode potentials and their significance in corrosion processes is discussed in detail in Section 2 of this chapter.

1.6 Solubility and precipitation reactions

An aqueous solution is a homogeneous mixture of water (the *solvent*) and any substance (the *solute*) dissolved in it. In general, there is a limit to the amount of solute that will dissolve at any given temperature the amount being different for each solute. When this limit has been reached, and no more solute will dissolve, the solution is said to be *saturated* and the quantity of the solute that has dissolved is called its *solubility*.

The solubilities of compounds vary over a wide range, some being very soluble and others only slightly soluble. Furthermore, the solubility of a compound can change quite dramatically in the temperature ranges prevailing in steam raising plant.

It is essential to understand the solubility behaviour of compounds that might be present in feedwater and boiler water to ensure that unsatisfactory conditions do not arise. If the solubility of a compound decreases with increasing temperature (i.e., it has a retrograde temperature/solubility relationship) then the solubility limit may be exceeded as the temperature rises. The excess solute initially in solution will now separate from the solution as a solid; it will precipitate and may form a hard deposit (a *scale*) on heat transfer surfaces. This could lead to progressive and possibly serious loss of heat transfer in that section of the plant.

The converse is also true. Depending on the behaviour of a particular compound, changes in temperature can result in the dissolution of a precipitated scale.

The solubility behaviour of aqueous solutions of chemical compounds present in steam raising plant is therefore important. In Table 2.4 the solubility, at 25°C, of a number of these compounds is listed and in Fig 2.2 the changes in solubility with temperature of a number of these compounds is illustrated. The quite varied behaviour of the different compounds is very evident and it is often difficult to predict the solubility at one temperature from data at other temperatures.

1.7 Solubility product

The compounds listed in Table 2.4 are essentially all strong electrolytes, i.e., when dissolved in water they are converted, very largely or completely, into cations and anions:

$$BA \leftrightarrow B^+ + A^-$$
$$\text{cation} \quad \text{anion}$$

In this situation interactions of ions formed from different dissolved compounds can lead to the precipitation of a third compound. An understanding of these interactions is therefore important.

As already described, a saturated solution of a pure compound will contain a fixed concentration of that compound depending on the prevailing temperature. Experimental work has shown that at a given temperature the mathematical product of the concentrations of the dissolved ions is a constant. This constant

TABLE 2.4

Solubility of some important compounds in water at 25°C

Compound	Chemical formula	Solubility g/dm^3
Calcium bicarbonate	$Ca(HCO_3)_2$	0.091
Calcium carbonate	$CaCO_3$	0.0153
Calcium hydroxide	$Ca(OH)_2$	1.59
Calcium phosphate	$Ca_3(PO_4)_2$	0.181
Calcium sulphate	$CaSO_4$	2.09
Magnesium carbonate	$MgCO_3$	0.169
Magnesium hydroxide	$Mg(OH)_2$	0.008
Potassium phosphate	K_3PO_4	900
Sodium phosphate	Na_3PO_4	15.5
Sodium chloride	$NaCl$	362
Sodium hydroxide	$NaOH$	1140
Sodium sulphate	Na_2SO_4	300
Sodium sulphite	Na_2SO_3	315

is called the *solubility product* and is normally given the symbol K_s and is defined as:

$$K_s = (B^+) \times (A^-)$$

where (B^+) = concentration of B^+ in g.ion/dm^3

(A^-) = concentration of A^- in g.ion/dm^3

Therefore when solutions of electrolytes are mixed together, interactions of ions formed from different dissolved compounds can lead to precipitation of a third compound. This is best illustrated with a practical example. A solution of calcium sulphate containing 1 g/dm^3 $CaSO_4$ (a strong electrolyte) will have 7.34×10^{-3} g.ion/dm^3 Ca^{2+} and 7.34 g.ion/dm^3 SO_4^{2-}. A solution of sodium carbonate containing 10 g/dm^3 Na_2CO_3 will have 1.88×10^{-1} g.ion/dm^3 Na^+ and 0.94×10^{-1} g.ion/dm^3 CO_3^{2-}. If, at 25°C, 1 dm^3 of each of these solutions is mixed together then the final volume will be 2 dm^3 and will contain:

3.67×10^{-3} g.ion/dm^3 Ca^{2+}

3.67×10^{-3} g.ion/dm^3 SO_4^{2-}

9.4×10^{-2} g.ion/dm^3 Na^+

4.7×10^{-2} g.ion/dm^3 CO_3^{2-}

Now, from Table 2.4 it can be calculated that the solubility product of $CaCO_3$ at 25°C is 2.34×10^{-8}.

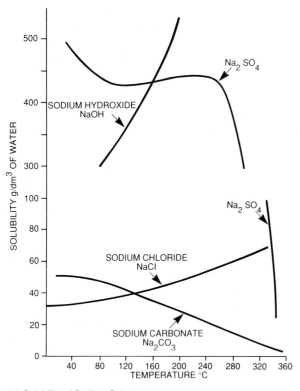

(a) Solubility of Sodium Salts

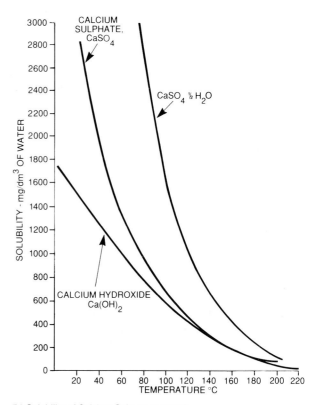

(b) Solubility of Calcium Salts up to 220°C

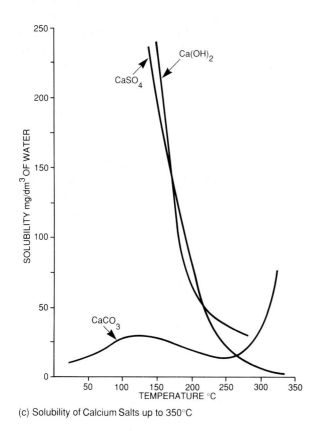

(c) Solubility of Calcium Salts up to 350°C

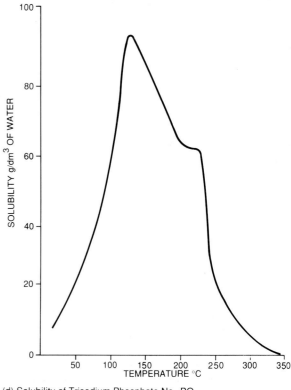

(d) Solubility of Trisodium Phosphate Na₃PO₄

FIG. 2.2 The effects of temperature on solubility of various compounds

In the foregoing solution the mathematical product of $(Ca^{2+}) \times (CO^{2-})$ is considerably higher than this value, i.e.,

$$(3.67 \times 10^{-3}) \times (4.7 \times 10^{-2}) = 1.72 \times 10^{-4}$$

The mixed solution is therefore *super saturated* with calcium carbonate and the excess over the solubility limit will precipitate. The overall reactions are therefore:

$$Na_2CO_3 \rightarrow 2\,Na^+ + CO_3{}^{2-} \rightarrow CaCO_3$$
$$CaSO_4 \rightarrow SO_4{}^{2-} + Ca^{2+} \quad \text{Precipitates}$$

Reactions of this type are frequently used in the pre-treatment of water, so that compounds which might precipitate and form scales in the steam raising plant are removed before the water is fed to the plant.

2 The corrosion of metals [3,4,5]

In nature, metals are usually found as relatively stable compounds (e.g., oxides, carbonates and silicates) which are converted into metal by a process of chemical reduction, for example, iron from iron oxide via a blast furnace, or aluminium from aluminium oxide in an electric furnace. Considerable amounts of energy have to be expended to bring about these chemical reactions. Consequently most commonly used metals are reactive and the process of corrosion may be defined as the reversion of the metal to a more stable, oxidised, state.

In general, the greater the amount of energy used to produce the metal from its ore, the greater is the tendency for the metal to return to a more stable form by corrosion. Under normal circumstances gold, which is mined as the free metal, does not corrode, whereas aluminium and magnesium, which require large quantities of electrical energy for their manufacture, have a strong tendency to corrode.

All metals are liable to corrode when the environment favours conversion of the metal to oxide by reaction with oxygen or a source of oxygen such as water, or into a metallic salt by reaction with salts or acids.

The initial effects of such a reaction is usually to produce a thin layer of oxide or salt on the metal surface. In some cases the oxide or salt film results in a state of 'passivity', preventing further corrosive action. For this to occur the film on the metal must be stable, adhere firmly to the metal and entirely cover the surface with a non-porous coating. These requirements are met if the metal oxide or salt film is insoluble in the fluid in contact with the metal and is self-repairing when damaged. If the film formed does not have these properties it will not protect the underlying metal and destructive corrosion may ensue.

In moist air, aluminium and chromium readily form protective oxide coatings, whereas the oxide formed on low alloy steel surfaces is generally porous and does not prevent further corrosion. The reddish brown coatings of rust which form on steel may in fact accelerate the rate of attack.

In most environments, the corrosion of metals is essentially an electrochemical process and the following sections discuss the principles of aqueous corrosion in those terms.

2.1 An electrochemical cell

A piece of metal, such as zinc, carries no overall electric charge. If, however, it is placed in an electrolyte solution, zinc sulphate say, some of the zinc metal dissolves to form zinc ions, leaving electrons on the metal. Thus a potential difference will exist between the zinc metal and the zinc sulphate solution, the level of which will depend on the concentration of zinc ions in the solution. The potential between the metal and a solution of its ions of standard strength (1 g.ion/dm^3) is called the *standard electrode potential* for the metal. In the case of zinc, it has a value of 0.76 volts.

Similarly, a copper electrode assumes a standard electrode potential with respect to a corresponding standard solution containing copper ions. This has a value of 0.34 volts. Such single electrode potentials are sometimes called *half cell* potentials (see Section 2.3 of this chapter).

If the zinc and copper electrodes are now joined externally, as in a simple type of Daniell cell (Fig 2.3), a current will flow, the zinc becoming the negative electrode (anode) and the copper the positive electrode (cathode) of the cell. The zinc dissolves (or corrodes) to form Zn^{2+} ions, and releases electrons, which pass via the external circuit to the copper where they are consumed in converting Cu^{2+} ions in the solution into Cu metal which deposits on the cathode:

$$Zn - 2e^- \rightarrow Zn^{2+} \text{ (anodic reaction – oxidation)}$$
$$Cu^{2+} + 2e^- \rightarrow Cu \quad \text{(cathodic reaction – reduction)}$$

The rates of reaction at the anode and cathode are electrochemically equivalent and the extent of dissolution of the zinc anode and deposition of copper on the cathode are in proportion to the quantity of electricity that has flowed through the cell.

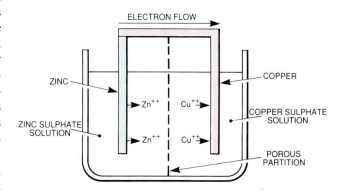

FIG. 2.3 A Daniell cell

2.2 A simple corrosion cell

In corrosion processes it is convenient to think of an anodic process as producing electrons and a cathodic process as consuming electrons in a simple cell.

If the solutions of zinc sulphate and copper sulphate in the Daniell cell are replaced by a solution of sodium chloride, a current will still flow, zinc ions again passing into solution and releasing electrons which are consumed at the copper electrode. However, the cathodic process will not be the same as in the Daniell cell, as the concentration of copper ions in solution is negligible. The only positive ions (cations) in the solution are sodium ions and hydrogen ions from the water. As corrosion proceeds, a third cation — zinc ions — will appear in the solution, from corrosion of the zinc anode. Of the cations present, the sodium cannot be deposited on the cathode because this would require a potential far higher than that which the zinc attains in this cell. The cathodic deposition of zinc ions in solution at the anode would imply that the cell was in equilibrium, which is not the case as corrosion of the zinc continues. Therefore, discharge of H^+ ions to produce hydrogen gas may be the only possible cathodic reaction in this particular system:

$$Zn - 2e^- \rightarrow Zn^{2+}$$
$$2H^+ + 2e^- \rightarrow 2H$$

followed by: $2H \rightarrow H_2$ (hydrogen gas)

However, if oxygen is present there is another possible cathodic process, namely, the reduction of oxygen gas (dissolved in the salt solution), to hydroxyl ions:

$$2Zn - 4e^- \rightarrow 2Zn^{2+}$$
$$O_2 + 2H_2O + 4e^- \rightarrow 4OH^-$$

The overall reaction would be:

$$2Zn + O_2 + 2H_2O \rightarrow 2Zn^{2+} + 4OH^-$$

The solution would become alkaline and may precipitate zinc hydroxide, $Zn(OH)_2$.

For corrosion occurring in approximately neutral, aerated solutions, the 'oxygen absorption reaction' shown immediately above is the predominant cathodic reaction.

In Fig 2.4 the Daniell cell has been modified to show a simple corrosion cell. In the solution of sodium chloride are immersed a strip of copper (the cathode) and a strip of iron (the anode), joined with a copper wire. As the reactions proceed the solution adjacent to the copper cathode becomes enriched in OH^- ions and the solution near the anode becomes enriched in Fe^{2+} ions. Both the anodic and cathodic reaction products are soluble but the ferrous and hydroxyl ions diffuse towards one another and form a precipitate of ferrous hydroxide somewhere between the anode and cathode of the cell. As the corrosion product is

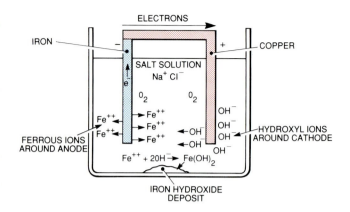

FIG. 2.4 A simple corrosion cell in aerated sodium chloride solution

not formed in contact with either electrode, there is no tendency for either the anodic or cathodic process to be stifled and corrosion of the iron continues.

If the solution in this bimetallic corrosion cell is de-aerated and the atmosphere above the cell contains no oxygen, then the hydrogen evolution reaction will become the cathodic process. The base metal of the couple (iron) will continue to function as an anode, and will corrode. However, if the solution is neutral or alkaline, the hydrogen ion concentration is very small, consequently the rate of the cathodic reaction is limited and tends to become 'polarised' — an effect which is discussed later. As the rates of the anodic and cathodic reactions must be equal, the rate of the anodic reaction will also be slowed down, and the corrosion of the base metal will be retarded.

In acid solutions, with higher concentrations of hydrogen ions, the hydrogen evolution reaction takes place very readily, the rates of corrosion of the iron anode can be very high and largely controlled by the anodic processes.

2.3 Electrode potentials and polarity of bimetallic couples

As already described, when a metal is placed in a solution, a potential exists between the metal and its own ions in the solution. If the potentials are determined for a number of metals, under standard conditions (i.e., the solution containing 1 g.ion/dm³ of the metal ions), then it is possible to construct a table of standard electrode potentials. To provide a datum with which the electrode potentials of metals may be compared, the standard electrode potential of hydrogen is taken as zero. For this purpose the standard hydrogen electrode consists of a platinum electrode in an acid solution containing 1 g.ion/dm³ of hydrogen ions, its surface being coated with hydrogen bubbles. The results obtained are given in Table 2.5.

The reactive metals, which have a strong tendency to form ions and compounds, are at the negative or 'base' end of the table; for example, sodium and po-

TABLE 2.5

Standard electrode potentials

Element	Electrode reactions	Standard electrode potential, (Eo) volts at 25°C on hydrogen scale
Noble end		
Gold	$Au = Au^+ + e^-$	$+1.68$
Platinum	$Pt = Pt^{2+} + 2e^-$	$+1.20$
Silver	$Ag = Ag^+ + e^-$	$+0.800$
Copper	$Cu = Cu^+ + e^-$	$+0.522$
Copper	$Cu = Cu^{2+} + 2e^-$	$+0.345$
Hydrogen	$H_2 = 2H^+ + 2e^-$	0.000
Lead	$Pb = Pb^{2+} + 2e^-$	-0.126
Tin	$Sn = Sn^{2+} + 2e^-$	-0.136
Nickel	$Ni = Ni^{2+} + 2e^-$	-0.250
Cobalt	$Co = Co^{2+} + 2e^-$	-0.277
Iron	$Fe = Fe^{2+} + 2e^-$	-0.440
Zinc	$Zn = Zn^{2+} + 2e^-$	-0.762
Manganese	$Mn = Mn^{2+} + 2e^-$	-1.05
Titanium	$Ti = Ti^{3+} + 3e^-$	-1.63
Aluminium	$Al = Al^{3+} + 3e^-$	-1.67
Magnesium	$Mg = Mg^{2+} + 2e^-$	-2.34
Sodium	$Na = Na^+ + e^-$	-2.712
Calcium	$Ca = Ca^{2+} + 2e^-$	-2.87
Potassium	$K = K^+ + e^-$	-2.922
Base end		

tassium. Conversely, the non-reactive metals such as gold and platinum, are at the positive or 'noble' end of the series.

If two metals in electrical contact are immersed in an electrolyte solution containing 1 g.ion/dm^3 of both metal ions, it is possible to predict from Table 2.5 the potential of the resulting cell and which of the two metals would function as the noble, cathodic member, and which as the base or anodic member of the cell. For the Daniell cell, it can be deduced correctly that the potential of the cell is:

$$+0.34 \text{ V} - (-0.76 \text{ V}) = +1.10 \text{ V}$$
$$Cu^{2+}/Cu \quad Zn^{2+}/Zn$$

the algebraic difference of potential between the $Cu^{2+}/$Cu electrode and the Zn^{2+}/Zn electrode both with respect to the hydrogen electrode. It can also be predicted that the zinc will corrode and copper will be deposited at the cathode.

However, the figures given above are Standard Electrode Potentials. If the concentration of metal ions in solution is reduced, the potential of the electrode moves in a base direction, the shift in potential being 29 mV for every tenfold decrease in concentration of the metal ions.

It is therefore possible to calculate the potential of a cell in which the electrolyte solution contains differing concentrations of the metal ions.

2.4 Galvanic corrosion

In Section 2.2 of this chapter, a simple corrosion cell of iron and copper in sodium chloride solution was discussed. The metallic iron in the cell was converted by corrosion into oxides resembling the ores from which the metal was first extracted. The corrosion of the less noble member of a pair of metals which are jointed together is called *galvanic* corrosion or *bimetallic* corrosion. However, the electrode potentials series can only be a guide in indicating relative polarities of components of bimetallic couples in solutions of their own ions, and not in other electrolyte solutions.

For example, if the zinc/copper couple were immersed in sodium chloride solution the solution would initially be free from copper and zinc ions, and the electrode potential series would be of little use in indicating the actual potential difference available to drive current round the circuit. Furthermore, the series could fail to predict the polarity of the electrodes if one of the metals is covered with a thin stable oxide film. For a zinc/aluminium couple, Table 2.5 predicts that zinc should be the cathode, but in practice the aluminium with a stable oxide film acts as the cathode, and the zinc as the anode.

For practical purposes, it is usual to take pairs of metals and/or alloys, and determine by observation which is anodic and which is cathodic in the solution under consideration. From the results a table can be drawn up, giving a *galvanic series* for the particular environment. Table 2.6 shows such a series for metals

TABLE 2.6

Galvanic series of metals and alloys in sea water

> *Noble end*
> Titanium
> Monel (67% Ni, 30% Cu, 1.2% Fe)
> *Passive stainless steels
> Inconel (80% Ni, 13% Cu, 6.5% Fe)
> Nickel
> Copper/nickel 70/30
> Copper/nickel 90/10
> Aluminium bronze
> Copper
> Alpha brass (70% Cu, 30% Zn)
> Aluminium brass
> Muntz metal (60% Cu, 40% Zn)
> Tin
> Lead
> *Active stainless steels
> Cast iron
> Mild steel
> Aluminium
> Zinc
> Magnesium
> *Base end*

* Passive stainless steel is covered with an oxide film and resists corrosion. Active stainless steel is of the same composition, but the oxide film has been destroyed and the alloy behaves in a much less noble manner.

165

in sea water. The actual electrode potentials are not included, as they are dependent on the particular conditions to which the metals are exposed.

The effects of galvanic corrosion are often serious, especially in power stations using estuarine or sea water for cooling, and this aspect of corrosion is discussed further in Chapter 3.

Comparison of Tables 2.5 and 2.6 shows that the order of the metals in the two series is not the same. Aluminium and nickel are less base in sea water than is indicated by the electrode potential series, and titanium, which is at the base end of the potential series, behaves like a noble metal as a result of the formation of protective oxide films.

2.5   The effect of polarisation on corrosion rates

The difference in potential between the anode and cathode in a corrosion cell is the driving force of the corrosion process and the rate of corrosion will be determined by the current flowing round the circuit. Any factor that increases the resistance of the circuit or decreases the potential difference will lead to a decrease in the current flowing, and therefore, decrease in corrosion rate.

For its continued propagation the corrosion process requires the removal of the corrosion products from corroding metal surface into the electrolyte solution. If this does not occur the electrode potentials and corrosion currents will fall and the metal is said to be *polarised*.

If an Fe/Zn couple is placed in an acid solution, rapid evolution of hydrogen would occur at the Fe electrode with attendant corrosion of the Zn anode. However, if this bimetallic couple is placed in a neutral or slightly alkaline de-aerated electrolyte solution, the hydrogen is formed very slowly on the Fe surface and can cover it with gas bubbles. This will halt the flow of ions to and from the solution, limiting the corrosion current. This effect is known as *cathodic polarisation* and is the most commonly occurring form of polarisation in practical electrochemical cells.

If however the electrolyte solution contains dissolved oxygen, as described earlier, the reaction:

$$O_2 + 2H_2O + 4e^- \rightarrow 4(OH)^-$$

could be the dominant cathodic process. This would prevent the accumulation of hydrogen bubbles on the Fe surface and the corrosion processes could proceed. In this role, oxygen is acting as a de-polariser.

When corrosion products accumulate at the anode surface and hydrogen at the cathode, both anodic and cathodic polarisation will occur. The potential of the cathode will become less positive and that of the anode less negative. The potential of the corrosion cell will then be reduced together with the corrosion current and thus the corrosion rate (Fig 2.5). The extent to which

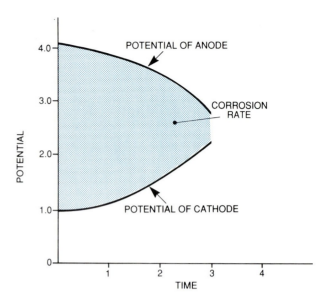

FIG. 2.5   The reduction of cell potential and corrosion rate due to polarisation of the electrodes

these processes take place will depend on the metals concerned and the composition of the electrolyte solution, and in the extreme case, the potential of the cell will fall to zero and the corrosion cease.

A practical use of the subject matter discussed in this section is the *cathodic protection* of metals by wasting or sacrificial anodes, or by means of an impressed current. Consider the couple, iron/brass, found in power station condensers, with sea water as the electrolyte. From the galvanic series Table 2.6, the iron will corrode. If, however, an auxiliary anode capable of readily providing a supply of electrons is joined to the couple as shown in Fig 2.6, the corrosion of the iron will be much reduced. The sacrificial auxiliary electrode (zinc or magnesium) is lower in the galvanic series than the iron and will corrode preferentially.

In practice, the most satisfactory form of cathodic protection is obtained by using an external source of

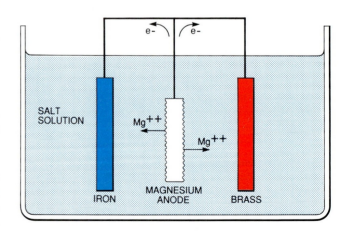

FIG. 2.6   Cathodic protection of iron/brass couple using an auxiliary magnesium anode

direct current in place of the sacrificial anode, with the negative pole of the DC source connected to the metal to be protected, and the positive pole to an inert electrode (usually platinised titanium), which is immersed in the electrolyte.

2.6 The corrosion of single metals

In the previous sections, corrosion has been discussed in terms of dissimilar metals. However, it is not necessary for dissimilar metals to be in contact for corrosion to occur. The most familiar example of single metal corrosion is the rusting of iron and steel objects in the atmosphere or underwater. It can be shown experimentally that this corrosion is electrochemical, indicating that anodic and cathodic areas must be present in the single metal.

If a piece of steel covered with a protective oxide film is immersed in dilute hydrochloric acid, the corrosion of the metal will initially be prevented by the oxide layer. However, if the steel is lightly abraded or scratched, some of this oxide will be removed and, when placed in the acid solution, corrosion will occur where the metal has been exposed with the retained oxide layer acting as a cathodic area for the formation of hydrogen (Fig 2.7).

In this example the anodic areas were produced by abrading the oxide film. Even without this treatment, corrosion would have occurred in time. Metals, after exposure to air or water, carry oxide films, but at some points the film may be cracked, porous or thin, and at these points it will be easier for metal ions to leave the metal lattice (i.e., corrode) than at areas where the film is intact. The mechanism is similar to that shown in Fig 2.7, except that the cathodic reaction is the formation of hydroxyl ions and hydrogen is not evolved. The formation of anodic and cathodic areas is also promoted by a lack of homogeneity in the metal, such as inclusions, surface imperfections, orientations of the grain structure or localised stresses and may also be a result of variations in the local environment.

2.7 Differential aeration

With metals forming protective oxide films the presence of dissolved oxygen will enhance oxide formation making the metal cathodic. Areas of the metal where the oxygen concentrations are higher will be more cathodic than areas where the oxygen concentrations are lower. Consequently, areas where the dissolved oxygen concentrations are low will become anodic and may corrode.

This effect is known as *differential aeration* and was first described by U. R. Evans, a pioneer in the scientific investigation of corrosion. The fact that currents flow between a well aerated electrode and a less aerated one of the same metal can readily be demonstrated.

Corrosion due to differential aeration is a commonplace and damaging process, for example, in unprotected structures, in boiler tubes and in water tanks. It is also of practical importance in the corrosion of condenser tubes or in other situations where deposits can impede access of oxygen to the underlying metal surface and lead to pitting corrosion and failure. A classic example of this type of corrosion is shown in Fig 2.8. Two pieces of steel are to be welded together, the joint being in contact with liquid. Within the sleeve joint oxygen replenishment is difficult and corrosion will take place in the crevice, unless the bulk liquid is also maintained at a low oxygen concentration.

2.8 Factors affecting corrosion rates

Anodic and cathodic reactions proceed in step with each other, but the overall rate of electrochemical corrosion may be controlled by either the anodic or the cathodic reaction rate.

For many base metals such as iron in aqueous solution, the rate of the cathodic reaction is controlled by a limited supply of oxygen and determines the rate of corrosion of the metal. If the supply of oxygen to the cathode is increased, the rate of cathodic reaction is raised and the anodic corrosion reaction rate correspondingly increases. If the cathodic area is constant and the area of the anode increased the total amount of corrosion would be about the same, but the amount of corrosion per unit area of the larger anodic area would be decreased, so the intensity of corrosion would be reduced.

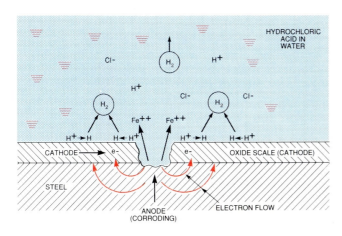

Fig. 2.7 Corrosion of steel in hydrochloric acid at a break in the oxide film

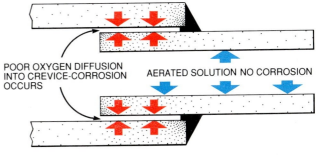

Fig. 2.8 Crevice corrosion due to differential aeration in steel pipework carrying an electrolyte solution

Conversely, if the situation arises where large cathodic areas are coupled to small anodic area in an electrolyte, attack on the anodes will be intense. For example, if steel bolts are used to fasten copper or brass components, the bolts could suffer very severe corrosion. Similarly, if steel tube or plate with an incomplete or cracked oxide film (e.g., mill scale) is immersed in water, rapid localised rusting (corrosion) will occur. The mill scale will act as a large cathodic area and the cracks in the film will expose a metal surface that will become small active anodic areas. Rapid pitting corrosion can then take place at these points.

The removal of oxygen from solutions in which metal is corroding means that the oxygen consuming reaction ($O_2 + 2H_2O + 4e^- \rightarrow 4OH^-$) cannot take place and the only available cathodic reaction is the reduction of hydrogen ions ($H^+ + e^- \rightarrow H$ and $H + H \rightarrow H_2$). In neutral and alkaline solutions, the concentration of hydrogen ions is small, therefore the cathodic process will take place at a very low rate, and the anodic process will be equally slow. Thus de-aeration of solutions will effectively reduce the corrosion rate to very low values, hence the common practice of using alkaline, de-aerated boiler feedwater to control corrosion of the boiler feed system plant.

It is evident that corrosion of steam raising and power plant can be minimised by careful consideration of the factors discussed here, during the design of plant and in the selection of operating water chemistries. However, other processes can result in the corrosive attack of plant materials, particularly at high temperatures; these are discussed in detail in later sections.

3 Boiler water treatment and steam purity

To maintain boilers and turbines at a high level of availability and efficiency, the chemical control of water and steam purity is aimed at the prevention of:

- Corrosion in feed, boiler and steam systems.
- Scale and deposit formation on heat transfer surfaces.
- Deposition and corrosion in turbines.

In this section some of the basic chemistry involved is discussed before proceeding to the detailed water treatment procedures in Section 4 of this chapter.

3.1 Formation of protective oxide films

As discussed earlier, when metals are in contact with water, corrosion of the metal may take place depending on the conditions prevailing. The readiness with which a metal corrodes depends on its standard electrode potential, the base metal being much more susceptible to corrosion than the noble metals. However, in practice there are particular cases, e.g., titanium, aluminium, where a base metal does not undergo extensive corrosion because a protective oxide film is formed on

the metal surface effectively stifling further corrosion. Steam/water circuits in power plant depend on such films for their integrity and the prime aim of the water treatment regimes developed is to produce and maintain conditions where these protective oxide films are formed and are stable.

When mild steel is immersed in water corrosion can readily occur, the rate depending on the pH and oxygen content of the water — high pH values and low oxygen levels giving the lowest corrosion rates. The effects of acid and alkaline solutions on the corrosion rate of mild steel at 310°C are illustrated in Fig 2.9, and quite clearly minimum corrosion rates are obtained at pH values (measured at 25°C) of 9–12. However, the results presented show that serious corrosion will occur in highly alkaline as well as under acid conditions — hence to obtain minimum corrosion rates it is essential to control the pH within set limits.

At the temperatures that prevail in high pressure steam plant, clean mild steel surfaces readily react with water to form, through a series of reactions, Fe_3O_4, (magnetite) and/or Fe_2O_3 (haematite). The first reaction is the formation of Fe^{2+}:

$$Fe + 2H_2O \rightarrow Fe^{2+} + 2OH^- + H_2\uparrow$$
(metal)

Further reactions then take place to form Fe_3O_4/Fe_2O_3:

$$3Fe^{2+} + 4H_2O \rightarrow Fe_3O_4 + 4H_2\uparrow$$
$$2Fe^{2+} + 3H_2O \rightarrow Fe_2O_3 + 3H_2\uparrow$$

In de-aerated water treated with alkali at high temperatures, the formation of magnetite from Fe^{2+} takes place very rapidly (the Schikorr Reaction) and the magnetite is formed as a dense, almost impervious, oxide layer firmly adhering to the metal surface. As the oxide forms it becomes a physical barrier between the metal and the water. Further attack of the metal is then primarily controlled by liquid diffusion through micro pores in the oxide film, and to a lesser extent by the diffusion of Fe^{2+}, Fe^{3+} and O^{2-} ions and electrons (Fig 2.10).

These diffusion rates are very low and are progressively reduced even further as the oxide film thickens. Consequently the rate of corrosion of the mild steel is quickly reduced to a very low level with correspondingly slow rates of thickening of the oxide film. For example, after some 15 000–20 000 hours operation at 350°C, the dense oxide film produced on mild steel boiler tube in good quality boiler water will only be some 20–30 μm thick.

During the formation of the protective oxide film, particularly during the early stages, not all the Fe^{2+} and Fe^{3+} ions formed are incorporated directly into the growing oxide film. Some of these ions diffuse out into the bulk water. However, the solubilities of Fe_3O_4 and Fe_2O_3 in boiler and feedwater are very low and when the reactions to form these oxides occur, the

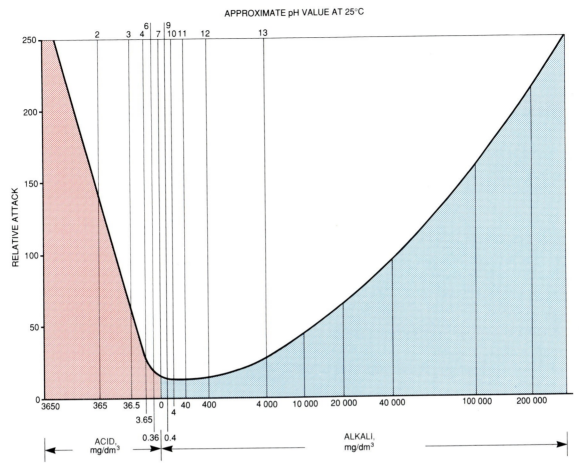

APPROXIMATE pH VALUE AT 25°C

FIG. 2.9 Attack on mild steel by acid and alkali at 310°C

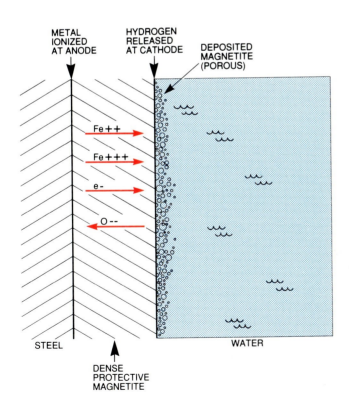

FIG. 2.10 Counter current mechanism of protective
magnetite growth

oxides precipitate as a rather porous granular layer on the outer surface of the adherent oxide (Fig 2.10). An illustration of a typical 'duplex' oxide on mild steel is shown in Fig 2.11.

The foregoing discussion has centred on the reactions of mild steel in contact with water, but similar reactions take place with other alloys. For example, if a high Cr content alloy is used, Cr^{3+} ions are formed in addition to Fe^{2+} and Fe^{3+} and become incorporated into the protective oxide film, the oxide formed (spinel) having a structure very similar to that of magnetite (Fig 2.12). However, other reactions can also take place that result in the formation of a very thin but highly protective layer of chromic oxide (Cr_2O_3) between the magnetite layer and the metal surface. When dry steam is in contact with metal surfaces a corresponding series of reactions again result in the formation of a protective oxide layer. In this case,

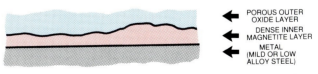

FIG. 2.11 Protective oxide layer on mild steel (duplex)

169

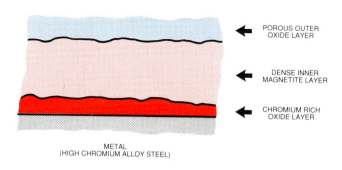

POROUS OUTER OXIDE LAYER

DENSE INNER MAGNETITE LAYER

CHROMIUM RICH OXIDE LAYER

METAL
(HIGH CHROMIUM ALLOY STEEL)

FIG. 2.12 Protective oxide layer on chromium steel (triplex)

however, no liquid is present and the metal oxides have a very low solubility in steam. Consequently all the metal corroded becomes incorporated into the oxide layer and growth of the layer is controlled by ionic diffusion in the oxide lattice.

3.2 Sources of alkalinity for boiler waters

The need to maintain alkaline conditions in boiler waters to prevent corrosion was established at an early stage in pressure steam plant development. The most readily available alkali to use was caustic soda (NaOH) or soda ash (Na_2CO_3), which was converted to NaOH under boiler conditions. This was satisfactory in low pressure plants, but with the move to higher pressures and higher heat flux, problems were encountered with boiler tube corrosion. At the caustic soda levels then being used to treat boiler water, the results presented in Fig 2.9 would suggest that corrosion rates should be very low. However, it was found that in the high heat flux zones of these boilers, very high localised concentrations of alkali were formed at tube surfaces by a process referred to as *hide out*. This effect may be observed with any soluble component of the boiler water to a lesser or greater degree, and the conditions influencing it are discussed in more detail in a later section.

In an attempt to combat the undesirable effect of caustic soda, sodium phosphates were introduced as an alternative source of alkalinity. Trisodium phosphate (Na_3PO_4) and disodium phosphate (Na_2HPO_4) can both effectively produce alkalinity by hydrolysis in water:

$$Na_3PO_4 + H_2O \leftrightarrow Na_2HPO_4 + NaOH$$
$$Na_2HPO_4 + 2H_2O \leftrightarrow NaH_2PO_4 + NaOH$$

The important difference between this source of alkalinity and that produced by free caustic soda lies in the reversibility of the above reactions. If acid species appear in the boiler water they will be neutralised by the sodium hydroxide:

$$HCl + NaOH \rightarrow NaCl + H_2O$$

On the other hand, if higher concentrations of sodium phosphate/sodium hydroxide are formed in the water

films at tube surfaces, the reactions move to the left effectively removing NaOH from the solution and limiting the maximum pH that can be produced. This led to the development of the *co-ordinated phosphate* method of boiler water treatment, in which phosphate concentration and pH are controlled by using mixtures of disodium and trisodium phosphate, such that the boiler waters had a satisfactory pH but effectively only contained very low concentrations of sodium hydroxide [6].

Figures 2.13, 2.14 and 2.15 illustrate this procedure. The curves represent the pH of a solution of Na_3PO_4. Above the line the water sample would contain a mixture of NaOH and Na_3PO_4 and below the line a mixture of Na_3PO_4 and Na_2HPO_4. The target is to operate just below the line by treating the boiler water with a suitable mixture of sodium phosphates.

These procedures have been used very successfully in the intermediate and high pressure range plants up to about 160 bar. However, at the higher pressures, considerable difficulties can be encountered in controlling the phosphate and pH values to the required limits. This situation arises because of the relatively low solubility of sodium phosphates at high temperatures which leads to a very marked tendency to hide out. This situation is exacerbated by further hydrolysis reactions that can result in sodium phosphates hiding out but leaving free caustic soda in the boiler water. Under these conditions satisfactory control of boiler water chemistry can be very difficult to maintain.

In CEGB plant, sodium phosphate is now only used in a limited number of lower pressure (60 bar) boilers. Sodium hydroxide is much more widely used in boilers operating at a range of pressures up to 160 bar. The increasing tendency for hide out, particularly prevalent in high pressure, high heat-flux boilers, is controlled by using lower concentrations of added alkali (see Section 3.3 of this chapter). Co-ordinated phosphate treatment is still favoured in many countries with, again, recommended dosing levels decreasing with increasing operating pressure with relatively narrow ranges of permitted pH and alkalinity values.

In some boilers, the tendency to hide out is so marked that it is practically impossible to control the levels of solid alkalis in the boiler waters and even the use of low levels of caustic soda can lead to boiler tube corrosion. This situation arises most frequently in boilers having relatively poor circulation and zones of high heat-flux and has led to the introduction of the *all volatile treatment* (AVT). In this case, the only alkalising agents used are ammonia, hydrazine or, occasionally, organic amines which will not concentrate up at tube surfaces because of their relatively high steam volatility. At boiler water temperatures (up to 355°C), these compounds are very much weaker alkalising agents than caustic soda or sodium phosphates and therefore the tolerable levels of potential acid impurities in the boiler water are correspondingly very much lower. In the event of serious ingress of impurities into

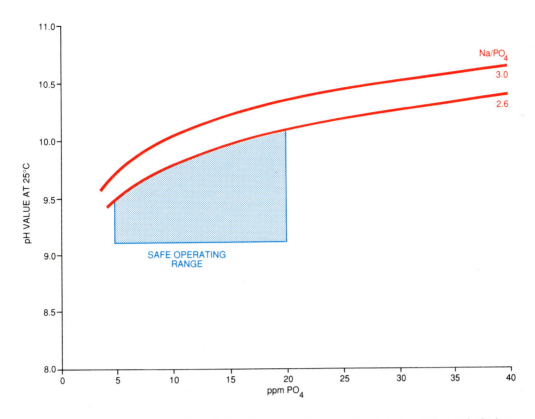

FIG. 2.13 Recommended co-ordinated phosphate curve for operation between 69 bar and 124 bar

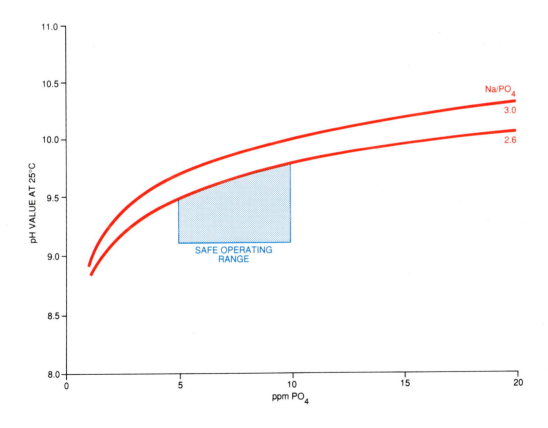

FIG. 2.14 Recommended co-ordinated phosphate curve for operation between 124 bar and 179 bar

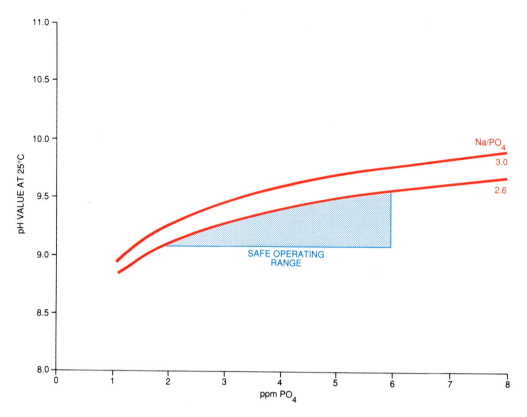

FIG. 2.15 Recommended co-ordinated phosphate curve for operation between 179 bar and 200 bar

the boiler water, it is common practice to inject low concentrations of sodium hydroxide/phosphate for a limited period until control of boiler water chemistry can be restored. With careful control of these impurity levels the AVT regime has been successfully used for 15–20 years in a considerable number of large, high pressure boilers.

3.3 Chemical treatment of boiler water — drum boilers

From the foregoing discussion it is clear that if safe, reliable, corrosion-free operation of high pressure drum boilers is to be achieved, it is necessary to operate with a boiler water chemical composition that will produce and maintain a stable protective oxide film on the tube surfaces. This can be achieved by using a water that simultaneously contains:

- Low, controlled levels of alkali.

- Very low levels of oxygen.

- Low levels of potential acid forming impurities such as Cl^-.

With increasing experience in operating such boilers, a better understanding of the necessary levels of alkali and the tolerable levels of impurities has led to the development of standard procedures for the chemical treatment of boiler waters.

These procedures take account of the pressure at which the plant operates, the boiler design (i.e., hori-

zontal or vertical boiler tubes), the maximum heat flux levels that prevail (and hence the tendency for salts to hide out) and the required purity of steam. As the bulk of impurities in boiler water are introduced with the feedwater, very rigorous control of feedwater quality is also necessary. Consequently, standards for the chemical treatment and purity of feedwater have also been developed and are discussed in detail in a later section.

The chemical regimes used in the range of drum boilers operated by the CEGB are given in Table 2.7 which details the boiler water and steam quality targets used in CEGB plant [7]. The targets given are for conventional, fossil-fuelled boilers operating at 60–160 bar, including high heat-flux oil-fired plant. The table also includes details of the boiler water treatments for the magnox nuclear plant which have horizontal, serpentine tube boilers with assisted circulation operating at up to 60 bar.

The table illustrates the features discussed above:

- The boiler water is made alkaline with modest levels of sodium hydroxide (up to 12 mg/dm^3) or trisodium phosphate. In the high pressure, high heat-flux oil-fired plant the targets are more stringent (0.75 mg/dm^3 NaOH).

- The levels of chloride are carefully controlled with the alkali levels being such as to neutralise any acid chloride produced. This is achieved by operating with a specified minimum of alkalinity in the boiler water, which is increased as the sodium chloride levels

TABLE 2.7

Summary of targets for drum-type boilers
The targets (expressed as μg/kg unless otherwise stated) relate specifically to steady state operation
and are set at levels considered to be achievable on a well-maintained plant for 95% of the daily running.
Any departures from these values should be treated as abnormal and should be investigated.

Parameter	Boiler class					Type of analysis
	60 bar	100 bar	160 bar (coal)	160 bar (oil)	Magnox (Nuclear)	
Boiler water						
Non-volatile alkali treatment						
Chloride (NaCl)	≤ 6000	≤ 4000	≤ 2000	≤ 500	≤ 2000	M/P
Silica (SiO_2)	≤ 5000	≤ 1500	≤ 200	≤ 200	≤ 5000	M/P
Sulphate (SO_4)	Local decision					M
Sodium hydroxide	1.5 × NaCl					M
(minimum)	(5000)	(5000)	(2000)	(500)	(2000)	
Sodium phosphate	Local decision				≥ 1000 (as NaOH)	M
All-volatile alkali treatment						
Chloride (NaCl)		≤ 200	≤ 200			M/P
Silica (SiO_2)		≤ 1500	≤ 200			M/P
Sulphate (SO_4)	Local decision					M
Saturated steam						
Silica (SiO_2)	≤ 20	≤ 20	≤ 20	≤ 20	≤ 20	M/P
Sodium (Na)	≤ 20	≤ 10*	≤ 10*	≤ 10*	≤ 20	M/P

*Target of ≤ 5 μg/kg when austenitic alloys are used

M — manual sampling   P — permanently installed instrument

increase, to maintain a mol ratio of NaOH to NaCl of ≮ 1.5:1. The boiler is then blown down to maintain the NaCl concentration within the target value.

• The permitted levels of alkali and impurities are steadily reduced as the plant operating pressures increase, as heat fluxes increase, and in boilers having horizontal, serpentine rather than vertical boiler tubes.

• When operating with weakly basic volatile alkalis such as ammonia, permitted impurity levels are very low.

These standards are applied in some 200 boilers operated at pressures up to 160 bar and in recent years careful operation to these standards has resulted in only a very small proportion of tube failures stemming from corrosion of the waterside of the boiler tubes.

The targets set out in Table 2.7, are to be used when boilers are operating at steady high loads. When plant is starting up (hot or cold) or operating at low loads, difficulties can be encountered particularly in meeting the oxygen target in feedwater. This in turn could lead to increased levels of oxygen in boiler water and an enhanced risk of boiler tube corrosion. Long term experience with plant operating at < 100 bar has shown that in general no serious problems should result from

initial operations outside the targets, provided that the oxygen in feedwater is reduced to target levels within two hours of synchronisation.

At higher boiler pressures, i.e., >100 bar, there is considered to be an increased risk of boiler tube corrosion if excessive oxygen levels in boiler water are allowed to persist for extended periods of operation.

To minimise this risk, the targets in Table 2.7 have been amended to cover the short term (up to 2 hours) excursions in oxygen levels that could occur at start-up and for longer term (up to 100 hours) low load operation.

Table 2.8 indicates that oxygen levels of up to 2000 μg/kg[3] can be tolerated for up to 2 hours if boiler water NaCl levels are within the targets given in Table 2.7 for the various boiler water chemistry regimes. However, if higher oxygen levels are being experienced then the boiler water NaCl levels should be reduced, by blowdown, as indicated in the table.

Table 2.9 shows that for more extended periods of low load operation it is necessary to operate at significantly lower levels of oxygen, or considerably reduce the boiler water NaCl levels.

3.4 Saturated/superheated steam purity

It is necessary to control impurity levels in steam, particularly sodium, as high levels could result in the

173

TABLE 2.8

*Boiler water chloride limits (μg/kg NaCl) for conditions
where oxygen targets are exceeded (> 100 bar)*

Boiler water chemistry	Boiler water dissolved oxygen, μg/kg			
	200	500	1000	2000
AVT	200	200	200	100
Sodium hydroxide 500–750 μg/kg NaOH	500	500	250	100
Sodium hydroxide 2000–3000 μg/kg NaOH	2000	600	350	200

In the absence of any boiler water dissolved oxygen measurements, the peak feedwater (economiser inlet) value as boiler operating temperature is approached, is assumed to apply. This table is applicable for the first 2 hours of operation.

TABLE 2.9

*Boiler water chloride targets (μg/kg NaCl) as a function of
oxygen concentration for low load operation*

Boiler water treatment	Boiler water dissolved oxygen, μg/kg			
	10	25	50	100
AVT	200	200	200	100
Sodium hydroxide 500–750 μg/kg NaOH	500	500	250	250
Sodium hydroxide 2000–3000 μg/kg NaOH	2000	1000	1000	1000
Sodium hydroxide/phosphate 2000–3000 μg/kg NaOH	2000	1000	1000	1000

In the absence of boiler water dissolved oxygen measurements, the peak feedwater (economiser inlet) value is assumed to apply. This table is applicable for up to 100 hours' low load operation.

deposition of sodium hydroxide in the turbine leading to stress corrosion failures (see Section 9 of this chapter) [8,9].

The targets for saturated steam, also given in the Table 2.7, have been developed from a combination of an understanding of the physio-chemical reactions occurring within the boilers, superheaters and turbines, and from operating experience [7]. The target levels for sodium salts in boiler water have, therefore, to be set at such a level that a combination of steam volatility of the salts and carry-over of boiler water would not result in the steam containing unacceptable levels of sodium.

The problems associated with excessive levels of SiO_2 in steam are discussed in Section 9.2 of this chapter. Silica is relatively volatile in high pressure steam, the volatility increasing with increasing pressure. Therefore, to meet the target of $\leqslant 20$ μg/kg SiO_2 in steam, the permitted maximum level in the boiler water is decreased with increasing operating pressure.

4 Feedwater treatment — drum boilers

Feedwater is the major source of soluble and insoluble impurities entering the boiler and therefore the

principal aim of feedwater treatment and monitoring is to minimise the levels of such impurities.

Insoluble impurities are very largely metal oxides derived from corrosion of the plant itself, particularly the materials of construction of the condenser and feed systems. Such corrosion is minimised by maintaining the feedwater at an alkaline pH and by removing dissolved oxygen down to very low concentrations. Control of the corrosion of feed system materials will also help to minimise the risk of corrosion failures in other sections of the plant (e.g., boilers).

Soluble impurities in feedwater most frequently arise from in-leakage of cooling water from condensers and other coolers, but can also result from poor quality make-up water. To ensure that the levels of such impurities are controlled, the concentrations in feedwater are periodically and/or continuously monitored such that in the event of unacceptable levels of contamination, steps can quickly be taken to correct the situation.

The feedwater treatment regimes used by the CEGB are typical of those used in most countries for high pressure drum boilers and are given in detail in Table 2.10 [1,7,10]. The individual parameters included in the table are discussed in the following sections.

4.1 Control of insoluble impurities in feedwater

4.1.1 Feedwater pH — the use of volatile alkalis

As illustrated in an earlier section, by making the pH of feedwater slightly alkaline the corrosion rates of ferrous alloys can be minimised. Where a system also contains copper-based alloys similar benefits can be obtained, but if the pH levels used are too high, particularly in the presence of oxygen, then corrosion of these alloys can be accelerated.

Consequently, in condensate and feed systems containing both ferrous and copper-based alloys the most satisfactory pH range to use is 8.5–9.2. In the absence of the copper alloys, corrosion of ferrous components may be further reduced by raising the pH to 9.4 or above.

In a drum boiler, any non-volatile salts present in the feedwater will rapidly increase in concentration in the boiler water to a level determined by the amount of blowdown being employed. The presence of such salts in feedwater could also lead to contamination of steam where spray desuperheaters are used. Consequently only volatile alkalis are used for feedwater pH control. The approach has several advantages, in particular:

- There is no excessive accumulation of alkali in the boiler water, the volatile alkali passing out of the boiler with the steam.

- As condensates are formed in the lower pressure section of the turbine, in the condensers and on the steam side of the feedheaters, the alkali is immediately available in these parts of the steam/water

TABLE 2.10

Feedwater targets for drum-type boilers
The targets relate specifically to steady state operation and are set at levels considered to be achievable on
a well maintained plant for 95% of the daily running. Any departures from these values should be treated as
abnormal and should be investigated.

Parameter	Boiler class					Type of analysis (see key)	Location (see key)
	60 bar	100 bar	160 bar (coal)	160 bar (oil)	Magnox (Nuclear)		
Feedwater							
Conductivity μS/cm at 25°C	*Plant decision to achieve boiler water targets					P	EPD
Sodium (Na)	*Plant decision					M/P	EPD
Sulphate (SO$_4$)	*Plant decision					M	EPD
Dissolved O$_2$ μg/kg	$\leqslant 50$	$\leqslant 50$	$\leqslant 50$	$\leqslant 50$	$\leqslant 50$	P	EPD
	$\leqslant 5$	$\leqslant 5$	$\leqslant 5$	$\leqslant 5$	$\leqslant 5$	P	BI
Oil, μg/kg	$\leqslant 200$	$\leqslant 200$	$\leqslant 200$	$\leqslant 200$	$\leqslant 200$	M/P	EPD
Hydrazine (N$_2$H$_4$), μg/kg	2 × Dissolved O$_2$					M/P	DAO
Ammonia (NH$_3$), μg/kg	$\leqslant 800$ ($\leqslant 1000$ when copper alloys absent)					M/P	DAO
pH (at 25°C)	8.8–9.2 (8.8–9.4 when copper alloys absent)					P	DAO
Total metals	Plant decision					M	BI
Spraywater							
Sodium (Na)	*Plant decision to meet Na target in steam					M/P	

Key: EPD — Extraction pump discharge   DAO — De-aerator outlet   BI — Boiler inlet
      M — Manual sampling   P — Permanently installed continuous monitor
      *Plant decision — the target values are based on specific plant operating experience such that boiler water
      chemistry and steam quality are satisfactorily controlled.

system, raising the pH of the condensates and hence suppressing corrosion.

Ammonia (NH$_3$) is the most widely used volatile alkali because of its low cost, ready availability and its stability at high temperatures.

However, ammonia does have some disadvantages in that it is very volatile in steam and it is a relatively weak alkali at high temperatures. Consequently initial condensates formed at high temperatures, for example, in feedheaters, may have relatively low pH values resulting in only limited protection from corrosion [12].

Other volatile alkalis, organic amines, have therefore been used to control feedwater pH in particular applications, the most widely used being:

- Morpholine.

- Cyclohexylamine.

- Diethyl amino ethanol.

- Amino methyl propanol.

- Methoxy propylamine.

In particular circumstances these amines can be superior to ammonia, being less volatile and so maintaining the pH of feedwater at higher temperatures and producing higher pH levels in condensates [13]. However, these compounds have limited thermal stability and are not normally used in plant operating with superheated steam temperatures above about 450°C

[14]. If significant decomposition of the amines occurs, not only will there be a loss of alkali from the system, but the breakdown usually produces organic acids and carbon dioxide that could enhance feed system and turbine corrosion. Whilst the use of volatile alkalis has many advantages in minimising the overall corrosion of the steam/water circuits, care has to be taken to avoid the formation of excessive concentrations of alkalis, in particular condensates. For example, plants have suffered severe corrosion of copper-based alloy condenser tubes in the non-condensable gas off-take zones of vacuum condensers. This results from condensates in this zone containing a 500–1000-fold increase in ammonia concentration over bulk feedwater/ condensate concentrations. The severity of the attack can usually be dramatically reduced by the use of alternative, more resistant condenser tube material, or by spraying bulk condensate into the affected zone.

4.1.2 Monitoring and control of feedwater pH

In all plants there will be some losses of water and steam, and the contained alkalising agent, from the steam/water circuits. Furthermore, there will also be losses of volatile alkalis from the circuit with the non-condensable gas removed from the condensers. It is therefore normal practice regularly or continuously to monitor the pH and/or direct conductivity of bulk condensate at the extraction pump discharge, or feedwater at boiler inlet, to ensure that the correct chemical conditions are being maintained. Figure 2.16 shows

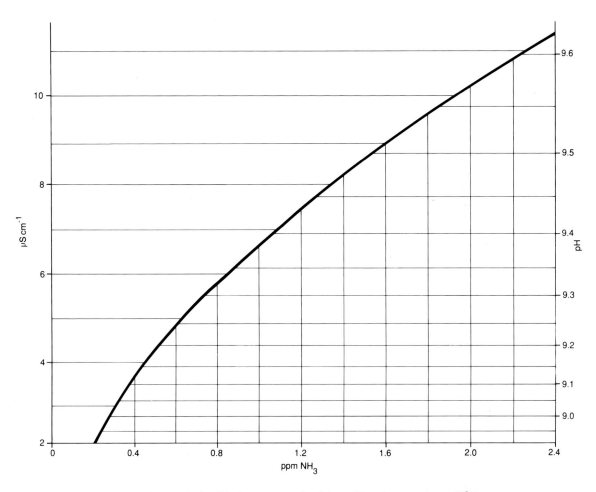

FIG. 2.16 Relationship between conductivity, pH and ammonia at 25°C

the relationship between pH, direct conductivity and ammonia concentrations in water at 25°C. Similar plots are available for the organic amines.

To make up for losses from the system, additional quantities of the alkali being used are injected into the feed system, normally in the low pressure sections, to maintain the required pH. In some plants where ammonia is used, the required make-up is generated by the decomposition of hydrazine (N_2H_4) injected as a de-oxygenating agent.

4.1.3 Oxygen control

To minimise corrosion in feed systems it is necessary to maintain low dissolved oxygen levels in condensate and feedwater (Table 2.10). Very low levels of oxygen in feedwater are necessary to avoid possible corrosion of economiser tubing and to minimise the risk of on-load corrosion of boiler tubing (see Section 6 of this chapter). The oxygen levels are controlled by a combination of physical and chemical de-aeration techniques.

Physical de-aeration

Physical de-aeration methods are frequently employed at two points in the condensate/feed system. In most

power plant the main condensers are operated under vacuum, and at the temperature and pressure prevailing, the equilibrium solubility of oxygen in the condensate is relatively low (e.g., <50 μg/kg O_2). If further de-aeration is required, this can be achieved in the condenser by the injection of live steam into the bulk condensate in the bottom of the condenser, producing a 'steam stripping' effect.

Alternatively, a further pressure de-aeration stage is frequently introduced in the low pressure section of the feed system when the feedwater temperature has been raised to 110–150°C. Again steam stripping is used either by injecting live steam into the feedwater or by spraying the feedwater through an atmosphere of steam. The steam, carrying the gas extracted from the feedwater, is then vented to the main condensers.

Using these de-aeration techniques, the oxygen level in feedwater in most modern plants can readily be reduced to <10 μg/kg and frequently to <2 μg/kg.

Oxygen may enter the feed system with make-up water from a water treatment plant or reserve feedwater tanks as, in many instances, these reserves of water are fully aerated. However, it is common practice for the water to be drawn into the feed system via the main condensers where de-aeration takes place. In some make-up water treatment plants, the oxygen levels of this water are substantially reduced by passing it through a vacuum degasser before it enters the feed system. A further alternative procedure is to 'sparge' nitrogen gas into the water in the reserve feedwater tanks to strip out the dissolved oxygen, and maintain the stored water at a low oxygen level.

Chemical de-aeration

Although the physical de-aeration techniques are very effective in reducing dissolved oxygen levels in feedwater, as an additional precaution, the last traces of oxygen are removed by injecting a chemical oxygen scavenging agent into the system. As already noted, in modern high pressure boiler plant it is essential to minimise the solid content of feedwater and hence chemical de-oxygenating agents should not produce any residual solid salts in boiler water. The most widely used agent is hydrazine (N_2H_4) [15]. This has the advantages of being steam volatile, slightly basic and reacting readily to remove even traces of dissolved oxygen at temperatures above $150°C$ producing only volatile or gaseous products, i.e.,

$$N_2H_4 + O_2 \rightarrow N_2 + H_2O$$

At higher temperature hydrazine itself decomposes to produce ammonia and nitrogen, i.e.,

$$3N_2H_4 \rightarrow 4NH_3 + N_2$$

As indicated earlier, the decomposition of N_2H_4 can be utilised to supply additional NH_3 to the steam/water circuits for pH control of feedwater.

A number of other organic compounds will react with residual dissolved oxygen at elevated temperature and can be used as de-oxygenating agents, for example:

- Organic amines used for pH control.

- Carbohydrazide.

- Ascorbic acid.

- Diethyl hydroxylamine.

- Erythrytol.

However, the breakdown of these organic compounds could lead to the formation of substantial quantities of carbon dioxide or organic acids which, at high concentrations, could increase the risk of corrosion in turbines and feed systems.

Catalytic de-aeration

One major disadvantage of most chemical de-oxygenating agents is their relatively slow reaction rates at temperatures below $100°C$, limiting their value in protecting the lower pressure sections of the feed system. Recent studies have led to the development of alternative procedures effective in reducing dissolved oxygen concentrations from the mg/kg level to $<10\ \mu g/kg$ rapidly at ambient temperatures. These techniques could be particularly useful for de-oxygenating make-up water or boiler feedwater during cold start-up of plants [16]. Two processes are now in plant use:

(a) *De-oxygenation using N_2H_4 and active carbon* [17,18]. Hydrazine will react rapidly with dissolved oxygen in water at ambient temperatures when in contact with active carbon. The reaction is rather more rapid at pH values above 9.0. To obtain a very high level of de-oxygenation, a slight excess of hydrazine is injected into the water which is then passed through a column of granulated active carbon or a bed of powdered active carbon. With contact times of approximately 30 s for the column and 4–6 s for the bed, the oxygen levels can be reduced from saturation ($\simeq 8$ mg/kg) to $<10\ \mu g/kg$. The de-oxygenated water then contains the dissolved nitrogen produced by the interaction of oxygen and hydrazine, together with some ionic impurities leached from the active carbon, particularly silica. It is therefore necessary to determine the extent of impurity leaching and, if necessary, remove the ionic impurities with an ion-exchange bed downstream of the carbon column/bed.

(b) *De-oxygenation using hydrazine or hydrogen and a palladium catalyst bed* A similar de-oxygenation process has been developed in which the water containing hydrazine or injected hydrogen is passed through a bed packed with a palladium-containing catalyst. The catalyst used is an ion-exchange resin with palladium metal precipitated into the pores of the bead structure. In the presence of a 40% excess of hydrogen or a 100% excess of hydrazine, dissolved oxygen levels are reduced from 4 mg/kg to $<10\ \mu g/kg$ with a contact time in the catalyst bed of some 20 seconds. The products of the reactions are water and nitrogen, although there is some evidence that very small quantities of palladium are released into the treated water when a new catalyst bed is first put into service.

4.2 Soluble ionic impurities in feedwater

4.2.1 Sources of impurities

The soluble impurity levels in boiler water, particularly chlorides, must be carefully controlled to avoid corrosion of boiler tubing (see Section 6 of this chapter). In most plants, virtually all soluble boiler water impurities are carried into the boiler in the feedwater; it is therefore essential to limit the concentrations of these impurities in feedwater. The actual concentrations of chlorides, sulphates, etc., in feedwater that can be tolerated will depend on the type and operating pressure of the plant and the boiler water chemical regime being used (Table 2.7). Excessive levels will lead to difficulties in controlling boiler water composition and even modest levels will require frequent or continuous boiler water blowdown with associated losses of heat and pure water. The extent to which such procedures can be used may also be severely restricted by the maximum rate of blowdown available on the particular plant.

The major source of these impurities in the feedwater is in-leakage of low quality water from the cooling water side of the main condensers or other cooling systems. When contamination is detected, action should be taken as quickly as possible to identify the source of the leak and take corrective action.

A further important source of impurities can be the make-up water added to the system. Good quality make-up water should have a conductivity of $<0.2\ \mu S/$cm and contain only a few $\mu g/kg$ of Cl^-, SO_4^{2-}, Na^+ and SiO_2, and this is generally achieved. However, malfunction of the water treatment plants used to produce the make-up water could quickly lead to a serious deterioration in make-up water quality and the passage of significant levels of ionic materials into the feedwater. It is therefore very important to monitor make-up water carefully and maintain good quality. A particular difficulty encountered with some raw waters, is that they contain non-reactive silica and non-ionic organic compounds that are not removed by ion-exchange processes and can lead to a rapid build up of SiO_2 and/or Cl^- and SO_4^{2-} in boiler waters. When such a situation arises special water treatment procedures may be required.

Impurities can also be introduced into both feedwater and boiler water with the alkalis injected into the system to control pH. Care should therefore be taken to ensure that only a satisfactory quality of dosing chemical is used, e.g., only use ammonia or hydrazine containing very low levels of Cl^-.

4.2.2 Sampling and monitoring

In general, feedwater monitoring for soluble impurities is carried out at the main condenser extraction pump discharge to detect any in-leakage of cooling water close to the source. However, in plant where impurities can also enter the feedwater downstream of the condenser, for example, in poor quality reserve feedwater, it is a common practice to have additional monitoring at the boiler/economiser inlet (see Section 12 of this chapter).

The levels of anionic impurities in feedwater are monitored continuously using 'after cation exchange' conductivity (K_H) which measures the conductivity of the anions present as free acids. This determination has the advantages of being simple and reliable, with a relatively high sensitivity to mineral acid anions such as Cl^- and SO_4^{2-}.

It is also common practice now to measure the Na^+ content of condensate and feedwater continuously using reliable and sensitive instruments based on ion selective electrodes. In most instances cooling waters and poor quality make-up water will contain Na^+, Cl^- or SO_4^{2-}, hence the combined monitoring of Na^+ and K_H is a very effective method of detecting

and measuring inleakage of cooling water or unsatisfactory make-up water.

Insoluble impurities in feedwater

As indicated, the insoluble debris in feedwater (mainly metal oxides) is minimised by controlling the pH and O_2 levels in feedwater and steam condensates. To determine the effectiveness of the feedwater conditioning, the levels of individual metals at boiler/economiser inlet are determined regularly (e.g., Fe, Cu, Ni), see Section 12 of this chapter. The special sampling requirements for obtaining satisfactory water samples for debris analyses are also given in detail in that section.

If the main condenser tubes are fabricated from copper-based alloys, it is also good practice to monitor the copper levels in condensate at the main condenser extraction pump discharge. The results may give early indications of corrosion problems on the steam side of the condensers. Furthermore, if changes are made to the feedwater chemical treatment regime, the copper levels in condensate will indicate whether or not the change has led to enhanced condenser corrosion.

Additional sampling for monitoring corrosion debris at other points in the feed system can also be of value in detecting specific problem areas, for example, an increase in debris levels could indicate corrosion problems whereas a decrease would suggest that deposition is occurring.

In most operating plants, the control of feedwater pH within the limits set out in Table 2.10 will lead to satisfactory control of debris levels. However, during the start-up of plant after a period off-load (i.e., plant maintenance) or following sharp changes of feedwater flow rates, the debris levels may be very much higher than normal. To prevent this debris passing forward to the boilers would require the putting to drain of large quantities of water. Consequently it is now frequently the practice to filter this water using cartridge or precoat filters that are capable of removing a very high percentage of the debris, thus allowing the plant start-up to proceed. Magnet filters, that very effectively remove magnetite (Fe_3O_4), are also used.

5   Feedwater chemistry — once-through boilers

5.1   Feedwater treatment regimes [7,10,11]

Once-through boilers are inherently different from drum boilers in that 100% of the feedwater entering the boiler tubes is evaporated and superheated in a single pass through the boiler, compared with some 10-25% in a drum boiler system (Fig 2.17). Consequently all soluble salts and insoluble debris in the feedwater will either deposit in the boiler/superheater tubes or be carried out of the boiler with the steam. To minimise the risk of boiler tube corrosion, avoid excessive debris deposition on boiler tube surfaces and to produce steam of the required purity, the impurity levels in feedwater

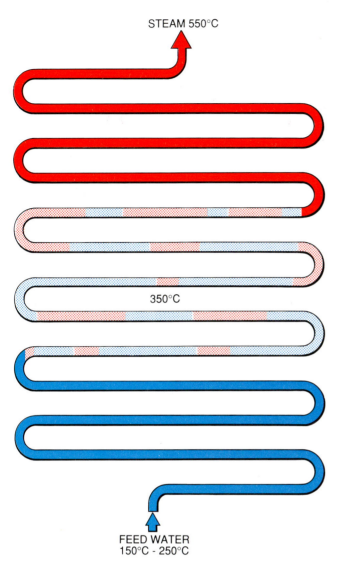

STEAM 550°C

350°C

FEED WATER
150°C - 250°C

Fig. 2.17 Once-through boiler

have to be rigorously controlled. This is normally achieved by:

- Using only volatile alkali for pH control.

- Installing full or part-flow condensate purification plant to minimise soluble potentially aggressive impurities in feedwater.

- Where necessary, using feedwater filtration to control insoluble debris to very low levels.

Several different feedwater treatment regimes have been developed for use in once-through boiler systems, involving different pH and oxygen targets. The details of these regimes are set out in Table 2.11 and the application of each particular treatment will depend, to some extent, on the operating pressure of the plant and the materials of construction of the feed system and boilers.

5.1.1 Low oxygen/high pH treatment

The feedwater treatment is essentially the same as that detailed for high pressure drum boilers using ammonia and hydrazine (Table 2.10), but with more stringent limits on feedwater impurity levels. This regime can be used in plants containing copper-based, ferritic and austenitic alloys. However, the feedwater quality must be controlled to the set targets or boiler tube corrosion, discussed in Section 7 of this chapter, may occur.

5.1.2 High oxygen/low conductivity treatment

If mild steel were exposed to water at pH = 7.0, relatively rapid attack would take place. However, some years ago it was established that if the water was essentially free from mineral anion impurities (e.g., Cl^-, SO_4^{2-}) and contained 200–300 $\mu g/kg$ O_2, a stable protective magnetite/haematite (Fe_3O_4/Fe_2O_3) layer was formed on the steel surface. This led to the development of a feedwater treatment regime for use in high pressure once-through boilers in which the feed and condensate systems contained no cuprous alloys apart from condenser tubes.

A particular advantage of this form of treatment is that no alkalis are added to control feedwater pH.

In addition to rigorously controlling the ionic impurity levels in the feedwater, it is also necessary to maintain the effective oxygen concentration by the injection of oxygen gas or hydrogen peroxide (H_2O_2). Both sources of oxygen appear to be effective but there have been reports of feedwater containing hydrogen peroxide attacking stellite components in valves and pumps.

This feedwater treatment has been widely used in once-through high pressure fossil-fuelled plant. However, there are some limitations:

- In low pH water, high oxygen levels (i.e., 200–300 $\mu g/kg$ O_2) can lead to enhanced attack on copper alloy feedheater tubing and result in excessive copper levels in feedwater. In modern high pressure plant this problem is avoided by the use of only ferrous alloys in feed system construction. This is particularly important in plant operating at supercritical pressure. Copper oxides have a relatively high solubility in supercritical steam but rapidly deposit in the turbine. Such deposits can lead to serious problems in turbine operation.

- The protective oxide layer formed on the metal surfaces is stabilised by a constant supply of oxygen in the feedwater. This requirement is met by maintaining the required oxygen level in the feedwater and by maintaining a sufficiently high flow rate in the feed system. If these requirements are not met, enhanced corrosion of the feed system will occur with much increased iron concentrations in feedwater.

5.1.3 Combined treatment

Many of the once-through plants built some 15–20 years ago still have copper-based alloy feedheater systems and the so-called 'combined treatment' was developed to allow a form of 'high oxygen/low conductivity treat-

TABLE 2.11

Primary targets for once-through boilers — feedwater at boiler inlet

Parameter and units	Type of analysis	Low pressure	High pressure subcritical and supercritical		
pH at 25°C	Permanent instrument	9.2–9.6	> 9.3	8.4–8.7	> 7.0
Oxygen (μg/kg O_2)	Permanent instrument	< 5	< 5	150	250
Hydrazine (μg/kg N_2H_4)	Permanent instrument	2 × O_2 (min 10)	2 × O_2 (min 10)	–	–
Ammonia (μg/kg NH_3)	As required to meet pH at 25°C				
Conductivity – direct (μS/cm at 25°C)	Permanent instrument	$\geqslant$ 4.3	$\geqslant$ 5.5	0.8–1.4	< 0.2
Conductivity (after cation exchange) (μS/cm at 25°C)	Permanent instrument (*Note 1*)	< 0.1	< 0.08	< 0.1	< 0.1
Sodium (μg/kg Na)	Permanent instrument	< 5	< 2	< 5	–
Iron (μg/kg Fe) Copper (μg/kg Cu)	Manual analysis	< 25	< 5 < 2	< 10 < 3	< 10 < 3
Reactive silica (μg/kg SiO_2)	Manual analysis	< 20	< 20	< 10	< 10

Note 1   Individual anions (e.g., Cl^- or SO_4^{2-}) determined as required to check conductivity values.

ment' to be used in these plants. Detailed investigations showed that the corrosion of copper alloys could be dramatically reduced by controlling the pH of the feedwater to 8.0–8.5, even in the presence of 200 μg/kg O_2, without any adverse effects on the ferrous alloys (Fig 2.18) [19,20].

This treatment has many of the advantages of the 'high oxygen/low conductivity' regime. The quantities of volatile alkali (normally ammonia) used to control the feedwater pH are relatively low (i.e., 50–100 μg/kg NH_3) and still allow long cycle times on the condensate purification plants between regenerations. This form of feedwater treatment has been adopted in a number of once-through plants in preference to the low conductivity regime.

5.2 Application to operating plant

From the foregoing discussion, it is clear that careful consideration must be given to plant design and materials of construction in selecting the most suitable feedwater treatment regime to be used. It is now the practice to select the feedwater regime at the design stage, avoiding the use of unsuitable materials. For example, feed and boiler systems are constructed free from copper alloys so that the plant can be operated

using the 'high oxygen/low conductivity' or 'combined' regimes as preferred.

The need to meet the stringent feedwater purity targets has also been stressed. To this end, monitoring of the condensate/feed system is carried out at various points and additional quality targets (e.g., specific limits on Na^+, Cl^- and SO_4^{2-}) may be introduced. This enables problems in plant operation to be quickly identified.

These points are illustrated in the feedwater targets set for high pressure, once-through boilers of the advanced gas-cooled reactor operated in the UK (Table 2.12). The boilers operate at 170 bar with a final superheat temperature of 550°C. The feed systems are free from copper alloys and the boiler/superheater tubes are constructed from a range of ferrous alloys, from mild steel to austenitic stainless steel. The decision was made to operate the boilers using a low oxygen feedwater regime with very close control of Na^+, Cl^- and SO_4^{2-} concentrations. However, in early testing it was found that in one particular boiler design, mild steel pipework at the boiler inlet was susceptible to erosion-corrosion in areas local to a flow control ferrule. Consequently a modified feedwater treatment was developed, involving the use of modest levels of dissolved oxygen (i.e., $\leqslant$ 25 μg/kg), to counter this problem. Higher

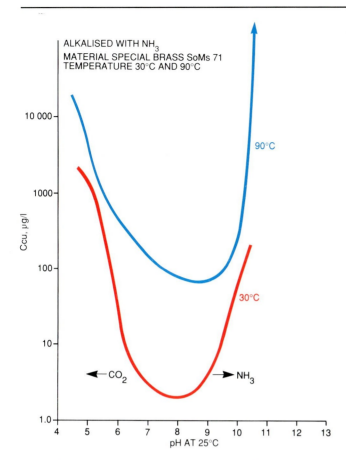

FIG. 2.18 Dependence of corrosion of brass on
pH value when alkalising with ammonium hydroxide

levels of hydrazine were also used to ensure complete
removal of this oxygen before the evaporative zone
of the boiler tubes, to avoid any increased risk of boiler
tube corrosion. In later plant designs, this section of
the boiler inlet tubing has been fabricated from alloys
resistant to erosion-corrosion, and these boilers are
operated with the low oxygen feedwater regime.

6 Boiler tube corrosion — drum boilers

When discussing the sources of alkalinity used to
control boiler water pH, the phenomenon of 'hide out',
i.e., the loss of dissolved solids from the bulk boiler
water, was highlighted and the problems that could
stem from its occurrence in the high heat flux zones
of the boiler tubes. The possible formation of potentially aggressive concentrations of boiler water salts is
reflected in the upper limits of boiler water impurities
and solid alkalis detailed in Table 2.7.

Hide out is characterised by the steady loss of
solutes from the boiler water when the boiler is on
relatively high load, the solutes reappearing when the
load is sharply reduced or when the boiler is taken off
load. To a lesser or greater degree, this effect can be
observed with all soluble species in the boiler water,
but is more pronounced with salts that have a low
solubility at boiler water temperatures and in boilers
operating at particularly high heat-fluxes (e.g., large
oil-fired boilers).

The loss of solute from boiler water has been observed over many years but only more recently has a
proper understanding of the mechanism been developed.
As already indicated, the phenomenon is frequently
observed with high heat-flux boilers but can also be
found in drum boilers with low/moderate heat-flux if
water circulation rates are low. It is enhanced by the
presence of porous oxide layers on tube surfaces or
where crevices are present. Within the pores of the
oxide or the crevices 'wick boiling' can take place
(Fig 2.19). Initially the pores and crevices are flooded
with bulk boiler water containing dissolved salts and
alkali. When boiling occurs the steam is ejected from
the pores leaving behind a concentrated solution. As
the pores approach dryness, the volume of steam being
produced falls away and the pores re-flood with boiler
water and the cycle is repeated. This cycle will repeat
indefinitely whilst the heat-flux is applied to that particular section of the boiler tube, resulting in the deposition of less soluble salts (Na_3PO_4, $NaCl$, $CaSO_4$)
or the formation of very concentrated solutions of the
more soluble salts or alkalis (e.g., $NaOH$). It has been
estimated that concentration factors as high as 10^4–10^5
can occur, i.e., a boiler water containing 2 mg/kg
$NaOH$ could produce, in the oxide pores, a solution
containing 20 wt% $NaOH$. The concentration factor
(cf) can be estimated from an equation of the form:

$$cf = \exp \frac{qs}{\alpha \varrho_w L D_w}$$

where D_w = diffusivity of salt in water
  L = latent heat of evaporation
  q = heat flux per unit area of inside wall
  s = depth of deposit
  α = porosity of deposit
  ϱ_w = density of water

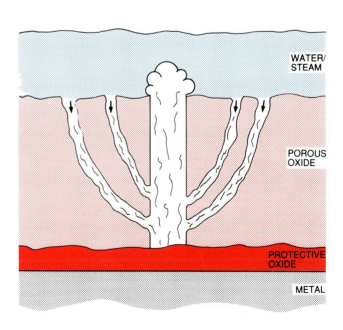

FIG. 2.19 Wick boiling

TABLE 2.12
Summary of primary targets for the steam/water circuit of AGR once-through boilers
The primary targets relate specifically to steady state operation and are set at levels considered to be achievable
on a well maintained plant for 95% of the daily running. Any departures from these values should be treated
as abnormal and should be investigated and corrective action taken as necessary.

Determined	Sampling position	Recommended minimum frequency of analysis	Primary target		Shutdown
			Low oxygen	Oxygen dosed	
pH at 25°C	BI	Continuous	> 9.3	> 9.0	
NH_3, µg/kg	BI	Daily	Greater than 300 Dependent on operating pH		
Conductivity — direct, µS/cm at 25°C	BI BO EPD CPPO	Continuous Continuous Continuous Continuous	> 2.8 > 2.8 > 2.8 < 0.08	Dependent on operating pH	1.0
N_2H_4, µg/kg	BI	Continuous	1.5 × dissolved O_2 concentration with a minimum of:		
			10	30	
Dissolved O_2, µg/kg	BI	Continuous	< 5.0	Not greater than 25	> 500 Time dependent
	EPD	Continuous	< 50		
Na, µg/kg	BI BO CPPO	Continuous Continuous Continuous	< 2.0 < 2.0 < 2.0		> 200
Conductivity — after cation, µS/cm at 25°C	BI BO EPD CPPO	Continuous Continuous Continuous Continuous	< 0.08 < 0.08 < 0.30 < 0.08		3.0
Cl, µg/kg	BI CPPO	Twice per week Continuous	< 2.0 < 2.0		> 150
SO_4, µg/kg	BI CPPO	Twice per week Weekly	< 2.0 < 2.0		> 100
SiO_2 — reactive, µg/kg	BO CPPO	Weekly Weekly	< 20 < 5		
Fe, µg/kg	BI	Monthly	< 5		
Cu/Ti, µg/kg	BI	Monthly	< 2		

Key: CPPO — Condensate polishing plant outlet   DAO — De-aerator outlet
     BI — Boiler inlet   EPD — Extraction pump discharge
     BO — Boiler outlet

The equation highlights the enhanced corrosion risk associated with increasing oxide thickness on the boiler tube surface. The oxide will gradually thicken with increasing periods of operation; for example, in large boilers operating at 160 bar, the oxide growth rate is approximately 15 µm per 10 000 hours. To minimise any resulting risk it is now common practice to clean the boiler tubes chemically when the oxide thickness has increased to 50–100 µm.

The actual quantities of salts involved are generally very much smaller than would be required to produce significant scaling of the tube surfaces and so affect heat transfer, and the hiding out of neutral salts (e.g., Na_2SO_4, $CaSO_4$ or SiO_2) are not considered to pose any risk to boiler integrity. These salts appear to de-posit in the pores and crevices without damaging the protective oxide films on the metal surfaces.

If, however, the salts present in the boiler water when concentrated-up can produce an acid solution or a strongly alkaline solution, there is then the possibility of the protective oxide film being attacked and destroyed leading to serious corrosion of the underlying metal. In operating plant, examples of both acid and alkaline corrosion have been found and resulted in boiler tube failures, but, in general, boiler tube steels are more resistant to concentrated alkaline solutions than concentrated acids (Fig 2.9). Both forms of attack can be avoided by adhering to the boiler water targets set out in Table 2.7. As hide out also involves crevices or porous oxide deposits, the effects can be minimised respectively by design or by regular chemical cleaning of boiler tubes to limit the accumulation of deposits.

6.1 Acid chloride corrosion [3]

As indicated, both concentrated acid and concentrated alkaline solutions can lead to serious boiler tube corrosion, but the two conditions differ very markedly in the way in which the tube is attacked and this affects both the rate of attack and mode of failure.

Detailed consideration of the boiler water treatment regimes given in Table 2.7 will show that the chloride levels are closely controlled and that the aim is to maintain a modest excess of alkali, assuming that all the chloride present could be converted to acid.

The formation of acid can occur in two ways:

(a) When contaminants leaking into the boiler water (via feedwater and condenser leakage) contain hydrolysable chlorides such as magnesium chloride which can produce hydrochloric acid, e.g.,

$$MgCl_2 + H_2O \rightarrow MgO + 2HCl$$

This is a particular risk in plants using sea water as cooling water because of its high hydrolysable chloride content; 14% of the total chlorides are magnesium and calcium chloride.

(b) By interaction of neutral sodium chloride with an oxidising agent. The process by which this occurs is not well defined but experimental rig and plant evidence indicates that high oxygen levels in boiler water can produce such effects. There is also evidence that oxides of metals that can readily change valency under boiler water conditions, e.g., copper oxide, can produce a marked increase in the ease with which 'acid chloride' attack is initiated. In the high heat-flux zones the acid solutions concentrate in the porous oxides, attacking the protective oxide and then the underlying tube metal:

$$Fe_3O_4 + HCl \rightarrow FeCl_2 + FeCl_3 + H_2O \quad (2.1)$$
$$Fe + 2HCl \rightarrow FeCl_2 + H_2 \quad (2.2)$$

Under operating conditions the iron chlorides are not stable and as they diffuse away from the corrosion site Reaction (2.1) is reversed leading to the re-precipitation of Fe_3O_4 and the reformation of HCl. The acid is then available to attack more iron oxides or the unprotected tube metal.

The reaction sequence produces a rapid thickening of the iron oxides layers which in turn increase the effects of hide out. The net effect is rapid attack of the boiler tubing with the formation of thick layers of iron oxides.

This cyclic process is illustrated in Fig 2.20: the form of the corrosion products being very characteristic of this type of attack. The re-precipitated iron oxides form a dense, laminated, but non-protective

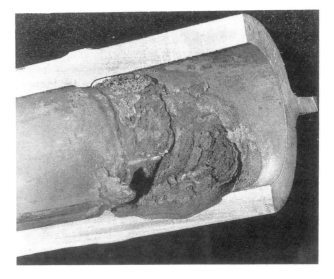

(a)

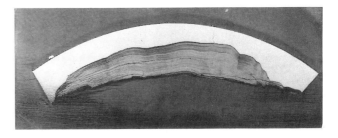

(b)

FIG. 2.20 Corrosion scab-laminated iron oxides
(a) Characteristic acid chloride attack on boiler tube
(b) Cross-section of corrosion products showing marked laminations

cap over the corrosion site, whilst the cyclic corrosion process takes place at the tube surface. A further frequently observed effect is hydrogen embrittlement or the decarburisation of the tube metal as a result of the hydrogen formed in the corrosion process converting the carbon in the steel into methane (CH_4) (Fig 2.21). Acid chloride attack can result in corrosion rates of several mm/year with the extent of decarburisation being more pronounced when high rates of corrosion have occurred. If very extensive decarburisation does take place then the ductile strength of the steel tubing is dramatically reduced and 'thick section failures' can occur. These can involve relatively large pieces being blown out of the tubing in a catastrophic failure (Fig 2.22), when only a fraction of the tube wall thickness has been consumed by the corrosive attack.

Because of the form of the corrosion product cap over the corrosion site and the cyclic corrosion reactions, the process, once established, can continue even if the boiler water composition is adjusted to have an excess of alkali as given in Table 2.7. When the boiler is taken off-load, the very restricted access of the bulk boiler water to the corrosion site may prevent the dilution of the concentrated aggressive chloride solutions, with the consequence that when the boiler is returned

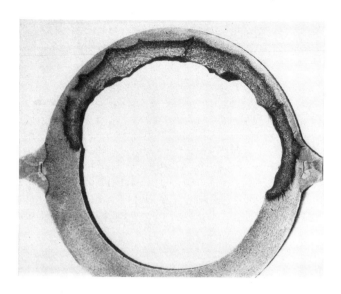

FIG. 2.21 Decarburised section of boiler tube

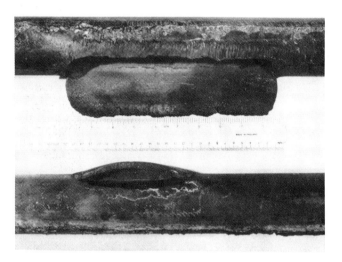

FIG. 2.22 Thick section failure due to decarburisation

to service the corrosion process restarts. It has been demonstrated that to arrest this form of corrosion it is necessary to hot soak or operate at low load with a boiler water containing a substantial excess of NaOH, which slowly penetrates into the corrosion areas and neutralises the acid chlorides.

Once this form of damage has become established in a boiler, it is necessary to remove all the corrosion products to ensure against a recurrence of the problem. This is best achieved by cutting out and replacing all the damaged areas of tubing. This is particularly important where hydrogen decarburisation has occurred as there is a high risk of further failures in affected tubing when the boiler is returned to load following tube replacement. The boiler is chemically cleaned to remove less obvious areas of corrosion products and accumulated porous oxides that could enhance concen-

tration of boiler water salts. This procedure is usually successful in halting the attack.

The chemical cleaning removes all the oxide layers from the tubing and subsequent operation with good boiler water allows adherent, protective oxides to reform.

As may be expected, this form of attack has appeared quite frequently at stations using sea water for cooling. However, it has also occurred at inland stations where control of boiler water chlorides has not been satisfactory and/or high oxygen levels in feedwater have persisted during boiler start-ups. The conditions leading to acid chloride attack can very readily occur in boilers being operated using AVT for boiler water pH control and great care has to be taken to maintain the required low levels of chloride in the boiler water. If difficulties are encountered in controlling chloride levels, solid alkali additions can be made to the boiler water for limited periods to allow the chloride concentrations to be reduced to within the normal operating targets.

6.2 Caustic attack

In common with all other soluble boiler water salts, sodium hydroxide can be concentrated-up by a hide out mechanism and, having a high solubility, can give rise to very concentrated solutions. Iron oxides have a significant solubility in such solutions and the protective oxide film is, effectively, dissolved away leading to the high corrosion rates indicated in Fig 2.9. In contrast to the situation with acid chloride attack, the dissolved oxides are carried away into the bulk boiler water leaving very little corrosion product in the corroding area. The overall effect is to produce relatively clean corrosion pits in the tube surface (Fig 2.23 'caustic gouging') although on occasion characteristic loosely packed Fe_3O_4 platelets are observed (Fig 2.24). Owing to the relatively ready escape of corrosion products from the corroding site, including any hydrogen produced, the underlying metal does not suffer decarburisation. Moreover, when the boiler is brought off-load, the concentrated caustic soda solutions are quickly diluted by bulk boiler water and the corrosion process is halted.

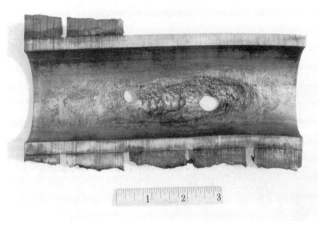

FIG. 2.23 Caustic corrosion gouging (Electric Power Research Institute)

FIG. 2.24 Metal oxide interface showing magnetite platelets (× 300)

This form of attack has also been observed in boilers operating with a sodium phosphate treatment regime. As indicated earlier, the sodium phosphates can hydrolyse to produce sodium hydroxide. If these reactions occur at the same time as sodium phosphate hide out, the overall result is a boiler water enriched in sodium hydroxide. The buffer action of the sodium phosphate is lost and caustic attack of the boiler tubes could result.

6.3 Stress corrosion cracking [21]

Another form of tube failures resulting from the localised concentration of dissolved salts in the evaporation sections is stress corrosion cracking. In addition to relatively high salt concentrations, stress corrosion requires the tube metal to be in a highly stressed condition. The stresses can be the result of operating conditions, residual stress in the tubing or stress resulting from structural constraints.

This type of failure is much less frequently observed than acid chloride or caustic attack, but can be more catastrophic as tube penetration can occur very rapidly. Although stress corrosion failures have been observed in mild steel, they occur more frequently in high strength alloys such as austenitic steels.

The attack takes the form of narrow, branching cracks that can penetrate through the tube wall via trans-

granular or intergranular pathways (Fig 2.25). Stress corrosion can be produced in acid or alkaline solutions, but reported incidents have more frequently involved concentrated alkali such as sodium hydroxide.

6.4 Scale formation

A 'scale' is a deposited layer of slightly soluble salt formed on a heat transfer surface when the solubility limit of the salt is exceeded. This can occur when boiler water, containing salts with retrograde temperature coefficients of solubility, comes into contact with boiler tube surfaces having temperatures substantially higher than the bulk boiler water. It can also occur when the prevailing boiler water chemistry leads to conversion of a soluble to a less soluble compound.

In well-run high pressure boiler plant, scale formation should not occur if good quality make-up water is used and the boiler water chemistry targets, set out in Table 2.7, are maintained. However, an understanding of the processes leading to scale formation is of value.

The formation of scale on a boiler tube would result in a loss of heat transfer and, consequently, a loss of boiler output. Moreover, if the thickness of the scale were allowed to increase it could result in serious over-

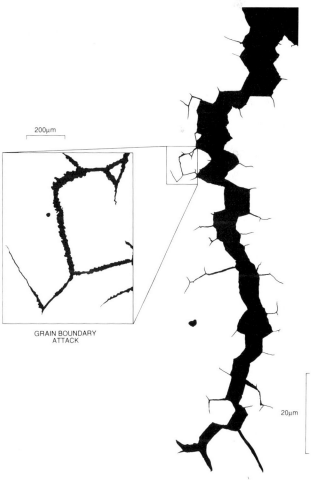

200μm

GRAIN BOUNDARY
ATTACK

20μm

FIG. 2.25 Intergranular stress corrosion cracking

185

heating and possibly failure of the boiler tube without any associated corrosion.

Scales very commonly consist of calcium and/or magnesium combined with sulphate, carbonate or phosphate and may also contain silica (SiO_2). These impurities enter the boiler as low level impurities in feedwater, largely as soluble salts, and slowly accumulate in the boiler water. If this process is allowed to continue, the solubility limits of a particular salt will be exceeded and that salt will then deposit. Precipitation may take place in the bulk boiler water, but as these salts have a retrograde temperature/solubility relationship, the deposition of the salt will take place preferentially on the heat transfer surfaces that are at a higher temperature than the bulk boiler water. Once a substantial scale has been formed on a boiler tube surface it can often only be removed by chemical cleaning of the boiler.

A particular advantage of using sodium phosphate as a source of boiler water alkalinity is that when ingress of calcium or magnesium salts occurs, the phosphate present rapidly converts them into relatively insoluble calcium/magnesium phosphates in the bulk boiler water. The salts precipitate as a loose flocculent material, rather than a scale, and can be removed from the boiler by blowing down. This is particularly useful in lower pressure boilers where the feedwater may not be of the highest purity.

7   Boiler tube corrosion and deposition — once-through boilers

7.1   Boiler tube corrosion by feedwater impurities

As indicated in Section 6, in a once-through boiler all the soluble impurities and debris present in the boiler feedwater will be retained in the boiler tubes or pass out of the boiler with the steam. In the evaporation zone of the boiler, the soluble impurities in the feedwater will concentrate in the residual liquid phase, until the concentration of the impurity in the steam equals its concentration in the feedwater, i.e., a dynamic equilibrium. Depending on the pressure the plant operates at, concentration factors can be as high as $10^5 - 10^7$. If the levels of impurities in the feedwater are excessive then, at that stage, the concentrated solution may be strong enough to attack the boiler tubing. Depending on the impurities present and the plant materials of construction, this could then lead to acid or alkaline corrosion or stress corrosion cracking as discussed previously in Section 6.

High concentrations of impurities would normally only be experienced in the final section of the evaporative zone. However, if porous oxide deposits are present over substantial areas of the tube surface, then hide out could occur with the formation of aggressive concentrations of impurities over much larger areas of the boiler tubing. It is therefore essential to maintain the very high quality of feedwater, as set out in Tables 2.11 and 2.12, to minimise the risk of corrosion. In plants where these standards have been achieved, the only salts found to accumulate in boiler tubes are low levels of $Ca/MgSO_4$, i.e., salts that have an extremely low steam solubility and are not aggressive.

When operational problems result in excessive quantities of impurities being fed to the boilers, it is good practice to take the boiler out of service and wash out the retained salts as soon as possible after the incident to minimise the corrosion risk.

Once-through boilers operating with a very high make-up rate (i.e., where steam is used in industrial processes and not condensed and recycled) have suffered corrosion from relatively low level impurities present in the make-up water in a non-ionic form and not removed in the water treatment or condensate polishing plants. Natural waters have been found to contain low levels of organic compounds, which on heating to $300-350°C$ in the boilers decompose to form sulphate, chloride and phosphate. If sufficient quantities of these compounds are entering the boilers then corrosion of boiler tubes can result [23]. Therefore, in operating high make-up plants, detailed and frequent monitoring of make-up water quality is essential.

7.2   Two-phase erosion-corrosion

Erosion-corrosion is a form of flow enhanced corrosion and can occur in regions of high turbulence on high liquid velocities, even when the feedwater/boiler water chemistry appears to be satisfactory. The high liquid velocities prevent the build-up of a sound protective oxide layer and consequently the underlying metal continues to corrode, with rates of several mm/year being observed. The problem can appear in single or two-phase (steam/water) sections of the plant and most frequently involves mild or low alloy steels. Two-phase erosion-corrosion has arisen in relatively low pressure (50 bar) once-through boilers, with mild steel serpentine boiler tubes. In these plants feedwater treatment was 'low oxygen/high pH' with pH control using ammonia. In the evaporation zone of the boiler tubes, the high relative volatility of ammonia, combined with its poor basicity at $260°C$, led to the residual liquid phase having a much lower pH than the feedwater (Fig 2.26). In the serpentine bends, phase separation occurred and the residual liquid phase was swept rapidly round the bends by the high velocity steam. The combination of low pH and high liquid velocities resulted in serious erosion-corrosion attack of the bends, with subsequent tube failures [13,23].

Where this situation has arisen in boilers with poor accessibility and it has not been possible to replace the mild steel tubing with more corrosion-resistant materials, the agent used for control of feedwater pH has been changed from ammonia to organic amines. The amines used (e.g., morpholine, 2-amino-2-methyl-propanol), are stronger bases than ammonia at the

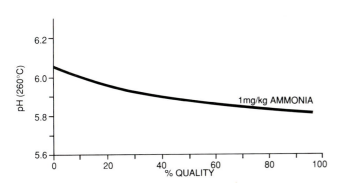

FIG. 2.26 pH versus steam quality in ammoniated boiler water at 260°C

operating temperatures, and have a much lower steam volatility. Consequently in the evaporation zone, the pH of the residual liquid films is maintained at a significantly higher value (see Fig 2.27) and the rates of metal wastage have been substantially reduced to rates consistent with required plant lifetime.

7.3 Debris deposition in boiler tubes

Metal oxides in the feedwater of once-through boilers will very largely deposit in the boiler tubes with some limited solubility in steam.

In high pressure plant, excessive deposition of debris on boiler tube surfaces can cause problems by increasing the risk of corrosion and by restricting boiler tube flows. This latter effect can occur to such an extent that boiler output is restricted because the boiler feed pumps cannot overcome the increased pressure drop in the boiler tubes. When this situation arises it is neces-

sary to take the boiler out of service and remove the deposited oxide by chemical cleaning the boiler.

In supercritical boilers, cuprous oxide in particular has a relatively high solubility in the supercritical fluid, which can then lead to serious deposition of copper and cuprous oxide on turbine valves and turbine blades. It is therefore particularly important that copper levels in feedwater to supercritical boilers are closely controlled. This can usually be achieved by avoiding the use of copper alloys in feed system construction, particularly in feedheaters, and by a careful choice of feedwater chemistry [24].

8 Boiler water chemical treatment for the recirculating steam generators used with pressurised water reactors

The treatment regimes adopted for the feedwater and steam generator water in pressurised water reactor (PWR) plant is essentially the same as described in the earlier sections of this chapter. However, because of certain features of steam generator design, serious problems have arisen in a number of plants and special consideration has to be given to the steam generator water chemistry regime used [25,26].

The steam generators operate at some 50–60 bar and initial water chemistries used paralleled those in use in conventional plant worldwide, with particular emphasis on the use of sodium phosphates in recirculating steam generators. The early designs of steam generators contained large numbers of crevices which were potential concentration sites, both at the main tube plates and also where the steam generator tubes passed through tube support plates. Insufficient control of feedwater pH and oxygen levels in early op-

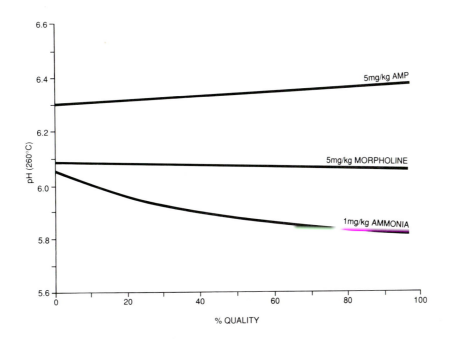

FIG. 2.27 pH versus steam quality in boiler water alkalised with amines at 260°C

eration led to considerable quantities of iron and copper oxides being carried into the steam generator and forming layers of sludge on tube plates and tube support plates. The combination of sludge and crevices resulted in extensive hide-out of boiler water salts and serious corrosion of the Inconel steam generator tubing, which in some cases has necessitated the complete replacement of steam generators.

Steps were then taken to improve the boiler water chemistries being used with a change to an 'all volatile alkali' form of treatment, very similar to that detailed in Table 2.7. This resulted in a substantial reduction in the sludge accumulation in the steam generators, but hide-out of salts still occurred in the tube to tube support plate crevices. 'Acid-chloride' type on-load corrosion of the mild steel tube support plates took place and the thickening laminated layers of magnetite so formed, crushed the steam generator tubing, a process described as *denting*.

Problems have also been encountered with forms of stress corrosion cracking in the bend areas of tubes in the steam generators and in the tube to tubeplate crevices.

In recent designs of steam generators more drastic steps have been taken to alleviate the problems, namely:

- Tube plates and tube support plates are designed in such a way as to minimise crevices.

- The support plates are fabricated from more-corrosion-resistant alloy steels.

- The feed systems are being constructed free from copper alloys.

- The steam generator water chemistries are now even more stringent than those used in conventional boilers operating on all volatile treatment regimes (Table 2.7).

- Feedwater quality has been much improved with low debris, ionic impurity and dissolved oxygen levels (Table 2.13).

- Some plants are now utilising amines such as morpholine for pH control to maintain high effective pH values at higher temperatures in all sections of the steam/water circuits.

The present practices for feedwater and steam generator water treatment at PWR stations are detailed more fully in Volume J.

9 Steam purity

9.1 Sources of impurities in steam

It will be noted in Table 2.7 that upper limits are set for sodium salts (Na^+) and silica (SiO_2) in steam. It is important that these limits are not exceeded, as a range of problems can result from deposition of sodium salts and silica in superheaters, reheaters and turbines.

TABLE 2.13

Typical secondary circuit water chemistry guidelines for on-load operation
Final feedwater

	Target	Expected value
Control parameters		
Direct conductivity at 25°C $\mu S/cm$	5.5 – 11	5.5 – 11
Cation conductivity at 25°C $\mu S/cm$	< 0.1	< 0.08
pH	9.3 – 9.6	9.3 – 9.4
Dissolved oxygen $\mu g/kg$	< 5	< 1
Hydrazine $\mu g/kg$	$3 \times O_2$ ≮ 20	20
Sodium $\mu g/kg$	< 2	< 0.5
Diagnostic parameters		
Ammonia $\mu g/kg$	750 – 2200	750 – 1050
Total iron $\mu g/kg$	< 5	< 5

Steam generator blowdown		
	Target	Expected value
Control parameters		
Cation conductivity at 25°C $\mu S/cm$	< 0.8	< 0.8
Sodium $\mu g/kg$	< 20	< 5
Chloride $\mu g/kg$	< 20	< 5
Sulphate $\mu g/kg$	< 20	< 5
Silica $\mu g/kg$	< 300	
Diagnostic parameters		
pH	9.0 – 9.5	9.0 – 9.5
Total iron $\mu g/kg$		<1000

In general terms the solubility of salts and silica in steam increases with increasing steam density, i.e., solubility increases with increasing pressure at constant temperature but decreases with increasing temperature at constant pressure.

The partition coefficient between steam and water (Kp) is defined as:

$$Kp = \frac{\text{concentration of solute in steam}}{\text{concentration of solute in water}}$$

and the overall relationship between Kp and steam density is of the form:

$$Kp = (\varrho_s/\varrho_e)^\eta$$

where ϱ_s = steam density and ϱ_e = water density under set conditions of temperature and pressure. Of particular interest is the partition under saturation conditions, and the changes in Kp (saturated) with pressure may be presented in the form of a ray diagram as shown in Fig 2.28.

If the boiler water impurity levels and sodium hydroxide limits given in Table 2.7 are not exceeded, then the equilibrium SiO_2 concentration in steam should

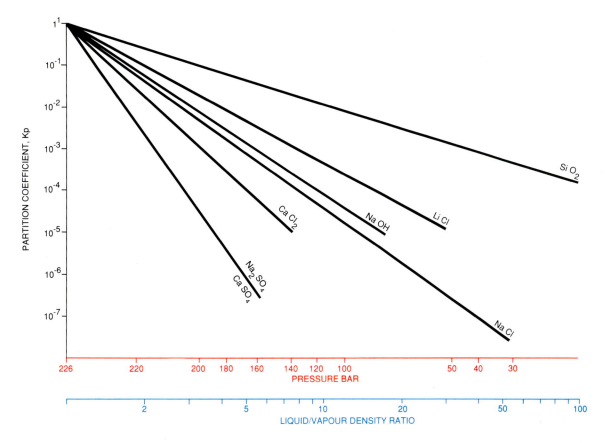

FIG. 2.28 Steam solubility ('ray diagram')

be < 20 $\mu g/kg$ and the Na^+ concentrations should be significantly lower than the set limits.

However, this is an ideal situation and impurities can enter the steam in at least three ways:

- By evaporation from the boiler water, as discussed.

- By entrainment of boiler water droplets in the saturated steam (i.e., carry-over), the salts then going forward into the superheater and turbine.

- As impurities present in feedwater used in desuperheater sprays or by leakage of boiler water into superheated steam in tubular desuperheaters.

Therefore in addition to controlling the boiler water chemistry, it is also necessary to determine the SiO_2 and Na^+ levels in saturated and/or superheated steam to ensure that the required steam purity is being maintained.

9.2 Problems arising from high impurity levels in steam

Excessive concentrations of SiO_2 or Na^+ in steam can lead to a number of problems ranging from reduced turbine performance to serious corrosion of superheater, reheaters and turbines. Some of these problems are discussed in the following sections and further information, particularly on deposition of salts

in turbines, can be found in reports of a number of detailed studies that have been carried out [9].

9.2.1 Silica in steam

As boiler plant operating pressures were increased it was observed, first in the United States, that silica (SiO_2) deposits could form on turbine blades, which in the more serious cases led to a significant loss of output. It was then necessary to take the turbine out of service to remove the deposits to restore generating capacity.

On investigation it was found that silica had a much higher partition coefficient than say, sodium chloride (see Fig 2.28), and consequently appeared in steam at much higher concentrations for a given concentration in boiler water. As this steam passed through the turbine, the sharp falls in steam density resulted in the steam becoming super-saturated, with SiO_2 then depositing from the steam onto the turbine blades.

A close study of the changes in SiO_2 solubility in steam over the range of temperatures and pressures encountered in operating turbines, showed that if the SiO_2 content of the superheated steam fed to the turbine was < 20 $\mu g/kg$, blade deposition would not occur at any stage of the HP, IP, or LP turbine. This figure is now very widely accepted as the safe upper limit for SiO_2 in steam (see Table 2.7).

The partition coefficient for SiO_2 was found to be substantially reduced by the presence of solid alkalis, particularly NaOH in the boiler water, probably due to the formation of silicates. From this data, and data obtained from boiler water containing no solid alkalis, working curves have been prepared indicating the maximum concentration of SiO_2 in boiler water that can be tolerated if the steam is to contain less than 20 $\mu g/kg$ SiO_2 (Fig 2.29). These graphs have been used to set the targets for SiO_2 in boiler water (Table 2.7).

When only minor deposition of SiO_2 has occurred, usually in the turbine final stages, it will normally be washed off the turbine blades by the condensate formed in the turbine during shutdown or start-up. However, when more extensive deposits have been formed it may be necessary to wash the turbine blades when the turbine is off-load; using warm, dilute alkali solution to re-dissolve the SiO_2 effectively.

Under normal operating conditions the SiO_2 levels in boiler water can be controlled by modest blowdown. However, if excessive ingress of SiO_2 is allowed to occur its control may require much more extensive blowdown with significant losses of heat and water from the system. This situation can readily arise when a boiler is being returned to service after an overhaul or outage for boiler tube repairs, as a result of silica-rich debris getting into the waterside of the boiler (e.g., insulating material, coal ash). In this case the boilers can be safely operated for restricted periods with SiO_2 in steam levels up to 50 $\mu g/kg$. However, if high SiO_2 levels persist, the boiler should be operated at reduced

pressure and load until the concentration in the boiler water has been reduced to within the set targets.

A further significant source of SiO_2 in boiler water in some cases, is material entering the system with the make-up water. In a demineralisation plant, soluble silicates in the raw water supply will be efficiently removed by ion exchange and very low levels of SiO_2 will be present in the demineralised water. Some waters, however, contain quantities of very finely divided solid particles of SiO_2-rich minerals such as clays or quartz (referred to as 'non-reactive' SiO_2) which are not efficiently removed by the normal water treatment processes and are carried forward with the make-up water. At boiler temperatures and pressure these compounds are converted into normal 'reactive' SiO_2 and can volatilise into the steam. Non-reactive SiO_2 will not be detected by the simple analytical techniques used to determine soluble SiO_2. If their presence is suspected from boiler water analysis, special analytical procedures should be used that determine 'total' SiO_2, i.e., both normal soluble silicates and the non-reactive forms.

9.2.2 Sodium salts in steam

As discussed earlier and illustrated in Fig 2.28, the equilibrium concentrations of sodium salts in steam are relatively low and should not result in the deposition of salts in superheaters and turbine in quantities that could lead to corrosion problems. However, even at these low levels some deposition/absorption of sodium onto the surfaces of superheater tubing does occur, the sodium becoming incorporated into the metal oxide layers without damaging the protective nature of these oxides.

If excessive levels of sodium are persistently present in steam, a number of problems could result depending on the actual levels of impurities present and their nature, i.e., whether it is predominantly sodium hydroxide or natural salts such as sodium sulphate.

Sodium hydroxide in austenitic superheaters — stress corrosion

The substantial increase in steam temperature produced in the superheaters results in a significant reduction in steam density and an accompanying reduction in the solubility of salts in the steam. This could lead to the steam becoming super-saturated and the contained salts being deposited on tube surfaces. If sodium is present in steam as sodium hydroxide, because of its very high solubility, this deposition could lead to the formation of concentrated solutions of sodium hydroxide on tube surfaces. At the temperatures that prevail in superheaters these solutions would be very corrosive and in austenitic tubing could result in rapid and catastrophic failures from stress corrosion cracking.

Stress corrosion of austenitic and high chromium content ferritic alloys is a well documented phenomenon, but in spite of extensive investigations the mechanism and range of this form of attack have not been fully elucidated. For stress corrosion to occur the metal

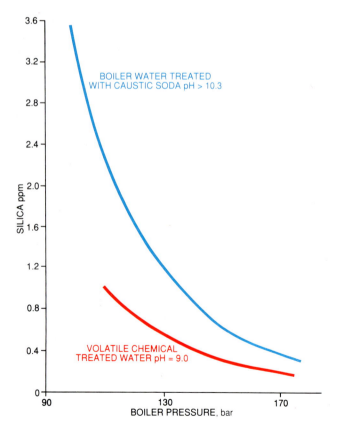

FIG. 2.29 Levels of silica in boiler water at different pressures to ensure not more than 20 $\mu g/kg$ in steam

has to be in contact with a relatively aggressive environment, e.g., concentrated solutions of NaOH or NaCl, particularly in the presence of oxygen. In addition, the area of the metal at risk has also to be in a stressed condition, for example, residual stresses from fabrication or stresses arising from operating conditions.

The situation is further complicated by the corrosion taking place in two different forms:

● Transgranular cracking, in which the stress corrosion crack advances into the bulk metal across the metal grains.

● Intergranular attack where the corrosion advances down grain boundaries.

The latter form of attack is more frequently observed in so called 'sensitised' material, in which in-service or fabrication heat treatment has led to a modification of the alloy with precipitation of chromium-rich metal carbides in the grain boundary areas. This precipitation leads to a depletion of chromium in the metal immediately adjacent to the grain boundaries, which is then more susceptible to corrosive attack by aggressive salt/sodium hydroxide solutions.

When conditions favour stress corrosion, the rates of penetration of the metal can be very rapid indeed. Considerable care should therefore be taken to avoid gross contamination of steam, particularly by sodium hydroxide.

Deposition of sodium salts in turbines

As superheated steam passes through the HP turbine, there are rapid decreases in temperature and pressure which produce a significant fall in steam density. This results in a sharp reduction in the solubility of solutes in the steam and the possibility of deposition on the metal surface in the turbine. The deposition of sodium hydroxide or sodium chloride could, as in superheater deposition, lead to the formation of aggressive environments and the risk of stress corrosion failures.

The deposition of sodium sulphate would not be expected to result in an enhanced risk of corrosion in turbines because it would be present as a dry neutral salt. However, it can contribute to corrosion problems in reheaters as indicated below.

In general because of the sharp fall in steam density on passing through the turbine, deposition could occur at significantly lower concentrations of salts in steam than would be necessary to produce superheater deposits. Therefore there is a risk of turbine corrosion from deposition if Na^+ levels in steam persistently exceed the set limit but without gross contamination of steam occurring.

In the HP turbine, deposition is generally slight and limited to salts with very low solubility in steam, e.g., sodium sulphate. In the lower pressure stages, more soluble salts can be progressively deposited and there have been a significant number of incidents of sodium hydroxide and sodium chloride deposition which, in some cases, have led to corrosion failures [9].

Sodium sulphate deposition in reheaters

Figure 2.28 indicates that the equilibrium concentration of sodium sulphate in saturated steam is very low, and can therefore be disproportionately increased by quite low levels of boiler water carry-over. As the steam leaves the HP turbine and passes through the reheaters, the sharp decreases in steam density and hence solubility continue. This effect has resulted in the deposition of quite significant quantities of sodium sulphate in reheaters from steam containing relatively low levels of impurities (i.e., $1-2$ $\mu g/kg$ Na_2SO_4) [27]. The deposited sodium sulphate, a neutral salt, is essentially innocuous and there is no evidence of on-load corrosion resulting from this source. However, when the unit is taken off-load and allowed to cool down, condensation will frequently occur in the reheater leading to the re-solution of the deposited salts. If the reheaters are of the pendent design the condensate formed will then drain down to the bottom bends taking with it the sodium sulphate in solution. (Sodium sulphate deposited in HP turbines will also form salt solutions in off-load periods which could subsequently be carried forward into the reheaters.)

On return to service this sodium sulphate solution will re-evaporate leaving the dry salt in the bottom bends. This cycle, of on-load deposition, off-load re-solution and accumulation in the bottom bends of the reheaters, occurs each time the unit is put into and taken out of service, and may ultimately lead to quite high concentrations of sodium sulphate in the solutions formed off-load. When air is admitted to the reheater, pitting attack of the reheater tubes can then take place, enhanced by the conductivity of the sodium sulphate solution, and lead to widespread reheater tube failures (Fig 2.30). There is also some evidence that aerated solutions of sodium sulphate can lead to low temperature stress corrosion of austenitic alloy tubing.

A number of steps can be taken to avoid such problems, the most direct being to limit the concentration of sodium and sulphate in steam by operating with boiler waters as pure as practicable and avoiding carry-over. If, nevertheless, the formation of such deposits is suspected, then the alternative approach is to prevent the formation of condensates when the units are taken out of service. In modern, large turbines this can be achieved by using the forced air cooling system to displace the residual steam/water vapour out of the turbine and reheater before condensation can occur. If the unit is to be out of service for a significant period of time, the turbine and reheater should be stored in dry air to prevent the subsequent absorption of moisture and the possible formation of sodium sulphate solutions (see Section 11 of this chapter).

10 Corrosion in condensate and feed systems

Corrosion does occur in condensate and feed systems but, in general, the consequences are not as potentially

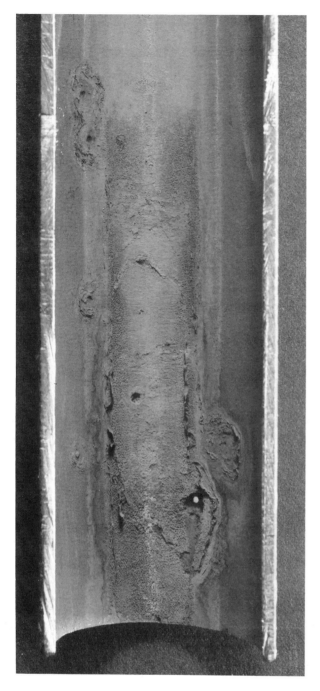

FIG. 2.30 Reheater corrosion due to salt deposition

serious as corrosion in boilers. Problems have arisen from two main causes:

- Lack of satisfactory control of feedwater chemistry, particularly when bringing plant back into service.

- Design features that lead to corrosion even when operating with good quality feedwater.

10.1 Corrosion resulting from poor quality feedwater

Proper control of pH and oxygen levels in condensate and feedwater will minimise corrosion of the plant

and also minimise debris levels in the feedwater passing forward to the boilers. The degree of control required varies with the materials of construction of the system.

10.1.1 Systems containing copper-based alloys

When copper-based alloys are present in feed and condensate systems, close control of pH and oxygen levels are required to avoid corrosion (see Table 2.10). Excessively high or low values of pH, particularly in the presence of high oxygen levels, will result in an increase in general corrosion rates and it is essential to minimise these effects. This is particularly important in the feed systems of once-through boilers.

Problems most commonly arise during plant start-up when control of pH and oxygen levels can be difficult. This is illustrated by instances of corrosive attack on copper-based alloy feedheater tubing (Fig 2.31). During the early stages of the start-up, the feedwater to the boilers can be relatively high in oxygen and carbon dioxide resulting from air ingress into the system, and relatively low feed system temperatures limiting the degree of deoxygenation that can be achieved in the de-aerator. When this feedwater enters the boiler, the oxygen and carbon dioxide are preferentially removed with the steam in the drum and pass into the turbine and then into the steam side of the feedheating system. Under these operating conditions, the initial condensates formed on the steam side of the feedheater tubes are relatively low in ammonia (and pH) and high in oxygen and carbon dioxide. This combination can lead to a much enhanced attack of the copper-based alloy tubing, with an associated risk of tube failures. As drains from the heaters are normally returned to the feed system, such corrosion also results in high levels of copper in feedwater being fed to the boilers.

The initial corrosion reaction is $4Cu + O_2 \rightarrow 2Cu_2O$ and frequently the steam side of the feedheater tubing is found covered in relatively thick layers of loose non-protective copper oxides. In cupro-nickel feedheaters, particularly those with lower nickel content (e.g., 90/10, Cu/Ni), a secondary reaction can take place leading to further corrosion of the tubing, i.e.,

$$Cu_2O + 2Ni \rightarrow 2NiO + 2Cu$$

In this process, referred to as *exfoliation corrosion*, the copper metal forms very thin films on the outside of the nickel oxide layer that readily become detached and are carried away with the heater drains into the feedwater.

10.1.2 Systems free of copper-based alloys

Where the condensate and feed systems are constructed from mild steel, alloy steels and titanium, condensate/feedwater pH levels of greater than 9.2 will give enhanced protection to the system and modest levels of oxygen, i.e., up to 50 μg/kg, can also assist the development of stable protective oxide films. However, care

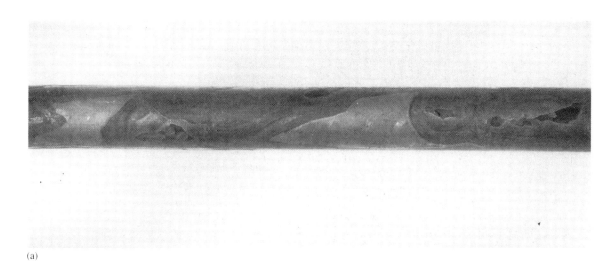

(a)

(b)

(c)

FIG. 2.31 Copper alloy feedheater tube corrosion
(a) Meniscal corrosion contours on an otherwise unaffected surface. Note the exfoliation of corrosion product
(b) Relief effect on steam side surface at bottom of the element
(c) Well defined transition boundaries and on the lower section a groove on the outer surface coincident with a tubeplate
(see also colour photograph between pp 208 and 209)

must be taken to ensure that the oxygen in feedwater does not appear in the boiler water and lead to an increased risk of 'on-load' corrosion. This can be achieved by having additional quantities of chemical deoxygenating agents (e.g., hydrazine) present in the feedwater when high oxygen levels are encountered (see Table 2.10).

Low feedwater pH values (< 8.80) will increase the overall corrosion rates with a corresponding increase in iron oxide levels in the feedwater. However, these corrosion rates are still relatively low and would not threaten the integrity of the system unless adverse design features were present.

10.2 Corrosion problems resulting from feed system design features

Corrosion has occurred in condensate and feed systems, operating with good quality feedwater, as a consequence of hydrodynamic features unwittingly designed into the systems. Such effects are well illustrated by considering two particular problems, one of which has affected copper-based alloy condenser tubes, and the other, various mild steel components in feed systems.

10.2.1 Condenser tube corrosion

Corrosion of copper alloy condenser tubing (e.g., admiralty brass) due to high ammonia levels has occurred in the non-condensable gas off-take zones of main condensers, even when the bulk feedwater/condensate has been of good quality with satisfactory ammonia levels. In this zone of the condenser, unusual condensation conditions prevail such that with steam entering the condenser containing some 0.5–1.0 mg/kg ammonia, local condensates are formed containing some 200–300 mg/kg ammonia. In the presence of the relatively high oxygen levels that prevail in the non-condensable gases, this concentration can lead to a vigorous corrosive attack of the condenser which results in tube failures.

The problem has been resolved in several ways:

- Using condenser tubes of more resistant material (i.e., 70/30 cupro-nickel or stainless steel, instead of brass).

- By the introduction of bulk condensate sprays in the vulnerable area of the condenser, preventing the formation of high ammonia concentrations.

- Designing the non-condensable gas off-take zone in a way that avoids the hydrodynamic conditions that lead to the problem.

The last approach is that used in large, modern power plants.

10.2.2 Erosion-corrosion of mild steel feed system components

Erosion-corrosion, a form of flow-enhanced corrosion of mild and low alloy steels, has affected a large

number of plant feed systems [28]. It has probably resulted from the increased flow rates and increased water velocities used in large power plants, together with the need to use low oxygen feedwater for high pressure drum boilers.

When a mild steel surface is exposed to water the initial reaction is:

$$Fe + 2H_2O \rightarrow Fe^{2+} + 2OH^- + H_2$$

which is then followed by:

$$3Fe^{2+} + 4H_2O \rightarrow Fe_3O_4 \text{ (magnetite)} + 4H_2$$

with the magnetite forming a protective oxide layer on the metal surface.

However, the rate of this second reaction is very dependent on temperature, pH and oxygen concentrations. In low temperature, low oxygen feedwater this reaction proceeds relatively slowly. In areas of high flow rate or local turbulence, the product of the first corrosion reaction (i.e., Fe^{2+}) is carried away from the metal surface before the second reaction can take place. Consequently the protective oxide layer does not develop and the underlying metal can continue to corrode at a relatively high rate (e.g., 2–4 mm/year loss of tube wall).

This form of attack has been found in the single-phase sections of the feed system (i.e., when only water is present) at bends, in turbulent zones downstream of flow control orifices and at mild steel feedheater tube inlets (Fig 2.32). It has also been found in two-phase sections of feed systems, e.g., where steam and water are present on the bled steam side of feedheaters.

Fig. 2.32 Single-phase erosion-corrosion of mild steel

The corroded areas all show the characteristic features of this form of attack; large numbers of smooth, scalloped pits, covered with a shiny black oxide.

In the single-phase situation, the rate of corrosion varies quite sharply with temperature, with a peak rate in the region of 130–180°C (Fig 2.33). The rate also varies with the pH of the feedwater (Fig 2.34) and is also very dependent on the oxygen content.

In two-phase sections of plant, erosion-corrosion has been found at temperatures up to 260–280°C and this is probably partly due to the preferential loss of ammonia (or other volatile alkali) into the

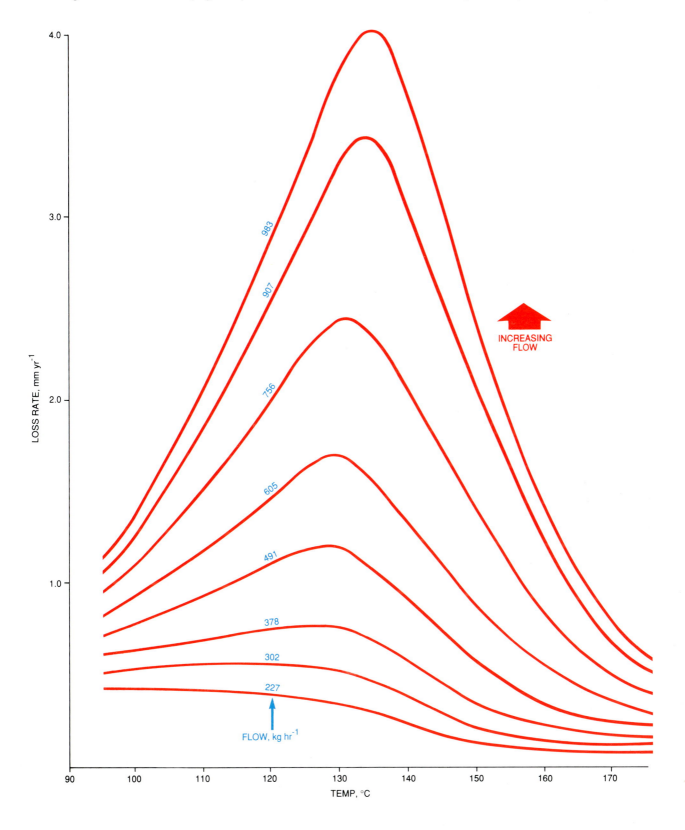

FIG. 2.33 Flow temperature dependence of erosion-corrosion rates

steam, leaving the residual water at a relatively low pH.

This form of corrosion has caused serious problems in a number of plants and several approaches have been used to correct the situation:

(a) Erosion-corrosion particularly affects mild steel. It has been found that the presence of low levels of Cr (i.e., $\geqslant 1\%$) in the steel are sufficient to reduce the rate of attack dramatically. Consequently areas of plant where this form of attack may occur are now being fabricated in 1% Cr or $2\frac{1}{4}\%$ Cr steels.

(b) Figure 2.34 indicates that increasing the feedwater $pH_{(25)}$ to 9.6 or above will significantly reduce but not completely halt the attack. This approach has been used with considerable success in feed systems free from copper-based alloys. When using volatile alkalis in two-phase situations the problems of maintaining higher water pH values are particularly difficult as:

● Volatile alkalis tend to become weaker bases at higher temperatures.

● Most volatile alkalis at present used in feedwater treatment, tend to fractionate preferentially into the steam phase, leaving the water short of alkali (see Section 5 of this chapter).

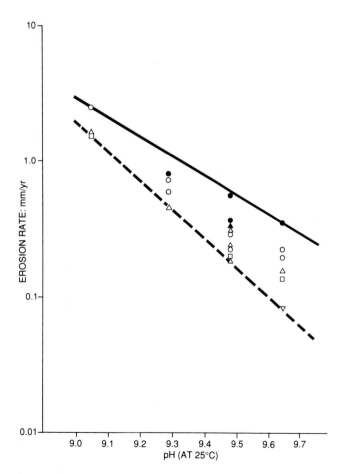

FIG. 2.34 Erosion-corrosion rate versus pH at 140°C

Both these problems have been overcome in the steaming economiser sections of some drum boilers by recirculating a small proportion of solid alkali-dosed boiler water back to the economiser inlet. The presence of the involatile solid alkali has successfully maintained a high pH in the liquid phase of the two-phase regions and halted the erosion-corrosion without increasing the levels of solid alkali in the boiler water.

(c) As indicated, erosion-corrosion appears to be a result of the slow formation of magnetite (Fe_3O_4) or haematite (Fe_2O_3) at relatively low temperatures (i.e., $< 280°C$) in low oxygen feedwater. These reactions can be greatly accelerated by introducing relatively low levels of oxygen into the feedwater (see Section 5 of this chapter). However, care must be taken to ensure that this oxygen is removed before the feedwater enters the boiler drum to avoid an increased risk of on-load corrosion. This is achieved by dosing additional quantities of de-oxygenating agent (e.g., hydrazine) to remove the oxygen in the higher temperature sections of the feed system and economiser.

11 Standby storage of steam/water circuits

Until a few years ago, plant storage had been a low priority activity in the overall strategy of power plant operations. More recently it has been increasingly recognised that to maintain high plant availability, it is essential to use good storage techniques during off-load periods [29].

Whilst the plant is in service, the on-load boiler water and feedwater treatment protects the steam/water circuit materials from corrosion by controlling oxygen concentrations, pH and impurity levels in the steam and water. However, when plant is taken out of service and cooled down, air may have to be admitted and a different situation prevails. The relatively low levels of alkalinity used in the circuits for on-load chemical control, will be rapidly neutralised by the carbon dioxide present in the air, producing a low pH solution, and the dissolved oxygen concentrations will rapidly increase. Under these conditions, mild and low alloy steels will readily suffer from corrosive attack, these effects being enhanced by any Cl^- or SO_4^{2-} present in the water. This was illustrated earlier when discussing the problem of reheater corrosion. A similar problem can arise when plant is drained down for maintenance work, leaving small pools and droplets of water on tube surfaces. A pool or droplet of water with a significant conductivity, probably containing some Cl^- and/or SO_4^{2-} in a differential aeration situation, represents an ideal system for producing off-load corrosion (Fig 2.35). The electrochemical cell set up by this system has the effect of accelerating the reaction:

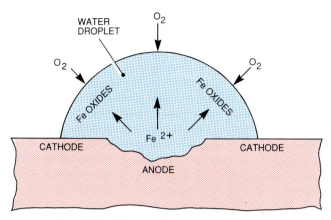

FIG. 2.35 Differential aeration corrosion at water droplet

$$Fe_{(metal)} \rightarrow Fe^{2+} {}_{(soluble)}$$

in the low oxygen zone beneath the water surface. On diffusing away from the metal, the Fe^{2+} species will oxidise further to produce magnetite and haematite which at low temperatures precipitate in a non-protective form. With a continuous supply of oxygen from the air in contact with the water, this reaction sequence can continue indefinitely. A special instance of this situation is where the metal surface is partly immersed in water, leading to the frequently observed preferential 'water line' attack, the resultant corrosion usually taking the form of steep-sided pits just below the water surface. On returning the plant to service the pits may be cleaned out, but, on redraining in a subsequent outage, they can become prime sites for further off-load attack.

Depending on the contaminants in the retained water and the plant operating conditions, it is also possible that corrosion, initiated off-load, may continue when plant is returned to service. This would be particularly true if the off-load pitting occurred in areas of high heat flux when the plant was on load.

A further feature of this problem is that although much or all of the corrosion occurs during off-load periods if not detected, the actual tube failures normally occur on-load as a result of the increased temperatures and pressures.

To avoid such problems, it is necessary to have well developed storage procedures and to have available the necessary additional equipment and plant modifications for their satisfactory application.

11.1 Plant storage procedures

Plant storage procedures should be used at all stages of the plant's operating life, including commissioning.

In selecting a storage regime, account must be taken of the limitations and demands imposed by the design of plant and the requirements for introduction/return to service. The major points to be considered are:

- Duration of off-load period.

- Required speed of return to service.

- Whether or not the plant can be, or is to be, drained during the outage.

- Availability of supplies of good quality water for refilling emptied plant.

- Materials of construction.

- Availability of specialised storage equipment.

For a particular plant, the final choice of storage procedure will require a balance of the above requirements, some of which may conflict with each other. A typical 'decision tree' used to assist in making such a choice is given in Fig 2.36.

11.1.1 Short/medium term storage plant — draining not required

Nitrogen capping of boilers

A frequent post-operational storage situation is that in which boiler plant is to be out of service for a short/medium period, possibly up to six weeks, and the plant does not have to be drained. In this case the preferred storage system is that referred to as *nitrogen capping*, which involves the injection of nitrogen into the steam spaces of the boiler to prevent air ingress as the temperature and pressure fall. The normal procedure would be to take the boiler off-load, allow the pressure to fall to almost atmospheric by natural cooling, and then apply a low pressure oxygen-free nitrogen supply to the only open air cock on the system. As the steam condenses, nitrogen is drawn into the plant, thus preventing the entry of oxygen and carbon dioxide. For satisfactory protection against corrosion it is necessary to hold the oxygen content of the nitrogen in the plant to less than 1 vol.% O_2. This can readily be achieved with well designed equipment.

The nitrogen can be supplied via a permanently installed system with suitable isolation at the air cock and a low pressure relief valve on the nitrogen supply side. Alternatively, in a simpler system, the nitrogen is supplied via a quick-action gas coupling on the selected air cock, the connection being made when plant pressure has just reached atmospheric. In applying this technique it is essential that the safety aspects of using a high pressure gas supply are satisfactorily covered.

As with any plant protection regime, to be successful it must be used correctly. If the plant leaks badly in the steam or water spaces then difficulties can arise from excessive consumption of nitrogen, or the need to keep injecting aerated water into the system to maintain water levels.

Nitrogen capping has the clear advantage that the boiler is not drained and the contained water remains close to that of normal boiler water or feedwater in composition. This means that the plant can very rapidly be returned to load, if required. The effectiveness of the procedure has been confirmed by boiler water

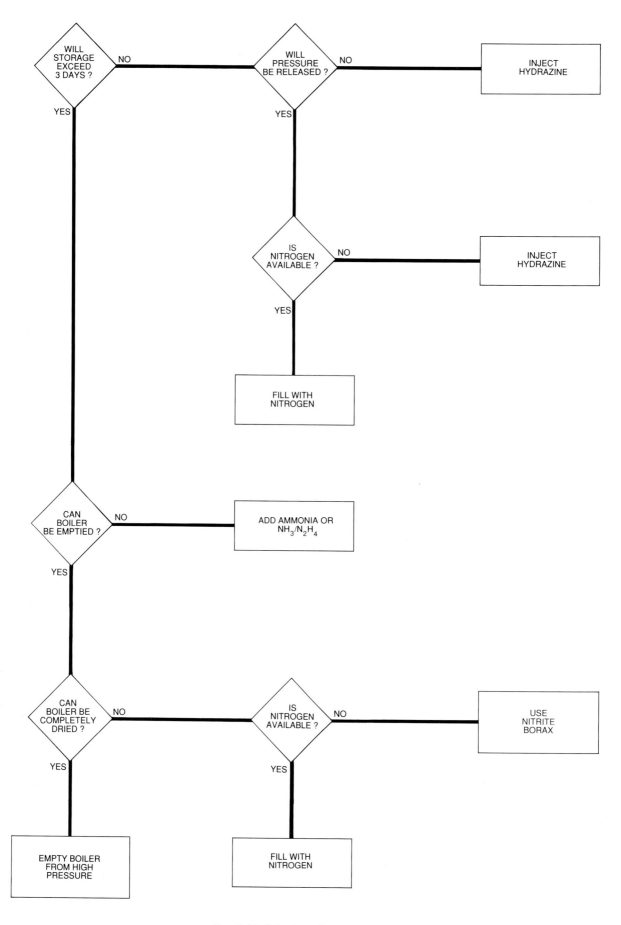

Fig. 2.36 Selection of storage processes

analyses and from the condition of test specimens inserted into boiler drums. This is the most satisfactory storage procedure that does not involve draining of the plant.

Wet storage of boilers

(a) *Using ammonia (NH₃) and hydrazine (N₂H₄)* An alternative storage procedure is to fill the cold boiler to the air cocks with a suitable wet storage solution. The favoured method is to drain the boiler down and refill with a solution of NH_3 or NH_3/N_2H_4. The former should contain about 300 PPM NH_3 and have a pH of above 10.5; the latter would normally contain 50 PPM NH_3 and 50 PPM N_2H_4, with a pH of greater than 10.0. Ideally, both solutions should contain less than 2 PPM NaCl, although there is some plant experience to suggest that much higher levels of NaCl can be tolerated.

If suitable dosing pumps and boiler circulation systems are available, these storage solutions can be generated from normal boiler water without the need to drain and refill the plant.

Solution of these compositions will protect steel components for very long periods, even in initially fully-aerated water. However, for continued protection it is necessary to check and maintain the pH at the required level. The storage solution should be regularly circulated and analysed, and if short of alkalinity, more NH_3 or NH_3/N_2H_4 should be injected or the plant should be drained down again and refilled with a solution of the correct composition.

A successful method of applying this procedure is to overfill the boiler and have a small reservoir of storage solution connected to the only open air cock. This has the effect of maintaining a reserve of storage solution such that leakage from the plant will quickly be evident, whilst further air ingress is avoided.

Wet storage is a method of storage which requires rather more attention than other procedures. It also has several other shortcomings:

● The plant has to be drained and refilled before returning to service, to remove the excessive amounts of ammonia and hydrazine.

● In cold weather, it may be necessary to supply some heat to the plant to avoid freezing.

● High concentrations of NH_3 should not be used to store sections of the plant containing copper based alloys.

(b) *Using borax/nitrite* An alternative wet storage solution is one containing 200 PPM sodium borate/200 PPM sodium nitrite, and this particular composition is satisfactory for storage of mixed metal plant items. The borate/nitrite approach has a much broader potential application and can be used with most alloys to be found in a power plant system. The procedures are discussed in detail as follows.

Feed systems and turbines

As far as feed systems and turbines are concerned, no specific protection is normally applied in the short to medium term. The storage of these sections of the plant would be complex and it is difficult to justify extensive procedures for limited periods. In general, feed systems are stored full of normally-dosed feedwater, water-wedged if possible. Turbines would receive no special treatment apart from natural ventilation. However, where there is a ready supply of oil-free dry compressed air (e.g., forced air cooling), then the drying out of the turbine and condenser can be assisted by passing air through the plant. Drying of condenser and LP turbines can also be accelerated by drawing vacuum on the condenser.

11.1.2 Short/medium term outage — plant to be drained

Dry storage — all sections of the steam/water circuit

In many instances it is necessary to drain off-load plant for plant overhauls, tube repairs, etc. In this situation the preferred option would be to use a dry storage procedure where possible. The plant would be blown empty from as high a pressure as practicable (possibly up to 30 bar) and, once the plant is empty, all vents would be rapidly opened to promote air circulation and to evaporate any retained water. This procedure has limitations in that the residual heat in the plant is frequently insufficient to complete the dry-out. There is also a serious risk of condensation occurring when the plant goes cold, therefore, if possible, the drying should be assisted by the circulation of dry air through the plant or by the use of low level heat.

The main danger areas for retention of water are pendant tubes or horizontal sections of tubing or vessels (e.g., economisers, superheaters). In boilers, considerable accumulated wastage has also been observed in wall tube headers in plant using the blowdown emptying technique without the use of supplementary drying procedures, due to water retention in the headers.

If a satisfactory dry storage system can be established, it is by far the most convenient and flexible procedure to be used where plant has to be opened to the atmosphere for prolonged periods.

A particular disadvantage of dry storage is the need to refill the plant before it can be returned to service, which can make considerable demands on water supplies and cause a significant delay in re-commissioning. A further disadvantage is that the feed systems of such boilers are normally stored full of water and 'passing' valves may result in ingress of water into the dry sections of the plant.

Fill and drain procedures — boiler and feed systems

As already indicated, complete drying of steam raising plant can be difficult and an alternative approach is a 'fill and drain' procedure using a solution containing 200 PPM sodium borate/200 PPM sodium nitrite. This approach accepts that there will be a significant quantity of residual solution left within the plant but containing non-volatile inhibiting chemicals. Extensive testing has demonstrated that corrosion will not take place for at least six weeks on exposure to the atmosphere under these conditions. The full procedure is to drain the plant, hot or cold, refill with the borate/nitrite solution, circulate and drain again. This procedure has the advantage that if the moisture evaporates, the inhibiting chemicals and the protection they give are not lost.

Sodium nitrite is an anodic inhibitor and it is essential that the concentration of nitrite is sufficiently high to produce the required inhibition. If the concentration is too low, there is a danger of actually accelerating the corrosion processes. The target should be to maintain an absolute minimum of 100 PPM sodium nitrite, and for longer term storage, concentrations of up to 1000 PPM have been used successfully.

The most satisfactory method of applying this technique is to prepare the inhibitor solution in an external tank and then pump into the plant. Extra tanks, or selected reserve feedwater tanks can be used for holding the borate/nitrite solution when not required. This arrangement has the advantages of having the solution immediately available when required and, when drained from the plant, the solution can be returned to the holding tanks, thus avoiding problems of effluent discharge, and reducing chemical costs. As already noted, a further advantage with this particular 'fill and drain' procedure is that the solution can be safely used in mixed metal sections of the plant.

The major disadvantage is that before returning the plant to service, it is necessary to drain and rinse thoroughly to reduce residual nitrite to acceptable concentrations, which can usually be achieved by completely filling and draining the plant twice with good quality water. This can make heavy demands on make-up water supplies. It is also advisable to avoid this solution getting into the superheater sections unless there is complete confidence that the residual borate/nitrite can be effectively removed by flushing or rinsing before return to service.

Nitrogen filling — boilers and feed systems

Another approach involves draining the plant, with oxygen-free nitrogen being drawn into the spaces created, thus replacing water with nitrogen and excluding air. Comprehensive nitrogen supply systems have been installed for the storage of boilers and feed systems, particularly in nuclear plant, and successful longer term storage has been achieved. As with nitrogen capping discussed previously, it is necessary to maintain the oxygen content of the nitrogen to < 1 vol.% O_2. This requires regular analysis of the nitrogen in the plant as gas purging is necessary if oxygen levels begin to rise.

This method of storage can involve the use of quite large volumes of nitrogen. When the plant is returned to service this nitrogen may be displaced into buildings surrounding the plant and particular care should be taken to ensure that this does not create a hazard for plant operators.

11.1.3 Long term storage

The long term storage of plant (i.e., for periods greater than two months) is obviously a much more complex situation as consideration has to be given to all sections of the plant including electrical and mechanical equipment, as well as the pressure circuit. As already indicated, 'wet storage', using ammonia solutions and possibly nitrogen filling, can be used for storage over quite extended periods. However, the first choice for the storage of boiler, turbines, condensers and feed systems would be 'dry storage' with dry air circulation.

When the plant is taken out of service boilers and feed systems should be blown down as soon as possible and header caps and air vents opened to promote air circulation. Turbines and reheaters can be blown free of residual steam using forced air cooling systems if available. Special air drier units are then used to circulate dry air through all sections of the plant continuously throughout the period of storage (Fig 2.37). When the plant is initially blown empty, it is very probable that small volumes of water containing impurities such as Cl^- and SO_4^{2-} will be retained in the plant. Therefore the relative humidity in the plant must be kept at < 30% to prevent retention of water by hygroscopic salts to maintain a dewpoint 10–15°C below the lowest temperature in the plant being stored.

Dry air is preferred to warm air circulation as it is more effective in drying out plant, is an easier system to sustain over long periods and uses much less energy to produce the desired effects. To apply this procedure successfully requires careful consideration for the most effective positioning of the driers and to have available suitable connection points to the plant. In practice, this procedure has proved very effective in storing plant over prolonged periods off-load.

In some plants considerable success has been achieved using very much simpler, if less reliable systems, e.g., by combining high pressure blowdown with the introduction of trays or sacks of drying agents, such as soda lime, or silica gel. It is necessary to replace the desiccants at regular intervals but, with care, satisfactory storage over extended periods has been achieved. When the storage period ends, it is essential to ensure that all the desiccant is removed from the plant before it is refilled and returned to service.

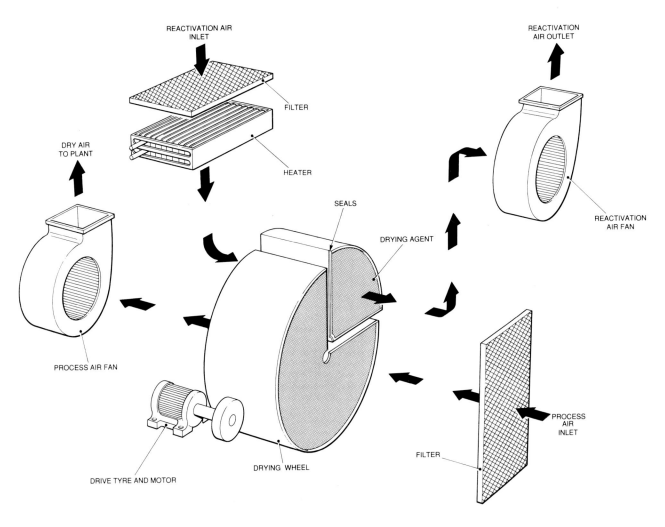

REACTIVATION AIR
INLET

FILTER

DRY AIR
TO PLANT

HEATER

SEALS

DRYING AGENT

REACTIVATION
AIR OUTLET

REACTIVATION
AIR FAN

PROCESS AIR FAN

PROCESS
AIR
INLET

DRIVE TYRE AND MOTOR

DRYING WHEEL

FILTER

FIG. 2.37 Dehumidified air supply system

12 Chemical monitoring — steam/water circuit

12.1 The need for detailed chemical monitoring

For the close chemical control needed to protect steam/ water circuits of power plant, detailed information on the prevailing chemical conditions is essential. In recent years, manual sampling and laboratory analysis have given way to the extensive use of on-line instrumentation for most, if not all, the important chemical measurements. This approach benefits the operator by providing him with continuous rather than intermittent information on chemical conditions in the circuit, and has the added advantage of reducing the risk of sample contamination inherent in manual sampling and analysis.

At modern power stations chemical measuring instrumentation is employed:

- To monitor the chemical conditions in the steam/ water circuit to ensure that satisfactory chemical control is being maintained and to alert operators to malfunctions.

- To monitor the operation and performance of plant installed to facilitate chemical control, e.g., the water treatment, condensate polishing plants and chemical dosing plant.

- To control the chemistry of auxiliary circuits, e.g., stator cooling circuit.

- To facilitate chemical investigations when unexpected problems arise.

The chemical data provided by on-line instrumentation can be displayed locally or relayed elsewhere, e.g., to the control room. The chemical conditions can then be judged against the criteria laid down for plant operation and, if necessary, appropriate corrective actions taken. More recently the operator's task has been simplified by the use of computers for the processing and presentation of chemical data and for raising alarms. This approach has been extended to provide automatic control of plant, notably the regeneration of water treatment and condensate polishing plant resins. More comprehensive systems of automatic chemical control have been developed, particularly for control of boiler water chemistry [30]. However, it is essential that such systems are based on reliable instrumentation and well developed control philosophies.

With the continuing development of more on-line chemical instrumentation and versatile data processing techniques, manual methods of analysis have been relegated to a supporting role at most modern power stations. They are not, therefore, included here. Further information on methods currently in use is available in a number of publications [31].

12.2 Monitoring requirements for steam/water circuit chemistry

The chemical monitoring required for steam/water circuits will vary according to the particular type of plant considered. However, the principal sampling locations and parameters to be measured are shown in Figs 2.38 and 2.39. They are:

- *Make-up water treatment plant* Measurements of direct and after-cation conductivity (see Section 12.4.2 of this chapter), reactive silica and sodium to monitor the performance of individual stages of the plant, to give warning of resin bed exhaustion and ensure the required outlet water quality is being maintained.

- *Condenser* Direct conductivity measurements by means of probes installed at various locations within the condenser shell or, alternatively, in indi-

vidual hot-wells to assist in the location of condenser leaks.

- *Condensate extraction pump discharge* Direct and after-cation conductivity measurements, supplemented by sodium monitoring, to provide warning of condenser leakage. Dissolved oxygen measurements are also required in order to establish the adequacy of oxygen removal at the condenser.

- *Condensate polishing plant outlet* Direct and after-cation conductivity, reactive silica and sodium measurements are required to monitor the performance of the plant and the need for resin bed regeneration. Additionally, the measurement of chloride and sulphate in the outlet water has assumed increasing importance.

- *Downstream of chemical dosing points* pH supplemented by ammonia and hydrazine measurements to provide a check on the adequacy of dosing.

- *De-aerator* For testing purposes it is necessary to have the facility to sample and monitor oxygen at both the inlet and outlet of the de-aerator to ensure satisfactory performance.

- *Final feed* Comprehensive chemical monitoring of feedwater just before it enters the boiler, particularly on once-through systems, to provide a final check on quality and acceptability. Measurements include those for direct and after-cation conductivity, dissolved oxygen, pH, sodium, chloride, sulphate

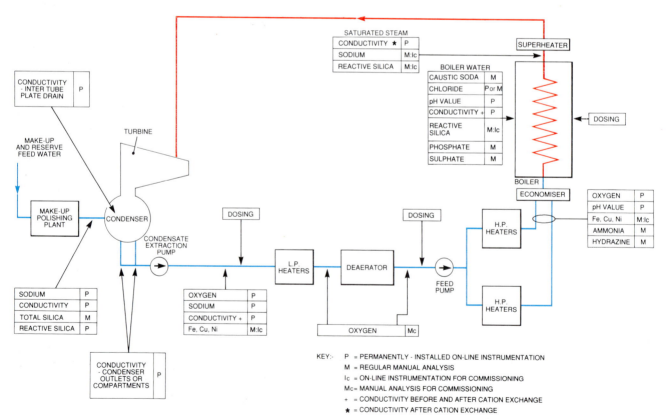

FIG. 2.38 Location of principal sampling and dosing points for drum-type boiler systems

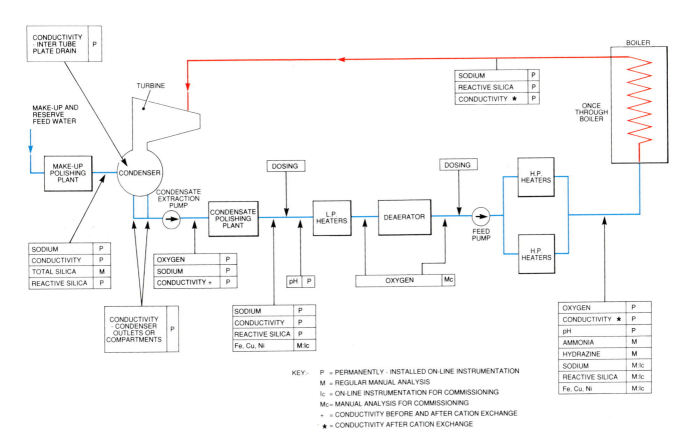

KEY:-
- P = PERMANENTLY - INSTALLED ON-LINE INSTRUMENTATION
- M = REGULAR MANUAL ANALYSIS
- Ic = ON-LINE INSTRUMENTATION FOR COMMISSIONING
- Mc = MANUAL ANALYSIS FOR COMMISSIONING
- + = CONDUCTIVITY BEFORE AND AFTER CATION EXCHANGE
- ★ = CONDUCTIVITY AFTER CATION EXCHANGE

FIG. 2.39 Location of principal sampling and dosing points for once-through boiler systems

and total iron. Manual methods are widely used to monitor iron because of the complexity of continuous monitoring equipment.

- *Boiler water* The monitoring of boiler water quality, as such, applies only in the case of drum-type boilers. Direct and after-cation conductivity, pH, chloride, sodium and reactive silica measurements are required to confirm that the correct boiler water conditions are being maintained.

- *Steam* Measurements of direct and after-cation conductivity, sodium and reactive silica should be carried out on saturated steam, superheated steam or both, to ensure that criteria based on the need to minimise salt deposition in the superheaters, re-heaters or turbine are being met.

Although this section concentrates on the sampling and analysis of the main steam/water circuits, most of the principles and techniques to be described are also valid for other power station water circuits where similar requirements exist. For example, in the stator cooling water systems, conductivity measurements are used routinely to provide a check on purity. In some nuclear power plant there are reactor ancillary cooling water systems and reactor pressure vessel cooling water systems where, in addition to conductivity and pH measurements, more specific analysis may be required, e.g., for carbon dioxide in gas-cooled reactor systems.

12.3 Sampling and sample conditioning [32,33,34]

It is essential to obtain a sample for analysis which is representative of the fluid in the system being monitored. Problems are most likely to arise when more than one phase is present, for example, special consideration is necessary when sampling particulates in water or droplets in saturated steam. The design and location of a sampling device must therefore take account of the fluid conditions at the sampling point. Where on-line measuring instrumentation is involved, these considerations must include the transport and conditioning of the sample prior to its presentation to the instrumentation. It is also essential that sampling and sample conditioning systems have minimal effect on the chemical nature of the sample.

Sampling equipment generally has fallen into two categories:

- Sampling equipment in which the sample flow rate and pressure are controlled by valves (fast samplers).

- Sampling equipment in which the sample flow rate and pressure are controlled solely by the length and bore of capillary tubing which forms an integral part of the sampler (capillary samplers).

Sampling equipment typical of the first category is shown in Fig 2.40 and is intended for abstracting water

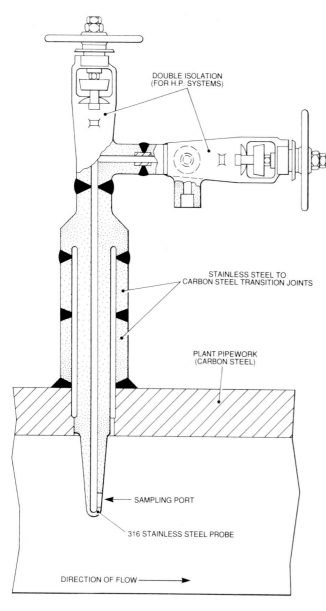

DOUBLE ISOLATION
(FOR H.P. SYSTEMS)

STAINLESS STEEL TO
CARBON STEEL TRANSITION JOINTS

PLANT PIPEWORK
(CARBON STEEL)

SAMPLING PORT

316 STAINLESS STEEL PROBE

DIRECTION OF FLOW ⟶

FIG. 2.40 Sampler for use where sample flow rate and
pressure are controlled by valves

samples for the monitoring of species in solution. Normally it is installed in pipework with its sampling port facing away from the direction of flow to minimise the ingress of particulates which may deposit in sample lines. To obtain a representative sample which includes particulates, a capillary sampler such as that shown in Fig 2.41 is more appropriate. This allows samples to be taken 'iso-kinetically', i.e., with sample flow rates adjusted so that the water velocity at the entry point of the probe equals that in the pipework being sampled, giving an accurate sample of the particulate material present. The use of a capillary to control the flow avoids the need for valves, which can become sites for particulate deposition. Unfortunately, capillary sampling systems only give satisfactory samples over a narrow range of water flow rates and are also prone to blockage. They are therefore only used

when accurate monitoring of particulates is important. The tendency on power stations is to use the fast sampler shown in Fig 2.40 as an alternative for this purpose, but with its sample port turned through 180° so that it faces into the direction of flow in a section of pipework where the flow is turbulent. With this approach, errors due to particulate deposition in the sample lines can be reduced by keeping the linear velocities in the sample line above 1 m/s, thereby ensuring turbulent conditions (Reynolds number greater than 3000). Nevertheless, it must be noted that the results of total iron and similar measurements on samples obtained in this way may be biased low compared with those obtained using a capillary sampling system [33].

The problems of sampling a multiphase system also arise with saturated and superheated steam. Again, iso-kinetic sampling is desirable, but deposition in sample lines once the steam has been condensed can be minimised by having high sample flow rates. Pressure and flow rate of the sample are frequently controlled by valves as an alternative to capillaries. When sampling steam, the cooling water requirements can also be considerable; 1 kg/min of superheated steam sample requires some 60 kW of heat to be dissipated. This could require a cooling water flow rate of 50 to 60 kg/min. Clearly, steam sample flow rates must be kept to a minimum consistent with the needs to maintain iso-kinetic conditions at the sampling point and the measuring instrumentation supplied with sufficient sample.

All samples must be transported and 'conditioned' with respect to temperature and pressure prior to presentation to the on-line instrumentation. In modern plant, it is common practice to route samples taken from the plant to instrumentation grouped in purpose built rooms, sometimes over considerable distances (e.g., 200 m). This is only of real consequence when significant changes can take place in a sample during transport (e.g., with hydrazine and oxygen).

Figure 2.42 shows a typical arrangement of sampling and sample conditioning equipment for use with on-line chemical measuring instrumentation. Particular importance is attached to the close control of the final sample temperature since this very often gives more reliable results than compensating for temperature variations at the instruments themselves. For most chemical measurements 25°C is chosen as the sample reference temperature.

12.4 Chemical measuring instrumentation

12.4.1 Choice and installation of equipment

To ensure satisfactory chemical control and operator confidence, all chemical instrumentation installed at a power station should have high reliability and availability. Four factors can contribute towards this:

● The intrinsic quality of the instrumentation.

● Its suitability for power station use.

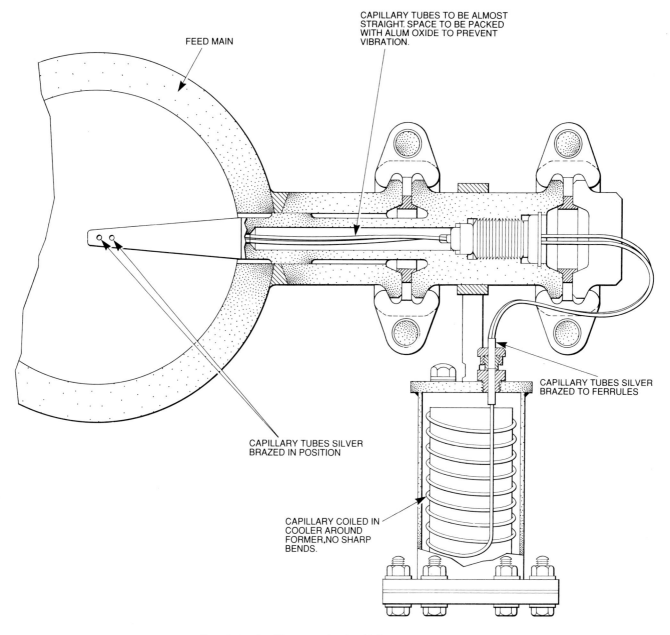

FEED MAIN

CAPILLARY TUBES TO BE ALMOST STRAIGHT. SPACE TO BE PACKED WITH ALUM OXIDE TO PREVENT VIBRATION.

CAPILLARY TUBES SILVER BRAZED TO FERRULES

CAPILLARY TUBES SILVER BRAZED IN POSITION

CAPILLARY COILED IN COOLER AROUND FORMER, NO SHARP BENDS.

FIG. 2.41 Capillary sampler for high pressure systems

- Correct installation.

- Regular and efficient calibration and maintenance.

To ensure that the chemical instruments installed meet these requirements, it is desirable to have some form of instrument evaluation/approval for power station use.

The satisfactory performance of all chemical instrumentation is very dependent upon its correct installation. Experience has shown that such equipment benefits from being housed in purpose built rooms isolated from the onerous environment of the plant [35]. Best modern practice is to combine this with the centralisation of the instrumentation and the sample conditioning equipment. One consequence can be the need for long sample lines, however, experience has shown that this is not a problem provided sample transit times are kept below 10 minutes. Where oxygen and hydrazine measurements are involved, sample temperatures are reduced at source to below 100°C to minimise any reaction and consequential loss in the sample line. Only in instances where very accurate measurements are required, especially those involving particulates, is it necessary for sample lines to be kept short and instrumentation to be sited local to the sampling point. This tends to apply more to specific short term investigations.

A typical chemical instrumentation installation at a modern power station is shown in Fig 2.43.

12.4.2 Measurement of electrical conductivity

The measurement of electrical conductivity is the simplest, most reliable and most useful of the measure-

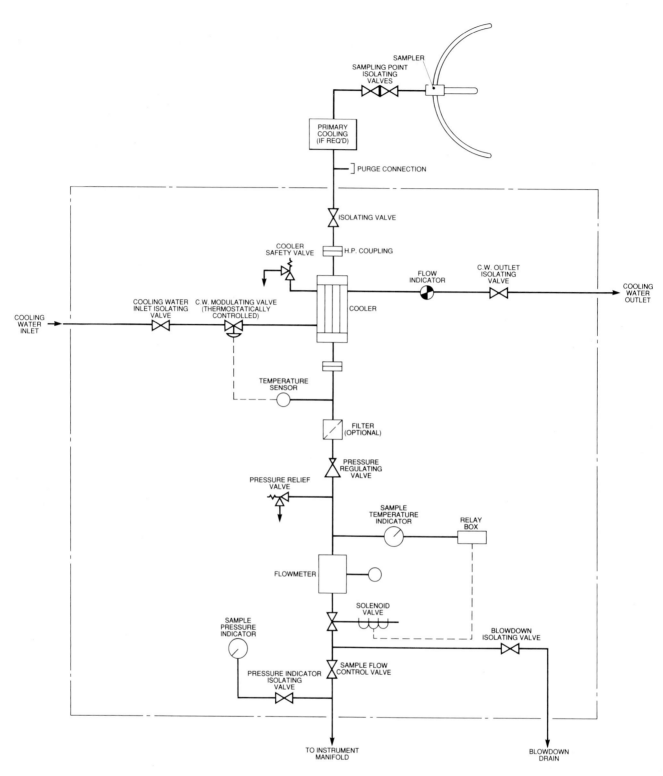

FIG. 2.42 Typical arrangement of equipment for sample conditioning

ments carried out on power station waters although, strictly speaking, it is a physical parameter. All salt impurities and conditioning chemicals such as hydrazine and ammonia, when dissolved in water, dissociate to form *ions* which are capable of conducting electricity. The conductivity of a water is therefore indicative of its purity and increases with the concentration of soluble ionic species. It is because of the latter that

expression of the results as conductivity units is preferred to its reciprocal measured in ohms. The unit of conductivity is the Siemen (S). This is defined as the conductivity of a solution which when placed in a cell having two 1 cm^2 electrodes, 1 cm apart, produces a cell resistance of 1 Ω. Because of the low conductivity of water samples from power plant, the unit of conductivity widely used is the μS.

FIG. 2.43 A typical chemical instrumentation installation

Electrical conductivity is measured by passing a small alternating current through a known volume of the sample contained in a cell, an example of which is shown in Fig 2.44. This simple 'flow line' cell comprises three ring-like electrodes spaced equally within an epoxy resin moulding which is accurately bored and thus contains a precisely determined volume of sample. The cell is threaded each end for mounting (vertically) in the sample line. Conduction through the sample takes place between the central electrode and the two outer electrodes. Electrical conduction is confined entirely within the cell where it is uninfluenced by the presence of adjoining metal pipework. The resistance of the sample contained within the cell, and hence the conductivity, is determined using an AC Wheatstone bridge circuit.

Cells can be manufactured to different dimensions in order to provide a range of 'cell constants' from which the most appropriate for a particular application can be chosen. The cell constant (k) has the units of

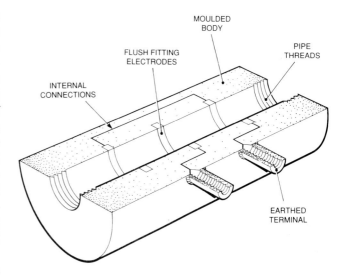

FIG. 2.44 Cutaway view of simple flow line conductivity cell (ABB Kent-Taylor)

207

L^{-1} and for a cell with planar electrodes it is defined as the ratio of path length (l) to cross-sectional area (a) of the electrodes, i.e., k = l/a.

The cell chosen for a particular application should have a constant which gives an easily measurable value of resistance between the electrodes, i.e., within the range 10^2 to 10^6 Ω. If the resistance (R) is given by:

$$R = \frac{k}{K} 10^6$$

where K is the conductivity of the sample (in μS/cm), then it can be seen that for a conductivity of 0.1 μS/cm a cell constant of 0.1 or less is required to bring the resistance into the acceptable measuring range. At higher conductivities, e.g., 10^4 μS/cm, then a cell constant of 1.0 or 10.0 is more appropriate. The use of a cell with an inappropriate constant can lead to errors in conductivity readings due to electrical lead and electrode-solution resistances becoming significant.

Cells are manufactured to close dimensional tolerances and for most purposes there is no need for further calibration. Where necessary, however, this can be carried out by making measurements using an electrolyte of known specific conductance. The electrolyte chosen for this purpose is aqueous potassium chloride, of which the specific conductance at various concentrations is known accurately.

The conductivity of aqueous solutions is dependent upon temperature, increasing as the temperature of the solution increases. In order to obtain accurate measurements it is necessary either to control the temperature of the sample within tight limits ($\pm 2^\circ$C) or, alternatively, to correct automatically the measured conductivity to that which would pertain at a chosen reference temperature. The temperature chosen in both instances is the internationally accepted standard temperature of 25°C.

For normal applications, automatic temperature correction is accomplished by means of a temperature sensing element in intimate contact with the solution contained within the conductivity cell. The sensing element is connected in such a manner that it opposes and precisely cancels the temperature-dependent conductivity deviations both above and below the solution conductivity at 25°C. Over a restricted temperature range, the conductivity change can be considered as a linear function of temperature and many conductivity meters incorporate a linear temperature correction which has the following form:

$$K_{Tr} = \frac{K_T}{1 + \alpha (T - Tr)}$$

where K_{Tr} and K_T are the conductivities at the reference temperature (Tr) and the measurement temperature (T) and α is the temperature coefficient. For solution conductivities above 1 μS/cm, a temperature coefficient between 0.015 and 0.025 per °C is usually

applied. Below 1 μS/cm, however, the method of applying automatic temperature correction becomes more complicated. This is because the conductivity versus temperature change is made up of two components. The first component, due to the impurities present (e.g., sodium ions and chloride ions), generally has a temperature coefficient of about 0.02 per °C. The second is from the conductivity of hydrogen and hydroxyl ions formed from the dissociation of water itself, which show a significantly higher temperature coefficient. As the conductivity of pure water (0.055 μS/cm at 25°C) is approached, the latter effect predominates and correction for this must be applied separately. Figure 2.45 shows the observed temperature variation of the conductivity of pure water compared with applied temperature coefficients of 0.02 and 0.04 per °C. When monitoring the conductivity of relatively pure water, it is therefore essential that the instruments used have the correct type of temperature compensation.

To improve the sensitivity of conductivity measurements in detecting anionic impurities such as chlorides and sulphates, it is common practice to arrange for measurements to be made after the sample has been passed through a column of strongly acidic cation exchange resin in the hydrogen form. This has two effects:

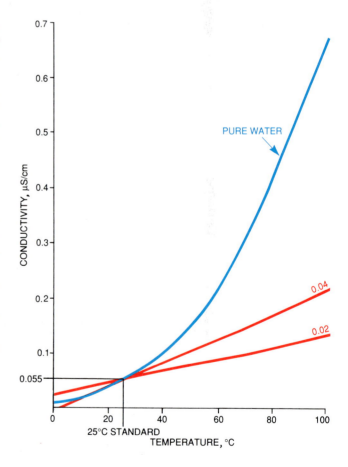

FIG. 2.45 Variation of conductivity with temperature for pure water with applied temperature coefficients of 0.02 and 0.04 per °C

FIG. 1.19   A Terex bowl scraper

FIG. 1.20   Typical large CEGB coal stock

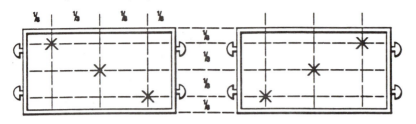 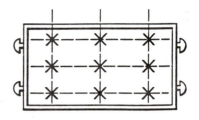

3 INCREMENTS PER WAGON (ALTERNATIVELY)   **9 INCREMENTS PER WAGON**

FIG. 1.23   Auger sampling from wagons

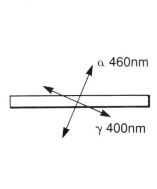

α 460nm

γ 400nm

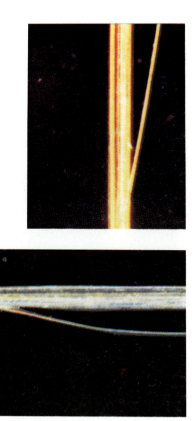

Amosite (Grunerite)

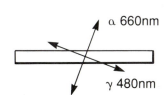

α 660nm

γ 480nm

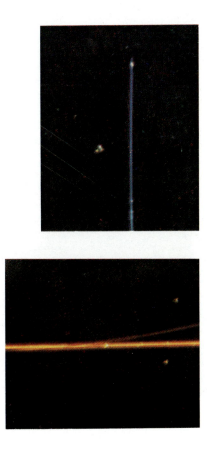

Amosite (Cummingtonite)

Fig. 1.89 Appearance of asbestos fibres in a
microscope (as per Fig 1.88)

FIG. 1.97 Operator with personal dust sampler and portable gas detector

FIG. 1.98   The MIRAN long pathlength infra-red analyser in use

(a)

(b)

(c)

Fig. 2.31   Copper alloy feedheater tube corrosion
(a) Meniscal corrosion contours on an otherwise unaffected surface. Note the exfoliation of corrosion product
(b) Relief effect on stream side surface at bottom of the element
(c) Well defined transition boundaries and on the lower section a groove on the outer surface coincident with a tubeplate

Fig. 4.12 Photograph showing initial surface of tube sections selected, for chemical cleaning tests

Fig. 4.13   Photograph showing the tube sections in Fig. 4.12 after chemical cleaning test
Top row: cleaned in 3% ammoniated citric acid, pH 3.5 Bottom row: cleaned in 5% w/w hydrochloric acid

- Conditioning chemicals such as ammonia and hydrazine and all other cations are removed. This reduces the background conductivity.

- Salts such as chlorides and sulphates are converted into their corresponding acids which significantly increases their respective contributions to the conductivity of the sample.

This measurement is known as *after-cation conductivity*, and is very frequently used as an on-line measurement of the overall anionic impurity levels in a sample.

In practice, a two-column, two-cell system such as that shown in Fig 2.46 is employed. Cell 2 measures the after-cation conductivity of the sample but, if differing from Cell 3, also indicates when Column 1 is exhausting. Cell 3 continues in service to provide a valid measurement at its outlet. When the exhausted resin column is being replaced, Column 1 and Column 2 are interchanged, the new column being put in the Column 2 position.

12.4.3 Chemical measurements using ion-selective electrodes

A number of chemical measurements, including those for pH, sodium, chloride and ammonia can be made with 'ion-selective' electrode systems, i.e., using electrodes which respond to changes in concentration (or, more accurately, activity) of particular ionic species in solution. These electrode systems are potentiometric devices and when connected with a suitable reference electrode to form an electrochemical cell give voltage outputs related to the activity of the determined ion according to the Nernst equation:

$$E = E_0 + 2.303 \frac{RT}{zF} \log a_x$$

where:
- E = potential output from the measuring electrode
- E_0 = a constant potential determined by the reference electrode and internal construction of the measuring electrode
- T = temperature of the sample, K
- z = charge on the ion
- F = Faraday constant
- R = gas constant
- a_x = activity of ion x in solution

If T remains constant the equation may be more simply expressed as:

$$E - E_0 \propto \log a_x$$

i.e., the voltage output, is proportional to the logarithm of the ion activity. Activity (a_x) and concentration (C_x) are related thus:

$$a_x = \gamma C_x$$

where γ is the activity coefficient and is dependent on the total ionic strength of the solution. If the total ionic strength is held constant then:

$$E - E_0 \propto \log C_x$$

i.e., the voltage output, becomes proportional to the logarithm of the ion concentration.

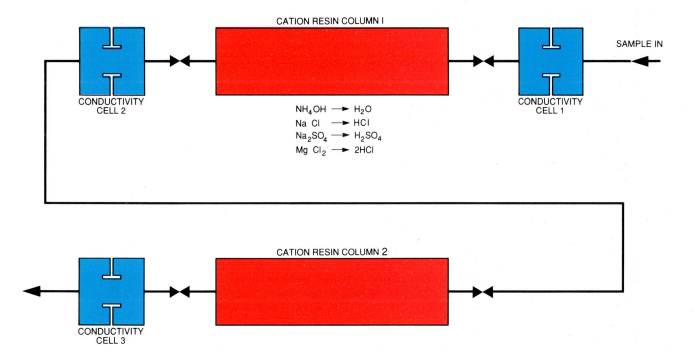

FIG. 2.46 Arrangement of before and after cation conductivity measurements

209

Except for pH measurement, the ionic strength should be maintained at a constant level by addition of an 'inert' electrolyte at high concentration. Small differences between ionic strengths of samples and standards then become insignificant compared with the strength of the added electrolyte. In some applications, notably the measurement of sodium ion, the ion-selective electrode may also be pH dependent and a buffer solution can be used in place of the inert electrolyte to serve the dual role of pH buffer/ionic strength adjuster.

There are four basic types of ion-selective electrode as shown in Fig 2.47. They are:

- *Glass electrodes* These have the sensing membrane made of special glass, often bulb-shaped to give minimum electrical resistance. The surface of the glass

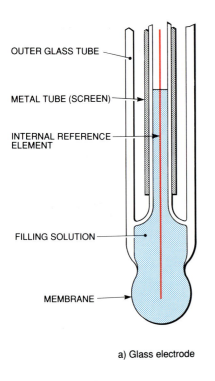

a) Glass electrode

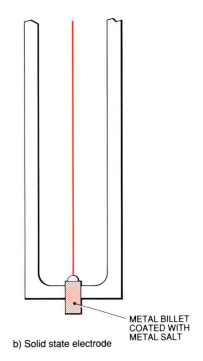

b) Solid state electrode

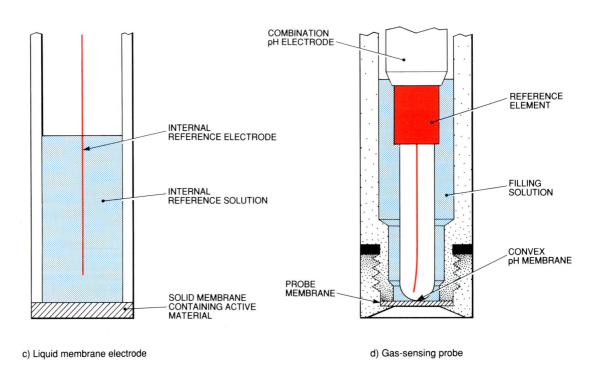

c) Liquid membrane electrode

d) Gas-sensing probe

FIG. 2.47 Types of ion-selective electrodes

acts as a cation exchanger usually exchanging sodium or lithium for the ion being sensed. The resulting change in the activity of these species in the membrane produces a change in the electrode potential. This type of electrode is used for measurements including pH (i.e., hydrogen ion concentrations) sodium and ammonia.

- *Solid-state electrodes* These incorporate a sparingly-soluble inorganic salt membrane, an example of which is the chloride electrode which has a sensing pellet based on an intimate mixture of mercuric sulphide and mercurous chloride.

- *Liquid-membrane electrodes* These have an inert support material with a cation or anion exchanger absorbed or impregnated onto it. The sample comes into contact with the active exchange material and is sensed by an ion-exchange mechanism. Calcium and nitrate electrodes are of this type.

- *Gas sensing electrodes* These comprise a plastic body having at one end a hydrophobic gas-permeable membrane. Gases such as ammonia or carbon dioxide released from the sample by adjustment of the pH permeate the membrane where they subsequently dissolve in the internal electrolyte, changing its pH. This change, sensed by a pH electrode, is a function of the activity or concentration of ammonia or carbon dioxide in the original sample. Electrodes of this type incorporate their own reference element inside the body and do not need an external reference electrode.

The most widely used reference electrode in ion-selective systems is the calomel electrode shown in Fig 2.48. It comprises a mercury/mercurous chloride element immersed in 3.8 M potassium chloride salt bridge solution. When used with on-line instruments it is mounted together with the ion-selective electrode in a flow cell through which the sample is passed. The ion-selective electrode precedes the reference electrode.

Conditioning of the sample, e.g., with respect to pH or ionic strength, is carried out before the sample reaches the flow cell. In the case of sodium measurements one convenient means of achieving the required pH is to permeate alkaline vapour, e.g., ammonia or diethylamine, through the walls of a short length of silicon tubing incorporated in the sample line to the instrument. Another method is to allow the gas to diffuse into the sample from another vessel. Both methods eliminate the risk of sample contamination which could occur with direct additions of buffer solution. Peristaltic pumps are usually chosen as a means of feeding the sample to the flow cell because they also reduce the risk of sample contamination.

A typical flow diagram for an ion-selective measuring system, in this case for sodium, is shown in Fig 2.49.

Where the ion-selective electrode responds according to the theoretical equation, single-point calibration is usually sufficient for satisfactory operation of the in-

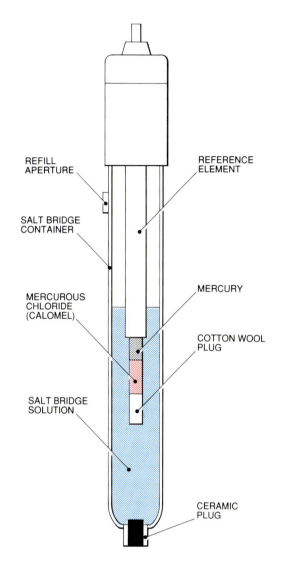

FIG. 2.48 Calomel reference electrode

struments over a very wide concentration range. The notable exception to this, however, is the solid state chloride electrode which departs from this behaviour at low concentration, i.e., as the chloride concentrations arising from the solubility of the sparingly-soluble chloride incorporated into the electrode becomes significant in relation to the chloride concentrations in the sample. As a minimum, a two-point calibration is required in this instance.

There are specific problems associated with the calibration of instruments for measuring pH in low conductivity waters. The first is that conventional standard pH solutions are of much higher ionic strength than the samples and this can lead to errors. One practical method of overcoming this difficulty is to calibrate instruments by using flowing dilute ammonia solutions of similar ionic strength to the samples. From the measured conductivity of the solutions, the ammonia concentration and corresponding pH can be accurately calculated and the instrument calibrated accordingly. For this to be successful, however, the ammonia solutions must be free from ionic impurities.

211

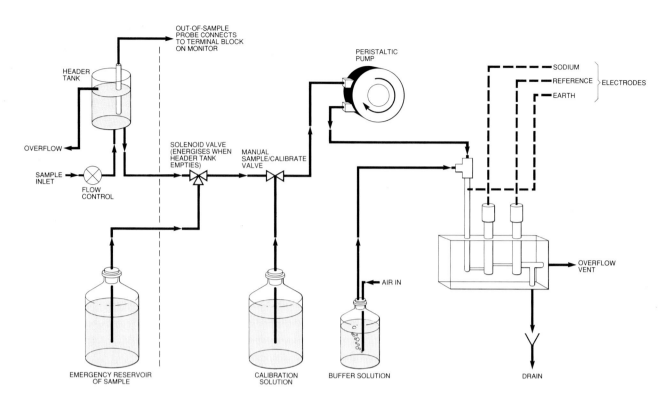

FIG. 2.49 Flow diagram for a sodium analyser (ABB Kent-Taylor)

Other difficulties which may be encountered with pH measurement in low conductivity waters are:

● Sensitivity to sample flow rate; flow rates must be continuous and reasonably constant.

● Errors due to electro-kinetic effects. Stainless steel sample lines and components must be used to ensure complete screening and allow the system to be earthed.

● Sensitivity to sample and buffer solution temperatures.

The problems experienced with compensating for sample temperature are similar to those associated with the measurement of low conductivities. Two effects must be considered: the temperature coefficient of the electrode output and the temperature variation exhibited by the sample pH itself. Instrumentation should have full temperature compensation facilities but it is often better to arrange for the sample (and standard solution) temperature to be closely controlled, i.e., within $\pm 2°C$ of a given working temperature. As with conductivity, the internationally accepted reference temperature for pH measurements is $25°C$.

12.4.4 Colorimetric analysers

Some chemical measuring instruments, notably those for the determination of reactive silica, iron and high level chloride are based upon colorimetric methods of analysis. The species to be measured is made to undergo a chemical reaction to produce a coloured solution, the intensity of which can be related to its concentration. The instruments are designed to add, automatically, the requisite quantities of reagents to a measured discrete volume of sample, or to inject the reagent at the required rate into a continuously flowing sample. Subsequent to the development of the colour, the solution is passed to a measuring cell. Light of a wavelength for maximum absorption by the coloured solution is passed through the measuring cell to fall onto a photocell or photomultiplier arrangement, and the amount of light absorbed is measured. The concentration of determinand in the sample is obtained by comparison of this reading with measurements of light absorption for solutions of known concentrations reacted in an identical way. In place of standard solutions of known concentration, colour filters are sometimes used for instrument calibration purposes.

A flow scheme of a typical instrument of this type used for the measurement of reactive silica is shown in Fig 2.50. In this example, the colour is produced by the reaction of silica in solution with ammonium molybdate and a reducing agent, such as ascorbic acid, to form molybdenum blue. Sample and reagents are drawn continuously into the instrument by means of a multichannel peristaltic pump. Reagent addition takes place sequentially with appropriate mixing between stages and the reacted sample then passes to the measuring cell. The light absorbed by the reacted sample is measured and first corrected for that absorbed by an unreacted sample in the reference cell which provides a sample 'blank') before being converted for display as silica concentration. Calibration is carried out by periodically introducing solutions of known silica

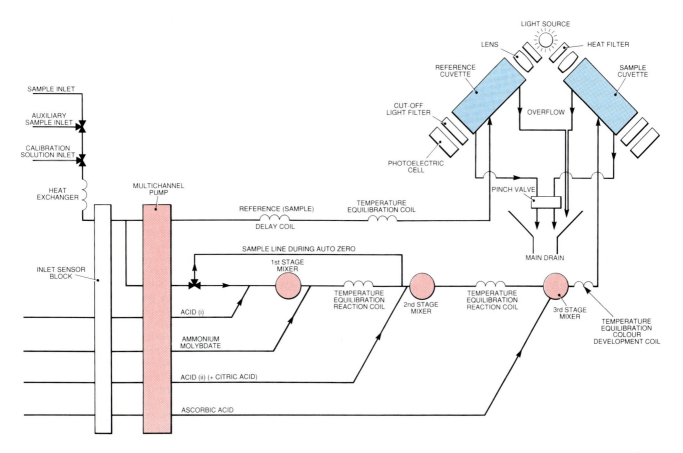

FIG. 2.50 Flow diagram for a silica analyser (ABB Kent-Taylor)

concentration to the instrument in place of the sample, either manually or as part of an automatic cycle. Alternatively, use may be made of standard colour filters to calibrate the instrument. However, the use of standard solutions is the preferred procedure.

The flow circuits of other instruments of this type are basically the same, but using different reagents. Further stages in the reaction process may also be needed. For example, a digestion stage is included in instruments for measurements of total iron to ensure complete solubilisation of particulates in the sample prior to reaction. The measuring principle, however, remains the same.

12.4.5 Measurements using electrochemical cells

Most on-line analysers for the measurements of dissolved oxygen and hydrazine in power station waters are now based on the use of electrochemical cells.

One of the most widely used cells for dissolved oxygen measurements is the Mackereth-type galvanic sensor shown in Fig 2.51. It comprises a central lead anode and an outer silver, or gold, cathode immersed in an alkaline electrolyte. The cell is separated from the sample by a gas-permeable membrane, e.g., silicone rubber or PTFE, through which oxygen in the sample can diffuse to be reduced at the cathode. This is illustrated in Fig 2.52.

The reaction at the metal cathode is:

$$O_2 + 2H_2O + 4e^- \rightarrow 4OH^-$$

whilst the corresponding reaction at the anode is:

$$Pb \rightarrow Pb^{2+} + 2e^-$$

The hydroxyl ions from the cathodic reaction migrate through the electrolyte to the lead anode where they combine with the lead ions (Pb^{2+}) to form lead hydroxide. A current output is obtained from the cell proportional to the amount of oxygen reaching the cathode. This, in turn, is proportional to the partial pressure of oxygen in the sample since this provides the driving force for diffusion of oxygen through the membrane. The current output can thus be related to the oxygen concentration of the sample. Single-point calibration is usually sufficient with this type of sensor.

With galvanic cells of this type the lead anode is consumed at a rate determined by the amount of oxygen reaching the cathode. The cells, therefore, have a finite life and need replacing at intervals depending upon the oxygen concentrations to which they are exposed.

Electrochemical cells for the measurement of hydrazine are similar in principle to those for dissolved oxygen although some require an external potential to generate the current. A typical cell is shown in Fig 2.53. It comprises an outer anode of platinum wire, which is allowed to come into direct contact with the sample,

213

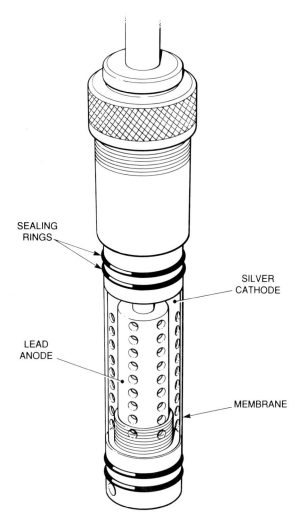

SEALING RINGS

SILVER CATHODE

LEAD ANODE

MEMBRANE

FIG. 2.51 Mackereth-type oxygen sensor construction

and a silver/silver-oxide central cathode. The electrodes are separated by a porous ceramic layer impregnated with a silver oxide gel. Hydrazine is reduced at the anode:

$$N_2H_4 + 4OH^- \rightarrow N_2 + 4H_2O + 4e^-$$

The corresponding cathodic reaction is:

$$4e^- + 2Ag_2O + 2H_2O \rightarrow 4Ag + 4OH^-$$

The current which flows is directly related to the concentration of hydrazine in the sample and independent of other chemicals usually found in feed and boiler waters. It does, however, respond to hydroxyl ions and the accurate performance of this type of cell requires the sample pH to be controlled within close limits (e.g., by addition of a buffer).

12.4.6 Ion chromatography

Ion chromatography (IC) is a technique still at the development stage and therefore does not yet lend itself readily to automatic continuous monitoring. Instruments presently available are essentially for laboratory use. However, the technique has assumed increasing importance in recent years as a means (sometimes the only means) of measuring the very low concentrations of impurities present in boiler feedwater. In current practice, ion chromatographs are used on power stations to provide information supplementary to that obtained from the installed on-line instrumentation. However, as it is essential to avoid sample contamination, they are often sited alongside the other instrument in the chemical instrumentation rooms. Whilst giving very useful information, ion chromatographs frequently require considerable care and attention to keep them operational.

IC is essentially a development from High Pressure Liquid Chromatography (HPLC) — a widely used technique for the separation of non-volatile components in solution. In the case of IC, it is ions in solution which are separated and then measured. A schematic diagram of an ion chromatograph is shown in Fig 2.54. An appropriate ionic solution, termed the eluant, is used to separate the ionic species in a discrete sample of the water to be analysed on a special, low capacity ion-exchange column operating at pressures up to 45 bar. The sample is introduced into the flowing eluant stream at the head of the column by means of a sampling valve and carried into the column. As the separated ions are eluted from the column, there is a change in the conductivity of the eluant which can be detected and measured using an appropriate conductivity cell and meter. The output of the conductivity meter during the analysis therefore displays a number of 'peaks', each of which corresponds to a particular ion. This is termed the *chromatogram*. A typical chromatogram for a simple inorganic anion analysis is shown in Fig 2.55. The various peaks, areas or heights are a quantitative measure of the concentration of each specific ion. The instrument is calibrated using solutions of known concentration.

The sensitivity of the conductivity meter to the eluted ions can be very much improved by the use of an ion-exchange 'suppressor', which effectively eliminates the high background conductivity signal produced by the eluant against which the conductivity of the eluted ions must be observed. In the case of anion analysis, eluants commonly used are appropriate mixed solutions of sodium hydroxide, sodium bicarbonate and sodium carbonate. The eluant leaving the separator column is passed through a further column of high capacity cation exchange resin in the hydrogen form to remove sodium ions and convert the eluant to low conductivity carbonic acid. At the same time, the anion salts are converted to their corresponding acids, thus increasing considerably their conductivity and the sensitivity of the instrument.

For cation analysis, dilute mineral acids such as hydrochloric acid are generally used as eluants and therefore an anion exchange suppressor in the hydroxide form is used to convert the hydrochloric acid to water.

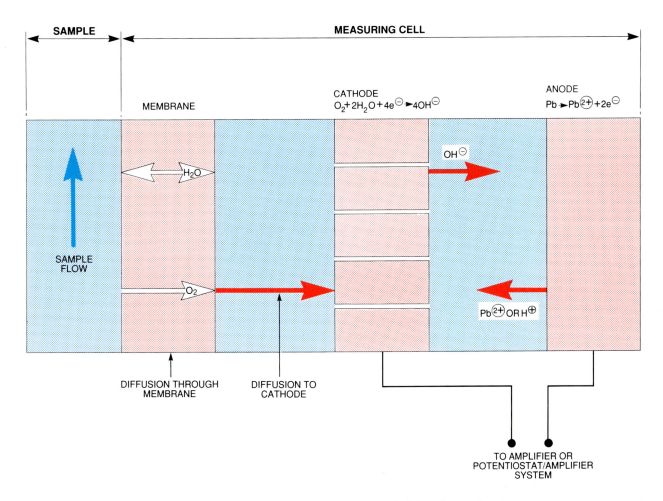

FIG. 2.52 Mackereth-type oxygen sensor principles of operation

Where the concentration of the ionic species of interest is particularly low (i.e., 0.1–10 μg/kg), a pre-concentrator column must be employed. This is a small column of ion-exchange resin of the same type as in the separator column through which a large volume of the same can be passed (e.g., 50–100 ml). The cations or anions absorbed onto the resin are then released by using an appropriate eluant and introduced directly as a concentrated sample onto the separator column.

Ion chromatography is a technique particularly useful for the measurement of sodium, sulphate and chloride in power station waters at concentrations as low as 0.1 μg/kg.

12.4.7 Chemical data processing

As in most areas of power station operation, computers have become an important tool in collecting, processing and storing data, particularly that provided by the on-line chemical measuring instrumentation. The benefits to be derived are considerable:

- More convenient access to a comprehensive range of recorded chemical measurements for both operational and diagnostic use.

- A convenient means of processing and displaying information so as to allow easy assessment of chemical conditions.

- During fault conditions, the use of algorithms can assist the operator in diagnosing faults and selecting the most appropriate corrective action.

- A means of compressing and storing data so as to provide historical records for retrospective examination.

- An improved means of validating data, identifying and diagnosing instrumentation faults.

- A means of improving automatic chemical control.

Computers used in this context can provide a more complete picture of the chemical conditions in the plant, assisting the operator in maintaining the close chemical control demanded by modern power station plant.

The introduction and development of computers for this purpose have closely mirrored that in other areas of power station operation and the various systems available for use in modern plant are adequately dealt with elsewhere (e.g., Volume F). On new stations the

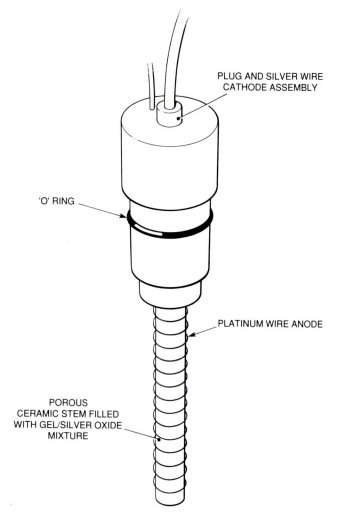

FIG. 2.53 Hydrazine sensor

chemical data processor is usually an integral part of the hierarchal computer system for the station as a whole. Facilities for appropriate alarms and trend display are usually included. It is also usual to provide for both short and long term storage of data, with summarised chemistry logs available in printed form.

Although the use of computers for chemical data processing could be extended to give full automatic chemical control, this has only been attempted to a limited extent because the understanding of the response characteristics of plant to perturbations in water chemistry is often incomplete. In general, the systems adopted initiate alarms to the operator and give appropriate instructions on the corrective action to be taken.

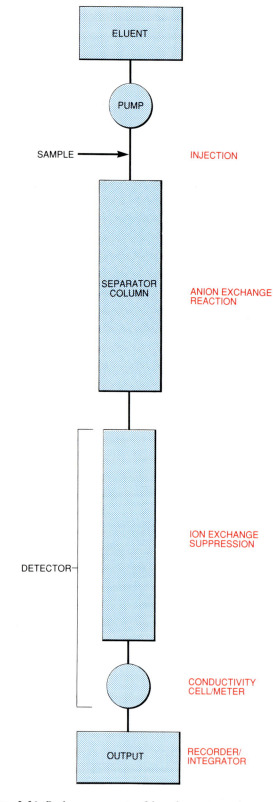

FIG. 2.54 Basic components of ion chromatograph

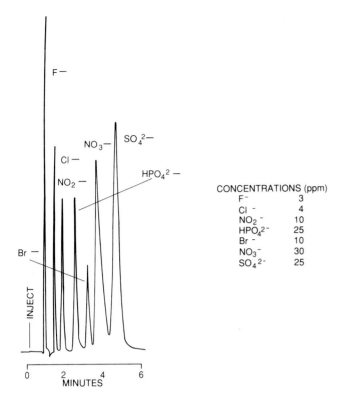

CONCENTRATIONS (ppm)

F⁻	3
Cl⁻	4
NO₂⁻	10
HPO₄²⁻	25
Br⁻	10
NO₃⁻	30
SO₄²⁻	25

FIG. 2.55 Typical ion chromatogram for simple anion separation

13 References

[1] Atkins, P. W.: Physical Chemistry: Oxford University Press

[2] Moore, W. J.: Physical Chemistry: Longmans, London

[3] Manual of Investigations and Correction of Boiler Tube Failures; Research Project 1890–1 Final Report April 1985, Ref CS–3945; EPRI: 1985

[4] Evans, U. E.: Introduction to Metallic Corrosion: Arnold: 1981

[5] Uhlig, H. H.: Corrosion and Corrosion Control, an Introduction to Corrosion Science and Engineering: Wiley: 1985

[6] Goodstine, S. L.: Recommended Water Chemistry Practices for High Pressure Drum Type Boilers; Proceedings Fossil Plant Water Chemistry — Paper 3: EPRI Ref CS–4950: 1985

[7] GOM 72: Chemical Control of Steam/water Circuits of Drum Type and Once-through Boilers (Part 1, Issue 5, April 1983; Part 2 Issue 6, September 1985; Part 3, Issue 7, September 1985: CEGB: 1983

[8] Solubility of Corrosive Salts in Steam: Research Project 969 — Final Report June 1984, Ref NP–3559: EPRI: 1984

[9] Lindinger, R. J. and Curran, R. M.: Corrosion Experience in Large Steam Turbines: Joint ASME/IEEE Power Generation Conf. October 1981, ASME Paper 81–JPGC-Pwr–32: 1981

[10] Bursik, A.: VGB Guidelines on Boiler Feedwater, Boiler Water and Steam of Water Tube Boilers: 45th Int. Water Conf., Pittsburgh USA; Paper 1WC–84–116: 1984

[11] Interim Guidelines on Fossil Plant Cycle Chemistry: Project 2712 1, Final Report, Ref CS4629: June 1986

[12] Wetton, E. A. M. and Lewis G. G.: High Temperature pH Calculations for Solutions of Volatile Amines; Water Chemistry of Nuclear Reactor Systems–4, Paper 99: BNES London: 1986

[13] Penfold, D., et al: The Chemical Control of Two-phase Erosion-Corrosion via Once-through Boilers: Water Chemistry of Nuclear Reactor Systems — 4, Paper 97: BNES London 1986

[14] Douvois, V. Lambert, I. Desmoulin, D. and Nordmann, F.: Laboratory and Plant Investigations on Decomposition Products of Morpholine in Secondary Systems of French PWRs: Water Chemistry of Nuclear Systems — 4, Paper 97: BNES London: 1986

[15] Dalgoord, S. B. and Sandford, M. O.: Review of Hydrazine/ Oxygen Reaction Kinetics: Materials Performance, pp 32–38: April 1982

[16] Kelly, S. F. M. and Aspden J. D.: The Control of Dissolved Oxygen in Make-up Water for Cyclic Operating Power Plant: Proc. American Power Conf. 47, pp 1042–1047: 1985

[17] Dickersen, R. C. and Miller W. S.: US Patent 4 556 492: December 1985

[18] Carmen, C.: Oxygen Removal via Activated Carbon Pre-coats: 23rd Annual Liberty Bell Corrosion Course: 1985

[19] Effertz, P. H., et al: Combined Oxygen/Ammonia Conditioning of Water-steam Circuits in Units with Once-through Boilers: Des Maschinenshaden, 51 (No. 3) pp 97–109: 1978

[20] Effertz, P. H. Hider, R. and Resch, G.: State and Prospects of the Oxygen/Ammonia Treatment of Feedwater: Water Chemistry of Nuclear Reactor Systems — 3, Proc. pp 285–300: BNES London: 1983

[21] Parkins, R. N.: Stress Corrosion Spectrum: Br. Corr. J.7, pp 15–28: January 1972

[22] Hochmuller, K.: Problems in High Pressure Boilers Caused by organic Matter in Feedwater: 43rd Int. Water Conf. Pittsburgh, USA; Paper IWC — 82–38: 1982

[23] Palomero, C. F. Pha, E. and Garbett, K.: An Evaluation of Two-phase Erosion-Corrosion Damage at Vandellos Nuclear Power Station: Proc. Water Chemistry of Nuclear Reactor Systems — 3, pp 235–242: BNES London: 1983

[24] Pocock, F. J.: The Babcock and Wilcox Co.: Water Chemistry Guidelines for Once-through Steam Generators and Drum Boilers: Proc. Fossil Plant Water Chemistry Symposium — Paper 2, Ref CS4950: EPRI: 1985

[25] Battaglia, P. T. Rootham, M. W. and Wootten, M. J.: PWR Secondary System Water Chemistry Guidelines and Their Application: Proc. 46th Int. Water Conf., Pittsburgh, USA pp 396–401: 1985

[26] PWR Secondary Systems Water Chemistry Guidelines: Revision 1, Report No. NP 5056 SR EPRI: March 1987

[27] Bates, A. J.: Off-load Corrosion in the Reheater Banks at Didcot Power Station: Report No. SSD/SW/81/N149: CEGB London: 1981

[28] Bates A. J. et al: Single-phase Erosion-Corrosion Research Programme: Nuclear Energy 25, pp 361 CEGB: 1986

[29] GOM 100: The Storage of the Steam/water Circuits of Generating Plant, Issue 3: October 1984

[30] Computer-based Chemical Data Processing Systems: CM/COP/06, Issue 1: CEGB: August 1986

[31] Water Analysis Equipment: Standard No. 500123, Issue 4, CEGB: April 1989

[32] Steam and Water Sampling — Sample Conditioning Systems: Standard No. 500136, Issue 1: CEGB: November 1986

[33] The Sampling of Steam and Water for Chemical Monitoring and Analysis: CM/COP/07, Issue 1: CEGB: February 1988

[34] Steam and Water Sampling — Samplers and Coolers: Standard No. 764501, Issue 5: CEGB: April 1988

[35] Chemical Measuring Equipment Rooms: CM/COP/04, Issue 1: CEGB: December 1982

CHAPTER 3

Water treatment plant and cooling water systems

1 Introduction

Large quantities of water need to be provided for power generation and the raw water supply to a power station is a critical factor in the selection of a site. Raw water supplies are required for three main purposes:

● To provide make-up water to the water/steam cycle.

● For cooling of the steam for re-use as condensate.

● General domestic use.

Inevitably the water supplies available will vary in composition from site to site; consequently, a variety of treatment techniques has to be employed in the power industry. Moreover, more than one type of raw water may be used at a particular site, e.g., towns water as the make-up supply and sea water for cooling purposes.

The make-up water to the modern high pressure fossil-fuelled boilers and nuclear reactors now employed has to be of extremely high quality with virtually complete removal of salts and gases. A number of techniques may be used in the initial treatment of a raw water, but invariably the final stage is an ion-exchange process.

Note, however, that the use of a high quality make-up water alone may not be sufficient for certain types and designs of plant, notably, nuclear reactors, and an additional stage of treatment of the total condensate has to be applied to produce a treated water that is virtually comprised solely of hydrogen and oxygen. This stage of treatment, again using ion-exchange, is known as condensate purification or polishing.

Cooling water treatment is less sophisticated and is usually confined to the addition of chemicals and/or the use of mechanical techniques to minimise the formation of inorganic and biological deposits in order to maintain satisfactory heat transfer conditions within the cooling circuitry.

This chapter is devoted to providing the background principles involved in the various water treatment techniques used in the power industry, together with details of the plants employed and the operating experiences encountered.

2 Water supplies to power stations

2.1 Types of water

Before discussing water treatment it is important to be aware of the various types of water that may have to be treated. Water can be broadly classified into five types. Table 3.1 gives typical analyses of each type.

2.1.1 Deep well water

This is rain water that has percolated through layers of various strata until it enters an underground aquifer, usually consisting of inert gravel and sand. During its passage, water will have dissolved elements charac-

TABLE 3.1

Typical analyses of various types of water

Constituent	Deep well water	Upland surface water	Clean river water	Industrial river water	Sea water
Alkalinity mg kg^{-1} CaCO$_3$	200	10	100	150	150
Calcium mg kg^{-1} CaCO$_3$	90	12	100	250	1000
Magnesium mg kg^{-1} CaCO$_3$	120	5	50	100	5500
Sodium mg kg^{-1} CaCO$_3$	20	8	50	160	24 000
Chloride mg kg^{-1} CaCO$_3$	10	10	50	130	27 500
Sulphate mg kg^{-1} CaCO$_3$	20	5	50	230	2500
Total organic carbon C	0.2	5	8	15	—

teristic of the strata through which it passes. Frequently limestone and dolomite rocks are present so that the water will contain predominantly calcium and magnesium salts and a high proportion will be present as alkalinity due to the free carbon dioxide in the leaching water. Any organic matter picked up from the top layers is invariably filtered or adsorbed by the strata beneath so that such water is virtually free from organic matter.

2.1.2 Upland surface water

This is water that does not percolate through mineral matter, but has run over impervious rock and through layers of peat. Consequently, it will contain little inorganic matter and is soft, but will contain dissolved and colloidal organic species from the peat (humic and fulvic acids). Such water will collect in reservoirs and lakes in areas such as Wales, Scotland and the Lake District.

2.1.3 Clean rivers

Much of this water will be from upland surface sources and therefore will again contain a significant level of natural organic compounds, both soluble and colloidal. However, depending on the strata over which it has passed and any other discharge entering, it will contain a higher level of dissolved inorganic matter than the upland surface water. Obviously, in any body of moving water, insoluble silt can be present and this will contain both inorganic and organic elements.

2.1.4 Industrial river water

This essentially is re-used water and contains the effluents from human activities, i.e., sewage and industrial wastes. The inorganic constituents of treated sewage will be higher than the original drinking water consumed mainly due to increases in sodium salts. Sewage also adds another class of organic species to the water. Industrial effluent can contain inorganic constituents such as alkalis and acids in addition to a vast range of organic compounds, e.g., proteins, dyes, oils, detergents, agricultural and pharmaceutical chemicals. It follows from this that a river of this type will usually be high in both inorganic and organic constituents and that suspended and colloidal organic and inorganic material will be present.

2.1.5 Sea or estuary water

These sources are included for completeness, but their use in the power industry is restricted mainly to cooling at coastal stations and for the use of sea water for the production of chlorine on-site for sterilisation of the cooling water (see Section 10.4.2 of this chapter). Previously, some power stations used sea water evaporators as a source of fresh water but this process has largely been discontinued.

2.1.6 Choice of raw water

Towns main water could be any of the types (see Sections 2.1.1–2.1.4 of this chapter) listed and will have been treated to comply with the regulations governing the supply of drinking water, thus it will be free from suspended solids and will have been chlorinated. In the past this source of supply was very frequently used for providing make-up water, but more recently supplies taken directly from rivers and on-site boreholes are increasingly being used. This has usually been done in terms of cost and availability of supply.

A decision based on cost comparisons of various raw water sources is difficult due to the number of factors involved. The costs of towns water have been increasing rapidly over the past 10 years and this directly affects the cost of providing make-up water. River water is still of negligible cost relative to towns supply, but its quality will frequently be poorer so that the chemical costs of treatment will be higher. Moreover, to convert a system from towns main to an alternative source will require frequent additional capital to provide pre-treatment facilities so that overall we have to consider:

Towns mains	*Alternative supply*
High cost of supply	Very low cost of supply
Low chemical costs	Higher chemical costs
No additional capital costs	Additional capital costs

Depending largely on the quality of the alternative source of water, cost could favour either of the systems. Clearly, if the alternative source is shown to be overall cheaper in addition to being continuously available, the decision is relatively easy. However, there will also be situations where the towns main water is so good in quality relative to alternative sources that it will give the overall lowest cost for producing make-up water. There will then be a need to consider the more nebulous question of the costs likely to be involved if the towns water were not available. A number of power stations have changed from towns main to river water for producing make-up water with an overall cost saving and the safeguarding of the supply in the event of a drought.

2.1.7 Chemical composition

Expression of concentration

In the analysis of water, concentrations of dissolved salts may be expressed in several ways.

In the UK power industry concentration is expressed as milligram per kilogram (mg kg^{-1}) of the concentration of the ion or associated molecule in water, e.g., mg kg^{-1} as chloride (Cl) and mg kg^{-1} as sodium chloride (NaCl).

In the water industry litre (l) is generally used in preference to kg.

In general water treatment practice, concentration is still frequently expressed as mg kg^{-1} or mgl^{-1} as the calcium carbonate equivalent. The equivalent weight of calcium carbonate is 50 (the molecular weight of CaCO$_3$ is 100 and carbonic acid is dibasic) and therefore

40 mg kg^{-1} chloride (Cl), say, is equivalent to $40 \times 50/35.5 = 56.3$ mg kg^{-1} as calcium carbonate ($CaCO_3$).

Another form of expressing concentration is as a milli-equivalent per litre (meq l^{-1}) and in this case 40 mg kg^{-1} chloride $= 40/35.5 = 1.13$ meq l^{-1}.

An equivalent mineral acidity, abbreviated to EMA, may be quoted in water analyses. This is a measure of the concentration of the strong anions, sulphate, chloride and nitrate present in the water.

The major constituents of all natural waters consist of the salts of sodium, potassium, calcium and magnesium associated with bicarbonate, sulphate; chloride and nitrate ions. Other constituents present in lower concentrations are silica, organic matter, fluoride, ammonia, iron and manganese. Silica occurs mainly as soluble ionised 'reactive' silica. Non-reactive silica can on occasions also be present. This is a form of silica that does not react in the standard test for silica and may comprise polymeric silica, clays and quartz which appear to exist in solution but are non-ionised. The non-reactive silicon can be found by determining the total silica present using either an alkaline extraction or hydrofluoric acid to render the non-reactive silica reactive. The non-reactive silica can then be obtained by difference. Phosphates may occur as simple ions or in polymeric form.

2.2 Potential effects of impurities in water

In general, the main potential adverse effects of the impurities present in raw water are to promote corrosion and scaling of plant items. Additionally, some substances can affect the efficient operation of the water treatment processes used to obtain satisfactory feed and boiler make-up water.

2.2.1 Calcium and magnesium salts

Calcium and magnesium bicarbonates will deposit carbonate scales if water is heated. Calcium sulphate also has a limited solubility and if the water is concentrated the solubility may be exceeded with resultant deposits. It is mainly in the cooling circuits of power plant that such deposits are likely to arise and the concentration factors of recirculating cooling waters have to be controlled by purging from the system to prevent deposition. Carbon dioxide is generated when bicarbonates decompose and this can result in locally acidic concentrations with an enhanced risk of corrosion occurring. Precipitation of calcium carbonate can also occur when alkaline agents are added to cooling water. For example, this can occur when sodium hypochlorite is used to control biological fouling in cooling circuits.

2.2.2 Sodium

Any salts present in water represent a potential threat of corrosion. Sodium salts such as chloride and sulphate have to be controlled to minimise corrosion of the boiler and turbine sections of the plant. Although small additions of sodium hydroxide are used to mini-

mise general corrosion in drum boilers, carryover into the superheater and turbine sections has to be avoided to avoid stress corrosion.

In once-through boilers for nuclear reactors where austenitic superheaters are fitted, the risk of stress corrosion from any chloride or hydroxide present would be very high. Accordingly, extremely low concentrations of sodium have to be maintained in the water/steam circuit.

2.2.3 Silica

Silica has a relatively high solubility in high pressure steam relative to other inorganic compounds. During passage through the turbine the pressure decreases and the solubility of the silica decreases and can deposit on turbine blades reducing the turbine efficiency. For this reason, silica concentrations in boiler waters have to be controlled to strict limits.

2.2.4 Organic matter

The main problem arising from organic matter in the raw water is fouling of anion exchange resins (see Section 4.4.1 of this chapter) and additional stages of treatment may have to be included in water treatment plant to minimise this effect.

Additionally, organo-chlorides may be present in the raw water and these being non-ionic are not removed by ion-exchange. Breakdown of such compounds with temperature occurs in the water/steam circuit liberating chloride ions and this can affect the chemical control within the water/steam circuit.

2.2.5 Suspended solids

These have to be removed prior to any ion-exchange or reverse osmosis process. Consequently, such processes have to be preceded by filters and/or clarification plant (see Sections 3.1 and 3.2 of this chapter).

2.2.6 Phosphate

In the presence of calcium, insoluble calcium phosphates can be deposited from raw water. The main factors influencing deposition are concentration, temperature and pH. In recirculating cooling systems using raw water containing phosphate, deposition occurs and this has to be controlled by operating procedures, the introduction of Taprogge plant and the addition of chemicals (see Section 10.1.2 of this chapter).

Polyphosphates may also be present in raw water. These can act in a similar manner to organic matter on anion exchange resins giving rise to fouling and difficulties in obtaining the required quality and quantity of water from such plants.

2.3 Quantities required

The quantity of water required for boiler make-up ranges from 1–4% of the maximum continuous rated steam flow. The demand for make-up is highest when stations are commissioning and during two-shift operation. The typical 3% design make-up capacity for a four-unit 2000 MW coal or oil fired station is 4600 m^3/day. The supply requirements are increased by the regeneration losses of the make-up treatment plant which can be up to 20% of the net output for high dissolved solids waters, increasing the total input to 5500 m^3/day. Adding the requirements for domestic water raises the total demand for raw water to about 6000 m^3/day.

The essential requirement for cooling services is an abundant supply of water and, for example, a modern 2000 MW station with a once-through cooling system, requires a flow of about 6.5×10^6 m^3/day. There are no rivers in the UK with a minimum fresh water flow of this magnitude; the largest river in the UK may, at times, have a fresh water flow of less than 0.854×10^6 m^3/day. Consequently, for inland stations, recirculating cooling systems incorporating cooling towers are installed. In this case the cooling water intake required will be about 200 000 m^3/day.

2.4 Water quality standards and treatment required

2.4.1 Make-up water

Only minute traces of impurities are allowable in the make-up water for modern high pressure boiler plant. A typical specification will call for the following guaranteed quality:

Conductivity before and after passage through a cation exchange column in the hydrogen form	< 0.10 μS cm^{-1} (at 25°C)
Sodium	< 15 μg kg^{-1}
Silica	< 20 μg kg^{-1}

It can be seen that quality of water required is close to that of absolute pure water which has a conductivity of 0.056 μS cm^{-1} at 25°C.

Other parameters may be added to the specification depending on site conditions. For example, a guaranteed total organic content (TOC) may be specified. This would usually be set at < 0.1 mg kg^{-1}.

The extent of treatment required will depend on the quality of the available water supply. If there is a significant concentration of suspended solids present, the initial stage of treatment will be clarification and filtration. This will then be followed by plant to remove the bulk of the dissolved solids. Ion exchange, evaporation, reverse osmosis or electrodialysis may be used, but ion-exchange is used most frequently, mainly because the majority of waters have dissolved solids contents less than 500 mg kg^{-1} so that the potential advantages of using reverse osmosis or electrodialysis on high dissolved solids waters cannot be realised. Evaporation is less commonly used than hitherto.

The final stage of treatment to achieve the purity specification given above is always by an ion-exchange process.

On occasions an additional section of plant may be installed specifically to remove organic matter. Ion-exchange, oxidation and adsorption by carbon have all been used.

2.4.2 Cooling water

Most waters can be treated to make them suitable for cooling services, the purpose of any treatment employed being to minimise the risk of fouling or corrosion of the heat exchange surfaces.

Inland power stations employ recirculating cooling water systems and cooling towers. Once-through cooling systems are used for power stations sited on large estuaries or at coastal locations where large quantities of water are readily available.

With recirculating cooling water systems, scaling problems become more severe due to concentration of the salts present in the water and these aspects are considered in greater detail in Section 10.1 of this chapter.

3 Pre-treatment of raw water

Many natural waters are coloured, turbid, and contain suspended solids (silt, clay, etc.) and organic matter. If such substances are not removed, using a primary stage of treatment the subsequent water treatment processes would be adversely affected, the severity of the effect depending on the type of process. For example, evaporator plant, although requiring water that has been pre-treated, would tolerate some ingress of suspended solids, whereas a deionisation plant using ion-exchange would be severely affected in terms of both fouling of the ion-exchange materials and the development of high pressure losses. In the case of reverse osmosis the membranes used cannot tolerate any suspended or colloidal solids and complete removal is essential (see Section 6.3.2 of this chapter). Pre-treatment processes cannot remove all the organic matter present in raw water. The aim is to remove as much as possible of the types of organic matter that would affect the later stages of treatment. The organic matter will consist of a large number of compounds with variable molecular weights and it will be present in soluble and colloidal form. The pre-treatment should remove all of the colloidal organic matter and a proportion of the higher molecular weight species.

If non-reactive silica is present some may be removed by the pre-treatment processes but generally the effectiveness of removal will be poor.

The usual form of pre-treatment in the power industry is clarification followed by filtration. However, filtration alone may be sufficient to pre-treat a raw water containing only a small quantity of suspended solids.

In general river water will always require to be pre-treated whereas water companies' supplies will have been treated at source, if necessary, for drinking purposes and will not usually require a further treatment.

3.1 Clarification

Clarification is the term used to encompass coagulation, flocculation and sedimentation reactions which are involved in the removal of suspended and colloidal matter from water.

In coagulation, chemicals providing metal ions, such as aluminium sulphate and ferric chlorides, are added usually by positive displacement variable stroke metering pumps to the raw water in the range 5–50 mg kg^{-1}.

The particles present in water are electrically charged (negatively in most natural waters) and it is the repulsive effects of such charges which prevents the particles forming aggregates which would be large enough to settle out. Now when a coagulant is added to water, e.g., aluminium sulphate, a chemical reaction occurs:

$$Al_2(SO_4)_3 + 3Ca(HCO_3)_2 =$$
$$2Al(OH)_3 + 3CaSO_4 + 6CO_2$$

The electrical character of the $Al(OH)_3$ produced is complex and a number of species of differing charge may be obtained. However, it is known that positively charged particles and ions will be present, and these will tend to neutralise the excess negative charge of the fine particles. The repulsive forces between particles will therefore be reduced so that there is a tendency for particles to join together forming masses that are sufficiently heavy to settle rapidly. In addition to charge neutralisation, metal hydroxides such as $Al(OH)_3$ are:

(a) Capable of enmeshing fine particles.

(b) Can form bridges between particles.

It is possible for the particles to lose their individual characteristics so that the system becomes basically a suspension of $Al(OH)_3$ in water. Flocculation of this suspension occurs, the heavy agglomerate ($Al(OH)_3$ + particles, etc.) form and a rapid sedimentation then takes place.

There are a number of factors which can affect the coagulation and flocculation processes, e.g., pH, temperature, nature of particles present in the raw water, and the chemical composition of the raw water. The most important factor is pH, and there is usually an optimum pH value at which most efficient coagulation/flocculation takes place. The pH range for satisfactory coagulation with $Al_2(SO_4)_3$ is 6.0–7.0 and in this range the soluble aluminium remaining after filtration would be 0.05–0.2 mg kg^{-1}. When waters having a pH outside this range have to be treated, acid or alkali is added with the $Al_2(SO_4)_3$ to ensure good coagulation and flocculation.

Polyelectrolytes may also be used as coagulating and flocculating agents either alone or combined with an inorganic coagulant. These are long-chained organic molecules with chemical groups attached to the chain. On adding to water the active groups become charged depending on the nature of the active group, either positively, negatively or with both so that an overall zero charge exists, and these are termed cationic, anionic and non-ionic polyelectrolytes, respectively.

Cationic polyelectrolytes are used as coagulants being capable of neutralising the negative charges present on suspended solids in natural waters. The other types are used solely as flocculant aids, for which the type of charge is unimportant, usually following an inorganic coagulant. For this purpose, the higher molecular weight polyelectrolyte products are employed.

Polyelectrolytes are more expensive than the inorganic based coagulants, but smaller quantities are required so that overall costs are comparable. They may be used alone or in combination with the inorganic agents.

Polyelectrolytes are usually applied as powders which must be added carefully to water to prepare a solution as the materials have a tendency to agglomerate and block the injection equipment. There is a need to provide sufficient turbulance at the dosing point to ensure rapid mixing with the water being treated. Special designs of automatic dosing equipment have been developed to dilute a strong solution to a more dilute and finally inject into the main stream.

Various types of clarification tanks are shown in Fig 3.1. Basically, a clarification plant consists of a coagulation section designed to provide intimate mixing of the raw water and coagulant used. At this stage it is often preferable to provide rapid agitation to effect charge neutralisation. Immediately coagulation starts a more gentle agitation is required to encourage flocculation (the gathering together of coagulated particles). The water then flows upwards at a reducing velocity so that the floc sinks rapidly to the bottom of the reaction tank. The fluidised bed formed by the accumulation of floc is called a floc blanket and this is the factor on which clarification efficiency is primarily dependent. A simple floc blanket tank has a vertical parallel-walled upper section with a flat or hopper-shaped base. The upward flow at reducing velocity assists flocculation to occur and the large floc particles remain in suspension within the tank. As floc accumulates the volume of the blanket rises and this then has to be controlled by removing solids from the system.

The mechanism of clarification within the floc blanket tanks is complex but involves flocculation, entrapment and sedimentation. The contact time of water within the blanket is usually greater than one hour. The blanket can be considered as a filter through which the small particles rising cannot pass. The pores are the voids between the larger particles which comprise the bulk of the floc blanket. The efficiency of entrapment will be affected by the spacing of the suspended particles which in turn is dependent on floc quality, water velocity, and particle size, range and distribution.

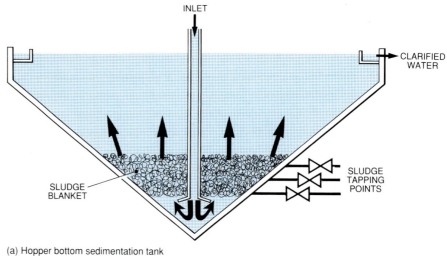

(a) Hopper bottom sedimentation tank

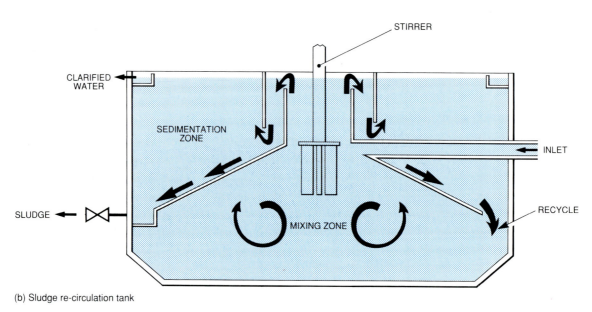

(b) Sludge re-circulation tank

FIG. 3.1 Clarification tanks

The following need to be considered in the design and operation of clarification plant:

- The most effective chemical treatment, e.g., coagulant type and quantity from laboratory tests taking into account the variations in the raw water composition.

- Concentration, particle size distribution and surface charge of suspended and colloidal solids.

- pH conditions for the clarification process.

- Coagulant mixing rate and retention time.

- Contact time through the sludge blanket.

- The upflow velocity limits for the maintenance of an acceptable water quality making allowance for seasonal effects (notably temperature).

- Blanket floc concentration. This is a useful method of assessing the ability of the blanket to remove floc and remain stable.

Usually by a suitable choice of coagulant and design of plant it is possible to achieve good clarification of most natural water, and, in addition to removing suspended matter, the plant removes some of the organic matter which is present in many raw waters. In general, 30–70% is removed, but better figures than this can be obtained in plants designed specifically to remove organic matter.

3.2 Filtration

Water leaving the clarification stage will contain mainly traces of floc as suspended material. During normal operation the concentration of suspended solids entering the filtration stage should be less than 5 mg kg^{-1} suspended solids. Perturbations in the sludge blanket level caused by changes in flow rate or changes in the nature of suspended solids and the composition of the raw water entering the clarifier leading to poor coagulation and flocculation, can result in excessive quantities of floc, etc., leaving the clarifier section. The most frequent problem arising in the operation of clarifiers is the carry-over of floc, resulting in overloading of the filters with a consequent loss of quality and the need for increased cleaning of the filter media.

Sand or anthracite are invariably used as filter media in water treatment and the filtration properties are defined by:

- Size.
- Shape of particle.
- Density.

The normal filtration medium used is a bed of closely size-graded sand with a particle size of about 1 mm, and rapid filtration is used with water velocity through the bed of 5–15 m/h. The filter vessels are normally vertical and cylindrical in shape. The sand layer is about 1 m deep and is supported either on gravel or on a flat plate fitted either with nozzles or a header and lateral collection system for the filtered water, and as distributors for the cleaning water. Pumps are usually employed to pass the water through the filter media under pressure, but gravity filters have also been used. Various types of filter are shown in Fig 3.2.

In operation, suspended solids collect in the top layers of the filter bed; consequently there is a progressive increase in head loss across the unit. At constant flow rate, the graph of head loss against time is parabolic thus once the filter begins to block there is a rapid rise in head loss and the filter has to be taken out of service for cleaning.

Filters are cleaned by backwashing with an upflow of water or a combination of air and water. This loosens the compacted bed and the solids are transported to the waste outlet at the top of the unit.

Other types of filter less frequently used in the power industry are:

(a) *Multi-media filters* To improve the capacity of the bed for retaining solids two filter media can be used. In dual-media or multi-media filters a layer of anthracite is placed on top of a layer of sand. The anthracite is of larger particle size to improve the dirt holding capacity of the bed and because it has a lower density than sand it remains as the top layer after backwashing.

(b) *Up-flow filters* In up-flow filters the water enters at the bottom of the filter bed and flows up through layers of sand of decreasing particle size, thus improving the dirt holding capacity of the bed. To prevent the bed being carried upwards a grid of vertical plates is positioned across the unit at the top of the bed. Because flow surges can tend to cause the bed to lift, it is important to provide a means of maintaining a reasonably constant flow through such filters.

Following clarification and filtration the water has to be of satisfactory quality for the next stage of treatment. A typical water quality specification for the clarification process preceding an ion-exchange would be:

Suspended solids

2.0 mg kg^{-1} at filter inlet, i.e., after flocculation,

0.2 mg kg^{-1} at filter outlet

as measured by filtration through a 0.45 μm filter dried and weighed.

Organic matter

The clarification treatment should decrease the organic content of the raw water by at least 50%. This will need to be demonstrated by analysis of a filtered sample before and after treatment using one of the following methods:

- Total organic carbon.
- UV absorption.
- Permanganate value — either 4 hours at 26°C or 30 mins at 100°C.

Residual aluminium or iron

Outlet from the filter should contain < 0.2 mg kg^{-1} of the metal ion.

Colour Hazen value < 5

It may also be useful to specify the sludge concentration required to obtain optimum performance of the clarifier. As a guide the sludge concentration should be about 15% vol/vol as determined by measurement of settlement for 30 min of a 100 ml sample in a 100 ml cylinder and expressing the results as the % proportion of the sample volume occupied by the settled sludge.

4 Ion-exchange

More than 100 years ago it was recognised that natural solids could exhibit ion-exchange properties, the first

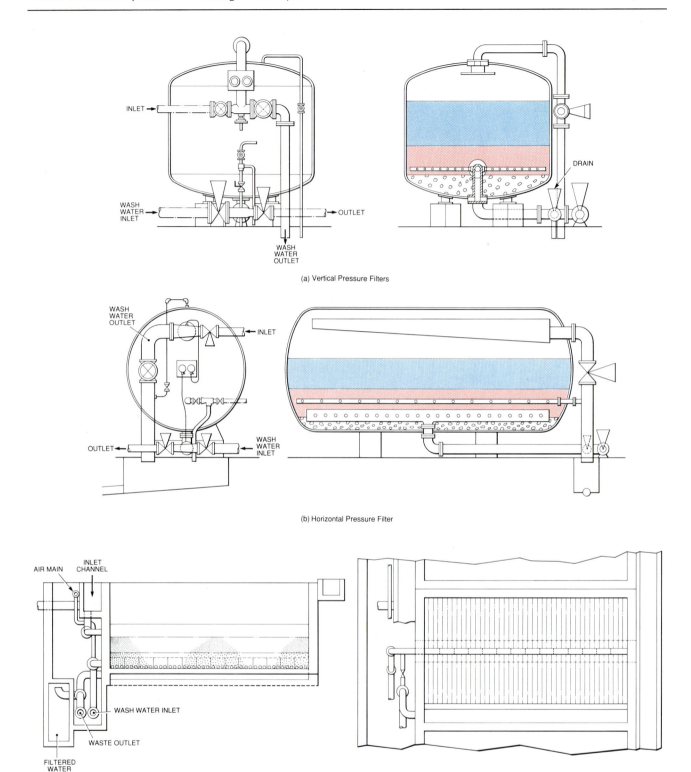

(a) Vertical Pressure Filters

(b) Horizontal Pressure Filter

(c) Rapid Gravity Filter

FIG. 3.2 Types of filter (Dewplan Ltd)

examples being soils and rocks (M. S. Thompson 1845). Subsequently, in 1935, synthetic ion-exchange materials were produced using the polymers then available (Adams and Holmes).

In 1944 the first polystyrene based resins were produced (D'Alelio), and this is still the major ion-exchange polymer used. More recently resins based on the use of acrylic polymers have been developed.

Today ion-exchange materials are employed in a wide variety of applications, ranging from pharmaceutical products to water treatment processes.

In water treatment the three main characteristics required from solid ion-exchangers are:

● Insoluble, but permeated by water.

● An ability to exchange ions with those in solution.

- To allow a flow of water through a bed of the material with an acceptably low resistance, yet, when necessary, acting as a filter medium.

The ion-exchange reaction can be exemplified by the removal of sodium chloride from water as follows.

A cation resin having properties equivalent to a strong acid can exchange hydrogen ions for the sodium ions associated with sodium chloride to produce hydrochloric acid

$$R_H + NaCl \leftrightarrow R_{Na} + HCl$$

If this product is then treated with an anion resin having the properties of a base, the acid is neutralised with the production of pure water

$$R_{OH} + HCl \leftrightarrow R_{Cl} + H_2O$$

In this way other salts giving ions such as calcium, magnesium, sulphate and nitrate can be removed from a raw water.

4.1 Preparation of synthetic ion-exchange resins

Information published by Purolite Ltd on the production of resins has been extensively used in preparing this section, and this is acknowledged [1].

These resins are produced in two major stages:

- The preparation of the appropriate polymer.

- The addition of the required ion-exchange functional group.

The polymer, which is usually a copolymer of styrene and divinyl benzene, can be prepared either as a gelular material in which the pores depend on imbibed water or with discrete larger pores; this latter type is usually referred to as macroporous.

Polymerisation reaction

The divinylbenzene (DVB) acts as a cross-linking agent and the quantity used can be varied mainly to produce resins with differing physical characteristics, this in turn can influence the chemical properties, notably, selectivity.

The copolymerisation is carried out to produce spherical beads by suspension polymerisation.

4.1.1 Gelular resin

Styrene has a very low solubility in water; consequently, when styrene, DVB and a catalyst are stirred in water, droplets of the monomer form. The water usually contains a variety of protective colloids or suspension agents to assist the formation of these droplets and then assist their stabilisation during polymerisation.

At the end of the reaction the polymer will be in the form of firm spherical beads. The size of beads

formed is a function of the stirrer speed, the geometry of the reaction vessel and the stabilising agents used.

Some heat is required to start the polymerisation process to produce free radicals from the catalyst; the reaction is then exothermic. The rate of polymerisation depends on temperature and the type of catalyst used. With a benzoyl peroxide catalyst a temperature of 80–90°C is used. The initial rate of conversion of monomer mixture to polymer is relatively slow, but at around a 10% conversion a 'gel point' is reached and the reaction rate increases and the particles change to a gelular insoluble material which has a large proportion of unreacted monomer. As the reaction proceeds, the beads contain less monomer and more polymer so that the beads harden, although they remain in the form of a gelular material.

The beads are then washed free from suspending agents and gently dried. A proportion of the beads will be excessively coarse or fine and need to be discarded. This could be done after the introduction of the functional groups, but it is preferable to screen before the activation stage.

4.1.2 Macroporous resin

Polystyrene is soluble in a variety of organic compounds one of which is styrene. The polystyrene cross-linked with DVB will not dissolve in styrene but will be swollen by it as polymerisation proceeds. The molecules of the styrene monomer are in close proximity to the growing polymer chains and hence the chains grow homogeneously and close together with only molecular sized pores in the gelular material.

If the polymer chains grow in an environment in which they are not compatible, a precipitated structure results. The agents used to create this effect are called 'porogens'. The essential properties of a porogen are that it should be fully miscible with liquid monomer, but that the solid polymer should be essentially insoluble in it or unswellable by it. Preferably, the porogen should also be relatively insoluble in water so that it will readily remain in the monomer phase in the aqueous dispersion. Examples of suitable porogens are the higher alcohols (butanol upwards) and aliphatic hydrocarbons.

As the polymerisation progresses, the growing polymer chains find themselves in an environment which is becoming less concentrated in the styrene with which they are compatible, and more concentrated in the porogen with which they are not. A stage is reached where the chain structure precipitates. The porogen quantity must clearly be controlled so that the precipitated structure still forms an interconnecting polymer network, but it now contains larger holes or 'pores' which at this stage are occupied by the porogen. At the end of the polymerisation reaction the porogen is removed, most conveniently by distillation, leaving discrete pores which are not dependent on water and will not collapse if the resin is dried. Light is refracted at the numerous internal phase boundaries giving the resin its characteristic opaque appearance. Figure 3.3 illustrates the formation of a macroporous bead.

The precise quantity of porogen depends on the type of porogen used and the purpose for which the resin is designed, but it will normally be within the region of 30–40% by volume of the monomer phase. Consequently, the yield of polymer per batch from the polymerisation vessel is considerably lower than with gel beads, accounting for the higher costs of macroporous resins. The activation steps which follow are essentially the same as those for gel resins.

4.1.3 Acrylic type polymers

Suspension polymerisation is used to produce the beads and DVB is used as the cross-linking agent. The acrylic polymers used are acrylate, methacrylate or acrylonitrile.

4.1.4 Introduction of the functional ion-exchange groupings

Strongly acidic cation resin

The polystyrene-DVB copolymer is reacted with sulphuric acid (see Fig 3.4). Under the action of strong sulphuric acid (>85%) at temperatures of 100°C or more, the aromatic ring will undergo a substitution reaction wherein a sulphonic acid group will replace a hydrogen atom. The reaction normally only proceeds as far as mono-substitution and, due to stearic and electronic influences, no substitution occurs in a DVB molecule.

The reaction is often carried out in the presence of a suitable swelling solvent. Since aromatic hydrocarbons would themselves sulphonate under these conditions, chlorinated hydrocarbons are conventionally used. The solvent permits a more even diffusion of

sulphuric acid into the polymer, resulting in a completely smooth transparent resin bead. Sulphonated without solvent, the resin takes on a reticulated, orange-peel appearance. Since the pores themselves in macroporous polymers allow more freedom of access to the sulphuric acid, solvent is rarely used in the sulphonation of these types. The solvent ultimately distils and is recovered as the sulphonation temperature is raised to 100°C.

This sulphonated polymer is all that is required chemically to make a cation resin, but at the end of the polymerisation reaction it is in an environment of strong sulphuric acid. Direct mixing with water would evolve a large amount of heat, and, in the case of gel beads, extensive bead breakage. To minimise these effects, the resin is 'hydrated' by stepwise addition of reducing strengths of sulphuric acid until eventually it can be washed directly with water. After thorough washing and boiling to remove any residual traces of solvent, the resin can straight away be packed as a hydrogen form product.

Weakly acidic cation resins

This involves the introduction of an active carboxylic grouping. This is difficult to obtain from a polystyrene matrix, but it can be done relatively easily by hydrolysis of acrylate, methacrylate or acrylonitrile co-polymers (see Fig 3.5).

Anion resins — polystyrene polymer matrix

The manufacture of anion resins is more complicated than cation. Complete mono-substitution of the sulphonic acid group can be virtually guaranteed provided minimum acid strength and temperature are exceeded. The anion activation process comprises two discrete stages, and the first of these, *chloromethylation*, is

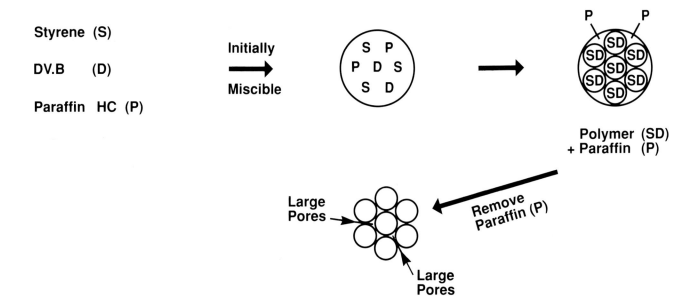

FIG. 3.3 Preparation of porous resin

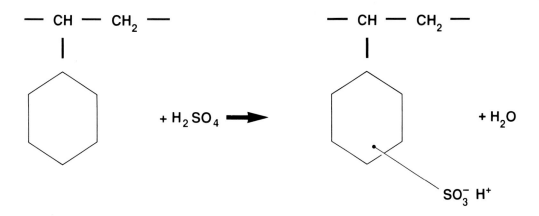

FIG. 3.4 Sulphonation of polystyrene

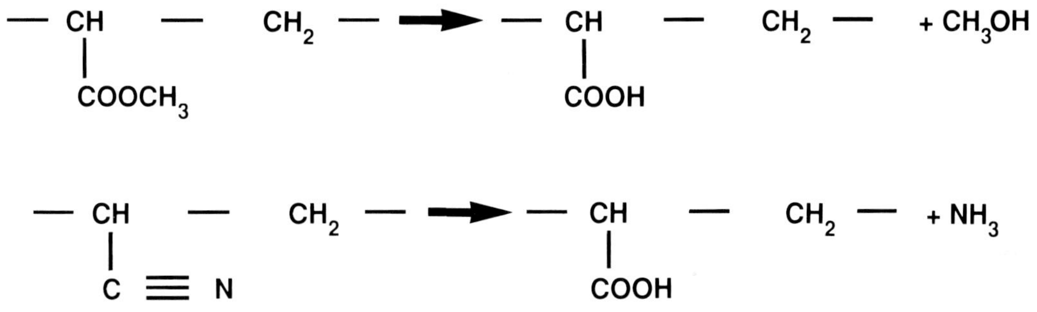

FIG. 3.5 Hydrolysis of acrylate

subject to much more variability, and complete mono-substitution is rarely achieved.

Chloromethylation

A chloromethyl group is substituted into the aromatic ring using the reagent methyl chloromethyl ether (CME) (see Fig 3.6).

This reaction is catalysed by Friedel-Crafts reagents, such as zinc chloride, aluminium chloride, ferric chloride or stannic chloride. Process temperatures are much lower than with the sulphonation reaction — between ambient and 60°C. Up to 50 different equilibrium reactions have been postulated by various research workers in the field, so that increased variability of the process is not hard to understand.

An additional complication is that the newly introduced, pendant chloromethyl group is itself capable of substituting in a neighbouring, hitherto unreacted aromatic ring. This introduces a form of *secondary cross-linking*, which tightens up the polymer matrix and reduces the overall maximum capacity of the resin, since two aromatic rings are now substituted without producing an 'active' group.

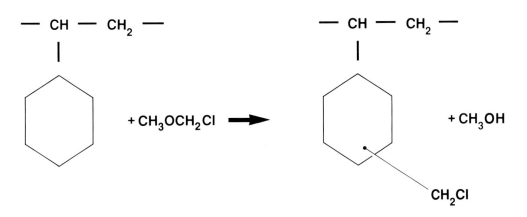

FIG. 3.6 Chloromethylation

Amination

The second stage of the anion resin production process is more straightforward, and involves the reaction of the still hydrophobic, pendant chloromethyl group with a suitable amine to form the amino group which is the basis of the anion exchange resin. Three such amines are conventionally used, the choice being dependent upon the desired functionality of the anion resin — trimethylamine (TMA), dimethylethanolamine (DMEA) and dimethylamine (DMA); see Fig 3.7.

Resins [A] in Fig 3.7 are known as *strong base anion type I*. Resins [B] in Fig 3.7 are known as *strong base anion type II*. Resins [C] in Fig 3.7 are *weak base anion*. The production of equivalent kinds of type I and type II anion resins is absolutely identical up to the end of the chloromethylation stage, the properties of the final product being determined entirely by the choice of amine.

Anion resins — acrylic polymer matrix

Again, the anion processes are more complex than the cation. Cross-linked polymethyl acrylate is reacted with a bifunctional amine, usually dimethylaminopropylamine, making use of an 'aminolysis' reaction between its primary amine function and the pendant acrylic ester. This directly gives a weak base anion resin by the function of the dimethylamino group.

$$- CH - CH_2 -$$
$$|$$
$$COOCH_3$$
$$\qquad + NH_2 (CH_2)_3 N(CH_3)_2 \rightarrow$$

$$- CH - CH_2 -$$
$$|$$
$$CONH (CH_2)_3 N(CH_3)_2 + CH_3OH$$

Strong base resins are not made by using a different amine, but by quaternisation of the weak base material, using methyl chloride.

$$R(CH_2)_3 N(CH_3)_2 + CH_3Cl \rightarrow$$
$$R(CH_2)_3 N^+(CH_3)_2 Cl^-$$

This is a *strong base type I* resin.

4.1.5 General properties in respect to deionisation processes

Strongly acidic cation resin

The general chemical properties are those of a strong acid and as such the resin when operating in the hydrogen form is capable of splitting strong and weak salts to produce the corresponding strong and weak acid. When operating in the sodium form it can exchange calcium and magnesium ions and may therefore be used to soften water.

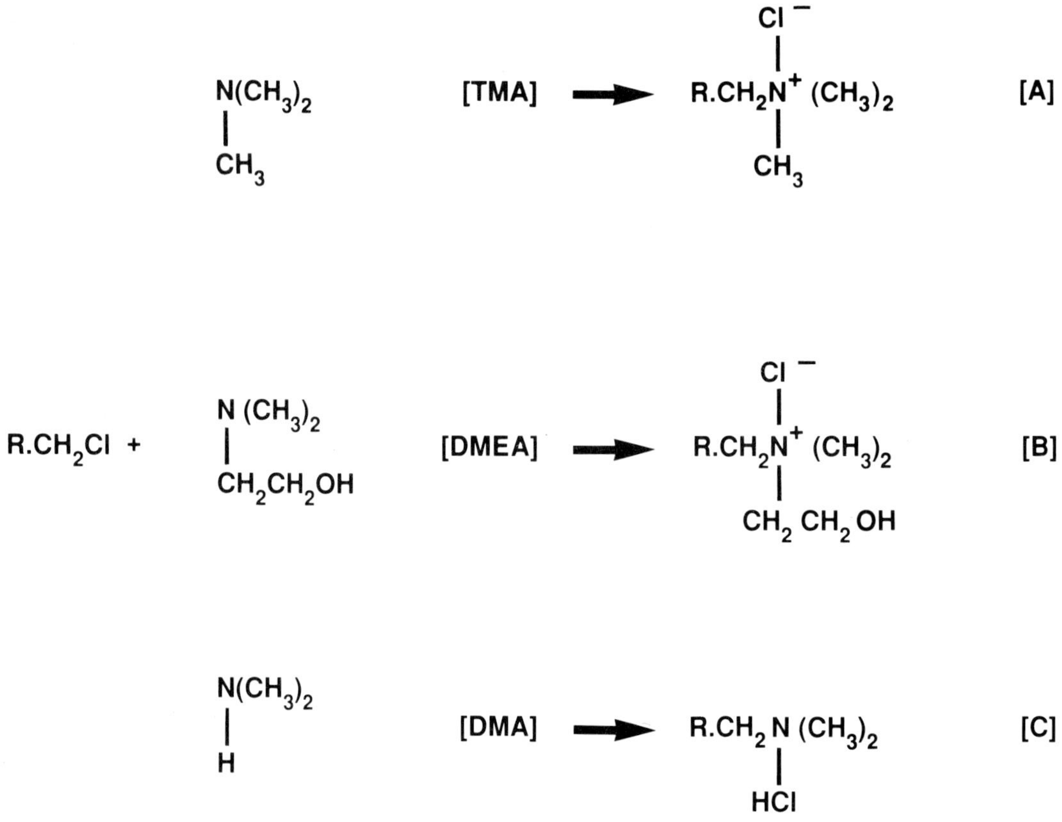

FIG. 3.7 Amination

Specific properties of the resin depend on the degree of cross-linking. Physical strength, i.e., the resistance to attrition and osmotic shock, is usually improved as cross-linking is increased, whereas the operating capacity will be reduced. The optimum cross-linking for general use of 8% divynyl benzene, but for additional strength 10–12% DVB resins are produced. Macroporous resins with DVB contents of up to 20% are produced for use under arduous conditions.

Weakly acidic cation resin

These resins are of lower equivalent weight than the strongly acidic resins and hence have higher total capacities.

The resin has a very high affinity for hydrogen ions. This arises from its structure and active grouping and can be compared chemically with a weak acid. The carboxylic grouping also has a very high affinity for calcium ions; accordingly, its main use in water treatment is to remove cations equivalent to the alkalinity of the raw water being treated.

Overall, the resins can be regenerated very easily and stoichiometric levels of acid can be employed.

Strongly basic anion resin

By definition this type of resin has a very low affinity for the hydroxyl ion. The lower the basicity the higher the affinity so that type II resin is more easily regenerated to hydroxyl form than type I.

Weakly basic resin

The resin has a high affinity for hydroxyl ions and can therefore be regenerated very effectively. However, its low basicity prevents the removal of weak acids such as silica or carbonic acid. Salts cannot be split by the resin and the reaction of the resin is that of neutralisation of mineral acids arising from a preceeding strongly acidic cation resin.

4.2 Ion-exchange reactions

In this section a brief outline is provided of the principles of the ion-exchange reactions taking place in the treatment of water.

It is useful to consider the ion-exchange resins as insoluble salts and we have already indicated that resins having strong and weak acid and base functional groups are available.

The reversible exchange reaction will be an equilibrium process involving:

$$\text{Resin X} + \text{Y} \leftrightarrow \text{Resin Y} + \text{X}$$

where X is a cation or anion attached to the resin and Y is a cation or anion in solution. If we regard the uptake of Y as the forward reaction this represents the treatment stage of a process whilst the application of an excess of X to the resin Y is the reverse reaction

and is equivalent to the regeneration stage. The reaction in any one direction cannot be 100% complete and, for example, in the forward reaction some of the Y ions will not be removed.

The equilibrium is represented by:

$$K = \frac{[R_y] [X]}{[R_x] [Y]}$$

where K is the equilibrium constant or selectivity constant for the particular resin used.

It follows that the highest removal of Y by R_x will occur when the resin is completely in the X form. As the ion-exchange reactions in water treatment take place in columns or beds of ion-exchange resin it is important to have the resin as high as possible in the X form to obtain a high quality product. In practice this can be achieved by ensuring that resin near the outlet from a bed is highly regenerated (see Section 5.4.2 of this chapter on counter flow regeneration). It should be noted that in using resin in a column the concentration of R_x is maintained by replenishment as the water passes through the column. This drives the reaction from left to right, i.e., in the forward direction.

Ion-exchange technology is primarily concerned with promoting the forward and reverse reactions of the reaction given above at minimum cost.

How do ions enter an ion-exchange resin?

In the case of a strongly acidic cation resin the positively charged cations in solution are attracted to the negatively charged sulphonate ions, whereas the negatively charged anions are repelled.

Electroneutrality is required at all times and there is a balance of charge of the ions entering and leaving a resin bead. The exchange is not restricted to the surface of a bead and migration of ions throughout a bead can occur. The exchange is a five stage process:

- Diffusion of ions from the solution to the surface of the bead.

- Diffusion of ions into the bead.

- Exchange of ions within the bead.

- Diffusion of the exchanged ion through the bead to the surface.

- Diffusion of the exchanging ion from the surface to the bulk solution.

The selectivity of a resin for different ions is controlled by electrostatic forces and depends on:

- The size of the ion and its charge (valency).

- The size of the hydrated ion; this is inversely proportional to the naked radii, for example, the naked lithium ion is half the size of a potassium ion, but the hydrated potassium ion is half the size of a lithium ion.

- The smaller the hydrated radii the greater the selectivity of the resin for ions of the same valency.

  For cations selectivity is $Rb^+ > K^+ > Na^+ > H^+$ for strongly acidic cation resin

  For anions the order is $HSO_4^- > NO_3 > Cl > OH$ for strongly basic anion resin

- The higher the valency the greater the selectivity, e.g., $Ca > Na$ and $SO_4 > Cl$.

Overall, it is the charge density of the ion which controls selectivity.

The rate of exchange is governed by two main factors:

- The diffusion across a static film of water at the surface of the bead (film diffusion).
- Diffusion within the bead (particle diffusion).

For the strongly acidic and strongly basic resins which are highly ionised, the rates of reaction within a bead are very high and in the treatment of normal waters which are relatively of low dissolved solids concentration *film* diffusion will control the rate.

Under high concentration conditions such as exist during regeneration, the rates of exchange, both across the film and within the bead, are high, but the rate controlling step is that of *particle* diffusion.

The weakly basic and acidic resins react more slowly than the strong types. This is because the resins are less ionic in nature and the exchange reaction involves the making and breaking of co-valent bonds and the rate controlling stage is within the bead (particle diffusion) even during the treatment stage.

The rate of exchange will also depend on the size of the ion involved, its valency and chemical structure. At the relatively low flow rates employed in make-up water treatment plants differences in the rate of exchange of inorganic ions are rarely observed, but the slow reaction rates of large organic ions can be detected.

Condensate purification plants operate at much higher flow rates (see Section 9 of this chapter) and under these conditions a difference between the rate of exchange of sulphate and chloride ions is evident.

The rate of exchange increases as the particle size of a resin bead decreases due to increased surface area and rate of diffusion. However, in a bed of resin as particle size is decreased the pressure drop across a bed will increase and hence for practical purposes there is a limit to which particle size can be reduced. It is usual to employ resins that have a bead size range of 0.3–1.2 mm.

4.3 Exhaustion and regeneration of ion-exchange resin

Under *co-flow* conditions, the water and regenerant solutions flow in the same direction during the exhaustion and regeneration stages, respectively.

In operating a co-flow ion-exchange plant, the ion-exchange units consist of a vertical column of resin through which the water is usually passed in a downward direction.

In order to understand the principles of operation of the ion-exchange column, it is useful to consider a cation exchange column with the resin initially in the hydrogen form.

The selectivity coefficients of a cation resin for calcium, magnesium and sodium exchange when operating in hydrogen form are:

$$\frac{Ca}{K} > \frac{Mg}{K} > \frac{Na}{K}$$
$$H \qquad H \qquad H$$

Hence, during the exhaustion stage calcium and magnesium are exchanged preferentially as the water passes through the bed so that it consists of layers of resin in different ionic forms as illustrated in Fig 3.8.

If, on regeneration, acid is passed through the bed in the same direction as the water flowed during the exhaustion stage, calcium, magnesium and sodium ions will be displaced down the column by hydrogen ions. At the end of a regeneration stage, the condition of the resin will be as shown in Fig 3.8, with the upper layers of resin in the hydrogen form but, at the outlet from the bed, sodium form resin predominates unless gross quantities of acid are used. During the subsequent exhaustion stage, water of high hydrogen ion content will be contacting resin of relatively high sodium content at the outlet from the bed. This is an unfavourable equilibrium and sodium will be displaced and appear in the treated water. This will continue for the early part of the cycle with the 'leakage' of sodium gradually decreasing. The sodium concentration present will then plateau for a period and during this time the sodium entering with the inlet water will be exchanged at the top of the bed with the exchange zone gradually moving down the column. Ultimately, the zone will reach the bottom of the column and the bed will again be in the condition shown in Fig 3.8, and there will be a gradual increase in the sodium content of the treated water. When a predetermined concentration is reached the bed is deemed to be exhausted.

It follows that if the exhaustion and regenerant flows are in opposite directions, following a regeneration, water leaving the unit would be in equilibrium with the most highly regenerated resin and extremely low levels of sodium leakage will be obtained throughout the exhaustion cycle up to the breakpoint. Figure 3.9 illustrates this point in terms of the conductivity obtained from a cation-anion system. This technique is known as 'counter flow' operation. A more complicated unit is required to permit this mode of operation, but a variety of designs is available and these are discussed in Section 5.4 of this chapter.

4.3.1 Capacity

An ion-exchange resin has a finite total capacity which depends on the total number of ion-exchange sites

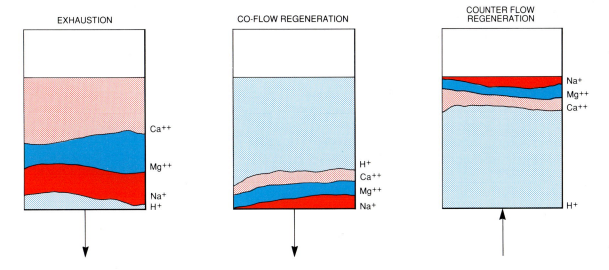

FIG. 3.8 Exchange of cations in service and regeneration

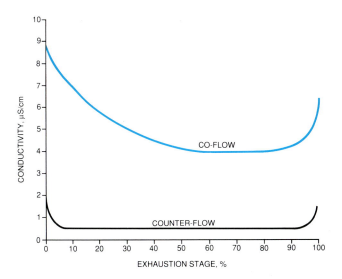

FIG. 3.9 Conductivity following cation-anion with
co-flow and counter flow regeneration

present. However, it is the operating capacity which is the more important parameter.

Operating capacity is the output obtained from a bed of resin up to a pre-determined breakpoint. For cation resin this will be set by the sodium leakage, but practically, this can be determined by a decrease in the conductivity of the water leaving a cation unit, or an increase in the conductivity of water leaving the following strongly basic anion unit.

The operating capacity of a cation resin depends on several factors:

(a) Ease of removal of exchanged ions.

(b) Strength and quantity of acid used.

(c) Type of anions present in the water being treated.

The ease of removal of cations is the major factor and this depends on their relative selectivities and we have already indicated that the order of selectivity is Ca > Mg > Na. Hence, for ease of removal the order is reversed, i.e., Na > Mg > Ca, and for a given usage of acid the operating capacity will increase as the proportion of sodium in the raw water increases. In the case of (b), the capacity will obviously increase with an increase in the quantity of acid used. However, the strength of acid is also important in that it reflects the concentration of hydrogen ions present during the regeneration. In the reaction:

$$R_2\,Ca + 2H^+ \leftrightarrow 2R_H + Ca$$

increasing the hydrogen ion concentration promotes the forward reaction so that the regeneration is favoured. The strength of acid has to be related to plant factors such as distribution within the unit, ease of handling and effects on the materials of construction.

In the UK power industry, sulphuric acid is usually employed because it is cheaper than the most obvious alternative, hydrochloric acid. With most waters, sulphuric acid has to be used at undesirably low concentrations (from equilibrium considerations) to avoid precipitating calcium sulphate. The actual strengths used vary from 1.5–3.0% depending on the proportion of calcium present in the raw water, as does the operating capacity.

The third factor affecting the operating capacity is the proportion of alkalinity present in the raw water. Alkalinity represents the bicarbonate (HCO_3^-) content. During the exhaustion stage alkalinity is removed as follows:

$$2\,R_H + Ca\,(HCO_3)^2 \leftrightarrow R_2\,Ca + 2H_2\,CO_3 \quad (3.1)$$

If we compare this with the removal of calcium chloride as by:

$$2R_H + CaCl_2 \leftrightarrow R_2\,Ca + 2HCl \quad (3.2)$$

Now in Equation (3.1) the carbonic acid produced is weakly dissociated, i.e., the concentration of hydrogen ions present is low so that the backward reaction is not promoted. Whereas in Equation (3.2), hydrochloric acid is produced which is highly dissociated and a high level of hydrogen ions will be present so that the backward reaction will be promoted and consequently the position of the equilibrium is less favoured for the removal of calcium. The uptake of cations is therefore more effective as the proportion of alkalinity present in the raw water increases and this allows greater depth of the resin bed to be used before exhaustion occurs and hence a higher operating capacity is obtained.

Ion-exchange resin manufacturers provide practically derived data showing the variation of operating capacity with the composition of the raw water and the quantity of acid used.

Flow rate during the exhaustion cycle can affect capacity and factors to allow for various flow conditions are also provided.

Figure 3.10 shows the type of data provided and Table 3.2 gives details of capacities attainable with differing levels of regeneration with various types of water.

TABLE 3.2

Capacity of cation resin

Quantity of regenerant g.H₂SO₄/1	Composition of raw water				Capacity g.CaCO₃/1
	%Na	%Ca	%Mg	Alkalinity	
48	40	35	25	50	25
64	45	30	25	80	33
80	20	75	5	80	34
96	60	0	40	30	42
112	83	8	9	95	53
128	100	0	0	10	54
144	61	26	13	72	46
160	7	9	84	53	50

The above has considered the behaviour of cation exchange resins during the exhaustion and regeneration cycle.

Anion exchange resins show a similar pattern of behaviour in respect to leakage and operating capacity. However, the effect of sodium leakage from the cation resin on a following anion exchange bed should be noted.

Any sodium ions leaving the cation resin will be associated with the anions entering the anion exchange section and the following reaction occurs:

$$R_{OH} + NaCl \leftrightarrow R_{Cl} + NaOH$$

The hydroxyl ions produced can affect the removal of weak anions particularly silica; consequently, the leakage of silica from an anion exchange bed is dependent on the sodium leakage from the cation unit.

The small quantities of ions leaving cation and anion exchange beds have then to be removed by the intro-

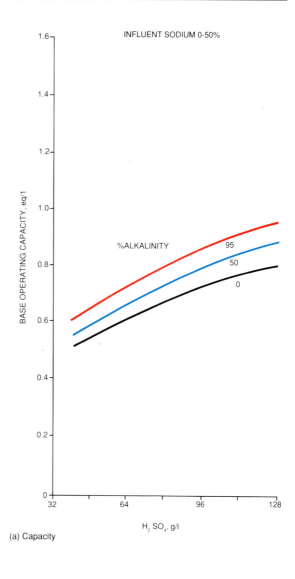

(a) Capacity

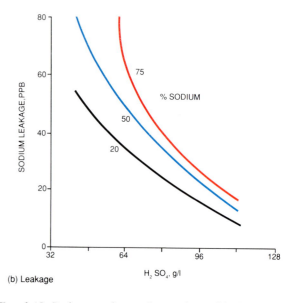

(b) Leakage

FIG. 3.10 Resin manufacturer's capacity and leakage data (Purolite Ltd)

duction of a final stage of treatment in a mixed bed unit (see Section 5 of this chapter).

4.4 Deterioration of ion-exchange resins

Ion-exchange resins will frequently give a number of years' service before deteriorating in performance, such that they have to be replaced.

Resin performance can be impaired by both physical and chemical adverse effects.

Physical breakdown

Resin beads can break down into smaller fragments by attrition, partly through mechanical damage and partly arising from the continual expansion and contraction that occurs during the service and regeneration cycles. If this occurs the main adverse effect is an increase in pressure drop across the bed. Some of the fines may be removed from the bed by backwashing but it may be necessary to skim the top of the bed and add new resin. Ultimately, complete replacement may become necessary.

Breakdown of the cation resin in a mixed bed makes separation from the anion resin more difficult, with a consequent risk of a poor quality water being obtained (see Section 5.4.5 of this chapter).

Chemical breakdown

Anion resins are inherently less stable than cation resins and, with time, losses in total and strongly basic capacity occur. The reactions involved are shown in Fig 3.11 and these follow a first order decay rate.

Figure 3.12 illustrates the changes that occur with time under make-up water conditions.

The rate of decay increases with a rise in temperature. Now, anion resins in the mixed beds used in condensate purification are exposed to water in the range 30–50°C. Consequently, a more rapid loss of active groups arises. This factor coupled with some fouling adversely affects the kinetic ability of the resin such that, at the very high flow rates used, the extremely high quality water required cannot be produced, and the capacity for coping with ingresses of cooling water can be seriously affected. The overall result is that anion resins in condensate purification plants currently have short working lives, ≈ 2 years. This problem is covered in more detail in Section 9 of this chapter.

Cation resins are much more stable and it is comparatively rare for any problems to arise from a loss of capacity even when the resins are subjected to temperatures in excess of 100°C.

Fouling

Fouling is the attachment to resins of materials which can adversely affect their performance. This can arise with all types of resin, but more serious fouling is usually associated with anion resins.

Cation resin is usually in the first unit of the treatment process, hence it is more likely to receive silt, iron oxides, aluminum hydroxide, etc., either from the raw water or from carry-over via a pre-treatment stage. Backwashing and the use of acid when operating with the resin in hydrogen form are generally able to minimise the problem, but greater difficulty can arise when the resin is used in the sodium form for softening water.

Cation resins can become fouled with calcium sulphate arising from regenerating with sulphuric acid,

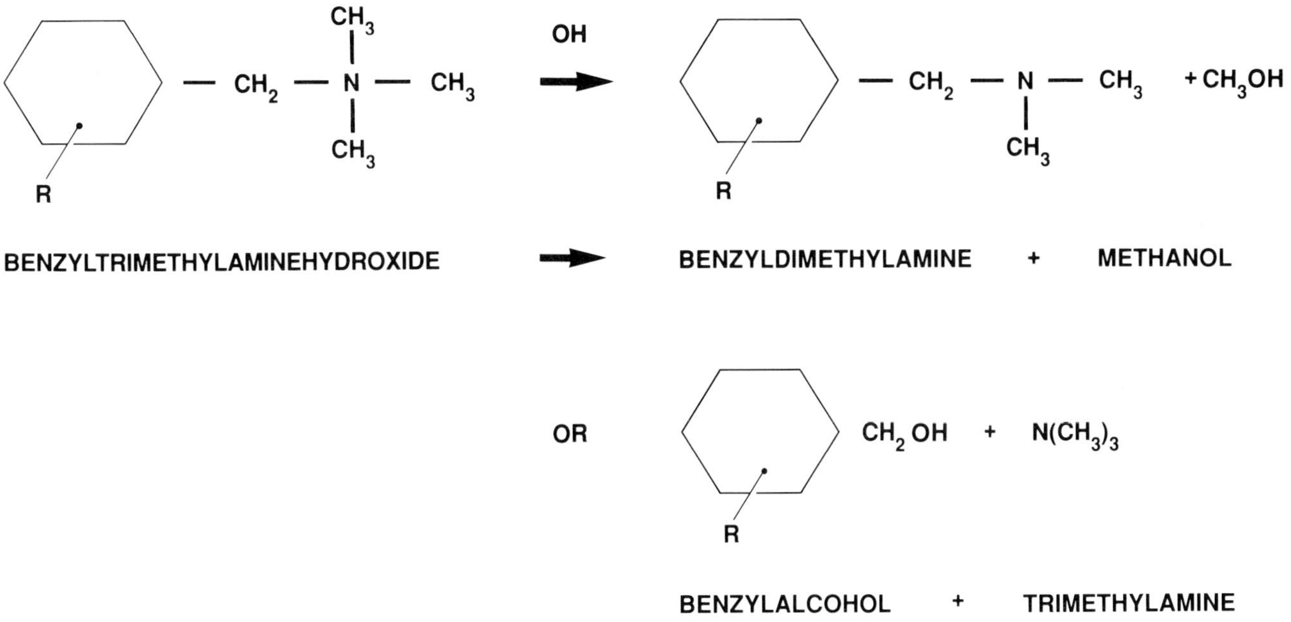

FIG. 3.11 Losses in total and strongly basic capacity

235

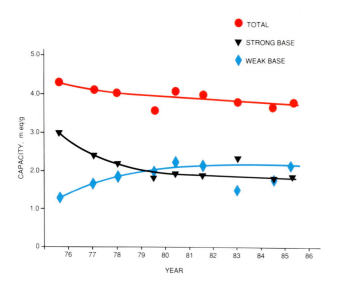

Fig. 3.12 Loss of exchange groups with time for an acrylic strong base resin

particularly if the concentration is too high and the flow too low during regeneration. The weakly acidic resins are particularly susceptible as they remove predominantly calcium ions from the raw water. Such resins are frequently used in a two-layered bed above strongly acidic cation resin, and when they become fouled their density increases and separate layers cannot be maintained. This results in a loss of operating capacity.

There is a need to remove organic matter from raw water, however, in so doing anion resins can be fouled. This is the most common form of chemical fouling that occurs in water treatment plants. Fouling of anion resins by organic matter and the closely related subject of the removal of organic matter have presented problems to the operators of plant since the advent of anion exchange resins. The subject is therefore considered in detail in the following sections.

4.4.1 Organic matter and anion resins

Methods of monitoring organic matter

Until recently, the methods of determining organic matter in water were time consuming and often empirical.

The development of the total organic carbon (TOC) in water analysers has enabled the overall level of or-

ganic matter to be estimated rapidly and reliably. The analysers are capable of detecting differences as small as 0.03 mg kg^{-1} organic carbon.

Before the advent of total organic carbon analysers, potassium permanganate demand was the main method used for determining the overall level of organic matter in water. This is a determination of the degree to which organic matter in water is oxidised by an acidic potassium permanganate solution. Unfortunately, over the years a variety of conditions for carrying out the test has been used. For example, the time of contact of the water with the permanganate has been varied from 30 minutes to 240 minutes, with temperature conditions ranging from 20°C to 100°C. Additionally, results have been expressed in terms of oxygen absorbed (O$_2$) and as permanganate (KMnO$_4$); consequently, it has frequently been difficult to compare data produced by different workers.

Water is also analysed for its fulvic acid content. This determination is not specific for fulvic acids. Many species, e.g., unsaturated compounds and iron, will absorb at the wavelength used and would consequently be recorded as fulvic acid. In this method the sample is acidified to pH 1.0 and filtered to remove humic acids. The fulvic acids are then determined by UV absorption at 300 nm.

Table 3.3 shows results for samples taken from a plant treating an industrial river water and analysed by various methods for determining organic matter content [2]. The relationship between total organic carbon and potassium permanganate value change little throughout the demineralisation plant. In contrast, the total organic carbon and fulvic acid levels show that the relative proportions changed at various stages of the treatment. As expected, fulvic acids were removed more completely (94%) than organic matter in general (82%). The average level of fulvic acids in the make-up water was 0.3 mg kg^{-1} for a TOC level of 1.4 mg kg^{-1}.

The nature of organic compounds and fouling of anion resins

The fouling of strongly basic anion exchange resins, which has been reported extensively since their introduction in 1952, is most frequently observed with peaty waters. These waters are low in dissolved solids, but contain significant quantities of organic material of complex structure resulting from the decomposition of vegetable matter. Although the composition and

TABLE 3.3

Comparison of total organic carbon, potassium permanganate values and fulvic acid levels

Position	TOC mg kg^{-1}	Permanganate value (PV) mg KMnO$_4$ kg^{-1}	Ratio $\dfrac{PV}{TOC}$	Fulvic acid mg kg^{-1}	Ratio $\dfrac{fulvic}{TOC}$
Cation inlet	7.9	44.7	5.6	5.8	0.73
Anion outlet	1.6	8.1	5.1	0.4	0.25
Mixed bed outlet	1.4	6.8	4.9	0.3	0.21

amounts of the organic compounds are not known in detail, they are broadly characterised as polycarboxylic acids of high but variable molecular weight. They have been further arbitrarily grouped into fulvic acids which are soluble at pH 1 and humic acids which are not. Elemental analyses are shown in Table 3.4. They indicate that humic acids have less carboxylic groups than fulvic acid and that the complex molecules may also contain amine type basic groupings.

TABLE 3.4

Elemental analyses of humic and fulvic acids

Element %	Humic acid	Fulvic acid
Carbon	54.4	49.4
Hydrogen	3.6	1.5
Nitrogen	1.9	2.4
Oxygen	40.1	46.7

The chain length and aromaticity of organic compounds will largely dictate the selectivity of uptake by anion resin. As the chain length increases, the selectivity increases, but eventually a point is reached when the size of the organic molecule is too large to enter the resin (this will depend on the type of resin) and selectivity will then appear to decrease with chain length.

Note that although the selectivity of the resin for large organic ions will be high, the rate of uptake will be lower than for the smaller inorganic ionic species and the degree of removal achieved will be controlled by rate processes rather than the equilibrium control that predominates in the removal of inorganic ions under normal operating conditions.

The nature of organic fouling is still not fully understood. It may be a combination of chemical exchange and physical adsorption and involve more than one group on the molecule. Fouling due to simple mechanical blocking may also be a possibility, but should be minimised by the coagulation of suspended and colloidal organic matter followed by flocculation and filtration, which is usually incorporated as pre-treatment in high grade water treatment plants. Chemical exchange of humic and fulvic acids does take place on strongly basic resins and there is some release on regeneration. However, without a detailed knowledge of the number of compounds involved and their compositions, one can only speculate on whether some fractions are eluted completely and others completely retained, or whether such factors as depth of penetration into the bead and possible chemical modification are important in irreversible fouling.

Four important general points can be made with respect to the mechanism of organic fouling:

- It must be possible for the organic matter to attach itself to the resin.

- The resin will exhibit a high selectivity for the organic matter.

- The organic matter will not be completely removed by the normal regenerating agent.

- The degree of fouling experienced will be a function of the nature of the organic matter and the chemical and physical characteristics of the resin used.

The effects of organic fouling are:

(a) Reduced capacity.

(b) Increase in electrical conductivity of the treated water.

(c) Prolonged rinse requirements.

(d) Low pH of final treated water.

These mainly arise from a loss of active exchange sites and a decrease in the exchange rate. Obviously (a) and (c) will affect operating costs and, overall, the working life of an anion resin could be considerably curtailed. For example, resin beds have had to be replaced after only a few months use, whereas the expected life is around 5 years.

4.4.2 Treatment of waters containing organic matter

Two conflicting situations arise:

- The need to remove organics from the water.

- The need to remove organics from the resin to maintain performance.

Unfortunately, the resins most likely to remove the organic matter are likely to be the resins least likely to release the organic matter on regeneration. Furthermore, the organic species that have most difficulty entering the resin will become the ones that are most difficult to remove from the resin. The variety of organic compounds present in water coupled with the variations in operation of ion-exchange plant make it impossible to select a single type of resin as the 'best' option. However, there are general guidelines that can be given to minimise the effects of organic matter.

The choice of the treatment process will depend on the level of organic matter in the water and its type in ratio to the total ionic load. Usually, the greater the ratio of organic matter to the total ionic load the greater the risk of fouling of the anion resins.

The main treatments currently used on organic-bearing waters to produce water of low organic content and to minimise organic fouling of anion resins, are:

- Flocculation and filtration.

- Macroporous resins in chloride form.

- Acrylic anion resins.

- Chemical treatment of the fouled resin.

Flocculation and filtration preceding the ion-exchange section can remove 40–70% of the organic matter, the actual figure will depend on the type of water being treated. In this stage of treatment it is mainly the larger organic species that are removed.

The use of an additional ion-exchange unit using an anion resin in the chloride form ahead of the main deionisation plant may be used instead of, or in addition to, flocculation. A macroporous resin is used and the unit is often described as a 'scavenger' unit or organic trap. The percentage removed across such beds is about 70% and again, some of the larger organic species with a high propensity for fouling the following resins operating in hydroxyl form are removed.

Capacity data for organic matter uptake are limited; frequently the units are run to a fixed volume through-put, for example, 2500 bed volumes. Regeneration is then carried out in two stages using, firstly, 10% sodium chloride with 2% sodium hydroxide, preferably warm, and, secondly, using a small volume of 4% sodium chloride.

Overall, a 'scavenger' unit has advantages in ease of operation and cost relative to the frequently used alternate clarification plant, but the quality of the final treated water will be similar with either option. Table 3.5 shows typical results using these two different water treatment options on the same station treating an upland surface water [2].

TABLE 3.5

Scavenger versus flocculation on a surface water

Samples	Total organic carbon mg kg^{-1}	
	Plant A (Scavenger)	Plant B (Clarification)
Raw water	2.2	2.05
Scavenger or pre-treatment	0.70	1.3
Cation outlet	0.70	1.3
Anion outlet	0.25	0.25
Mixed bed outlet	0.25	0.20

Acrylic resins are now widely used on the main anion units of water treatment plants. They are avail-

able as type I strongly basic anion resin and have high operating capacities and are relatively resistant to organic fouling. This latter property arises from their lower selectivity for organic species compared to similar resins based on a polystyrene matrix.

A comparison of quality of water produced from acrylic and polystyrene based anion resins when used in the main anion section of a water treatment plant treating an industrial river water gave the results shown in Table 3.6 [2].

Even with new resins, differences in the quality of water from the mixed beds which followed the anion units was apparent, particularly in the lower level of total organic carbon (TOC) in the water from the stream using the acrylic resin. After an eight month period of operation the conductivity of water from the stream using the polystyrene resin had increased appreciably.

The polystyrene resin was replaced with acrylic resin after 3 years, subsequently, the acrylic resins continued to produce a similar quality water to that given in Table 3.6 for a ten year period.

This example shows the advantage of one type of resin over another, but also highlights a problem remaining that is unlikely to be overcome by the use of ion-exchange processes alone. That is, the removal of the residual TOC which was present even when the acrylic resin was used. Alternative, and probably expensive, stages of treatment would be needed on this water to produce a treated water of low TOC.

Organic fouling of anion resins can be alleviated by cleaning the resin with a brine/caustic solution. For best results a programme of regular cleans is required, say every 50 cycles, but this is frequently impractical due to the demand for treated water. The optimum concentration of the cleaning solution is 10% sodium chloride with 3–4% sodium hydroxide and soaking the resin in the solution for several hours will give the best results.

As indicated above, despite extensive treatment, organic matter can be present in make-up water and a small proportion may be present as organo-chloride. The subsequent decomposition of the organic material within the boilers can result in contamination of the system by organic acids, ionic chloride and carbon

TABLE 3.6

Acrylic versus polystyrene resin for organic removal

		Anion outlet		Mixed bed outlet	
		Acrylic resin	Polystyrene resin	Acrylic resin	Polystyrene resin
New resins	TOC mg kg^{-1}	0.7	1.2	0.5	0.9
	Conductivity μS cm^{-1}	2.3	0.8	0.08	0.12
After 8 months	TOC mg kg^{-1}	0.8	1.4	0.5	1.2
	Conductivity μS cm^{-1}	3.3	0.7	0.1	0.9

dioxide. This could have two effects, firstly, control of the feedwater conditions becomes more difficult and, secondly, the risk of corrosion occurring is increased. Qualitatively, the potential effects of this ingress of organic matter are known, but whether the quantities involved could be significant is still uncertain. The worst situation is likely to be an ingress of organo-chloride to a plant susceptible to corrosion.

Various additional treatments to produce make-up water of low TOC content have been investigated including the use of activated carbon, and oxidising agents, such as ozone and hydrogen peroxide. Of these, activated carbon would provide the most effective reduction in TOC and Table 3.7 shows the results obtained during investigational work [2]. However, the cost involved in the frequent changes of the material would be high and makes its use unattractive.

More recently, a photo-oxidation technique has been developed [3]. UV light is passed through the water to which hydrogen peroxide has been added and the optimum position for its use would be at the outlet from the main anion exchange unit. Significant removal of organo-chloride in addition to general organic matter can be obtained and the costs of treatment are likely to be significantly lower than alternative techniques.

The use of reverse osmosis in place of the main cation-anion stage in make-up water treatment plant (see Section 6 of this chapter) would allow a low residual TOC to be obtained, and this may be one of the factors influencing whether it is chosen in preference to ion-exchange for a particular application.

4.5 Ion-exchange resin testing — chemical tests

There are several reasons for testing ion-exchange materials.

There is a need to examine new resin supplied by the manufacturers to determine whether the material conforms to its specification. In the power industry contracts are let on the basis that the quality of the resin, in terms of its chemical and physical condition,

will conform to an agreed specification. Tables 3.8 and 3.24 give details of the standards required for use in make-up and condensate purification plants respectively.

It is important that material is tested before being used to determine that the contract conditions have been met. In this way, the use of poor quality materials which would have to be replaced in a short period is avoided. Additionally, the consequent risks of short-falls in the quantity and quality of water produced arising from inferior materials are reduced.

The previous section has indicated how deterioration in the performances of ion-exchange resins arises. By examining the resins at regular intervals the degree of deterioration can be assessed and an estimate of the actual and potential increase in operating costs arising can be made. This allows resin replacement programmes to be soundly based in terms of cost and potential risk to the plant.

Tests have been developed to determine the chemical and physical condition of an ion-exchange resin. Table 3.9 lists the types of tests which may be employed.

As stated previously anion resins are chemically less stable than anion resins, consequently, they are likely to be tested more frequently and a greater number of types of test may be used to assess their condition.

4.5.1 Capacity tests on cation resins

Total cation exchange capacity, C_T

This is determined by completely converting a known weight of the resin to the hydrogen form, by passing a considerable excess of 1N nitric acid through it. The regenerated resin after rinsing to remove excess acid, is then allowed to stand for several hours in contact with decinormal (0.1N) sodium hydroxide solution containing 10% (w/v) sodium chloride, and the reduction in sodium hydroxide concentration which results is determined by titration with standard acid solution. This reduction in hydroxide is equal to the amount of hydrogen ions released from the resin and hence the total exchange capacity of the resin.

TABLE 3.7

Performance of activated carbon for removal of organics

Volume treated BV	Total organic carbon mg kg^{-1}		
	Anion unit outlet	Mixed bed column outlet	Activated carbon outlet
432	3.2	0.8	0.4
864	2.7	0.7	0.2
2406	2.4	0.7	0.3
4558	3.8	0.6	0.2
8147	3.2	0.5	0.1
10 820	2.8	0.8	0.7
11 540	2.2	0.8	1.0

TABLE 3.8
Specifications for ion-exchange resins in make-up water treatment plants

	Single beds				Mixed beds*		
	Cation exchanger	Anion exchangers			Cation exchanger	Anion exchangers	
Polymer matrix type	Styrene gelular strongly acidic	Styrene gelular strongly basic (Type 1)	Styrene macroporous strongly basic (Type 1)	Acrylic gelular strongly basic (Type 1)	Styrene gelular strongly acidic	Styrene gelular strongly basic (Type 1)	Styrene macroporous strongly basic (Type 1)
Total exchange capacity (eq/litre in ionic form)	≮ 1.9 (H$^+$ form)	≮ 1.2 (Cl' form)	≮ 1.0 (Cl' form)	≮ 1.2 (Cl' form)	≮ 1.9 (H$^+$ form)	≮ 1.2 (Cl' form)	≮ 1.0 (Cl' form)
Bead integrity — Whole crackfree Beads, %	≮ 95	≮ 95	≮ 95	≮ 95	≮ 95	≮ 95	≮ 95
Broken beads, %	≯ 2	≯ 2	≯ 2	≯ 2	≯ 2	≯ 2	≯ 2
Physical strength Breakdown after OSA test, %	≮ 10	≮ 20	≮ 10	Test not applicable	≮ 10	≮ 20	≮ 10
Number of OSA test cycles	25	500	500		25	25	500
Bead size distribution (Diameter in ionic form)	(H$^+$ form)	(Cl' form)	(Cl' form)	(Cl' form)	(H$^+$ form)	(Cl' form)	(Cl' form)
Low flow rate (<30 m/h linear flow) — Range (min-max), mm	0.3 – 1.2	0.3 – 1.2	0.3 – 1.2	0.3 – 1.4	0.6 – 1.2 *(All flow rates)*	0.4 – 1.0 *(All flow rates)*	0.4 – 1.0 *(All flow rates)*
Harmonic mean, mm	0.60 – 0.85	0.60 – 0.85	0.60 – 0.85	0.75 – 1.0	0.65 – 0.90	0.55 – 0.75	0.55 – 0.75
Undersize, mm	≯ 1% < 0.3	≯ 1% < 0.3	≯ 1% < 0.3	≯ 1% < 0.3	≯ 1% < 0.6	≯ 2% < 0.4	≯ 2% < 0.4
Oversize, mm	≯ 5% > 1.2	≯ 5% > 1.2	≯ 5% > 1.2	≯ 5% > 1.4	≯ 5% > 1.2	≯ 2% > 1.0	≯ 2% > 1.0
High flow rate (>30 m/h linear flow) — Range (min-max), mm	0.3 – 0.4	0.3 – 1.4	0.3 – 1.4	0.3 – 1.4			
Harmonic mean, mm	0.75 – 1.0	0.75 – 1.0	0.75 – 1.0	0.75 – 1.0			
Undersize, mm	≯ 1% < 0.3	≯ 1% < 0.3	≯ 1% < 0.3	≯ 1% < 0.3			
Oversize, mm	≯ 5% > 1.4	≯ 5% > 1.4	≯ 5% > 1.4	≯ 5% > 1.4			
Ionic form as supplied	H$^+$ or Na$^+$	Cl'	Cl'	Cl'	H$^+$ or Na$^+$	Cl'	Cl'

* For two resin systems

TABLE 3.9

*Examination and testing
of ion-exchange resins*

Techniques and tests

Total exchange capacity
Individual group capacity
Operational capacity
Extraction of foulants
Identification of foulants
Kinetic leakage
Osmotic shock/attrition
Visual microscopy
Moisture and density
Bead size distribution

$$C_T = (V_1 N_1 - 4N_2 t_1)/W(100 - M) \times 100 \text{ meq g}^{-1}$$
of dry resin

where V_1 = volume of 10% sodium chloride/0.1N sodium hydroxide

N_1 = normality of the decinormal sodium hydroxide solution

N_2 = normality of the standard acid

t_1 = titre, ml

W = weight of moist resin in hydrogen forms, g

M = % moisture content of the resin

Note: meq = milli-equivalents

Strongly acidic exchange capacity

This is determined by passing 1 litre of 4% (w/v) sodium sulphate solution through a known weight of the fully regenerated resin and titrating a 100 ml aliquot of the resulting solution with standard sodium hydroxide solution.

The weakly acidic capacity is obtained by difference.

The chemical capacity under plant conditions (operational capacity) can be determined as described later for anion resins, but using sodium hydroxide to exhaust the resins rather than sulphuric acid and measuring the volume throughput to a 30 μS cm^{-1} conductivity break.

Cation resins are extremely stable and frequently only the total exchange capacity will be carried out.

4.5.2 Capacity tests on anion resins

The determination of the strong and weak groupings is particularly important for anion resins because of their relatively low stability.

Weakly basic capacity C_W

This is determined by completely converting the resin to the chloride form, passing 1250 ml of 1% (v/v) ammonia solution through a known weight of the resin in this form and collecting the first 1000 ml of effluent in one receiver and the remaining 250 ml fraction in a second receiver. The chloride content is then determined by titrating 100 ml aliquots of the fresh 1% (v/v) ammonia solution and the first and second fractions, in turn, against standard silver nitrate solution. The ammonia solution elutes the chloride from all weakly basic groups plus some of the strongly basic groups and the value of chloride content in the second fraction is used to estimate the amount of chloride derived from strong groups, which is contained in the first fraction. Then:

$$C_W = 1000N(t_1 - t_2)/W(100 - M) \text{ meq g}^{-1}$$
of dry resin

where N = normality of the silver nitrate solution

t_1 = titre in ml, of the first fraction aliquot

t_2 = titre in ml, of the second fraction aliquot

W = weight of moist resin taken, g

M = % moisture content of the resin

Total basic capacity C_T

1000 ml of 4% sodium sulphate solution is then passed through the same resin and the effluent collected. The chloride content is then determined as before on a 100 ml aliquot of the effluent and n 100 ml of the fresh 4% (wv) sodium sulphate solution. Then:

$$C_T = 1000N(t_1 - t_3) + (t_2 - t_3/4) + (t_4 - t_5)/W(100 - M)$$

$$C_T = 250N(4t_1 + t_2 - 5t_3 + 4t_4 - 4t_5)/W(100 - M) \text{ meq g}^{-1}$$
of dry resin

where t_3 = titre in ml, of the fresh 1% (v/v) ammonia solution

t_4 = titre in ml of the effluent after passing sodium sulphate

t_5 = titre in ml of the fresh 4% (w/v) sodium sulphate

Strongly basic capacity, C_S

Strongly basic capacity is obtained by difference, i.e., $C_S - C_T - C_W$ meq g^{-1} of dry resin

A modified procedure can also be used in which the total and strongly basic salt-splitting capacity is determined and the weakly basic capacity is calculated from the difference. Similar values for total capacity are obtained with either method, but by carrying out a salt-splitting capacity within the second method a direct estimate of strong base capacity is obtained. This overcomes the difficulty of making an allowance for any chloride eluted from the strongly basic groups when ammonium hydroxide is used.

As a result the latter method gives slightly higher results for the strongly basic capacity.

Typical values for capacities determined by the modified procedure for a styrene type I strongly basic resin are: total capacity 4.4 meq g^{-1}, strongly basic capacity 4.3 meq g^{-1} and weakly basic capacity 0.1 meq g^{-1}.

4.5.3 Operational capacity

The methods given below are for anion resins as the test is more likely to be required for this type of resin. However, the same principles apply to testing cation resins.

The operational capacity can be determined by passing a standard strength sulphuric acid solution through the resin at fixed flow rates until a set end point is reached. The apparatus used, employs two test columns operating at different flow rates. The resins are regenerated with sodium hydroxide at 128 g l^{-1}, rinsed to a conductivity of 30 μS cm^{-1} and then exhausted with a 342 mg kg^{-1} sulphuric acid solution containing a trace of silicic acid.

One column operates at 18 bed volumes/h to show the kinetic performance of the resins. Both columns are run to an end point of 0.2 mg kg^{-1} SiO$_2$ until steady conditions are reached.

A preferable, but more time consuming approach, is to use columns of resin regenerated under conditions which simulate the actual plant conditions under which the resin will be used.

4.5.4 Identification of foulants

A wide variety of metallic elements, such as iron, aluminium, calcium, barium, copper and nickel can be present on cation resin. Fouling can arise from a single element, e.g., calcium, or from the collective effect of metals such as iron, copper and nickel.

Extraction techniques using hot solutions of acids can be used to remove the contaminants and extracts can then be examined using an atomic absorption technique.

The removal of organic foulants from anion resins is difficult, but for investigational purposes an aqueous alcoholic caustic brine solution is the most effective method. The organic nature of the extract can then be indicated by absorption at 300 nm. It is also useful on occasions to omit the alcohol from the extraction stage, the organic matter extracted can then be determined by a total organic carbon determination and/or using a permanganate oxygen absorption technique. This extractant solution can more readily be used under plant conditions, hence the test can be used to assess the possible effectiveness of a plant clean.

4.5.5 Kinetic test

The kinetic abilities of resins are important under the high flow rate conditions which obtain in condensate purification plants.

A kinetic test method has been developed within the power industry and details on its use are given in Section 9 of this chapter [4].

4.6 Ion-exchange resin testing — physical tests

4.6.1 Bead size distribution

Bead size distribution is important in determining the pressure drop characteristics of resin and therefore there is a need to check this parameter on both new and used resins. Additionally, in testing the physical strength of resin a knowledge of bead size is required to assess the degree of breakdown after applying a chemical and physical shock to the resin.

A HIAC particle size analyser could be used as an alternative to wet sieving. In this instrument a collimated beam of light shines through optical windows across a fluid stream in which the particles of ion-exchange resin to be measured are suspended. The light passing through the cell falls upon a photodiode, the output from which is amplified and measured. When a particle passes through the cell a shadow is cast on the diode due to that portion of the incident light which impinges on the particle either reflected or refracted. This shadow is proportional to the cross-sectional area of the particle. The reduction in electrical output from the photodiode equivalent to this extinction area activates a counter in the measurement channel between whose millivolt threshold settings the value lies. The instrument has to be calibrated with particles of known size to be able to correlate millivolts with microns.

Obviously, the capital cost of the HIAC instrument is much greater than using the manual wet sieving technique, but it provides a much shorter and simpler analysis. Moreover, potentially more accurate and reproducible results can be obtained. These advantages are particularly important in view of the trends in water treatment in respect to physical breakdown, and the demand for resins of more uniform size.

4.6.2 Osmotic shock and attrition test

This test was developed within the power industry to assess the ability of resins to withstand plant conditions using an accelerated test procedure [5].

During service, resins are subjected to both physical and chemical stresses. For example, compressive forces arise from the resistance of a bed of resin to flow and sudden changes in flow create an impact effect. Abrasion can arise when the resins are backwashed or transferred along pipelines from one vessel to another, such as occurs during the regeneration of resins used in condensate purification plant (see Section 9 of this chapter). Regeneration of the resins with relatively concentrated chemical solutions produces osmotic effects and changes in volume both leading to a chemically induced stress condition.

The above effects are introduced in the test apparatus by subjecting the resin to a sequence of regeneration, rinse, exhaustion and rinse stages. The regeneration and exhaustion stages introduce the main osmotic shock effect, whilst during each rinse stage physical stress is introduced by impacting the resin against a mesh screen. The sequence is repeated automatically for a predetermined number of cycles, and any resin which passes the screen during the test is collected to provide a quantitative measure of breakdown. Figure 3.13 shows the apparatus in which two samples of resin can be tested simultaneously. Details of the procedure are as follows.

Firstly, beads less than 0.5 mm diameter are removed from samples of cation and anion resin as measured in the sodium and chloride form, respectively. The bead size distribution is then measured using the HIAC particle size analyser mentioned previously.

A 2.5 ml representative sample is added to the glass column in the test apparatus, which is 610 mm in length and 10 mm diameter with 0.5 mm mesh sieve material fitted across each end.

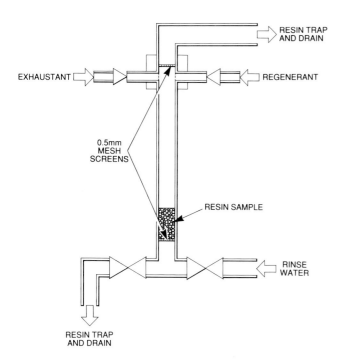

FIG. 3.13 Osmotic shock apparatus

The resin is then subjected to the four sequential test phases and for strongly basic anion resin these would be:

(a) Downward flow of 40 ml of 15% w/v sodium hydroxide solution at 1 ml/second, a 20-second soak and a 10-second drainage period.

(b) Backflow and rinse with water from a pump operating at 6 bar such that the resin is impacted against the top screen and held there for 10 seconds. The resin is then allowed to fall to the bottom of the tube and is drained for 25 seconds.

(c) As for (a) but sodium chloride is used to exhaust the anion resin.

(d) As for (b).

A single cycle takes 4 minutes and the cycles are either continued until measurable breakdown occurs or fixed depending on the objective of the specific test. Usually 25 cycles are performed on a gelular resin, 500 cycles on a macroporous resin.

The degree of breakdown is measured by determining the percentage of beads which are less than 0.5 mm using the HIAC bead size distribution data. The % reduction in average bead size is also calculated. There is a linear correlation between these two parameters of measurement of breakdown, as shown in Fig 3.14. The standard deviation of the breakdown to <0.5 mm is 2.5 to 3.5% at the 25–30% breakdown level, and for the reduction in average bead size the standard deviation is 1.5–2% at the corresponding 15.20% level.

Typical results from the test over a recent five year period are shown in Table 3.10.

The results show the general greater strength of macroporous resins compared with the gelular materials, with the weakest form of resin being the gelular strongly basic type.

Note that these tests simulate the normal conditions likely to be present. However, recently 20% DVB macroporous resins in condensate purification plant treating a condensate containing morpholine and AMP (2 amino 2 methyl propanol) have been shown to be physically less stable than gelular material (see Section 9.8.2 of this chapter).

TABLE 3.10

Results from osmotic shock/attrition tests

Description	Gelular (25 cycles)		Macroporous (500 cycles)	
	Cation (SA)*	Anion (SB)**	Cation (SA)	Anion (SB)
Breakdown to < 0.5 mm, %	3	24	2	3
Reduction in bead size, %	4	15	2	2
Variability (SD)	5	13	1	2
Number of samples	26	48	30	29

* Strongly acidic   ** Strongly basic type I

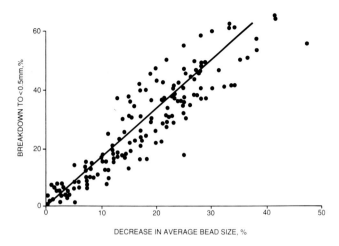

FIG. 3.14 Percentage breakdown versus decrease in bead size for gelular anion resins

4.6.3 Visual and electron microscopy

Visual microscopy is a useful tool for the examination of resin. Cracks and flaws in beads can be seen easily and this can provide useful data on the risk of physical breakdown of the resin. Indeed, the results can be expressed in terms of:

- The number of whole beads as a %.

- The number of cracked beads as a %.

Fouling of the internal structure of an acrylic weakly acidic cation by calcium sulphate can be observed, whilst general fouling by organic matter and the metal elements previously mentioned can also be detected.

The scanning electron microscope can be used where a greater magnification is required. This technique allows the internal and external structures of resins to be observed. This can be particularly useful when a new product is introduced by the manufacturer and its structure can be compared with established materials.

When the electron macroscope is used in conjunction with energy dispersive X-ray analysis the elemental analysis of the surfaces of resin beads can be determined, whilst the elemental composition and valency states can be obtained from X-ray photo-electron spectroscopy.

Other similar techniques such as X-ray diffraction and X-ray fluorescence can be used to identify foulants, such as the precipitated calcium sulphate periodically found within weakly acidic cation resin.

5 Deionisation processes for producing make-up water

5.1 Design

The type of deionisation plant installed on a power station will depend primarily on the following factors:

- The composition of the raw water.

- The quality of treated water required from the plant.

- The capital cost.

- The operating cost.

For today's high pressure plant the water quality required dictates that the final section of treatment is a mixed bed unit, but see Section 5.4.4 of this chapter in respect to the quality of treated water obtainable from counter flow regenerated systems.

Almost all of the plants will have strongly acid cation and strongly basic anion units, and a degasser to remove carbon dioxide.

The main additional options are units containing:

- A weakly acidic cation resin.

- A weakly basic anion resin.

- An anion resin in chloride form.

- Activated carbon.

In special situations an oxidation stage such as described in Section 4.4.2 of this chapter may be employed.

As indicated in Section 4.3.1 of this chapter, alkalinity can be removed from a water relatively easily, and for waters having a high alkalinity the plant will usually employ a weakly acidic cation resin and a degasser unit. This resin can be regenerated stoichiometrically and, hence, the operating cost is low and the carbon dioxide released from the alkalinity after cation exchange is removed in a degasser. This in turn reduces the ionic load on the following anion resin. For the majority of raw waters available, the removal of carbonates and bicarbonate present can be more efficiently achieved by the physical degassing of carbon dioxide than by absorption as carbonate ions on a strongly basic anion resin. In situations when the alkalinity of the raw water is low but the salt content is high, the use of a weakly basic anion resin prior to the strongly basic anion is favoured to obtain a more efficient use of the sodium hydroxide regenerant.

Figure 3.15 shows some of the types of plant currently used in the power industry.

5.1.1 Raw water analysis

It is imperative that accurate data are available on the composition of the water that is to be treated. As discussed in Section 2.1 of this chapter, the composition of waters in different areas varies widely and even the water from a specific source can vary considerably with time. Accordingly, analyses of water should be obtained over as long a period as is practical. Some attempt should also be made to assess the risk of any possible future adverse variations.

Particular attention needs to be paid to parameters which are difficult to change and to constituents likely to cause fouling of the anion exchange resins, e.g.,

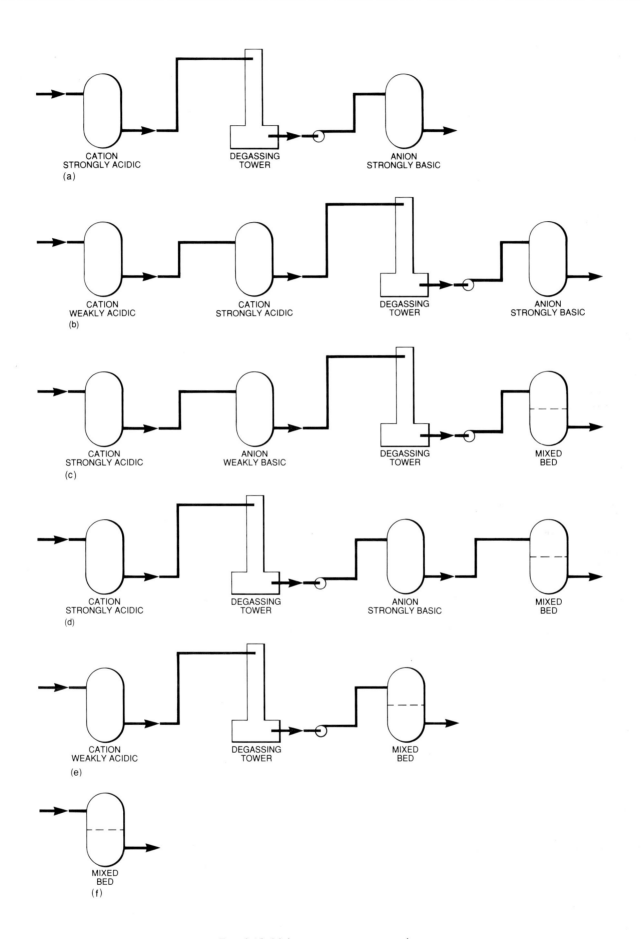

FIG. 3.15 Make-up water treatment plants

organic matter. Table 3.11 is an example of a detailed water analysis used in a specification for the supply of a make-up water treatment plant.

5.1.2 Design throughput

There are usually three main criteria determining the throughput of a plant, firstly, the output to be produced per day, secondly, the maximum output per hour to cope with short term high demands for treated water and, thirdly, the output required between regenerations of the ion-exchange beds. The outage time for regeneration has to be taken into account to ensure the required daily outputs are met.

The installation of a deionisation plant for power generation is an essential item of equipment and, therefore, a single stream does not provide sufficient safeguard in meeting the make-up water requirements at all times.

Plants are usually designed to operate continuously and this requires a multi-stream plant to cover for regeneration periods where one stream is off-line. As a guide to plant capacity the Littlebrook D plant serving 3×660 MW units is rated at 4.2% of the MCR steam flow, equivalent to a flow of 273 m^3/h or 6552 m^3/day.

More recently, plants have been designed with $2 \times 100\%$ streams. These are less expensive, in spite of the greater volume of resin required, due to their simpler construction and ease of operation.

The plants are also sized to ensure that 100% of the design daily output can be produced even if one stream is out of service.

TABLE 3.11

Specification design — raw water analysis

	Design	min	max
Raw water analysis			
pH	7.7	7.2	8.2
Equivalent mineral acidity, mg kg^{-1} CaCO$_3$	415	280	500
Total alkalinity, mg kg^{-1} CaCO$_3$	105	70	140
Total hardness, mg kg^{-1} CaCO$_3$	340	270	400
Calcium hardness, mg kg^{-1} CaCO$_3$	244	190	300
Magnesium hardness, mg kg^{-1} CaCO$_3$	96	75	110
Sodium, mg kg^{-1} CaCO$_3$	180	90	200
Potassium, mg kg^{-1} CaCO$_3$	–	–	–
Sulphate, mg kg^{-1} CaCO$_3$	260	190	340
Chloride, mg kg^{-1} CaCO$_3$	130	80	150
Nitrate, mg kg^{-1} CaCO$_3$	25	10	40
Iron, mg kg^{-1} Fe	0.1	0	0.18
Aluminium, mg kg^{-1} Al	< 0.01	0	0.1
Barium, mg kg^{-1} Ba	< 0.1	–	–
Ammonia (free and saline), mg kg^{-1} NH$_3$	1.0	0.4	1.4
Permanganate value, mg kg^{-1} KMnO$_4$ (30 min at 100°C)	50	30	60
Total organic carbon, mg kg^{-1} C	12	8	16
Reactive silica	7.0	5	10
Non-reactive silica, mg kg^{-1} SiO$_2$	–	–	–
Appearance		clear	
Suspended solids, mg kg^{-1}	30	5	500
Detergents (Manoxol OT), mg kg^{-1}	< 0.1	–	–
Phosphates, mg kg^{-1} Na$_3$PO$_4$		approximately 2.5	
Water quality at cation outlet			
Sodium, mg kg^{-1} Na		Not greater than 2	
Water quality at anion outlet			
Conductivity (at 25°C), μS cm^{-1}		Not greater than 20	
Silica, mg kg^{-1} SiO$_2$		Not greater than 0.2	
Final water quality (at MBU outlet)			
Conductivity (at 25°C), μS cm^{-1}		0.1	
Sodium, mg kg^{-1} Na		0.015	
Total silica, mg kg^{-1} SiO$_2$		0.02	
Organic matter (TOC), mg kg^{-1} C		< 0.2	
Dissolved oxygen, mg kg^{-1} O$_2$			

An important design consideration is the frequency of regeneration and therefore the length of run between regenerations. The run length must be sufficient to ensure that regeneration of the off-line stream can always be completed by the time the on-line stream exhausts. The choice of an excessively long run length increases the size and cost of the ion-exchange units and the regeneration system, and also increases the resin inventory.

Typical stream regeneration times are in the range 3–5 hours and the design run length about 8 hours, equivalent to no more than three regenerations per day. The design throughput must allow for the raw and treated water required during regenerations and the volume of waste water arising from deionisation plants depends on the composition of the raw water. Typical wastage figures range from 5–25% for low and high dissolved solids, waters, respectively.

5.1.3 Vessel sizing

Vessel sizes are based mainly on flow rate, resin capacity, exhaustion cycle time and potential pressure losses across the unit.

The plant designer will select the type of ion-exchange resin and the quantity of regenerant chemical needed to give the required water quality, the mode of regeneration and the arrangement of the ion-exchange units. Using data provided by the resin manufacturer, the ion-exchange capacity is determined and this fixes the resin volume required to obtain the design throughput between regenerations.

The designer will have to balance the use of low levels of regenerant chemicals which give an enhanced efficiency in terms of usage of chemicals, against higher levels of chemicals with lower efficiency but smaller resin volumes and reduced vessel size and cost.

A simplified design approach for a cation unit for a single stream of treatment would be as follows.

Assuming the unit has to produce 55 m^3/h of treated water, then when sizing the units, a flow rate of 10-bed volumes/hour through the bed is initially selected with a bed depth of 1.8 m.

Under these conditions, to produce the volume of water required, a surface area of 3 m^2 is needed. Hence a vessel of 1.97 m diameter would be required.

The volume of resin chosen on the basis of flow conditions then needs to relate to the quantity of water that will be produced between use and regeneration. Consider a water of composition:

	mg kg^{-1}	% of total cations
Sodium as CaCO$_3$	160	31
Calcium as CaCO$_3$	250	49
Magnesium as CaCO$_3$	100	20
Total cations	510	

with 40% of the total cations being present as alkalinity.

By using co-flow regeneration with sulphuric acid at a level of 80 g H_2SO_4/litre the capacity can be obtained from the resin manufacturers data. Under the conditions selected, this would be 33.5 g CaCO$_3$/litre of resin. Raw water will consume some of this capacity during rinsing, and if 7-bed volumes of rinse water are added this is equivalent to a load of 3.5 g CaCO$_3$/litre of resin. Hence the net operating capacity will be 30 g CaCO$_3$/litre of resin.

The concentration of cations removed from the water during the exhaustion cycle is 0.510 g/litre of water minus the leakage of cations, which from the data given will be 0.005 g CaCO$_3$/litre of water, which equals 0.505 g CaCO$_3$/litre of water. Therefore, 330 m^3 of water can be treated before the bed is exhausted and at a flow rate of 55 m^3/h, this would give a 6-hour exhaustion stage.

Obviously, by varying the flow rate, the optimum period between regenerations can be obtained. Note, however, that the flow rates used have to be compatible with pressure loss criteria for the bed of resin.

A similar approach would be adopted for the sizing of anion units.

In the case of the final mixed bed stage of treatment, the ionic load will normally be very low so that the unit is usually designed on flow considerations rather than the capacity of the resins. Generally, this leads to a smaller unit, operating at a higher flow rate than the preceding cation and anion units.

Vessel heights are selected on the basis of the overall height to be 50–100% greater than the depth of the resin. This is to provide sufficient freeboard for backwashing the resins prior to a regeneration.

5.2 Construction of deionisation units

The main components in an ion-exchange water treatment plant are the ion-exchange units, the regenerant handling facility, storage of chemicals, the control system and the effluent neutralisation and disposal section. Most plants also have degassers for carbon dioxide removal after the cation exchange stage and, occasionally, vacuum de-aerators may be included. Other components such as pumps, pipework and control valves are not unique to water treatment plants.

The main technique used in ion-exchange is the fixed bed, alternatives being batch, fluidised or suspended beds and continuous systems.

Most commonly, the water to be treated flows down a compacted bed of resin and the process continues until there is an increasing concentration of ions not removed by the ion-exchange resin, as described in Section 4.3 of this chapter. At this stage, in older plants, regeneration is carried out usually in a downward direction (co-flow), but increasingly the regenerant is passed upwards (counter-flow) to obtain an improved quality of treated water with a relatively low usage of regenerant chemicals.

Details are given in the following sections of the construction of the units and the regeneration procedures used on make-up water treatment plants.

5.2.1 Strongly acidic cation and strongly basic anion units

The bulk of dissolved solids present in a raw water are removed within these two first ion-exchange beds.

A typical unit will consist of a cylindrical steel pressure vessel with dished ends which is internally rubber-lined. Internal equipment is fitted to ensure a uniform flow distribution throughout the unit.

To obtain an efficient performance from the ion-exchange resins, water and regenerant distribution and collection systems are required. A single distributor may be used to handle the water inlet, backwash and regenerant flows, but frequently two distributors are provided, one for water and the other for regenerants.

A number of types of top distribution systems are in use as follows:

Tun dish This is the simplest arrangement used and water (or regenerant) cascades from the dish in a 'bell' type distribution.

Splash plate In an alternative design the inlet pipe is directed downwards against a splash-plate as in Fig 3.16. The distribution pattern is similar to that given by the tun dish type. For large diameter units, two or more splash-plate distributors are employed but careful design of the inlet pipework is necessary to ensure that the flow is evenly divided between them.

Wedge wire strainer This acts as both a filter and a distributor. It is cylindrical in shape and is mounted vertically as shown in Fig 3.16. It is fabricated from stainless steel. Water passes through vertical slots in the sides of the cylinder which are formed by wedge-shaped wire elements. The slots so formed, widen in the direction of the service flow and this helps to minimise clogging during operation. Large particles are trapped by the strainer but are flushed to waste when the flow direction is reversed during backwashing. Any small resin particles or foreign matter present in the resin bed can become securely lodged in the slots during backwashing, giving rise to restricted flow and uneven distribution. Because of this it is usually necessary to remove the strainer periodically for cleaning. If the slots are kept clean this type of strainer provides very even annular distribution.

Spider This is a more elaborate system and comprises a central hub type header from which radiate perforated distribution pipes, as shown in Fig 3.17.

Header and lateral These types should give the most uniform distribution as water is entering at a multiplicity of points. It comprises a central header pipe from which are fed a number of distributor laterals with holes or slots drilled along their length, similar to the spider arrangement. In some designs the laterals are covered with a fine plastic mesh or by a stainless steel mesh (see Fig 3.16). These wrapped laterals are

more prone to clogging, particularly by fine resin particles carried forward from a preceding unit. If clogging occurs, both flow and distribution can be affected and the clogged lateral will need to be replaced.

Bottom collection and distribution systems are fitted to *collect* the treated water and spent regenerant during co-flow operations, to *distribute* water for backwashing of the resin. In counter-flow operation the system will be used to *distribute* regenerants.

Two main types are used:

● Header and lateral with nozzles fitted.

● A flat or slightly dished nozzle plate.

In the header and lateral system, the laterals are usually spaced at equal distances along the header and are of varying length to fit the circular area of the vessel. The nozzles are fitted at intervals along the lateral arm and are usually situated on the underside of the lateral (see Fig 3.18).

In the nozzle plate design the plate acts as a false bottom to the unit and supports the resin bed. Nozzles fitted in the plate permit the flow of water to or from the dished end of the vessel (see Fig 3.19).

The nozzles used in both designs are similar. Each nozzle will be capable of handling up to about 0.5 m^3/h of water under normal operating conditions. In practice the mean flow rate, based on the number of nozzles fitted and the flow rate, varies considerably in units of different design as shown in Table 3.12.

Several types of plastic have been used to make nozzles. The nozzles are of moulded construction and the collecting slots or orifices are usually formed in the mould. The dimensions of the collecting slits in the nozzles have to be sufficiently small to retain the resin bead but not to cause an excessive restriction to flow. A typical nozzle is shown in Fig 3.20.

Special designs of distribution and collection systems are required for counter-flow operation of cation and anion units and this aspect is covered in Section 5.4 of this chapter.

Serious problems can arise if resin escapes from a vessel. For example, resin leaving a cation unit will enter the degasser and the following anion unit. Failure of a plant component, usually a nozzle, lateral or header in the final stage of treatment, could result in cation and anion resins entering the steam/water circuit. If this occurs under high temperature conditions which prevail the cation resin will break down to give acidic products which are non-volatile, whereas anion resin, as described in Section 4.4 of this chapter, breaks down at relatively low temperatures to give volatile alkaline products. The overall result is that the boiler water becomes acid with a high risk of rapid corrosion occurring, whereas the steam and condensate are more alkaline which is not so serious, but some effect could arise on non-ferrous components in the system. For these reasons all ion-exchange units are fitted with resin traps on the outlet pipework. These are fine strainers designed to retain ion-exchange beads which have to

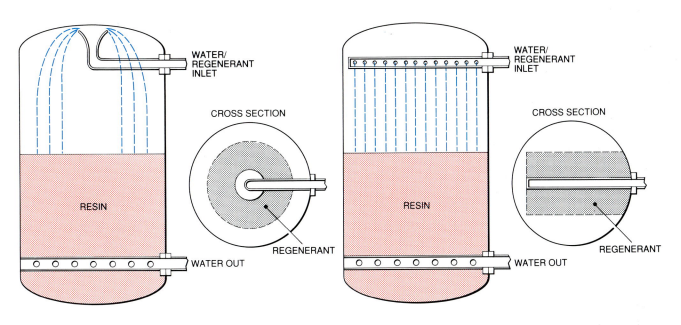

(a) Tun-Dish Distributor

(b) Perforated Pipe Distributor

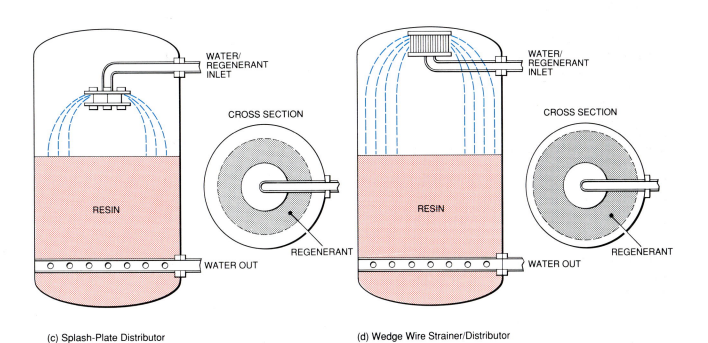

(c) Splash-Plate Distributor

(d) Wedge Wire Strainer/Distributor

FIG. 3.16 Inlet water and regenerant distributers

be capable of withstanding the maximum pressure of the upstream pumps.

5.2.2 Weakly acidic cation resin units

With waters of high alkalinity content a weakly acid resin may be installed to reduce operating costs. This may be used in a separate vessel or within a vessel containing the strongly acid resin which reduces the capital cost of the plant.

The weak resins are less dense than the strong resins and form a separate layer above the strongly acidic cation. The beds may be regenerated in either a co-flow or counter-flow mode. The counter-flow system is to

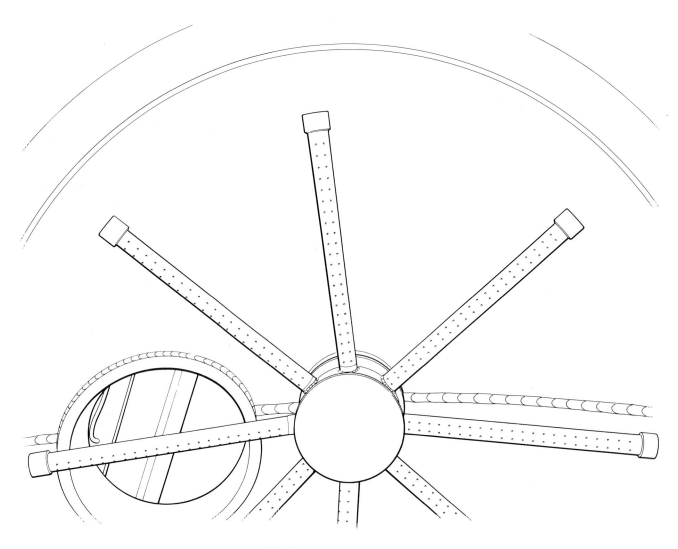

FIG. 3.17 Spider type distribution system

TABLE 3.12
The performance of nozzle collector elements

Nozzle	Type of unit and collector system	Unit flowrate m^3/h	Number of nozzles/unit	Flowrate/nozzle m^3/h
A	Mixed bed bottom collector/distributor	13.6	17	0.80
B	Cation unit bottom collector/distributor	22.7	69	0.33
C	Cation unit bottom collector/distributor	22.7	88	0.26
D	Anion unit bottom collector/distributor	22.7	88	0.26
E	Mixed bed bottom collector/distributor	22.7	41	0.56
F	Cation unit bottom collector/distributor	56.8	150	0.38
G	Mixed bed intermediate collector/distributor	56.8	64	0.89
H	Mixed bed bottom collector/distributor	56.8	109	0.52

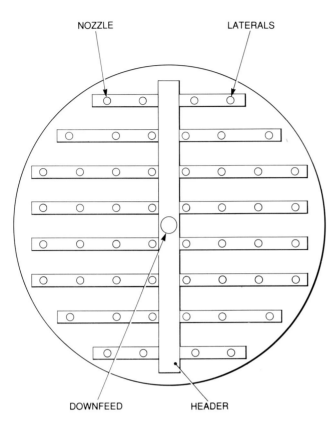

FIG. 3.18 Typical mid and bottom header and lateral collector system

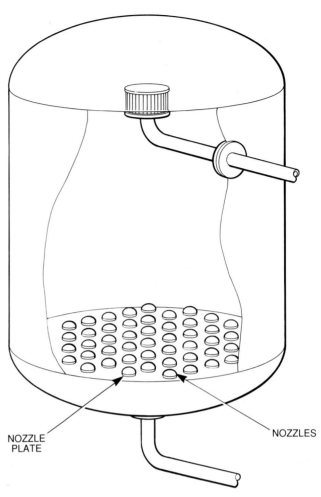

FIG. 3.19 Nozzle plate collection/distribution system

be preferred as the fresh acid contacts the strongly acidic resin first and the partially-spent acid reaching the weakly acidic resin is still capable of giving a satisfactory regeneration.

The vessels and internals are as described for standard cation and anion vessels.

5.2.3 Weakly basic anion resin units

As above, the resin can be used in a separate vessel or within a vessel containing strongly basic resin.

The use of weakly basic resins has decreased since the advent of type I acrylic strongly basic resins having good regeneration efficiencies. More recently, resins with both weakly and strongly basic groupings within the beads have been produced. This overcomes some of the problems of layered beds and still has the economic advantage of being used in a single vessel.

The construction and components used are the same as for the units considered earlier.

5.2.4 Mixed bed unit

Mixed beds are usually smaller than the main cation and anion units and the service flow rates are higher. However, the flow rate through this final section of the plant becomes more critical as anion resin in the main section and the mixed bed ages, or is fouled; for this reason, design flow rates are usually restricted to less than 30-bed volumes per hour.

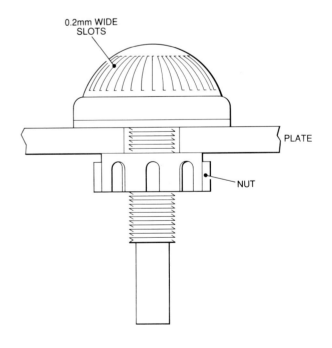

FIG. 3.20 Typical nozzle

251

As this is a single unit containing cation and anion resin, the construction is more complicated than the other vessels in the stream. This arises from the need to regenerate two types of resin with different chemicals. As a result an additional distribution/collection system has to be incorporated at the interface which exists between the cation and anion resins when they are separated prior to regeneration. This is of similar construction to those described previously (see Fig 3.18). As the mixed bed is the final unit in the water treatment stream, it is essential that high quality distribution/collection systems are installed to obtain the extremely high quality water which is demanded at this stage.

5.2.5 Degassers

When raw water is passed through a cation resin any alkalinity present reacts to produce carbon dioxide. This could be removed by the following anion resin, but additional resin capacity would have to be provided and there would be an increased caustic soda usage. Carbon dioxide can be removed more cheaply using a physical degasser.

Air blow scrubber towers consist of vertical towers packed with polypropylene saddles through which the water flows downwards whilst air is passed counterflow in an upward direction, see Fig 3.21. The packing height is typically 2–3 m and the specific flow rate is in the range 50–100 m/h. The air flow is usually 20–25 times the water flow and under these conditions the carbon dioxide content can be reduced to <5 mg kg^{-1}.

A storage sump for the degassed water is provided at the bottom of the tower from which water is drawn for the next stage of treatment and for rinse purposes during regenerations of the various resins.

5.3 Chemicals and storage facilities

Large quantities of chemicals are used on modern power stations and extensive storage facilities are included to ensure that the risk of chemicals being unavailable because of industrial disputes or adverse climatic conditions is minimised. The principal reagents used for regenerating ion-exchange resins are sulphuric acid, sodium hydroxide and occasionally hydrochloric acid.

Sulphuric acid is purchased at 96% strength and caustic soda as a 47% solution. Hydrochloric acid is supplied as a 28% to 32% solution, the lower strength being normally supplied during summer months to reduce the nuisance from hydrochloric acid vapour.

Chemicals are delivered by road tanker in quantities up to 20 tonnes. Transfer to the storage tanks is by means of air pressure applied to the road tanker barrel.

The storage tanks are usually pressure vessels designed for a pressure of 2.1 bar gauge. For the maximum storage capacity, non-pressure tanks may be used but these require larger size vents.

Storage tanks for sulphuric acid and caustic soda are unlined steel vessels. Sulphuric acid attacks steel when diluted below 70% and because the acid readily

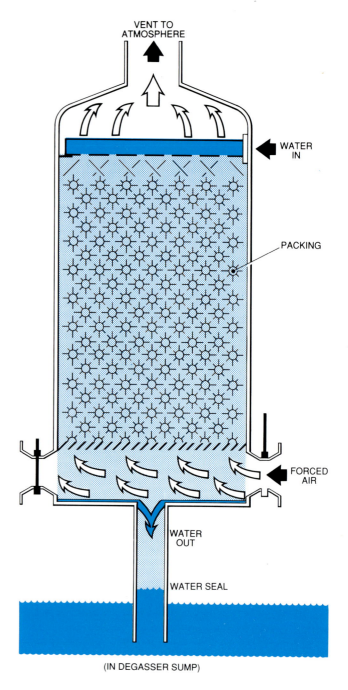

(IN DEGASSER SUMP)

Fig. 3.21 Degasser tower (Dewplan Ltd)

absorbs moisture from the air; acid at the surface can become diluted during prolonged storage. It is desirable to ensure that the acid level does not remain constant and in some instances a purge flow of dry air is provided. A 6 mm corrosion allowance is specified. Caustic soda is stored in unlined steel tanks which are trace heated and lagged. As there is a risk of stress corrosion cracking at temperatures above about 50°C, the tanks are usually stress-relieved. Hydrochloric acid is very corrosive towards mild steel and all steel surfaces must be protected with a suitable lining such as rubber.

The storage tanks are surrounded by bund walls which form a collecting tank sized to contain the whole contents of the storage vessel in the event of a major leak. Storage tanks for 47% sodium hydroxide are maintained at a temperature of about 25°C to prevent freezing. There is little danger of 96% sulphuric acid freezing but 98% acid freezes at above 0°C. There is no risk of hydrochloric acid freezing. Figures 3.22 and 3.23 show freezing point curves for caustic soda and sulphuric acid.

5.4 Regeneration procedures and equipment

5.4.1 Chemical addition

The concentrated chemicals as supplied have to be diluted to provide an adequate volume for contact with the resin. When high strength sulphuric acid is mixed with water heat is generated and therefore the dilution is carried out in two stages. In this case a dilution tank is installed in which the initial acid is diluted from 90% to 10–20%. As discussed in Section 4.3 of this chapter, relatively low concentrations of sulphuric acid have to be used to avoid precipitation of calcium sul-

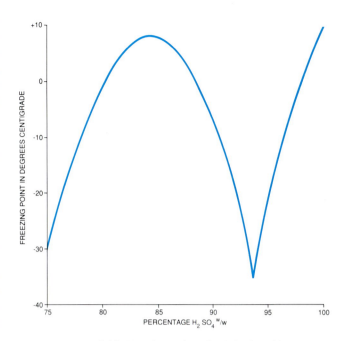

FIG. 3.23 Freezing point of sulphuric acid

phate. The second stage of dilution is therefore designed to produce an acid strength in the range 1–3%.

A typical dilution system for sulphuric acid is shown in Fig 3.24. Sulphuric acid is drawn up by vacuum from the bulk tank into a high level measure. Vacuum is produced using a water-powered ejector and a high level barometric loop is provided to prevent water being drawn up into the measure or acid being pulled over into the water system. The acid then flows down at a set rate into the dilution tank where it is added to a continuous flow of water to produce the 20% solution. Dilution of sulphuric acid produces a considerable amount of heat. Hence the dilution tanks are usually rubber-lined steel or lined glass reinforced plastic construction and are designed to ensure rapid mixing to prevent overheating. The diluted acid is drawn from the tank by means of a water ejector which serves to dilute acid further to the required strength for regeneration.

Less dilution is required in the regeneration of anion resin with caustic soda. A diluted strength of 4–5% is used. The dilution tank may therefore be omitted.

The vessels and pipework containing concentrated caustic soda are again trace heated and lagged to prevent freezing.

5.4.2 Regeneration techniques

There are two main types of regeneration used: designated co-flow and counter-flow.

In co-flow regeneration the chemicals are passed through the bed in the same direction as the service water flow, whereas in counter-flow they are passed through the bed in the opposite direction to the service flow. In the UK power industry, invariably service flow is downwards and regenerant flow upwards, but

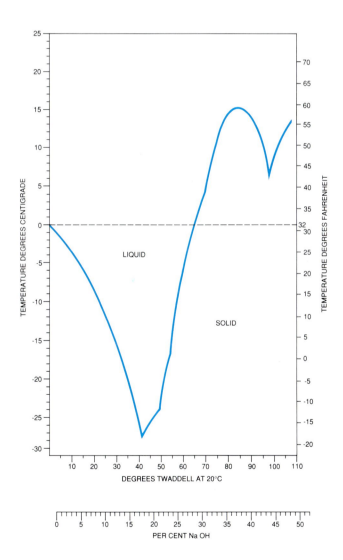

FIG. 3.22 Freezing point of caustic soda solutions

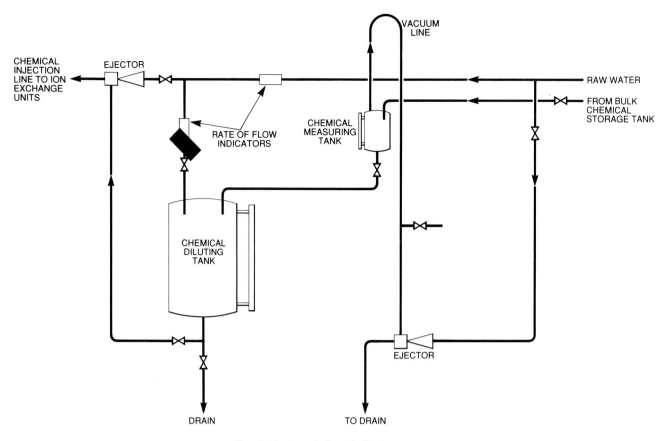

FIG. 3.24 A typical acid dilution system

designs using service flow upwards and regenerant flow downwards are available and are widely used in Europe.

The key factor in counter-flow regeneration is that the water being treated passes through the most highly regenerated resin immediately before leaving the unit. As discussed in Section 4.2 of this chapter, this gives a favourable equilibrium reaction and a very high quality treated water can be obtained.

With co-flow regeneration, the opposite applies and this gives a much poorer quality water in terms of leakage of unwanted ions in the treated water.

It follows that an essential requirement for a counter-flow operation using an up-flow of regenerant is that the bed should not rise and mix.

Several designs of units have been adopted to prevent this happening as follows:

Air hold-down (Fig 3.25) The resin bed is held in compaction by a crust of dry resin above the waste regenerant collector system which is positioned 100–200 mm below the top of the bed. This lateral can be subjected to severe stress from the piston effect of a slight movement of the compacted bed and therefore needs to be of high strength. Frequently stainless steel is used for the header and laterals and the lateral tubes are wrapped with a stainless steel mesh (see Fig 3.26). The top and bottom of the laterals are perforated, the top holes collecting the downflowing air and the

bottom holes the upflowing effluent. Each regeneration is preceded by a surface flush of the top layers of resin via the buried header and lateral and leaving through the backwash outlet, in this way, the bottom of the bed is undisturbed which is important in maintaining a high quality treated water.

Regenerant enters the vessel via the nozzle plate and passes upwards and out of the buried collector. An upward displacement rinse follows and the final rinse is then carried out downwards.

Water hold-down The design is similar to the above, but water is passed from the top of the vessel downwards through the shallow layer of resin and leaves by the submerged regenerant collector.

Packed bed (Fig 3.27) This system can be designed to operate either using downflow service and upflow regeneration or upflow service and downflow regeneration.

Plants installed in the UK power industry have usually been of the former type. In these systems nozzle plates are fitted at the top and bottom of the vessel, and the space between them is filled with resin plus inert plastic beads on top of the resin, which can absorb the shrinking and swelling of the resin beads during the regeneration and exhaustion cycles.

The regeneration procedure is similar to the other systems. However, to clean the resin the inert material has to be removed from the vessel. The resin is then

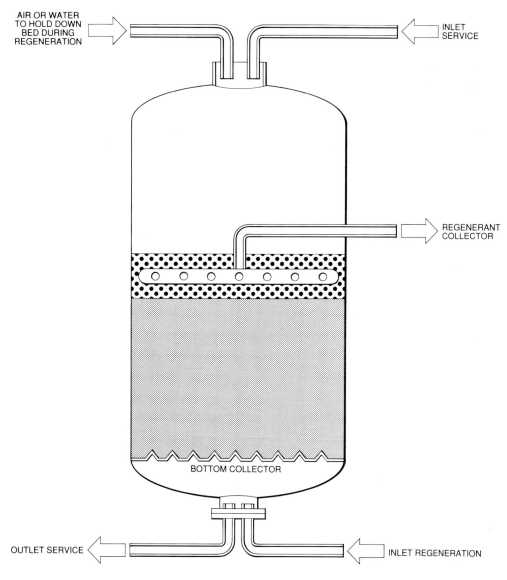

AIR OR WATER
TO HOLD DOWN
BED DURING
REGENERATION

INLET
SERVICE

REGENERANT
COLLECTOR

BOTTOM COLLECTOR

OUTLET SERVICE

INLET REGENERATION

FIG. 3.25 Counter-flow regeneration air and water hold-down system

cleaned by backwashing and the inert material returned to the vessel. A potential advantage of this system is that it may be possible to convert existing co-flow systems to counter-flow without major modifications.

In the upflow service, downflow regenerations systems, the resin bed is again contained between two flat nozzle plates filling all but a small space at the top. During the service cycle the bed is lifted as a plug and the top layers are held firmly against the top nozzle plate. The remainder of the bed can be allowed to fluidise without affecting the quality of the water produced. On regenerating downwards the resin drops back against the bottom nozzle plates so that the regenerant passes through a settled bed, hence there is no risk of the bed being disturbed during the passage of regenerant. As the vessel is almost full of resin, only a small quantity of water is used in the displacement and final rinsing of the bed.

In-situ backwashing to clean the resin would not be effective since the bed expansion available is minimal

and the resin has to be removed from the vessel to a separate unit for backwashing. It follows that the raw water needs to be free from suspended matter to minimise the number of time consuming cleaning cycles.

5.4.3 Strongly acidic cation unit regeneration

Cation units being the first in line are more likely to receive suspended matter which is then filtered out, so it is important that prior to regeneration good backwash facilities are available on this unit. That is, the passage of water upflow through the bed. The bed needs to be expanded by about 50% to ensure good clean-up and to do this a backwash flow rate of approximately 15 m/h is required for a sulphonic cation resin.

Sulphuric acid is usually employed to regenerate the cation resin, the main reason for its use being that it is cheaper than the alternative, hydrochloric acid. Hydrochloric acid would have the advantages of giving a higher capacity for a given quantity of regenerant

FIG. 3.26 Top lateral construction for a counter-flow air hold-down system

and sodium leakage would also be lower. This arises from the fact that sulphuric acid has to be used at low concentration (2%) to avoid calcium sulphate precipitation, whereas hydrochloric acid can be employed up to 7% strength.

However, the capacity advantage is not sufficient to offset the difference in prices of the two agents and this coupled with the greater difficulty in handling hydrochloric acid (fume, etc.) leads to sulphuric acid being used in the vast majority of plants in the UK.

The three main methods of minimising calcium sulphate precipitation are to:

- Inject acid at low concentration.

- Inject at high flow rate.

- Use counter-flow regeneration.

Usually, therefore, plants are designed for the initial injection of a low concentration of H_2SO_4, say 1% at a relatively high flow rate. This displaces a high proportion of the calcium on the bed so that the strength of the acid which follows can be increased progressively and introduced at a lower rate.

Calcium sulphate precipitation under counter-flow conditions is less likely because at the end of a service cycle the bottom of a cation bed is predominantly in the sodium form. It is this fraction of the bed that the acid will first contact in counter-flow regeneration, so there is less time for the super-saturated calcium sulphate solution to precipitate before it leaves the unit.

In co-flow regeneration the dilute acid is admitted above the resin and flows downwards through the bed for a period of about 30 minutes. After regeneration the bed is rinsed by downwards flow to remove the excess acid, the rinse water being run to waste until the quality is satisfactory for the unit to be returned to service. The regenerant is prepared using raw water for dilution and the bed is rinsed with raw water.

For counter-flow regeneration a similar sequence of resin cleaning acid injection and rinsing takes place, but the procedures differ because of the more complicated design used.

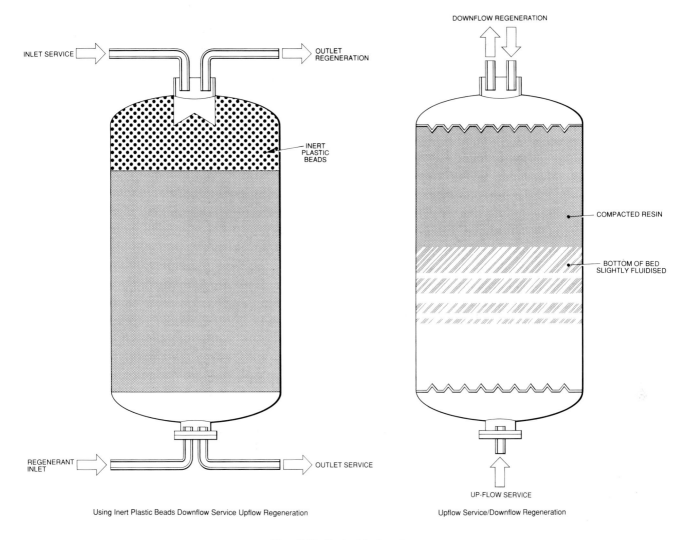

INLET SERVICE

OUTLET REGENERATION

INERT PLASTIC BEADS

REGENERANT INLET

OUTLET SERVICE

Using Inert Plastic Beads Downflow Service Upflow Regeneration

DOWNFLOW REGENERATION

COMPACTED RESIN

BOTTOM OF BED SLIGHTLY FLUIDISED

UP-FLOW SERVICE

Upflow Service/Downflow Regeneration

FIG. 3.27 Packed bed system

Details of a typical sequence of events for an air hold system are:

(a) A surface flush using filtered water is applied via the regenerant outlet lateral at a flow rate of $18 \ m^3 m^{-2} h^{-1}$ to lift the surface of the bed and thereby clean off any deposited materials from the resin. This loosens the resin around the top lateral and prevents the possibility of hydraulic damage from the impact of a tightly packed resin bed.

(b) The bed is allowed to settle for 3 minutes and then the water in the vessel is drained down via the top lateral system and this takes 7 minutes.

(c) Upward injection of the 1.5% acid at a flow rate of $10 \ m^3 m^{-2} h^{-1}$ is carried out for 23 minutes. This is then followed by the injection of 3% acid at a flow rate of $11.5 \ m^3 m^{-2} h^{-1}$ for 11 minutes.

(d) An upward rinse (using water from the degasser sump) is carried out at a flow rate of $9 \ m^3 m^{-2} h^{-1}$ for 25 minutes.

(e) The vessel is then refilled with water. This takes 6 minutes.

(f) A final downward rinse of the bed is applied with filtered water at a flow rate of $9 \ m^3 m^{-2} h^{-1}$ for 5 minutes.

(g) The bed is then ready for return to service.

Note that in (d) the rinse water used is decationised and this prevents contamination of the bottom of the bed which would occur if raw water is used.

5.4.4 Regeneration of anion units

No difficulties should arise in the regeneration of anion resins with sodium hydroxide provided the regenerant is distributed uniformly, has adequate contact time with the resin and is rinsed with decationised water. The regeneration procedure is similar to that used for the cation regeneration.

Optimum conditions for the regeneration of strongly basic anion resin are:

257

Concentration of NaOH	4%
Flow rate	2 bed volumes/h
Rinse flow rate	2 bed volumes/h for 1 bed volume and then 12 bed volumes/h
Rinse volume	6–12 bed volumes

Temperatures can be important in the removal of silica from anion resin, particularly if silica ages and polymerises within the bed so that the caustic soda is acting as a solvent rather than an exchanging ion. Under these conditions the use of warm caustic is beneficial.

At this stage it is of interest to compare the quality of water that would be obtained from the cation-anion removal stage using counter-flow and co-flow regeneration.

Operating with a high solids water with a sodium concentration of 30% of the total cations gave the following results, using counter-flow regeneration:

Conductivity	$\mu S\ cm^{-1}$	0.6
Sodium	$mg\ kg^{-1}$	0.03
Silica	$mg\ kg^{-1}$	0.02

A plant operating on the same water under comparable conditions, but using a co-flow regeneration system, would be expected to produce a treated water of the following quality:

Conductivity	$\mu S\ cm^{-1}$	30
Sodium	$mg\ kg^{-1}$	3.0
Silica	$mg\ kg^{-1}$	0.2

In favourable circumstances, on a low sodium content water, it has been possible to produce treated water of as high a quality as that obtainable from a mixed bed. Under these conditions the mixed beds were taken out of service with a consequent saving in operating costs.

5.4.5 Mixed bed regeneration

The use of two different types of resin in a single unit makes regeneration much more complicated. Figure 3.28 and Table 3.13 show the sequence of events, whilst Table 3.14 gives details of the regeneration conditions.

The first stage is to separate the two resins into two discrete layers. The cation and anion resins will have different densities, so that the initial backwashing causes the resins to separate, the lighter anion resin forming the top layer. To facilitate regeneration of the resins, a central combined collector and distributor is provided at the interface of the two layers. Caustic soda is introduced at the top and the spent regenerant runs to waste from the central collector. The rinse water follows the same path. A buffer water flow upwards through the cation resin may be used

TABLE 3.13

Typical mixed bed regeneration sequence

Stage	Operation
1	Backwash to separate resins
2	Inject anion resin regenerant
3	Rinse anion resin
4	Inject cation resin regenerant
5	Rinse cation resin
6	Drain bed
7	Air mix the resins
8	Allow bed to settle
9	Refill unit
10	Rinse.

TABLE 3.14

Typical regeneration conditions for resins in a mixed bed

	Anion	Cation
Regeneration level, gl^{-1} resin	64	64
Regenerant concentration wt%	4	5
Regeneration injection time, h	0.5	0.5
Regeneration flow rate, BV h^{-1}	3.2	2.6
Rinse time, h	0.5	0.5
Rinse volume (bed volume)	5	5
Regenerant	NaOH	H_2SO_4

to minimise cross-contamination. Acid is then introduced at the central distributor, passes down through the cation resin and is run to waste from the bottom of the unit. A buffer flows downwards through the anion layer with help to minimise cross-contamination. The resin is then rinsed and afterwards re-mixed by passing low pressure air upwards through the bed, this is followed by a final rinse.

Difficulties in consistently obtaining the required quality from mixed beds ($< 0.1\ \mu S\ cm^{-1}$ conductivity) do arise from time to time. Surveys of plant and investigational work on mixed beds supplied to the power industry [6] have shown a number of design and operating faults, mainly in respect of the regeneration equipment and procedures; they also show that most of the impurities in the treated water result from regenerants left within the mixed bed resins as opposed to failure of the resins to remove influent ions.

In the regeneration procedure there is a risk of acid contacting the anion resin and caustic soda contacting the cation resin. It is the former type of contamination that gives most problems on make-up water treatment mixed beds, and the symptoms are observed on the return to service of a freshly regenerated bed when it may require a greatly extended rinse before producing water of the specified quality. This treated water will have a high conductivity and sodium content and in severe cases the condition may continue for several days. The increased level of sodium has often been

Typical mixed bed sectioned view

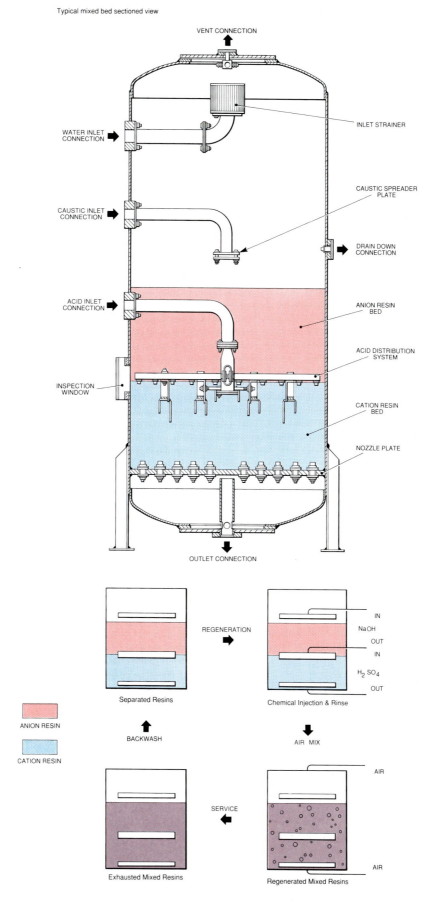

FIG. 3.28 Mixed bed unit and sequence of regeneration procedure

259

attributed to leakage from either cation resin that has become contaminated by sodium hydroxide or from a slow release of sodium from a regenerated, organically fouled resin. Both these sources would produce an alkaline rinse water, but mixed beds suffering a prolonged final rinse invariably have a predominantly acidic rinse water, the major acidic anion being sulphate.

Condensate purification mixed beds have to produce a very much higher quality water than that for make-up supply; consequently, such effects could be very serious. Further details on this problem will therefore be given in Section 9 of this chapter.

How can such cross-contamination arising on the regeneration of make-up mixed beds be minimised? A number of factors are involved as follows:

- Thorough mixing may not completely eliminate the effects of acid contaminated anion resin, but is, nevertheless, an important factor and can best be achieved by having an adequate depth of water above the bed prior to mixing to ensure total fluidisation of the bed when the mixing air is admitted.

- Maintaining a good separation of anion and cation resins prior to chemical regeneration.

- Ensuring that the anion/cation resin interface is maintained above the acid injection or collection centre lateral.

- Ensuring that any cation resin losses are replaced.

- Maintaining a downward buffer flow of water through the anion resin during cation resin regeneration and rinsing.

Table 3.15 gives guidelines for the regeneration parameters and design features to achieve a good chemical regeneration with minimal contamination of the anion resin by acid.

5.5 Effluent treatment

It is generally necessary to ensure that the water treatment plant effluent is neutralised prior to discharge. Effluent discharge pH limits are in the range pH 5–10 or pH 6–9 depending on local discharge conditions.

Water treatment plants are designed such that effluent from the cation unit is mixed with effluent from the anion unit. For most waters, when using a degasser to remove carbon dioxide, there would be an excess of acid over alkali in the effluent if the required levels of chemicals to regenerate the resins satisfactorily were used. However, the additional sodium hydroxide that would have to be added to the effluent is passed through the anion resin. This will enhance regeneration, but the bulk of the alkali will pass into the effluent sump and provide the necessary neutralisation.

It follows that if the plant could be run without the need to produce a neutral effluent the operating costs could be reduced. It may be possible to use an acid effluent for other purposes on site providing there are

TABLE 3.15
Mixed bed regeneration guidelines

Regeneration stage	Guideline parameter
Backwash Flow rate:	$8-10 \ m^3 \ m^{-2} \ h^{-1}$
Duration:	10 minutes
Anion resin regeneration	
Regenerant injection time:	20 minutes minimum
Displacement rinse:	> 1 bed volume at regenerant injection rate
Total rinse:	> 4 bed volumes
Cation resin regeneration	
Regenerant injection:	Downwards from underslung slotted button type centre lateral with buffer flow through anion resin
Injection time:	15 minutes minimum
Displacement rinse:	> 1 bed volume at regenerant injection rate
Total rinse	> 4 bed volumes
Resin mixing	
Air flow rate time:	$> 1.0 \ m^3 \ m^{-2} \ min^{-1}$
	10 minutes
	Plus the addition of sufficient backwash water to raise the free water level by 25 mm

no adverse environmental effects, e.g., to reduce the pH of cooling water.

Effluent from the stream regeneration is collected in a lined concrete sump. The contents are mixed by recirculation and the pH monitored. Acid or caustic is added as required to bring the pH within the acceptable band. Figure 3.29 shows a typical effluent treatment system.

5.6 Instrumentation and control

Some idea of the complexity of the instrumentation and control required for modern demineralisation and condensate polishing plants can be gained from Fig 3.30 which shows the make-up plant control panel and the condensate polishing plant regenerators at Heysham *1* nuclear power station.

The instrumentation and controls have two main purposes. Firstly, to control the plant and monitor the performance during normal operation and thus determine when regeneration is required and, secondly, to enable regeneration to be carried out efficiently.

During normal operation the following instrumentation is used:

Flow meters These are usually orifice plate type and provide both the flow rate through the various units and the integrated flow. Integrated flow measurement is necessary because many units are run to a fixed throughput between regenerations, it also enables resin

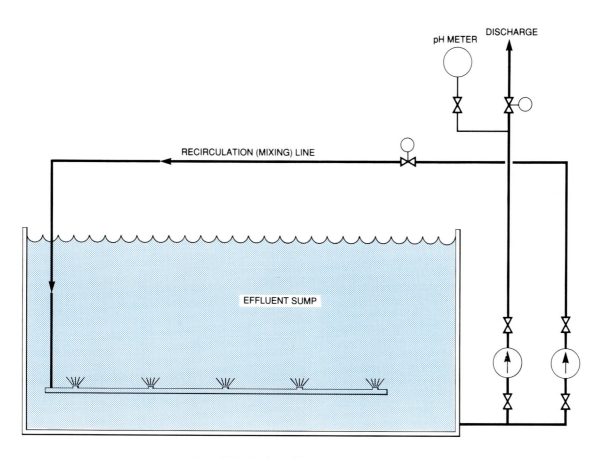

FIG. 3.29 Typical effluent treatment system

performance to be checked. Flow metering is also required for correct setting of regeneration flows.

Pressure gauges These are provided at pump discharge and at the inlet and outlet of all ion-exchange units, filters and resin traps. They indicate when resins are becoming fouled with suspended solids, when breakdown of resin into smaller particles is occurring, when filters require backwashing or recoating and when resin is escaping from the units to block the resin traps. Differential pressure gauges are also provided for some of these applications.

Conductivity meters Conductivity meters are the most simple and reliable means of detecting the purity of water throughout ion-exchange plants. As a resin bed exhausts, the conductivity will move towards the conductivity of the influent water. Thus for the three main types of unit the change in conductivity as each resin exhausts are:

Cation unit — fall in conductivity

Anion unit — rise in conductivity

Mixed bed unit — rise in conductivity

For cation units the fall in conductivity is usually detected by monitoring the differential conductivity of a sample taken from the vessel outlet and a sample taken from just above the bottom of the bed.

For mixed bed units and condensate polishing plants where the conductivity is very close to that of pure water, a special type of conductivity monitor is required which accurately compensates for the change in conductivity with water temperature.

Conductivity meters may also be used for monitoring the regeneration processes and are sometimes used for checking the strength of acid and caustic regenerant solutions.

Silica meters Silica only influences the conductivity of water to a small extent and for anion units the increase in silica will precede the rise in conductivity. Silica meters are therefore frequently employed to monitor water quality at the outlet of anion and mixed bed units.

Sodium monitors Although differential conductivity may be used to monitor cation unit breakthrough, sodium monitors provide a much more accurate and sensitive indication of slip from the cation units. Sodium monitors are also used for checking final mixed bed quality.

pH monitors pH monitoring is used mainly for the control of the pre-treatment plants before the ion-exchange sections. Here the pH meter controls the rate of addition of coagulent or acid/alkali to optimise the effectiveness of the clarification process. pH mea-

261

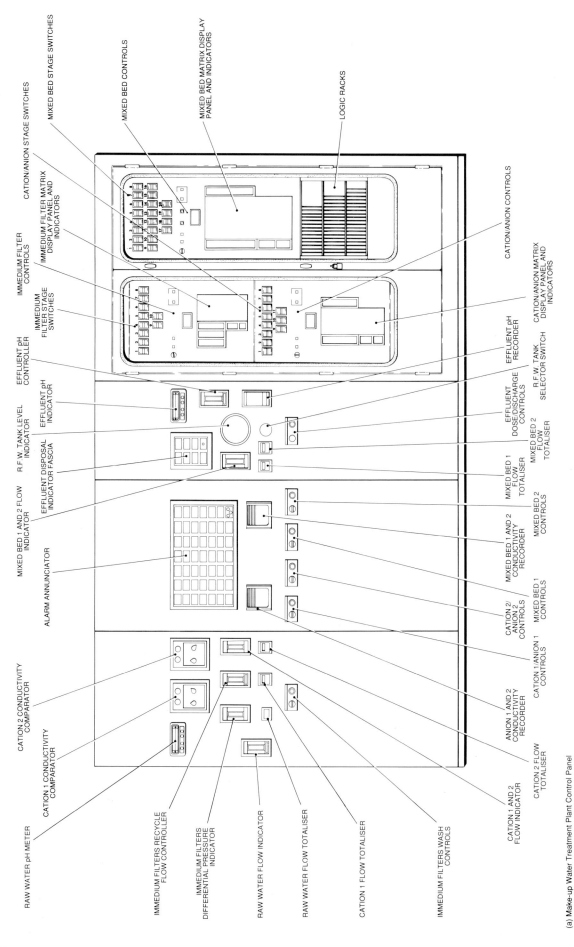

RAW WATER pH METER

CATION 2 CONDUCTIVITY COMPARATOR

CATION 1 CONDUCTIVITY COMPARATOR

MIXED BED 1 AND 2 FLOW INDICATOR

ALARM ANNUNCIATOR

R.F.W. TANK LEVEL INDICATOR

EFFLUENT pH CONTROLLER

EFFLUENT pH INDICATOR

EFFLUENT DISPOSAL INDICATOR FASCIA

IMMEDIUM FILTERS RECYCLE FLOW CONTROLLER

IMMEDIUM FILTERS DIFFERENTIAL PRESSURE INDICATOR

RAW WATER FLOW INDICATOR

RAW WATER FLOW TOTALISER

CATION 1 FLOW TOTALISER

IMMEDIUM FILTERS WASH CONTROLS

CATION/ANION STAGE SWITCHES

MIXED BED STAGE SWITCHES

MIXED BED CONTROLS

MIXED BED MATRIX DISPLAY PANEL AND INDICATORS

LOGIC RACKS

IMMEDIUM FILTER CONTROLS

IMMEDIUM FILTER MATRIX DISPLAY PANEL AND INDICATORS

IMMEDIUM FILTER STAGE SWITCHES

CATION/ANION CONTROLS

CATION/ANION MATRIX DISPLAY PANEL AND INDICATORS

EFFLUENT pH RECORDER

EFFLUENT DOSE/DISCHARGE CONTROLS

R.F.W. TANK SELECTOR SWITCH

MIXED BED 2 FLOW TOTALISER

MIXED BED 1 FLOW TOTALISER

MIXED BED 1 AND 2 CONDUCTIVITY RECORDER

MIXED BED 1 AND 2 CONTROLS

CATION 2/ ANION 2 CONTROLS

MIXED BED 1 CONTROLS

CATION 1/ANION 1 CONTROLS

ANION 1 AND 2 CONDUCTIVITY RECORDER

CATION 2 FLOW TOTALISER

CATION 1 AND 2 FLOW INDICATOR

FIG. 3.30 Make-up water treatment and condensate purification plant control panels

(a) **Make-up** Water Treatment Plant Control Panel

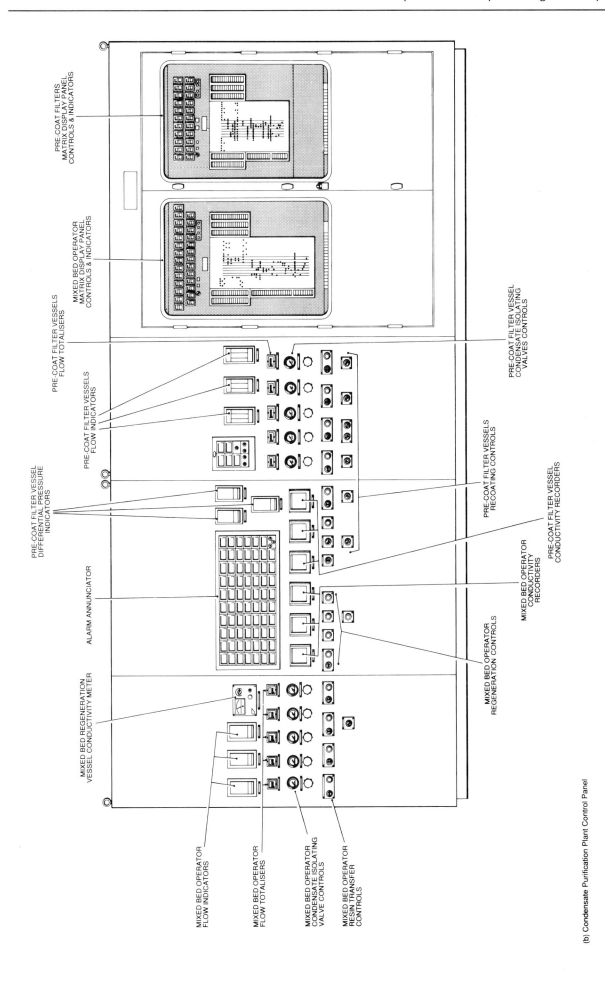

PRE-COAT FILTERS
MATRIX DISPLAY PANEL
CONTROLS & INDICATORS

MIXED BED OPERATOR
MATRIX DISPLAY PANEL
CONTROLS & INDICATORS

PRE-COAT FILTER VESSELS
FLOW TOTALISERS

PRE-COAT FILTER VESSELS
FLOW INDICATORS

PRE-COAT FILTER VESSEL
DIFFERENTIAL PRESSURE
INDICATORS

ALARM ANNUNCIATOR

MIXED BED REGENERATION
VESSEL CONDUCTIVITY METER

MIXED BED OPERATOR
FLOW INDICATORS

MIXED BED OPERATOR
FLOW TOTALISERS

MIXED BED OPERATOR
CONDENSATE ISOLATING
VALVE CONTROLS

MIXED BED OPERATOR
RESIN TRANSFER
CONTROLS

PRE-COAT FILTER VESSEL
CONDENSATE ISOLATING
VALVES CONTROLS

PRE-COAT FILTER VESSELS
RECOATING CONTROLS

PRE-COAT FILTER VESSEL
CONDUCTIVITY RECORDERS

MIXED BED OPERATOR
CONDUCTIVITY
RECORDERS

MIXED BED OPERATOR
REGENERATION CONTROLS

(b) Condensate Purification Plant Control Panel

FIG. 3.30 (cont'd) Make-up water treatment and condensate purification plant control panels

surement is also used for controlling the effluent neutralisation and discharge from the ion-exchange units.

Level gauges All sumps and tanks including bulk chemical storage tanks are fitted with level gauges or contents indicators.

Plant control Make-up and condensate polishing plants are usually controlled automatically. The regeneration process consists of a number of timed stages performed in a fixed sequence. Cam timers or solid state relay sequence controllers are used but, increasingly, microprocessor-based programmable logic controllers are being employed.

Make-up plant stream regenerations may be initiated manually on receipt of an alarm from plant instrumentation or automatically at a fixed throughput initiated by the flow integrators or from conductivity or silica meters at anion outlet. Mixed-beds are normally regenerated at a set fraction of the cation/anion regeneration frequency, for example, at every 10 cation/anion regenerations.

Once initiated the regeneration sequence proceeds without further operator intervention. First the standby stream is rinsed and put on line then the exhausted unit is taken off-line, regenerated, rinsed and placed on standby. The effluent is automatically neutralised and discharged.

6 Reverse osmosis for make-up water treatment

Osmosis occurs when two solutions of different concentrations are separated from one another by a membrane which is permeable to the solvent but impermeable to the solute. Solvent flows from the dilute to the concentrated solution until, at equilibrium, the chemical potential of the solvent is equal on both sides of the membrane — just as the chemical potential of a substance is the same in two co-existing phases of a species, for example, liquid and vapour. The well defined pressure which just prevents such flow is called the osmotic pressure, and is directly proportional to the concentration of the solute. For a concentration difference of dissolved solids across a membrane of 1000 mg kg^{-1} the osmotic pressure will be about 0.8 bar. Note that osmotic pressure is a property of the solution and does not depend on the nature of a membrane that is ideally semi-permeable or on the mechanism of operation of the membrane.

If a pressure greater than the osmotic pressure is applied to the concentrated solution, the solvent can be forced through the membrane leaving the dissolved substances behind.

This method for purifying water is termed *reverse osmosis* (see Fig 3.31). In practice, to produce purified water in sufficient quantities for power station applications, operating pressures of 20–40 bar are needed in addition to the osmotic pressure, in order to overcome the frictional resistance of the membrane. The rate at which the solvent is separated from the concentrated solution and the efficiency of removal of the solutes are determined by the characteristics of the membrane.

The separation of the solutes from the solvent in front of the membrane leaves a concentrated solution there, increasing the concentration difference of solutes across the membrane and therefore increasing the osmotic pressure. For a constant applied pressure the flow rate of purified water will decrease. In order to lessen this effect, turbulence promoters can be used to mix the solution upstream of the membrane.

As indicated in Section 2 of this chapter, the increasing cost and scarcity of towns water has led to the need to consider treating lower grade water which is usually available to most power stations. In general this water will have a higher dissolved solids content than towns water and may contain organic compounds which will inhibit the performance of ion-exchange resins. Other costs associated with ion-exchange such as plant items and the chemicals used have also been increasing and, therefore, other possible water treatment techniques have to be considered.

Reverse osmosis is a possible alternative technique for the initial stage of treatment to remove most of the dissolved solids present in a water. The water treated by reverse osmosis would then be passed through an ion-exchange mixed bed to achieve the final quality required for make-up purposes.

6.1 Membranes

A membrane is usually considered as being a thin sheet of a plastic material of about 100 μm thickness and this forms a partial physical barrier between two fluids such that permeation of some of the species present in the fluids can occur. A number of membranes can be produced with differing permeabilities. In reverse osmosis, the membrane needs to be of very low porosity to reject the inorganic ions found in water and typically the cut off is 0.0001 μm.

Three main types of membrane are used, namely, cellulose acetate, polyamide (nylon type) and a composite membrane comprising a thin film of polyamide of low porosity for the removal of salts on top of a substrate polymer such as polysulphone.

The cellulose acetate and polyamide are asymmetric, i.e., made in a one stage process, whereas in the composite membrane the thinner active polymer is deposited on the surface of the substrate polymer in a two stage process. This technique allows very thin active layers to be produced and as a result the layer can be of very low porosity with high salt rejection capability, and yet have a flux similar to that of the thicker, but more porous active layer on an asymmetric membrane.

Cellulose acetate and composite membranes have usually been employed in sheet form, whilst the polyamide is used primarily as hollow fibres.

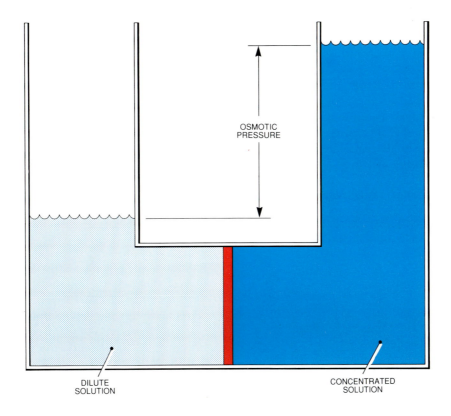

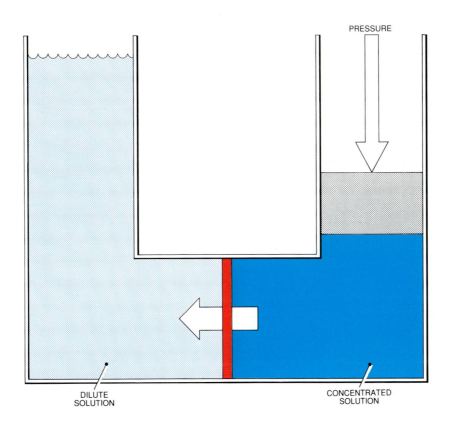

FIG. 3.31 Reverse osmosis

Typically the fibres are of 80 μm external and 40 μm internal diameter and have an outer active dense skin of 0.1–1.0 μm thickness. The fibre structure has the advantage of providing a large surface area within a small volume.

The membranes must be able to withstand the operating pressure whilst allowing the purified water to pass through with minimal resistance.

In practice, this is usually achieved by supporting the membranes in various configurations within a pressure vessel, the whole being known as a module. These modules have been developed to obtain high surface areas of membranes within a small vessel volume.

6.2 Removal of salts and organic matter from water

6.2.1 General

The rate of water flow and the passage of salt through the membrane are the most important properties.

The water flow Q_w is defined by:

$$Q_w = K_w(\Delta P - \Delta\pi)A/t$$

where K_w = permeability for water

ΔP = pressure difference across membrane

$\Delta\pi$ = osmotic pressure difference across membrane

A = area

t = thickness

The rate of salt passage is defined by:

$$Q_s = K_s (\Delta C)A/t$$

where Q_s = rate of flow of salt through the membrane

K_s = permeability for salt

ΔC = concentration difference across membrane

It can be seen that by increasing the pressure, the water flow increases at constant salt flow thus giving a lower salt concentration in the product water.

The recovery or conversion is the fraction of the feed flow which passes through the membrane. At a conversion of 75% for example, the ratio of product flow to reject flow is 3:1 and the concentration of dissolved salts in the reject water approximately four times the inlet concentration.

The 'salt passage' is the ratio of salt concentration in the product stream or permeate to the salt concentration in the feed thus:

$$SP = C_p/C_f \times 100\%$$

where SP = % salt passage

C_p = salt concentration in product

C_f = salt concentration in feed

The term salt rejection is normally used to define salt removal performance. Salt rejection is equal to 100% minus the salt passage.

Reverse osmosis membranes reject larger and divalent ions (such as Mg, Ca and SO_4) more effectively than smaller monovalent ions such as Na and Cl, typically > 95% compared with 85% so that in practice the main impurities in permeate are sodium and chloride. The rejection efficiency for silica is approximately 80% and dissolved gases such as CO_2 and oxygen are not removed.

6.2.2 Removal of organic matter

The degree of removal (rejection) of organic matter by reverse osmosis membranes depends on a number of factors. The most important factor being the type of condition of the organic matter present. Organic species in water can range from the totally soluble through to colloidal particles and suspended matter.

Paradoxically, the colloidal and suspended components will be most easily removed, but at the same time will contribute significantly to the degree of fouling of the membrane surface.

In general the removal of truly soluble organic compounds depends on:

● The molecular weight — the higher the molecular weight the greater the rejection.

● The degree of ionisation — the greater the ionisation the higher the rejection.

Table 3.16 gives typical % rejection figures for a number of organic compounds.

The type of membrane used will also affect the degree of removal obtained, particularly with the low molecular weight species. For example, it has been shown that for polar low molecular weight organic compounds such as alcohols, aldehydes, ketones and phenols, better rejections (but still low, relative to high molecular weight species) are obtained with polyamide membranes relative to cellulose acetate membranes.

TABLE 3.16

Removal of organic matter by reverse osmosis

Component	% Rejection
Humic acid	> 99
Tannic acid	90 – 95
Alkyl benzene sulphonate	90
Lindane	80 – 90
Acetone	40 – 70
Acetic acid	20 – 50
Phenol	15 – 20
Chlorophenol	50
Methanol	5 – 20
Ethanol	20 – 60
Propanol	40 – 90

Fortunately, most organic matter present in raw waters to be treated for industrial purposes is derived from decaying vegetation and effluents giving rise to large organic molecules, with some of the organic matter being present in colloidal form so that high removal efficiencies are likely to be obtained.

6.3 Deterioration of reverse osmosis membranes

Reverse osmosis membranes can deteriorate in service due to two main effects; hydrolysis and fouling.

6.3.1 Hydrolysis

The production rate (flux) can increase with time, but there is a decrease in the quality of water produced, i.e., leakage of salts. This is due to hydrolysis and in the case of cellulose acetate membranes can be caused by the water being treated having a pH greater than 8. Polyamide membranes are not so susceptible to pH and can operate up to pH 11.0. A similar effect can arise from bacterial attack, but this is not so common as the gradual deterioration due to hydrolysis.

To minimise hydrolysis the pH should be in the range 4–6 and the temperature should be low. Obviously, in practice there has to be a compromise between minimising hydrolysis and maximising the efficiency of treating the water, e.g., an increase in temperature increases the flux rate and hence lowers the cost of producing treated water, albeit with a shorter membrane life.

6.3.2 Fouling

As water flows through the membrane the concentration of salts in the surface film increases. If there are sparingly soluble salts present there is a risk that 'scaling' of the membrane will occur. For this reason it is necessary to ensure that the velocity over the surface is sufficiently high and is uniform.

As with ion-exchange resins, the mechanism of fouling of membranes is still not fully understood. Obviously, in a physical process such as reverse osmosis the prime cause of fouling will be the entrapment of material on the surface of the membrane. Again, as with ion-exchange it is possible that membranes 'age' such that a reduced output is obtained, e.g., compaction of the membrane material itself may arise.

Irreversible fouling, therefore, is probably the result of an accumulation of solid material at the membrane surface which cannot be removed by physical and chemical methods, and an ageing process which could involve both physical (compaction) and chemical effects (degradation of the membrane surface). The role of organic matter in this process is not fully established, but investigational work on the treatment of sewage by reverse osmosis indicated that the two main factors influencing the rate of fouling were particles less than 5 μm, and the dissolved organic components of the sewage. It is possible that at the membrane surface, where maximum concentration occurs, organic matter can be precipitated and is relatively easily compressed into the surface of the membrane rather than being swept away.

Reversible fouling is the accumulation of material which can be relatively easily removed from the membrane. In this case inorganic components can be a major factor causing scaling and deposits of oxides of iron, aluminium and silica.

To minimise scaling, the raw water can be softened or the pH controlled such that the solubilities of substances such as calcium carbonate, calcium sulphate and magnesium hydroxide, etc., are not exceeded. Prefiltration is applied to minimise the ingress of the oxides mentioned above, but unless a very fine filter is used (at a high cost) small particles frequently colloidal in nature will be present and give rise to fouling. As in the case of irreversible fouling, organic matter is likely to be a factor influencing reversible fouling, and both colloidal and the initially soluble organic matter could contribute to the fouling process.

Biological fouling arising from the growth of slimes on the surface of the membrane can also occur. Disinfectants can be applied to the incoming raw water. However, for the polyamide membranes the chlorine residual should not exceed 0.1 mg/l, whereas the cellulose acetate systems will tolerate up to 1 mg/l.

Alternative sterilising agents are ozone, hydrogen peroxide, peracetic acid and formaldehyde.

Sterilising is unlikely to be effective in preventing fouling by non-biological organic matter which usually represents most of the organic matter in the raw water.

It should be emphasised that reverse osmosis membranes have a relatively short working life of 1–5 years. The deterioration is likely to arise from a combination of factors:

- Physical stresses.
- Chemical degradation.
- Ageing.
- Fouling.

It is difficult to quantify the contributions of these factors. However, in respect to fouling it is likely to be due to the adherance of colloidal matter on the surface, and this can be present in the incoming water or produced at the high concentration zone which arises at the surface of the membrane.

An empirical test has been devised to determine the potential of a raw water for fouling the membrane and is called the silt density index or SDI. The SDI is measured by monitoring the rate of plugging when feedwater at 2 bar pressure is passed through a 0.45 μm filter.

This is done by passing a given volume of water through the filter and noting the time taken. The water is allowed to continue to pass through the filter for a defined period (5 or 15 minutes), and the timing of the passage of the given volume of water is repeated.

$$SDI = [1 - (t1/t2)/T] \times 100$$

where t1 is the filtration time for the first passage of water

t2 is the filtration time for the second passage of water

T is the period between readings

The SDI recommended by membrane manufacturers is usually < 4 for spiral wound and < 3 for the hollow fibre type of membrane

6.3.3 Cleaning of membranes

To combat fouling, membranes are cleaned both physically and chemically.

Physical

High velocity water flushing is usually employed by releasing the pressure at the downstream end of the plant and can be made automatic to operate on a time cycle.

Air has also been employed to generate high turbulence conditions at the membrane surface.

Unfortunately, the spiral wound and hollow fibre elements most frequently used, for economic reasons, are the most difficult to clean. It follows that there is a need to pre-treat the raw water to a very high standard if an adequate service life from the membranes is to be obtained.

Chemical

A variety of chemical agents has been used to clean membranes to restore the output from reverse osmosis plant with varying degrees of success. Table 3.17 shows the range of agents employed [7].

These agents have been applied singly and in various combinations. For specific fouling conditions such as by iron the choice is simple, i.e., an acid, with citric

TABLE 3.17

Cleaning agents for reverse osmosis membranes

Class	Chemical compound
Acids	1–2% citric acid 2–4% hydrochloric acid 2–4% sulphuric acid
Alkalis	1% sodium hydroxide
Phosphates	2% sodium hexametaphosphate 2% sodium triphosphate
Oxidising agents	hydrogen peroxide peracetic acid perborates
Disinfectants	1% formaldehyde
Detergents	Variety — usually at 0.2–0.4 w/v strength

acid likely to be most effective. For aluminium oxide and silica, alkaline washing should be effective.

In practice, a single component is unlikely to be the cause of fouling and therefore a combination of the properties of the agents is required to attack a deposit comprising mixed metal oxides, siliceous material and organic matter — note, however, that any chemical cleaning technique has to be compatible with the material from which the membrane is constructed.

The so called biological detergents can contain several of the above agents, 'Ariel' for example is an alkaline anionic detergent containing sodium perborate. This agent has been used to combat fouling on reverse osmosis systems treating surface waters and sewage effluent, both of which would be expected to give deposits containing appreciable quantities of organic matter.

With any cleaning operation it is always better to start before serious accumulation of fouling material occurs, and to maintain regular application of the cleaning agent.

6.4 Reverse osmosis plant

6.4.1 General

A typical reverse osmosis plant consists of the following items:

- Pre-treatment including acid dosing for pH control and dosing of scale control additives.

- High pressure pumps which may be high speed centrifugal, multi-stage centrifugal or reciprocating type.

- The reverse osmosis membranes. The membranes or permeators are usually connected in a series/parallel arrangement such that the reject from the earlier stages is used as the feed to the later stages. This increases the plant conversion.

- A pressure regulating valve, this is used to maintain the necessary reject flow and control the membrane inlet pressure.

- The post-treatment system, this usually includes a degasser to remove carbon dioxide formed when acid is used for pH control.

The majority of permeators used are either spirally wound sheet or hollow fibres.

Spiral wound

The cellulose acetate, or thin film composite membrane, and a backing material are sealed together down one edge and the whole is rolled into a spiral around a central perforated product collector tube. The raw water enters at one end of the spiral and passes axially through it, the product water passing through the membrane and via the porous backing spirals inward to the perforated collection pipe (see Fig 3.32).

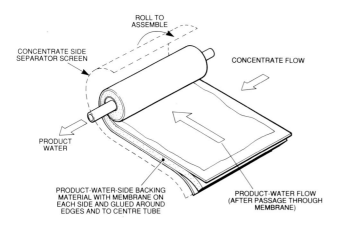

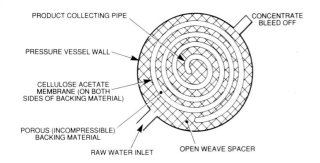

FIG. 3.32 Spiral-wound reverse osmosis module

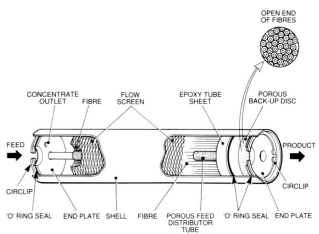

FIG. 3.33 Hollow fibre reverse osmosis module

Hollow fibre

Fine polyamide capillaries about the thickness of human hair are arranged in a U-shaped bundle inside a pressure tube. The ends of the bundle are embedded in an epoxy tube sheet which is machined to expose the open ends of the capillaries. Polyamide has a lower permeability than cellulose acetate but this is offset by the greater surface area per unit volume. Feedwater enters at the centre of the shell and flows outwards. Permeate flows radially inwards through the walls of the fibres and discharges from the product tube sheet. The construction is shown in Fig 3.33.

6.4.2 Experience in the UK power industry

Relatively few reverse osmosis plants have been installed because good quality raw water with total solids < 500 mg kg^{-1} has generally been available and hence the use of ion-exchange has been the preferred option for providing make-up water, for example, the total installed capacity of reverse osmosis plant is approximately 4000 m^3/day, which is less than 5% of the make-up requirements of the industry.

The largest plants are at Hartlepool nuclear power station [8]. Here a 44 m^3/h plant was installed in 1978 and a second plant was commissioned in 1985 making the total capacity 104 m^3/h.

The first plant was designed to compensate for an increase in total dissolved solids of the towns supply such that the output from the existing ion-exchange plant was decreasing to an unacceptable extent.

The Hartlepool water supply is derived from deep wells. It contains negligible levels of suspended solids (SDI < 2) and organics and is therefore ideally suited to reverse osmosis treatment.

As shown in Fig 3.34, a proportion of the towns main water feed to the deionisation plant is treated and returned to produce a blend of reverse osmosis product water and towns water to give a composition similar to that of the original towns water specification.

The towns water is boosted to about 4 bar pressure, dosed with 5–10 mg kg^{-1} scale inhibitor and with sulphuric acid to reduce the pH to the range 4.5–5.5. The pre-treated feedwater is filtered through 25 micron cartridge filters and then pressurised to 28–36 bar by the multistage centrifugal high pressure pumps. The 78 reverse osmosis modules are housed in 13 horizontal stainless steel pressure tubes, 0.2 m diameter by 6 m long each containing six spiral wound cellulose acetate elements connected in the series.

The pressure tubes are arranged in a series/parallel flow configuration in three stages consisting of seven, four and two tubes. Figure 3.34 shows the flow arrangement. Water is pumped to the first seven modules connected in parallel, concentrate from these modules then passes as feed for four further modules and the concentrate from these is the feed to the final two modules. Using this technique a raw water utilisation of 80% is achieved.

The 20% reject water (concentrate) flows through a control valve to waste. The product water flows to the degasser for CO_2 removal and from the sump is pumped back to the feed line to the make-up plant.

The plant operated for six years on the original set of membranes and produced 700 000 m^3 of treated water to the quality shown in Table 3.18. After this time the product water quality started to deteriorate to an unacceptable level due to hydrolysis of the membrane, and replacement membranes were therefore fitted.

A number of problems arose during the commissioning and early operation of the plant. Some of these

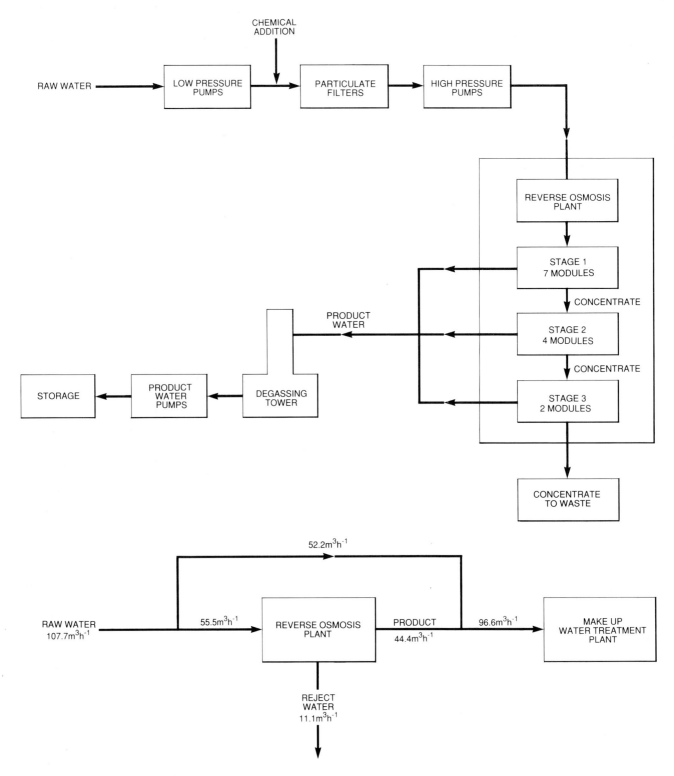

FIG. 3.34 Reverse osmosis plant — block diagram

were malfunctions of components such as the acid dosing pump, valves, etc., which could readily be corrected. Membrane telescoping occurred during the commissioning period when full pressure was applied to the membranes at start-up. The membrane arrangement is such that no movement in the pressure tubes can occur at the centre of the membrane due to the rigidity of the product collection tube. The brine seals on the

outer surface are less restrained allowing a telescoping effect to occur. This can cause damage to the membrane with possible failure of the glue line and contamination of the product water. The risk of damage was overcome by installing a slow start valve which reduced the pressure shock on the membranes at start-up.

Leaking of end cap and module O-ring seals on the assemblies used to interconnect membrane product

TABLE 3.18

Water analyses from the reverse osmosis plant at Hartlepool power station

Analysis	Pre-treated town mains water	Product water	Reject water
Conductivity μS cm^{-1}	1560	145	6050
Total hardness mg kg^{-1} CaCO$_3$	560	30	2700
Sodium mg kg^{-1} Na	100	15	600
Sulphate mg kg^{-1} SO$_4$	455	15	2300
Chloride mg kg^{-1} Cl	180	23	800

tubes caused a decrease in water quality. This was rectified by dismantling the pressure tubes and fitting new seals, but it was a time consuming and labour intensive activity.

However, over the six year period which followed, the plant performed satisfactorily, and when further water treatment capacity was required, a second reverse osmosis plant was installed having an output of 60 m^3/h. Spiral wound cellulose acetate membranes were again used with six elements in a module using a ten to four module array.

6.5 Future applications of reverse osmosis

The operating cost of reverse osmosis is mainly dependent on the pumping cost and the cost of replacement membranes and these factors represent about 85% of the operating cost, but it is only slightly affected by the concentration of dissolved salts in the raw water. In comparison, for ion-exchange plants the main operating cost is for regenerant chemicals and this relates directly to the dissolved solids of the raw water.

The main application is likely to continue to be for the partial treatment of raw water to ion-exchange plants where changes in supply result in the design capacity of the plant being exceeded, and/or where space is at a premium. Application to new installations will require a careful evaluation of the comparative costs of reverse osmosis plus ion-exchange and ion-exchange alone.

In the UK there are not many water supplies as favourable to reverse osmosis as at Hartlepool. However, as the cost of clean, good quality water accelerates, cheap, but dirty water may need to be used for the raw supply.

Waters of this type, e.g., industrial river water, have not been used in the power industry as feed to reverse osmosis plant, but investigational work on two industrial rivers, the Aire and Trent, has clearly demonstrated the need for extensive pre-treatment, and the requirement is likely to be more stringent

than that required for ion-exchange under comparable conditions.

A comparison of the advantages and disadvantages of reverse osmosis relative to ion-exchange is given in Table 3.19.

TABLE 3.19

Advantages and disadvantages of reverse osmosis relative to ion-exchange

Advantages of reverse osmosis	Disadvantages of reverse osmosis
Can be used on waters of higher dissolved solids	Higher capital cost on waters of low dissolved solids
Lower operating costs	High membrane replacement costs
Higher percentage removal of organic matter and colloidal silica	Lower percentage removal of inorganic salts
Lower chemical usage and hence environmentally more acceptable	Higher susceptibility to fouling and hence raw water may need extensive pre-treatment
Lower space requirement	

7 Electrodialysis for make-up water treatment

7.1 General

Electrodialysis is a process in which water flows between alternative cationic and anionic ion-exchange membranes between two electrodes. When a direct current is applied, cations in the water, being positively charged, will move towards the anode. Now cations can only pass through the cation membrane and anions can only pass through the anion membrane; consequently the salts present in the water will be concentrated and depleted in alternate compartments as shown in Fig 3.35. The problem with electrodialysis is that the membranes readily foul as the result of precipitation processes and the efficiency of separation of the salts decreases. In electrodialysis reversal (EDR), the electrical polarity is reversed at regular intervals so that the concentrated and depleted flow paths are interchanged. This is done continually, typically at 20 minute intervals. The reversal direction of ion movement is intended to release the potentially scale-forming ions and any colloidal matter fouling the membranes, and hence a major problem associated with straight electrodialysis should be alleviated.

The manufacturers state that this will allow operation without any continuous chemical treatment provided the Langelier Index (see Section 10.1.1 of this chapter) of the concentrate is +2.2 or less, and the calcium sulphate concentration is less than 1.75 times the saturation value.

The ion-exchange components of the membranes are similar to those used in ion-exchange resins, i.e., copolymers of polystyrene and divinyl benzene containing sulphonate and quaternary ammonium ion-exchange groupings on the cationic and anionic membranes, respectively.

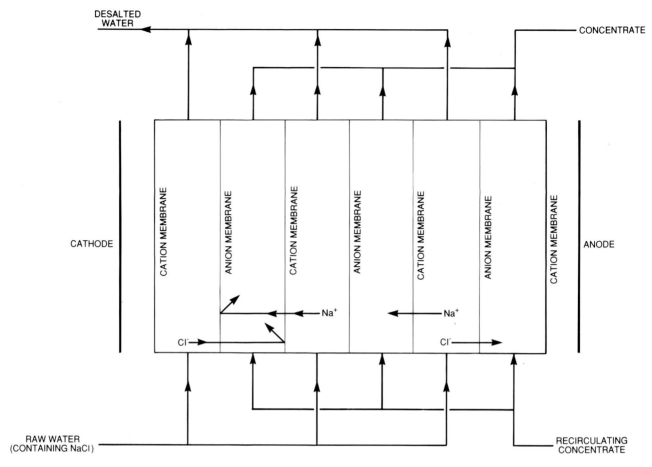

FIG. 3.35 Electrodialysis unit

Each cell pair comprises a cation transfer membrane, a diluate spacer, an anion transfer membrane and a concentrate spacer. The water flows across the membrane surface where the ions transfer through the membrane under the influence of the applied potential.

A membrane stack comprises cell pairs between two platinised electrodes and typically there will be 300–500 cell pairs per stack. A single hydraulic pass through a stack will produce a level of removal of salts which depends on a number of factors such as temperature, flow rate, applied voltage and composition of the water. The degree of removal will range from 30 to 60% per pass or 'stage' of treatment. A number of stages will be used to obtain the required quality of treated water.

In smaller EDR units several hydraulic and electrical stages may be used within a membrane stack, but for larger capacity plants the stacks are likely to be single stage and to obtain the required quality several stacks are operated in series. As in the case of reverse osmosis, electrodialysis is a possible technique for an initial stage of treatment of raw water for the removal of a high proportion of the dissolved solids present. The treated water will then need to be passed through a deionisation plant using resins to obtain a final quality of water suitable for make-up purposes.

EDR is a relatively recent innovation; consequently very few full-scale plants have been installed compared with reverse osmosis systems. However, Cordina and Curry [9] have reported experience gained from a full-scale installation on a make-up water treatment plant and they were able to compare EDR performance with a reverse osmosis plant operating under identical conditions. The general conclusions were that reverse osmosis can produce a higher quality water, particularly in respect to silica and organic matter, but the membranes are more sensitive to suspended and colloidal solids in the raw water than is the case with EDR. Although complex, the operation of EDR was considered to be easier than for reverse osmosis, being particularly easy to start-up. Operating costs of the two systems were found to be very similar.

Experience with EDR in the power industry is limited to pilot plant tests carried out at Ferrybridge power station and a full-scale plant recently installed at Drakelow power station. Details of the experience gained are given below.

7.2   Electrodialysis pilot plant tests at Ferrybridge power station

An Ionic Incorporated Aquamite III electrodialysis plant was operated for 100 days on several types of raw water and the performance of the plant was assessed in terms of salt rejection, water conversion and resistance to fouling.

The EDR unit comprised 200 vertically-stacked cell pairs, and there were six hydraulic and two electrical stages within the stack, as shown in Fig 3.36.

Typical operating conditions were:

Pressure, bar	Inlet	3.7
	Outlet	0.2
Flow rate, 1 min^{-1}	Inlet	24
	Concentrate waste	4
	Product	20
Voltage, V	Stage 1	70
	Stage 2	63
Current, A	Stage 1	1.6
	Stage 2	0.8

Initially, the plant was operated for 1250 hours on Ferrybridge towns water which is a clean supply containing minimal levels of suspended solids and organic matter.

The salt rejection obtained was as shown in Table 3.20. During this period of operation measurements of stack pressure, current, voltage and product flow rate remained constant, demonstrating that no fouling of the stack was taking place.

In the second period of operation the River Aire was used as the raw supply. This is a moorland water with a relatively high suspended solids and organic matter content, and the dissolved solids concentration is more than twice that of the towns water.

The water was pre-treated using ferric chloride and a cationic polyelectrolyte prior to filtration, but this failed to reduce the Silt Density Index (see Section 6.3.2 of this chapter) to < 20, which was above the normal limit of 15 set by the manufacturers.

Using this poorer quality raw water, it was necessary to chemically-clean the stack weekly to maintain output. The clean consisted of a sequential application of

TABLE 3.20

Quality of water obtained from an EDR plant treating Ferrybridge towns water and River Aire water

Parameter	% Rejection	
	Ferrybridge towns water	River Aire
Conductivity μS cm^{-1}	90	88
Total alkalinity CaCO$_3$	86	85
Chloride mg kg^{-1} Cl	92	93
Sulphate mg kg^{-1} SO$_4$	97	91
Nitrate mg kg^{-1} NO$_3$	93	95
Calcium mg kg^{-1} Ca	94	94
Magnesium mg kg^{-1} Mg	92	93
Sodium mg kg^{-1} Na	62	88
Potassium mg kg^{-1} K	83	90
Total organic carbon mg kg^{-1} C	66	50

hydrochloric acid and alkaline brine via the clean in place (CIP) system incorporated into Ionics EDR plants. During the test of 1000 hours' duration, the mean water conversion was 70% and the % rejection of salts was similar to that obtained with towns water as shown in Table 3.20.

Overall, the EDR plant tested was shown to be capable of removing > 90% of dissolved solids from both of the raw waters used. The removal of organic matter is significantly less than inorganic salts and the level achievable will depend on the type of organic matter present.

The electrical power requirements for pumping and providing the DC stack supply when treating the towns water were 1 kWh/m^3 and 0.2 kWh/m^3 of treated water, respectively, and for river water the figures were 1.2 kWh/m^3 and 0.3 kWh/m^3.

7.3 Electrodialysis plant at Drakelow power station

Until recently the raw water supply to the *C* station make-up deionisation plants was a towns supply, with a smaller usage of River Trent water. Towns water had become increasingly expensive and currently costs 40 p/m^3. A further disadvantage of using towns water was that it contained a small, but significant, quantity of organochloride arising from the chlorination process at source. These compounds are not removed by ion-exchange processes (including EDR); consequently they pass into the water/steam circuit where they breakdown by hydrolysis to give ionic chloride. This results in an increased blowdown demand from the boilers to maintain the required boiler water quality to minimise corrosion.

These disadvantages in the use of towns water led to a general consideration of the make-up supplies on the station with the objective of minimising the costs involved by changing to a cheap, alternative raw water supply.

River Trent water is cheap, but contains about twice the level of dissolved solids of the towns water, and hence there was a need to install an additional treatment stage before the water could be fed to the existing deionisation plant.

The pre-treatment requirements for an EDR plant should be less stringent than for reverse osmosis, the other main possible alternative form of treatment. This meant that with EDR an existing flocculation and filtration plant could be used, whereas a reverse osmosis plant would have required a new pre-treatment system. Consequently, a considerable capital saving would arise from installing EDR.

Studies were carried out to optimise the make-up supply using an EDR plant. This resulted in a decision to install a unit to remove 75% of the total dissolved solids in clarified River Trent water. The specification for the quality of the water to be treated by EDR is given in Table 3.21.

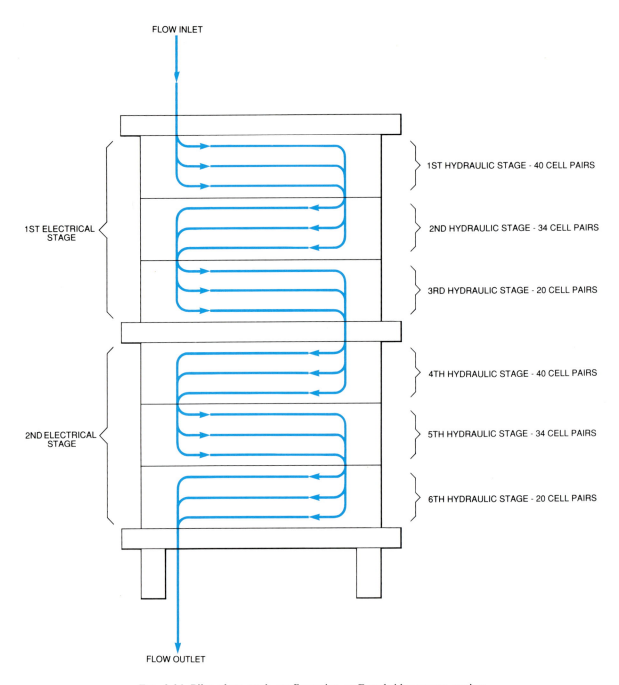

FLOW INLET

1ST ELECTRICAL STAGE

2ND ELECTRICAL STAGE

1ST HYDRAULIC STAGE - 40 CELL PAIRS

2ND HYDRAULIC STAGE - 34 CELL PAIRS

3RD HYDRAULIC STAGE - 20 CELL PAIRS

4TH HYDRAULIC STAGE - 40 CELL PAIRS

5TH HYDRAULIC STAGE - 34 CELL PAIRS

6TH HYDRAULIC STAGE - 20 CELL PAIRS

FLOW OUTLET

FIG. 3.36 Pilot plant stack configuration at Ferrybridge power station

Subsequently, an Ionics Ltd, Aquamite XX6/2 plant was purchased. This was designed to produce 50 m³/h of treated water with a removal of 75% of the total dissolved solids at a temperature of 10°C. The plant also had to produce 80% of treated water from the raw supply.

The plant comprises two streams, each with three stacks (having single hydraulic and electrical stages) in series, with each stack containing 500 cell pairs. The specified power consumption was 0.9 kWh/m³. At each stage 36% of the total dissolved solids is removed to give the overall 75% salt removal (see Fig 3.37).

The main components in addition to the cell assemblies are feed and concentrate stream pumps, a 10 μm cartridge filter fitted immediately before the cell array, valves, a gas separator tank and blower, a 45 kVA rectifier and a control panel. Instrumentation includes pressure gauges and a pressure differential gauge, flow indicator for the diluate and concentrate streams, a conductivity meter for diluate quality measurement and a temperature indicator.

A clean in place (CIP) equipment is installed for chemical cleaning of the membranes. This uses the concentrate recirculation pump with appropriate fittings to a chemical stage tank.

The EDR plant has to be taken out of service when the membranes require cleaning. The frequency of cleaning will depend on the quality of water in

TABLE 3.21

Chemical compositions of the raw and treated water on the EDR plant at Drakelow power station

Temperature 12.9°C		Specified raw water composition	EDR inlet measured	EDR outlet measured	Removed by EDR
Conductivity	(μS/cm)	1120	1179/1105*	339/339*	71%/69%
					(70%/68%) Δ
pH		7.5	7.4	7.5	—
Sodium	(mg kg^{-1} Na)	80	73	27	63%
Calcium	(mg kg^{-1} Ca)	106	89	20	78%
Magnesium	(mg kg^{-1} Mg)	28	20	4.9	75%
Total hardness	(mg kg^{-1} CaCO$_3$)	381	306	70	77%
Total alkalinity	(mg kg^{-1} CaCO$_3$)	184	111	44	60%
Sulphate	(mg kg^{-1} SO$_4$)	186 +	167	55	67%
Chloride	(mg kg^{-1} Cl)	120 +	127	22	83%
Nitrate	(mg kg^{-1} NO$_3$)	55	44	8	82%
EMA	(mg kg^{-1} CaCO$_3$)	380	315	100	68%
TOC	(mg kg^{-1})	—	6.2	4.9	21%
Silica	(mg kg^{-1} SiO$_2$)	8.4	6.4	4.3	33%
TDSΔ	(mg kg^{-1})	740	637	185	71%
Suspended solids ++	(mg kg^{-1})	3.0	0.43	—	—
Total iron ++	(mg kg^{-1})	1.0	0.15	0.22	—
Soluble iron ++	(mg kg^{-1})	0.3	0.02	0.05	—

*   RTC/Ionics conductivity meters Δ calculated

\+   After flocculation (addition of ferric sulphate)

++ After flocculation and filtration

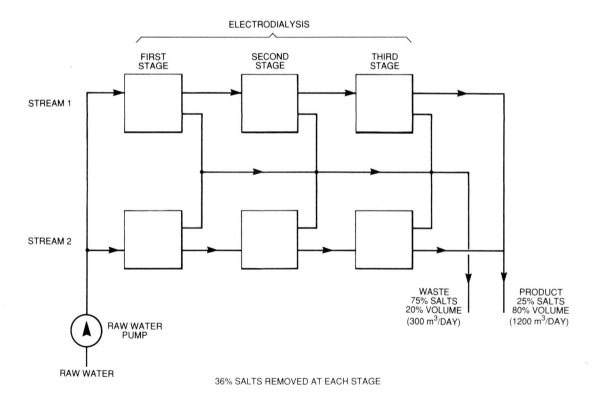

FIG. 3.37 The two-stream EDR plant each with three stages of salt removal, at Drakelow power station

terms of suspended and colloidal matter passing through the cells.

In addition, there is an electrode CIP process. This is carried out automatically during the normal operation of the EDR plant. Dilute acid is pumped through the electrode compartments for about an hour per day to remove any scales or deposits that have formed.

Figure 3.38 shows the general layout of the EDR plant.

The EDR plant has operated for a year and in general has performed satisfactorily. The results in Table 3.21 indicate the typical quality of water produced. The relative level of removal of individual ions was generally as expected, e.g., sodium removal less than calcium. Interestingly, a small proportion of silica was removed in addition to the main salts.

The main difficulty encountered in meeting the specified conditions was during the winter period when the plant could remove only 70% of the salts present at a temperature of 10°C.

The reason for this has not yet been fully established. At the time, the membrane cleaning technique comprised:

● Polarity reversal at 20 minute intervals.

● Monthly chemical cleaning via the (CIP) system using 3% hydrochloric acid.

Despite these techniques and with the suspended solids entering the plant being less than 0.5 mg kg^{-1}, there was evidence from pressure drop measurements that some fouling of the membranes was taking place. Additionally, the 10 μm filter cartridges were having to be changed more frequently than the 3-week period specified.

This experience indicates the importance of pretreatment to the successful operation of EDR. The nature and size of particles reaching the membranes may be critical, and these are likely to depend on the water source, the chemical coagulents used and the clarifier design.

Eventually, the plant was stripped down and it was evident that the membranes required a manual clean. This was done using scrubbing brushes. Subsequently, there was an increase in the effectiveness of salt removal, but by this time the temperature of the river water had increased so that the performance of the plant at 10°C with the membranes in a clean condition has not yet been established. As a result of this difficulty, weekly chemical cleaning was instigated and the next low temperature period is awaited to determine whether the membranes can now meet the specified 75% salt removal at 10°C. This may also establish how frequently the membranes have to be cleaned manually. The specified period was every two years, but clearly this has not been sufficient up to the present time.

The variation in river water temperatures complicates the operation of the plant, as frequent voltage adjustments have to be made to ensure that the required salt removal is maintained.

The major benefit from the installation has been the provision of cheaper make-up water and, in this respect, significant cost savings have been achieved. There have also been cost savings from reduced boiler blowdown requirements.

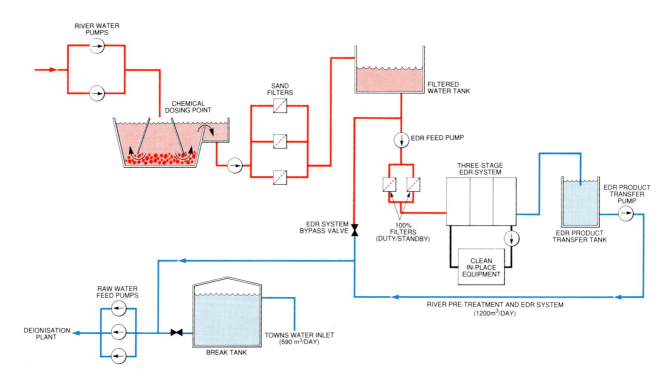

FIG. 3.38 General arrangement of an EDR plant at Drakelow power station

Overall, experience with EDR in the power industry has shown that it is most likely to be used when it enables expensive raw water to be replaced by a cheaper source. As with most membrane systems, the quality of the water entering in terms of particulate matter will be a critical factor determining the successful use of the technique. There has not been sufficient operational experience to assess the potential life of the membranes used, whose replacement cost is a substantial proportion of the total plant cost. In this respect, the source of the raw water and the form of the pre-treatment provided are likely to be significant factors.

8 Use of evaporators in the preparation of make-up water

In recent years, evaporators for providing make-up in the power industry have largely been superseded by ion-exchange processes; hence, only a brief description of the water treatment processes used will be given.

8.1 Treatment of raw water

When used in a make-up water treatment process, evaporation plant will play a role similar to that of reverse osmosis, that is, removing the bulk of salts present in the raw water, with further treatment being required by an ion-exchange stage to obtain the required final quality. Similarly, some pre-treatment of the water used as feed to evaporators has to be carried out.

Initially, a stage of clarification and/or filtration may be needed as described in Section 3 of this chapter. In the case of inland raw waters, this would then be followed by a softening process (removal of hardness salts), and base-exchange softening is normally used to minimise scaling within the evaporator.

8.1.1 Base-exchange softening

This is an ion-exchange process similar to that described in Section 4 of this chapter.

The scale-forming cations, calcium and magnesium, in the water are replaced by the sodium ion using a bed of strongly acidic cation resin.

Typical reactions are:

$$Ca(HCO_3)_2 + 2\ Resin\ (R)Na \leftrightarrow 2NaHCO_3 + R_2Ca$$
$$CaSO_4 \quad + 2RNa \quad\quad \leftrightarrow Na_2SO_4 \quad + R_2Ca$$
$$MgCl_2 \quad + 2RNa \quad\quad \leftrightarrow 2NaCl \quad\quad + R_2Mg$$

Brine is used to reverse these reactions when the bed is exhausted, as indicated by a breakthrough of hardness salts.

$$CaR_2 + 2NaCl \leftrightarrow 2RNa + CaCl_2$$

The residual hardness in the treated water will depend mainly on the level of regenerant used. For a water of average composition the treated water total hardness should be > 2 mg kg^{-1} as $CaCO_3$ when the resin is regenerated with 80 gl^{-1} of salt.

A base exchange softener is constructed similarly to the ion-exchange units described in Section 5.2 of this chapter, comprising a cylindrical steel pressure vessel with dished ends, a distribution system for the incoming water and brine, and a collection system including support for the resin bed. For economy, a graded gravel bed is frequently used for this purpose.

The salt solution is prepared in a simple salt saturator.

8.1.2 Use of sea water

Sea water cannot be softened economically; consequently, the only form of treatment given is the addition of sulphuric acid to remove the temporary hardness, $Ca(HCO_3)_2$, and convert it into permanent hardness, $CaSO_4$, with the liberation of carbon dioxide. This will reduce the propensity for calcium carbonate scaling in the evaporator, leaving calcium sulphate as the main scale forming constituent.

8.2 Operation of evaporators

8.2.1 General

The various types of evaporators used in power stations are described in Volume C, Chapter 3, hence the following paragraphs are concerned only with the chemical aspects of evaporator operation and maintenance.

The choice of bled steam or live steam evaporators is largely a question of economics; however, the characteristics of the water to be evaporated should influence the design of the evaporator employed.

When the water is softened by a base-exchange process the risk of scaling is reduced, but some additional chemical treatment within the evaporator is needed to give trouble-free operation.

In the case of sea water evaporators, experience in the power industry has shown that although the rate of scaling within the evaporator can be minimised by the treatment outlined above, the control of the treatment is not sufficiently precise and periodic cleaning is necessary. Moreover, serious corrosion often occurs in sea water evaporators, due to the strong electrolyte solution in the evaporator. Ferrous materials suffer the worst attack (stay bolts, evaporator shell, etc.) and regular maintenance is necessary. Additionally, in theory the evaporator distillate side of the evaporator complex (Fig 3.39) is isolated from the feedwater and the distillate only enters the feed system as make-up after further treatment in the mixed bed polishing units. However, feedwater contamination has occurred when due to defective water level control in the evaporator, salt water from the evaporator has carried over into the 'flash box' at point (A) in Fig 3.39 and thence, via the No 1 LP heater and drain cooler drains, into the condenser.

8.2.2 Scale prevention

Where scale deposition is likely to occur, the severity of scaling can often be minimised by adding certain

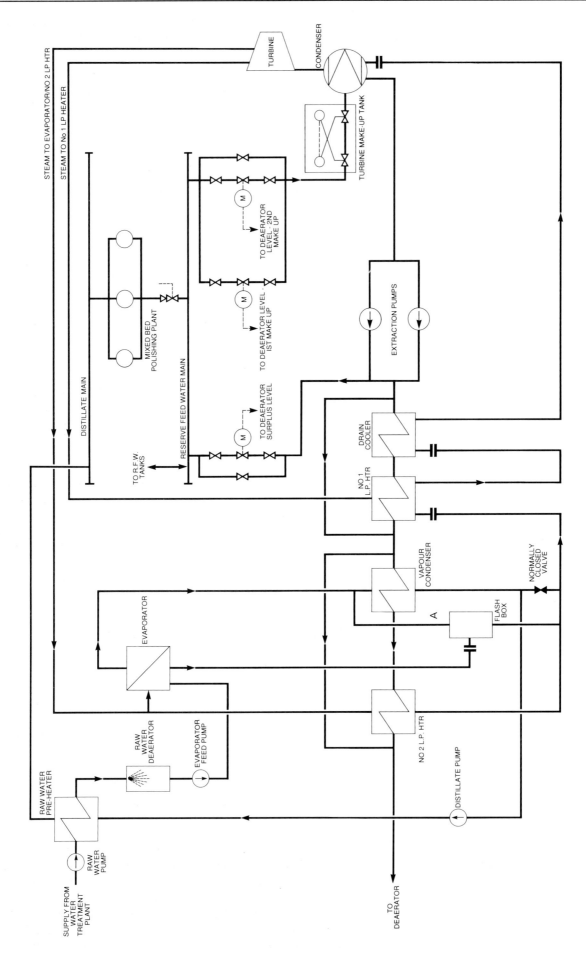

FIG. 3.39 Evaporator and associated plant — flow diagram

chemicals to the water before it enters the evaporator. The chemicals used act in one of three ways:

- By complexing calcium and magnesium in a soluble form and so preventing precipitation.

- By causing salts thrown out of solution to be precipitated as a harmless sludge, which is subsequently removed with the blowdown.

- By interfering with the crystallisation process, such that soft rather than hard deposits are formed on the heat exchange surfaces.

In the latter case only thin deposits are formed, which due to their nature do not materially affect heat transfer. Chemicals which have been used for evaporator treatment of the three types referred to above, include:

- Polymetaphosphate.

- Ethylenediamine tetra-acetic acid (EDTA).

- Phosphates.

- Organic polymers.

The choice of treatment chemical, if any, can only be determined by conducting plant trials. Different waters respond differently to the additives, as do evaporators of different design.

8.2.3 Blowdown

It is necessary to limit the concentration within the evaporator body to ensure that the less soluble constituents are not precipitated from solution and to prevent priming due to the concentrated solution of salts, which would lead to a deterioration in the quality of the distillate.

For most towns main supplies a 5% to 10% blowdown is adequate, whilst with sea water a blowdown of approximately 50% is necessary. This is usually carried out continuously, the blowdown rate being controlled by a fixed orifice. The chemist needs to regularly check the water in the evaporator body and arrange for additional blowdown if necessary.

8.2.4 Cleaning

Where, in spite of the best possible chemical treatments, scaling does occur, a stage will eventually be reached where heat transfer is materially affected and the evaporator will no longer produce the required output.

With coil type evaporators, cleaning can often be done manually, but with tubular evaporators this may be very time consuming and it is necessary to seek a chemical method of cleaning. If the scale contains 20% or more calcium carbonate, or other alkaline constituents, it is usually possible to remove it with dilute inhibited acid. If the scale is essentially calcium sulphate, acid cleaning is not very effective and although other cleaning agents such as EDTA have been used,

these are very expensive and slow in action so that manual cleaning is likely to be required.

8.2.5 Evaporator distillate quality

The quality of the evaporator distillate will depend upon:

- The concentration of dissolved salts in the water in the evaporator body.

- The foaming characteristics of the raw water.

- The design of the evaporator.

- The efficiency of the gas venting arrangements.

Where the raw water is softened towns supply water, or water of similar quality, the electrical conductivity of the distillate may vary from $0.8\ \mu S\ cm^{-1}$ to $3\ \mu S\ cm^{-1}$, the higher value being often due to inefficient venting of CO_2.

With sea water the following analysis is typical of the distillate quality:

Conductivity	$11.6\ \mu S\ cm^{-1}$
Silica	$0.011\ mg\ kg^{-1}$ as SiO_2
Ammonia	$0.22\ mg\ kg^{-1}$ as NH_3
Carbon dioxide	$6.4\ mg\ kg^{-1}$ as CO_2
pH	5.3

The maximum permissible dissolved salt concentration in the evaporator body is determined by the quality of the raw water and its scaling tendencies. For sea water, the conductivity of the water in the evaporator body will be in the range of 65 000 to 90 000 $\mu S\ cm^{-1}$ and for town water 6000 to 8000 $\mu S\ cm^{-1}$. It can be seen therefore that the amount of priming and carry-over that can be allowed is very small indeed.

9 Condensate purification

9.1 General

The term condensate purification (or polishing) is used to describe the treatment of condensed steam from turbines operating in the power industry.

There are several possible sources of soluble contaminants into the water/steam circuit of a power plant. Cooling water ingress to the condenser can arise despite the introduction of double tubeplates and the use of titanium condenser tubes, and this can be a particularly severe problem where sea water cooling is employed. For example, an ingress of less than 0.5 litre/hour of sea water into a 500 MW unit could result in an unacceptable quality of feedwater. On occasions, contamination of the system from the make-up water can occur. This could arise from mal-operation such that regeneration chemicals enter the system, or the plant may overrun during the service cycle. Additionally, complete removal of certain substances may not occur, particularly organic matter (see Section 4.4 of this chapter).

Insoluble impurities can also be present in condensate. Usually, these are corrosion products from the materials in the system, with copper and iron being the predominant species. However, under commissioning conditions and following outages for maintenance, a number of other substances, in particular silica from lagging and cementitious materials, may be present in insoluble form.

Any ingress of soluble and insoluble contaminants into the feedwater will increase the risk of corrosion within the boiler and turbine sections. There is also a risk of substances depositing within the turbine, which can result in a loss of output.

It follows from the foregoing that a condensate purification plant (CPP) needs to be capable of removing both soluble and insoluble contaminants from condensate.

The quality of treated water required to be produced by a CPP depends on the design of the power plant and the operating conditions. The greater the risk of corrosion and the subsequent economic penalties the higher the quality. This is exemplified by comparison of the specifications for the required water qualities from CPP installations at AGR and PWR nuclear plants shown in Table 3.22.

TABLE 3.22

Treated water quality required from condensate purification plant

	AGR	PWR
Sodium, μg kg^{-1}	< 2	< 0.3
Chloride, μg kg^{-1}	< 2	< 0.3
Sulphate, μg kg^{-1}	< 2	< 0.5
Conductivity, μS cm^{-1}	< 0.08	< 0.06

It can be seen that the limits for sodium, chloride and sulphate are an order of magnitude lower for the PWR system. Similar limits to those used on the AGR are applied when full-flow CPP is installed on fossil-fuelled stations and magnox reactors.

It follows, therefore, that for once-through boilers, both nuclear and fossil-fuelled, or high heat flux boilers such as at oil-fired coastal stations, the whole of the condensate flow is likely to be treated using an additional plant to that used to provide the make-up water supply.

Ion-exchange resins have the capability of removing soluble and insoluble contaminants hence this is the main section of a CPP, but an additional filtration facility may be included.

9.2 Partial condensate purification

Partial condensate purification facilities may be provided on large conventional fossil-fuelled plants using the final stage of the make-up water treatment plant to treat a proportion of the feed flow to the boiler for a short period. For high pressure drum type boilers the greatest risk of contamination is during the commissioning period and at subsequent start-ups when the level of insoluble corrosion debris is high. A higher level of soluble impurities will be present during start-up and these arise mainly from cooling water ingress during the off-load period and from the washing of sodium salts and silica from the turbine blades during the initial passage of wet steam.

To accommodate partial condensate purification, the final mixed beds in the make-up plant are increased in size. The arrangement is that water dumped on start-up is purified in the mixed beds together with reserve feedwater and can be recycled through the beds. As a further refinement the full-load condensate flow from one unit in a multi-unit station can be polished.

A typical arrangement is illustrated in Fig 3.40, which shows the make-up system at Littlebrook *D*. It will be noted that there are two make-up mixed beds and two polishing mixed beds of equal size and with a common regeneration system. The polishing mixed beds are used to recycle the reserve feedwater, thus all make-up to the condensers and to the reserve feed tanks is freshly polished water.

At start-up, the condensate section valve is closed and condensate is recovered to the inlet of the polishing mixed bed. The 30% make-up pump takes suction from the polished water main and pumps through the low pressure heaters up to the de-aerator.

Systems of this type are installed at many coal and oil-fired stations.

9.3 Full-flow condensate polishing plant — general design features

The purpose of the condensate polishing plant (CPP) is to treat the total condensate flow to enable boiler/feedwater quality standards to be maintained in the event of any impurity ingress into the condensate.

A wide variety of CPP has been used, but as will be seen from the examples listed below, the main treatment section comprises mixed beds.

Types of condensate purification plants:

● Mixed beds.

● Powdex → mixed beds.

● Mixed beds → Powdex.

● Cation units → mixed beds.

● Filters → mixed beds.

● Filters cation units → mixed beds.

● Cation unit anion unit → cation unit.

Details of the types of system used in the power industry will be given in he following sections. References [10,11,12,13] provide additional information.

A CPP has to treat large quantities of water, for example, the normal condensate flow from a 660 MW unit such as at Heysham 2 is 1733 m³h⁻¹. Hence, to minimise the capital cost of plant, the ion-exchange beds have to be operated at a high flow rate relative to make-up ion-exchange plant. High flow rates produce two main effects; a high differential pressure across a bed and a decrease in the time available for the ion-exchange reactions to occur. These factors will be considered in relation to design and operation in Section 9.8 of this chapter.

Protection of the feed and boiler circuits in power plants having full flow condensate purification will be provided by the addition of ammonia, to elevate the pH, and hydrazine, to remove any residual dissolved oxygen present after de-aeration.

The pH will be controlled in the range 8.8–9.6 with the actual pH being set by the particular site operating conditions. Normally, a pH range of 8.8–9.2 will protect the system and minimise the transport of metallic impurities from the feed circuit to the boiler. Occasionally, where there is a high risk of erosion/corrosion occurring, the pH may be raised to 9.4 or above. It should be noted that increasing the pH level can have a significant affect on the design and operation of CPP. For example, the ammonia concentration has to be increased by a factor of four to raise the pH from 9.2 to 9.6. Consequently, this increases the load onto the cation exchange resin when it is operated in the hydrogen form, and the regeneration frequency has to be increased.

Alternative amines such as morpholine and AMP (2amino–2methyl propanol), may afford better protection than ammonia, by providing improved pH conditions in the water phase of the two-phase (steam/water) regions of the boiler. Again, the use of these amines will affect the operation and performance of the CPP (see Section 9.8 of this chapter).

The use of hydrazine generally poses no problems to the CPP as the breakdown products are mainly ammonia from high temperature degradation and nitrogen from reaction with any oxygen in the water.

The time to exhaustion of a bed containing cation resin in the hydrogen form will obviously depend on the pH used, and the ability to sequence the regenerations is a main design consideration. In general, the regeneration frequency should not exceed once per day for the total station polishing plant. Under conditions where pH has to be controlled continuously at pH 9.4 and above, the frequency of regenerations of the mixed beds can be decreased by the inclusion of upstream cation units, or by operating the cation resin of the mixed bed in ammonium form.

9.3.1 Plant location and layout

To remove the maximum amount of corrosion products the polishing plant should be located where it will process the total boiler feed. In most power stations this would be after the de-aerator, but at this point the temperature is well above the maximum anion resin service temperature of 60°C. Polishing plants are therefore located in the LP feed train at either the condenser extraction pump discharge or the gland steam condenser outlet where the temperature is about 30–40°C. Thus a considerable proportion of the heater drains flow may by-pass the plant via the de-aerator, as a result, at many stations the polishing plant may treat only 60–70% of the total boiler feed flow.

For AGR stations, 100% condensate polishing is necessary at all times and the plants comprise 3 × 50% polisher vessels or 2 × 50% with a common standby unit. Details of some AGR condensate purification plants in respect to type and number of units employed are shown in Fig 3.41.

The requirement for polishing at coastal drum type units is mainly for control of condenser leakage. Consideration was given to the choice of a central polishing plant with condensate piped from all the units and to the provision of unit plant with a common regeneration system. The most economic solution has been to provide each turbine-generator with two 50% flow polishers in a shunt loop and hold a spare change of regenerated resin in the common regeneration plant. In the event of condenser leakage, it is usually possible to tolerate the increased impurity ingress during the resin change period of about one hour when polishing is reduced to 50%. Alternatively, resin transfer is carried out at a time when it is convenient to drop load.

Polishing plants for drum type boilers are not provided with standby regenerators since the units can operate without polishing in the event of regenerator failure.

For the magnox nuclear stations only partial polishing was initially provided on each unit with the ability to combine the plants and polish at 100% flow on one set in the event of condenser leakage. Because of the frequency of condenser leaks the plants were later uprated to treat 100% of the condensate flow on all turbine-generators.

9.4 Filtration

Insoluble corrosion products will be present in condensate and levels are particularly high during the early commissioning period when concentrations of ion as high as 2000 μg kg⁻¹ can be experienced. Thereafter, significant levels of debris are likely to be present only during start-ups, following a disturbance to the system. At this time, concentrations of 100–200 μg kg⁻¹ of iron may be present for a short period. Under normal load conditions total concentrations of iron and copper will usually be <5 μg kg⁻¹.

Early full flow CPP in the power industry used filters as the initial stage of treatment. The plant at Drakelow C included deep bed sand filters which could be pre-coated with cellulose fibre and 12 × 3 m diameter vessels had to be provided on each 370 MW unit.

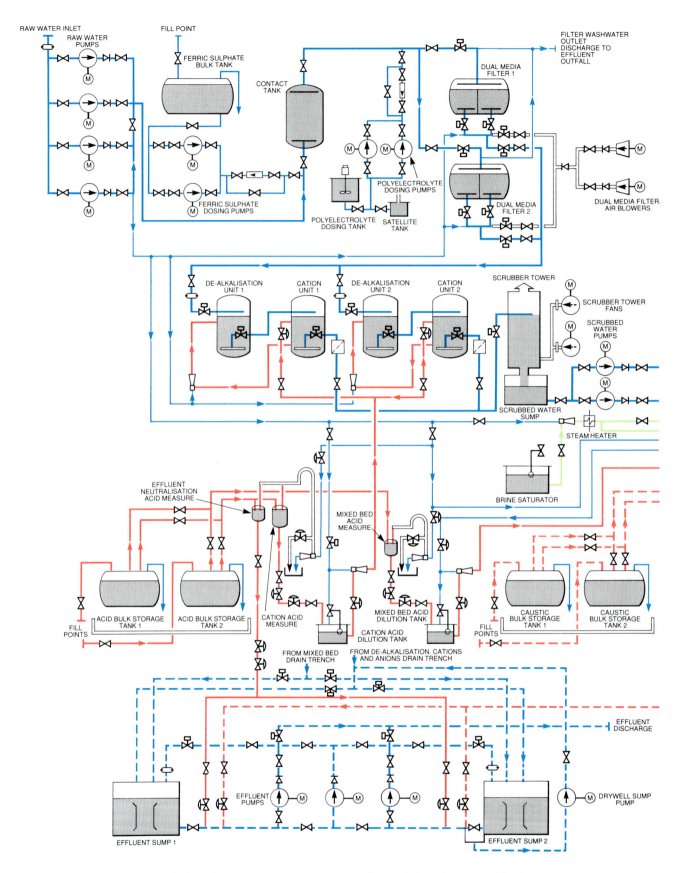

FIG. 3.40 Make-up water treatment system with partial condensate purification

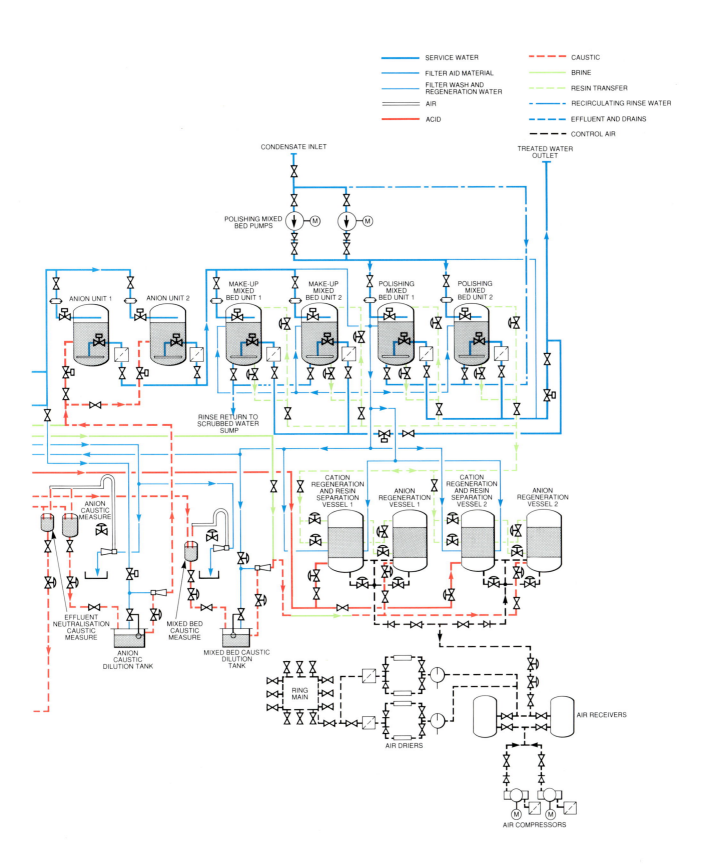

FIG. 3.40 (*cont'd*) Make-up water treatment system with partial condensate purification

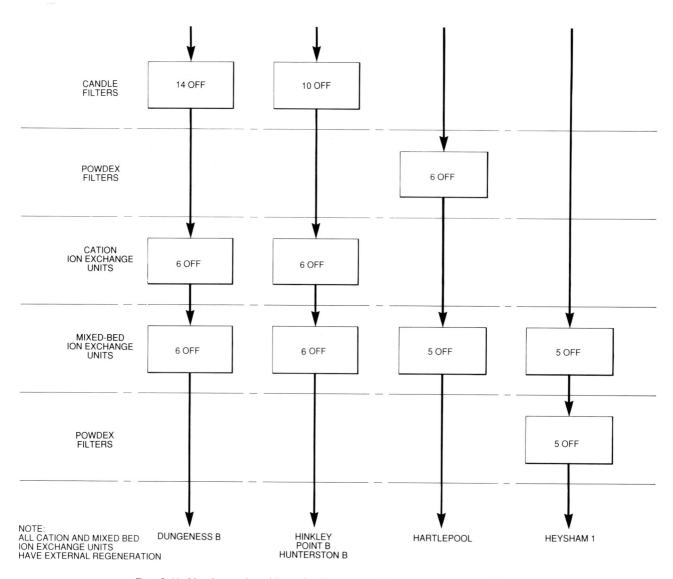

FIG. 3.41 Number and position of units in an AGR condensate purification plant

This type of plant is expensive and occupies a very large area.

9.4.1 Pre-coat filters

Other plants included pre-coat filters, usually with tubular stainless steel wire-wound filter candles, A typical pre-coat filter is shown in Fig 3.42. These filters were expensive, but occupied a much smaller space than deep bed types.

The filters are pre-coated by circulating a cellulose powder suspension through the elements and this deposits a layer of cellulose about 6 mm thick on the elements (see Section 9.8.4 of this chapter where the use of powdered ion-exchange is discussed). After a pre-service rinse the units are put on line and operate at a specific flow rate of about 8 m/h. The filters are taken out of service on reaching a high differential pressure and are backwashed using air and water, The backwash procedure involves compressing air in the filter head followed by a rapid opening of the body drain valve. The sudden expansion of the air provides a rapid flow to remove the pre-coat.

The pre-coat filters generally gave a good removal efficiency for suspended iron, but pre-coat removal was not always effective and this led to poor re-coating and the eventual need for mechanical cleaning of the elements. Consequently, the filters were frequently by-passed.

Meanwhile, investigational work had shown that cation and/or mixed beds could act as effective filters without being fouled by particulate matter. The efficiency of iron removal was shown to be in the range 75–85% and the pressure drop was shown to increase only slightly throughout the service run.

Removal of the iron by backwashing of the bed was occasionally found to be a problem but this could be overcome by development of techniques in which the particulates were removed through the bottom collector system rather than via the backwash when difficult conditions were encountered.

9.4.2 Magnetic filters

Magnetic filters have been tested at some UK power stations to assess their ability for removing particulate

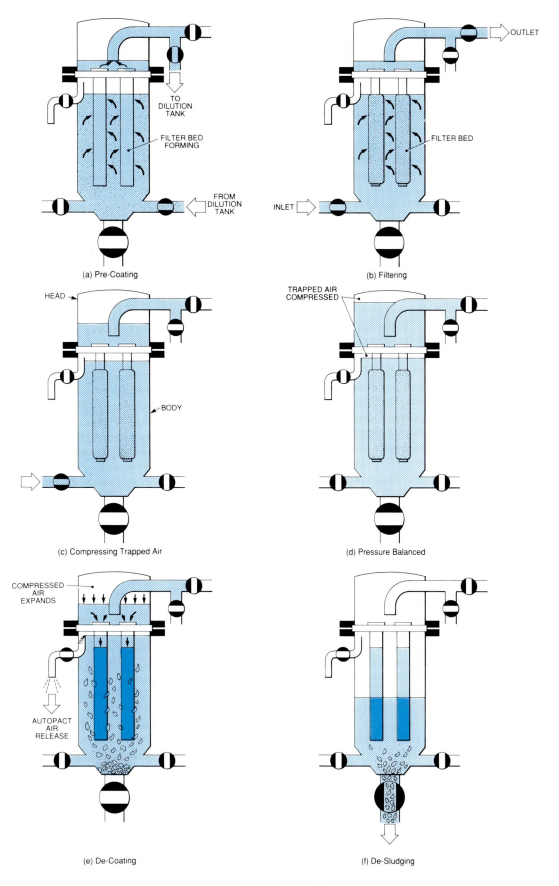

(a) Pre-Coating

(b) Filtering

(c) Compressing Trapped Air

(d) Pressure Balanced

(e) De-Coating

(f) De-Sludging

FIG. 3.42 Pre-coat filter

285

iron oxide from feedwater. They can be used to filter the water at the hot end of the feed train, i.e., after the boiler feed pumps, and hence the whole of the feed water can he filtered. The filters can also be used at other points in the feed train, e.g., heater drains.

Magnetic filters remove particulate ferromagnetic material and, to a limited degree, copper oxide. Magnetite (F_3O_4) should be removed effectively at an efficiency of about 90%, but haematite removal will be less effective. Hence, the overall efficiency will depend on the ratio of magnetite to haematite present as particulate matter, and variable results can be expected.

Permanent magnetic filters were tested at Wylfa nuclear power station to protect the once-through boilers. The filters had to be physically cleaned and did not provide a significant reduction in iron level, whereas an electromagnetic filter installed at Richborough power station for on-load cleaning was able to reduce the iron levels in feedwater, particularly at start-up.

Overall, the use of magnetic filters has not been pursued in the UK power industry. This is due to the strict control of chemical conditioning, and the installation of condensate purification plant at stations where extremely high quality feedwater is required.

Newer types of high gradient magnetic filters are now available which are claimed to give a much improved efficiency. The main possible application for this product would appear to be in the filtration of activated corrosion products in nuclear systems.

9.5 Mixed beds

Mixed beds in CPP work on the same basic principles as described for make-up plant (see Section 5 of this chapter). However, the operating demands are much more severe due to three main factors: the higher flow rate, the higher quality of treated water required, and the higher temperature conditions, each of which make the performance of the ion-exchange resins more difficult.

The high flow rate conditions dictate that shallow beds of large diameter are used to operate at acceptable pressure loss levels. These are unsuitable for in-situ regeneration of the cation and anion resins because of the difficulty in obtaining a uniform distribution of the regenerants. Additionally, in-situ regeneration would present a relatively high risk of direct contamination of the boiler feed water by regenerant procedure. Finally, the additional distribution and collection lateral system required at the interface of the two resins would impede the flow during the service cycle. For these reasons mixed beds in CPP are invariably provided with external regeneration facilities.

It follows that the mixed bed service vessel is similar in construction to the main cation and anion units in make-up plants. Table 3.23 indicates the range of the design conditions for the service mixed bed.

The beds are shallow compared with the diameter and this coupled with the high flow rate conditions

286

TABLE 3.23

Typical condensate purification plant design and operating conditions

Parameter	Mixed bed unit	Cation unit
Vessel diameter, m	1.5 – 3.5	1
Bed depth, m	1 – 1.5	1
Flow rate $m^{-3}\,m^{-2}\,h^{-1}$, mh^{-1}	*70 – 150	100 – 150
Bed volumes/hour	50 – 150	100 – 150
Cation regeneration Agent	H_2SO_4	H_2SO_4
Quantity, kgm^{-3}	80 – 200**	50 – 80**
Anion regeneration Agent	NaOH	
Quantity, kgm^{-3}	80 – 150	
Cation/anion resin ratio	1:1 to 2:1***	
Temperature	30 – 40°C	

* The latest UK plant (Sizewell PWR) will operate at 72 $m^{-3}\,m^{-2}$ h

** Acid usually thoroughfared via the mixed bed cation to a preceding cation bed

*** This ratio is used when a cation unit does not precede the mixed bed

makes uniform distribution of the inlet water and collection of the treated water more difficult.

Inlet systems include perforated tube headers and multiple splash plate designs as shown in Fig 3.16 and 3.43. The base of the unit is either a flat or dished nozzle plate (Fig 3.19) or a flat plate supported on a concrete fill with nozzles set in a header and lateral arrangement immediately above the plate. The flat nozzle plate provides a combination of good flow distribution and resin removal but is difficult to support when designed for a high pressure drop. With concrete underfill the flat nozzle plate is more suitable for the highest differential pressures, but the internal header system can impede the removal of resin for regeneration.

The dished nozzle plate is easier to support, gives excellent resin removal, but may give a less uniform flow due to the variation in bed depth across the unit.

It is important to be able to observe the top of the resin bed to check that the depth is correct and the surface is even, and to observe the bottom of the bed to ensure that resin removal to the regenerator is complete. For this purpose, sight glasses are provided at both these positions.

9.5.1 Regeneration procedures

At the end of the service cycle the mixed resin is hydraulically transferred from the service vessel to the regeneration plant. The transfer pipework extends to the base of the unit and is arranged to ensure complete removal of the resin.

Regeneration plant does not have to be located adjacent to the service units. Ion-exchange resins can

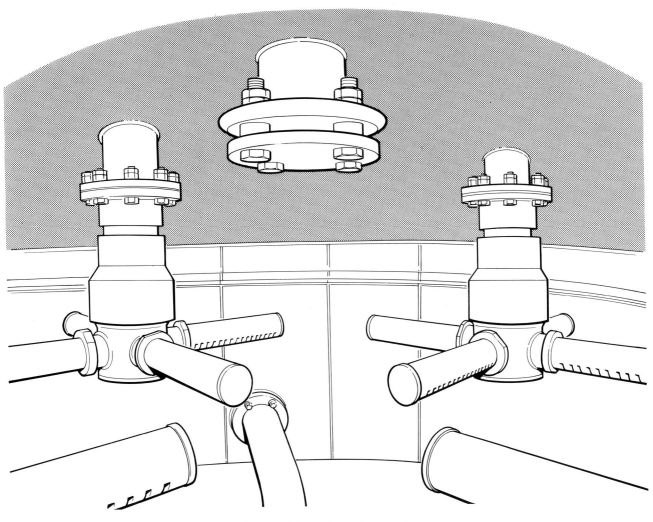

FIG. 3.43 Multi-cruciform plate distributor

be transported over considerable distances making it possible to locate the regeneration plant close to the make-up water treatment plant and the bulk chemical storage tanks. At Fawley power station, for example, resins are transferred over a distance of about 600 m.

In the regeneration plant the basic procedure is as follows and this applies irrespective of whether the cation exchange resin is to operate in the hydrogen or ammonium form:

- Backwashing and flushing to remove corrosion products filtered out during the run.

- Separation of the two resins.

- Regeneration of the cation resin with sulphuric acid followed by rinsing.

- Regeneration of the anion resin with sodium hydroxide followed by rinsing.

- Airmixing the resins.

- Final rinsing.

After final rinsing the resins are either returned to their original service vessel or transferred to a resin holding vessel to await transfer to the service vessel at the next regeneration.

The critical requirement in the regeneration procedure is to minimise cross-contamination of the resins by the regenerants, that is, the contamination of the cation resin by caustic soda, and of the anion resin by the sulphuric acid.

Slight contamination of the cation resin with sodium hydroxide is relatively unimportant when operating in the hydrogen form. In ammonium form operation (see Section 9.8.2 of this chapter), however, it is vitally important to have the minimum contamination in order to obtain a satisfactory quality of treated water in terms of sodium content.

Contamination of the anion resin with sulphuric acid can result in difficulty in obtaining the low sulphate concentration required in the treated water. The following equations illustrate how this arises:

(a) $RN^+(CH_3)_3\ OH^- + HSO_4^- \rightarrow$

$$RN(CH_3)_3\ HSO_4 + OH^-$$

(b) $2(RN^+(CH_3)_3HSO_4)^- + H_2O \rightarrow$

$$(RN^+(CH_3)_3)_2\ SO_4^{2-} + H_2SO_4$$

In reaction (a) the anion resin is converted to the bi-sulphate form; subsequently, when the resins are mixed and the final rinsing begins the bisulphate form is converted into the sulphate form with a release of sulphuric acid (reaction (b)). The sulphate then has to be removed by the bulk of the anion resin which will be in hydroxyl form. This can be done when the anion resin is new, but with any deterioration of the anion resin due to fouling/ageing can result in a very protracted period before satisfactory sulphate concentrations can be obtained.

The first requirement in relation to minimising cross-contamination is that the cation and anion resins are of suitable size and density to give a good separation on backwashing.

Ideally of course, anion resin of small particle size and low density would provide the best separation conditions. However, other factors such as pressure loss, physical strength, operating capacity and kinetic behaviour have to be satisfied so that a compromise

has to be made when specifying resins for CPP operating conditions.

The UK power industry's current specification for the properties of resins to be supplied for CPP use is given in Table 3.24.

A range of combinations of various types of cation and anion resins can be satisfactorily separated using the same particle size specification:

- Gel anion medium cross-linked with gel 10% DVB cation resin.

- Macroporous anion with gel 10% DVB cation.

- Macroporous anion with macroporous 12% DVB or 20% DVB.

The second requirement is to have regeneration equipment which, after the initial separation stage, will give minimum cross-contamination.

Several types of regenerators are used to regenerate the resins externally.

TABLE 3.24

Specifications for ion-exchange resins in condensate purification plants

	Single or mixed beds*			
	Cation exchangers		Anion exchangers	
Polymer matrix type	Styrene gelular strongly acidic	Styrene macroporous strongly acidic	Styrene gelular strongly basic (type I)	Styrene macroporous strongly basic (type I)
Total exchange capacity (eq/litre in ionic form)	$\not< 2.0$ (H^+ form)	$\not< 1.7$ (H^+ form)	$\not< 1.2$ (Cl^- form)	$\not< 1.0$ (Cl^- form)
Bead integrity Whole crackfree Beads, % Broken beads, %	$\not< 98$ $\not> 1$	$\not< 98$ $\not> 1$	$\not< 98$ $\not> 1$	$\not< 98$ $\not> 1$
Physical strength Breakdown after OSA test, %	$\not> 10$	$\not> 5$	$\not> 10$	$\not> 5$
Number of OSA test cycles	100	500	100	500
Bead size distribution (Diameter in ionic form)	(H^+ form)	(H^+ form)	(Cl^- form)	(Cl^- form)
Range (min-max), mm	0.6 – 1.0	0.6 – 1.0	0.4 – 0.85	0.4 – 0.85
Harmonic mean, mm	0.65 – 0.85	0.65 – 0.85	0.55 – 0.65	0.55 – 0.65
Undersize, mm	$\not> 1\%$ < 0.6	$\not> 1\%$ < 0.6	$\not> 2\%$ < 0.4	$\not> 2\%$ < 0.4
Oversize, mm	$\not> 5\%$ > 1.0	$\not> 5\%$ < 1.0	$\not> 1\%$ < 0.85	$\not> 1\%$ > 0.85
Bead density (kg/m^3 in ionic form)	$\not< 1180$ (H^+ form)	$\not< 1180$ (H^+ form)	$\not> 1120$ (SO_4'' form)	$\not> 1120$ (SO_4'' form)
Mass transfer coefficient (m/s) for sulphate exchange (1 mg/litre H_2SO_4) at 100 m/h linear flow rate	Test not applicable	Test not applicable	$\not< 2.0 \times 10^{-4}$	$\not< 2.0 \times 10^{-4}$
Ionic form as supplied	H^+ or NH_4^+	H^+ or NH_4^+	Cl^- SO_4'' or HCO_3'	Cl^- SO_4'' or HCO_3'

*For two resin systems

Single vessel

The regenerator is similar to an in-situ regenerated mixed bed as used in make-up water treatment plant but with the addition of resin inlet and outlet connections (a typical unit is shown in Fig 3.28). The vessel height must be sufficient to allow the resin to be expanded by about 100% during the backwashing stage. After backwashing, the resins are separated into two layers by means of a slow upward water flow. The interface between the resins is designed to be located at the position of the intermediate acid inlet/caustic outlet distributor. It is therefore most important that the correct resin charge volume is present and that the cation and anion resin volumes are in the design ratio. If the interface is not correctly located then either the cation resin will be contaminated by sodium hydroxide or the anion resin will be contaminated by sulphuric acid.

Even with correctly separated resins, cross-contamination can occur due to imperfect flow distribution at the interface collector and blocking water flows are used to minimise regenerant migration. During the caustic injection stage, for example, an upward water flow is maintained through the cation resin to minimise caustic migration down into the cation resin.

Separate cation and anion regeneration vessels

One of the first modifications to condensate purification plant regeneration systems was the use of separate cation and anion regeneration units (see Fig 3.44).

Because the resins are contained in separate vessels, each resin can be given the correct backwash rate consistent with the required degree of bed expansion and removal of crud is much easier. Although the process can avoid direct cross-contamination it is most important that the resins are accurately separated since any cation resin transferred with the anion will be completely converted to the sodium form and an anion remaining behind in the separator vessel will be completely converted to the sulphate form. The position of the interface has to be accurately defined and it is essential, therefore, that the volume of cation resin is constant, hence resin transfers from service units to regenerators must always be complete and the ratio of the two resins must remain the same. This was not

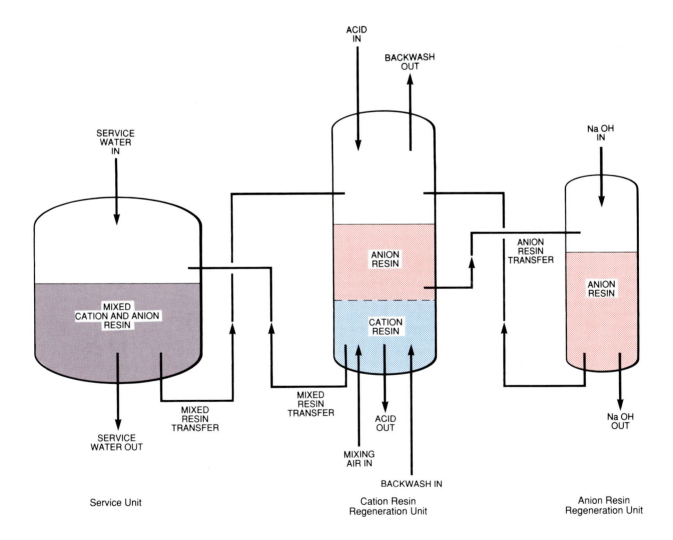

FIG. 3.44 Two-vessel regeneration

always possible to guarantee and, hence, cross-contamination has arisen with this system.

Conesep (NEI Thompson Kennicott) [14,15] uses two regeneration vessels, but in this case cation resin is transferred from the separator vessel to a smaller cation regenerator vessel (see Fig 3.45).

The mixed bed is transferred from the service vessel to the main separation unit which also serves as the anion regeneration vessel. On backwashing, the resins separate and the lower cation resin layer is then hydraulically transferred to a separate vessel simply by introducing water into the base of the Conesep vessel and forcing resin to flow out through a transfer pipeline. During this transfer the anion/cation resin interface descends smoothly through the vessel, the base of which is cone-shaped to compress the interface, and eventually enters the transfer pipeline. In this pipe it is sensed by conductimetric or optical means and the transfer terminated. On the original full scale installation this procedure has given cation in anion cross-contamination figures as low as 0.3% v/v. This level of cross-contamination is acceptable for plants aiming to produce water with sodium levels of less than 0.1 μg kg^{-1} when operating in H-OH mode.

Variations introduced since the design was first used include adjustments to the all important base design, the development of alternative interface sensing techniques and the introduction of a secondary separation stage. This involves a secondary classification, by backwashing, of the anion resin component after its regeneration with sodium hydroxide. This second separation relies on the fact that any cation resin beads in the anion resin are now in the more dense sodium form. Thus they are more easily separated from the anion resin which is not in the less dense OH form.

The small amount of cation resin recovered by this procedure is moved to an interface (holding) vessel and subsequently added to the next charge to be regenerated.

These improvements have particular benefit to operation in the ammonium or amine cycle when the cation resin in the mixed bed has to have a very low residual sodium content. Separation down to < 0.06% v/v of cation resin in anion resin has been achieved on full-scale operating plant.

Two other processes to alleviate cross-contamination were developed in the USA. *Ammonex* (Crane Cochrane) [16] was designed specifically for operating with the cation resin of the mixed bed in ammonium form. Two regenerator vessels are used. Following transfer from the service vessels separation is carried out by backwashing. The anion portion is then transferred to the second regenerator and in this stage the resin transfer point is usually biased towards the anion section in order to limit the quantity of cation transferred with the anion resin. The anion resin is then regenerated with sodium hydroxide and rinsed. Ammonium hydroxide is then passed through the bed to convert any cation resin present from the sodium to ammonium form. Meanwhile, the main cation section is regenerated with sulphuric acid and rinsed. The cation and anion resins are then mixed. Subsequently, the bed may be ammoniated either externally or during the service cycle.

In the *Seprex* (Ecodyne Graver) system the resins are transferred from the service vessel to the first of two regeneration vessels. The resins are then separated by backwashing and the anion portion is transferred to the second vessel. After draining water from the anion resin, a 16% solution of sodium hydroxide is applied, this enhances the separation by floating the anion resin from any traces of cation resin that have been carried over. The anion resin is then taken from the vessel to the third vessel for rinsing. The cation resin remaining is rinsed and transferred to the main

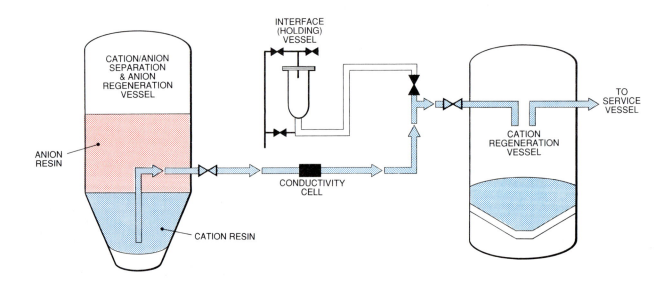

FIG. 3.45 Conesep regenerator (NEI-Thompson-Kennicott)

cation regeneration vessel for mixing with the next charge of cation resin to be regenerated.

The main cation bed is regenerated and rinsed in vessel one prior to the transferring to the third vessel for mixing with the anion resin in readiness for use.

Obviously, cross-contamination would not arise if mixed beds were not used to treat the condensate. *Tripol* (Permutit Ltd) [17] is an alternative system to the use of mixed beds. As shown in Fig 3.46, the resins are in three discrete layers of cation-anion-cation in the service vessel, and are transferred from the service vessel for regeneration.

The cation and anion resins are kept apart in the regeneration system, even to the extent of having separate resin movement lines so that the cross-contamination problem of the wrong regenerant on the resin is completely avoided. Resin ratios and volumes can be altered to suit particular operating conditions but typically the lead and trail cation are each 350 mm

deep, and the anion 500 mm deep. The linear flow rate is typically 200 m/h and the separated beds technique avoids the compaction which occurs in a deep mixed bed so that the headloss is about the same as a mixed bed condensate polishing plant running at 120 m/h.

9.5.2 Chemical ingress protection following a regeneration

External regeneration systems are much less likely to result in regenerant contamination of the polished water than are internally regenerated polishers but nevertheless there is some risk that insufficiently rinsed resins could be transferred to the service units and this contamination passed forward to the boilers.

Because of the high risk to the AGR boilers from caustic or acid contaminants, the chloride ingress conductivity monitoring and trip system at the polishing

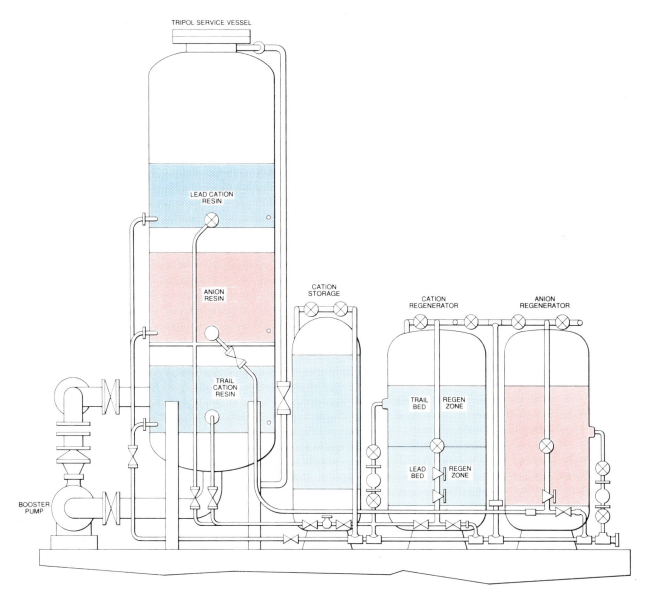

Fig. 3.46 The Tripol condensate purification plant (Permutit Ltd)

plant outlet is backed up by a polisher rinse check; the chemical ingress protection system.

The protection system is independent of the normal plant control system. The system is initiated when the acid or caustic valves open and ensures that the resin transfer valves are kept closed. The rinse flow and conductivity are monitored by at least two sets of flowmeters and conductivity probes. When it has been confirmed that the correct rinse quantity has flowed through the resin and that the conductivity is sufficiently low, the interlock on the resin transfer valve is released thus allowing resin to be transferred from the regenerator to the service units.

9.6 Cation and mixed bed plants

The design of the cation unit will be similar to that of the mixed bed service vessel and Table 3.23 gives general data on the size and operating conditions. The resin in the cation unit is transferred to an external regenerator vessel. Here the resin can be easily and quickly regenerated and no problems of cross-contamination arise.

With a preceding cation unit the more complicated mixed-bed regeneration can be performed less frequently and in the absence of condenser leaks the mixed beds can be regenerated after about every five cation bed regenerations.

When the cation and mixed bed are regenerated together, the acid from the cation unit is thoroughfared to the cation resin of the mixed bed.

9.7 Powdered ion-exchange/filtration units

Pre-coat filters have been adapted for use with powdered ion-exchange resin [18]. A thin film of powdered cation and anion resins is formed onto the outside of a hollow cylindrical element. A variety of designs of elements exist but usually they comprise of filters wound onto a perforated stainless steel former. The liquid to be treated flows inwards passing through the ion-exchange resin material and out to service (see Fig 3.47).

Chemically the resins are similar to those employed in deep bed ion-exchange. The particle size of the resin is approximately 30 μm and when the cation and anion resins are mixed, a bulky floc forms. The characteristics of the floc vary depending on the ratio of cation/anion resin used. To form the coat the mixed resin floc is passed on to the element and water is circulated until a satisfactory coat is obtained. The unit can then be put into service and operated until either a specified increase of pressure differential is reached or up to exhaustion of ion-exchange capacity. The dry weight of the resins used to form the filter coat is equivalent to 1 kg per m^2 of element surface and the usual service flow rate is 100-200 m^{-2}min^{-1}. The coat has a low resistance to flow, and typical

pressure differentials are 0.1-0.2 bar for a freshly pre-coated unit. The unit can be operated to exhaustion of ion-exchange capacity, or to a specified pressure differential (1.7 bar), if filtration is the main function.

When the end point is reached the spent resins are removed by passing a flow of water in the reverse direction to that employed during the service cycle.

The unit is then ready for re-coating and the spent resins being of no further value are usually discarded.

The powdered cation resin can be supplied in hydrogen and ammonium forms, whilst the strongly basic anion resin is supplied in hydroxyl form. Additionally, the powdered resins can be supplied already mixed in various cation/anion ratios and in the required ionic form. This is a distinct advantage in the preparation of pre-coats.

9.8 Ion-exchange processes in condensate polishing plant

A variety of ion-exchange resins has been used in CPP with varying degrees of success, largely dictated by the plant operating conditions. Resins have ranged from high DVB macroporous resins to low DVB gelular types, and the combinations of cation/anion in mixed beds have ranged from gel:gel to macroporous: macroporous.

For optimum performance the resins used in CPP should have the following properties:

- High resistance to osmotic shock and attrition (physical strength).

- High rates of reaction.

- High cation/anion separation characteristics.

- High capacities.

- High resistance to fouling.

- High thermal and oxidative stability.

- High uniformity of physical size.

- Low resistance to flow (pressure loss).

- Low organic leachable content.

Details of the specification for resins to be used in CPP are given in Table 3.24.

9.8.1 Cation resin — hydrogen form

The resin will remove the conditioning agent — usually ammonia — from condensate and this represents the main load. When a cation unit precedes a mixed bed operating in hydrogen forms, the cation to anion resin ratio in the mixed bed will usually be 1:1. In the absence of the preceding cation bed, a higher ratio of 2:1 cation:anion is frequently used.

The majority of plants operate with hydrogen form cation, and the quantity of ammonia that can be removed per cycle will usually dictate the frequency of

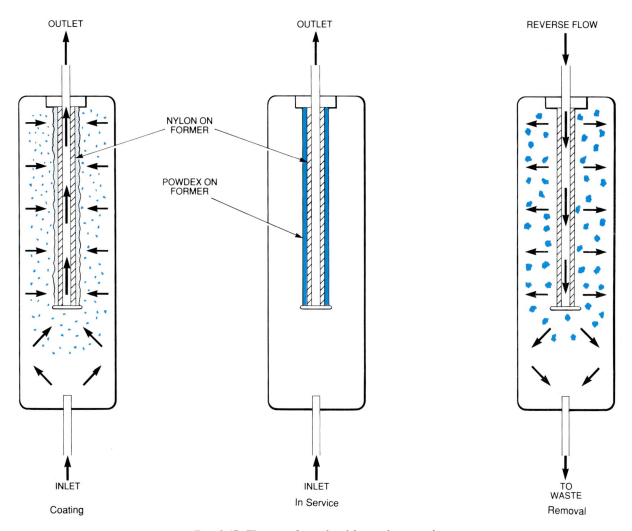

NYLON ON
FORMER

POWDEX ON
FORMER

OUTLET

OUTLET

REVERSE FLOW

INLET

INLET

TO
WASTE

Coating

In Service

Removal

FIG. 3.47 The use of powdered ion-exchange resin

regeneration. The capacity obtained from a cation resin will depend on, the level of regeneration, the breakpoint chosen and the type of resin used.

The total volume capacity of a gelular resin will normally be greater than an equivalent macroporous type, i.e., 2.00 eq/l and 1.7 eq/l, respectively. The levels of regeneration used on CPP are about 100–200 g/l H_2SO_4. Using 150 g/l H_2SO_4 the operating capacity from laboratory data should be 1.6 eq/l for gelular resin and 1.4 eq/l for the macroporous type. However, these results have rarely been achieved on CPP at UK power stations. Typically, capacities around 0.8 eq/l have been reported and the main reason for the discrepancy has been attributed to poor distribution of water through the beds.

The capacity of hydrogen form cation resin for sodium when treating condensate contaminated with sea water will be approximately in proportion to the ratio of NH_4:Na present, i.e., the total operating capacity will be as when removing ammonia only, but the quantity of ammonia removed will be a decreasing proportion of the capacity as the concentration of sodium present increases.

Strongly, acidic cation resin has a high selectivity for sodium relative to hydrogen; consequently, sodium can be removed to a very low level in a mixed bed (< 0.3 μg kg^{-1} Na).

Overall, from an ion-exchange viewpoint the cation-mixed bed system would be the preferred option relative to a mixed bed. The main advantages being:

• A higher quality water is produced.

• A higher capacity can be obtained from the anion component of the mixed bed.

• The system has a greater ability to cope with changes in the pH of condensate.

• Greater flexibility in the cation/anion ratio used in the mixed bed.

• By reducing the number of regenerations of the mixed beds, the potential mass of sulphate introduced into the system should be less.

• An additional filtration facility is provided which may also provide some protection to the mixed bed resins.

The disadvantages are: an increase in pressure loss across the plant, increased capital cost and a greater space requirement.

9.8.2 Cation resin — ammonium form

The major disadvantage of operating with the cation resin in the hydrogen form is that the uptake of the chemical conditioning agents leads to the mixed bed having to be regenerated frequently, and as discussed, this has attendant problems in respect to maintaining a satisfactory quality of treated water. By operating with the cation in ammonium form, removal of ammonia does not take place and the pH of the condensate is virtually unchanged through the bed. However, adventitious cations such as sodium and calcium can be exchanged on the resin by the following reaction:

$$R_{NH4} + Na^+ \rightarrow R_{Na} + NH_4^+$$

where R_{NH4} and R_{Na} are the ammonium and sodium forms of the resin, respectively.

The equilibrium or selectivity constant R_{NH4}^{Na} is represented by

$$K = \frac{[R_{Na}][NH_4^+]}{[R_{NH4}][Na^+]}$$

Two factors make operation more difficult in the ammonium form:

- The presence of ammonium ions.

- The uptake of sodium by the resin for a release of ammonium ions is an unfavourable equilibrium.

The presence of ammonium ions pushes the reaction from right to left and to nullify this effect the concentration of resin in sodium form (R_{Na}) has to be extremely low. For example, to obtain a concentration of 2 μg kg^{-1} of sodium in the treated water at pH 9.4 the concentration of R_{Na} in the cation resin has to be < 0.4%. For comparison, when operating in hydrogen form, the resin has to be < 30% to obtain the same quality, i.e., the equilibrium leakage of sodium when operating in the ammonium form is high relative to hydrogen form exchange. It follows that on regenerating an ammonium form mixed bed, there is a need to minimise cross-contamination of the cation resin by sodium hydroxide, in addition to removing any sodium absorbed during the service cycle, to obtain the low concentration of R_{Na} required.

A superior two-vessel regeneration system has to be used to obtain these stringent conditions.

The selectivity co-efficient $K_{NH4}^{Na} = 0.77$ is lower than $K_H^{Na} = 1.7$ therefore the resin will have a lower capacity for sodium than when operating in the hydrogen form. The equilibrium conditions for sodium/ammonium exchange lead to a longer zone of exchange than for sodium/hydrogen exchange and hence a lower capacity will result.

The actual capacity obtained will depend on the concentration of sodium in the water to be treated, the higher the concentration the greater the capacity.

The high pH conditions in the mixed bed will also adversely affect the exchange rate of the component anion resin. The zone of exchange is extended and therefore a higher kinetic leakage of chloride and sulphate ions and a lower capacity would be predicted for the anion resin relative to when it is used with a hydrogen form resin.

Ammonium-form operation has not been widely used in the UK power industry, but pilot plant and full-scale trials have indicated that the process can satisfactorily produce high quality water, and operate under condenser inleakage conditions.

Amine-form operation requires similar operation conditions and the morpholine form has been used successfully at Oldbury nuclear power station using the Conesep regeneration technique [19].

As with ammonium form systems, sulphuric acid is used to convert the resin to hydrogen form after exhaustion. The bed is then converted to the amine form by the uptake of the chemical conditioning agent during the service cycle.

It should be noted that experience with CPP treating condensate dosed with amines (morpholine at Oldbury power station and AMP at Wylva power station) has shown that there is a risk of physical breakdown of macroporous cation resins, and that the degree of breakdown is dependent on the DVB content of the resin. The higher the DVB content the greater the breakdown.

Laboratory tests showed that the breakdown results from osmotic forces which develop when an amine form resin is regenerated with acid, and this effect is enhanced by any contact between the resin and the sodium hydroxide used to regenerate the anion resin. The tests showed that 20–25% DVB macroporous resins were particularly susceptible to breakdown, whereas 10% DVB gelular and 12% macroporous resins were virtually unaffected [20].

Hydrogen form versus ammonium or amine form cation resin

The relative advantages of the two forms are:

Hydrogen form

- Resin separation conditions may not be so critical as with the ammonium form which requires superior regeneration facilities such as provided by Conesep, Ammonex, or Seprex.

- Should be able to meet the extremely high water quality requirements (0.3 μg kg^{-1} sodium chloride and 0.5 μg kg^{-1} sulphate) more easily.

- Higher capacity under condenser leak conditions.

- Kinetic leakage of anions will be lower.

Ammonium form

- The number of regenerations of the mixed bed will be reduced, hence the risk of sulphate ingress into condensate will be lower.

- There is saving in the condensate conditioning chemical and the resin regenerants.

The advantages of amine form will be similar to ammonium form, but there will be a greater saving in the conditioning agent costs because a high concentration has to be used to achieve a given pH compared with ammonia. A disadvantage is that the risk of physical breakdown will be higher than when operating in hydrogen or ammonium forms.

9.8.3 Anion resin

Polystyrene macroporous strongly basic resins have generally been used within the mixed beds of CPP in the UK. These have performed satisfactorily in terms of separation characteristics during regeneration, and have been shown to have adequate physical strength.

When the resin is new, a high quality of water can be consistently produced and treated water of chloride and sulphate concentration < 0.3 μg kg^{-1} can be consistently obtained. However, in a relatively short period of time, deterioration in the performance of the anion resin has been experienced. The main effects arising from the deterioration are an increase in chloride and sulphate leakage during condenser leaks and an increase in the time of sulphate rinse-out following a regeneration. These adverse factors have led to the need to replace anion resin more frequently than anticipated, and the life of the resin can be less than a year.

At the higher flow rates used in CPP, ion-exchange kinetics are of greater importance than under make-up plant conditions. However, even under the low flow rate conditions employed for make-up plants, adverse kinetic effects do occur from the irreversible absorption of organic matter onto anion resin (see Section 4.4.2 of this chapter). Inherently, anion resin is more likely to take up foulants and it is also chemically less stable than cation material, and, therefore, the problems encountered on CPP are considered to be associated with these factors.

Considerable investigational work has been carried out within the UK power industry on the deterioration of anion resins in CPP [21,22].

Ion-exchange kinetics at low concentrations are controlled by diffusion across a thin layer of static water that exists at the surface of a bead, and the transfer of ions follow a general first-order kinetic reaction. Based on this concept a number of practical equations have been used to study ion-exchange kinetics in packed beds.

Most information in the UK power industry has been obtained from the following mass transfer equation:

$$\log \frac{C}{Co} = \frac{MSZAR}{V} \text{ (after Frisch and Kunin 1960 [23])}$$

where
C = column outlet concentration
Co = column inlet concentration
M = mass transfer coefficient, ms^{-1}
S = specific surface area of resin beads, $m^2 m^{-3}$
Z = depth of exchange zone, m
A = cross-sectional area of test column, m^2
R = volume fraction of exchange material in bed
V = volumetric flow rate, $m^3 s^{-1}$

Now by using a suitable rig; A, R, Z and V can be fixed and C/Co and S can be measured.

This enables M, a mass transfer coefficient, to be determined which is independent of bead size distribution and bed geometry. The higher the value of M, the faster the rate of transfer. Note, however, that M will vary with flow rate.

This follows from $M = D/\delta$ where D is the diffusion coefficient for the exchanging ion and δ is the boundary layer thickness. The value of δ depends on flow rate, hence so does M, and it is important to obtain test column data based on flow rates representing conditions on plant; for comparative tests identical flow conditions should be used. Note that although the rate increases, the time of contact decreases and the overall mass transferred actually decreases.

The test rig used is shown in Fig 3.48.

The anion resin to be tested can be used either mixed with cation resin or as a single bed in a 50 mm diameter column, with a bed depth of 550 mm.

Deionised water maintained at $20 \pm 0.5°C$ is circulated through the test bed at a superficial flow rate of 100 mh^{-1} and the column influent is dosed with either chloride or sulphate to simulate condenser leakage and/or regeneration malfunction.

The column outlet is monitored continuously for conductivity and sodium ion concentration and samples are taken during each test for subsequent ion-chromatographic analysis for chloride or sulphate.

Typical figures obtained for mass transfer coefficients for new and used resins are given in Table 3.25.

The UK power industry now specify a mass transfer coefficient for the supply of new anion resins for CPP. This is based on sulphate exchange (1 mg kg^{-1} H_2SO_4) at 100 mh^{-1} linear flow and the mass transfer coefficient should not be less than 2.0×10^{-4} metres per second for new anion resin.

Using the kinetic test, data on coefficients have shown:

- The anion resins likely to be used in CPP all have similar mass transfer coefficients.

- The mass transfer coefficient increases with an increase in flow rate.

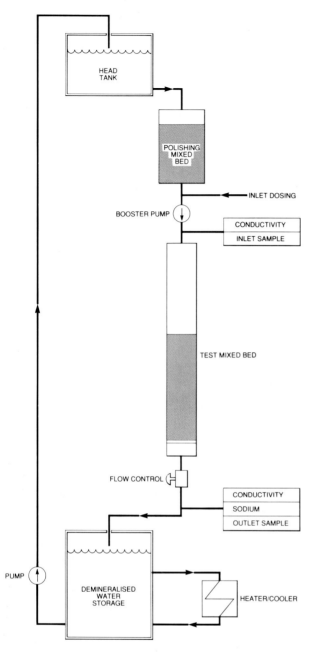

FIG. 3.48 Kinetic test column apparatus

TABLE 3.25

Mass transfer coefficients of new and fouled strongly basic anion resins

Typical mass transfer coefficients $\times 10^{-4}$ ms^{-1}				
Resin	Chloride exchange		Sulphate exchange	
	New	Fouled	New	Fouled
Polystyrene gel	2.3	2.1	2.15	1.9
Polystyrene macroporous	2.3	2.1	2.15	1.75
Acrylic gel	2.3	2.0	2.15	1.75

faster the kinetics the shorter the exchange zone and hence the higher the operating capacity and the quality of water produced.

With new resin a relatively short zone of exchange will be obtained, but as the resin ages and/or fouls, etc., reaction rate will decrease and the zone of exchange will increase. If, for example, a zone of exchange increases to 80% of the total bed depth, then, effectively, only 20% of the bed is available to take up contaminants and the bed will have a low operating capacity.

The length of the exchange zone depends on a number of factors:

- Velocity — it is $\alpha \sqrt{\text{(velocity)}}$ hence doubling the flow rate will give a 50% increase in the length of the zone exchange.

- Bead size — the larger the beads the longer the exchange zone.

- Concentration of influent anions — the lower the concentration the shorter the exchange zone.

- Type of ion — a longer zone of exchange for sulphate ions compared with chloride ions.

- pH — the higher the pH the slower the rate of exchange of anions, particularly sulphate. It follows that in an ammonium form bed where high pH conditions always exist, the zone of exchange will be longer than when operating the same bed in hydrogen form. Consequently, the capacity in ammonium form mixed bed for anion exchange would be expected to be lower than that of a hydrogen form system.

The foregoing have indicated how laboratory data from kinetic tests can be used to estimate the potential performance of CPP. It must be emphasised, however, that these data are likely to represent the best performance conditions whether the anion resin be new or old. In practice a number of plant factors will also play a role in determining both the quality of water achievable and the ability of anion resins, in particular, to cope with condenser leaks. These factors include flow distribution, regeneration, mixing and temperature, and they could dominate both the quality of water produced and the capacity obtained.

- The mass transfer coefficient for exchange of sulphate ions is lower than that obtained for chloride ion-exchange.

- Mass transfer coefficients of resins that have been in service for a period generally decrease, and the effect can be particularly severe for the exchange of sulphate ions.

- The mass transfer coefficient is pH dependent and for anion resin decreases in moving from acidic to alkaline conditions.

Now the above equation also allows the zone of exchange (the depth of bed required to achieve the required degree of exchange) to be determined. The

The mechanism of deterioration in anion kinetics has not yet been fully established. Factors likely to hinder the transfer of ions across the liquid boundary include fouling and this could arise from the presence of organic matter in the make-up water or from the resins themselves, e.g., organic sulphonates from cation resin have been shown to affect the anion kinetics. As indicated in Section 4.4 of this chapter, degradation of the active groups on anion resins occurs with time leading to a loss of total and strongly basic groupings. These changes would be expected to take place more rapidly at the surface of the resin, and hence the rate of transfer of ions by the film diffusion process would decrease.

The fouling/ageing process, resulting in poorer kinetics, has been the major problem encountered with the anion resins used in CPP, and the magnitude of the effect will depend on the specific flow rate. Hence, under conditions where the highest quality of water has to be produced, there may be an overall advantage, accepting the higher cost involved, in operating at a lower flow rate. This concept has been used in the design of the CPP mixed beds for the Sizewell *B* PWR station.

9.8.4 Powdered resins

Powdered resin filters have been shown to be more effective filters than cellulose powder filters, particularly for hydrated oxides. The ion-exchange capacity is however limited and, because of the high cost of the throw-away resin, the use of powdered resin is confined to treatment of condensate with only traces of contaminants.

In the CEGB, Powdex filters are operated at Hartlepool and Heysham *1* power stations. At both stations the powdered resin filters were intended to be on-line continuously and operated in the ammonia form with the mixed beds on standby. The installation of the chloride ingress trip system, which detects contamination by means of direct conductivity monitors, precludes ammonia form operation where there is a background conductivity due to ammonia and requires the mixed-beds to be on-line continuously.

At Hartlepool (see Fig 3.41), the filters operate in the ammonium form before the polishing mixed-beds and perform a filtration role to protect the mixed-bed resins. Because ion-exchange performance is unimportant a high ratio of cation to anion resin is used. The use of alternative pre-coat materials is being explored.

At Heysham *1* (see Fig 3.41), the filters come after the polishing mixed-beds and operate in the hydrogen form. Because they contain highly regenerated resin they are able to remove traces of contaminant slip, notably sulphate, from the mixed-bed polishers.

9.8.5 Performance of resins in a Tripol plant

Experience with this process in the UK is limited to investigational work on a unit capable of treating the flow of condensate from a 60 MW unit [24].

This work showed that operating with the cation resin in hydrogen form a very high quality of water could be obtained with sodium and chloride concentrations of 0.1 μg kg^{-1}. The sulphate content was higher at 0.6 μg kg^{-1} but this value represented conditions throughout a run including the period immediately following the return to service after a regeneration.

High capacities were obtained for the removal of ammonia by the cation resin. Using 220 kg H$_2$SO$_4$ per m^3 and operating to a 0.1 μS cm^{-1} conductivity breakpoint a capacity of 1.7 eql^{-1} could be obtained. This is significantly higher than is usually obtained from full scale mixed bed units, and is attributed to the efficient flow distribution on the Tripol plant used.

Operating in ammonium form at a pH of 9.6 under condenser inleakage conditions a quality of treated water of 0.5 μg kg^{-1} Na, 0.1 μg kg^{-1} Cl and 0.6 μg kg^{-1} SO$_4$ could be obtained. The capacity of the anion section during condenser leaks was approximately 30% of the total volume of the capacity of the macroporous resin used. Again, this is higher than has been obtained from full scale mixed bed plants and the good distribution characteristics of the relatively small test plant may be a contributory factor.

10 Cooling water

The main use of cooling water is to cool the exhaust steam from the turbine in a surface condenser of the shell and tube type. Condensation takes place in the shell as the steam contacts banks of tubes through which the cooling water is passing. The condensate falls to the bottom of the shell and re-enters the circuit as boiler feedwater.

In condensing the steam the cooling water is raised in temperature by approximately 10°C. It may then pass immediately back to the source of abstraction and this type or system is usually referred to as once-through/ direct cooling (Fig 3.49). Extremely large volumes of water are required to operate in this mode and therefore it is employed primarily at coastal power stations where sea water can be the cooling medium. The cooling water intake requirement will be about 5.5×10^6 m^3/ day for a 2000 MW station, which is more than the dry weather flow of most rivers in the UK. Consequently, for inland stations a tower cooling system is used (Fig 3.50) and, in this case, all the water is recirculated from the condenser to cooling towers where it is cooled for re-use by an induced flow of air against a downflow of water. After cooling, the water is returned to the condenser. A proportion, about 1% of the water recirculating, is lost continuously by evaporation from the cooling towers. For a 2000 MW station this represents a 55 000 m^3/day loss which has to be replaced by raw river water. The water vapour leaving the tower is free from salts; consequently, the water in the circuit is concentrated and the system has to be purged to maintain the concentration at a level to minimise the risk of scale forming on the heat exchange surfaces.

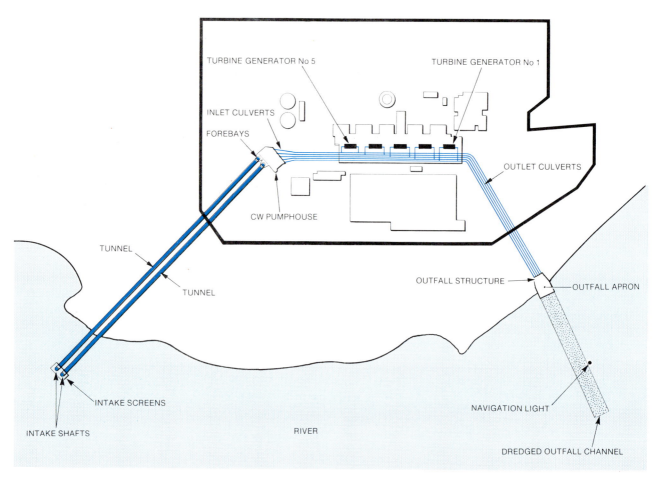

FIG. 3.49 Once-through/direct cooling water system

The concentration factor, i.e.,

$$\frac{\text{concentration of cooling water}}{\text{concentration of river water}}$$

is usually kept in the range 1.25–2.0. The actual value will depend on the composition of the river water and the operating regime. Water purged is replaced by river water. The best estimate of the concentration factor is obtained by determining the concentration of an ion such as sodium, which is soluble and cannot be deposited within the system.

Any fouling of the surface of the condenser tubes will result in a loss of heat transfer and a risk of a high cost penalty in respect of the production of power.

This section considers the chemical factors involved in plant operation to minimise the potential loss of heat transfer arising from scales, deposits and organic biological growth and slimes.

10.1   Scaling

Several ionic species initially soluble in the river water can, as the temperature and concentration of the cooling water increase, be precipitated from solution onto the heat transfer surfaces and counter measures have to be taken.

Technically, complete prevention could be easily applied, but in many cases it would not be cost effective, and at worst could cost more than that of the problem. For example, the scaling constituents could be completely removed from the influent water using ion-exchange or precipitation techniques. The capital and operating costs of such plant would be extremely high and in most cases they can be ruled out on this basis.

Careful consideration is needed to reach an optimum solution. Firstly, the potential effect of any scaling needs to be quantified to ensure that action is needed. Then, possible methods need to be assessed to determine their effectiveness and cost. Finally, an assessment of the degree of control required and maintenance of any counter-measures taken.

10.1.1   Calcium carbonate deposits

Calcium carbonate is frequently present in river water in appreciable quantities and if the water is concentrated, heated or subjected to contact with excess air the following reaction takes place:

$$Ca(HCO_3)_2 \leftrightarrow CaCO_3 + CO_2 + H_2O$$

The loss of carbon dioxide may lead to the solubility product of calcium being exceeded and hence the water is potentially scale forming.

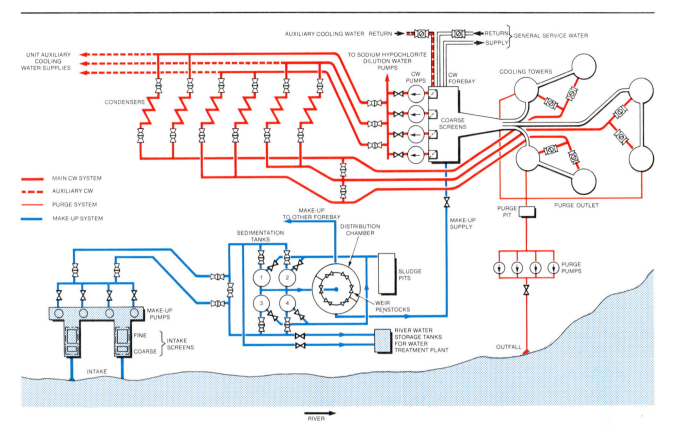

FIG. 3.50 Tower cooling water system

A fine balance exists between whether a water containing calcium carbonate is scaling or corrosive. In the presence of excess carbon dioxide it can be corrosive and when super-saturated it can form scales.

Langelier [25] developed a method of predicting the scaling or corrosive potential of a water from the chemical composition of the water, the pH and temperature.

The Langelier or Saturation Index = $pH_{(actual)} - pH_{(saturation)}$. $pH_{(saturation)}$ is defined as the pH the water would have if it had the same total alkalinity but was just saturated with respect to calcium carbonate.

$$pH_{(saturation)} = pK_2 - pK_s + pCa + pAlk$$

where p = $\log_{10}$, the reciprocal of the adjoining symbol

Ca = concentration of calcium as gram-equivalents/litre

Alk = alkalinity as gram-equivalents/litre

K_2 = $(H^+)(CO_3^{2-})/(HCO_3^-)$ which is the second dissociation constant of carbonic acid

K_s = $(Ca^+)(CO_3^{2-})/(CaCO_3)$ which is the dissociation constant of calcium carbonate

The quantity $(pK_2 - pK_s)$ varies with temperature and total ionic concentration and values for this quantity are given in Table 3.26, which is taken from 'Industrial Water Treatment Practice' by Hamer, Jackson and Thurston.

If $pH_{(determined)}$ is less than $pH_{(saturated)}$ the water is corrosive. If $pH_{(determined)}$ is greater than $pH_{(saturated)}$ the water is scale forming.

As the method is not precise and some degree of super-saturation can frequently occur without scale forming, it is usual to recommend that the water has a small positive index rather than risk corrosion.

Ryznar [26] proposed an empirical index $I_R = 2\ pHS - pH$, which in general appears to be more representative of the corrosive or scale forming nature of cooling water.

I_R < 6 the water is scale forming

I_R > 6 but < 7 the water is close to equilibrium

I_R > 7 the water is corrosive.

Calcium carbonate deposits in once-through cooling systems are seldom a problem. Carbon dioxide is not lost from the water and the time of contact with the system is short. Moreover, the cost of treating the large throughput of water would be prohibitive.

In a recirculating cooling system, if the make-up water has a high temporary hardness the cooling water will initially also have a high temporary hardness. As the water cascades from the cooling towers it will loose carbon dioxide to the atmosphere. The water will also concentrate and thus a water which initially con-

TABLE 3.26

Values of [pK$_2$ − pK$_s$]

As ionic strength	As TDS (PPM)	Values of (pK$_2$ − pK$_s$) at a temperature of				
		0°C	10°C	20°C	50°C	80°C
0.000	0	2.60	2.34	2.10	1.55	1.13
0.001	40	2.68	2.42	2.18	1.63	1.22
0.005	200	2.76	2.50	2.27	1.72	1.32
0.010	400	2.82	2.56	2.33	1.79	1.39
0.015	600	2.86	2.60	2.37	1.84	1.44
0.020	800	2.89	2.64	2.40	1.87	1.48

tained bicarbonate alkalinity and free carbon dioxide will then contain carbonate and bicarbonate alkalinity.

Where the carbonate ion concentration is sufficiently high, the solubility product of calcium carbonate may be exceeded and the water both in the tower and passing to the condensers is potentially scale forming; this may give rise to scale deposition in the warmer parts of the circuit. It is by no means certain that scale deposition will occur and only a small proportion of CEGB cooling tower stations have been troubled with calcium carbonate scales. The potential for scaling may be determined, as shown shown earlier, by estimation of the Langelier or Ryznar scaling indices. It is clear that for a given concentration ratio the potential for scaling will primarily depend on the calcium hardness and alkalinity of the make-up water. For an existing station pH$_{(actual)}$ can be measured; for a new installation pH$_{(actual)}$ can be estimated from the predicted alkalinity of the circulating water. If a scale forming tendency is evident, then dosing of acid or scale inhibiting chemicals may need to be considered. Further details are given in the following Sections 10.2.2 and 10.2.3. respectively.

10.1.2 Calcium phosphate deposits

A number of stations experience phosphate scaling and this has been particularly prevalent in the Midlands where cooling water is taken from the River Trent. Phosphate enters rivers from domestic sewage, industrial effluent and land drainage. The problems are most severe in hot, dry summers, but occur annually in varying degrees according to the season and water patterns.

To minimise the cost of preventive measures, an equation was developed within the Midland Region to predict the rate of scale build-up from readily available parameters with the object of applying treatment selectivity when the rate becomes significant [27].

The general form of the overall equation selected for data filling was:

$$\text{Rate of scaling} = kS \exp(-E/RT)$$

where E is the apparent activation energy for the complete process, i.e., transport, nucleation and growth of deposit.

$S = (C_s − C_o)/C_o$ which is the relative super-saturation, where C_s is the actual concentration of the participating species and C_o is the equilibrium concentration

k   would depend on the nature of the calcium phosphate deposited and β tri-calcium phosphate was chosen as data on the solubility product and variation with temperature was available.

The practical data were obtained from an operational plant taking cooling water from the River Trent. Cooling water outlet temperatures and relevant chemical parameters such as phosphate, calcium and pH were recorded at 10 minute and daily intervals for a period of 70 days. Relative super-saturation values could then be calculated for each 10 minute interval after interpolation of the chemical parameters. Supporting data was obtained from other Trentside stations by extracting temperatures logged for operational purposes. Corresponding cooling water analyses were not always available, therefore a grand mean of the summer data on water chemistry for all stations was used, i.e., 200 mg kg^{-1} Ca, 6 mg kg^{-1} PO$_4$ and pH 8.4.

Using this data, a semi-empirical equation was derived (see [27] for full details), which correlates the rate of calcium phosphate scaling of condensers with the concentrations of calcium and total ortho-phosphate in the cooling water, the pH and the condenser outlet temperature.

The equation which applies to regions of average heat flux in the condenser is:

Rate =

$1.0 \times 10^3 \, S \exp [50 \times 10^3 \, (1/298 − 1/T)]$ mg m^{-2}h^{-1}

where S = super-saturation index for β tri-calcium phosphate

T = bulk outlet water temperature in degrees K.

On the basis of this equation, the following index could be proposed:

$$\log R = F_1(T) + F_2(PO_4) + F_3(Ca) + F_4(pH)$$

where F_1, F_2, F_3 and F_4 are functions of the given parameters and Tables 3.27 to 3.30 give calculated results over a range of parameter values. Based on performance losses of condensers attributable to calcium phosphate scaling, an action limit of $R = 1.0$ mg m^{-2} is required and a suitable warning limit would be $R = 0.5$ mg m^{-2}h^{-1}. Note that when $R = 1$, $\log R = 0$ so that any value of $\log R > 0$ is unacceptable.

pH control is frequently used to prevent scaling and the index will give the pH conditions required. For example, a water containing calcium and phosphate concentrations of 150 mg kg^{-1} and 10 mg kg^{-1}, respectively, having a pH of 8.5 and operating at a temperature of 30°C will (from Tables 3.27 to 3.30; $F_1(T) = 0.879$, $F_2(PO_4) = 1.0$, $F_3(Ca) = 2.934$ and $F_4(pH) = 4.095$) give $\log R = 0.718$. This is unacceptably high, indicating that scaling will occur. To reduce $\log R$ to zero, $F_4(pH)$ has to be changed by 0.718 to -4.813. Table 3.30 shows that it is equivalent to a pH of 7.8. The pH of the cooling water therefore needs to be reduced from 8.5 to 7.8 if scaling is to be avoided.

Note also that $\log R = F_1(T) + 0.16$ and by referring to Table 3.27 it can be seen that the cooling water will give a positive scaling index only at a temperature of 26°C.

As the data used to derive the equation covered a relatively small range, the equation may not be valid over a wide range of water composition. Estimates of the ranges are 100–200 mg kg^{-1} Ca^{2+}, 3–10 mg kg^{-1} PO$_4$3$^-$, pH 7.5–9.0 and temperature 26–38°C.

TABLE 3.27

Scaling factor F_1 (temperature)

T(°C)	$F_1(T)$	T(°C)	$F_1(T)$
15.0	−3.022	28.0	0.358
16.0	−2.762	29.0	0.619
17.0	−2.503	30.0	0.879
18.0	−2.243	31.0	1.140
19.0	−1.983	32.0	1.401
20.0	−1.724	33.0	1.662
21.0	−1.464	34.0	1.923
22.0	−1.204	35.0	2.184
23.0	−0.944	36.0	2.445
24.0	−0.683	37.0	2.707
25.0	−0.423	38.0	2.968
26.0	−0.163	39.0	3.229
27.0	0.098	40.0	3.491

The equation was developed specifically for stations using River Trent water for cooling purposes. For other rivers whose use might lead to calcium phosphate fouling of heat exchangers, the same procedure could be applied in respect to the calcium and phosphate factors, but the temperature factor may need to be established empirically.

10.2 Prevention of condenser tube scaling

10.2.1 Taprogge

Taprogge is a physical method of continuously cleaning condenser tubes by the circulation of sponge rubber balls in the cooling water. The system is simple, see

TABLE 3.28

Scaling factor F_2 (phosphate)

$[PO_4^{3-}]$ mg kg^{-1}	$F_2(PO_4)$	$[PO_4^{3-}]$ mg kg^{-1}	$F_2(PO_4)$	$[PO_4^{3-}]$ mg kg^{-1}	$F_2(PO_4)$
1	0.000	6	0.778	11	1.041
2	0.301	7	0.845	12	1.079
3	0.477	8	0.903	13	1.114
4	0.602	9	0.954	14	1.146
5	0.699	10	1.000	15	1.176

TABLE 3.29

Scaling factor F_3 (calcium)

$[Ca^{2+}]$ mg kg^{-1}	$F_3(Ca)$	$[Ca^{2+}]$ mg kg^{-1}	$F_3(Ca)$	$[Ca^{2+}]$ mg kg^{-1}	$F_3(Ca)$
100.0	2.722	170.0	2.997	240.0	3.170
110.0	2.772	180.0	3.026	250.0	3.190
120.0	2.818	190.0	3.054	260.0	3.209
130.0	2.860	200.0	3.079	270.0	3.227
140.0	2.898	210.0	3.124	280.0	3.245
150.0	2.934	220.0	3.127	290.0	3.262
160.0	2.967	230.0	3.149	300.0	3.279

TABLE 3.30

Scaling factor F_4 (pH)

pH	F_4(pH) at [Ca^{2+}] mg kg^{-1}				
	100	150	200	250	300
7.0	−5.883	−5.904	−5.921	−5.936	−5.948
7.1	−5.732	−5.765	−5.773	−5.788	−5.802
7.2	−5.587	−5.611	−5.630	−5.646	−5.661
7.3	−5.447	−5.472	−5.493	−5.510	−5.525
7.4	−5.312	−5.339	−5.361	−5.379	−5.394
7.5	−5.183	−5.211	−5.233	−5.252	−5.269
7.6	−5.058	−5.087	−5.110	−5.130	−5.147
7.7	−4.937	−4.967	−4.991	−5.011	−5.029
7.8	−4.820	−4.851	−4.875	−4.896	−4.914
7.9	−4.705	−4.737	−4.762	−4.784	−4.802
8.0	−4.594	−4.626	−4.652	−4.673	−4.692
8.1	−4.484	−4.517	−4.543	−4.565	−4.584
8.2	−4.377	−4.410	−4.436	−4.458	−4.477
8.3	−4.270	−4.304	−4.331	−4.353	−4.372
8.4	−4.165	−4.199	−4.226	−4.249	−4.268
8.5	−4.061	−4.095	−4.122	−4.145	−4.165
8.6	−3.958	−3.992	−4.020	−4.042	−4.062
8.7	−3.856	−3.890	−3.917	−3.940	−3.960
8.8	−3.753	−3.788	−3.815	−3.838	−3.858

Fig 3.51; balls are injected at the inlet to the condenser via a pump and collected at the outlet of the condenser by a strainer, and then re-injected into the circuit. The abrasive action of the balls, which are slightly larger than the condenser tube diameter in the uncompressed state, is intended to maintain the heat transfer characteristics of the tube surfaces in a satisfactory condition. The number of balls normally used is sufficient for each tube to receive a ball every five minutes, assuming that the balls are uniformly distributed in the condenser.

The main justification for using Taprogge is to minimise deposition and scaling of the condenser tubes.

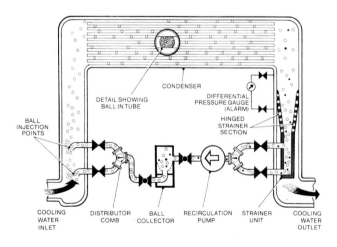

FIG. 3.51 Taprogge condenser tube cleaning system

However, there may also be a secondary advantage in a reduction in the quantity of chlorine used in the cooling circuit to combat biological growth.

Operation and maintenance

Operation is simple and mainly involves the changing of balls at regular intervals. At a number of stations a service agreement with Taprogge is used to provide replacement balls and general maintenance of the equipment.

The condenser design dictates the number of recirculating pumps used. For example, a condenser designed with four quadrants will employ four ball-circulating pumps.

Several types of ball are available ranging from soft sponge-like rubber, to carborundum-coated abrasive balls. The type actually used will depend on the cooling water conditions; the greater the risk of scaling, the more abrasive the ball required. Frequently, a mix of two different types of ball is used.

Two main procedures are adopted for changing the balls:

(a) After a set period of operation.

(b) Catching the balls, counting, examining and replacing any that are worn or lost. This will usually be carried out at a greater frequency than (a).

Table 3.31 gives details of ball usage at several stations where a risk of calcium phosphate scaling exists.

Balls lost enter the cooling circuit, but there is no evidence that serious difficulties have arisen from these stray balls. However, it is difficult to predict where they are likely to reside finally. It is conceivable that they eventually reach the river, but the number involved is likely to be small.

Various methods have been used to check the effectiveness of Taprogge installations for maintaining clean condenser tubes. These are:

● Use of condenser performance data.

● Visual examination of tubes removed from the condensers.

● Application of a chemical cleaning agent on a selected number of tubes followed by chemical analysis from which the quantity of scale/deposit present can be calculated.

● Heat transfer tests of condenser tubes.

The results obtained have shown that the tubes can generally be kept in a satisfactorily clean state, but occasionally, particularly during summer periods, scaling can arise and the tubes have then to be cleaned chemically.

Taprogge installations will not overcome the need for chlorination of cooling water, which will be required to combat tower weed and to protect auxiliary coolers from bacterial slimes. However, a reduction in the use

TABLE 3.31

Data on Taprogge ball usage

Station	Types of ball ball used %		No of balls used as % of tubes	Ball replacement balls/1000 hours/ condenser
	Abrasive	Plain		
A	40	60	10	4000 (500 MW)
B	25–50	75–50	7.5–10	6000 (500 MW)
C	50	50	7.5	6000 (500 MW)
D	10	90	7	3000 (200 MW)
E	0	100	6.6	1000 (200 MW)

of chlorine does arise and the quantity used is likely to be in the range of 10–30% of that required in the absence of Taprogge.

There is a risk of thinning of the condenser tubes by the abrasive action of the balls, leading ultimately to tube failure. However, there is little quantitative evidence available on the rate of loss of metal from the tube surfaces attributable to this cause. Experience to date suggests that any adverse effects would arise only in the long rather than the short or medium term.

10.2.2 Acid dosing

Acid dosing represents a chemical alternative to physical techniques for maintaining clean condenser tubes. The foregoing has indicated the effect of pH on the deposition of calcium phosphate and calcium carbonate, and therefore complete prevention of scaling can be obtained by operating at suitably low pH conditions.

However, almost all of the components in contact with the circulating cooling water are vulnerable if subjected to an acidic environment for extended periods, obviously mild steel would corrode. Similarly, concrete could be attacked by any acid; furthermore, even in neutral solutions, sulphate ions react with concrete and the rate of attack depends on the composition of the concrete, its compaction and the sulphate concentration. The combined effect of pH and sulphate concentration can lead to a very rapid attack.

There is a need therefore strictly to control both the pH and the sulphate concentration. The pH should not fall below 7.0 and the sulphate concentration should not exceed 600 mg kg^{-1}.

The choice of the acid for pH adjustment is dictated by environmental and economic factors. Sulphuric acid is relatively cheap and its use is environmentally acceptable.

Overall, acid dosing is a relatively low capital option but with high operating cost. Acid additions do not prevent bacterial slimes forming in condenser tubes so that a high chlorine dosage also has to be maintained. An advantage of acid dosing is that the whole of the cooling water is rendered non-scaling so that auxiliary coolers can also be protected from scaling.

Financial, operational, strategic and environmental are all factors involved when protection of plant from

scaling is considered. These need to be compared and contrasted prior to deciding whether a chemical or physical technique is used in a specific situation.

10.2.3 Chemical additives

A multiplicity of commercial additives are available for preventing scaling. The additives include polymaleic acid, polyacrylates, polymethacrylates, polycarboxylic acids, phosphonates, ethoxylated amines, non-ionic detergents and quaternary ammonium salts. They are used to modify the crystal structure of compounds precipitating from super-saturated solutions so that scaling does not occur.

Such compounds represent a low capital option being simply injected into the cooling water circuit. They are expensive, but are used at a relatively low dose rate, no bonded tank storage or monitoring equipment is required and overdosing is not likely to cause attack of metals or concrete.

A large number of factors can influence both the degree of scaling likely to arise and the performance of chemical additives in inhibiting such scale formation. Consequently, plant trials are required to establish the effectiveness of these agents.

Chemical additives have not been used widely in the power industry, partly because of their high cost and the large volumes of water that have to be treated. For example, a typical dose rate of 2 mg kg^{-1} can be 20% more expensive than using a much larger quantity of sulphuric acid to control pH conditions.

Rig and plant trials of additives for the prevention of calcium phosphate scaling have shown a wide range of effectiveness (from zero to 96%).

10.3 Removal of scale

Once scale has accumulated to any significant degree, it has to be removed. This is usually carried out chemically with acids, but there is also a limited use of physical methods. Chemical cleaning will result in some metal loss which inevitably restricts the total number of chemical cleans which can be carried out in the life of a station.

10.3.1 Acid cleaning

Condensers are normally acid cleaned using one of two methods; namely, off-load and reduced-load cleaning.

Off-load cleaning

The condenser is taken off-load, filled with a solution of inhibited 1.5% w/w acid and left to soak at ambient temperature for several hours. Sulphuric acid is usually employed, but hydrochloric acid has also been used.

Reduced-load cleaning

This method is occasionally used on high merit plant where to take a unit out of service would incur a considerable cost penalty.

Load is reduced to about 60–70% of the maximum continuous rating so that it is possible to isolate the condenser shell which is then filled with a 1–2% w/w solution of sulphuric acid and left to soak at a temperature of about 40°C. Plant tests have shown the efficiency of the cleaning process to be variable, being as low as 22% and averaging 50%. This is due to gas blanketing, especially in the centre of the condenser which is bowed to facilitate draining. The gas is generated by the action of the acid on the calcium carbonate invariably present.

10.3.2 Physical cleaning

Mechanical methods

The mechanical methods include the use of bulleting with rubber, nylon or twisted wire bullets. In general these methods do not remove scale satisfactorily, but help in removing silt prior to chemical cleaning.

Water jetting

High pressure water jetting at around 600 bar pressure is used on many stations. However, it is time consuming due to the large number of tubes involved, and is of little value to high merit plant. Results show that the efficiency of the process is dependent on the operator. Typically, the effectiveness of the technique can be as low as 30% in the absence of station supervision and is unlikely to exceed 85%, even with continuous supervision.

10.4 Control of biological fouling

Cooling water is obtained from rivers and the sea; consequently biological material will be present. Micro-organisms present will tend to grow and multiply on the components within the circuit forming slimes, algae and fungi; whilst in the case of sea water the growth of macro-organisms (mussels), will occur.

Micro-organisms can form films on heat transfer surfaces and the formation of slime on the inner surfaces of condenser tubes is a well known phenomenon. Algae can grow within the cooling towers leading to weed which tends to block screens, reduce air flow in the tower and produce streams of water rather than droplets so that efficiency of cooling may be decreased.

Fouling of circulating water culverts by mussels can arise where sea water is used for cooling. Single shells of mussels, which have grown after the fine screens installed at the inlet to the station, can enter the condenser and partially block the condenser tubes. The turbulence created leads to erosion/corrosion of tubes with subsequent failure and a consequent inleakage of cooling water into the condensate.

10.4.1 Chlorination

Chemical control of biological fouling is by the use of chlorine as a disinfectant.

There are two main methods of applying chlorine:

(a) As chlorine gas dissolved in water.

(b) As sodium hypochlorite solution.

For many years (a) was the sole method used in the power industry, but because of the potential risk from the storage and handling of the liquid chlorine, from which the gas is obtained, most stations now use sodium hypochlorite. Details of the production and handling of this chemical are given in Section 10.4.2 of this chapter.

Similar products arise when chlorine or sodium hypochlorite is added to water:

$$Cl_2 + H_2O \leftrightarrow HOCl + HCl$$
$$NaOCL + H_2O \leftrightarrow HOCl + NaOH$$

The hypochlorous acid (HOCl) formed then dissociates to give hypochlorite ions:

$$HOCl \leftrightarrow OCl^- + H^+$$

Now the equilibrium concentrations of HOCl and OCl^- depend on pH. The lower the pH the greater the level of the undissociated hypochlorous acid, which is a more effective biocide than the hypochlorite ion. At pH 7.4, the HOCl and OCl^- are of approximately equal concentration, whereas at pH 9.5, all the available chlorine is in the OCl^- form.

The presence of ammonia and organic matter in cooling water reduce the effective dose of chlorine added. The presence of ammonia leads to the formation of mon, di and tri chloramines, whilst organic matter can react with chlorine to give chloroform and other halomethanes. Note that the chloramines produced can still act as disinfectants whereas the haloforms cannot. In this respect, free available chlorine is defined as that existing in water as hypochlorous acid and hypochlorite ion, and combined available chlorine is the residual chlorine existing in water in chemical combination with ammonia or organic nitrogen compounds. The chlorine demand is defined as the difference between the amount of chlorine added to water and the amount of free available chlorine remaining at the end of a specified contact period.

At inland stations intermittent chlorination is carried out, during which each of the condensers and auxiliary coolers are chlorinated in turn for a period of about ten minutes and the whole cycle repeated every four hours. Normally, the dosing is adjusted to give a residual of 0.5 mg kg^{-1} chlorine at the condenser outlet for five to ten minutes.

At sea water stations where fouling by mussels is the major problem, the most practicable chlorination regime is to apply a continuous dose throughout the mussel settlement season. If the chlorination procedure is adequate for mussels then other organisms in the culverts are unlikely to present a problem.

Continuous chlorination should be carried out whilst the sea temperature is above 10°C, certainly during the period from April to November.

Continuous chlorination throughout the year, regardless of temperature, may be necessary in stations close to nursery mussel areas such as Morecambe Bay (Heysham *1* and *2* power stations). Chlorination should be controlled to provide an effective chlorine residual of 0.2 mg kg^{-1} at the condenser inlets. This normally requires an injection rate of about 1 mg kg^{-1}. Where systems are to be shutdown, the system should be dosed to a higher concentration prior to shutdown. Levels of 6 or 20 mg kg^{-1} are required depending on whether the shutdown time is less than or greater than three days.

The presence of algae biofouling within the cooling tower is due to filamentous green algae predominantly of cladophera species. This occurs mainly where the heat transfer pack is exposed to light between the pond sill and the lower skirt. The normal intermittent chlorination regimes can usually contain such growth, but in extreme cases an additional batch treatment using higher concentrations of chlorine may be required.

10.4.2 Chlorination plant

The majority of power stations in the UK now use sodium hypochlorite as the sterilising agent, and there are two main sources of supply:

(a) By tanker from chemical manufacturers.

(b) Production on site from the electrolysis of brine.

However, the latter approach is limited to coastal stations because of the need to import salt to inland stations, the high capital cost of the plant and its greater operational complexity relative to (a).

Use of imported sodium hypochlorite

A typical composition of sodium hypochlorite provided by chemical manufacturers is:

Available chlorine	15%
Sodium hydroxide	1–1.5%
Sodium carbonate	0.7–1.0%

It follows that on a weight basis the usage of sodium hypochlorite will be approximately seven times that of chlorine, and in terms of available chlorine the cost is about three times that of chlorine liquor. For a 1000 MW station operating without Taprogge, the usage of sodium hypochlorite may be up to a 1000 tonnes/annum.

Sodium hypochlorite is slightly caustic and is an oxiding agent and therefore needs to be handled with care, but it is not a dangerous chemical when maintained in an alkaline condition. It must not be allowed to contact acid, as free chlorine gas would be released. Most of the sodium hypochlorite dosing systems installed have been retrofitted to existing cooling water systems; using the distribution lines and motive water pumps previously employed for the injection of chlorine. New pumps are provided for injecting the sodium hypochlorite solution into the motive water line together with a new sequence controller. Figure 3.52 is a typical flow diagram of a sodium hypochlorite dosing system.

Electrochlorination

In electrochlorination plants, sea water is passed through electrolytic cells in which sodium chloride is converted to sodium hypochlorite and hydrogen.

The anodic reaction is

$$2Cl^- \rightarrow Cl_2 + 2e^-$$

and the cathodic reaction is

$$2Na^+ + 2e^- \rightarrow 2Na^-$$
$$2Na + 2H_2O \rightarrow 2NaOH + H_2$$

As the cathode and anode products are not separated the overall reaction is

$$2NaOH + Cl_2 \rightarrow NaOCL + NaCl + H_2$$

The cells for sea water electrolysis are of two main types, concentric tube and parallel plate.

The concentric tube cells are illustrated in Fig 3.53. The cell comprises inner and outer bipolar titanium tube electrodes, and the saline water flows down the annulus between the tubes. The anode areas are coated with platinum or platinum/iridium. The electrode gap is minimised to reduce the internal resistance and maintains a high velocity over the surfaces to prevent fouling. The cells are connected electrically in series and the water flows through a number of cells in a series/parallel arrangement. The parallel plate cells consisting of plastic housing contain arrays of closing space plate electrodes. The cathode material is either titanium or Hastalloy *C*. The titanium anodes are coated with platinum/iridium or precious metal oxides usually ruthenium oxide.

In general the oxide coatings give a satisfactory power consumption but are more restricted as regards operation at low temperatures.

Figure 3.54 shows a typical parallel plate cell arrangement. The electrode gap is of the order of 2.5 mm to maintain high velocities and low resistance. The hydroxide produced at the cathode increases pH and results in the gradual build up of insoluble magnesium hydroxide and calciferous deposits. These are removed by the application of regular hydrochloric acid cleaning, and permanent recirculatory acid cleaning systems are installed.

As sea water flows through the electrolysis cells the hypochlorite concentration increases. Efficiency reduces at higher concentrations and the normal operating range is from 1000 to 2000 mg kg^{-1} as chlorine equivalent.

Several other factors affect the efficiency of electrolysis expressed as kWh/kg chlorine equivalent, the

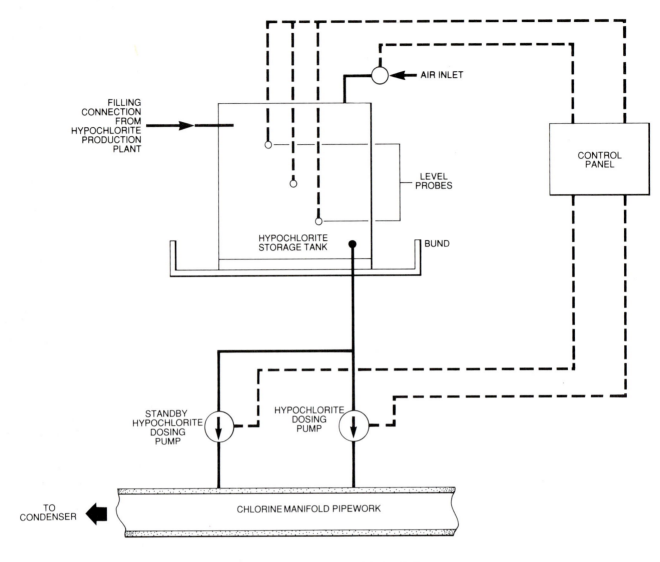

FIG. 3.52 Typical sodium hypochlorite dosing system

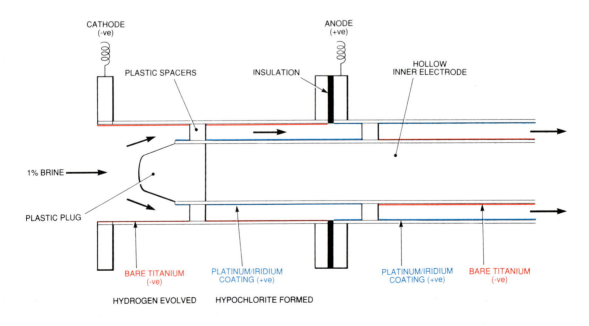

FIG. 3.53 Concentric tube electrolysis cells for producing sodium hypochlorite

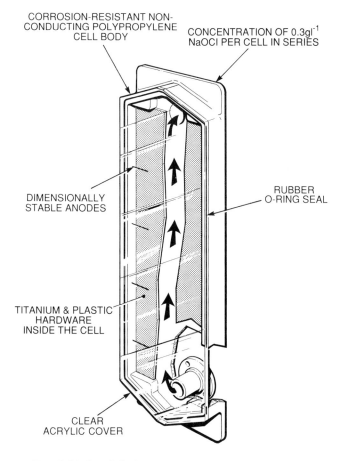

CORROSION-RESISTANT NON-CONDUCTING POLYPROPYLENE CELL BODY

CONCENTRATION OF 0.3gl⁻¹ NaOCl PER CELL IN SERIES

DIMENSIONALLY STABLE ANODES

RUBBER O-RING SEAL

TITANIUM & PLASTIC HARDWARE INSIDE THE CELL

CLEAR ACRYLIC COVER

FIG. 3.54 Parallel plate electrolysis cells for producing sodium hypochlorite (NEI-Thompson-Kennicott)

most important are salinity, temperature and current density:

- *Salinity* As salinity increases, the conductivity of the electrolyte increases giving a lower internal resistance and improved current efficiency. Cells are designed to operate in a range of salinities from 1.3–3.5% expressed as sodium chloride.

- *Temperature* As temperature increases, the conductivity increases giving an effect similar to an increase in salinity.

  At temperatures below 12°C, oxide coatings deteriorate and current density has to be reduced. Operation at temperatures lower than 5°C is not possible.

- *Current density* As current density, expressed as A/m², increases the overvoltage due to the internal resistance increases and the efficiency falls. The cell efficiency must be balanced against the cost of additional surface area. It is clear that operation at reduced output is best effected by reducing the overall cell voltage rather than by reducing the number of cells in service.

Plant performance is typically in the range 3.5–5 kWh/kg chlorine equivalent and it is important to monitor the plant so that the reason for any deterioration in performance can be ascertained.

During electrolysis large quantities of hydrogen are formed and must be removed safely. Some systems use air blowers to reduce the hydrogen concentration in the air flow to less than a quarter of the lower explosive limit. The safety procedure is to disengage the hydrogen from open-topped product tanks. This is a passive system which avoids problems with air flow failure interlocks.

Plant rooms are well ventilated to ensure that there is no possibility of a build up in hydrogen.

Figure 3.55 shows the arrangement of the sea water electrolysis plant at Blyth *A* and *B* power stations.

The plant is designed to produce 160 kg/h of chlorine equivalent. Sea water from the cooling water circuit is passed to a set of 3 × 50% booster pumps with two operational and one on standby. The water is then pumped through strainers which retain suspended solids down to a 500 μm size to protect the electrolysis cells. The water leaving the strainers is passed to four streams of electrolysis modules each capable of providing 33% of the hypochlorite required and each stream comprises 12 cells, i.e., a total of 48 cells. The flow through each stream is 31.9 m³/h, and the concentration of equivalent chlorine produced is 1250 mg kg⁻¹.

The sea water pumps have sufficient head to discharge the product after electrolysis to the top of a sodium hypochlorite storage tank. This tank is open-topped and of sufficient size to allow disentrainment of the hydrogen present to the atmosphere. The tank is outside and is above roof level so that the hydrogen released is widely and safely dispersed by natural ventilation.

Four sets of hydrogen gas detectors are fitted in the cell room. If the alarm on any of these is initiated, the incoming 415 V circuit-breakers are tripped and the whole plant is shut down.

'Sanilec' electrolysis cells are used (see Fig 3.54) which have Hastelloy *C* cathodes and expanded titanium metal anodes suitably coated with precious metal oxides (including ruthenium) to allow a cold water input. The cell is constructed from polypropylene with a clear acrylic plastic cover; this allows visual detection of any scaling of the cells.

The four transformers are 3.3 kV/415 V rated at 200 kVA, and are three-phase, 50 Hz air cooled giving approximately 415 volts at full load.

The life of the anode coating is guaranteed for 10 years under normal operating conditions. However, a major factor influencing the life obtained will be the time the plant operates with less than all four streams in service.

An acid flush system is incorporated to remove any scales which form within the cells. This is carried out by an entirely manual procedure and extreme care is taken to flush the modules thoroughly to remove completely the sodium hypochlorite before admitting the acid. A dilute solution of hydrochloric acid is used.

Manganese deposits can form on electrodialysis cells if the incoming sea water has a concentration of

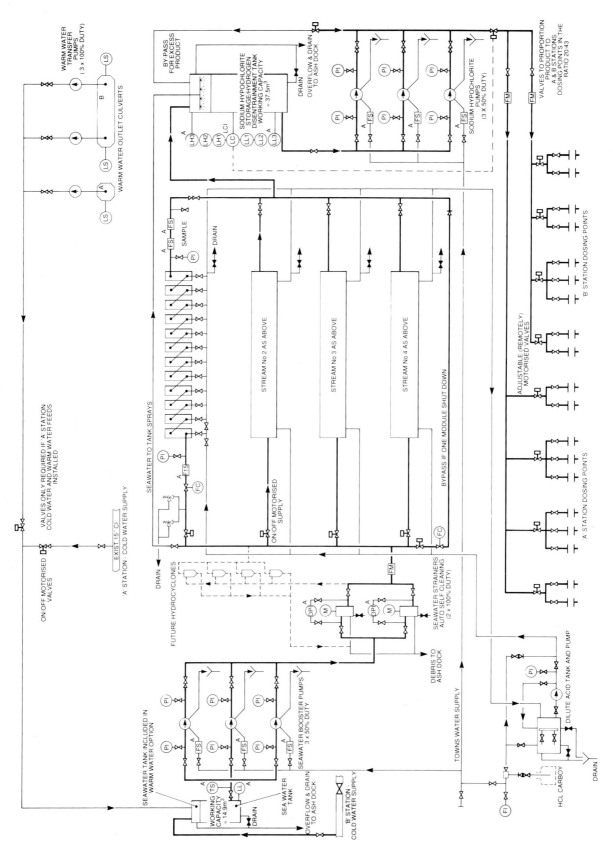

Fig. 3.55   Sea water electrolysis plant for the production of sodium hypochlorite at Blyth power station (NEI-Thompson-Kennicott)

> 20 μg kg^{-1} manganese. At Blyth, the electrical polarity of the cells is reversed on a timed cycle to minimise the accumulation of such deposits.

The normal operating regime is to run all four streams to produce 160 kg/h of equivalent chlorine. The water flow through each module is constant and preset at the module inlet. The current flowing through each module is preset to give the total quantity of sodium hypochlorite required. Subsequently, the setting is automatically maintained for changes in sea water salinity and temperature.

In the event of a stream being out of action, the power output from the transformer can be increased to ensure that full plant output is available from the remaining streams.

The operation of the plant is basically manual with safety interlocks and override functioning.

Water flow is established through the cells and is sensed by the main and back-up flow switches. At start-up there is an initial delay until the flow switches are energised before power is supplied to the cells. If a loss of flow occurs, this is detected by a module inlet/outlet flow switch and the stream will be shut down.

A fail level control system is provided on the sodium hypochlorite storage tanks. At high level the power is switched off immediately and then, after a short delay period to purge the system of hydrogen, the feedwater pumps are switched off. No automatic re-start facility is available. At low tank level the discharge pumps from the tanks are switched off and an alarm is indicated, but output from the system continues until the tank is full and the high level switch then initiates a plant shutdown.

From the detraining tanks the hypochlorite solution is pumped to cooling water intakes. Hypochlorite solution is extremely corrosive, hence rubber-lined steel or plastic pipework is used.

In comparing electrolytic chlorination plants with gas chlorine or hypochlorite injection systems it should be noted that the size of electrolytic systems is directly related to the peak design chlorine dose level, whereas this has little effect on the cost of dosing systems.

10.5 Entry of debris and particulate matter

The coarse screens at the cooling water station inlet arrest large pieces of debris, but many smaller particles and pieces of debris still enter. Additionally, dust and broken timber from the tower packing can be present. At many stations, tube plates have to be 'picked' to remove materials blocking off tubes. Some sea water and estuarine stations have used secondary screens within the water box and reduced throat inserts in the tubes to prevent clogging the tube plate and to prevent partial blocking of tubes that could cause local high velocities, thereby increasing the risk of erosion/corrosion or impingement attack at the entrance to the tubes.

The ingress of sand from the river water used at inland stations can also give rise to erosion and erosion/corrosion at tube inlets. Nylon inserts into the tube inlet and protective coatings on the tube plate for a short distance into the tubes have been used to combat this problem.

10.6 Cooling tower packings

For many years the only materials used as the heat exchange medium in cooling towers were timber or asbestos cement packing (see Fig 3.56). The wooden lattices cool the water by the splashing action through the pack, whereas asbestos cement packing is in the form of corrugated sheets stood on edge and the water flows over them in a continuous film. The latter is the more efficient as there is less resistance to the passage of air and hence the air has a higher velocity. At least six and often eight counterflow towers packed with these materials are required on a 2000 MW fossil-fuelled station; Fig 3.57 shows a typical cooling tower.

Wooden packings suffer from damage from a number of causes, i.e., salt crystallisation and dry and wet fungal rot. The fungal rot is alleviated by impregnating the new timber with solutions containing copper, chromium and arsenic salts. This has allowed a useful life of around twenty years to be obtained, but towards the end of the period minor collapses of timber occur and this causes general problems in the circulation of water through the plant.

Asbestos cement packings have tended to break up with time as the result of not being adequately supported.

Until recently, the normal practice when towers needed to be repacked was to use timber, which was cheaper and safer than asbestos. However, spiralling costs of timber and the associated construction and installation costs led to a consideration of pre-formed plastic packings made of unplasticised PVC.

In a plastic pack, corrugated PVC sheet is glued together to form square or V-shaped diagonal channels. Typically, the blocks are three metres square (see Fig 3.56).

Several types of packing are available and their efficiency depends on the spacing between the corrugated sheets. The most efficient packing has the closest spacing.

The use of plastic packs has been shown to give an improvement in cooling of 2–3°C, depending on the type of plastic pack and the operating conditions [28]. This is equivalent to a saving of around £2 × 10^6 per annum on a 2000 MW station. It has also been shown that when an installation does not achieve the above improvement the main reason has been poor water distribution. Subsequently the fitting of extra nozzles and pipework has enabled the design figures to be met. The main factors affecting the performance and life of plastic packs are scaling, biofouling, icing and structural deterioration.

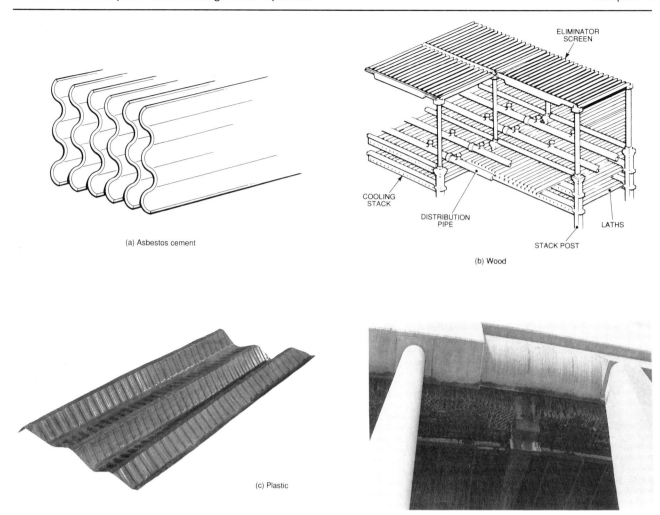

(a) Asbestos cement

(b) Wood

(c) Plastic

FIG. 3.56 Materials for cooling tower packings

Initially, several installations suffered from scaling with both calcium phosphate and calcium carbonate. This has been alleviated by pH-control using acid dosing as described in Section 10.2.2 of this chapter.

Biofouling is a major problem with plastic packs, the high surface area and temperature conditions are conducive to the growth of biofilms. These have been enhanced by the presence of silt in the river water supply and by a degree of scaling.

Increase in weight from fouling has been shown to comprise 80% biofouling, 10% silt and 10% scale. The biofouling rate is greatest for the high efficiency packs and as might be expected is seasonal, being highest during the summer and autumn. The actual rates are variable depending on site conditions and figures obtained from test-section weight measurements have given 0.01–0.6 kg/m^3/day for the high efficiency packs, but only up to 0.05 kg/m^3/day for the lower efficiency type [28].

Figure 3.58 shows how the performance of the packings deteriorates as the weight of foulant increases.

Treatment to avoid biofouling would be difficult and expensive, to reduce it to very low levels is likely to require the installation of filters and the use of high concentrations of disinfectants. Slug dosing of sodium hypochlorite can reduce the rate of fouling but does not remove existing deposits.

Icing of plastic packing could result in more serious structural failure than would be the case with wood or asbestos. Consequently, the ice prevention system which is a water ring main around the tower has to be maintained in a good condition.

Long term experience with plastic packing in cooling towers outside the UK power industry has indicated that physical degradation is minimal after 10 years' operation. It follows that the main factor determining the life of a pack will be the accumulation of fouling material which lowers the efficiency of cooling and creates an unacceptable load on the packing support structure.

At present the life is specified as ten years, made up of eight years at the mean fouling rate and two years of extreme fouling conditions equivalent to twice the mean rate due to two hot summers.

Packings are chosen on the basis that they will not accumulate a fouling load exceeding 350 kg m^{-3} over the specified life.

10.7 Health precautions in relation to cooling water

Cooling water is drawn from sources which contain potentially harmful organisms and, therefore, safe-

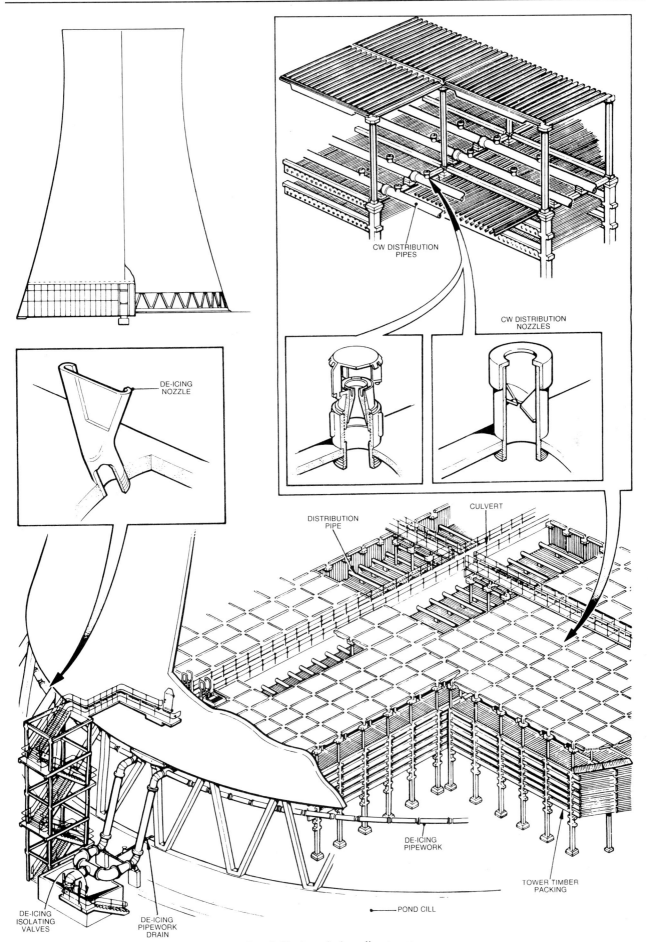

CW DISTRIBUTION
PIPES

CW DISTRIBUTION
NOZZLES

DE-ICING
NOZZLE

DISTRIBUTION
PIPE

CULVERT

DE-ICING
PIPEWORK

TOWER TIMBER
PACKING

POND CILL

DE-ICING
ISOLATING
VALVES

DE-ICING
PIPEWORK
DRAIN

FIG. 3.57 A typical cooling tower

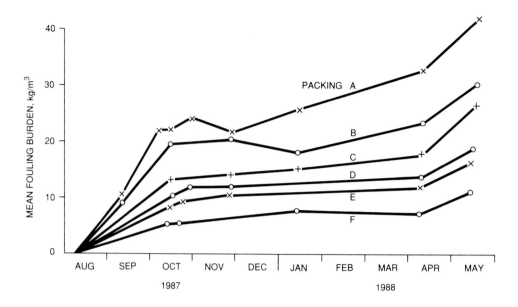

FIG. 3.58 Accumulation of biofouling on plastic packing

guards have to be applied to protect the health of personnel working on cooling water systems.

A code of practice exists to provide guidelines on protecting the various respiratory ailments, and eye damage. The two main respiratory diseases which could arise are:

- Allergy — hypersensitivity to inhaled matter of microbial origin.

- Infection — an invasive growth of micro-organisms in the respiratory tract.

The *allergic* response occurs soon after exposure and is thought to be due to inhalation of micro-organisms. The illness is influenza-like with general malaise, cough and shortness of breath, but it resolves without treatment in a few days. It has been referred to as 'Condenser cleaners' disease'.

Legionnaires' disease is an *infection* caused by the bacterium — Legionella pneumophila. It is a common organism found in a range of habitat including stagnant water, slime and silt. The possibility of an outbreak of Legionnaires' disease only occurs if the environmental conditions are such that the bacteria can multiply rapidly, and then only if a respirable aerosol is produced. The initial symptoms of the disease, following an incubation period of 2–10 days, consists of malaise, dry cough and headache followed by a fever similar to influenza. The illness usually resolves itself naturally after a few days, but can progress to pneumonia requiring treatment especially in those with established lung disease. Surveys carried out in the power industry have shown that there is only an intermittent presence of Legionella pneumophila in low numbers in power station cooling water systems. The numbers of organisms found are not considered to be

a health risk and the measures indicated in Table 3.32 provide satisfactory protection.

In addition to the above diseases, there is also a risk of eye damage from cleaning processes and this has to be safeguarded against.

The levels and frequency of chlorination normally used in main circulating water systems are intended to control biofouling for the protection of plant and to achieve a high level of plant performance. Chlorination and the use of on-load ball cleaning techniques are effective in reducing biofouling. However, when a

TABLE 3.32

Conditions requiring respiratory and eye protection during maintenance/cleaning of CW systems

Conditions	Suitable protective equipment
Pressure water jetting	Either: A respirator, ventilated visor/helmet
Bulleting	or
Condenser hosing	
Any other work which may create a spray or aerosol	A respiratory pneumatic face seal and goggles, box type
Condenser tube rodding	*Respirator, filtering face piece disposable
Condenser tube-end picking	
Cleaning cooling tower ponds	or
Entry into cooling towers in service	Respirator, pneumatic face seal
Fire fighting sprinkler systems — deluge testing	or
	Respirator, ventilated visor/helmet
Cleaning auxiliary water systems where entry is necessary	*Disposable respirators should not be used in conditions where the respirator may become standard

312

system is drained for cleaning, some algae and amoebas may remain in any residual silt and slime, and other microorganisms may also be present. It is for this reason that respiratory protective equipment has to be worn by personnel engaged in maintenance and cleaning processes.

A variety of methods of cleaning accumulated sludge, scale and corrosion products has been described in this section (10.3), and Table 3.32 gives details of the protective equipment which needs to be used when carrying out such procedures.

On occasions, major work may be needed on condensers where the use of respiratory protective equipment would be impractical. In this case, the system has to be drained down and refilled with towns main water, and drained again. Any sludge or slime remaining must be hosed away with towns main water by personnel wearing respiratory protection and stationed outside the condenser water box. The condenser then has to be allowed to dry out prior to entry.

In the case of cooling water auxiliaries, they should be flushed at an adequate flow rate for a period immediately before isolating and draining for cleaning and maintenance work.

As a general precaution, it is necessary to chlorinate plant prior to carrying out cleaning or maintenance work, ideally, the system should be filled with water, treated to give a residual free chlorine concentration of 5 mg kg^{-1} after mixing and recirculating for 24 hours. The system should then be flushed and refilled with clean water.

In addition to the chlorination of the main cooling water circuitry, specific items of plant may require individual continual dosing of chlorine. For example, small auxiliary cooling towers in which the induced air is discharged directly to atmosphere require regular chlorination to prevent bacterial and algal growth. A slug dose should be applied weekly to give a uniform free residual chlorine of 5 mg kg^{-1}.

11 References

[1] Golden, L. S.: The fundamentals of ion-exchange resin production: Purolite Ltd.

[2] Brown, J. Ray, N. J. and Jenkins, M. A.: Ion-exchange in the treatment of low grade water. 6th National Conference on Water, problems in power plants at Karlovy Vary, Czechoslovakia: 1977

[3] Tittle, K. Knowles, G. and Tyldesley, J. D.: The oxidative pretreatment of organic matter in natural waters. Ion-exchange technology: Ellis Harwood, Chichester: 1984

[4] Harries, R. R. and Ray, N. J.: Anion exchange in high flow rate mixed beds: Effluent and Water Treatment Journal 24, 131: April 1984

[5] Ball, M. Harries, R. R. and Pickering, W. W.: The physical strength of ion-exchange resins: 47th International Water Conference, Pittsburgh, Pennsylvania: 27–29 October 1986

[6] Harries, R. R. and Ray, N. J.: Acid leakage from mixed beds: Effluent and Water Treatment Journal: October 1978

[7] Fitchett, G. A. Ray, N. J. and de Whalley, C. H.: Reverse osmosis for water purification: CEGB research No.12: July 1981

[8] Fitchett, G. A.: The role of reverse osmosis in providing high purity water for power generation. The application of membrane processes in the production of high purity water: Dept. of Mechanical Engineering, University of Glasgow: March 1986.

[9] Cordina, O. J. and Curry, K.: Side by side comparison of reverse osmosis and EDR: Ultrapure Water, October 1987 pp 24–27: 1987

[10] Mills, G. R. and Bolton, H. R.: Design and operation of condensate polishing at CEGB Fawley: Effluent and Water Treatment Journal: February 1980

[11] Ray, N. J. Ball, M and Parry, D. J.: Condensate purification plant for nuclear systems: Proc. British Nuclear Engineering Society Meeting; Water chemistry: 11 paper 24: 1980

[12] Tittle, K.: Condensate polishing of AGR stations: Chemistry and Industry: February 1987

[13] Tonge, D.: PWR water treatment: IMechE C274/89: 1983

[14] Emmett, J. R. Grainger, P. M.: Ion-exchange mechanisms in condensate polishing: Proc. 40th International Water Conference, Pittsburgh; 1979

[15] Sadler, M. A., Bolton, H. R.: Regeneration of condensate polishing mixed resins — A review of 10 years experience with the Conesep regeneration system: 50th Annual Meeting International Water Conference: Pittsburgh: October 1989

[16] Crits, G. J.: Condensate polishing with the Ammonex procedure: Ion-exchange technology; Ellis Harwood, Chichester: 1984

[17] Smith, J. H. and Peplow, T. A.: The Tripol process — a new approach to ammonia cycle condensate polishing: Paper 26, condensate purification for nuclear systems; British Nuclear Energy Society international conference on the water chemistry of nuclear reactor systems, Bournemouth: October 1989

[18] Duff, J. H. Leoandsky, J. A.: 'Powdex' A new approach to condensate purification: 24th annual meeting, The American Power Conference, March 27–29th: 1962

[19] Sadler, M. A. Bates, J. C., Darvill, M. R.: Morpholine form condensate polishing at Oldbury on Severn nuclear power station: Ion-exchange for industry: Ellis Harwood, Chichester: 1988

[20] Harries, R. R. Bates, J. C. and Greene J. C.: Volatile amines in the steam/water circuit. The importance of resin choice for condensate polishing water chemistry of nuclear systems: 5th Proc. of the International Conference organised by the British Nuclear Energy Society of Chemistry, the European Nuclear Society and the Institute of Chemical Engineers, Bournemouth: October 1989

[21] Harries, R. R. and Tittle, K.: Deterioration of exchange kinetics in condensate purification plant. 4th Conference on water chemistry of nuclear reactor systems, Bournemouth: British Nuclear Engineering Society: 1986

[22] Harries, R. R.: Ion-exchange kinetics in ultra pure water systems: 3rd Ultrapure Water Symposium, Society for Chemical Industry, London: December 1989

[23] Frisch, N. W. and Kunin, R.: American Institute Chemical Engineers, J.6, 640,: 1960

[24] Ball, M. Jenkins, M. A. and Burrows, R. J.: Ammonium form cation resin operation of condensate purification plant: Ion-exchange technology: Ellis Harwood, Chichester: 1984

[25] Langelier, W. F.: Chemical equilibria in water: American Water Works Association, 38. 169: 1946

[26] Ryznar, J. S.: A new index for determining the mound of calcium carbonate scale formed by a water: American Water Works Association; 36, 472: 1944

[27] Kingerley, D. G. Rantell, A. and M. J. Willett. A rate equation for the scaling of condensers by calcium phosphate. First estimate from plant data: Power Industry Research 1. 17–28: 1981

[28] Johnson, T.: Plastic packings for large cooling towers: The Chemical Engineer: July 1990

C H A P T E R 4

Plant cleaning and inspection

1 Introduction

In the context of the cleaning and inspection of power station plant, this chapter is devoted principally to the discussion of chemical cleaning and associated factors.

Chemical cleaning is an important option in the overall strategy for maintaining the integrity of power plant in the face of corrosion and fouling. Cleaning processes of one sort and another may be employed at all stages in the life of plant: at works, during fabrication of the individual components, during and after site construction and thereafter at intervals during its operational life.

The steam/water circuits of modern power plant must be operated under extremely clean conditions if adverse consequences are to be avoided. A variety of contaminants, either in the working fluid or in the form of deposits on the metal surfaces, may give rise to a number of undesirable phenomena. Thus, solid debris from construction, if not removed before commissioning, may damage surfaces — particularly in the turbine. Solid contaminants may cause corrosion by direct attack on the surfaces, or indirectly by impeding the growth of protective oxides. In addition, dissolved material may lead to deposit formation and hence to impaired heat transfer.

Furthermore, the products of corrosion in the feed system transported into the boiler can accumulate on the waterside surfaces and form concentrations which lead to rapid on-load corrosion and possibly to overheating failures.

Basically, chemical cleaning is used to obviate such hazards or, where some deterioration is inevitable, to restore the efficiency of heat transfer equipment.

With conventional boilers, both before and after operation, waterside deposits/layers which are possibly no more than 50 μm thick are removed to prevent consequential corrosion, etc.

With heat exchangers, terminal temperature difference data provide the criteria for cleaning. Apart from the main condenser and feedheaters, the many coolers in the auxiliary circuit all need attention from time to time.

In current nuclear plant, the risks of corrosion and overheating are less than in fossil-fired boilers and the prime justification for cleaning is safety. As will be seen later, however, nuclear plant has its own particular problems and the demands have inspired novel solutions.

Cleaning may be required at a number of stages in the life of plant. At works, components are chemically cleaned during manufacture — in some cases on several occasions. When installed, power plant is subjected to a pre-commissioning clean. Again, in service, re-cleaning may be considered a necessary remedial measure after certain adverse occurrences, as well as being adopted as a routine preventive maintenance procedure.

In this chapter, the basic science of the cleaning processes available is considered to the extent necessary to provide a scientific understanding of the principles involved.

The processes dealt with here are:

- Degreasing.
- Oxide and scale removal.
- Inhibition against severe attack on the base metal.
- Passivation — which is employed both as a stage in its own right and sometimes as a temporary measure for storage when unforeseen events interrupt the cleaning procedure.

Finally it is emphasised that cleaning is not a substitute for, nor an alternative to, inadequate water treatment and corrosion prevention measures.

2 Fundamentals of cleaning processes

2.1 Degreasing

The removal of grease, oil and similar substances from plant or components after fabrication or operation is necessary to prevent the formation of agglomerates of crud and oil, with the consequent risk of blockages, inefficient operation and possible overheating.

Removal is achieved by using alkaline/wetting agent solutions such as 1000 mg/kg solutions of sodium hydroxide, trisodium phosphate or sodium silicate, at temperatures at/or above ambient depending on circumstances. In the alkali boil-out stage of boiler cleaning, for example, a solution of this kind is used and the boiler is fired to 40.0 bar for 24 hours.

In those instances where there is a risk of contributing to the stress-corrosion of austenitic stainless steel if caustic alkalis are used, organic emulsifying degreasants and wetting agents are employed. These materials contain in the same molecule both hydrophobic and hydrophilic groupings (e.g., sodium aryl sulphonates) to facilitate the removal of organic (hydrophobic) material in an aqueous environment.

Where *clean conditions* working is employed during fabrication and erection, the amount of oil/grease requiring removal is less than hitherto. It is usual therefore to use a wetting agent, which has the effect of reducing the surface tension of water by a considerable amount for a relatively low concentration of agent.

In post-service cleans, the need for degreasing varies considerably depending on whether there has been ingress of oil or organic matter during operation of the plant. In boilers, degreasing is seldom necessary. However, in condenser steamside cleaning, and more particularly in feedheater cleaning, oil/grease removal may be the principal reason for carrying out the clean.

Wetting of the surfaces being cleaned is always important, so wetting agents are incorporated into proprietary inhibitor formulations.

In extreme cases, neither caustic nor emulsifying solutions are adequate and it is necessary to resort to organic solvents to remove grease, e.g., halogenated hydrocarbons. Particularly difficult or intractable cases may require manual swabbing.

2.2   Oxide/scale removal

2.2.1   Acidic formulations

The main stage in most cleaning processes is the removal of oxides from the substrate metal, either those grown in situ or transported to the metal. Acids are widely used for this main stage and an extensive range is available, either alone or in mixtures with other acids and/or chemical agents. The selection of a formulation depends upon a number of factors such as cost, commercial availability, toxicity.

Mineral acids have been widely used for many years for cleaning mild steel; principally hydrochloric, sulphuric and phosphoric acids. Nitric and hydrofluoric acids have been employed for cleaning stainless steels, particularly at the manufacturers' works.

Increasingly, organic acids are used for power station plant. This is because, although more expensive than mineral acids and less efficient for oxide dissolution, they are safer to handle, less corrosive and (perhaps the most important technical advantage) they do not contribute to the stress-corrosion of austenitic stainless steel. However, their action is more temperature-sensitive and may need temperatures of 90–100°C for effective cleaning, compared with 65°C which is sufficient with hydrofluoric and hydrochloric acids.

Of the organic acids, citric is the most popular; but formic, hydroxy-acetic, oxalic and ethylene diamime tetra-acetic acid (EDTA) have all been used either individually or in mixtures.

For the removal of mill-scale, hydrochloric acid is typically used at 5% and citric acid at 3%, depending upon the amount of material to be removed. A hydrochloric acid concentration of 5% is equivalent to 2.4% iron, 3% citric acid concentration is equivalent to 0.9% iron. Where far higher or much lower potential iron burdens are estimated the acid concentrations employed can be adjusted accordingly. Citric acid can be used in a number of stages of substitution, and for the removal of oxide from mild steel the mono-ammonium salt is advantageous. The rate of dissolution of iron oxide is greater than in the acid alone, but the conditions are still sufficiently acidic to avoid the precipitation of ferric ammonium citrate.

Citric acid ($C_6F_8O_7$) is able to ionise with increasing pH through the three carboxylic groups (successively) and finally the hydroxyl group. Figure 4.1 shows the relative amounts of the various ionic form as a function of pH. Figure 4.2 shows the rate of iron dissolution superimposed on the theoretical curve for the monocitrate ion (extracted from Fig 4.1) both as a function of pH.

A number of other organic acids exhibit a similar trend as the acid strength of the carboxylic acid group varies only slightly in these.

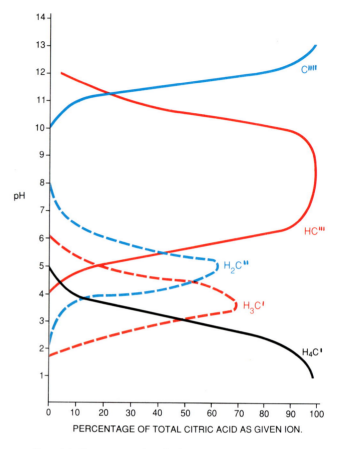

FIG. 4.1   Percentage of each citrate ion present versus pH

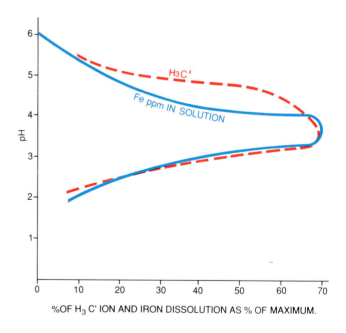

FIG. 4.2   Rate of iron dissolution from magnetite compared with the theoretical curve for the monocitrate ion (extracted from Fig 4.1) as a function of pH

In contrast to this, the rate of dissolution of iron oxide in mineral acids changes linearly with pH.

Ammonium bifluoride (ABF) or sometimes sodium fluoride (NaF) may be added to acid cleaning solutions.

Originally used to improve dissolution of silica, they were also found to aid iron removal.

Comparable data on the relative dissolution of silica from various forms of siliceous material are shown in Table 4.1.

Probably fluoride additions are helpful because iron is brought into solution more quickly, being then held there by the principal acid. ABF and NaF are used at about 0.5% concentration.

Fluorides are more effective in organic acids than mineral acids, probably owing to the relative ease of ionisation in the former, as the acid dissociation constant (Ka) for hydrofluoric acid and the first carboxy group of many organic acids is about the same (approximately 10^{-4}), whilst mineral acids with virtually 100% ionisation suppress dissolution of the fluoride.

The performance of acid formulations is influenced by concentration, flow and temperature. Within limits, the rate of dissolution of oxide is not the factor determining the concentration selected. The concentration is fixed by the amount of material to be taken up and held in solution (usually estimated).

Flow is mainly for reagent replenishment to ensure that cleaning is not stifled by reagent depletion at the interface. It is also necessary to avoid the accumulation of unduly high concentrations of corrosion products, e.g., ferric ion Fe^{3+} which would be corrosive. A rate of flow of 0.3 m/s is found to be generally satisfactory. A deposit intractable at this rate will dissolve no faster at higher rates, and higher rates of flow could lead to enhanced residual corrosion of cleaned metal.

An excess of acid over the contaminant (e.g., iron oxide) is maintained both to ensure that dissolution continues at an acceptable rate and to prevent precipitation of material already removed by ensuring that the saturation solubility limit is not exceeded.

Dissolution will cease if and when:

(a) The deposit is all dissolved.

(b) The remainder of the deposit is insoluble.

(c) The reagent is exhausted.

In practice, (a) is the objective and in power plant boiler cleaning a period of about 6 to 8 hours is normally adequate for the removal of oxide formed in service. However, if abnormally thick oxides are allowed to develop, dissolution may be hindered as in (b).

Hydrofluoric acid has been used throughout the world for chemical cleaning for some years and was accepted by the CEGB as a third option for iron oxide removal from boilers along with hydrochloric acid and citric acid based formulations. Its immediate advantage is that of capacity: 1% hydrofluoric acid is equivalent to 3% citric. So in certain unusual postoperational cleans, one fill of 2% hydrofluoric acid could be used instead of two fills of 3% citric acid or one fill of 5% hydrochloric acid. Hydrofluoric acid has some complexing ability as well as a straight acidic action, and also has the advantage over citric acid of being effective at 65°C rather than 95°C. It is not particularly good as a solvent for ferric oxide but will strongly complex the ferric ion once it is in solution. Its main drawback is one of safety because of its toxicity, but given proper precautions, experience has shown that it can be used effectively.

2.2.2 Chelating agents

Chelating agents have been used abroad for the prevention and removal of deposits in power plant, perhaps

TABLE 4.1

Dissolution of various forms of silica in typical stages of chemical cleaning procedures

Cleaning agents	Test/ duration hours	Silica, mg SiO_2/kg						
		Scale from LP boiler *	PF ash	Lagging	Used lagging	Sand	Amorphous	Quartz
5% HCl acid	6	52	222	215	215	28	48	3.5
5% HCl acid plus 1% fluoride	6	505	820	630	725	130	390	525
3% Ammoniated citric acid (pH 3.5)	6	90	210	260	210	30	100	Not detected
3% Ammoniated acid +1% fluoride	6	87	80	100	68	153	37	95
Copper strip solution	6	3.8	17.5	6.3	11.5	1.7	53	13.8
Alkali boil-out solution, i.e., 1000 PPM trisodium phosphate at 41.5 bar and 250°C	24	280	120	350	•	594	400	185

* This scale was a mixture of serpentine ($3MgO\ 2SiO_2, 2H_2O$) and apatite ($[Ca_3\ (PO_4)_2]_3\ Ca(OH)_2$)

• Determination omitted in view of similarity of other pairs of results for 'lagging' and 'used lagging'.

the most extensively applied being ethylene diamine tetra-acetic acid (EDTA). However, they have been employed only rarely for large operations in the CEGB.

EDTA has been used in acidic conditions, e.g., with citric and oxalic in condenser cleaning, or in similar mixtures for the off-load cleaning of boilers (USSR) or reactors (USA and UK). As EDTA is a much stronger chelating agent than any dicarboxylic acid, it may be that its main role is to impart to the mixture the optimum pH for dissolution.

EDTA was initially used as a cleaning agent as the tetra-sodium salt at 10 bar for 24 hours. Amongst the claimed advantages were safety of handling compared with acids and minimal corrosion, although an inhibitor is needed. It is not particularly cheap, however, and the justification for its use may depend on any reduction in outage costs, which may be achieved compared to conventional acid cleaning, since fewer stages are needed.

EDTA has been used for on and off-load cleaning, as either the sodium or ammonium salt. Whilst, as would be expected from its historic use as a chelant for calcium and magnesium, it is suitable for cleaning evaporators and condensers of hardness salts; it is also useful for removing iron oxide deposits. Its performance confirms theoretical predictions that both Fe_3O_4 and Fe_2O_3 are soluble over a wide pH range.

On-load usage of EDTA for oxide and scale control in the USA has been at up to pH 10 and 300°C with concentrations of up to 10 mg/kg being used. With concentrations of a few hundred mg/kg fed directly to the boiler drum rather than to the feed, and without adequate de-oxygenation, corrosion has occurred. The trisodium salt of nitrilo triacetic acid (NTA) has also been used as an on-load cleaning agent.

A comparative study showed NTA to be better than EDTA for preventing hardness scale up to 80 bar, but less so at 100 bar.

NTA is cheaper than EDTA, but the calcium salt has a retrograde solubility.

The use of EDTA and analogues both on and off-load is constrained by the thermal decomposition of both the parent substances, and that of their complexes with metallic ions. The extent and nature of degradation appears to depend upon, at least, both temperature and the presence or absence of dissolved oxygen. Decomposition is probably step-wise, some of the products being chelants themselves. Above about 280°C, magnetite is precipitated from solutions of alkaline EDTA. This effect has been proposed as a means of increasing the corrosion resistance of steel, as the oxide layer so produced is more compact than that formed by growth.

As noted above, thermodynamically both magnetite and haematite are soluble in alkaline, as well as acidic, EDTA solutions. There are, however, kinetic limitations and to obtain realistic cleaning rates the process temperature has to be increased to 150°C. This makes it convenient to apply to a boiler coming

off load. Under these conditions, however, the corrosion rate is unacceptably high so that the addition of inhibitors specially formulated for use in alkaline solutions is necessary.

A commercial version of alkaline EDTA cleaning is the alkaline copper removal (ACR) process, formerly termed Vertan 675 (USA) or VT 675 (UK). (The name is for patent reasons. The dissolution of iron oxides by these means is not patentable, but that of copper is.)

One of the claimed advantages of the process is that if it is applied soon after a boiler comes off load, the still-hot boiler water can be used for the process solution with a consequent saving in time.

In this process, both iron oxides and copper and its oxides are claimed to be removed from the boiler. It is a 'one-fill' process utilising a solution, typically, of about 5% of the tetra-ammonium salt of EDTA. This is the only chemical added, and the entire operation is carried out within the alkaline pH range, although the procedure is carried out in two stages.

First, the iron oxide removal phase is carried out. The $(NH_4)_4$ EDTA plus an inhibitor are introduced into the boiler and the boiler fired to attain about 150°C, and thereafter at intervals to induce circulation (unless the boiler is of the forced circulation type in which case a circulation pump may be used). Circulation of the reagent dissolves iron and copper. The iron forms stable soluble chelates but the copper, just as in conventional cleaning, plates-out on the steel surfaces within an hour.

The first stage is continued until a stable iron plateau is attained. The process solution is then cooled at 100°C and air blown through. This reacts with the ferrous chelate to form ferric chelate. In turn, the ferric chelate oxidises the 'plated-out' copper which consequently is then claimed to dissolve to form a chelate solution, termination of this second stage being indicated by the achievement of a copper plateau.

During copper removal, the steel surface becomes passivated in the prevailing alkaline oxidising conditions. In consequence, after completion of the ACR process, a separate (and time consuming) passivation stage should not be required. The boiler is merely flushed with hot deionised water before being made available for service.

Whilst used abroad, this type of process has not yet achieved widespread use in the UK. Once accepted, the case for its use would depend on the economics of its application, i.e., both process cost and outage time, against alternative processes. In principle, however, if a true 'one-fill' process could be achieved, it would allow the clean to be completed within a weekend and so, if desired, outside the critical (survey) path.

2.2.3 Chelants for possible continuous on-load future cleaning

In addition to the use of EDTA and substances having analogous properties, quite different chelants have been considered for application to power plant circuits with the aim of continuous on-load cleaning of once-through boilers. Ideally, chelants for this purpose would need to be steam-volatile, as would the uncomplexed reagent and any degradation or reaction product, and they would also need to be stable under all conditions encountered in the steam/water circuit. Materials of possible application are oxine (8-hydroxyquinoline) and catechol-(1,2 benzene diol). The complexes must also be so thermodynamically weak at low temperatures that they can be removed by the ion-exchange resin and recycled, otherwise the economics of their use are very unfavourable. As yet, there has been no large scale application of this method.

2.2.4 Nickel removal

In the waterside deposits of some boilers, the third most abundant metal, after iron and copper, is usually nickel. Although in some cases there is more nickel than copper, for example, where the HP feedheaters are monel rather than cupro-nickel. Nickel arises from the presence of alloys in the steam/water circuit and its removal during cleaning is of interest. Apart from nuclear applications, a specific stage for this purpose is not employed.

Investigations (Kingerley *et al.*) showed that nickel metal is removed by either hydrochloric or citric acids, but nickel oxides are only removed by hydrochloric acid. This is apart from any incidental *mechanical* removal as a result of dissolution of more-soluble components of the deposit matrix. Nickel thus appears in the iron removal stage and is not taken up by the 'copper strip' procedure.

2.2.5 Silica removal

Silica may be encountered in various forms as a result of deposits or from ingress of lagging or pulverised fuel ash (PFA). Its removal is important in view of the operational requirement to limit silica concentrations in boiler water, minimising its carryover and deposition on turbine blades and the consequential penalty of long periods of operation at reduced pressure. Indeed, the need to reduce the time for a unit to achieve full load after cleaning was one of the reasons for the introduction of chemical cleaning.

Ammonium bifluoride (ABF) was added to acids because of its claimed beneficial effect of silica removal. However, fluorides in acid solution also improve the removal of iron oxides.

The results of a study by Brown and Kingerley of the effect on the dissolution of a number of forms of silica, mainly in the context of pre-commissioning cleaning, are summarised in Table 4.1.

2.2.6 Stoichiometry

Iron, whether in the ferrous or ferric states, forms predominantly 1:1 complexes with citric acid. Variations of the predominant form of citrate ion with pH are shown in Fig 4.1. The use of mono-ammonium citrate, rather than the hydrogen form, is preferred to avoid the precipitation of an insoluble complex. For the same reason an excess of 0.5% of acid over the total iron species present should be maintained.

For complex formation of iron with both citrate and EDTA in alkaline solution, the general equation is:

$$Fe_3O_4 + Fe + 8H + 4\ chelant \rightarrow 4Fe\,(chelate) + 4H_2O$$

Complexes of ratios other than 1:1 may be formed under certain conditions, e.g., where one reactant is greatly in excess of the other, such as the dicitrate complex with copper.

2.2.7 Interaction of parameters

During cleaning processes, in addition to the main reactions of oxide dissolution, metal dissolution, and inhibition, certain interactions can occur which must be borne in mind.

The simplest of these is the effect of temperature. This normally follows Arrhenius, i.e., an exponential factor of $\exp(-E/RT)$ applies. The exponent may be greater for the metal-acid reaction than for oxide dissolution, and so determine the process temperature. Descaling of iron with sulphuric acid is an example of this compromise.

Flow is also an important parameter. Whilst some flow is required to avoid reagent depletion at the surface, if the rate is too high it adversely affects inhibitor efficiency and can lead to severe corrosion. For example, ferric ion corrosion is aggravated by high flow rate.

Finally, concentrations of acid and inhibitor are important parameters. The stronger the acid (or conversely the weaker the inhibitor) the greater will be the dissolution of metal and the extent of corrosion.

2.2.8 Mechanism of oxide removal and descaling

In removing hardness scale with acids, direct chemical dissolution applies, e.g., $CaCO_3 + 2HCl \rightarrow CaCl_2 + H_2O + CO_2$.

Similarly, in the case of mill scale on mild steel, FeO, this is so readily soluble that, again, simple chemical dissolution applies.

$$FeO + 2HCl \rightarrow FeCl_2 + H_2O$$

This may govern in-works or pre-service cleaning. The oxide layer in mill scale, successively from the metal outwards, may consist of wustite (FeO), magnetite (Fe_3O_4) and haematite (Fe_2O_3). Of these FeO is readily soluble in acids and attack along this interface may therefore detach the outer layers in solid form.

With the works cleaning of alloy steels the oxide layer is soluble in the usual cleaning acids, although with some difficulty, and oxidation to form more soluble species (e.g., chromic from chromous) is often required. For this purpose, nitric acid is used, fulfilling both requirements. It also passivates the substrate metal. A widely used formulation is a mixture of nitric and hydrofluoric acids, which holds the oxidised species in solution. In some cases, pre-treatment in alkaline permanganate is applied.

In the post-service cleaning of mild and low alloy steels, e.g., boilers and economisers, a different mechanism applies. The oxide layer is predominantly magnetite but usually with a number of other substances incorporated, principally copper and its oxides, which necessitate the introduction of a further stage of treatment. An essential difference between oxides formed in operation and those encountered in tube manufacture, is the absence in the former of wustite, which is unstable below about 580°C. In consequence, in post-service cleaning, a readily soluble inner-layer is absent. The mechanism is then one of reductive dissolution, for magnetite and for haematite:

$$Fe_3O_4 + 6H^+ + H_2 \rightarrow 3Fe^{2+} + 4H_2O$$

$$Fe_2O_3 + 4H^+ + H_2 \rightarrow 2Fe^{2+} + 3H_2O$$

Ferric ions act as cathodic reactants thereby causing loss of the substrate steel:

$$2Fe^{3+} + Fe \rightarrow 3Fe^{2+}$$

In certain instances, additional substances may be added to aid this process, e.g., formic acid, by providing an alternative reducing agent to steel.

With increasing content of alloying elements in steel, the oxide layer changes from the magnetite structure to a spinel form as the inner layer, but with an outer layer of magnetite. In certain situations, the greater thickness and solubility of the latter and the relatively thin layer of inner oxide has led to the view that partial cleaning may be sufficient, i.e., that the inner layer may be left in place.

Nevertheless, the formulations currently considered to be suitable for these materials in plant are still mainly based in the seemingly less aggressive citric acid, but with various additives to aid dissolution.

The cleaning of copper-based alloys alone merely to remove the corresponding oxides presents few problems as these are readily soluble in a range of acids. However, this simple requirement is rarely met. The most frequent need for the cleaning of copper-based alloys arises in the descaling of condensers on the water side to restore thermal efficiency following the accumulation of hardness salts and the consequential impedance of heat transfer. Almost invariably, the deposit contains enough calcium carbonate or phosphate to render it easily acid-soluble, so that the process presents few problems provided that due attention is paid to the possible need for inhibiting against corrosion of copper-based alloys, as well as steel.

A more difficult situation is encountered on the steam side of condensers. Heat transfer problems on the steam side of the tubes are usually slight but a few cases have arisen in which both oil deposits and oxides of iron, both haematite and magnetite, have been found together. The large volume of process solution needed in cleaning the steam side of a condenser of a 500 MW unit for this purpose also means that heating is not feasible, so that the process must be effective at ambient temperature. The removal of the organic matter requires treatment along the lines already discussed as a first step. The cleaning of the iron oxides from the foreign substrates depends upon their nature. If they are predominantely magnetite, then hydrochloric acid will probably be effective. If, however, they are haematite, more complicated formulations are required. A mixture of citric, oxalic and EDTA has been successfully employed on many occasions. According to the intractability and make-up of the deposit, these have been applied either in a mixture with, or as alternate stages to, a degreasing formulation.

An analogous situation is encountered in cleaning feedheaters. Again, whilst the inorganic component of the deposit is readily soluble in mineral acids any organic component may have to be removed first if the acidic stage is to be successful. Sequential or even reiterative treatment is therefore needed to achieve a satisfactory performance.

The descaling of auxiliary equipment, e.g., minor coolers, heat exchangers and evaporators is usually fairly straightforward, in principle. The deposits encountered, apart from the suspended matter deposited from solution, are mainly sufficiently soluble in common acids (e.g., hydrochloric) to present no undue problems, as are the relatively thin layers of oxidation product of the substrate metal.

2.2.9 Decontamination of nuclear plant

In the nuclear field, the principal reason for cleaning so far has been for safety rather than to restore efficiency or to pre-empt corrosion, although there is continuing discussion on the possible eventual need for the latter. Decontamination is the use of chemical means for the removal of out-of-core activity from nuclear reactor circuits.

Initially, decontaminations were carried out with acids or mixtures of acid, a common constituent being oxalic acid which was found to be advantageous for the alloy steel as the associated inhibited corrosion rates are very low.

Thus, mixtures of citric acid and oxalic acid have been used in the 'Citrox' process, inhibited with phenylthiourea, for decontamination of the stainless steel in pressurised water reactor circuits.

On the other hand, it may be noted that 'APHC' and 'APACE', two commonly used decontaminants, corroded mild steel at up to 20 μm/h without an inhibitor; inhibitors decreased this by 80%. Clearly, the amount of corrosion with some of these processes is undesirably high and this has led to a reconsideration of the fundamental nature of oxide dissolution processes.

Thermodynamically, taking as an example the reductive dissolution of magnetite as already shown:

$$Fe_3O_4 + 6H^+ + H_2 \rightarrow 3Fe^{2+} + 4H_2O$$

Then, dissolution is clearly aided by:

- High Fe^{2+} solubility which can be increased by addition of complexing agents.

- Reducing conditions.

- An appropriate pH.

Kinetically, whilst in conventional cleaning reliance was placed on oxide dissolution (which usually includes some reductive dissolution) in decontamination considerations, increased emphasis is placed on reduction of metal ions actually in the lattice:

$$M^{n+} + Ze \rightarrow M^{(n-Z)+}$$

to impair the stability of the oxide lattice.

The outcome of this approach is the use of the so-called LOMI reagents exemplified by V^{2+} tripicolinate and CO^{3+} trispyridyl.

Oxide dissolution rates with this type of reagent can be two to three orders of magnitude greater than those in citric or hydrochloric acid formulations.

More importantly, in the present context, corrosion rates are less than those found with conventional formulations (see Table 4.2). These formulations are very expensive to prepare and have only about 5% of the capacity of conventional cleaning formulations and their use is restricted to nuclear decontamination.

2.3 Inhibitors

The formulations employed for dissolving iron oxides and other deposits from the internal surface of power plant may also attack the substrate metal itself. Precautions therefore have to be taken to restrain this undesirable side effect by the use of inhibitors, i.e., substances which, when added to cleaning solutions, decrease the attack upon the metal.

In this section, the following aspects of inhibition are described:

- Types of substance with inhibitive properties.

- Testing of inhibitors.

- Factors affecting performance and typical results obtained.

2.3.1 Types of substance for inhibition in acid solution

Many substances have been found to have inhibitive effects on metallic corrosion in acid solution. The substances range from halide ions to high molecular weight materials such as starch, and in physical form from solids to gases. Organic compounds are useful, particularly those with inhibitive groupings containing Groups 5 and 6 elements of the periodic Table (N, P, As, O and S). The organic group may consist of or contain alkyl and/or aryl radicals, and saturated or unsaturated groups. Inhibitive properties are found empirically to vary along an homologous series and according to whether the alkyl chain, for example, is straight or branched.

Acetylene, if bubbled through an acidic solution, induces some degree of inhibition of steel. The related propargyl alcohol is contained in a number of commercial inhibitors.

The onium class of compounds (i.e., the substances formed by the protonation of certain compounds R_2As,

TABLE 4.2

Estimates of general corrosion during chemical cleaning with commercial inhibitors

Material	Cleaning process	Corrosion mg m^{-2}s^{-1}	μm/8 hrs equivalent	% inhibition
Mild steel	3% Ammoniated citric acid pH 3.5, inhibited, 90°C	0.66	2.5	99
1 Cr$\frac{1}{2}$Mo	3% Ammoniated citric acid pH 3.5, inhibited, 90°C	2.0	8.0	95
2 $\frac{1}{4}$Cr1Mo	3% Ammoniated citric acid pH 3.5, inhibited, 90°C	0.3–3	1–10	87–98
9 Cr1Mo	3% Ammoniated citric acid pH 3.5, inhibited, 90°C	1.0	4.0	95
12 CrMoV	3% Ammoniated citric acid pH 3.5, inhibited, 90°C	0.5	1.7	95
Austenitic 316	3% Ammoniated citric acid pH 3.5, inhibited, 90°C	< 0.1	0.36	99
Mild steel	5% Hydrochloric acid, inhibited 75°C	4	14.2	90
	1.5% Sulphuric acid, inhibited 40°C	0.3	1	85
	10% Na$_4$ EDTA, inhibited 150°C	0.3	1	90
Brass (70–30)	1.5% Hydrochloric acid, inhibited 25°C	10^{-3}	Not detectable	99
	1.5% Sulphuric acid, inhibited 40°C	3×10^{-3}	Not detectable	67
410	V^{2+}/picolinate/formate	1	4	Inhibitor
321	V^{2+}/picolinate/formate	10^{-3}	0.04	absent

R$_3$P, R$_3$N, R$_2$SO, (RNH)$_2$CS) include many with inhibitive properties.

In some cases, in addition to the intrinsic inhibitive property to a given substance, the product of its reaction or degradation also has some powers of inhibition. Thus, arsine from arsonium compounds and the polymerisation products of acetylene fall into this category.

2.3.2 Mixtures/proprietaries

For some purposes it is sufficient to use single substances as inhibitors. Where these are readily soluble, they can be supplied as the solid or liquid, as appropriate. Other inhibitors are not readily soluble and need to be formulated with surface active agents to disperse them in colloidal form. A number of commercial inhibitors fall into this category, possibly with further inhibitors to increase efficiency either additively or synergistically, together with emulsifying agents and a holding solvent.

Care in use must be exercised to avoid breaking the emulsion and so losing inhibitor performance, taking into account the effects of temperature, pH, solubility, etc.

Manufacturers' claims regarding performance should be taken only as a guide. If possible, tests should be conducted which closely simulate conditions required in service.

2.3.3 Inhibitor performance

Test procedures and results

The inhibitors approved by the CEGB for use in chemical cleaning are largely understood from long experience. However, occasions may arise when their use in unfamiliar circumstances is contemplated and prior to such application proving tests may be considered necessary.

In testing, the choice of substrate samples, their surface preparation, superficial dimensions, and the number of replicates are of importance. The material must be free from welds and stress, and be in the correct condition of heat treatment (unless these are variables for assessment).

A standard surface finish should be adopted, and samples should also be degreased before testing.

Damage appraisal falls into two categories: general corrosion and pitting corrosion. General corrosion is assessed by weight loss, either direct or indirect, e.g., metal in solution, and is used to calculate inhibitor efficiency (E) by the formula:

$$E = (U - I/U)\ 100$$

where U = weight loss without inhibitor

I = weight loss with inhibitor

Pitting, or localised corrosion, may be assessed in a variety of ways, e.g., by metallurgical sectioning, by direct counting — in some cases microscopically, by the microscopic measurement of pit depths, or by profile measurement.

Clearly, pitting can be expressed as an equivalent amount of thinning or penetration. Test duration is important — the apparent kinetics over a short period may give a misleading impression of the course of events over a longer period. Aeration, temperature and flow are important and must be closely controlled. The volume of test solution in relation to the surface area of the metal can also affect the result obtained, as can the composition of the solution, which may alter during the test. Heat flux also requires consideration as temperatures above a recommended level can, in extreme cases, modify adversely the effect of inhibitors.

For many purposes, inhibitors can be adequately assessed in the beaker tests subject to the above provisions, metal coupons being immersed in an appropriate solution for a suitable time and the general and pitting corrosion measured and calculated. By such means, a number of inhibitors can be screened quickly, or one inhibitor assessed under a range of conditions.

Following screening, more sophisticated tests may be desirable on the most promising formulation to simulate the system of interest. Where, for example, cleaning of pipework is the ultimate aim, rig tests in which a given formulation is circulated through a tube which constitutes the test coupon are more suitable. In this way, the vital factor of flow can be more closely controlled and a number of other parameters more satisfactorily monitored (e.g., surface area/volume).

2.3.4 Effect of fundamental parameters

A number of fundamental parameters influence the performance of inhibitors, e.g., temperature, acid concentration, inhibitor concentration and flow.

Temperature

With and without inhibitor, the corrosion rate of most metals in acidic solution increases with temperature according to Arrhenius.

The relative corrosion rates with and without inhibitor are such that inhibitor efficiency [E = (U − I/U) 100] increases with temperature, but the corrosion rate, which is the more important factor with respect to metal loss, also increases.

Inhibitor concentration

Increase in inhibitor concentration usually leads to an increase in inhibition but not linearly. Investigations show that in some cases at very low concentrations prior to the onset of inhibition, corrosion may be stimulated by the addition of an inhibitor.

The increase of inhibition with increase of inhibitor concentration may in some cases be interpreted quantitatively in terms of an adsorption isotherm, e.g., that of Langmuir.

Flow

Increase of flow rate may impair the efficiency of the inhibitor.

2.4 Copper removal

In the post-service chemical cleaning of boilers, there is the need to remove copper deposits, in addition to iron oxides. These arise from the transportation into the boiler of the corrosion products of the feed system, where the condenser and the low pressure heaters on most plants are often made from either brass or cupro-nickel. Any corrosion products, both from on- and off-load corrosion, may be subsequently transported to the boiler and so incorporated into the magnetite layer. They may be present as the metal and/or its oxides, and possibly in discrete layers.

Their presence can lead to a number of problems. They may participate in corrosion processes; interfere with the welding of tube inserts; be transported into the turbine with the eventual need to clean that part of the plant; or they may lead to overheating failures, after chemical cleaning, if not completely removed. During iron removal, copper is taken into solution and then quickly plates-out galvanically. The 'plating' is of poor quality but, if the subsequent 'copper stripping' stage is incomplete, some will remain. In subsequent operation, the plated layer of copper may become partly detached and leave an annulus between the tube and the copper. Steam fills this annulus and owing to its low thermal conductivity causes overheating of the boiler tube.

For all of these reasons, it is desirable to remove the copper deposits periodically.

The standard formulation for this purpose is 1% citric acid, plus ammonia to pH 9.5, plus an oxidant. Copper oxides are soluble in ammonia (to form the cupric or cuprous ammonium complexes), but copper metal is not and so requires prior oxidation to convert it to the oxide. A variety of oxidants has been used: sodium bromate, nitrite, persulphate, hydrogen peroxide, perforate, and even air. All of these are effective in varying degrees, but bromate is the most effective. It is conventionally employed at 0.5%.

Depending upon the copper content of the deposit, either one or two stages of copper removal are applied.

When the copper content of the base oxide exceeds 5%, two copper 'strip' stages are deemed necessary and they are applied in the *sandwich process* — one before and one after the iron removal stage. The first stage facilitates access of acid, to be used later, to the iron oxide which could otherwise be partially masked and the effectiveness of the stage compromised. When the copper content of the base oxide is less than 5%, only one copper strip is needed and it is usually convenient to carry out this after iron removal. This is because the introduction of 50% of the citric acid takes up any residual iron oxide which has either escaped the iron strip or results from after-rusting, and also

because the ammoniacal copper strip solution has limited passivating properties. Thus, on visual inspection, after the second copper strip, the internal surfaces look better than when no copper strip has been applied.

The capacity of the standard strip solution for copper is about 6.35 g/l. This is in line with the stoichiometry for the reaction:

$$3Cu + NaBrO_3 \rightarrow 3CuO + NaBr$$

In practice, very few copper strip solutions are required to take up more than 2 g/kg, and the majority are less than 1 g/kg. An estimate of the amount likely to be removed can be made from prior analysis of tube samples provided that the samples are representative of the whole boiler.

During copper removal, no inhibitor is required as the solution used takes up very small amounts of iron. In fact, the solution being alkaline has some passivating effect, and the presence of oxidants increases this tendency, owing to the formation of ferric ions which are less soluble than ferrous.

Effluent disposal of copper strip solutions is usually more important than that of iron.

2.5 Passivation

Passivation is an anti-corrosion measure carried out after chemical cleaning so that, in the interval before operation, or, in the case of post-service cleaning, before commissioning, the system will not suffer corrosion and undo the good done by the cleaning operation. There are various passivation procedures. Alkalis, such as sodium hydroxide or ammonia, feature in a number of processes. Again, phosphates of the alkali metals, which provide a controlled alkalinity, are often used. In addition substances such as sodium nitrite may be employed. Typically, these formulations are applied at some temperature above ambient, at about 60°C (or about 250°C for pressure passivation which simulates boiler operating conditions) for periods of about four to more than 24 hours.

These processes all have in common the production of a thin layer of an oxide of the substrate metal, predominantly iron in the present context, with the object of restraining further corrosion. The degree of success achieved depends upon a number of factors.

A layer of oxide on a substrate metal will offer some protection against corrosion provided that it is continuous, adherent and impervious. Clearly, it also needs to be insoluble in the environment. It may happen that a passivation procedure is carried out and the solution then drained. This will usually result in either 'wet atmospheric' conditions in the plant, or some localised immersion. In either case, it is quite likely that the oxide, although passive when formed, is not resistant to the changed conditions.

As with most chemical processes, passivation is accelerated by temperature and it has been shown to be of particular advantage to treat steel at 250°C. In this case, the alkalinity is provided by ammonia; an oxygen scavenger, hydrazine, is also added to react with any oxygen present in the air space above the liquid at the start of the process. There is little scope for hydrazine to participate to any useful extent during the process itself, as it then undergoes thermal decomposition. Above 250°C, the iron oxide formed during the process of passivation largely adheres to the substrate metal, and although some unattached material is also produced, the protection afforded is superior to that achievable at lower temperatures. Below 250°C, the iron oxide formed does not adhere to the metal, and affords less protection.

Above 250°C, the adherent layer is about 100 μm thick after 24 hours' treatment, with the growth apparently following parabolic kinetics, whilst the outer layer is far thicker, at about 0.5 mm, but porous. Increasing the duration of treatment to 36 hours, or the pressure to 66 bar, or the chemical concentrations, are not of practical benefit in view of the limited improvement obtained.

Once formed, the film produced by high temperature passivation is very resistant to corrosion for some weeks in moist atmospheric conditions. In subsequent operation, the passive layer is forced off, as newly grown material forms at the metal/oxide interface below the passive film.

Where pressure cannot be raised for the purposes of passivation, other (mostly less-effective) means must be employed. Hydrazine/ammonia treatments at atmospheric pressure are and have been used, and whilst they are effective so long as the metal is immersed in them, corrosion soon initiates after drainage, i.e., they are really storage solutions.

It will be apparent that on some occasions during a cleaning procedure interruptions may arise which impose the need for the plant to be temporarily protected against corrosion. The logic diagram for such situations is shown in Fig 4.3.

As noted in Section 2.4 of this chapter, the standard copper strip solution of citric acid/ammonia/oxidant has some passivating properties and this, or a formulation of comparable effect (e.g., incorporating alternative oxidising agents such as hydrogen peroxide), is applied in some cases. Most of the foregoing discussion applies to mild steel; low alloy steels, e.g., CrMo, rust more easily and are subject to intergranular attack in nitrite (a *dangerous inhibitor*). Stainless steels, provided they are not in reducing conditions, are self-passivating. Non-ferrous metals are less prone to corrosion and a separate passivating stage is rarely necessary for them after cleaning.

2.6 Storage

At a number of stages in its history, the components which make up a complete plant may have to be stored. During storage, there are requirements to prevent the ingress of adventitious impurities and also to provide protection against corrosion. Usually, this means atmospheric corrosion, although in some cases immersed corrosion also has to be guarded against. Such precautions are important as the extent of pitting corrosion of idle, inadequately protected plant can reach 1.5 mm/year and this can result in the need to replace the item so damaged.

The basic corrosion reactions, in the case of steel, are the anodic reaction:

$$Fe \rightarrow Fe^{2+} + 2e$$

balanced predominantly by the cathodic reactions:

$$H^+ + e \rightarrow \tfrac{1}{2}H_2 \quad \text{and/or}$$
$$\tfrac{1}{2}O_2 + H_2O \rightarrow 2OH^-$$

However, other cathodic reactions, e.g., copper deposition or ferric ion reduction, may participate. The hydrogen ion reduction reaction is increasingly severe with decreasing pH (increasing acidity), particularly below pH 3, whilst oxygen reduction is increased by aerated conditions. Both of these undesirable features are met in wet atmospheric conditions which may exist in unprotected plant during inadequate storage. In addition, atmospheric contaminants such as carbon dioxide and sulphur dioxide will dissolve in any moisture present on the surface and lower the pH with a consequent increase in the rate of corrosion.

Essentially, therefore, the basic methods for the protection of plant during storage, depend upon either excluding moisture, or adjusting the pH and oxygen concentrations to acceptable levels.

Regarding the first option, the complete exclusion of water as vapour is not necessary — if the relative humidity is kept below 50% at ambient temperature (more usefully expressed as 30% at 0°C), atmospheric corrosion is minimal. This condition can be achieved either by drying, e.g., with currents of warm and/or dry air, or, in the case of plant internals, by introducing vapour phase inhibitors (VPI). These materials are sufficiently volatile to diffuse and become attached to the metal surface and so inhibit corrosion. They are mainly used for the protection of tubes, etc., before fixing in the boiler.

In many cases, however, dry storage is not a feasible option, and the environment then has to be modified to control the pH and dissolved oxygen content. Bearing in mind the need to restrict the amounts of non-volatile solid materials in such power plant to avoid the need for draining and flushing during commissioning, pH adjustment is mainly done by means of ammonia, and dissolved oxygen control by hydrazine.

At the end of the storage period, it is necessary to ensure that the concentration of storage chemicals does not exceed the permitted concentrations for plant operation and it may be necessary to dump the storage solution and refill before operation.

During storage, adequate concentrations of reagents must be maintained at the surfaces to be protected and this requires that the solution be tested, re-dosed if necessary and the solution periodically agitated to ensure proper mixing.

Another solution used in certain storage applications is borax/nitrite. This system was applied primarily to economiser storage, but has some application in other areas such as boiler tubes. The formulation is used at concentrations of 200 mg/kg of each salt provided the steel surfaces have an oxide layer, or 1000 mg/kg if the surfaces are new or unoxidised. Again the solution is drained before operation. Nitrogen capping is sometimes also used as a means of restricting oxygen ingress, either in conjunction with wet storage or alone.

Finally, it should be noted that plant surfaces also require protection from corrosion during hydraulic testing. A suitable solution for this purpose is deionised water with ammonia added to give a pH of 9.3 to 9.4 if the plant is to be put into service immediately.

If not, a storage solution of borax/nitrite as described should be used for the hydraulic test and then left in the plant, until replaced by a solution suitable for operation.

After drainage, care should be taken that harmful conditions do not arise in the interval before plant operation. With good planning the interval may be very short, but otherwise, appropriate storage measures should be applied.

A distinction must be drawn in some cases between storage procedures for ferrous and non-ferrous alloys.

Ferrous systems can be treated with 50 mg/kg hydrazine at pH 10, given by ammonia, in deionised water (of which the chloride content must be less than 2 mg/kg as NaCl). If copper alloys are also present, the pH must not exceed 10.7. Copper alloys should preferably be stored dry. If wet storage is unavoidable, two options exist: either demineralised water with 50 mg/kg of hydrazine as N_2H_4 or nitrite/borax.

2.7 Cleaning by non-chemical methods

Non-chemical methods of cleaning are broadly as follows:

- Grit and shot-blasting.
- Water and aqua-jetting at various pressures.
- Grinding.
- Scatter-scaling.
- Fettling.
- Flushing and steam purging.
- Rodding of condenser tubes.
- Cleaning of turbine blades.
- Bulleting of condensers.
- Grit blasting of condensers.
- Water jetting of condensers and feedheaters.

Whilst these methods clearly fall largely outside the scope of chemistry, they are mentioned briefly because, as with any wet technique, there is always some risk of corrosion and in some instances, e.g., wet sand blasting, an anti-corrosion inhibitor is added.

3 Cleaning practice — general considerations

3.1 Managerial aspects

3.1.1 Logistics

Chemical cleaning operations require the commitment of considerable resources and these must be correctly assessed ahead of the event if the success of the undertaking is not to be jeopardised.

Experience has demonstrated the need for full consultation between the main plant contractors, cleaning contractors, design services, scientific services, headquarters and station staffs before pre-service cleaning: and between cleaning contractors, station staff and scientific services, prior to post-service cleaning. For example, in pre-service cleaning it will be necessary to decide whether or not clean conditions working have been effectively adhered to. If not, a prolonged, as opposed to a brief, clean of the erected plant may be needed. Such a decision will probably require the involvement of scientific staff to carry out the requisite investigational work. Also in post-service cleaning, e.g., of a boiler or a condenser, the decision as to whether a clean is required or not will depend on the condition of tubes. Ideally, samples should be removed from the plant for laboratory tests well ahead of the outage to facilitate planning.

The time from setting up to completion of the cleaning operation may be provisionally estimated at 100–200 hours. This is a significant proportion of a major survey time, which is typically scheduled at 40 days, and clearly the decision to clean or not must be made at the earliest possible date.

The major logistical considerations are as follows.

Water requirements demand careful programming. Large quantities of both softened towns water and deionised water will be required. For pre-service cleaning, the prior commissioning of water treatment plant, including the provision of adequate storage facilities, is essential for adequate supplies of treated water. In post-service applications, the complete procedure may call for up to 12 boiler capacities of deionised water. In addition, the rate of supply of suitable quality water at the critical stages needs assessing, bearing in mind the possible demands of operation of the other units on the station.

Chemicals in sufficient quantity for the expected cleaning procedure must clearly be available on site. In this context, the occasional requirement to repeat

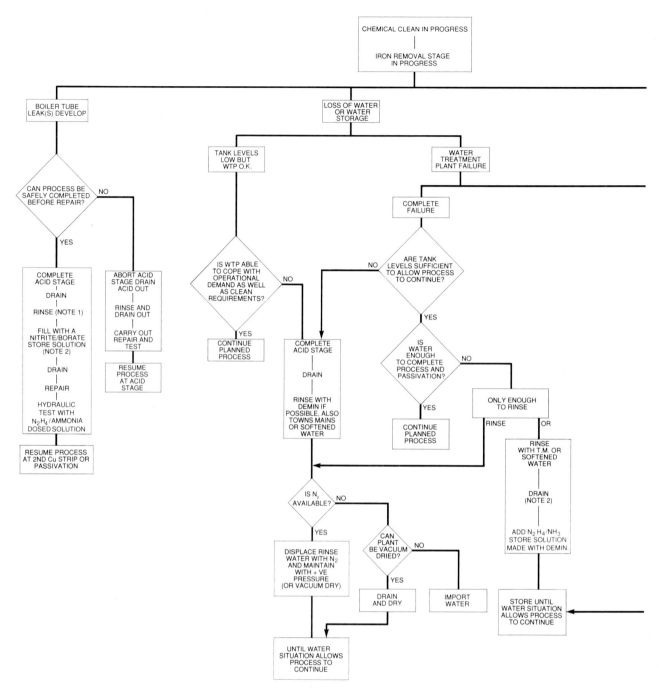

FIG. 4.3 Logic diagram for boiler storage where passivation has not been carried out (post-service cleaning)

a stage — not always a known factor before the event — must also be acknowledged, and appropriate provision made. Such situations can be catered for by having available on site enough acid and inhibitor for an additional or repeat iron oxide removal stage. The importance of this precaution cannot be overemphasised. The quantities of chemicals held at the depots of the cleaning contractors may be minimal, for reasons of economy. Consequently, if supplies run out at a station during a clean, more has to be obtained from the *manufacturers*, and the time taken to arrange delivery can cause disruptions to the planned outage at the station and affect the actual result of the clean itself.

Effluent disposal is controlled by statutory legislation, and it is important that the provisions of such legislation are observed. Where the option of tankering-away is selected, care should be taken to ensure that the tankerage is not only of the correct capacity but that it is available as required.

Given that appropriate quantities of water and chemicals are available on site it is also necessary to heat and circulate the cleaning formulations. If a package boiler has to be used for this purpose it should be delivered to site and demonstrated to be operational at least a week ahead of the target date for its required

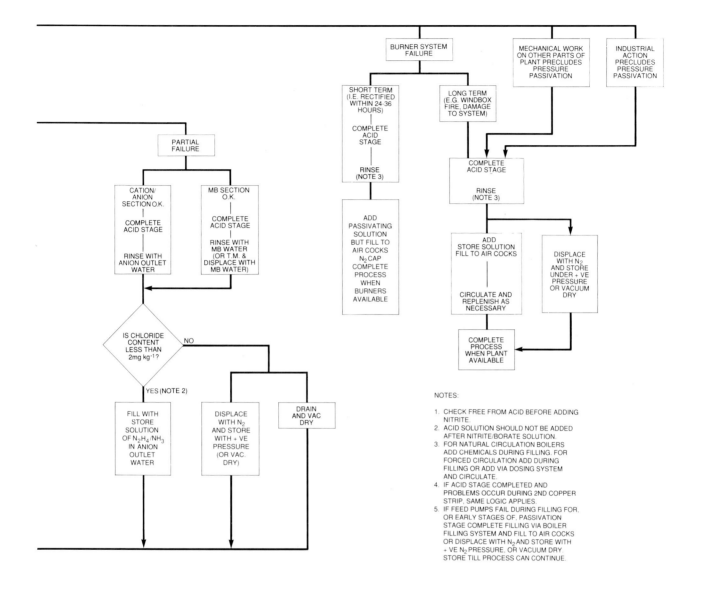

FIG. 4.3 (cont'd) Logic diagram for boiler storage where passivation has not been carried out (post-service cleaning)

use. Also pumps, whether electric or diesel, should be operational and compatible with local supplies and not have to rely on station maintenance effort, already committed elsewhere.

Where the motive power for circulation is gas — either air or nitrogen 'auto-circulation' — sufficient and reliable supplies should be provided. Finally, the appropriate manning levels must be established and arranged. They will usually be apportioned between the station staff and the contractor, but at certain crucial times, e.g., approaching the end of the stage when the element of decision as to whether or not to terminate a specific process is involved, the responsible

officer, or his delegated representative, should be present to ensure that the necessary quality of advice is available.

3.1.2 Effluent disposal

Chemical cleaning produces considerable quantities of spent liquors which have to be disposed of safely and with due regard for the environment. In addition to the actual cleaning solutions there will also be rinse solutions to consider.

There are three basic means employed for disposal of such liquors: tankering to a licensed disposal site, direct to ash lagoons, and incineration.

Tankering

Tankering is simple in principle, but requires careful planning in practice. Tanker capacity sufficient for the required volume has to be on site at the right time — not always easy to ensure in an activity as subject to delays as chemical cleaning. The tanker then travels to some approved and possibly remote disposal site where the liquors are dumped. The appropriate declarations of the quality of the effluent have to be made in advance of the clean. The main problem which arises with this method is the failure to have tankers available when needed, causing possible delays to the next stage of the clean.

Disposal in ash lagoons

The ash from coal-fired stations is usually disposed of in ash lagoons, being pumped there in the form of a water slurry and the water then decanted and returned to a river. As a 2000 MW power station burns, each day, some 20 000 tonnes of coal with a typical ash content of 15%, then this leads to the formation of about 3000 tonnes per day of PFA. Typically, the amount of water used is 4 million gallons per day.

In addition to their main purpose, ash lagoons provide a useful means for the occasional disposal of effluents from chemical cleans, using the inherent alkalinity of the ash to neutralise and precipitate heavy metals. So far, the water authorities have approved the discharge of such effluents into ash lagoons, provided that the final effluent conforms to the normal consent level. Otherwise, if there is any doubt about the potential quality, other means of disposal are specified, almost invariably tankering away (although in future incineration is available as an alternative).

The fates of the used liquors from chemical cleans of several boilers have been investigated, the findings show that disposal via ash lagoons is a satisfactory method — even for the effluents from copper strips which hitherto have been tankered away and have been more of a problem than effluents from iron oxide removal.

Both physical and chemical factors affect the quality of the effluents from ash lagoons in which chemical liquors have been disposed. Physical factors include dilution, flow rate, retention time (which depends upon geometry), density and temperature; chemical factors include neutralisation, precipitation, adsorption and settlement processes.

Chemically, the liquors discharged from a clean will vary according to the specific nature of the process as regards quality and in proportion to the size of the plant being cleaned as regards quantity.

The main reactions leading to improved effluent quality from an ash lagoon are:

- $FeCl_3 + 3OH^- \rightarrow Fe(OH)_3 + 3Cl^-$

- The dissociation of dilution of the various cuprammonium complexes and the precipitation of copper hydroxide.

The general formula of the complexes is:

$$Cu(NH_3)_\chi (H_2O)^{2+}_{6-\chi},$$

where χ can be 0 – 6 but the compound $Cu(NH_3)_4 (H_2O)_2$ is the most abundant.

Retention times could be increased by reducing the flow at the lagoon outlet for the duration of the clean. Studies (N. J. Ray and M. J. Jenkins, 1979) have shown that an ash lagoon with a retention time of 14 hours and a capacity of 8×10^4 m^3 is adequate to deal with the hydrochloric acid effluent from a 120 MW boiler.

A lagoon with 6×10^4 m^3 of standing water and a retention time of 10–20 hours is adequate for the disposal of citric acid from all stages of 120 and 200 MW boiler cleans to achieve acceptable outlet concentrations. The estimated retention time for a 500 MW boiler clean with a standing volume of 10×10^4 m^3 is 20 hours.

The studies concluded that dilution is a major factor in determining water quality leaving a lagoon. The degree of dilution depends on the volume of free water (which is a function of how full the lagoon is) and the extent of dispersion of the cleaning liquor.

Dispersion depends on the geometry of the lagoon and climatic conditions. In practical terms, the liquor will be dispersed into about 50% of the lagoon's *free volume*.

Incineration

Where ash lagoons are either unavailable, too small, or too full, incineration is an option for effluent disposal. In this method, certain liquors from the cleaning process are disposed of by combustion in the furnace of an operating boiler in fairly close proximity to the plant being cleaned.

Incineration has been proven for citric acid based solutions and is clearly applicable to EDTA-type formulations, and probably to hydrocarbon waste (e.g., from degreasing operations). Other formulations such as those based on hydrochloric acid, cannot be treated by incineration as they would be too corrosive to the boiler and lead to unacceptable atmospheric emissions.

In essence, incineration is effected by pumping the spent liquor from the plant which has been cleaned, e.g., in the case of a boiler, from the bottom header, and introducing it through spray lances into the combustion chamber of an operating boiler. No adverse effects on boiler operation has been found, nor would any be expected so long as the effluent injection rate is less than 5% by weight of the boiler firing rate. This corresponds to complete disposal in about 4 hours. Disposal time can be reduced by using two boilers, if possible.

Trials in the CEGB (W. Moore, 1982) confirmed the suitability of a point of injection just above top burner level. Pipework carrying the effluent must be capable of withstanding a minimum of 1.5 times the working pressure (normally less than 4 bar) and may be of steel or chemically-resistant rubber. The spray lances so far employed are of Type 316 stainless steel, each with a precision engineered nozzle. Up to four were inserted in each boiler to achieve the required distribution.

Atmospheric emissions were found to be acceptably low, but in UK prior approval of the use of the method at any site is required from the appropriate Inspector.

In addition to the normal safety precautions, the circuitry for the incineration should be pressure tested with water prior to the operation, and operatives should wear eye protection throughout the disposal.

Economically, incineration is not much more costly than lagoon disposal and is cheaper than tankering, although the location obviously influences the comparison with tankerage. The ratio at one site was 1:4 incineration to tankering.

3.2 Engineering aspects

3.2.1 Rinsing and flushing

Rinsing and flushing are necessary operations at a number of stages in cleaning operations. Rinsing is applied between stages to remove chemicals, as well as the substances taken into solution and suspension. The water quality to be employed and the criteria for completion of the rinsing process, e.g., in terms of conductivity or pH, are specified in the appropriate Appendices.

Water flushing (Appendix A) is used before and after cleaning to remove any contaminant which has entered the plant during installation, as well as any loose corrosion products and volatile corrosion inhibitors which have been used for protection of plant during storage and erection, and both chemicals and sludge after site cleaning processes.

Failure to remove such materials could have adverse effects.

Both the quantity and quality of the water used are important. Quantity governs the flow rate which can be sustained; it is important because much of the material to be removed is insoluble in water and must therefore be taken into suspension and swept away with the flushing water. If flushing is to be effective it is essential to achieve at least the velocity of flow which a given section of plant will experience in actual operation.

Care must be taken in certain cases, e.g., DC heaters, that the plant will support the weight of flushing water.

Regarding quality, deionised water should be the final fluid used, even if towns water is used for preliminary flushing to conserve supplies of higher quality water.

For austenitic stainless steels, all flushing must be carried out with deionised water to avoid any residual contamination which might lead to stress corrosion cracking. An exception to this requirement is where

flushing is to be followed by a further stage in an overall chemical cleaning process.

Again, unless flushing precedes a further stage in chemical cleaning, the plant should then be stored (either wet or dry, depending upon circumstances).

3.2.2 Steam purging

Steam purging is usually carried out immediately after the pre-service chemical clean of a unit to ensure that no debris remains in the system which might be carried out by the steam during operation and cause damage to plant such as turbine blading or safety valves (see Appendix B). Steam purging therefore consists of blowing superheated steam through the steam pipework in such a way as to exert a greater force on any debris than would be experienced under the maximum rated operating conditions. The object is to dislodge and remove any debris, e.g., weld beads, metal swarf. The efficiency of purging is quantified in terms of the *disturbance factor*, which is defined as the ratio of the force exerted on an object by the steam purging operation to that which would be exerted in operation at the maximum steaming rate of the boiler. A convenient criterion has been found to be that which is equivalent to a maximum continuous rating of 1.6.

Pipework is usually purged in two stages:

Stage 1 — Superheater and main steam pipes (see Fig 4.4)

Stage 2 — Superheater, main steam and reheater (see Fig 4.5)

A target plate, preferably of material similar to that of a turbine blade, is placed in the pipework temporarily erected at the outlet to enable the progress of steam purging to be monitored visually.

The target plate should be about 50 mm square and 19 mm thick with a surface milled smooth or lightly ground to remove any pit marks. It should be securely mounted in the path of the discharge pipework outlet at a distance such that the velocity of the steam flow is at least 240 m/s and not less than 100 diameters (internal) of the temporary pipework from any bend in the pipework. This arrangement simulates the effect of particles striking a turbine blade.

The temporary pipework should conform to the following requirements:

- The steam conditions in the permanent pipework being purged must produce the required disturbance factor.

- The discharge of steam plus any removed debris must be located so that equipment and/or personnel are not at risk.

- The temporary pipework must be as short as possible to reduce pressure losses.

- The design, installation and inspection of the temporary pipework and supports should conform to

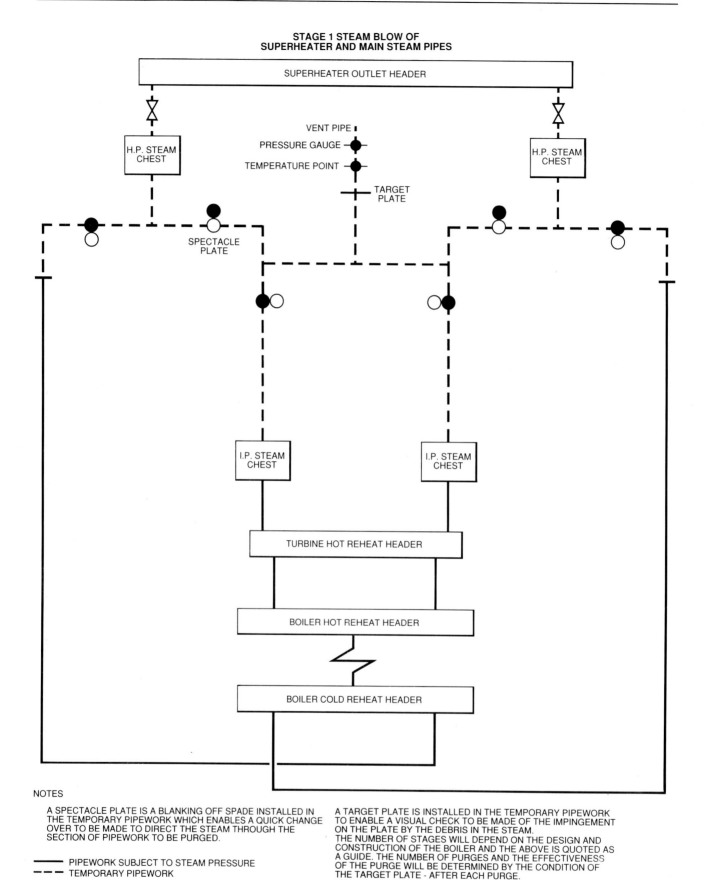

FIG. 4.4 Diagrammatic arrangement of pipework for a superheater purge

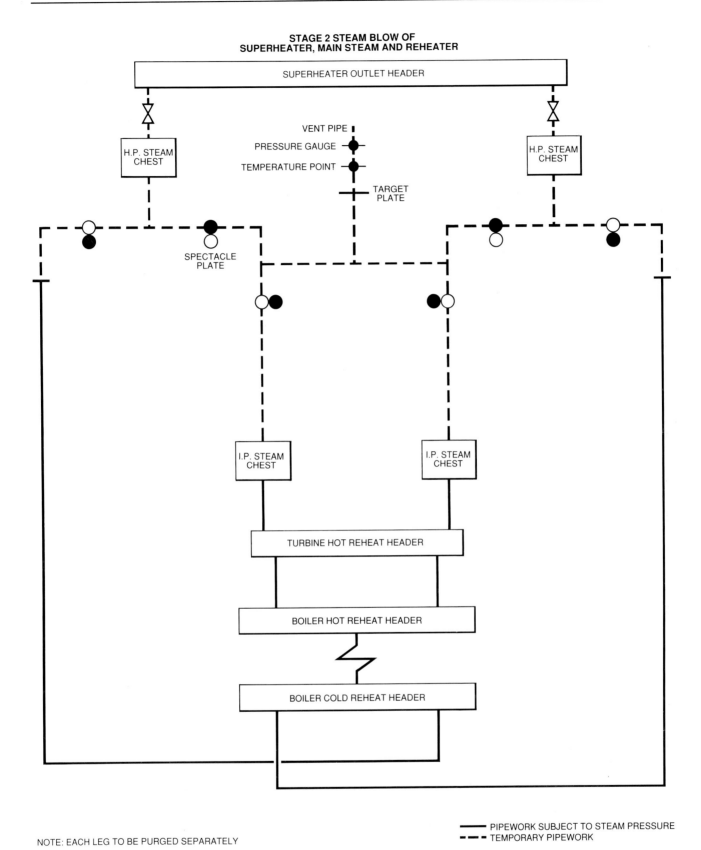

FIG. 4.5 Diagrammatic arrangement of pipework for a reheater purge

standards equivalent to those applicable to the permanent pipework.

● Vent piping should be securely anchored at the tail-piece. When steam is exhausted to atmosphere there is a large reaction created by the 'jet' effect. The thrust in the vent pipe will be transmitted to the building or supporting structure by the pipe supports; for a 500 MW boiler this is about 20 tonnes. Hence, in designing the pipe supports, factors such as temperature gradients, shock forces with possible shock loading cascade failure of supports, and the provision of ample margin for spring movement, should all be taken into account.

The disturbance factor is defined as:

$$Fd_b/Fd_n = (M_b/M_n)^2/(p_b/p_n)$$

where Fd_b and Fd_n are the drag factors under steam *blow* conditions and *normal* steel load conditions respectively

M_b and M_n are the corresponding mass flows

p_b and p_n are the corresponding densities.

Calculations in this context are facilitated by the method of Lapple, in which only a single pressure measurement at a point in the system is required, as long as the velocity at the point of discharge to atmosphere exceeds sonic (which occurs when the pressure in the final discharge pipe is greater than 1.3 bar).

Effective steam purging is a complex undertaking and each case requires individual consideration; CEGB Design Memorandum 068/25 deals with this subject. SPICO2 is a program which calculates the pressure and the temperature conditions for effective steam purging of the superheater, reheater and main steam pipework of a boiler; the disturbance factor, friction factors in the pipework; pressure drops across heat transfer stages and the thrust at the exit from the system.

The steam blowing procedure must be carefully planned and controlled with regard to technical and environmental aspects. For example, temperature differential limitations on drums, headers and permanent steam pipes must not be transgressed (the boiler manufacturer will supply the necessary data). The maximum feasible mass flow rate through a 200 MW boiler is about 10^6 kg/h. Above this figure the drum level would not be controllable. With drum temperatures at reduced loads, temperature gradients may cause drum distortion and, again, the manufacturer's advice must be sought.

During steam purging there is an increased risk of carry-over from the boiler as the pressure drops are higher than in normal operation. Solid alkalis, such as sodium hydroxide, should not therefore be used for boiler water treatment at this time. Instead, ammonia

with hydrazine (for oxygen scavenging) are employed, each at 50 mg/kg concentration, in both the initial boiler charge and subsequent make-up. In addition, the boiler water should not contain more than 0.1 mg/kg of chloride (as chlorine) or 0.2 mg/kg sodium (as sodium). Before starting the steam purge all cleaning chemicals should have been thoroughly flushed from the circuit.

Environmentally, the noise nuisance created by the purge should be minimised both by the use of silencers and by selecting the least inconvenient time of the week/day. People within the locality of the noise should be advised before the event; in this context, lights and sirens are normally used. Also personnel should use ear defenders, as well as observe the usual safety precautions.

Water flushing as an alternative to steam purging

The quantities of steam needed for steam purging are available from the reserve in the boiler drum in the case of drum boilers. Once-through boilers have no drum and therefore require the use of package boilers as an alternative source of steam. However, such measures are expensive and time consuming, and possibly not feasible owing to the size of the pipes concerned.

For such situations, it has been shown (R. R. Cranfield and J. Lawrence, 1973) that water flushing is an acceptable alternative to steam purging, provided that certain values of flow can be achieved, according to the size of the debris undergoing removal. Tables specifying the appropriate values are included in the paper.

3.2.3 Reagent circulation — requirements and methods

In all but the simplest cleaning applications, reagent circulation is essential. The purpose is to avoid reagent depletion at the surface, so that the cleaning process is not hindered, and also to prevent the local accumulation of substances entering the solution during cleaning, e.g., ferric ions which are themselves corrosive. The optimum flow rate is a compromise between the need to achieve a satisfactory rate of deposit dissolution and to avoid incurring the higher corrosion rates which accompany increasing flow rates even with inhibitor present. The target flow rate in most cases is 0.3 m/s (linked to a tube of 25 mm bore) with a maximum of five times this value. In practice, the target rate may be exceeded in some parts of the circuit with a different priority, e.g., through orifice plates. Experimental work has shown that flow rates of 1.5 m/s double the corrosion rate found at 0.3 m/s.

From the boiler capacity, the number of tubes and their bores, the circulation rate necessary to achieve the target flow rate, and hence the pumping requirements, can be estimated. For example, a 200 MW unit with a boiler capacity of 90 000 litres and 800 boiler tubes, each of 25 mm bore, 2 kg/s must be circulated to achieve the desired flow rate of 0.3 m/s. Such calculations can only provide an estimate; the rates achieved should be measured, and checks made to ensure that there are no stagnant areas.

Circulation is achieved by means of pumps or induced by the introduction of gas by diffusers (Fig 4.6) into a vertical tube (e.g., a downcomer in the case of a boiler), so that, as a result of the difference in density created in parallel circuits, liquid flows. Contractors' pumps may be electric or diesel, and standby capacity should be arranged to guard against failure.

Station pumps are sometimes employed for the circulation of cleaning formulations, e.g., boiler circulation pumps. Precautions should be taken in such cases to prevent ingress of cleaning solutions to the pump internals, as well as suspended matter which may scour the impellers. The pump manufacturers' advice must be sought on the effect that cleaning solutions may have on the component materials throughout the estimated plant lifetime.

With gas-induced circulation, nitrogen is used for acidic solutions, as air has been found to lead to unacceptable corrosion even in the presence of inhibitors. However, air is acceptable for the circulation of copper strip solutions and has some advantage in providing oxidising capacity supplementary to the oxidising agent (bromate) already present. Details of the circuitry required for the four principal modes of circulation are shown in Figs 4.7, 4.8. 4.9 and 4.10.

As already noted, air or nitrogen may be used for this purpose, depending on whether copper stripping or iron removal is being carried out.

In some circumstances, more than one means of reagent circulation may be required on the same clean, for example, in the case of a gas-induced boiler clean, a pump must be included in the circuit if the economiser is also to be cleaned.

The cleaning operation, as well as the prior rigging, is facilitated by the installation of purpose-built permanent connecting points (see Fig 4.11). The location of these depend on the recommendations of the plant manufacturer and DM 068/25.

The use of flexible pipework is not prohibited, but all temporary connections should be made in rigid materials wherever practicable. All pipework, flexible

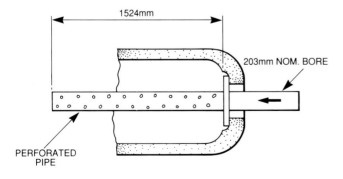

FIG. 4.6 Temporary connection for distributor drum

or rigid, should be sufficiently reliable for the whole of the chemical cleaning operation. Extra care is needed with flexible pipework, and suitably placed valves must be provided for isolation in case of failure.

Whilst cleaning is in progress, the plant should be vented to the outside of the building into an area inaccessible to personnel. Care should also be taken that no wind-borne spray from the vents can cause harm where it falls — including the vent from the constant head overflow in boiler cleaning.

Circuit integrity is clearly of importance and this must be established by filling with water before introducing any reagents.

The removal of drum internals and fittings is desirable during cleaning, for the purposes of reagent circulation and the removal of debris, but is not usually done because it is time consuming.

During cleaning, it is essential to prevent inadvertent ingress of certain reagents, notably halides or caustic alkali (actual or potential). It is also important to prevent cleaning formulations entering the superheater, particularly chlorides and caustic alkalis, where austenitic stainless steel superheaters are installed. To this end, the superheater is water-plugged before any circulation of cleaning formulations begins, and thereafter back-flushed (into the boiler) at intervals during the cleaning process, and between stages of the complete cleaning procedure.

In this context, the constant head overflow device fitted to the boiler drum (see Appendix C) is essential, and also a drum level device is necessary during cleaning.

3.3 Process control

3.3.1 Chemical aspects

Chemical addition

Chemical control must be imposed in respect of the various substances added to the circuit during cleaning operations, including the monitoring of chemical changes which occur during the process and which are indicators of progress, hazard, etc.

All chemicals should be added to the cleaning solution in such a way as to avoid harmful localised concentrations in the plant or mixing tanks. Addition should be at a constant and continuous rate whilst the solution is being circulated, until the required concentration is reached. Unduly high concentrations of acidic solutions could lead to unacceptably high corrosion rates whilst unnecessarily low rates of addition will prolong the process and could even result in an incomplete clean, bearing in mind that the solution temperature will begin falling once chemical addition has started because heat usually cannot be added to the solution by firing the boiler after this point.

During ammonia addition, care must be taken to avoid over-alkalisation as certain inhibitors tend to precipitate with increasing pH.

333

NOTES :
1. SUPERHEATER BACKFILLING WOULD BE CARRIED OUT USING THE FEED OR TOPPING
   PUMP DISCHARGING THROUGH ATTEMPERATER SPRAYS.
2. THE BOILER WOULD BE FILLED WITH WATER USING BOILER FEED OR TOPPING PUMP, CONTRACTORS PUMP CAN BE USED.
3. BLOW DOWN VESSEL AND LINES NOT SHOWN AND WOULD NOT BE USED.
4. ALL DIMENSIONS IN mm.

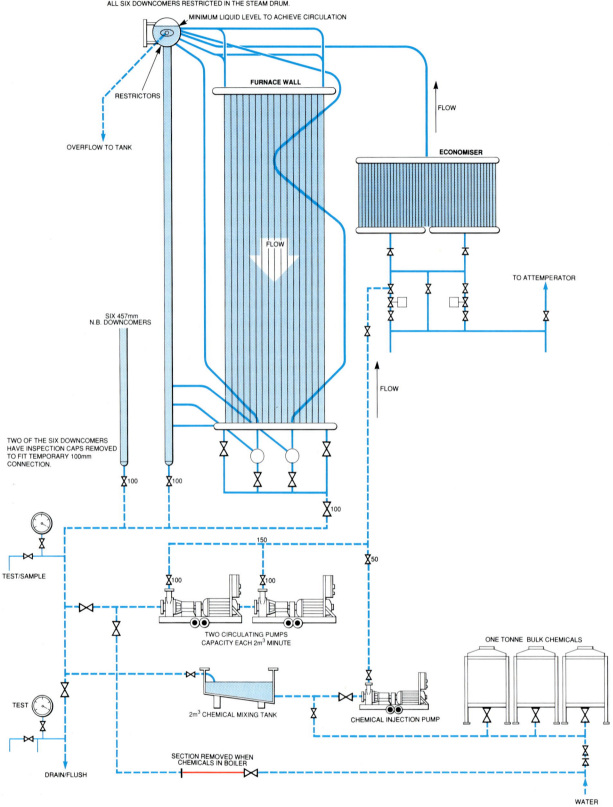

FIG. 4.7 Pump circulation of chemical cleaning solutions

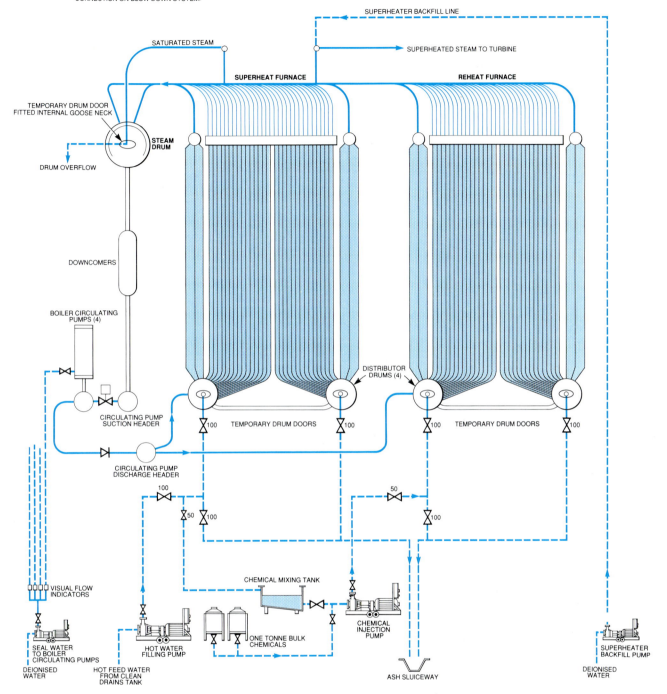

NOTES:
1. ONLY ONE PERMANENT BOILER CIRCULATING PUMP USED TO CIRCULATE CLEANING SOLUTIONS.
2. EITHER PUMP B OR C NORMALLY SELECTED.
3. BOILER FILLING PUMP USED TO FILL SYSTEM.
4. SYSTEM FINALLY DRAINED THROUGH TEMPORARY CONNECTION ON BLOW DOWN SYSTEM.

FIG. 4.8 A typical circuit for chemical cleaning in assisted circulation boilers

In copper removal during boiler cleaning, the process is carried out at 50°C. Frequently, the water is heated to a higher temperature before chemical addition to compensate for a decrease, but it should be noted that above 60°C the oxidant tends to react with the citric acid and is thus unavailable for its primary role in the cleaning process.

Inhibitor addition is dealt with in Section 3.3.4 of this chapter. Water quality is important for super-heater flushing. Contaminated water containing significant quantities of chloride has sometimes been found in the superheaters and reheaters of new drum-type boilers after erection, presumably as a result of the use of unsuitable water for hydraulic testing. To avoid the risk of stress corrosion of austenitic material from this source, especially during the pre-commissioning alkali boil-out, flushing of the superheaters and reheaters with deionised water should be carried out at

335

NOTES :
1. SUPERHEATER BACKFILLING WOULD BE CARRIED OUT USING THE FEED
   OR TOPPING PUMP DISCHARGING THROUGH ATTEMPERATOR SPRAYS.
2. THE BOILER WOULD BE FILLED WITH WATER USING BOILER FEED OR TOPPING PUMP. CONTRACTORS PUMP CAN BE USED.
3. BLOW DOWN VESSEL AND LINES NOT SHOWN AND WOULD NOT BE USED.

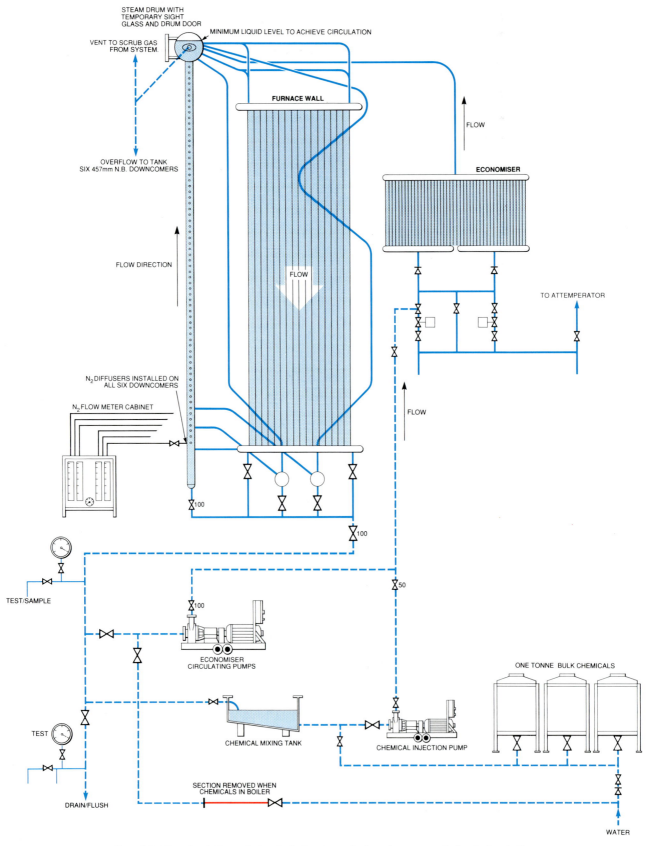

FIG. 4.9 Gas circulation of chemical cleaning solutions in a natural circulation boiler

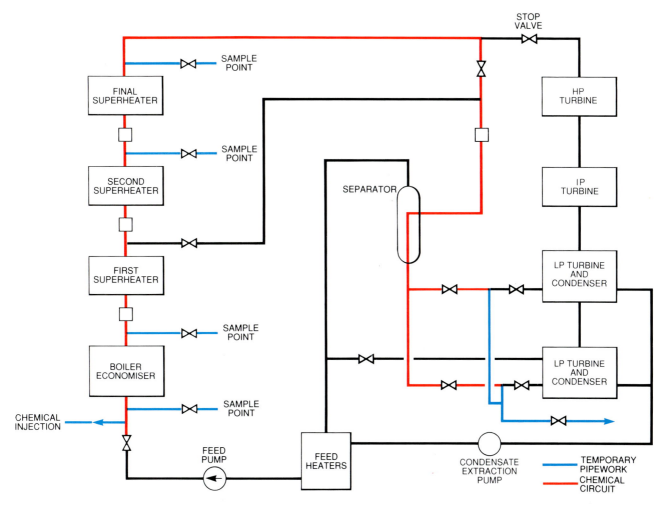

FIG. 4.10 Typical circuit for a post-service clean of a once-through boiler

the outset. Intermittent flushing should be employed until the conductivity of the effluent is below 5 μS/cm and the chloride concentration is below 1 mg/kg.

The control of chemical conditions is important not only for the cleaning stages as such, but also for the supplementary operations of rinsing, flushing, hydraulic testing and storage, either long or short. An occasional cause of corrosion is the unavoidably extended residence of a solution in plant (see Section 3.3.7 of this chapter).

During the clean, chemical monitoring is necessary both to ensure that the specified process conditions are maintained and to measure the extent of contaminant material removed as a guide to the progress of the cleaning operation. Thus in the case of a boiler clean, the concentrations of both total and free acid, as well as inhibitor, should be correct, and the total iron (and in some cases ferric iron) concentrations need to be monitored.

Methods of analysis to be used during chemical cleaning

The following methods are to be employed during various stages of the chemical clean. The sample must be filtered before any analysis is carried out.

pH Use Method 2 given in BS1427: 1962.

Concentration of hydrochloric acid or hydrofluoric acid Take 1 ml sample, add 50 ml of demineralised water. Add pH 4.5 indicator and titrate with 0.1N NaOH.

$$\text{mg/kg HCl} = \text{Titre} \times 0.365$$
$$\text{mg/kg HF} = \text{Titre} \times 0.2$$

Total citrate Take 2 ml sample, make up to 200 ml with distilled water and pass through a strongly acidic cation exchange column*. Discard the first 100 ml, and titrate the second 100 ml with 0.1N NaOH using a phenolphthalein indicator.

$$\% \text{ total citrate (as citric acid)} = \text{Titre} \times 0.64$$

*Note: The dimensions of a suitable ion-exchange column are: internal diameter, 15 mm; height, 200 mm. The resin is in the hydrogen form and the diluted sample is passed through the column at a rate of about 10 ml/minute.

Where sodium bromate has been used an allowance must be made for the presence of bromic acid after

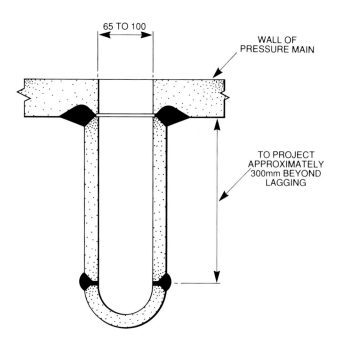

65 TO 100

WALL OF PRESSURE MAIN

TO PROJECT APPROXIMATELY 300mm BEYOND LAGGING

1. THE DIAMETER WILL DEPEND UPON THE NUMBER OF POINTS AT ANY POSITION AND SHOULD BE AGREED WITH THE BOILER-MAKER SO AS TO GIVE A SUITABLE CIRCULATION RATE.

2. THE CONNECTING POINT SHOULD BE DESIGNED TO B.S. 1113.

3. ALL CONNECTING POINTS SHOULD BE EASILY ACCESSIBLE.

FIG. 4.11 Design of permanent connecting points

cation exchange. For the addition of 0.5 per cent sodium bromate the calculation becomes:

% total citrate (as citric acid) = (Titre − 0.3) × 0.64

Free citric acid The free citric acid is obtained by deducting from the total citrate concentration the equivalent of the total iron found in solution:

% free citric acid =

(% total citrate) − (3.4 × % total iron)

Fluoride Determine using a fluoride selective electrode and an epoxy bodied reference electrode

Alkalinity Take 100 ml sample, add pH 4.5 indicator and titrate with 0.1N nitric acid

$$mg/kg\ NaOH = Titre \times 40$$

Sodium Sodium may be measured using flame photometry. Instruments giving a full scale deflection for 5 mg/kg or less of sodium should be satisfactory, and should be operated in accordance with the manufacturer's instructions.

Total iron in citric acid, hydrochloric acid or hydrofluoric acid

Dilute the sample to give a final iron concentration in the range 0–60 mg/kg (3% citric acid saturated

with iron will have an approximate iron concentration of 8500 mg/kg). For most analyses, a hundredfold dilution should be adequate. With hydrofluoric acid, dilute with 2% boric acid to protect glassware.

Determine the total iron by atomic absorption directly on the diluted sample or take 50 ml of the diluted sample, add 0.5 ml thioglycollic acid and 5 ml excess of 0.880 ammonia. Make up to 100 ml. Measure the colour using a 0.5 cm cuvette and a 605 filter. Read off the total iron concentration from a calibration graph prepared under identical conditions.

Ferric iron in citric acid or hydrochloric acid Take 25 ml sample and adjust the pH to 2.5 with approximately normal Na_2CO_3 solution.

Dilute the solution to 100 ml and add five drops of 5% sulphosalicylic acid solution. Titrate with 0.1M EDTA until the pink colour just disappears. Add a further five drops of indicator and continue the titration until the pink colour disappears.

$$mg/kg\ Fe^{3+} = Titre \times 223$$

Nickel The methods given in CERL 'Methods of Sampling and Analysis', Volume 1 should be used with suitable modifications to allow for the greater concentrations of nickel expected. Place 2 ml of the sample in a separating funnel, add four drops of 10% hydrogen peroxide solution and two drops of phenolphthalein solution. Then add 5 ml of 10% sodium potassium tartrate and mix the contents. Add 2.5N sodium hydroxide solution dropwise until the phenolphthalein just turns pink and then add 0.1N hydrochloric acid dropwise until the colour is just discharged. Add 5 ml of 0.15% furil-alpha-dioxime followed by 25 ml of 2N ammonia solution. Swirl the contents of the flask to mix, and add 15 ml of chloroform through a filter paper and measure the absorbance in a suitable size cuvette. Determine the nickel content from a calibration curve.

Methods of analysis to be used during post-service cleaning of condensers

The iron, copper, zinc and nickel content of samples of the chemical cleaning liquor can be determined by atomic absorption spectrophotometry using standard instrument operating conditions.

3.3.2 Temperature

Temperature control is important as it is necessary to achieve a compromise. If the temperature is too high, attack on the substrate metal will be excessive, and if too low, dissolution of oxides, scale, etc., will be inconveniently slow. A factor to be borne in mind is that in acid solutions heat cannot be put in once the reagents have been added. With citric acid formulations at about 3% strength, the water must be raised to about 95°C before the addition of acid and inhibitor. The process has then to be completed before the temperature falls below about 70°C, after which reaction rates become impractically low and, in addition, there is an increasing possibility of inhibitor precipitating from the solution.

In the case of hydrochloric acid formulations (about 5%) the *upper* temperature limit is lower, at about 70°C, but the inhibitors are not subject to precipitation as a result of the temperature falling. However, the scaling rate falls off with temperature, and for iron oxides a *lower* limit of 50°C applies.

In both cases, the period for an iron removal stage in a boiler clean will be 6–8 hours, unless a plateau in the control plot of iron versus time indicates that the process is complete sooner. Obviously, *ambient* conditions can influence the situation.

With copper removal in boiler cleaning, the process temperature is 50°C, the solution being pre-heated to about 60°C to give operational flexibility. In this case, a plateau is usually fairly quickly achieved, and falling temperatures are not a threat to completion. (Note, however, that copper removal may still be incomplete in some instances owing to the presence of unusually heavy deposits, and an additional stage may be needed.)

With the cleaning of condensers and feedheaters, temperature is rarely a problem. In condenser cleaning, in view of the large amount of solution involved, approaching 450 000 litres for a 500 MW unit condenser, the use of temperatures above ambient is not feasible. With feedheaters, the temperatures are low (50°C) and their maintenance presents no obstacle because of the relatively small volumes employed.

Heat is not put into a cleaning solution by firing the boiler once the reagents have been added, to avoid the risk of decomposition of reagents which could ensue if temperatures become locally elevated above the process target, as well as the greatly increasing risk of corrosion of the plant. There is, in consequence, the temptation to raise the temperature to an unduly high level before chemical addition.

With citric acid, where the target temperature for boiler cleaning is 95°C, inhibition will be lost at above 100°C when boiling occurs.

With hydrochloric acid, where the target is 70°C, temperatures above 75°C can lead to excessive corrosion, but the process still provides reasonable scale dissolution down to about 50°C, so temperature here is not quite so critical.

In the case of copper removal in boiler cleaning, the target temperature is 50°C. If however, the water is at above 60° when the chemicals are added, the oxidant (bromate) will be partly consumed in reaction with the citric acid, with a corresponding loss to the process.

In all of these instances, and more generally, it must be borne in mind that even though the boiler should not be fired once the chemicals have been added, in some cases steam from an external source may be used to increase the temperature.

3.3.3 Flow

In a number of cleaning processes, flow must be controlled to provide reagent replenishment, yet not be so great as to lead to excessive corrosion.

In boiler cleaning, as already noted, a suitable flow rate is about 0.3 m/s in a tube of 25 mm bore. A reagent which will be ineffective for oxide removal at this rate will be no better at higher rates. As regards corrosion, increasingly high flows result in relative loss of inhibition. Adequate inhibition is achieved up to about 2 m/s, after which increasing corrosion occurs.

Circulation in boiler cleaning is achieved by one of three means, i.e., internal circulation pump, external pump or by gas-induced flow. One of the two/three boiler circulation pumps used in operation is normally sufficient to achieve the required rate in a post-service boiler clean.

With an external pump, the required size is estimated knowing the system capacity, number of tubes concerned, and their bore.

In the case of gas-induced flow, the gas quantity required can be estimated and then introduced before finally controlling the rate by means of adjustments to the valves linked to in-line flow meters on the gas diffusers.

Whichever means is employed, however, checks are required. These include: the use of orifice plates strategically placed, injection of a tracer chemical (e.g., ammonia) and timing its transit around the circuit, as well as checking for temperature equilibration, or the use of thermal cameras to ensure that hot spots are absent.

In condenser cleaning, on the water side, the prime reason for flow is reagent replenishment, the deposit being mostly fairly soluble and not strongly susceptible to the effects of flow. The presence of carbonates results in local agitation of the solution by carbon dioxide evolution and, indeed, may lead to gas blockage of tubes.

In feedheater cleaning, although the use of pumps provides a positive flow, excessive deposits may cause local blockage and hence inadequate flow in the very locations where cleaning is most needed. It is for this reason that powerful flushing is required.

In steam side condenser cleaning, the solution flows over, rather than through the tubes, and flow is provided by means of an external pump without the constraint of the internal tube surfaces.

It should be remembered that without adequate flow, the samples taken from analysis may not be very effective guides to the progress of the clean.

Particular attention should be paid to potentially stagnant locations, e.g., instrument connections, to ensure that they are properly cleaned and flushed before being put back into service. Temperature measurements or the use of an infra-red camera may show if circulation is being achieved. Chemical monitoring may also provide some indication of adequate flow.

3.3.4 Inhibition

In the context of inhibition, process control in chemical cleaning has the twin objectives of limiting corrosion to acceptable levels and of ensuring the safety of personnel. The inhibitors approved for use on plant are given in Appendix D (Boilers) and Appendix E (Condensers).

For boilers, Armohib 28 is used for hydrochloric acid processes; Dodigen 95 or Stannine LTP for citric, sulphuric and related acids; and Dodigen 95 for hydrofluoric acid. This is the case for ferrous alloys, and also for condenser cleaning, where the mild steel or cast iron water boxes are the most vulnerable of the construction materials. In certain circumstances, it is also necessary to provide protection for the copper based alloys in condensers, and in those situations, 2-mercaptobenzotriazole (MBT) is also added.

In feedheater cleaning, whether the alloy is steel or cupro-nickel, Armohib 28 is the approved inhibitor for the hydrochloric acid (oxide removal) stage.

For hydrofluoric acid cleaning of boilers, Dodigen 95 was found to be technically suitable in the three plant trials carried out in CEGB.

Process control to ensure adequacy of inhibition is achieved by a number of means. Basically, the correct amount of the right inhibitor is added to the solution in the specified way, and other components of the process maintained in the range where they support acceptable inhibitor performance.

The approved inhibitor should be present under acidic conditions at all times to ensure that it is fully dissolved and not subsequently precipitated. In making up the solution it should always be added direct to an acidic solution and not be previously diluted with water, nor should it be added to water to which acid is added later. This requirement can be met by injecting the acid and the inhibitor simultaneously at the suction of the pump used for injecting chemicals into the cleaning circuit.

Some inhibitors have very limited solubility and it is particularly important that they are added to the cleaning solution in a controlled manner, incrementally with the acid injection. The inhibitor must be added at a constant rate over that period of time required for complete circulation of the cleaning solution. This can be achieved by the use of either a metering pump or a drip feed arrangement.

Once added, the concentration of the inhibitor can be determined by analysis, but this is usually done retrospectively in view of the length of time required.

The *wire wool* test has traditionally been used for the assessment of the adequacy of inhibition of *pickling* (chemical descaling) solutions, but it must be emphasised that it is subject to a number of uncertainties. It depends upon hydrogen bubbles in a corroding solu-

tion becoming entrapped in a ball of wire wool and so imparting greater buoyancy than in a well-inhibited solution; i.e., if the ball floats, inhibition is inadequate; if the ball sinks, then inhibition is acceptable. However, empirically, a number of factors such as temperature, duration of reaction, etc., all affect the results obtained.

In addition, considering the test in the light of the principles of inhibition discussed in Section 2.3 of this chapter, then clearly the only cathodic reaction catered for by this test is the evolution of hydrogen. Ferric ion corrosion, which in some circumstances may be a significant cause of metal loss, is without effect on the wire wool. For these reasons, therefore, this should be used as no more than an adjunct to other means of process control.

Electrochemical methods are feasible in principle, although none has yet been developed to the stage where it is suitable for routine use.

3.3.5 Inspection

Pre-service cleaning

Inspection in pre-service cleaning is initially *visual*. Much depends upon the efficiency of the clean conditions regime prior to plant installation. A decision may be required as to whether the individual component can be cleaned alone, at works or on site, and then stored, or if it is preferable to include it in the clean of a larger circuit.

Inaccessible areas will require the use of methods such as introscopes, or tube cameras. It may also be possible to assess the state of surfaces by applying the 'Sellotape' method, which depends upon pressing the sticky side of tape to the surface in question and estimating the quantity of material adhering to it after removal.

Unusually deep pits or zones of severe damage (which may pre-date cleaning), will require specialist comment and even rejection and replacement of a component.

Post-service cleaning

Again, initial examination is visual. Some indication of the likely degree of contamination and a consequent case for cleaning should be provided by the operational history of the plant. Thus, a boiler may have undergone on-load corrosion, or have been subjected to adverse water chemistry for some time prior to the decision to clean. Also, a heat exchanger in need of cleaning will have been performing unacceptably. In these cases, sample examination before deciding on the need to clean, and the method, is strongly advised; sample examination after cleaning will provide comparative data on which to assess the success achieved.

In addition, introscope and TV camera surveys may again be useful. Attention should be given particularly to those areas known, historically or by prior inspection, to be affected or vulnerable.

3.3.6 Plant storage during an interrupted clean

Disruptions of the planned cleaning procedures can be envisaged for various reasons, such as water shortage, plant breakdown or industrial action. In the case of such occurrences, there are methods of storing the partially-cleaned plant to protect it against corrosion until cleaning can be resumed. The methods employed are based in principle on those already described in Section 2.6 of this chapter. Their application according to the circumstances of the specific interruption, are given in the logic diagram presented as Fig 4.3. It should be noted that these protective procedures are applicable to both pre- and post-service cleans.

3.3.7 Some effects of incorrect cleaning

Departure from good cleaning practice may lead to adverse effects in a number of ways, and these may only become apparent during or after the clean. The use of hydrochloric acid to clean steel containing heat affected zones can lead to pitting of the order of 1 mm deep, and has been the subject of a special study. To avoid such occurrences, hydrochloric acid should be excluded from use on plant where there is any possibility of some heat affected zones remaining owing to incomplete heat treatment. This precaution normally is confined to new plant, as for any boiler with a history of cleaning with hydrochloric acid and freedom from this effect there is no point in changing course.

The types of attack cited are equally possible in pre- or post-service cleaning, although it must be emphasised that such occurrences are quite rare in any kind of cleaning.

Some forms of corrosion have been reported which are confined to post-service practice. Thus, subsurface cavitation can result from the cleaning of boiler tubes already suffering from hydrogen damage caused by severe on-load corrosion, A suggested interpretation (C. Settle, *et al*, 1966) was that the corrosion product acted as a semi-permeable membrane and permitted passage of hydrogen ions (but not the inhibitor) to the metal substrate, weakened already by hydrogen damage, so that corrosion was unduly severe as a result. Subsurface cavitation should never be allowed to occur at all. Tubes suffering from on-load corrosion to the extent that they exhibit hydrogen damage should be replaced before the boiler is cleaned as they are probably irretrievably weakened, and any associated corrosion scabs are not susceptible to chemical treatment.

There are no reported cases of subsurface cavitation with citric acid.

The possible adverse effects of repeated cleaning have been discussed by J. Rice, 1961. Damage of this order would not be envisaged within modern plant which has been regularly cleaned using properly inhibited formulations, and, in addition, where cleaning had not been used as inadequate compensation for poor water-treatment.

In addition to the corrosion of equipment during cleaning, if any residual chemicals are still present after cleaning they clearly present a hazard when operation is resumed. Not uncommonly, after cleaning with hydrochloric acid, the feedwater and boiler water are found to be unacceptably high in chloride on recommissioning. The cause may be chemicals retained in 'dead legs' in the circuit, or from iron chloride formed on the roof of the boiler drum during cleaning (the drum is not full during cleaning); Armohib 28 is less volatile than hydrochloric acid, so that corrosion could be more marked there than anywhere else in the boiler. However, it could not conceivably be sufficient to affect the integrity of the drum significantly. Damage due to cleaning is not to be confused with the removal of deposits from long established pits which then become visible.

In condenser cleaning to remove combined oil and oxide, the solvent employed damages the plastic plugs used routinely to repair and seal off leaking tubes, and the plugs need to be replaced after cleaning is complete.

4 Cleaning practice — specific areas of application

4.1 Works and off-site cleaning and protection of plant

Whilst it is not the intention to discuss here the cleaning and protection of plant off-site, it should be noted that these aspects have received considerable attention. Improved operational availability of plant is to be sought by the adoption of suitable cleaning and protection procedures during manufacture and installation. The first of these subjects is covered in CEGB Code of Practice 098/15 'The Cleaning and Protection of Power Station Plant' and the second by ECP 328 'The Storage of the Water/Steam circuits of Generating Plant.'

4.2 Pre-service cleaning

4.2.1 Objectives

The objectives of pre-service chemical cleaning are to enable the user to commission and operate plant with the internal surfaces of the water-steam circuit as free as practicable from all foreign matter, such as adventitious metal oxides, grease, oil and dirt, and to establish a satisfactory basis for the protective magnetite film to grow on the plant internal surfaces. In particular, it is intended:

- To help ensure that the relevant chemical control requirements are obtained with the minimum delay and restriction of load.

- To remove debris and thick oxide accumulated during fabrication or construction.

4.2.2 Former UK practice

Initially, the pre-service cleaning of power plant was mechanical, merely to remove extraneous matter such as mill scale, possibly with some chipping or grinding away of attached pieces. The first chemical treatment employed was an alkaline process for degreasing and, in one form, this survives to the present day. In time, the advantages to be derived from the use of acids for pickling, in terms of speed of descaling and the cleaning of inaccessible zones, were recognised, but there were early drawbacks in the form of the increased corrosion, which led to the introduction of inhibitors into cleaning practice.

Power plant cleaning increased in scope until, in about 1963, the first complete system clean in the CEGB was carried out at Rugeley A power station. There was then a gradual decrease in the extent of cleaning of any new plant. It was found to be feasible to clean feedheaters, for example, at works, keep them in a preserved state at works and during transport to site, and finally on site until they were connected into the circuit ready for operation.

Similarly, it was found to be practicable to omit the pre-service clean of the superheater and reheaters (as was first demonstrated at Fawley power station) and to rely solely on the steam purging operation. Both of these economies in the pre-service cleaning procedure were only feasible because of the introduction of a clean conditions working regime.

At present, therefore, instead of an overall cleaning procedure which includes a number of separate sub-circuits, it should be possible to carry out a pre-service clean of only the boiler and economiser. However, if there is a substantial shortfall in the efficiency of clean conditions working, more extensive cleaning of the erected plant may still be advisable.

4.2.3 Criteria for chemical cleaning

Steam/water circuit

As a result of the application of clean conditions working, the condensate and feed systems of drum boilers, superheaters and reheaters should not require pre-service chemical cleaning. Instead, they should be either water-flushed (Appendix A) or steam-purged (Appendix B). The furnaces of drum type boilers still require to be cleaned, but hydrochloric acid should no longer be necessary unless there has been an appreciable departure from clean conditions working (see Appendix F).

Turbine-generator oil systems

An extremely high degree of cleanliness is essential in turbine-generator oil systems to ensure trouble free operation. This is particularly so in the turbine governing circuit to ensure not only satisfactory response during operation, but also safety in case of emergency.

Chemical cleaning as such is not normally applied to these systems, but if necessary the conventional methods of deposit removal used elsewhere in the plant could easily be adapted to turbine-generator systems. It is, however, essential that a high degree of cleanliness be maintained during assembly and erection, and that adequate filtration, both temporary and permanent, be provided during commissioning of the set. Cleaning and protection procedures are necessary for every operation from an early stage of manufacture.

Essentially, the procedures should stipulate how the pipework and various components of the circuit need to be cleaned at works — either mechanically and/or chemically — and then treated with either temporary or permanent protectives as appropriate.

Cleaning, both after erection and before commissioning, and subsequently following overhaul, is then carried out by means of oil flushing; the bearings meanwhile being protected by means of strategically placed filters.

4.3 Post-service cleaning

4.3.1 Objectives

The objectives of post-service chemical cleaning are essentially the same as those stated in Section 4.2. for pre-service cleaning, but with an additional consideration. In post-service cleaning it is also necessary to remove debris and thick deposits which have accumulated during operation and maintenance and which, if left untreated, may lead to overheating and possibly to on-load corrosion.

4.3.2 Boilers

General

Plant which has been in service may contain scale and deposits arising from operation of the plant itself, which can bring about adverse effects if not removed. Whilst in normal circumstances the waterside oxides on the furnace wall tubes typically accumulate at a rate of about 2 μm/year (in the range of 0.5 to 5 μm/year), rates up to 100 times these may occur if on-load corrosion is experienced. The factors leading to such occurrences are discussed in Chapter 2, but for the present purposes we may note briefly that on-load corrosion is brought about by some abnormalities in water chemistry, operation, and, in some cases, design.

The immediate consequences are unusually thick deposits of iron oxides, contaminated (according to the particular local circumstances) by copper oxides, copper metal, the oxides of nickel and zinc, and possibly hardness salts. The longer term consequences — tube thinning and/or embrittlement — however, have adverse implications for plant integrity. It must be emphasised that if on-load corrosion occurs, the application of a number of remedial measures must be considered as well as chemical cleaning, e.g., upgrading of the water treatment (both internally and externally to the boiler), de-aeration, tube surgery and replacement, and modifications to the operational procedures. Chemical cleaning can only help to restore the situation, and if other factors are not tackled the problem will remain. The indications that on-load corrosion may have occurred, or is an increasing risk, are given by the following symptoms:

- A change in the operational chemistry, e.g., the occurrence of *hide-out* or incidence of persistent condenser leakage.

- Large amounts of debris in the boiler or feed system as revealed by periodic examination, either visually or by NDT, or by tube removal and sectioning.

- In once-through boilers, where increased pressure drop indicates the presence of unacceptably large deposits.

- On-load corrosion may actually lead to sudden failures without any of the foregoing symptoms and may necessitate extensive re-tubing. A corollary to this is the case where there is the need for extensive re-tubing owing to fireside corrosion and that the associated ingress of adventitious impurities is so great that the boiler merits a clean.

These symptoms refer largely to conventional boilers, both drum and once-through.

The case for routine cleaning

Even when the foregoing symptoms do not occur, there is a case for cleaning periodically. This view is based on the fact that, whilst cleaning is capable of dissolving the normally-grown magnetite and the overlying deposited debris, it cannot remove scales due to severe or persistent corrosion. Unduly-thick porous oxides may well concentrate the normally innocuous bulk boiler water to an unacceptably corrosive level, and so lead to even more severe oxide thickening and/or hydrogen damage — failures within about 100 hours are not unknown. In addition, very thick deposits may result in single overheating failures. The aim of periodic cleaning, therefore, is to pre-empt the development of such conditions, and such cleans are initiated on the basis either of oxide thickness or of time, i.e., the period of operation since the last clean.

Oxide thickness as a guide to the need for cleaning

The view that waterside oxide thickness can provide the most reliable criterion for cleaning is now widely accepted. For this purpose, measurements should be made on samples of tubing, preferably taken from the high heat-flux zones of the boiler or those known to be prone to a high rate of deposit accumulation.

Data from a wide range of plant operating at 100 bar and 160 bar (see Table 4.3) indicate that rates of oxide growth of about 1 μm/1000 hours represent the best conditions. Greater rates are therefore considered to present an increased risk of corrosion and it is advised that when some figure is substantially exceeded, say

50 μm (i.e., equivalent to 50 000 hours' operation in the best case but less (pro rata) in others), then the plant should be considered for cleaning. Two-shift operation (unless locally established data show otherwise) may lead to more rapid deposit accumulation than the base load operation and so to the need for cleaning after correspondingly fewer hours. In the absence of tube samples, routine cleans should be carried out on a time basis.

The above considerations are based mainly on fossil-fuelled boiler plant. With nuclear plant (secondary circuit) the accent is on 'ultra pure' water and it is hoped that the need to clean will not arise. Sample tubes are not accessible and the waterside oxide thickness will have to be estimated on the basis of rig data and operational history. However, chemical cleaning is not expected before at least 150 000 hours operation.

Practical aspects

Post-service cleaning is applied to a variety of designs and makes of boiler. Examples of deposit thickness and composition are given in Table 4.4. The results of examination of, say, five tube samples from the high heat-flux zone will enable a decision to be made as to whether the boiler should be cleaned at once or if a clean should be delayed subject to the results of monitoring the waterside oxide.

Analysis of the waterside oxide for composition provides some qualitative guidance to the type of clean required with respect to iron and copper removal. The range of deposit compositions encountered in the CEGB led predominantly to concentrations in the cleaning solutions of about 0.3 to 0.6% of iron and from zero to 0.15% of copper. Typical tubes from a high pressure boiler before and after cleaning are shown in Figs 4.12. and 4.13 respectively. The economiser is not invariably included in the circuit when a boiler is cleaned, unless the preferred mode of circulation (e.g., external pump) makes it more difficult to omit than to include. However, an economiser should not normally be allowed to operate for more than about 10 years without chemical cleaning. This is to minimise the risk of deposits from the waterside of the economiser being transported to the boiler in the form of slugs, leading to overheating and tube failures. An example of this showed that the deposits consisted of metallic oxides mixed with oil. The condition of the economiser waterside oxide should be monitored as a guide both to when a clean is due and to the extra amount of reagent needed over and above that required for the boiler alone.

There are three process options for post-service boiler cleaning. Guidance on which process to select is given in Fig 4.14.

Examples of the procedures for post-commissioning cleaning of drum-type boilers and the post-service cleaning of once-through boilers are given in Appendices G and H respectively.

TABLE 4.3

Data on boiler tube water side oxides

Station/boiler	Approximate running hours since last chemical clean	Scale thickness, µm — Fireside Total	Fireside Inner	Fireside Outer	Remote Total	Remote Inner	Remote Outer	Spectrographic analysis — Fe	Cu	Ni	Ca	Mg	Chemical regime	Remarks
Agecroft No 4	52 450	60			35	10	25	21.0	35.3	7.0	0.3	0.5	Zero solids	Chemical cleaning advised
Castle Donington No 1	13 023	55	10	45				*	8.2	0.1	2.8	0.2	Caustic soda	
Castle Donington No 1	17 450	21	10	11				*	5.6	1.1	1.0	0.6	Caustic soda	
Castle Donington No 2	6446	13	11	2	17	9	8	*	11.2	5.9	6.4	0.4	Caustic soda	Evaluations undertaken to
Castle Donington No 3	10 900	8			15	7	8	*	3.4	0.4	3.6	0.1	Caustic soda	determine routine cleaning
Castle Donington No 4	8125	27	7	20				*					Caustic soda	policy
Castle Donington No 5	37 800	50						*	9.4	11.0	1.0	0.8	Caustic soda	
Castle Donington No 6	–	45						*	8.9	3.1	0.7	0.1	Caustic soda	Thicker oxide probably due to
Castle Donington No 6	9580	70	15	30	35			*	14.0	5.4	0.3	0.2	Caustic soda	two-shifting
Cottam No 1	24 508	73	11–17	1	(15 samples)			*	1.4	2.7	0.01	0.7	Zero solids	Evaluation for cleaning following CW ingress
Cottam No 3	5327	12–18			15			*	0.3	0.4	0.4	0.5	Zero solids	Evaluation for cleaning
Cottam No 3	17 327	30			15	7		*	0.5	0.3	0.4	0.4	Zero solids	Evaluation for cleaning
Cottam No 4	20 000	110			17	7	10						Zero solids	
Drakelow No 6	–	30	21	18	15	9	6	28	32	16.7	0.1	0.2	Caustic soda	
Hams Hall C No 1	9006	16	10	6									Caustic soda/ sodium phosphate	
Hams Hall C No 2	32 218	107												
High Marnham No 1	–	25	15	10				63	2.0	2.2	0.1	0.1	Zero solids	Evaluation for routine cleaning
High Marnham No 1	18 473	33		(5 samples)	24			*	4.3	5.0	0.1	0.1	Zero solids	Evaluation for routine cleaning
High Marnham No 2	–	13						60	0.8	7.9	0.4	0.4	Zero solids	Evaluation for routine cleaning
High Marnham No 2	12 823	25		(16 samples)	11			*	1.3	10.7	0.01	0.2	Zero solids	Assessment of adequacy of sampling
High Marnham No 3	15 455	11						*	4.5	0.1	0.6	0.5	Zero solids	Evaluation for routine cleaning
High Marnham No 4	17 518	17	7	10	15	7	8	50	3.0	20.0	0.1	0.5	Zero solids	Evaluation for routine cleaning
High Marnham No 4	25 025	27	7	20				*	4.3	4.9	0.7	0.2	Zero solids	Evaluation for routine cleaning
High Marnham No 5	12 181	22	7	15				*					Zero solids	Evaluation for routine cleaning
High Marnham No 5	23 200	55	20	35	30	10	20	*	3.8	13.0	0.6	0.4	Zero solids	Evaluation for routine cleaning
High Marnham No 5	23 260	55	20	35	35	10	25	*	2.1	16.0	0.3	0.2	Zero solids	Evaluation for routine cleaning

* = Majority

FIG. 4.12   Photograph showing initial surface of tube sections selected, for chemical cleaning tests (see also colour photograph between pp 208 and 209)

FIG. 4.13 Photograph showing the tube sections in Fig 4.12 after chemical cleaning tests
Top row: cleaned in 3% ammoniated citric acid, pH 3.5 Bottom row: cleaned in 5% w/w hydrochloric acid
(see also colour photograph between pp 208 and 209)

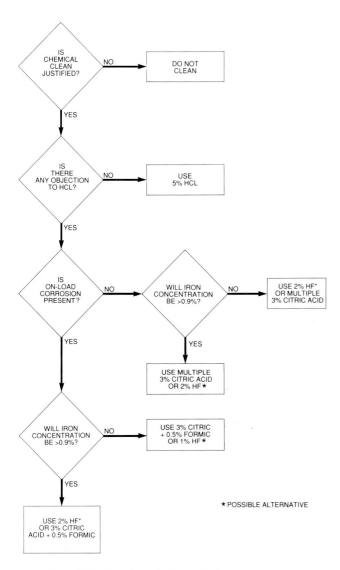

FIG. 4.14 Selection of chemical cleaning processes for boilers

4.3.3 Condensers — cooling water side

General

Cleaning is an alternative to or a supplement to preventive measures. The deposits usually contain some carbonate (Table 4.4 from Willett, 1977) which by preferential dissolution detaches other less-soluble material so that reagent exhaustion is rarely a problem.

In most power stations, the need to chemically clean condensers arises only exceptionally. At some stations outside the CEGB, particularly on the Continent, usually with recirculatory cooling systems operating at high concentration factors (> 2), it has been necessary for many years to try to prevent and/or remove waterside scale caused mainly by the concentration of hardness salts in the cooling water. Similar problems can arise quite suddenly at previously unaffected locations, as was shown in CEGB in 1975 and 1976 when the hot dry summers led to rapid and severe condenser scaling, particularly with calcium phosphate, and the consequent need for cleaning.

In some cases the Taprogge system is used for the prevention and removal of scale. In this process, sponge rubber balls are circulated periodically and this may largely eliminate the need for chemical removal of scale.

Condenser cleaning problems prompted much investigational work into various modes of cleaning and the associated question of the protection of plant during acid-cleaning by the use of inhibitors. The findings, together with the lessons of field experience, form the basis of the CEGB's currently approved methods.

Objective of cleaning

The objective of cleaning is to remove material impeding heat transfer and so restore the thermal efficiency. The potential losses are high. A loss of 1 mbar on a 500 MW unit costs £60 000 per annum (1989) and this can arise from 5 μm of deposit on condenser tube walls.

Criterion for cleaning

The criterion for assessment of the need for cleaning is an unacceptable fall in condenser efficiency. If it can be judged unambiguously that the losses are due to waterside scaling as opposed to air in-leakage, for instance, then cleaning should be considered by comparing the cost of the clean with the losses. Some supplementary guidance can be obtained by cleaning individual tubes.

Available cleaning processes

There are three distinct methods for chemical cleaning of condensers, all of which depend on the use of acid to dissolve the scale. The methods available are distinguished as *off-load*, *reduced-load* and *on-load* and full details are given in Appendices I, J and K.

In the off-load method, the acid solution is introduced into the condenser and circulated until monitoring shows scale dissolution to be effectively complete. The reduced-load method differs in that it is applied to one shell of the condenser, isolated for the purpose and effectively off-load, whilst the other is still operating.

In the on-load method, the pH of the cooling water entering the condenser is briefly decreased to about 2–2.5 by acid dosage. The buffering capacity of the cooling water, together with the large volume of the system, restrict the change of pH elsewhere in the circuit.

For the condensers associated with once-through boilers only the off-load method should be used, to avoid any possible hazard from leaks during cleaning.

Choice of acid

Whilst in principle any of a number of acids could be used to remove the scales found in condensers, hydrochloric and sulphuric acids are recommended as, in addition to being cheap and readily available, they

347

TABLE 4.4

Analysis of solid samples of condenser water side deposits

Sample description	Loose tube deposit	Tube plate	Pond sediment	Condenser tube scale		
				inlet	centre	outlet
Aluminium, as % Al$_2$O$_3$	9.8*	0.4	8.0*	0.6	0.4	0.6
Antimony						
Arsenic						
Bismuth						
Boron						
Calcium*, as %CaO	10.2	29.9	3.9	35.8	38.1	38.9
Carbon* (total), as %C	7.3	3.7	9.6	4.9	4.3	3.0
Chromium, as %Cr$_2$O$_3$	0.1	0.1	0.1	0.1	0.1	0.1
Copper, as %CuO	0.3	0.1	0.1	0.6	0.6	0.3
Iron*, as %Fe$_2$O$_3$	5.6	13.9	6.7	1.9	1.4	0.7
Lead, as %PbO	0.1	0.1	0.1	0.1	0.1	0.1
Magnesium, as %MgO	1.6	1.6	1.0	2.9	3.2	4.2
Manganese*, as %MnO	3.6	2.3	3.9	6.2	12.4	1.6
Molybdenum, as %MO$_2$	0.1	0.1	0.1	0.1	0.1	0.1
Nickel, as %NiO	0.3	0.2	0.2	0.4	0.4	0.25
Phosphorous*, as %P$_2$O$_5$	0.9	6.9	20.5	0.5	28.2	29.9
Silicon, as %SiO$_2$	39.4*	0.4	49.4*	0.6	0.4	0.9
Sodium, as %No$_2$O	0.1	0.1	0.1	0.1	0.1	0.1
Sulphur*, as %SO$_3$	1.7	0.8	1.0	ND	TR	TR
Tin, as %SnO$_2$	0.1	0.1	0.1	0.1	0.1	0.1
Titanium, as %TiO$_2$	0.3	0.1	0.3	0.1	0.1	0.1
Vanadium, as %V$_2$O$_5$	0.1	0.1	0.1	0.1	0.1	0.1
Zinc, as %ZnO	0.1	1.9	0.1	0.1	1.0	1.0
Loss on ignition, %	21.8	28.6	20.3	23.8	23.4	22.5
Carbonate*, as %CO$_2$	2.4	5.0	2.1	6.0	6.2	5.6
Organic carbon, as %C	6.6	2.3	9.0	3.3	2.6	1.5

* Figure obtained by chemical methods, AA, etc.

ND = Not detected

TR = Trace

have been used satisfactorily in the investigational studies of this subject. Of the two, hydrochloric acid is recommended only when the whole condenser is isolated, i.e., in an off-load clean. Where there is any possibility of contamination of the feedwater by the cleaning solution, as in the reduced-load or on-load methods, then sulphuric acid is advised.

Inhibition

The three basic modes of cleaning impose quite different requirements for inhibition and these are determined by the materials of construction of the plant and by the estimated long term frequency of cleaning. Investigations have shown how, by the intelligent use of inhibitors, acceptable conditions can be achieved in respect of general and pitting corrosion of water boxes and tubes, as well as galvanic attack at junctions and copper deposition from the clearing solution. The attack on copper-coated deposition from the clearing solution is considered to be the major corrosion risk during the chemical cleaning of condensers.

The findings were that 2-mercaptobenzothiazole (MBT) is an effective inhibitor for both brass and

steel in cold 1.5% w/w hydrochloric acid or hot (40°C) 1.5% w/w sulphuric acid, provided a suitable surfactant is used to ensure solubility. The use of either MBT or Armohib 28 suppresses copper deposition on the water box and also counteracts galvanic corrosion.

The degree of risk to condenser integrity is related to the credible frequency of cleaning. Recommendations for inhibitor requirements are given in Appendix E.

Selection of cleaning process

To aid in the selection of a suitable cleaning process a logic diagram is given in Fig 4.15.

Plant considerations

Terminal points Suitable permanent terminal points should be made as branch connections, e.g., 150 mm bore to the water box drains downstream of the drain valves so that rigging can be carried out on-load and subsequent flushing can take place without delay.

Circuitry Where appropriate, the water boxes should be connected in series into a circuit containing a tank

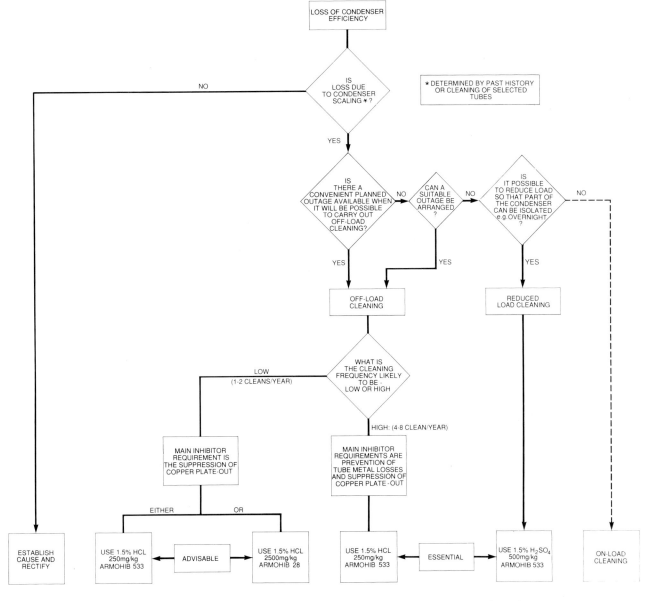

FIG. 4.15 Selection of a chemical cleaning process for the cooling water side of condensers

and pump, the water box vents being fitted with temporary connections back to the chemical mixing tank.

Protection of plant after cleaning   In contrast to boiler plant, there are no special requirements for the protection of condensers after cleaning provided that the cleaning solution has been flushed away adequately.

4.3.4   Condensers — steam side

General

Experience has shown that accumulated deposits on the steam side of condensers can cause significant efficiency losses. At one station in particular, removal of steam side deposits led to improvements in condenser back pressure of between 6 and 12 mbar for the units cleaned, and similar levels of deposits have been observed in a number of other stations. However,

cleaning of condenser steam sides is expensive and so the potential benefit has to be carefully assessed.

Objective of cleaning

The objective of cleaning is to remove material impeding heat transfer and so restore thermal efficiency.

Criterion for cleaning

The criterion for cleaning is an unacceptable fall in condenser efficiency. If it can be shown that the losses, or a sufficient proportion, are due to steam side fouling as opposed to air inleakage or waterside scaling, then steam side cleaning should be considered.

The potential financial benefit of cleaning is best judged on the basis of estimates of the fouling burden and the associated resistance to heat transfer.

The fouling burden and its composition on individual tubes can be assessed chemically by determining

349

the amount and nature of the deposit on a given area. Tubes extracted from the condenser for this purpose must always be removed by cutting between sagging plate positions and removing the sections from the steam side. Extracting tubes through the tube plate will lead to dislodgement of the steam side deposit.

In-situ deposit thickness measurements on accessible tubes will give an indication of the extent of fouling. Deposit thickness can be measured non-destructively using commercially available impedance measuring instrumentation.

Fouling resistance data can be obtained by externally cleaning selected tubes, in situ, with emery cloth and comparing the cooling water rise for pairs of dirty and clean tubes. On stations where 'Taprogge' is not installed to keep water side tube surfaces clean, due allowance should be made for water side losses when using this technique.

Fouling resistance data can also be obtained directly from heat transfer tests carried out on extracted segments of tubing.

Available cleaning processes

Currently, there is only one basic cleaning process for which satisfactory field experience is available. It can be used in two variants, one for deposits which are oil-free and one for deposits which contain oil.

Cleaning is carried out at ambient temperatures and cleaning times are typically 40 hours.

Details of each method are given in Appendices L and M.

Practical aspects

Studies have shown that steam side deposits which foul condensers are principally a porous mixture of hydrated iron oxides (nominally $Fe_2O_3 \cdot 2H_2O$) together with the corrosion products tenorite (CuO) and zincite (ZnO) formed in service. In some instances oil has been found incorporated into the deposit.

Usually a clean will only be justified when the deposit contains oil, etc., as well as metal oxides. Table 4.5 gives the composition and thickness of some actual examples (Furlong, et al, 1982).

It should be noted that even a small amount of oil will render the deposit hydrophobic, making removal by

aqueous-based systems impracticable. Therefore prior to cleaning, the oil content of the deposit should be determined as this will influence the final formulation of the cleaning solvent and the process requirements. Oil content should be determined by solvent extraction from segments of individual tubes and the metal oxide burden found by laboratory cleaning tests. Great care should be taken to ensure that representative tube samples are used for these tests. Guidance on the selection of a cleaning process is given in Appendix N.

Inhibitors

Extremely low base metal corrosion rates are a feature of the available processes and hence the use of corrosion inhibitors is not necessary.

The plating-out of copper on to ferrous substrates is not a significant problem, so inhibitors for its suppression are unnecessary.

Water requirements

During the clean there will be a requirement for large quantities of high quality water for preparing solutions and rinsing. The use of water having a significant calcium content is prohibited as it may cause the precipitation of calcium oxalate and a subsequent rise in solution pH.

Although the clean is carried out at ambient temperature, if possible a supply of high quality hot water (e.g., blowdown) should be available to assist and hasten the dissolution of oxalic acid crystals when the cleaning solution is being prepared.

Temporary holding tanks

Sufficient temporary holding tanks should be provided to hold the pre-mixed solution prior to transferring it to the condenser, and to hold the effluent at the end of the clean.

Mixing procedure

The mixing procedure and the quality of the tetrasodium EDTA as received are particularly important to ensure complete dispersal of the surfactant and that the pH is at the optimum for the process of oxide dissolution (4.5 for 0.08 mm NA_4 EDTA and 0.08M oxilic acid). If necessary, formic acid may be used for pH adjustment.

Circulation and agitation

Throughout the clean, the solution should be circulated by external pumps and agitated by nitrogen sparging to ensure good mixing. Air agitation must not be used, as this causes corrosion of the condenser.

Duration of clean

The clean should be allowed to proceed until analysis shows iron uptake has ceased or the total dissolved metal ions are equivalent to 85% of the available capacity.

TABLE 4.5

Condenser steam side — measured fouling burdens on individual tubes

Station	Ratcliffe				Willington	Ironbridge
Unit	1	2	3	4	B5	B1
Fe, gm^{-1}	19.4	24.4	43	130	70	34.9
Cu, gm^{-1}	8.8	13.4	9.2	36.5		16.9
Zn, gm^{-1}	9.4	2.3	1.85	15.3		5.4
Oil, gm^{-1}	4.4	6.7	83	7.9		0.75
μm	50	60	110	65–100	110	65

Experience has shown that provided the correct pH is attained, cleaning is completed in about 40 hours.

Effluents

Once the process is complete, the spent solution should be discharged to the holding tanks prior to disposal.

Engineering aspects

Care must be taken to ensure that the condenser steam side is completely isolated from the rest of the steam/water circuit prior to cleaning.

If air extraction pumps are used to draw off degreasant vapour, traps must be inserted to ensure that the pumps do not draw over any of the cleaning liquor.

A temporary sight glass should be provided to indicate the level of solution in the condenser steam side, and a vent/overflow should be provided just above the anticipated maximum solution level.

Tube plugs

Plastic tube plugs fabricated in Alkathene, a proprietary polythene, are not completely resistant to either of the approved degreasants and may fail either during a clean or on return to service. Such plugs should be replaced with brass plugs prior to cleaning.

Analytical methods

During the course of the clean it will be necessary to monitor at regular intervals for dissolved iron, copper and zinc, together with nickel in the case of stations where cupro-nickel tubes are fitted.

4.3.5 Feedheaters

General

Pre-service cleaning of feedheaters was formerly carried out using an acid/neutralisation/passivation sequence much the same as that currently available for boilers. However, the introduction of 'at works' cleaning, followed by preservation of the heater by means of vapour phase inhibition or dehumidification, largely superseded on-site cleaning which was then only carried out to remedy occasional shortfalls in the performance of the newer technique.

Where *post-service* chemical cleaning of feedheaters is required solely for the removal of inorganic constituents, this should be accomplished by employing appropriate stages referred to in Appendix G. The situation becomes more complicated, however, when organic contamination is present due to the ingress of turbine lubricating oil into the feedwater. In passing through the feed system, the oil appears to undergo progressive degradation so that whilst in the LP heaters it is easily removed by an alkali/detergent treatment, the deposits in the HP system may comprise an intractable mixture of degraded/carbonised organic matter and metallic oxides, and resist this treatment. In consequence a process was developed which is detailed in Appendix P.

Objective

As for condensers, the objective of feedheater cleaning is the restoration of the thermal performance of the feedheater by removal of contaminant material from the tubes.

Criterion

The criterion for cleaning is the loss of heater efficiency due to waterside fouling. Examination of removed tube samples would provide a useful check, as would any history of oil ingress into the feed system.

LP feedheaters

Deposits of an oily nature present in the LP feedheaters can be readily removed by using one of the many proprietary alkali/degreasant treatments which are available.

HP feedheaters

The processes currently available for post-service cleaning of HP feed heaters include:

• A chemical process incorporating degreasant and acid stages designed to remove both organic and inorganic contaminants.

• A physical process using high velocity water jetting.

However, the high velocity water flushing is not applicable to all types of feedheating plant; and cannot be generally recommended at the present time. Therefore the only process which can be recommended is that given in Appendix P, although it should be recognised that experience of this process and its application is limited and due attention should be given to the practical considerations that follow.

Practical aspects

Deposits on feedheaters appear, on the limited evidence available, to vary widely both from unit to unit and, on a given unit, along the feed train. In many cases, only inorganic oxide is present, partly grown, partly deposited. This usually presents no problem with regard to heat transfer and the question of cleaning does not arise.

Where oil fouling is present, it will typically be worst on the hottest HP heater, and progressively less on the previous heaters.

However, in severe cases, the entire train will be contaminated. The difficulty of removal is due to the mixed inorganic/organic nature which has been firmly 'baked-on' over a period of time, so an iterative procedure has been developed and applied.

Plant to be included

The process given in Appendix P has been applied to both full HP feed heater trains and single HP feedheaters. The former can present problems in ensuring correct and complete circulation in all areas of

351

the plant being cleaned and therefore it may be preferable both technically and economically to clean HP feedheaters singly or in pairs, if practicable, particularly if the contamination is localised.

Examination of tubes prior to cleaning

The nature of deposits present on the HP feedheater surfaces will vary considerably according to the source and extent of contaminants. The process given in Appendix P has been formulated to deal with a wide combination of likely contaminants. However, wherever possible, tube samples should be removed for laboratory trials to assess the effectiveness of the process, prior to its application.

Prevention of ingress of cleaning solution to other plant items

Care must be taken to minimise the risk to other sections of the plant by secure isolation from the boiler plant or disconnection, as would also be appropriate in the case of the condenser.

Care must be taken to ensure that all the cleaning solution has been removed by extensive flushing before the plant which has been cleaned is reconnected to the condensing or boiler plant, and this means dealing with each tube individually. Particular attention should be paid to the boiler water analysis when returning the plant to service, to confirm the absence of cleaning solution contamination.

4.3.6 Turbine blades

The purpose of cleaning turbine blades is to restore turbine efficiency. Turbine blade deposits arise from a variety of causes, notably silica in various forms, some of which are removed by blade washing of the LP turbines when the set comes on load. However, in some circumstances copper deposits form on blades, particularly of HP turbines, owing to the solubility of copper compounds in steam. This is most marked on super-critical units, and special precautions, aimed at limiting ingress of copper to the system, are taken to minimise these deposits in such plant. The effect has also been observed on sub-critical turbines.

As far as possible, preventive measures should be employed. If nevertheless deposition has occurred, mechanical methods of removal should be used. When these have been exhausted, chemical cleaning remains as an option. This was applied successfully to the fixed blades during a survey of Willington No 5 unit.

The formulation used contained a reducing agent to reduce the oxides to the metal, together with a complexing agent. The deposit analysis was largely cupric oxide, and hydrazine and EDTA were found to be suitable as a first stage. This led to some copper metal being plated on to the blades and diaphragms, so that subsequent oxidation, employing hydrogen peroxide and EDTA, again as the complexant, was needed.

Generally, as much deposit as is practicable should be removed by mechanical means from accessible areas before the application of a chemical clean.

4.3.7 Evaporators

Where still used, evaporators are prone to scaling, despite dosage during operation. Depending upon the solids content of the evaporator feedwater, there may be sufficient carbonate content in the scale to render simple (inhibited) acid treatment viable. Otherwise, where calcium sulphate and magnesium silicates are present as a high proportion of the scale, exposure to EDTA may be necessary.

4.3.8 Turbine lubricating oil system coolers

The cleaning of these coolers is dealt with on the assumption that corrosion has not occurred to an unacceptable degree. Treatment is therefore limited to flushing the oil side with oil to remove suspended matter.

Where this assumption is not justified, chemical cleaning may be desirable. For this purpose, some formulations used for other purposes may be applied. Prior checking is necessary to ensure that system materials are not at risk and that appropriate inhibitors are used.

4.3.9 Other plant items

Superheaters and reheaters are not normally cleaned for the purpose of removing iron oxide during their service life (with the exception of once-through boilers in which the superheater and reheater are automatically included when the evaporative surfaces are cleaned).

If any abnormal requirement to remove iron oxide deposits from any of these plant items should arise, it could probably be met by employing a modification of one of the other processes, but specialist advice should be sought.

However, certain superheaters and reheaters are prone to spalling of steam-grown oxide and, to reduce the risk of particulate material damaging the turbine, chemical cleaning has been advocated.

Apart from this, the phenomenon of off-load reheater corrosion is aggravated by salt deposits — mainly sodium sulphate in the reheater, which arises from steam carryover (and possibly other routes). This may be alleviated by periodic water-washing during an off-load survey, the deposits typically found being readily water-soluble. To minimise associated corrosion of the wetted surfaces, ammonia and hydrazine are added to the water used for washing, but otherwise inhibitors as such are not needed.

5 Cleaning strategies

Whilst inhibitors can usually be found with efficiencies of over 90%, this should not be used as a pretext for cleaning too often, or for relying on cleaning in lieu of good chemical control of the accumulation of products of corrosion or fouling. Ideally, a combined cleaning/

operations strategy should be worked out for the plant's estimated lifetime.

For example, an assessment of the future cleaning requirements of boilers (J. Brown, D. G. Kingerley and M. J. Longster, 1978), pointed out that the cumulative losses included those due to operationally-formed oxide as well as metal consumed by residual corrosion during the cleaning process itself. However, in a well controlled plant, cleaning at suitable intervals, say every four years, could be accomplished within the design corrosion allowance.

In other cases, the basis of the prediction may differ in some respects. For example, in the cleaning of the steam condensers to combat water side scaling, the analogous formula includes a component for the build-up of operationally-formed oxide according to a parabolic law, rather than the effectively linear kinetics found in most (but not all) cases to apply in boiler practice.

In the context of combating fouling of heat exchangers, the range of options available including preventive measures and mechanical cleaning, with chemical cleaning used only on isolated occasions, there is no problem unless fouling is so severe as to necessitate cleaning many times each year.

For AGR boilers, additional criteria apply over and above those for fossil-fuelled plant. These include; flow imbalance between boilers, flow imbalance between tubes, corrosion under thick oxides in the evaporator, and reduction in heat transfer owing to the presence of deposits. These factors are allowed for when applying the cumulative loss concept to AGR boilers.

In practice, the difficulties associated with cleaning the AGR boilers have led to stringent and expensive measures to avoid it. For example, the use of ultra-pure water, the maintenance of a specified cation/anion balance, and rig studies to establish the location of and conditions arising in the dry-out zone.

6 Chemical cleaning contractors

It is usual to employ contractors to carry out chemical cleaning unless the extent of the cleaning process is relatively small and of short duration, for example:

- Auxiliary heat exchangers (small coolers).

- Condensers of turbines up to 100 MW in the off-load mode.

- On-load cleans of any duration and not requiring extensive manning.

Contractors should be vetted before being engaged for a first time. Thereafter, their performance should be reviewed from time to time.

Factors to be taken into account in assessing a potential or existing contractor are: the calibre and number of personnel; the type and amount of equipment; their experience in UK and abroad, particularly on large scale plant; the location and capability of their depots to deal with those jobs which are performed off-site. It is also necessary to ensure that a contractor's resources are sufficient to deal with jobs which might be delayed, and perhaps clash with other commitments. Apart from technical requirements, the financial standing of contractors should also be examined.

7 Safety

7.1 General

This section deals with the precautions necessary for the safe handling, storage and use of chemicals, including inhibitors, and also describes the procedures for entry into plant after cleaning and disposal of residues.

The chemicals used in cleaning processes are potentially hazardous. The hazards, together with the storage requirements, are listed in Appendix Q. The officer responsible for the safety aspects of cleaning should be familiar with the appropriate documentation covering the location in question. In addition, suitable protective equipment is listed in Appendix R and emergency provisions and first aid treatment is dealt with in Appendix S.

The safety precautions include measures to deal with the potential eye-toxicity of certain inhibitors. Investigational work (J. Brown and D. G. Kingerley, 1970) has indicated that this toxicity was due to the formation of a short-lived reaction product (a dialkyl carbodimide) when symmetrical dialkyl thioureas are used as inhibitors. If it is necessary to enter confined spaces on plant when this substance may still be present, then full protective clothing, including a positive pressure air-hood must be worn. However, providing that certain precautions are taken regarding inhibitor addition and the plant is adequately ventilated after cleaning, and also that the temperature has fallen to below 30°C, the use of breathing apparatus is not now considered to be necessary.

7.2 Responsible officer

The officer responsible for the overall safety aspects of chemical cleaning should be the project site manager in the case of pre-commissioning cleaning and the station manager where post-commissioning cleaning is being carried out. In both circumstances the station chemist, or his nominated deputy, acts as adviser to the officer responsible for safety and should be the supervising officer to ensure that the safety precautions are strictly observed and to ensure that suitable hygiene and first aid facilities are available. The supervising officer consults with the station's medical service with respect to first aid facilities. Supervision of the required safety precautions by the station chemist should also include supervising the chemical cleaning contractor's employees; any failure to comply with the safety requirements should be notified immediately to the project site manager or station manager as appropriate.

The supervising officer or his deputy should be present at the scene of operations whenever chemicals are being received or are in the cleaning system or during subsequent inspections. The station chemist should ensure that chemicals are introduced by the contractor in the correct manner to the cleaning circuit and that the ventilation and temperature requirements before entry into the plant for inspection or working have been complied with.

7.3 General precautions and personal hygiene

Wherever and whenever there is danger of contact with chemicals, all personnel should:

- Be warned of the nature of the potential hazards and the necessary precautions and be instructed in the correct use of protective clothing and equipment by the supervising officer.

- Observe a high standard of personal hygiene avoiding contact with chemicals, their vapours or contaminated residues.

- Not eat, drink, smoke or inhale substances unnecessarily.

- Avoid wiping nose, eyes or face other than with clean paper tissues.

- Place all debris and chemical residues removed from the plant in clearly labelled polythene bags for subsequent disposal.

- On leaving the plant remove their protective clothing for cleaning before re-use, and clean themselves and their equipment using disposable materials.

- Wash themselves thoroughly as soon as possible after leaving the operational area.

7.4 Storage of chemicals

Chemicals should always be stored in a cool, well-ventilated and secure building which, as far as possible, is free from fire risk. Because of their particular properties and in some cases their incompatibility, some chemicals call for special storage precautions; these are given in Appendix Q. If any chemical is to be used which is not already listed, the advice of the medical adviser should be sought and the appropriate authorities notified so that the information in Appendix Q can be updated.

Chemical containers should be clearly labelled at all times. The supervising officer should ensure that notices detailing the action to be taken in the event of spillage, the emergency provisions and first aid treatment, are prominently displayed in the chemical storage and operating area. The notice should state the location of equipment necessary to deal with spillage or personal contamination. An adequate water supply should be available to deal with any leakage or spillage of chemicals. See Appendix T for details regarding hydrofluoric acid.

Stock control should be implemented so that there is a sequential turnover of chemicals.

The chemical cleaning contractor should provide suitable equipment to deal with spillage or personal contamination. This equipment should be readily available and properly maintained.

7.5 Handling chemicals

Manual handling of chemicals should be avoided as far as practicable. Wherever possible, mechanical aids should be used for transporting chemicals and for dispensing and transferring chemicals from containers. See Appendix T where hydrofluoric acid is involved.

The appropriate protective clothing and equipment specified in Appendix R should be worn when handling chemicals.

Spillage, splashing or the creation of dust or accumulation of vapours should be avoided as far as is reasonably practicable.

7.6 Precautions during the cleaning process

The chemical handling, mixing and temporarily hazardous areas should be roped off, cleared of extraneous matter, and have proper warning notices erected and authorised by the supervising officer. Persons not normally concerned with the cleaning process should be excluded from the area.

Whenever chemical solutions are in the plant the entire circuit, both temporary and permanent, should be periodically examined for leaks. Any leakage should be stopped as quickly as possible and the contaminated area thoroughly sluiced down. If the supervising officer deems the leak to be dangerous to personnel, the process should be suspended and the affected circuit drained and flushed prior to repair.

Appropriate protective clothing should be worn while examining for and dealing with leaks (refer to Appendix R for specific chemicals and Appendix T where hydrofluoric acid is concerned).

Entry into any confined space into which solutions or vapour can leak from the cleaning circuit during the process should be avoided as far as is reasonably practicable. If entry into such spaces must be made, the precautions given in the following Section 7.7 should be observed.

Whilst chemical cleaning operations are in progress, venting of the boiler, including the vent from the constant head overflow, should be to the outside of the building into a situation inaccessible to personnel and well away from other vents or intakes, to avoid the possibility of vapour re-entering the building.

7.7 Approach to plant (excluding body entry) after any stage of chemical cleaning

When the plant is to be opened up *after* the passivation stage or after any acid stage *during* the process, the following precautions shall be observed.

The chemical cleaning circuits should be drained to the approved disposal point taking care to minimise spillage, splashing of solutions or accumulation of vapour. If chemicals are drained into a holding vessel care should be taken that mixing of chemicals from different stages does not occur in the vessel. It is particularly important to avoid:

(a) Acidifying solutions containing nitrite or bromate.

(b) Mixing strong acids and alkalis.

(c) Mixing strong oxidising and reducing agents.

Contact with vapour emerging from drains and vents should be avoided as far as is reasonably practicable. Where this is unavoidable the precautions given in the next Section 7.8 should be observed. Vapours issuing from plant openings should be dispersed by blowing clean air through the access point using the best possible ventilation to obtain a rapid dilution. The breaking open of access points should be carried out by personnel wearing an approved type of coverall and gloves, and breathing apparatus incorporating a hood to give a full protection to the head. Suitable equipment is listed in Appendix R.

Contact with surfaces which are or have been wetted by chemicals should be avoided as far as possible.

Components which are contaminated with chemicals should be placed on impervious sheets for cleaning prior to replacement.

Personnel should not expose themselves to risk by entering plant areas without observing the following precautions.

7.8 Entry of personnel into plant which has been chemically cleaned

The following precautions are necessary if entry is to be made into plant areas after the passivation or acid stages of the cleaning process.

Entry to the plant should be restricted to the minimum number of people necessary.

During the time that personnel are inside the plant, a standby person should be positioned at the point of entry.

Before entering plant after chemical cleaning there shall be a minimum delay of three hours after opening all entry points. During this period ventilation of the plant should be carried out at a rate of at least 10 air changes per hour using a filtered air supply. It is essential that this three hour delay period is observed, even if a higher ventilation rate is adopted. Ventilation of the plant at a rate of at least 10 air changes per hour should be continued whilst personnel are in the plant area for inspection, or to carry out reinstatement work such as removal of flow restrictors.

Additionally, before personnel enter the plant area, the temperature should be allowed to fall below 30°C, if weather conditions permit.

When the safety conditions have been fully complied with, personnel entering the plant should wear a coverall, PVC gloves and carefully fitted eye-goggles as given in Appendix R. Where such conditions have not been fully complied with personnel entering the plant should wear a coverall, PVC gloves, rubber boots, breathing apparatus and hood attachment giving full protection to the head as given in Appendix R.

Removal of debris or chemical residues from the plant should be carried out so as to avoid contamination of external surfaces.

Ideally, the residue should be scooped up into impervious containers and passed out to persons wearing gloves and goggles (as given in Appendix R).

7.9 Exits of personnel and materials from plant containing hazardous residues

After chemical cleaning, debris and chemical residues should be placed in clearly labelled containers within the plant and then disposed of into a system where considerable dilution can be achieved. Equipment used for handling chemicals should be cleaned after use by also washing into a disposal system where considerable dilution can be achieved.

Personnel protective clothing should be used for the duration of the work period and then thoroughly cleaned or discarded; the operator then thoroughly washing the exposed parts of the body.

7.10 Disposal of chemical waste

Disposal of all surplus solutions and chemical waste should be by an approved route in line with National legislation and/or local regulations.

Appendix A

Water flushing

Flushing of plant with water will be required for the following purposes:

- To remove any contaminant which has entered the plant during erection and any loose corrosion products.

- To remove volatile corrosion inhibitors if used for protection of plant during storage and erection.

- To remove chemicals and residual sludge after site cleaning processes.

Notes:

1 This procedure may be unsuitable for large vessels, e.g., DC heaters, because of weight support limitations.
2 After water flushing has been completed, the plant should either be dried out or put in wet storage unless the flushing immediately precedes a further stage of cleaning.
3 Water flushing should be carried out using deionised water if austenitic components are present or for once-through boilers. If any other plant is given a preliminary flush with towns main water, it should be thoroughly flushed out afterwards with deionised water, except where it is to be followed by a chemical cleaning process.

Appendix B

Steam purging

B1 Introduction

Effective steam purging requires extensive temporary pipework and involves complex theoretical considerations which have to be studied for each individual design. It is known that to be fully effective it is necessary to achieve a scouting action at least equivalent to that of full-load steam flow. Design Memorandum 068/25, 'Steam Purging', gives a method for calculating sizes of pipework, together with the necessary chart. Since the issue of this generation design memorandum, there is now a computer program available for carrying out the required calculations. Practical guidance, based on experience of steam purging at many stations is also included in the design memorandum.

B2 Steam blowing

During steam blowing there is an increased risk of water carryover since the pressure drop which occurs is higher than in normal operation. To minimise the risk of stress corrosion, sodium hydroxide should not be added to the boiler water during steam blowing. The boiler should initially be filled with 50 mg/kg hydrazine and 50 mg/kg ammonia. No further additions of hydrazine are required, but 50 mg/kg hydrazine and 50 mg/kg ammonia should be added to the make-up water to the boiler to react with dissolved oxygen. Before commencing the steam blowing process, it should be proved that all cleaning chemicals have been thoroughly flushed from the boiler water circuit. The boiler water should not contain more than 0.1 mg/kg chloride (as Cl^-) or 0.2 mg/kg sodium (as Na^+).

Appendix C

Constant head overflow and temporary gauge glass arrangements

C1   Constant head overflow

In addition to water plugging and back-flushing the superheater, a constant head overflow device should be fitted to the drum to ensure that chemical cleaning solutions do not enter the superheater. A suitable arrangement is shown in Fig 4.16.

The overflow device may be fitted using a temporary drum door. It is not envisaged that the overflow would be required to take full flow of the chemical circulation pumps and a 100 mm line should be adequate. Particular care will be necessary when initially adding the acid to the water in the boiler so as not to exceed the working level. In this context, a temporary gauge glass extending the full height of the drum should be provided.

In the event of a stoppage of the acid circulation pump, the level in the drum may rise due to water draining back from the tubes. This factor should be taken into consideration in determining the height of the overflow, which should be positioned 75–100 mm above the required working level.

The valve shown in Fig 4.16 should be open at all times except when it is required to rinse the upper surfaces of the drum. Before such rinsing is carried out, it should be ensured that the solution contains less than 2 mg/kg chloride.

Provision should be made at the overflow outlet for the boiler to be vented outside the building, in a position inaccessible to personnel, and for any liquid overflowing to be led to a convenient sump where it may be neutralised if necessary. This may be achieved by leading the overflow outlet into a small tank from which the liquid can be drained and the vapour vented outside the buildings.

C2   Temporary gauge glass

A temporary gauge glass is required since the normal means of level indication cannot be used for several reasons, these are:

- The temporary level during cleaning is higher than the maximum of the permanent indicator.

- The cleaning chemicals cannot be rinsed out adequately from the permanent indicators without a risk of corrosion.

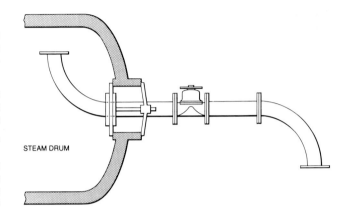

FIG. 4.16 Diagrammatic arrangement of constant head overflow in boiler drum

- The difficulty of reading the level on an otherwise out-of-service boiler by an operator stationed locally to the permanent glass.

- Increased use of Hydrastep water level gauge.

Experience shows that the boiler drum becomes slightly pressurised during temperature raising and, with an open-ended temporary gauge glass, a false indication of drum level is obtained. Both links of the indicating column must be connected to the drum. It is therefore advised that permanent tapping points are provided for drum level indication during chemical cleaning. Each tapping point should be fitted with double-isolating valves and a blanked flange suitable for full boiler pressure. The lower tapping point preferably should be from a line other than the permanent gauge glass impulse line, e.g., the boiler dosing or sampling line. The top tapping point could be teed into a drum air release line before the normal air valve.

Appendix D

CEGB approved inhibitors for boilers

The inhibitors which are approved for use during acid cleaning processes and the recommended concentrations are:

- For 5% w/w hydrochloric acid, use 0.25% w/w Armohib 28.

- For 1 or 2% w/w hydrofluoric acid, use 0.1% w/w Dodigen 95 (Note 1).

- For 3% w/w citric acid, use 0.05% w/w Stannine LTP or preferably 0.05% w/w Dodigen 95 (Note 2).

The inhibitor concentrations are based on the manufacturer's recommendations and have been shown to be technically acceptable in tests carried out by the CEGB.

The approved inhibitors should be present under acid conditions at all times to ensure that they are fully dissolved and not subsequently precipitated. They should always be added direct to an acid

solution and not be previously diluted with water or added to water to which the acid is added later.

The safety requirements for the addition of inhibitors to the cleaning solution should be strictly observed.

Storage and handling of the approved inhibitors should be in accordance with Appendices Q and R.

Notes:

1   With hydrofluoric acid, the material should be brought onto site as a dilute solution (typically 15% w/w) to which the inhibitor has already been added, to eliminate the need to dilute concentrated solutions before injecting into the boiler.

2   Dodigen 95 is preferred because its chemical composition is such that it is incapable of producing the hydrolysis product which has been associated with eye toxicity in the use of Stannine LTP.

Appendix E

CEGB approved inhibitors for condensers

If condensers are to be cleaned frequently (i.e., more than once per year), it is necessary to use inhibited hydrochloric acid or inhibited sulphuric acid.

The inhibitors which are approved for use during condenser acid cleaning processes and the recommended concentrations are:

- For 1.5% w/w hydrochloric acid, use 0.25% w/w Armohib 28 or 0.025% w/w Armohib 533 (see *Note*).

- For 1.5% w/w sulphuric acid, use 0.05% w/w Armohib 533 (see *Note*).

The inhibitor concentrations are based on the manufacturer's recommendations.

The approved inhibitors should be present under acid conditions at all times to ensure that they are fully dissolved and not subsequently precipitated. They should always be added direct to an acid solution and not previously diluted with water or added to water to which the acid is added later.

The safety requirements for the addition of inhibitors to the cleaning solution should be strictly observed.

Storage and handling of the approved inhibitors should be in accordance with Appendices Q and R.

For ferrous metals specifically

Hydrochloric acid inhibited with Armohib 28 or Armohib 533 or sulphuric acid inhibited with Armohib 533 may be used.

For ferrous metals and brass together

Hydrochloric acid or sulphuric acid inhibited with Armohib 533 is advised.

Note: Armohib 533 is a commercial formulation of MBT.

Appendix F

Pre-service cleaning of drum-type boilers and economisers (excluding superheaters and reheaters)

The following stages constitute a complete clean and must be carried out consecutively and with minimum delay.

Precautions must be taken to ensure that the chemical cleaning solutions used in Stages 3 and 5–8 do not enter the superheater. During Stage 3 the water level should not be taken above normal working level. During Stages 5–8, the superheater should be filled with deionised water to form a water plug and back-flushing should be carried out periodically. In addition, during Stages 5–8, a constant head overflow device should be fitted to the drum (see Appendix C):

Stage 1 Fill and drain the boiler as far as possible using water of a quality at least equivalent to a towns main supply, to remove foreign matter.

Stage 2 Flush the superheaters and reheaters intermittently using deionised water until the water leaving the superheater and reheater has a conductivity of less than 5 μS/cm and a chloride content of less than 1 mg/kg chloride.

Stage 3 Fill the boiler to working level with towns main demineralised water containing 1000 mg/kg trisodium phosphate. Raise the pressure to 40 bar with intermittent blowing down for 30 seconds every hour. The final pressure should be reached in 8–10 hours.
Maintain the boiler at 40 bar pressure with the specified chemical concentration for at least 24 hours. Drain while the boiler is still as hot as practicable and, if possible, whilst under pressure.

Stage 4 Open the boiler drum and distribution drums (if any) for examination and remove any loose deposits.

Stage 5 Circulate through the plant a solution in deionised water containing:
(a) 3% w/w citric acid, an approved inhibitor (Appendix D) and ammonia to give pH 3.5–4.0; or,
(b) 1% w/w hydrofluoric acid and an approved inhibitor (Appendix D).
If (a), the solution should initially be circulated at not less than 90°C and circulation should be continued until the iron content is constant (*Note 1*). The concentration of free citric acid should be maintained at not less than 0.5% w/w during circulation.
If (b), circulate at 55°C until the iron concentration is constant (*Note 1*). The concentration of free hydrofluoric acid should be maintained at not less than 0.2% w/w during circulation (see *Note 2*).

Stage 6 Raise the pH to 5.0 by careful addition of ammonia added slowly during circulation to avoid high pH values (*Note 3*).

Stage 7 Drain the boiler and when using ammoniated citric acid, rinse with a solution containing 0.2% w/w citric acid and ammonia to give pH 3.5–4.0.

Stage 8 Flush the plant with deionised water until pH remains above 6.0.

Stage 9 Fill the plant with deionised water containing 50 mg/kg hydrazine and 50 mg/kg ammonia.
Maintain a minimum pH of 10.0.
Raise the pressure to 40 bar and maintain for not less than 24 hours. No further additions of hydrazine to the boiler are required since this has been shown to be unnecessary. However, 50 mg/kg hydrazine and 50 mg/kg ammonia should be added to the make-up water to the boiler to react with dissolved oxygen.

Stage 10 On completion of Stage 9, blow the boiler empty at the highest possible pressure, but in any event not less than 4 bar.

Notes:

1 Experience shows that, providing the system has been lagged, the temperature should not have fallen below 75°C before cleaning has been completed when using ammoniated citric acid or 50°C with hydrofluoric acid.

2 Hydrochloric acid is not recommended

3 The reasons for raising pH to 5.0 are:
- To minimise corrosion of the plant after cleaning but before displacement of acid.
- To avoid corrosion in discharge points and culverts.

Appendix G

Post-commissioning cleaning of drum-type boilers and economisers

The following stages constitute a complete clean and must be carried out consecutively and with minimum delay (*Note 1*).

Precautions must be taken to ensure that the chemical solutions do not enter the superheater. The superheater should be filled with deionised water to form a water plug and back-flushing should be carried out periodically. In addition, a constant head overflow device should be fitted to the drum (see Appendix C).

Stage 1   In some circumstances this copper removal stage may be unnecessary (*Note 2*) and in others it may be carried out at reduced strength (say 50%) of the solution specified below.

Otherwise, fill the boiler with a solution containing 1% w/w citric acid, 0.5% w/w sodium bromate and ammonia to give a pH of 9.5.

Approximately 0.5% 0.880 ammonia will be required, the bulk of which should be added before the citric acid is introduced. The solution should be made in deionised water and circulated at a temperature of 50°C until the concentration of copper is constant (approximately 4–6 hours). Drain the boiler and backflush the superheaters with deionised water.

Stage 2   A choice has to be made in any given case between hydrochloric, hydrofluoric and ammoniated citric acids for iron oxide removal. Then the boiler should be filled with a solution of:

(a) 5% w/w hydrochloric acid (Fig 4.14), 0.5% w/w ammonium bifluoride and an approved inhibitor (Appendix D); or,

(b) 1 or 2% w/w hydrofluoric acid (Fig 4.14) and an approved inhibitor (Appendix D); or,

(c) 3% w/w citric acid, an approved inhibitor (Appendix D) and ammonia to give pH 3.5–4.0.

If (a), circulate at 65°C (*Note 3*) until the iron and nickel concentrations are constant (normally 6–8 hours).

If (b), circulate at 55°C (*Note 3*) until the iron concentration is constant (normally 3–4 hours).

If (c), the solution should initially be circulated at 90°C (*Note 3*) and continued until the iron concentration is constant.

Stage 3   Drain the boiler and when using hydrochloric or ammoniated citric acid rinse with a solution containing 0.2% w/w citric acid and ammonia to give a pH of 3.5–4.0.

Stage 4   Copper removal will almost invariably be required at this stage unless the water/steam circuit contains no copper alloys. The standard formulation and procedure is given in Stage 1.

The full strength solution, however, should be used only when the boiler is known to be heavily contaminated with copper and its oxides.

Where it is judged that copper contamination is not heavy, the 50% solution should be employed.

Stage 5   Drain the boiler and thoroughly backflush the superheaters with deionised water which has conductivity less than 0.5 μS/cm. Rinse the boiler with deionised water to remove the cleaning solution.

Stage 6   Refill the boiler with deionised water and add 50 mg/kg hydrazine and 50 mg/kg ammonia, maintaining a minimum pH of 10.0. Raise the pressure to 40 bar and maintain for at least 24 hours. No further additions of hydrazine to the boiler are required to maintain a reserve in the boiler water, since this has been shown to be unnecessary. However, 50 mg/kg hydrazine and 50 mg/kg ammonia should be added to the make-up water to the boiler to react with dissolved oxygen.

Stage 7   On completion of the passivation process, the boiler should be blown empty at the highest possible pressure, but in any event not less than 4 bar.

Notes:

1   No alkali boil-out is generally necessary for the post-commissioning cleaning of boilers, on the assumption that oil and siliceous materials will not be present. Where this assumption is unjustified, Stage 1 should be preceded by an alkali boil-out as specified in Appendix F, Stages 3 and 4.

2   The capacity for copper of the formulation given in Stage 1 is 6.3 Cu/kg. Only rarely is more than 30% of this capacity required. Depending upon local judgement as to the amount of copper in any boiler, it may be possible to omit Stage 1.

3   Cleaning rates fall with decreasing temperature. Experience shows that typical temperatures at the end of the iron removal stage are 70°C with citric acid, 60°C with hydrochloric acid and 50°C with hydrofluoric acid.

Appendix H

Post-service cleaning of once-through boilers including superheaters and reheaters

The following stages constitute a complete clean and must be carried out consecutively and with minimum delay.

Stage 1 Flush with deionised water at a velocity greater than 1.5 m/s in all sections of the plant.

Stage 2 Circulate a solution in deionised water containing:
- 3% w/w citric acid (*Note 2*) and 0.5% w/w formic acid (*Note 6*).
- An approved inhibitor (see Appendix D).
- Ammonia to give pH 3.5–4.0.

The solution shall be initially circulated at not less than 90°C and continued until the iron content is constant. The concentration of free citric acid should be maintained at not less than 0.5% w/w during circulation.

Stage 3 Raise the pH to 5.0 by careful addition of ammonia; added slowly during circulation to avoid local high pH values (*Note 3*).

Stage 4 Drain and rinse with a solution in deionised water containing 0.2% w/w citric acid and ammonia to give pH 3.5–4.0.

Stage 5 Flush with deionised water at a velocity greater than 1.5 m/s until the pH remains above 6.0.

Stage 6 Circulate a solution in deionised water containing 300 mg/kg hydrazine and 50 mg/kg ammonia.

Raise the temperature of the solution to at least 90°C and circulate for not less than 24 hours. Further additions of hydrazine should be made if necessary to maintain a concentration of not less than 25 mg/kg.

Stage 7 On completion of Stage 6, drain the plant at the highest possible temperature.

Notes:

1 In the post-commissioning cleaning of once-through boilers, it will not normally be necessary to include the feed system. Where copper is present, a copper removal stage as specified in Appendix G Stage 1 should be introduced between Stages 4 and 5.

2 The chloride content of the citric acid used should be such that a 3% w/w solution does not contain more than 2 mg/kg chloride as Cl^-.

3 The reasons for raising the pH to 5.0 are:
- To minimise corrosion of the plant after cleaning but before displacement of acid and subsequent flushing.
- To avoid corrosion in discharge points and culverts.

4 After Stage 7, the plant should be dried out or wet-stored.

5 In the event of the need to clean condensate or feed systems after service, the processes required could be based on this Appendix.

6 Formic acid is required to obtain an acceptable rate of reaction for dissolving the outer layer of duplex oxides from superheaters.

Appendix I

Hydrochloric acid process for the off-load post-service cleaning of condensers

The condenser should be made available for cleaning with all plant isolation, depressurising and draining complete.

Circulate at ambient temperature a solution containing 1.5% w/w hydrochloric acid and an approved inhibitor (see Appendix E), until the calcium concentration is constant and an excess of at least 0.5% w/w of acid still remains. Open the vents at intervals to release carbon dioxide liberated by the descaling process.

On completion of cleaning, drain the condenser and pass normal cooling water flow for 30 minutes. Then drain and inspect.

Appendix J

Sulphuric acid process for reduced-load post-service cleaning of condensers (excluding units with once-through boilers)

The part-condenser to be cleaned should be made available with appropriate plant isolated and drained.

Circulate at ambient temperature a solution containing 1.5% w/w sulphuric acid and an approved inhibitor (see Appendix E), until the calcium content is constant and an excess of at least 0.5% w/w of acid still remains. The temperature of the cleaning solution will increase as it takes up heat from the system up to a maximum of about 40°C, depending upon the particular plant.

Open the vents at intervals to release any carbon dioxide liberated. On completion of cleaning, drain the condenser and return to service.

Appendix K

Sulphuric acid process for on-load post-service cleaning of condensers (excluding units with once-through boilers)

As experience of on-load descaling of condensers is successful but limited, this method should be considered only when off-load or reduced-load cleaning cannot be applied.

The method is, therefore, given only in outline and detailed guidance should be sought before undertaking an on-load clean.

The process consists of adding sulphuric acid to reduce the pH to 2–2.5 locally within the condenser, yet restricting the total sulphate concentration of the cooling water to less than 700 mg/kg SO_4 and maintaining the bulk pH above 6.0 to minimise the risk to the concrete and metals in the cooling water circuit.

It has been found acceptable, at some locations, to introduce the acid via the chlorination system, with the additional advantage that the isolating valves facilitate dosage of individual passes of the condenser.

For any system; the practical details of on-load cleaning depend upon the system volume, the severity of scaling, the water analysis and the individual condenser design. The water quality is kept within the required limits of sulphate concentration and pH by varying the duration of acid addition and by purging when necessary.

An assessment of the probable effect of dosing on a given cooling water chemistry is advised where on-load cleaning is being considered.

Appendix L

Cleaning method for removal of oil-free deposits from the steam side of condensers (*Note 1*)

The condenser should be made available for cleaning with all plant isolation, depressurising and draining complete.

Circulate at ambient temperature a solution containing 0.08–0.13M EDTA (added as the Na_4 EDTA salt), 0.08 oxalic acid, and formic acid to give a pH of 4.5, until the iron concentration is constant or until the total metal ion concentration is 85% of the available capacity, when further dissolution rates are minimal.

On completion of cleaning, drain the condenser, rinse and inspect.

Note 1: Deposits containing less than 5% oil.

Appendix M

Cleaning method for removal of oil-containing deposits from the steam side of condensers (*Note 1*)

The condenser should be made available for cleaning with all plant isolation, depressurising and draining complete.

Circulate at ambient temperature a solution containing:
- 0.08–0.13M EDTA (added as the Na_4 EDTA salt) and 0.08M oxalic acid.
- Up to 10% v/v approved degreasant (see *Note 2*).
- 1% w/v Ethomeen S25.
- Formic acid to give pH of 4.5.

Circulate until the iron concentration is constant or until the total metal ion concentration is 85% of the available capacity, when further dissolution rates are minimal.

On completion of cleaning, drain the condenser, rinse and inspect.

Notes:

1 Deposits containing more than 5% oil.
2 Commercial available degreasants suitable for use in the steam side cleaning of condensers are based on C9 Naptha. As supplied, the degreasants normally contain C9 Naptha nonylphenyl surfactants and dodecylbenzene sulphonate surfactants.

A number of commercially available degreasants have been evaluated and two have been approved for use in this application. These are Industrial Polyclens and BannerSolve H536.

It should be noted that with both these degreasants it is necessary to add the additional surfactant, Ethomeen S25, to obtain the required emulsion stability for the duration of the clean.

Appendix N

Selection of cleaning process for condensers

N1 Selection of cleaning process

Let Cu = average copper burden, g/m^2

Fe = average iron burden, g/m^2

Ni = average nickel burden, g/m^2
(applicable only if cupro-nickel tubes are fitted)

Zn = average zinc burden, g/m^2
Cu, Fe, Ni and Zn are determined by chemical cleaning tests

V = volume of solution required to fill the condenser steam side, litres

A = total surface area of condenser tubes, m^2

B = total metal ion burden, moles

$$B = A\left(\frac{Cu}{63.5} + \frac{Fe}{56} + \frac{Ni}{59} + \frac{Zn}{65.4}\right)$$

If $B \leqslant 0.048V$ use 0.08M EDTA

$B > 0.048V$ use up to 0.13M EDT
(required concentration = 1.73/V)

$B > 0.078$ obtain scientific advice

The requirement for a degreasant can be established by solvent extraction and comparative cleaning tests. Normally, even the presence of a small quantity of oil will necessitate the use of degreasant. It is extremely important when determining fouling burden to ensure that representative tube samples are used.

When establishing the volume of solution required to fill the steam side of the condenser, consideration should be given (subject to the design of the condenser) to blanking off 'dead' areas or filling with gas bags thereby reducing solution requirements.

Appendix P

Degreasant/hydrochloric acid process for the post-service cleaning of HP feedheaters

Stage 1 Circulate a solution of a 50% v/v 'Applied Chemical 4–43' (see *Note 2*), at a temperature of 80–90°C for 10 hours.

Stage 2 Flush to waste at a flow equivalent to full feed flow until a conductivity of <2000 μS/cm is recorded.

Stage 3 Circulate a solution of 5% w/w hydrochloric acid inhibited with 0.25% w/w Armohib 28 at a temperature of 75°C for 2 hours.

Stage 4 Flush to waste at a flow equivalent to full feed flow until a conductivity of <2000 μS/cm is reached.

Stage 5 Repeat Stage 3.

Stage 6 Repeat flush to waste at a flow equivalent to full feed flow until a conductivity of <5 μS/cm is reached.

Notes:

1 If the HP feedheaters being cleaned are heavily fouled, it may be necessary to repeat all stages.

2 An equivalent alkaline detergent may be used if Applied Chemical 4–43 is not available.

Appendix Q

Summary of main chemical hazards and special storage requirements

Chemical	OEL (*Note 1*)	Main hazards	Special storage requirements
Acetic acid	10 PPM	Corrosive, irritant can cause severe burns. Flammable.	Store away from chromic and nitric acids.
Chromic acid	0.05 mg/m^3 (as Cr)	Can cause severe burns. Can cause violent explosion in contact with reducing agents.	Store away from oxidising agents and any fire risk.
Citric acid		Powder and solution can cause eye injury.	Store away from oxidising agents.
Formic acid	5 PPM	Corrosive. Can cause severe burns.	Store in a well ventilated area.
Hydrochloric acid	5 PPM	Corrosive and irritant.	Store in a well ventilated area.
Hydrofluoric acid (see also Appendix T)	3 PPM	Toxic and corrosive. Can cause severe burns.	Store in a well ventilated area.
Nitric acid	2 PPM	Oxidising and corrosive. Can cause severe burns.	Store away from acetic acid ammonia and combustible materials.
Sulphamic acid		Emits toxic fumes when heated. Irritant.	Store away from heat.
Sulphuric acid	1 mg/m^3	Extremely irritant and corrosive. Causes severe burns.	Store away from alkalis.
Ammonium bifluoride		Corrosive. Reacts with mineral acids to produce HF gas. Highly irritating to skin, eyes and nose.	Store away from acids. Fire risk.
Sodium fluoride	2.5 mg/m^3	Highly toxic.	Store away from acids and combustible materials.
Sodium hydroxide (caustic soda)	2 mg/m^3 (*Note 2*)	Highly corrosive. Severe irritant. Dust or mist can cause irritation to upper respiratory tract.	Store away from acids.
Potassium hydroxide (caustic potash)	2 mg/m^3	Highly corrosive. Severe irritant. Inhalation of dust or mist causes intense irritation.	Store away from acids.
Dipotasium hydrogen phosphate		Produces toxic, irritant fumes when heated.	
Disodium hydrogen phosphate		Produces toxic, irritant fumes when heated.	
Sodium carbonate (soda ash)		Skin irritant. Dust or mist irritating to upper respiratory tract.	Store away from acids.
Tripotassium phosphate		Skin irritant. If heated emits highly toxic fumes.	
Trisodium phosphate		Skin irritant. If heated emits highly toxic fumes.	
Ammonia	25 PPM	Corrosive. Severe irritant.	Store away from acids and/or bromate.
Hydrazine	0.1 PPM	Strongly caustic. Irritant. May cause damage to liver and kidneys.	Store separately.
Hydrogen peroxide	1 PPM	Oxidising and corrosive. Can cause severe damage to eyes and skin. May explode in contact with dust.	Store separately and away from all combustible materials in a cool area.
Potassium or sodium bromate	0.05 mg/m^3 (as Br)	Oxidising agent. Eye/skin irritant. With acid solutions toxic gases are produced. Possible explosion hazard with ammonia and contact with organic materials.	Store away from acids, combustible materials and ammonia.
Potassium or sodium chromate		Eye/skin irritant. With acid solutions toxic gases can be produced. Possible explosion hazard on contact with organic materials.	Store away from acids and combustible materials.

363

Appendix Q

Summary of main chemical hazards and special storage requirements (*Cont'd*)

Chemical	OEL (*Note 1*)	Main hazards	Special storage requirements
Potassium or sodium dichromate		Eye/skin irritant. With acid solutions toxic gases can be produced. Possible explosion hazard on contact with organic materials.	Store away from acids and combustible materials.
Potassium or sodium nitrite		Eye/skin irritant. With acid solutions nitrous fumes are produced. Possible explosion hazard on contact with organic materials.	Store away from acids and combustible materials.
Potassium or sodium permanganate		Eye/skin irritant. With acid solutions toxic gas may be produced. Possible explosion hazard on contact with organic materials.	Store away from acids and combustible materials.
Inhibitors (Armohib 28, Armohib 533 Stannine LPT Dodigen 95, 2 mercaptobenzo thiazole, etc.)		Can cause skin and eye irritation (some severe). Some combustible/toxic/corrosive.	
Wetting agents (Lissapol, Teepol, etc.)	—	May cause eye and skin irritation.	
Applied chemicals 4–43	—	Can cause skin and eye irritation; if swallowed, gastro-intestinal irritation.	Store away from sources of heat or strong acids.
C9 Naphtha degreasant	—	Flammable.	Store away from fire risk in a well ventilated area.
Oxalic acid	1 mg/m^3	Corrosive. Avoid breathing dust or vapour.	Store away from oxidising agents.
Tetra sodium ethylene diamine tetra acetate	—	Concentrated solution is caustic. Skin and eye irritant.	Store away from acid.

Notes:

1 OEL = Occupational exposure limit

Appendix R

Summary of the protective equipment to be used when handling chemicals

Conditions	Protective equipment
Normal handling of the following: Ammonia Ammonium bifluoride Applied chemicals 4–43 Armohib 28 C9 Naphtha degreasant Citric acid Diaminetetra acetic acid Ethomeen S25 Formic acid Hydrazine Hydrochloric acid Hydrofluoric acid Lissapol C Lissapol N Oxalic acid Sodium bromate Sodium hydroxide Sodium nitrite Stannine LTP Sulphuric acid Tetra sodium eythylene Trisodium phosphate	● Chemical safety goggles ● PVC apron ● PVC gloves ● Rubber boots (moulded) In addition, for particulate chemicals: ● Dust respirator In addition for formic acid, hydrochloric acid and sulphuric acid, ● Visor
For entry into plant (a) After chemical cleaning in which the requirements for inhibitor additions, and temperature have been fully complied with	● Chemical safety goggles ● One-piece cotton/polyester coverall with hood or one-piece disposable overall in PE-coated Tyrek ● PVC gloves ● Rubber boots (moulded)
(b) For handling debris and chemical residues removed from chemically cleaned plant	As for (a)
(c) After chemical cleaning in which the requirements for inhibitor additions, temperature and ventilation have not been fully complied with	● Compressed air line breathing apparatus ● One-piece cotton/polyester coverall with hood or one-piece disposable overall in PE-coated Tyrek ● PVC gloves ● Rubber boots (moulded)
When breaking open access points after chemical cleaning	● Compressed air line breathing apparatus (if entry also required). or ● Open-circuit self-contained breathing apparatus (if entry not required). ● One-piece cotton/polyester coverall with hood or one-piece disposable overall in PE-coated Tyrek. ● PVC gloves ● Rubber boots (moulded)
Where contact with vapour emerging from drains or vents during a chemical process is unavoidable	● Compressed air line breathing apparatus ● One-piece cotton/polyester coverall with hood or one-piece disposable overall in PE-coated Tyrek ● PVC gloves ● Rubber boots (moulded)
(d) Where severe spillage or escape of the following chemicals has taken place: ● Sodium bromate ● Nitrous fumes ● Stannine LTP ● Armohib 28	● Open-circuit self-contained breathing apparatus ● Compressed air-line breathing apparatus ● PVC gloves ● Rubber boots (moulded) ● One-piece cotton/polyester coverall with hood or one-piece disposable overall in PE-coated Tyrek (alternatively, PVC chemical splash suit)

Appendix S

Emergency provisions and first aid treatment

Emergency provisions

Flushing and washing water supplies

(a) Ample supplies of tepid flushing and washing water shall be provided at all possible points of discharge, spillage or escape of chemicals.

(b) Adequate provision shall be made for emergency treatment of the eyes, comprising eyewash bottles, located conveniently to places where discharge, spillage or escape of chemicals can occur.

First aid room

A suitable first aid treatment room with outside telephone facilities shall be provided within reasonable distance of the place where the chemicals are being used.

Store room

A suitable room shall be provided for housing the protective clothing and apparatus required for emergency use.

First aid treatment

Injury	Chemicals	Treatment
Splashes to the eye	All chemicals	• Immediately flood the eye with water. To be effective the eye must be opened. After a quick preliminary swill to wash away fluid around the eye, the eyelids should be pushed, apart using the thumb and index finger of the left hand. The casualty will probably not be able to open the eye himself because of painful spasms. If an eyewash bottle is used the jet should not be directed at the front of the eye but should be directed in from the side, so that flow is over the surface of the eye. • Irrigation should be continued for 5–10 minutes, after which the casualty should be taken to the first aid room. • After thorough irrigation the eye should be covered with a pad; the patient should be referred for medical opinion.
Irritation of the skin	All chemicals	If signs of skin irritation occur the person should be removed from contact and referred for medical opinion. In the event of splashing of the skin with chemicals the affected area should be washed thoroughly avoiding spreading contamination to the face and eyes.
Gassing	All chemicals	Remove person to fresh air, remove contaminated clothing, cover with blanket and keep person still and under observation. Refer for medical opinion.
	Ammonia or nitrogen. Bromine from Sodium bromate. Nitrogen dioxide from sodium nitrite.	Additional to above, if breathing is distressed give oxygen. If breathing fails, give artificial respiration, summon doctor to site.

Appendix T

The special safety case for hydrofluoric acid

T1 Introduction

If allowed to contact the skin or eyes, aqueous solutions of hydrofluoric acid can cause burns, the severity increasing with exposure and concentration of acid (40% or stronger acid will cause severe burns). The onset of pain following exposure may be delayed up to 24 hours. If not treated, the condition can result in extensive and permanent damage which may involve the underlying bone. Accompanying the pain there may be a visible reddening of the damaged skin. Such solutions are rapidly absorbed through the skin into the blood stream and may give rise to acute fluoride poisoning.

Additionally, solutions of hydrofluoric acid containing more than 40% w/w acid give off acid fumes which are extremely irritating and toxic. The recommended threshold limit value for hydrofluoric acid vapour is three parts per million by volume. At this concentration its presence is just detectable. Inhalation of the fumes leads to internal absorption again with the risk of acute fluoride poisoning.

Solutions of hydrofluoric acid in excess of 40% w/w free acid have a significant heat of dilution and addition of water to such solutions may result in violent eruptions. Dilution of strong or even intermediate strength acids should therefore always be achieved by adding the acid to water (as in the well known case of sulphuric acid). However, it can be delivered in dilute form to reduce hazard and, at actual cleaning strength, 2%, is 'relatively' safe to handle.

T2 Services and facilities required for the chemical clean

The *local hospital* and local plant nursing officer must be notified of the intention to use hydrofluoric acid well before the clean commences. The site first aid team should also be informed of the clean and, if necessary, made aware of currently approved methods for the treatment of burns and other injuries from hydrofluoric acid.

T3 First aid equipment and procedures

In addition to normal chemical first aid and rinsing equipment, the following special first aid equipment shall be on-site prior to commencing operations:

(a) A fluoride treatment kit.

(b) Sufficient number of tubes of calcium gluconate gel to enable all personnel involved to be issued with a tube.

(c) Sufficient copies of the appropriate first aid instructions to enable all personnel involved to be issued with a copy.

The first aid procedures for the treatment of exposure to hydrofluoric acid are given later in this appendix.

T4 Protective clothing

Personnel intending to handle hydrofluoric acid solutions stronger than 10% shall be equipped with the following protective clothing:

- Rubber boots with steel toecaps.
- PVC oversuit with hood.
- PVC gauntlet gloves.
- Full face perspex shield.

The PVC oversuits are provided with elasticated wristbands which seal onto the gauntlet gloves and a hood which is fitted with a drawstring enabling it to seal onto the face shield.

Breathing apparatus (not Puretha respirators) shall be at hand for dealing with spillages of strong solutions. After use but before removal by the wearer, the protective clothing should be decontaminated by hosing down with water.

T5 Supply of hydrofluoric acid

Hydrofluoric acid is normally supplied as concentrated solutions containing between 40 and 70% w/w HF, which can cause serious burns if spilled or splashed. Risks of serious accidents must be minimised by arranging for the supply of hydrofluoric acid in more diluted form. A strength of 1–2% w/w is appropriate for boiler cleaning. The acid should be supplied in bulk containers to minimise the number of transfers involved.

T6 Reception of intermediate strength acid

The bulk containers should be received into a location selected to give the minimum practicable transfer distance to the chemical dosing tank. The area should be temporarily bunded with sand bags to contain spillages and sufficient lime must he available to neutralise all the acid from a single container.

Access of personnel to the reception area should be restricted to those equipped with the protective clothing listed in Appendix R.

T7 Dilution and injection of the acid

The intermediate strength acid must be transferred from the bulk container to the chemical dosing tank, where it is immediately diluted to near working strength by either suction or gravity feeding only. The transfer shall be completed without exposing the acid to the atmosphere, the chemical dosing tank being covered throughout the injection and cleaning procedure. Transfer pipes should be flushed through with water after dilution is complete.

During the dilution stage, access should be restricted to one or two personnel wearing the full protective clothing with face masks; with the breathing apparatus at hand for emergencies. One operator in full protective clothing should stand by outside the bunded area ready to assist in emergencies.

T8 Progress of boiler clean

Autocirculation using nitrogen is the preferred method of inducing circulation. All external pipework and mass flow rates should be minimised to reduce the risks of leaks outside the boiler casing. The boiler should be externally vented using a simple gas/liquid separator to remove any hydrofluoric acid spray or foam. It should be noted that at working strength (1–2% w/w hydrofluoric acid) the risks of fluoride poisoning are small. Full protective clothing is nevertheless still required for all personnel working close to the boiler.

T9 Effluent treatment

It may be necessary to treat the effluent before final disposal. Some stations possess large ash lagoons with dwell times suitable for disposal and all that may be required is to encourage precipitation of the insoluble calcium fluoride by mixing a 50–100% excess of strong calcium chloride solution with the spent acid as it is drained from the boiler. It is important to ensure good mixing at this stage. Antifoam may also be required to minimise foaming in open channels during the disposal.

In the absence of ash lagoons, the effluent should be tankered away to a suitable disposal site.

T10 First aid kit for treatment of exposure to hydrofluoric acid

The complete 'kit' contains the essentials for both first aid and medical treatment of hydrofluoric acid burns, etc., and could be useful and convenient for use in transport, on site surgery supply and accompanying the patient to hospital.

Appendix T

The special safety case for hydrofluoric acid (*Cont'd*)

Each kit should contain:

2 × small jars or tubes (about 25 g) calcium gluconate gel 2.5%	*
2 × 1 kg pots calcium gluconate gel 2.5%	
2 × tubes 20 Sandocal tablets (Sandoz) each containing 400 mg Ca (20 meq.Ca2$^+$) and 20 mg ascorbic acid About 12 sealed pack(s) sterile medicated swabs	*
1 × sealed instructions — First Aid	*
1 × sealed instructions — Medical Treatment	*
1 × 20 ml hypodermic syringe in sterile pack	+
1 × 10 ml hypodermic syringe in sterile pack	+
5 × 10 ml sterile solution of calcium gluconate 10% (Sandoz or equivalent)	+
5 × 54 ml sterile solution of calcium gluconate 10% (Sandoz or equivalent)	+

Items marked with an asterisk (*) should prove adequate for the first aid treatment of hydrofluoric acid burns, etc.
When medical treatment could be involved, then the additional articles marked (+) will be required.

T11 First aid measures for use with hydrofluoric acid

When it is known or suspected that hydrofluoric acid has come into contact with the skin or eyes or if the vapour has been inhaled, the following first aid measures mst be implemented IMMEDIATELY.

Splashes on skin
- Remove the affected person from exposure.
- Drench the contaminated area with water and remove contaminated clothing whilst continuing water drench.
- Wash the bare skin for at least one minute; longer when the flesh has been penetrated.
- Apply calcium gluconate gel liberally to the affected area and massage in. Continue the application for 15 minutes after the pain has subsided and refer to hospital by ambulance.
- If the affected area is greater than 160 cm^2 give by mouth six calcium gluconate (Sandocal) tablets in water.
- Refer to hospital by ambulance.

Splashes in eye
- Irrigate eye with isotonic saline or water for at least 15 minutes.
- Refer to hospital by ambulance.

Inhalation of vapour
- Remove the affected person from exposure.
- Give by mouth six calcium gluconate (Sandocal) tablets in water, repeated at two hourly intervals.
- Refer to hospital by ambulance.
- Loosen clothing and place patient in a comfortable position.
- Keep patient warm.

Ingestion of acid solution
- Give repeated large doses of calcium gluconate (Sandocal) tablets — 6 tablets per dose, repeated at two hourly intervals.
- Do not induce vomiting, but give copious quantities of fluid to drink (water or milk).
- Refer to hospital by ambulance.

Volume E
Metallurgy

CHAPTER 5

Introduction to Metallurgy

The modern power station can be expected to have an operational life of at least forty calendar years. For reasons of economy and safety it is important to ensure the structural integrity of the major engineering components over that service lifetime. An understanding of material behaviour and of metals in particular is an essential requirement in meeting this objective.

Chapter 6 discusses steel, the most important of these materials, in some depth, but the role of non-ferrous alloys and non-metallic materials such as plastics and ceramics in power plant is recognised by discussion of their properties and applications in Chapters 7 and 8 respectively.

Chapter 9 examines the basis of material selection at the design stage for fossil-fuelled boilers, nuclear reactors and turbine-generators and discusses the production methods used to fabricate major components. One of the most important fabrication techniques is welding and Chapter 10 reviews the welding processes in current use, their metallurgical consequence and the various types of defects that can be associated with weldments.

The assurance of plant component integrity, including welds, has been greatly improved in recent years by the developments made in non-destructive testing methods and these are considered fully in Chapter 11.

As previously indicated, the operational life of power stations is being extended to times that often lie beyond those considered at the time of design. It is therefore of increasing importance to be able to judge the probable lifetime of components. Chapter 12 gives attention to this issue by discussing defect analysis both within and beyond the originally proposed design parameters.

The concluding Chapter 13 considers environmental effects in power plant operation that affect component integrity. This is discussed in terms of corrosion processes arising from boiler furnace and reactor gases and of failure mechanisms of stress corrosion and corrosion fatigue.

The technological advances that have been made in advancing power plant steam conditions over the past fifty years have depended greatly upon the availability of metals and alloys capable of withstanding high pressures and temperatures for extended periods of time.

In parallel with these improvements, there has been a much better awareness of structure/property relationships so that designers of plant components have been able to achieve more advanced and more economic designs, given a sound knowledge of material behaviour.

It is most important to consider the various mechanisms by which failure of components may occur. *Failure* in this context can be defined as the development of a defect that can grow to such a size as to make a specific component inoperable. Defects in material below this level can be contained, either by monitoring to ensure that a critical size is not reached or by carrying out a repair of the affected region. This type of approach has been greatly aided by the development of fracture mechanics, a quantitative treatment of the relationship between materials, tolerable defect sizes, the environment to which the material is exposed and the engineering requirement expressed in terms of stress, time and temperature.

Potential failure mechanisms at lower temperatures (up to about 200°C) are brittle (fast) fracture and fatigue. At higher temperatures (above about 400°C) failure mechanisms of interest include fatigue, both low and high frequency cycle, creep/fatigue, creep and corrosion of various types. Since the majority of power plant operates at temperatures between 500°C and 600°C, there is naturally a greater interest in failure mechanisms in that temperature range than at lower temperatures. However, the latter aspect cannot be ignored as some vital components such as low pressure turbine rotors and PWR pressure vessels must be shown to possess adequate levels of fracture toughness and thus be immune to brittle, fast-fracture.

Power plant is expected to operate for very long periods, possibly up to 300 000 hours, and this requires all time-dependent metallurgical processes to be properly understood. This comprises the customary database for creep and creep/fatigue, augmented by data more specific to the operational regimes of particular plant items. Examples of this include corrosion effects brought about by particular environments whether it be the flue gases of a coal or oil-fired boiler, or the CO_2 in a gas-cooled nuclear reactor and thermal cycles reflecting the start-up and shutdown conditions of some power plant components.

Understandably, much of this technical data has to be derived from a combination of laboratory tests, plant trials and feedback of plant operational experience. One aspect that the latter has demonstrated over recent years has been the importance that welded fabrication can have upon the property behaviour of materials. This has led, on occasion, to a number of practical problems with cracked components leading to expensive repair and maintenance procedures. For these reasons, considerable attention has had to be given to the selection of welding processes, their control during manufacture and the effect such processes have on material performance, particularly at elevated temperatures. There have been major advances in welding technology in recent years that bear directly upon these issues and these are considered in Chapter 10.

Equally, there is a growing demand to obtain positive guarantees of component integrity, both as manufactured and during operational service. This has led to increased attention being given to non-destructive testing techniques with notable advances in techniques of detection employing ultrasonics and eddy current methods. These complement the more traditional, well established visual, magnetic-particle and radiographic techniques of examination. It is suggested that the reader treats the developments in welding (Chapter 10) and non-destructive testing (Chapter 11) as interrelated technologies that aid one another in providing engineering components of higher quality and integrity. In both areas, modern electronics and computer availability are being used to ensure permanent records of processes that in themselves are being made less dependent on human error.

Power stations have been designed, historically, for operational lives of about 150 000 hours. The growing realisation that future gains in thermal efficiency are likely to be small has been paralleled by an enhanced emphasis on plant availability and reliability and the possibilities of a much extended operational life for plant. Target lives of about 300 000 hours are now being considered as readily attainable for modern power plant designs.

As a result of this switch in emphasis by utilities, a demand is arising for thorough technical assessment of critical plant components already in service to establish a basis for their continued safe operation. Metallurgical knowledge of time-dependent damage mechanisms has a vital role to play in making judgements affecting either refurbishment or replacement of time expired components. This aspect is treated in relevant detail in Chapter 12 by examining the methods of defect analysis beyond the originally postulated design life.

Since power station heat producing sources, either fossil-fuelled boilers or water/gas-cooled nuclear reactors, present a variety of external environments to the metal heat transfer surfaces it is not surprising that considerable attention has had to be given to the ensuing processes of corrosion. These aspects have

governed material selection in many instances, leading to the choice of steels with a sufficiently high level of chromium content to resist corrosive attack. It is worth noting, however, that corrosion mechanisms have provided a technological temperature barrier that it has not proved economically worthwhile to surmount. Examples are limits provided by oil high in vanadium content, coals high in sulphur and/or chlorine content and *breakaway* corrosion of carbon steel by CO_2 gas at temperatures above 400°C. These aspects are discussed in some depth in Chapter 13.

Historically, there have been two thrusts in power plant design development with which metallurgy has been closely involved. One has been to achieve a goal of enhanced thermal efficiency and hence lower fuel costs for a given electrical output, by allowing the use of higher temperatures and pressures in the steam cycle. The other has been to increase the size of the power producing unit in order to achieve economies in construction and operational costs.

The pattern of steam temperatures and pressures for central station power plant over recent years is shown in Table 5.1. Whilst there has been a continuing rise in the pressure parameter, there has been little tendency to increase steam temperatures much beyond 565°C. This levelling off reflects the fundamental difficulty, for fossil-fuelled plant, of overcoming in an economic manner the problems of boiler fireside corrosion due to the products of coal or oil combustion. Indeed, in the case of oil-fired boilers, the upper steam temperature limit is closer to 540°C due to the corrosive attack from vanadium products of oil combustion. Additionally, temperatures much beyond 580°C are

TABLE 5.1

*Plant sizes and steam conditions
versus year of commissioning*

Unit size MW	Steam conditions bar / °C / °C			Year first unit commissioned
30	41	454	NR	1941
60	62	482	NR	1952
100	103	566	NR	1956
100	103	524	518	1957
120	103	538	538	1958
200	159	566	566	1959
275	159	566	566	1962
300	159	566	566	1963
350	159	566	566	1964
375	241	*593	566	1965
500	159	566	566	1965
500	159	538	538	1968
660	159	566	566	1973
660	159	538	538	1976

Note:	NR denotes Non-reheat.
	Steam conditions — inlet steam pressure / inlet temperature / reheat temperature.
	*Supercritical steam conditions.

close to the limit for large ferritic steel forgings that are required in the high pressure rotor of a large modern turbine-generator. Their substitution by austenitic steel forgings, whilst not impossible, poses a number of serious technical challenges.

The present position therefore in respect of thermal efficiency gain is that further advances are only merited by the prospect of high fuel costs and thus are of interest to those countries that either lack an indigenous fuel supply or find its importation a costly exercise. In metallurgical terms this means that the materials discussed in subsequent chapters of this book in general span operational temperatures up to about 600°C, their respective chemical compositions being selected to meet mechanical property and corrosion resistance requirements for specific applications in boilers, reactors and turbines.

The second thrust in power plant design, increase in the size of the heat and power producing units, is also illustrated in Table 5.1. In more recent years the rate of growth in unit size has declined, for several reasons. From the point of view of security of supply from an electrical power network, it is undesirable to have any one power producing unit with an output greater than about 10% of the total electrical system. In consequence, unit sizes much greater than about 1000 MW have limited appeal on a worldwide market. Secondly, the issue of unit size has been influenced by the more recent trend towards building light-water reactors rather than advanced design fossil-fuelled plant. These have proved attractive to utilities by offering lower 'ownership cost' despite the lower thermal efficiency of their steam cycle and shifting the balance of cost towards construction and away from operation. Finally, in metallurgical terms, any continuing increase in unit size would require the manufacture of larger, massive, turbine rotors and there can be technical difficulties in achieving the mechanical properties demanded by designers in such large volumes of steel. This affects both high temperature creep strength and low temperature fracture toughness.

The succeeding chapters examine the role of metals in power plant technology in a logical manner. They reflect a relative decline in the introduction of new alloys coupled with an awareness that longer service lives demand a more thorough evaluation of those material properties that underpin structural integrity assessments.

CHAPTER 6

Materials behaviour

1 Modern steelmaking technology

The intensive application over the last decade of analytical techniques to component failure in service and to defects generated during solidification and fabrication has led to the progressive realisation that many of the quality characteristics of steels that ensure reliability over a long life are established at the melting, refining and ingot casting stages of production. A greater emphasis is therefore increasingly being placed on steelmaking considerations in specifications for a wide range of applications and especially for critical components of power and chemical plant. Obvious examples are rotor forgings and steels for pressure vessels operating at both low and high temperature.

The demand for improved qualities of steel in bulk quantities to meet more stringent product requirements has led the steelmaker to re-assess his raw material supplies and process routes to meet these demands at minimum cost. The rapid downward trend in sulphur levels to improve the properties of plate and pipe-line steels is being followed by reduced specification maxima for phosphorus, nitrogen and oxygen with targets for P + S + N + O of 50 PPM (0.005%) being predicted`in Japan. Sulphur and phosphorus levels below 0.005% are already routinely achieved throughout Europe. Tighter ranges for deoxidant, residual and microalloying elements are also increasingly specified.

The means of producing economically competitive tonnage steels at this level of cleanliness and purity have arisen by exploiting the understanding of the chemistry of slag and steelmaking reactions derived from operation of the now obsolete Siemens-Martin *open hearth furnace*. The chemistry of *open hearth steelmaking* is complex and has been covered extensively in the literature. The evolution of steelmaking processes since the *open hearth* is therefore now reviewed in terms only of the most relevant features for fine composition control which have led to improved properties.

Iron is present in the natural ore as a mixture of oxides. Reduction to liquid iron is achieved in the *blast furnace* by reaction with carbon and limestone. The resulting hot metal is rich in carbon, manganese, phosphorus, silicon and sulphur, all of which contribute to the brittle nature of the cast iron. The vastly superior properties of steels result from the reduction in the levels of these elements during the refining stages of the steelmaking process. The raw materials for modern steel furnaces are therefore hot or cold metal from the *blast furnace* and/or cold scrap.

Steelmaking processes are essentially the removal of the excess carbon, silicon, manganese, sulphur and phosphorus by chemical reaction, principally oxidation, under a slag whereby the excesses are transferred to the slag. The primary reaction is the removal of carbon to the specified level by oxidation:

$$C \text{ (in Fe)} + O \text{ (in Fe)} = CO \text{ (gas)}$$

The reaction promotes a vigorous boiling of the steel, the 'carbon boil'. This is followed by additions of the necessary levels of C, Si, Mn, Al and alloying elements (e.g., Cr, Mo, Ti, V, W) to meet the specification levels. Element transfer to slags depends on oxidation reactions, for example:

$$Mn \text{ (in Fe)} + O \text{ (in Fe)} = MnO \text{ (slag)}$$
$$2P \text{ (Fe)} + 5O \text{ (Fe)} = P_2O_5 \text{ (slag)}$$
$$Si \text{ (Fe)} + 2O \text{ (Fe)} = SiO_2 \text{ (slag)}$$

The oxygen is replenished by diffusion from the slag. Oxygen potential can be increased by adding iron oxide in the form of millscale or iron ore, or by direct injection of oxygen into the metal.

Sulphur removal, however, is favoured during the later stages of refining by reducing rather than oxidising conditions:

$$S \text{ (Fe)} + O^{2-} \text{ (slag)} = S^{2-} \text{ (slag)} + O \text{ (Fe)}$$

The reaction is pushed from left to right by increasing the oxygen ion potential (basicity) of the slag.

1.1 Converter processes

1.1.1 The basic oxygen steel process

Reaction rates in the stagnant open hearth furnaces were very slow (10–20 tonnes/h). This led to the development of converter processes in which the reaction rate is accelerated (to >500 tonnes/h in the larger vessels) by blowing air or, with the advent of tonnage oxygen after the war, oxygen through the hot metal from the base of the converter (bottom blown) or from lances onto the metal surface (top blown). The Basic Bessemer or Thomas converter process allowed the phosphoric limonite ($2Fe_2O_3 \cdot 3H_2O$) ore fields of France and Germany to be exploited and became the major steelmaking process in Europe before being superseded by the oxygen blown methods in the 1960s. The most widely used of all current processes is now the LD (Linz/Donawitz) or BOS (Basic Oxygen Steel) process which produces 500 million tonnes, or 70% of the world's steel production, annually. The process was developed to convert the high phosphorus European irons by top-rejecting powdered lime through the oxygen lance (Fig 6.1).

Bottom blown converters are comparatively new and as yet account for only 60 million tonnes per annum (approximately). Higher injection rates and shorter tap-to-tap times are claimed than for BOS with higher scrap consumption (oxy-fuel burners) and very low minimum carbon levels. Basic slags favour the removal of both sulphur and phosphorus and have been generally adopted in modern converters, together with dolomite refractory linings. Only 70% of the charged sulphur can be removed in the BOS process, however, because of its highly oxidising nature. Phosphorus removal is favoured by low temperatures and high dissolved oxygen. High phosphorus irons (0.4–2.0%) are therefore treated by flushing off the initial slag (the LD-AC process) which is quickly saturated since phosphorus recovery is rapid at the low temperatures prevailing early in the blow, and adding a second slag to improve the removal rate as the carbon content falls and the bath oxidises.

Converters essentially refine hot metal from the blast furnace but approximately 30% of scrap can be accommodated.

1.2 The basic electric arc furnace

The increasing availability of steel and iron scrap, together with precise temperature control and clean melting in a low sulphur atmosphere, led to the widespread adoption of the BEA furnace for the melting of quality carbon and alloy steels (Fig 6.2). The fur-

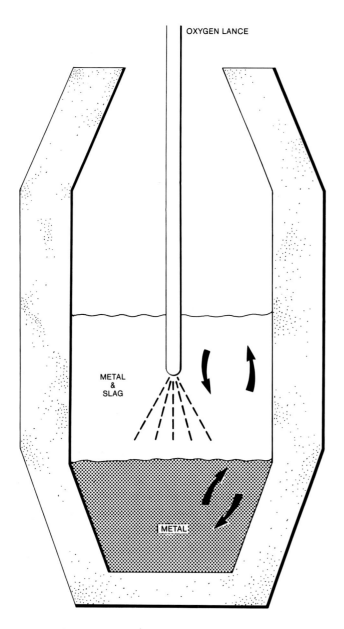

FIG. 6.1 The basic oxygen converter
Refining rates are accelerated by blowing oxygen
and powdered lime onto the surface of the
molten iron (top blown). Modern vessels have up to
450 tonnes capacity.

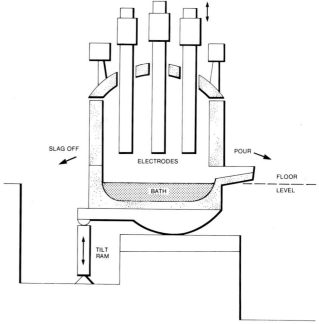

FIG. 6.2 The basic electric arc furnace
The process offers rapid, clean melting in a low sulphur
atmosphere and is especially attractive for quality steel
alloys. Oxygen lancing follows melting.

carbon to the bath. Sulphur levels can be further
reduced by a second reducing slag but this is time
consuming and can be avoided by a low sulphur
charge.

Final deoxidation prior to pouring is achieved, as in
the BOS processes, by additions of mixed deoxidants
such as ferromanganese, ferrosilicon and aluminium.
Ferro alloys are added to bring the steel into specification and the bath temperature is quickly increased
for tapping.

1.3 Secondary refining processes

Low sulphur and phosphorus levels (0.001–0.003%)
can be rapidly achieved in the BEA furnace by the
double slag process. This 'Refurnacing' is a viable
option for smaller ingots, but for ingots requiring steel
from three to five or even more furnaces a multitude
of secondary refining processes based either on holding
ladles or separate refining vessels has evolved. These
combine the ability to economically expand the capability of existing BEA or BOS facilities by accumulating
several melts with improved composition and tap temperature control.

1.3.1 Ladle processes

These bring about the final adjustments in analysis and
temperature which, although small, have a profound
effect on casting, working of the ingot and properties
of the final product. Refining processes carried out
in a ladle served by a BEA or BOS furnace can be
classified as in one of the following ways.

nace is charged with clean scrap, steel or pig iron as
available and melting is carried out as quickly as possible. Limestone is added to produce a refining slag
and carbon is added in the form of anthracite or
broken electrodes. Refining is generally accelerated by
blowing oxygen through the molten steel to promote
rapid decarburisation. The carbon is removed as CO
resulting in the 'carbon boil' which promotes good
slag/metal mixing and eliminates temperature and
concentration gradients by stirring the bath. Si, Mn,
P and Cr are also removed by oxidation. Sulphur is
best removed after the oxygen blow by removing the
highly oxidised CaO-FeO first slag and forming a
reducing slag by adding calcined lime, fluorspar and

Static processes

The liquid steel is treated with a synthetic slag, with or without vacuum or argon stirring (Fig 6.3). A gentle stream of nitrogen or argon bubbles, introduced through a porous conical plug at the ladle base, rises through the steel producing a gentle stirring action which reduces temperature and composition variations. Inclusions in the steel are taken to the surface and entrained in the synthetic slag added after the furnace slag has been prevented from entering the ladle. Oxygen and nitrogen removal is encouraged by an argon stream but effective removal rates require a combination of vacuum and stirring (by gas, mechanical or electrical induction), since the ferrostatic head in a stagnant ladle restricts degassing.

The benefits of efficient degassing on steel properties are such that around a quarter of all production is now degassed in some way. The process is relatively cheap but limited by the high refractory consumption, since the steel must be tapped hotter, and the additional time in the arc furnace to effect the initial refining operation.

Ladle injection processes

A powdered reactive material (Ca, Mg, rare earths) is fed deeply into the steel by means of a lance, usually with argon gas as carrier (Fig 6.4). Injection achieves maximum exposure of the reactive material to the liquid steel, enhanced further by the stirring action of the gas stream. The processes are designed to desulphurise, deoxidise and modify inclusion morphologies at minimal temperature loss and provide enough stirring to maximise inclusion removal to the slag.

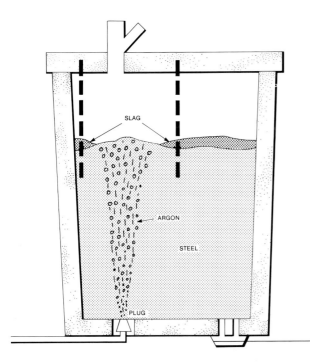

FIG. 6.3 Ladle stirring with a stream of nitrogen or argon bubbles

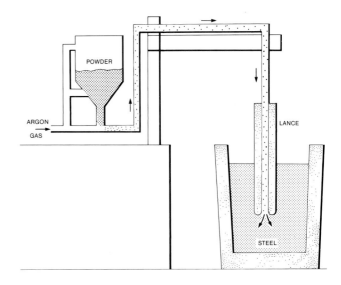

FIG. 6.4 Ladle injection system
The lance is submerged to near the bottom of the ladle to ensure stirring and an evenly-distributed reaction.

Ladle holding furnaces

A supplemental heating facility means that the temperature of the steel provided to the ladle is not critical and extremely close tolerances on composition and pouring temperature can consequently be met. The degree of refining and alloying achieved in the arc furnace thus becomes relatively unimportant. When combined with vacuum, oxygen injection, gas or induction stirring and additional facilities, the system becomes extremely versatile. Because of the supplemental heating, additional slags can be used to realise very high degrees of desulphurisation. With a vacuum facility, a high degree of deoxidation can be accomplished with carbon (Vacuum carbon deoxidation, see Section 1.5 of this chapter) without the use of strong deoxidisers such as aluminium and silicon, whose inclusions can have detrimental effects on steel properties. Refining can be achieved under vacuum with argon stirring at faster rates than those in the reducing stage of modern, fast melt down BEA furnaces, leading to a significant increase in productivity at lower capital cost. It is possible to produce very large ingots by combining several successive BEA furnace heats since ladle heating rates can be engineered to exceed the steel cooling rate. Ladle furnaces can be used with BEA or BOS furnaces for primary steelmaking, provided adequate dephosphorisation is carried out in the BOS furnace, the choice depending on the availability of hot metal, scrap and electric power. The choice of ladle process is mainly an economic one since all ladle refining methods produce steel of comparable quality.

1.4 Refining vessel processes

The argon oxygen decarburisation (AOD) process, widely used for stainless steel production because of its excellent chromium recovery (80% world production) is becoming a popular means of producing low alloy steels. The *gas refining arc furnace* is a similar process but incorporates heating and more vigorous blowing.

1.4.1 Argon oxygen decarburisation

The steel is melted, usually in a BEA furnace, and then decarburised and refined after transfer to the AOD vessel.

In the standard BEA furnace rapid decarburisation is accomplished by lancing the bath with oxygen. As carbon is removed from the bath, however, the equilibrium chromium level is reduced to only 3–4% at 1800°C, any excess rapidly oxidising out. The 18% Cr required for stainless steel production can then only be introduced by adding large amounts of ferrochromium immediately prior to teeming, which cools the bath significantly. Only small amounts of stainless scrap can therefore be used in the BEA furnace and high superheats (1900°C) are required, leading to high refractory costs. The AOD process overcomes the problem by progressively replacing the oxygen flow with argon as the carbon is oxidised. This reduces the partial pressure of CO and hence the rates of oxidation of carbon and chromium such that chromium is retained in the bath while carbon is removed to low levels. This means that the charge can be all stainless steel scrap, which is the lowest cost form of chromium.

The gas mixture is injected through under bath tuyères located in the side wall or base. No external heat source is supplied, the required heat being generated from controlled oxidation. Excellent bath mixing and slag/metal contact is secured by the gas injection with very high resultant reaction rates (desulphurisation < 0.005% in less than 5 minutes). As the argon proportion increases in the later stages low residual levels of O, N and H result, an advantage where 'vacuum stream degassing' is not installed. Dephosphorisation is best accomplished during melting since it does not occur under the conditions necessary for deoxidation and desulphurisation.

1.4.2 Gas refining arc furnace

The process combines the heating and argon bubbling abilities of the ladle refining furnace with the vigorous gas injection system of the AOD, but in a simple furnace converter vessel with a bottom tap hole. Heating and argon bubbling take place in the vertical position, but for high volume argon or oxygen blowing the vessel is tilted to submerge the tuyères. The close control of temperature and composition, use of synthetic slags and powder injection through the tuyères make the process suitable for a wide range of low and alloy steels.

1.5 Deoxidation

To avoid the possibility of CO evolution and associated porosity in the ingot, additions of strong deoxidising elements are added to the steel prior to teeming. These remove the remanent oxygen as insoluble solid or liquid compounds. Oxide inclusions in the finished product are however invariably detrimental to mechanical properties and there has been an increasing demand to reduce oxide volume fractions in modern steels. Deoxidation can be carried out in the melting furnace but, where secondary processes are adopted, some or all additions are more usually made in the holding ladle or refining vessel.

1.5.1 Vacuum carbon deoxidation (VCD)

Carbon itself is a powerful deoxidant, its effectiveness increasing rapidly as the partial pressure of CO over the metal decreases:

C (in liquid steel) + O (in liquid steel) = CO (gas)

As the pressure of CO decreases, the reaction moves rapidly to the right. A low pressure of CO can be achieved either by bubbling argon, in which the CO is diluted, or by submitting the steel to a vacuum, for example, the equilibrium concentration of dissolved oxygen in solution for 0.2% C at 1600°C and 0.01 atmospheric pressure (1.01 kPa) is only 1 PPM. The advent of vacuum furnaces and argon facilities has thus led to a trend to deoxidise as far as possible with carbon, thereby minimising the levels of deoxidising additions and associated inclusions.

Vacuum carbon deoxidation is more effective than the traditional method of adding silicon and can therefore replace the latter and avoid the introduction of detrimental silicate inclusions. Silicon in VCD steels is thus reduced from about 0.25% to ≃ 0.05%. VCD has become standard practice with NiCrMoV steels for LP turbine rotors and for some generator rotors.

1.5.2 Vacuum degassing

Residual gases in steel, especially hydrogen, have been principal causes of fabrication defects. Exposure of the liquid steel to vacuum is however successful in achieving innocuous gas levels and is now applied to over a quarter of all steel production.

1.5.3 Ladle degassing

To remove H and N effectively and economically in the ladle, a combination of vacuum and stirring by gas, mechanical or electrical induction is necessary to overcome the effects of the ferrostatic head. Metal/vacuum interaction can be facilitated by using a separate circulating vacuum chamber. Modern varieties use a twin-legged system and gas injection to assist metal circulation.

1.5.4 Stream degassing

Several methods exploit the exposure of a large surface area to achieve effective degassing during pouring as the stream breaks up into droplets.

Tap or stream degassing involves the use of a separate pony ladle from which the steel is drawn into the tapping ladle by the vacuum. For heavy forgings, vacuum stream degassing during pouring into the mould is the most popular method of hydrogen control. The hot-topped ingot mold sits inside the vacuum-tight chamber, the roof of which connects with the pony ladle (Fig 6.5). The stream of steel is exposed to the vacuum as it enters the mould, resulting in < 1 PPM H pickup. Exposure to hydrogen sources (especially moisture) during and after treatment is thereby almost eliminated.

1.6 Remelting processes

General improvements in mechanical properties arising from reductions in the volume of inclusions (oxides and sulphides) and in the levels of residual elements arise if an electrode is directionally solidified. The principal processes are vacuum arc remelting (VAR) and electroslag remelting (ESR), both of which utilise electric power to slowly remelt a solid electrode into a water cooled crucible to produce a resolidified ingot.

VAR uses an electric DC arc at 25–40 V in vacuum (Fig 6.6). As melting proceeds, the electrode is lowered

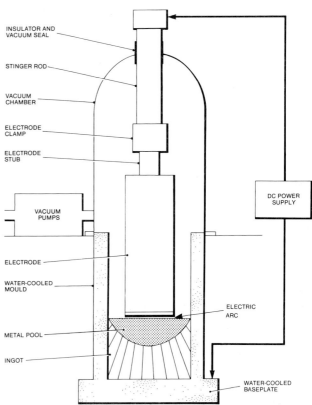

FIG. 6.6 Schematic illustration of a typical vacuum arc remelting (VAR) furnace

by an automatic control system which maintains arc stability, arc length and molten pool depth, each of which is crucial to controlled progressive solidification. High gas content of the electrode can lead to glow discharge phenomena and add to the gas evolution associated with the carbon deoxidation reaction. Many VAR producers employ vacuum induction melted (VIM) or vacuum degassed electrodes to avoid these problems (the VIMVAR process).

ESR uses line or low frequency AC in air to remelt the electrode(s) under a chemically active protective slag (Fig 6.7). The slag temperature is raised well above that of the melting point of the electrode(s) thus enhancing slag/metal reactions as the metallic droplets from the melting electrode(s) fall through the slag. The ESR ingot has lower inclusion volume, reduced micro and macro segregation, less porosity and less pipe than conventionally cast ingots.

Both processes improve the metallurgical (dendrite) structure of the ingot. ESR permits the production of shaped castings, such as slabs, from multiple electrodes whereas VAR is normally limited to round sections from a single electrode. VAR only slightly modifies the electrode composition but removes hydrogen, whereas ESR can substantially refine, especially for sulphur, but is subject to hydrogen pickup. The heaviest ingots that can currently be produced by VAR and ESR weigh approximately 55 t and 180 t respectively. ESR can

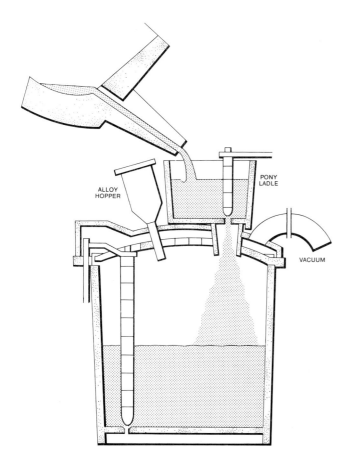

FIG. 6.5 Tap or stream degassing
As the steel is drawn from the pony ladle into the tapping ladle the stream physically breaks into droplets, thereby exposing a large surface area to the vacuum

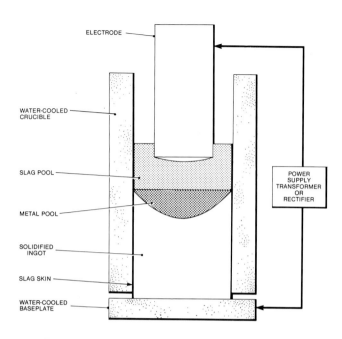

FIG. 6.7 Schematic illustration of a basic electroslag
remelting (ESR) furnace

thus be used for HP and IP (but not LP) rotors, and
has been used successfully for 12 Cr HP rotors in
which segregation has presented quality problems.

1.7   Residual elements

Residual elements not removed by oxidation or in
the slag during the steelmaking process accumulate
as scrap is recycled. Most scrap charged to the arc
furnace is merchant scrap (consumer durable goods)
exhibiting unpredictable variations in composition.
Residual levels, especially of Sn, Cu and Ni can be
controlled at increased cost by scrap selection and
dilution with low residual scrap, for example, virgin
scrap from the blast furnace/LD operation, blast furnace iron or directly-reduced iron-ore pellets.

Secondary refining, degassing, deoxidation and remelting methods (Fig 6.8) have made possible a very
high degree of control over the levels of specified

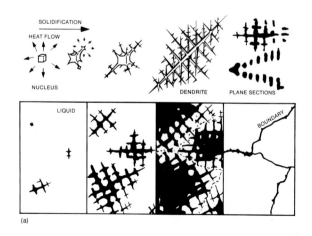

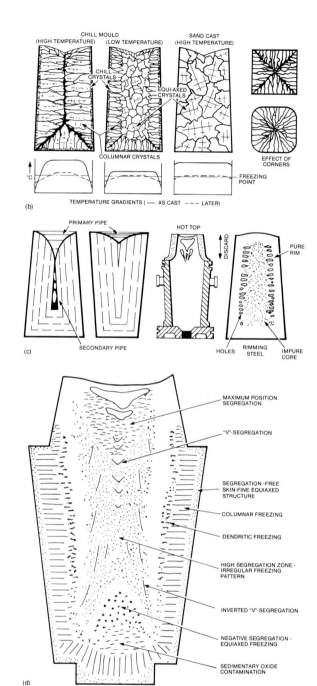

FIG. 6.8 Ingot microstructures
(a) All large castings and weld metals consist of grains
whose boundaries are defined when the growing fir tree
crystallites (dendrites) impinge during cooling.
(b) Dependence of distribution within an ingot of chill
and columnar crystals on temperature gradient.
(c) Primary and secondary contraction cavities can be
eliminated by maintaining a reservoir of liquid metal at
the top of the ingot (hot top) to feed the contraction
channels during solidification.
(d) Zones of segregation within an ingot resulting from
rejection of lower melting point constituents ahead of
the freezing interface.

elements, residuals, process related elements (O and N)
and non-metallic inclusions. The following levels can

now be routinely achieved in production steelmaking (weight %):

P 0.002, S 0.002, Si 0.02, Al 0.005 maximum, Sn 0.002, As 0.002, Sb 0.001. H PPM 1.0, O PPM 25, N PPM 50.

These represent a reduction by a factor of $\geqslant 10$ over typical levels two decades ago. The consequences for power plant components have been significant improvements in the consistency of product integrity resulting directly from improved inherent resistance to defect initiation and propagation over the whole range of castings, forgings, rolled and extruded products and weldments. Pressure vessel and rotor steels, for example, now have much improved toughness, leading to critical defect sizes for fast fracture which are unlikely to go undetected in operation. Weldments in steels produced by modern methods similarly exhibit enhanced resistance to weld metal solidification cracking and to cracking during cooling, subsequent post-weld heat treatment for residual stress relief and service operation.

2 Microstructure/composition/property relationships of steels

The mechanical behaviour of steels is very significantly influenced by both microstructure and composition. For example, the stress to fracture plain carbon steel in 100 000 hours at 500°C can be increased by more than a factor of 5 by alloying with not more than a half-weight percentage each of chromium, molybdenum and vanadium. The ductility of the same alloy in the bainitic condition at 700°C is reduced however from more than 20% elongation to below 2% by increasing the grain size from 20 to 100 microns, as occurs in welding. Physical metallurgy seeks to understand the reasons for the many beneficial and detrimental effects on mechanical properties of variations in composition and microstructure in order that the often conflicting property requirements for specific applications can be optimised and maximum benefit derived from steels for power plant applications. Specifications defining the permitted ranges of composition and microstructure (heat treatment) have therefore become increasingly stringent as the factors responsible for inadequate performance have been identified. Specific examples are given in Section 4.1 of this chapter, where some failure mechanisms identified by modern analytical methods and their consequences for plant components are treated in more detail. The following brief summary of the metallurgy of power plant steels provides some of the basic scientific background to the initial limits on specifications set for composition and microstructure for specific applications, highlighting the more relevant areas.

2.1 Solidification microstructures

After completion of the steelmaking operation, the molten metal is poured into either refractory moulds for the production of shaped castings, or into a cast iron mould for the production of ingots for subsequent fabrication into forged billets, bars, sheets or piping.

A wide variety of structures can exist in cast metals, including as-deposited weld metals, depending on the relative rates of nucleation and growth of the solid crystallites formed within the solidifying liquid. This is evident in the crystal structure produced across a solidifying ingot in a metal mould where the layer next to the mould wall is cooled very rapidly. The temperature falls well below the freezing point (undercooling) resulting in a high rate of nucleation, and a fine grain sized layer of metal results in this the chill zone. The crystals have similar dimensions in each direction, i.e., are equi-axed. As the zone layer thickens, the temperature gradient, associated undercooling and nucleation rate all reduce rapidly and crystal growth down the temperature gradient now dominates nucleation. Long columnar crystals having a characteristic fir tree morphology (dendrites) grow towards the mould centre (Fig 6.8). Here both temperature gradient and undercooling are slight and coarse, equi-axed crystals nucleate and grow ahead of the dendrite interface. If the liquid steel is superheated before pouring however (i.e., heated well above the melting point) the chill zone may be remelted and disappear. When freezing begins again, cooling is slower and only a few nuclei form, leading to the formation of very coarse columnar crystals. If the mould is water cooled, the columnar dendrites can be encouraged to grow to the ingot centre. If the mould is of thin section, however, the fine chill crystals may extend across the entire width; this giving good mechanical properties. Similar effects can be encouraged by adding solutes and nucleating agents to increase the nucleation rate for a lower temperature gradient.

The contraction cavity at the top of an ingot can be minimised by fitting a thermally-insulated refractory or exothermically-lined feeder head to maintain a reservoir of liquid (hot top, Fig 6.8). Shrinkage between dendrites is also important on a microscopic scale. This can lead to fine scale porosity unless precautions are taken to ensure that the interdendritic liquid channels are continuously fed with liquid metal, thereby avoiding contracting volumes of semi-solidified material isolated by the solid dendrites.

2.1.1 Segregation

Segregation refers to all non-uniformities of composition in an ingot or casting. Many types of segregation of elements and inclusions can occur during solidification, usually with detrimental consequences for mechanical properties. Lower melting point constituents tend to be rejected ahead of the freezing interface and concentrate therefore in the centre of the casting. If the constituents are trapped between impinging dendrites a form of microscopic segregation of the lower melting point constituents to the outer surfaces of the dendrite arms results. This is termed 'coring'.

2.1.2 Crystallographic structure, grain boundaries and the allotropy of iron

The initial grain structure of a metal is established during solidification as the growing solid crystallites impinge. Within each grain the atoms are arranged in a characteristic three dimensional pattern or crystal structure. Although many crystal structures exist, those of the common metals conform to one of three types:

- Body-centred cubic (BCC): for example, chromium, molybdenum, vanadium and alpha- or delta-iron.

- Face-centred cubic (FCC): for example, nickel, manganese and gamma-iron.

- Close-packed hexagonal (CPH): for example, zinc, magnesium and cadmium.

Although the arrangement of atoms within each grain is defined by the crystal structure of the metal, the grains have different orientations with respect to each other and a boundary therefore exists between adjacent grains, known as the grain boundary, within which the atomic arrangement is relatively disordered. The disorder extends over a distance of only 2–3 atom layers. The properties of the grain boundaries are fundamentally different from the grains and this has important consequences for the behaviour of metals. The higher energy associated with the disordered structure facilitates decohesion relative to matrix atoms, which lie closer together and are therefore more strongly bonded. Grain boundaries also become preferred sites for particle nucleation and growth and for segregation of impurities, either of which may further weaken the atomic bonding and induce intercrystalline failure (see Section 4 of this chapter).

Some elements are able to exist in more than one crystallographic form. Each form is termed an 'allotrope' or 'allotropic modification'. Iron itself exists in two forms of the cubic pattern, namely body-centred cubic and face-centred cubic (Fig 6.9). The body-centred cubic form exists between room temperature and 910°C, and between 1400°C and the melting point at 1539°C. The lower temperature form is known as 'alpha iron' and the higher temperature form as 'delta iron'. An aggregate of grains of this crystallographic structure is termed 'ferrite', the alpha form being 'alpha-ferrite' and the delta form 'delta-ferrite'. The face-centred cubic form exists between 910°C and 1400°C and is referred to as 'gamma-iron'. An aggregate of grains of this crystallographic structure is termed 'austenite'.

The two differing crystalline structures exhibit different physical properties and, to illustrate two of the more important ones, austenite has a coefficient of thermal expansion approximately 1.5 times that of ferrite and has a lower thermal conductivity.

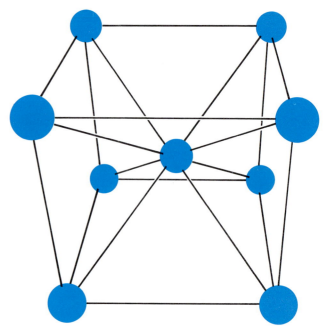

(a) Body-centred cubic

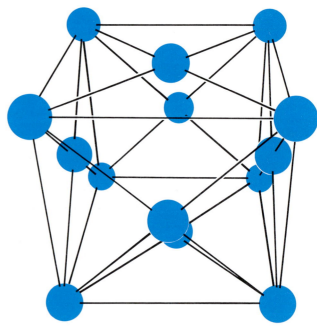

(b) Face-centred cubic

FIG. 6.9 Crystal structure of iron
(a) The atomic lattice of ferrite
(body-centred cubic, alpha iron) and
(b) austenite (face-centred cubic, gamma iron).
The close packing of atoms in the austenite lattice leads to a significant volume contraction as ferrite transforms to austenite on heating. The reverse occurs on cooling and this can lead to the generation of high stress levels within the material, especially where temperature gradients are severe, as for example in weldments. The stresses may persist to ambient when they are termed 'residual' stresses.

If a piece of iron is heated, its crystallographic structure changes from alpha-iron to gamma-iron at 910°C, from gamma-iron to delta-iron at 1400°C, and it melts at 1539°C, the reverse occurring on cooling. A further change occurs at a temperature of 769°C, when alpha-iron loses its magnetism; above this temperature iron is non-magnetic. The changes in crystal structure which occur at 910°C and 1400°C are accompanied by changes in volume, and on heating a contraction occurs at 910°C and an expansion at 1400°C. The reverse occurs on cooling through these temperatures.

2.1.3 Allotropic transformation under equilibrium conditions by nucleation and growth (civil transformation)

Iron and plain carbon steels transform from one allotropic form to another by a 'civil' process of nucleation and growth, even at relatively fast heating and cooling rates. As the steel is heated between the ferrite and austenite fields, nuclei of austenite, each a perfect crystal of independent orientation, begin to appear at the ferrite grain boundary triple points (junctions of three grains), Fig 6.10. Further nuclei follow along the grain boundaries and in the ferrite matrix. As the temperature is raised the austenite nuclei grow and impinge to define the austenite grain boundary network. The initial austenite grain size is thus often significantly smaller than the ferrite grain size. Raising the temperature further into the austenite field results in rapid grain growth, the smaller grains being consumed by the larger in order to minimise interface area, in analogous manner to the growth behaviour of soap bubbles. At 1400°C for pure iron the reverse process occurs and a new ferritic grain structure is established. The delta ferrite grains bear no known relation to the lower temperature alpha network. Cooling reverses the sequence as FCC austenite crystals nucleate and grow to form a new gamma iron boundary network, frozen during cooling before the final transformation by nucleation and growth of ferrite establishes the ambient boundary network (Fig 6.11). The transformations are termed civil because of the way in which the atoms change allegiance across the expanding boundary in a relatively unco-ordinated manner.

2.1.4 Transformation microstructures of steel

As mentioned previously, the alloying of iron with carbon is the basis of all steel making. Additions of up to 0.008%C have no marked effect on the structural changes which occur on heating or cooling, since the iron will contain this amount of carbon in solution at temperatures up to its melting point. The only effect is to cause the structural changes to occur over a small temperature range rather than at one specific temperature. Increasing the carbon above 0.008%, whilst it can be retained in solution in austenite at elevated temperature, results in its being rejected from solution on transformation to ferrite in the form of particles

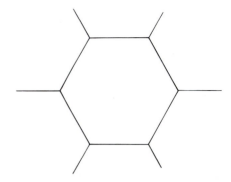

(a) Coarse grained ferrite structure

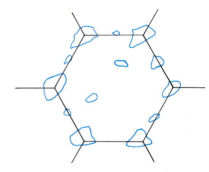

(b) Nucleation of austenite on heating above Ac₁

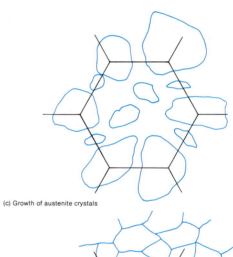

(c) Growth of austenite crystals

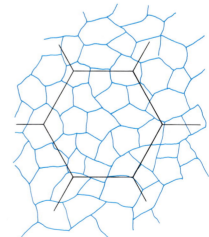

(d) Fine austenite grain structure at ⩾Ac₃

FIG. 6.10 Nucleation, growth and coalescence of new crystals changes the grain boundary network and frequently refines the grain size, as iron changes its allotropic form during heating and cooling.

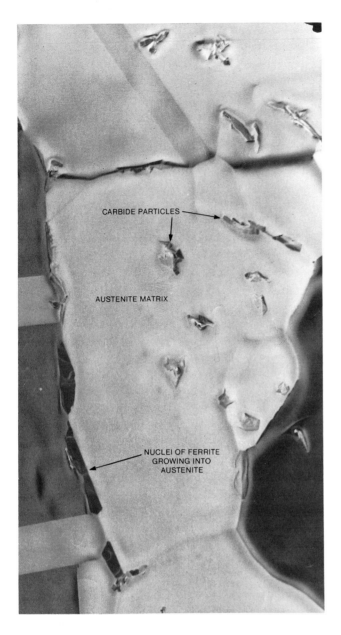

FIG. 6.11 Direct observation by photo emission
electron microscopy of nucleation of ferrite nuclei
on an austenite grain boundary during cooling of
a low alloy steel

of iron carbide, which are interspersed along the grain boundaries and within the crystals of ferrite (inter and intra crystalline particle nucleation). This is the basis of the strengthening of steels and is termed dispersion strengthening. The nature, size, spacing and distribution of carbides in ferrite is a dominant feature in determining both the low and high temperature properties of steels (toughness, tensile and creep strength, creep ductility, etc.) and is controlled principally by the types of alloying elements and the cooling rate during transformation.

Other effects of increasing carbon contents are:

● To depress progressively the melting point from 1539°C and to introduce a wider temperature range

over which melting or solidifying occurs instead of a clearly defined melting point.

● To raise progressively the transformation temperature at which the gamma-iron changes to delta-iron, from 1400°C to 1492°C at 0.10% carbon, above which the transformation temperature decreases with increasing carbon. In addition, the transformation then occurs over a temperature range instead of at a clearly defined temperature.

● To depress progressively the transformation temperature at which gamma-iron changes to alpha-iron, from 910°C to 723°C at 0.80% carbon. Increasing the carbon content above 0.80% raises the transformation temperature to 1130°C at 2.0% carbon. The transformation also occurs over a temperature range rather than at a clearly defined temperature.

All these changes in structure of necessity take time to occur, and whilst those at high temperature occur quite rapidly it will be appreciated that those occurring at low temperatures are very sluggish and long periods of time are necessary before completion.

If we now construct a graphic diagram, plotting the temperatures of the structural change against carbon content, with pure iron at the left-hand edge, a diagram of the form shown in Fig 6.12 is obtained. This represents equilibrium conditions and is known as the iron/carbon (or more exactly, as applied to steels, the iron/iron-carbide) equilibrium diagram. The point J on the diagram represents a complex solidification mechanism called a peritectic reaction, but this is of little significance to the final properties of steels and will not be discussed here. The point C represents another complex solidification mechanism called a eutectic reaction which will be referred to briefly later in discussing cast iron. Point S refers to a similar mechanism occurring in this instance in the solid state and called a eutectoid reaction.

The region bounded by points, N, H and A represents the solid solution of carbon in delta iron (ferrite), the region N-J-E-S-G represents the solid solution of carbon in gamma-iron (austenite) and the region G-P-Q represents the solid solution of carbon in alpha-iron. These illustrate the variation in solubility of carbon in iron at all temperatures up to melting. The point S denotes the minimum temperature and the carbon content at which austenite can exist and below this temperature any austenite is immediately transformed to a constituent called 'pearlite'. This is an intimate mixture of ferrite and iron carbide (or 'cementite', Fe_3C) consisting of alternate layers of ferrite and iron carbide which, when viewed under a microscope, has a lustrous appearance similar to 'mother-of-pearl'.

At room temperature the structure of carbon steels up to 0.80%C consists of a mixture of grains of ferrite and pearlite increasing with carbon content until at 0.80% carbon the structure is wholly pearlite. Above

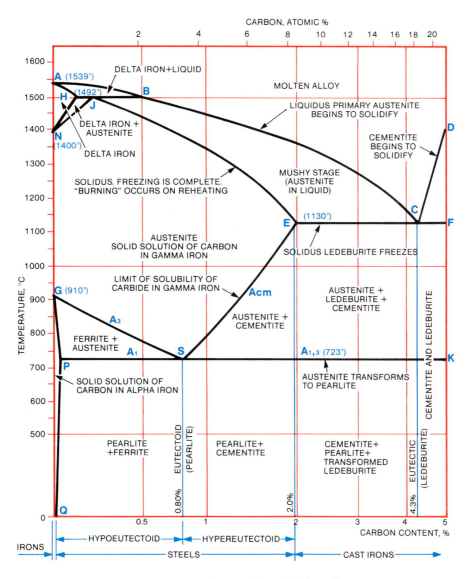

FIG. 6.12 The iron/iron-carbide equilibrium diagram

0.80% C, steels consist of grains of pearlite with small islands of iron carbide with no ferrite other than that contained in the pearlite. These steels are used for cutting tools and are not normally required to be welded. Figure 6.13 is a simplified form of the 'steel' portion of the iron/iron-carbide equilibrium diagram, and illustrates the appearance of the ferrite/pearlite, pearlite, pearlite/cementite structures, when viewed through a low power optical microscope, as the carbon content varies from 0 to 1.5%.

The important practical effect of increasing carbon in steels is an increase in the hardness and tensile properties, and it is possible to design steels to obtain specific hardnesses or tensile strengths by selecting the carbon content. The hardness of both carbon steels and alloy steels is largely dependent on the carbon content. Carbon also has a pronounced effect on the weldability of steels, and it is always desirable to keep the carbon content as low as possible in steels which are required to be welded to avoid the formation

of hard, brittle constituents in the weldment heat affected zone and weld metal.

Heating a eutectoid steel, that is one containing 0.80% carbon at room temperature and composed exclusively of pearlite, causes a transformation in its structure as the temperature passes through the point S on the diagram, at 723°C. This occurs by nucleation and growth of austenite, the cementite particles going rapidly into solution as they are intersected by the growing austenite nuclei. Under practical conditions of heating the formation of austenite requires a certain length of time and takes place, in the case of plain carbon steels, within a temperature range of 5° to 10°C.

Similarly with steels containing more or less than 0.80% carbon, which are designated hyper- or hypo-eutectoid steels respectively, the transformation into austenite occurs when the temperature passes the line PSK on the diagram. The transformation temperature is commonly referred to as the A_1 point. If the temperature rises beyond this point the excess ferrite

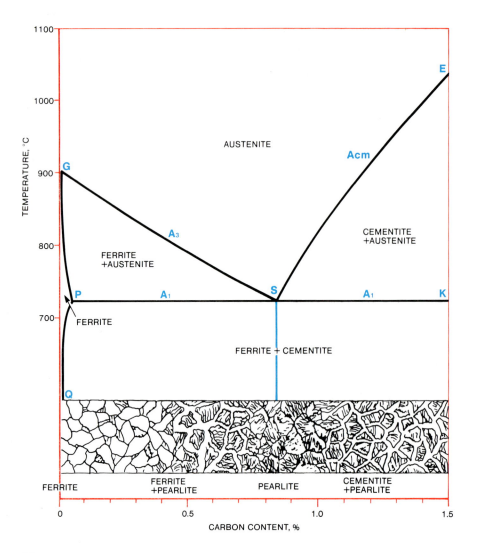

(a)

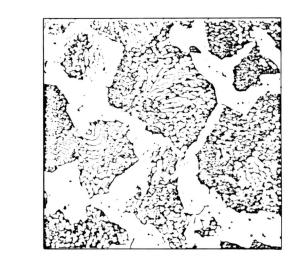

(b)

FIG. 6.13 Simplified iron/iron-carbide equilibrium diagram
(a) Change in microstructures of steels slowly transformed from austenite to ferrite
as the carbon content varies from 0 to 1.5 wt%.
(b) Detail of ferrite/pearlite microstructure of a 0.4% carbon steel at higher magnification.

or cementite, as the case may be, dissolves in the austenite. Lines GS and SE in the diagram mark the completion of the transformation according to the carbon content of the steel and are known briefly as either the A_3 and the A_{cm} points, respectively. Beyond these critical temperatures, the steel has a purely austenitic structure. It should be noted that whereas practical requirements of steel generally call for a fine grain structure, the grain size of austenite increases with the temperature and duration of the temperature.

If a plain carbon steel is slowly cooled from above its critical temperature (A_3, Fig 6.12), that is, from the austenite phase region, the transformations are reversed. In steels containing less than 0.80% carbon, ferrite nucleates and grows as the temperature falls through the range delineated by the lines GS and PS in the figure. Meanwhile, since ferrite contains little carbon in solution, the carbon content of the residual austenite progressively increases by long range diffusion until at the A_1 temperature the structure consists of ferrite of composition P(0.025%C) and austenite of composition S(0.8%C). Further extraction of heat causes the austenite to transform at constant temperature to pearlite (ferrite + cementite). Thus at temperatures just below A_1 the structure consists of grains of ferrite and pearlite, the relative proportions of which depend upon the carbon content of the steel. Further cooling to room temperature produces no significant change in structure although it may be noted that the carbon content of the ferrite is reduced still further (line PQ) by precipitation of more carbide.

The transformation of steels containing more than 0.8% carbon (hyper-eutectoid steels) is very similar except that the austenite initially rejects the carbon rich phase, cementite (Fe_3C) containing about 6.8%C, whilst cooling in the range between lines SE and SK, the carbon content of the residual austenite gradually falling to the eutectoid value. Further removal of heat causes transformation of the residual austenite to pearlite. Thus the structure consists of iron carbide (cementite) and pearlite, the proportions, again, depending upon the carbon content of the steel.

If the steel is exactly of eutectoid composition (0.8%C) it will, of course, transform to pearlite at the A_1 temperature without prior precipitation of ferrite or cementite.

2.1.5 Effects of cooling rate

The events described above occur if the steel is cooled slowly under equilibrium conditions (hence the term 'equilibrium diagram'). However, the attainment of equilibrium conditions is governed by the ability of the various atoms present to move, that is to diffuse and arrange themselves in the position required by the crystal lattices of the new phases. This diffusion is dependent in turn upon the temperature and the time available for the necessary movements to be accomplished. The higher the temperature and longer the

time the easier is the attainment of equilibrium. In plain carbon steels the structures produced by annealing (heating followed by slow cooling) are reasonably well represented by the equilibrium diagram. At faster rates of cooling, however, equilibrium is not approached and significant differences in structure arise. Indeed, if the rate of cooling is sufficiently rapid, the transformations which depend upon diffusion may be suppressed and other modes of transformation occur (to bainite or martensite). It is this ability to change the structure by varying the rate of cooling which forms the basis of the heat treatment of steel for mechanical property control.

The first effect of a higher rate of cooling than the equilibrium rate is to lower the temperature at which transformation occurs, and an understanding of the different structures which may be produced can be gained by considering the processes which would occur if the steel were cooled instantaneously to a temperature below the equilibrium transformation temperature, and then held at constant temperature until the transformation were complete. This can be simulated experimentally by rapidly quenching specimens of the steel from the austenite phase region to the transformation temperature and holding them isothermally to determine the initiation, progress and completion of the transformation. The structural changes may be followed by monitoring the dimensional changes of the specimen (dilatometry). These reflect the allotropic changes since austenite has a smaller volume than the transformation products. Alternatively specimens may be rapidly quenched after prescribed periods of isothermal treatment to retain the transient structure for microscopical examination of the extent of transformation and the nature of the product. From information of this sort it is possible to construct a time-temperature-transformation (TTT) diagram of the type shown in Fig 6.14. TTT diagrams (or continuous cooling diagrams) have been defined and are readily available for all power plant steels.

At temperatures just below A_1, the transformation from austenite to pearlite takes an appreciable time to initiate and complete, and the product is lamellar pearlite. As the temperature is lowered the time to initiate transformation shortens and the product is pearlite of increasing fineness, which at temperatures approaching 550°C cannot be resolved into its lamellar constituents using the optical microscope.

Further decrease in transformation temperature causes a lengthening of the incubation period and a change in structure of the product to a form known as 'bainite' (in honour of E. C. Bain, an American metallurgist who discovered the phase). These differences in incubation period reflect the competing influences of the increase in thermodynamic driving force as the temperature is lowered below the equilibrium value (A_1) and the increasing difficulty of diffusion of atoms to form the new phases which result in the characteristic C curve behaviour.

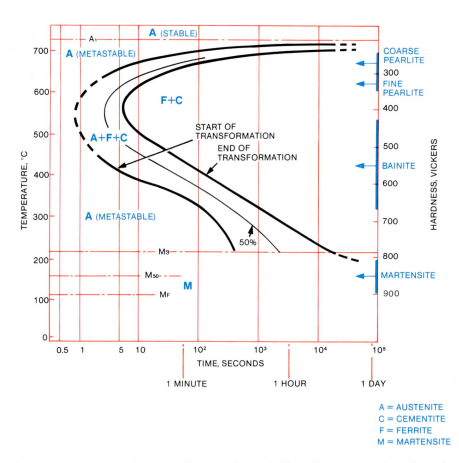

FIG. 6.14 Typical isothermal transformation diagram of a steel (Time, Temperature, Transformation or TTT)

2.1.6 Crystallographic nature of bainite and martensite

In common with the transformation to pearlite, ferrite in a bainite transformation nucleates initially at the austenite grain boundaries but thereafter grows by the formation of long needles which frequently traverse the austenite grain and branch along crystallographic planes. Pearlite formation is suppressed by the inadequate carbon diffusivity and instead the carbon is segregated ahead of the ferrite/austenite interface until large carbide particles are nucleated in the carbon-rich areas between impinging needles. This is upper bainite. As the transformation temperature is lowered, or cooling rate increased, the needles become finer, more well defined and even more crystallographically-oriented with the austenite matrix and fine carbide particles nucleate within as well as between the plates to form lower bainite.

If the temperature is lowered sufficiently, the diffusion-controlled nucleation and growth modes of transformation are suppressed completely and the austenite transforms by a diffusionless process in which the crystal lattice effectively shears to a new crystallographic configuration known as 'martensite'. This phase is morphologically similar to bainite in the optical microscope but has a tetragonal crystal structure and contains carbon in super-saturated solid solution. The formation of martensite, unlike pearlite and bainite, is dependent

not on time but on temperature. Thus there is a temperature at which the first traces of martensite will form, known as the martensite start temperature (M_s). Further transformation will only occur if the temperature is lowered and will be complete at some temperature known as the 'martensite finish temperature' (M_f).

Bainitic and martensitic transformations are frequently referred to as military transformations because of the co-operative, regimented way in which the atoms change allegiance from the austenite matrix to the ferrite needles. The M_s temperature is between 150° and 350°C and the M_f temperature about 100°C lower.

The above account has considered only the transformation of austenite below the critical temperature, and in all but steels of precisely eutectoid composition this will be preceded and accompanied by precipitation of ferrite (pro-eutectoid ferrite) in hypo-eutectoid steels or cementite in the case of hyper-eutectoid steels. Lines indicating the formation of these phases can also be depicted on the TTT diagram.

In practice the transformation of a steel normally occurs under continuous cooling conditions rather than isothermally but similar processes are involved.

2.2 Tempering

In the martensitic condition the steel is hard and brittle, as shown by the comparison of steels of different carbon content in the slowly cooled (annealed) and rapidly cooled (quenched) conditions tabulated below.

	Carbon content, %					
	0.1	0.3	0.5	0.7	0.9	1.2
	Hardness, VPN					
Slow-cooled	109	121	170	229	248	285
Water-quenched	390	640	810	920	920	920

Steels are rarely used in the hard but brittle martensitic condition, but high strength coupled with improved ductility can be obtained by tempering, i.e., reheating the martensitic structure at an appropriate temperature below the critical temperature (A_1). This procedure of hardening followed by tempering is fundamental to the heat treatment of steels.

Martensite, as pointed out previously, is a non-equilibrium structure and the reheating or tempering treatment will tend to restore the material to its equilibrium phase constitution of ferrite plus carbide. Thus if a steel is tempered at progressively higher temperature the first effect is to relieve the high internal stresses which characterise the martensite structure. The carbon in super-saturated solid solution then commences to precipitate as carbide, in this case in the form of discrete particles rather than the lamellar form associated with its formation as pearlite. As the temperature of tempering is raised and the time increased, the carbides grow in quantity and size and the matrix becomes depleted in carbon. These changes are accompanied by a reduction in hardness and strength, and by an increase in ductility. If tempering is continued, the structure ultimately consists of approximately spherical particles of carbide in a relatively soft matrix of ferrite. By judicious choice of tempering conditions the desired mechanical properties may be obtained.

The lamellar form of carbide in pearlite may also be spheroidised by long term heating at temperatures approaching the lower critical temperature (the A_1), that is, 500° to 720°C. In this case the driving force for spheroidisation is the reduction of interfacial energy between the carbide phase and the matrix — analogous to the spherical shape of a soap bubble, which is caused by surface tension effects — but the rate of approach to the equilibrium condition is determined by the diffusion of atoms which, in turn, depends on the temperature and duration of heating. Given a knowledge of the original structure, it is possible to assess the thermal history of a specimen from the degree of spheroidisation which has occurred. In this way it is possible to demonstrate that a high temperature component, such as a superheater tube, has been overheated and this is useful in the investigation of plant failures.

2.2.1 The prior austenite grain boundary (PAGB)

A fundamental distinction in behaviour of the austenite boundary exists between civil and military transformations. In civil transformations the ferrite nuclei formed on the austenite triple points and boundaries grow more or less equally into the adjoining grains. The austenite boundary is thereby totally destroyed. In military transformations the bainite or martensite needles (plates), because of their strong crystallographic growth nature, grow into only one of the adjoining austenite grains if boundary nucleated, or are confined within the parent austenite grain if matrix nucleated. The austenite boundaries are never traversed by the growing ferrite and each austenite grain in effect transforms in isolation to preserve the original austenite boundary network. Since it is now a boundary network in ferrite rather than austenite it is referred to as the prior austenite grain boundary. It is of special significance in weldments and castings transformed to bainite or in any component in which the peak temperature has permitted austenite grain growth to occur before transformation. Subsequent ageing or creep deformation can severely embrittle the PAGB of coarse-grained structures and this is a common source of premature failure. The effect is now well understood and precautions taken to avoid its consequences (see Section 4 of this chapter).

2.2.2 Recovery, recrystallisation and renormalising

The mechanical and thermal processes involved in the primary fabrication of plant components frequently result in the formation of other than the optimum microstructure for the desired mechanical properties. The examples of coarse-grained bainitic microstructures in weldments and coarse columnar grains in castings have already been cited. Other examples are the heavily deformed structures resulting from cold working where the grain structure is severely elongated in the rolling or forging direction. Cold working leads to a massive increase in the dislocation density of the material which impedes further crystal slip and thus produces a large increase in tensile strength, but reduces toughness and ductility. The cold worked state is fairly stable mechanically and can persist indefinitely at low temperatures. Heating at relatively low temperatures leads to a reduction in the dislocation density and the toughness improves for little loss of strength (recovery). Heating at between 450°C and 700°C however leads to a drastic softening of cold worked steels due to a process known as recrystallisation. This is similar to an allotropic transformation in that it proceeds by the nucleation, growth and coalescence of new, strain-free grains to form a relatively fine-grained and equi-axed microstructure of ferrite. The driving force is however the reduction in the stored strain energy associated with the dislocations rather than the free energy change of the allotropic transformation.

This primary recrystallisation is complete when all the original worked grains have been replaced. At higher temperatures in highly textured materials the primary recrystallised grains may be replaced by a new wave of recrystallisation. This is termed 'secondary recrystallisation'. Dynamic recrystallisation of austenite is the basis of the hot working processes where ingots or billets are given massive deformations by comparatively small applied stresses. The time required for recrystallisation sets a limit to the speed of the hot working operation and very rapid operations need appropriately higher temperatures. Alloys become more difficult to hot work however as the level of addition increases because of the strong influence of substitutional elements in raising the recrystallisation temperature.

2.3 Normalising

The optimum microstructural state for many applications is that of a uniform fine-grained structure of dispersion strengthened ferrite/pearlite, bainite or tempered martensite. The fine grain size is usually beneficial for ductility at both high and low temperatures and can be induced by heating to the minimum temperature in the austenite field consistent with adequate dissolution of the strengthening particles. This minimises austenite grain growth, ensuring a high nucleation rate on subsequent transformation to ferrite during cooling and the formation of the desired fine ferrite grain structure. Slow cooling in a furnace is referred to as annealing (a term also widely used for any softening treatment by heat). Natural cooling from a low austenite temperature is known as a normalising heat treatment.

2.3.1 Renormalising

The mechanical properties of the microstructural constituents of many parent and weld metals progressively deteriorate under the combined action of stress and temperature during service. The matrix strength is often reduced as the strengthening carbide dispersion coarsens and segregation and localised deformation frequently weaken the grain boundaries. Properties can however often be recovered if the component can be retransformed to austenite and cooled at the appropriate rate, termed renormalising. As the material recrystallises, the damaged boundaries become isolated either within the new austenite grain structure, which persists if the subsequent transformation is to bainite, or within the new ferrite/pearlite aggregate of grains resulting if the cooling rate is slower or the steel less highly alloyed. In either case the final grain size can be controlled by adopting the minimum austenitising temperature and time consistent with adequate dissolution and represipitation of the strengthening carbide dispersion. For low alloy steels this would be 950°C for several hours.

Renormalising, often combined with a subsequent temper to further stabilise the structure, is specified for a wide range of applications to ensure that the entire component, together with any detrimental microstructures resulting from the fabrication process (including those associated with welding), is transformed to a uniformly fine grained aggregate. The designer is thereby assured of uniform properties throughout the structure.

2.4 Effects of alloying

2.4.1 Hardenability

Plain carbon steels, as outlined above, can be dramatically strengthened by rapidly cooling austenite to form a bainitic or martensitic microstructure. Practical limitations of cooling rate control however severely limit the realisation of the full potential strength of plain carbon steels in plant components of heavy section. The cooling rate remote from the surface is restricted by the thermal conductivity of the steel such that, even for high quenching rates at the surface, the rate not far below the surface will be insufficient to form bainite or martensite and the result is a component with a hard martensitic exterior and a relatively soft core of pearlite. Attempts to increase the cooling rate across the entire section by, for example, water spraying or oil quenching generate very high thermal stresses at the surface with a risk of cracking. The ease with which transformation to martensite is obtained over the entire cross-section of the component is termed the 'hardenability' of the steel and all plain carbon steels are of poor hardenability. This difficulty can be overcome by adding alloying elements to the steel which suppress the nucleation and growth mechanism of pearlite formation by reducing diffusion rates. The civil transformation is thus postponed to longer times and lower temperatures, i.e., the curves in the TTT diagram (Fig 6.14) are pushed lower and to the right and the M_S and M_F temperatures are changed. It is not then necessary to resort to *drastically fast methods of quenching* and thorough hardening can be achieved in thick sections by oil quenching or even by natural air cooling (air hardening). An important example is the use of nickel to achieve full section hardenability in rotor steels.

As well as their effects on times to transform, alloying elements also have a powerful influence on the temperature ranges over which ferrite and austenite are stable and the shape of the iron carbon equilibrium diagram is accordingly changed. Elements are classified into groups of austenite or ferrite stabilisers, depending on whether they expand or contract the temperature range over which austenite is stable. In the former group are carbon, manganese, copper, nickel and cobalt, while silver, chromium, molybdenum, tungsten, niobium, vanadium and titanium extend the alpha and delta ferrite ranges. The effect of austenite stabilisers on hardenability, discussed above, is initially to lower the transformation temperature, thus inducing formation of bainite or martensite at slower cooling rates, but at higher alloying levels the transformation is suppressed below ambient and the steel remains austenitic.

2.4.2 Solid solution and dispersion strengthening

The tendency of alloying elements to open or close the gamma loop in the iron/carbon diagram is a consequence of their dissolution in the iron matrix. The atoms of the alloying elements substitute for those of iron within the austenite matrix and, because of the atomic size difference, the ferrite matrix produced on transformation is elastically strained. Dislocation motion is consequently impeded and the alloy becomes stronger. This is known as solid solution hardening and is an important component of the strength of alloy steels containing Cr, Ni, Si, Mn, Mo or Cu. Molybdenum and chromium are especially important since their strengthening effects persist to high temperatures and confer a significant element of creep resistance. Further strength can be conferred by the presence of a dispersion of very fine particles spaced sufficiently closely to impede dislocation motion, known as dispersion strengthening. Dispersions can be produced in several ways.

In age or temper hardening the dispersion and associated strength develops only during the ageing or tempering treatment of a super-saturated solid solution produced by a quench. This differs therefore from martensite and bainite hardening which occurs during the quench. Super-saturated solid solutions are produced for alloy systems exhibiting the phase diagram characteristics shown in Fig 6.15. The alloy can be homogenised as a single phase (α) by heating to and holding at temperature T_1. After this 'solution treatment' the alloy is cooled to the two phase region, generally to room temperature, sufficiently quickly to prevent separation of the solute. This gives a super-saturated solid solution. Hardening results when the alloy is aged at some low temperature, or tempered at a higher temperature to accelerate the process, when the β phase separates in a finely dispersed form. Typical dispersion phases in steels are carbides and nitrides, but intermetallic particles also exist. Tempering of bainites and martensites results in a rapid softening of the ferrite as the *as quenched* dislocation density reduces. The resulting loss of strength is compensated in part by the precipitation of fine carbides, and the process can continue to very long times during service (tens of thousands of hours). Carbide forming elements profoundly affect the tempering reaction and are used, in addition to improving hardenability, to delay softening and improve toughness. Some elements, especially molybdenum and vanadium produce such an intense dispersion that the hardness increases during tempering above 550°C before falling at longer times. This is termed 'secondary hardening'. Molybdenum, vanadium and chromium carbides are also highly resistant to coarsening. Low and medium alloy creep-

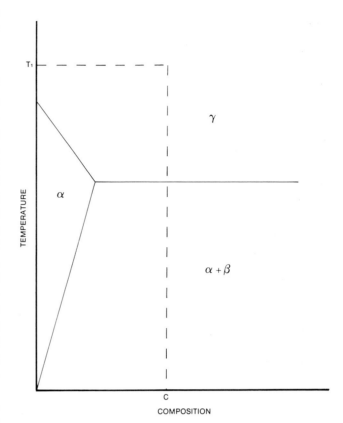

FIG. 6.15 Principle of dispersion hardening

resistant steels owe much of their long term high temperature strength to this effect.

An additional strengthening mode, also due to fine carbide dispersions, occurs in normalised low alloy steels, particularly those containing vanadium, and is known as 'interphase precipitation'. Precipitation in this case occurs during the austenite to ferrite civil transformation as sheets of particles on the α/γ interface. The growing particles momentarily pin the growing α/γ boundary which then breaks away to a new parallel position before being pinned again by the growing V_4C_3 particles. Under the continuous cooling conditions of normalised steels the particle size continually decreases as the temperature falls and diffusion distances reduce. The process can be observed dynamically in the photo emission electron microscope (Fig 6.16) or statically in thin foils in the transmission electron microscope.

2.5 Stainless steel

Chromium, although in itself a ferrite and carbide former, has a side effect of making the structural changes very sluggish, which makes the suppression of the austenite to ferrite change in heat treatment easier to achieve. It is, therefore, extensively used in steels to be hardened and tempered. A further important property of chromium, particularly marked when present in quantities above about 2%, is to improve resistance to corrosion and oxidation.

Resistance to corrosion and oxidation of steel depends on the film of oxide formed on its surface and in carbon and many low alloy steels, this oxide film offers little or no resistance to atmospheric corrosion. At elevated temperatures these steels have good resistance to oxidation in air or flue gas up to 575°C (1065°F) but above this the rate of oxidation increases rapidly. The presence of chromium however, in excess of about 5%, promotes the formation of a more protective oxide film and although 5% is insufficient to obtain useful resistance to atmospheric and aqueous corrosion, it is enough to improve the oxidation resistance up to about 600°C. Further increasing the chromium content produces a more resistant oxide film and at 13%, satisfactory resistance to mild corrosion media, such as wet steam, is achieved. Applications of this type of steel are steam turbine blades, propeller and pump shafts, impellers and water turbine runners. Increasing the chromium content above 13% produces improved resistance to more corrosive media and at 28% chromium, satisfactory oxidation resistance at 1100°C can be obtained.

Steels with chromium contents of up to 14%, provided the carbon content is above 0.12%, can still be hardened and tempered. With chromium greater than 14% however, because of its strong ferrite forming tendencies, the alpha- and delta-ferrite regions join together at the expense of austenite, with the result that such steels are wholly ferritic up to their melting point and cannot be hardened by heat treatment. Such steels tend to have a very coarse crystal structure and are then relatively brittle at room temperature and difficult to fabricate. They are not, however, brittle at elevated temperature and find application in furnace parts such as grate bars, clinker dams and dampers, generally in the form of casings.

In order to be able to utilise the good corrosion-resisting properties of these high chromium steels and at the same time attain satisfactory engineering properties, it is necessary to re-establish the austenite region. This can be done by adding nickel and with an 18% chromium steel the addition of about 2% of nickel does this and produces a steel which can be hardened and tempered. This is the well known steel (En 57) which is widely used for pump shafts in the marine field. Maintaining a chromium level of 18%, the addition of increasing amounts of nickel extends the re-established austenite region until at 8% nickel the temperature of the change from austenite to ferrite is depressed below room temperature and the structure at room temperature consists of grains of austenite. These steels are termed austenitic and include the well known 18/8 stainless steel.

Since the austenite-ferrite change on which hardening and tempering are dependent is suppressed below room temperature, these austenitic steels are similar to the high chromium ferritic steels in that they cannot be hardened by normal heat treatment. They are different from the ferritic high chromium steels in that they are extremely ductile, and ideally suited for deep pressings and similar applications. In addition, since they are austenitic they are non-magnetic and have a high coefficient of thermal expansion and low thermal conductivity, the combination of which renders them susceptible to thermal fatigue cracking during plant cycling.

Although these steels are not hardenable and have relatively low tensile strength at room temperature, they do have exceptionally good elevated temperature tensile properties which, when combined with their good corrosion resistance, suit them to applications demanding this combination. These include superheater tubing and steam piping where the metal temperatures are in excess of 550°C, gas turbine components and numerous types of pressure vessels employed in the chemical and allied industries.

The bulk of the austenitic steels produced are of the chromium and nickel type and other elements such as titanium, niobium, molybdenum, copper, tungsten, cobalt and aluminium may be added to impart special properties. A common example found in modern power stations is the superheater tube and steam pipe austenitic steel containing 18% chromium, 10 to 12% nickel and $2\frac{1}{2}$% molybdenum.

The 'Schaeffler' diagram, Fig 6.17, illustrates the phases present in stainless steel weld metals of various compositions, expressed as total equivalent chromium and nickel contents, and provides a useful reference to the following categories of stainless steels.

2.5.1 Magnetic

Martensitic, for example, turbine blades and cutlery. Ferritic, for example, corrosion-resistant decoration (motor car trim, etc.) and high temperature furnace parts. The magnetic types contain essentially chromium as the major alloying element; they are heat-treatable if the carbon content exceeds approximately 0.12% and may be welded, but the more hardenable types require pre- and post-weld heat treatments as outlined for the low alloy weld and parent metals. (Martensite areas on diagram.) At high chromium and low nickel equivalent values the structure is always fully ferrite and generally coarse-grained and brittle at ambient temperatures. It cannot be refined by post-weld heat treatment.

FIG. 6.16 Typical carbide dispersions found in steels observed precipitating dynamically by photo emission electron microscopy. The plate is one of a sequence, taken as austenite transforms to ferrite and carbide in a low alloy steel containing vanadium and molybdenum. Carbides form as large dendrites on the austenite boundaries and specimen surface, as needles of molybdenum carbide (Mo_2C) and as sheets of vanadium carbide (V_4C_3) particles along the advancing austenite/ferrite interface (interphase precipitation). Magnification 5000 diameters.

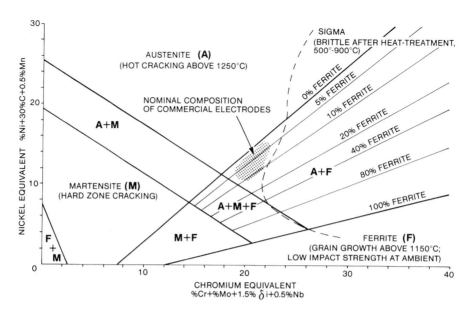

FIG. 6.17 Constitutional diagram for stainless steel weld metal (after Schaeffler)

2.5.2 Non-magnetic

Austenitic, for example, superheater tubes and hangers, steam pipes and sootblower nozzles. The non-magnetic types contain sufficient nickel or other austeniser in addition to chromium to retain the austenite structure at room temperature. Therefore, no hardening occurs due to fast rates of cooling and welding does not generally require any pre- or post-heating (Austenite with up to 20% ferrite areas on diagram).

Possible disadvantages with these steels when subject to welding processes and/or corrosive environments or high temperature conditions are:

- *Hot cracking of weld metal* — austenitic weld metals exhibit a tendency to hot cracking unless a small amount of ferrite is present. Normally this is permitted but is often a compromise if ferrite is undesirable, as is the case when high temperature strength is required. A convex-shaped weld bead is less prone to cracking than a concave bead and basic low hydrogen coatings are superior to rutile (titanium oxide) coatings.

- *Liquation cracking* — the presence of liquation cracks in that part of the heat-affected zone immediately adjacent to the weld metal. This is, in effect, hot cracking of the parent plate and all the austenitic stainless steels in current use for steam piping are potentially susceptible. For this reason the entire surface of all welds in these steels and that portion of the parent plate next to the weld are dressed after welding and final heat treatment, and subjected to careful non-destructive examination.

- *Susceptibility to weld decay* — a phenomenon involving preferential precipitation of chromium carbides at the grain boundaries, involving denudation of chromium and subsequent reduced cor-

rosion resistance at the grain boundaries. Carbide precipitation occurs within the temperature range 500–700°C, and this temperature range must be produced at some position in the parent material when it is welded, hence the term 'weld decay'. Weld decay may be obviated by reducing the carbon content to a maximum of 0.03% or by fixing the carbon with a strong carbide-former such as niobium or titanium so preventing the formation of chromium carbides. The latter practice is more common and the steels are then known as 'stabilised' steels.

- *General embrittlement* — due to precipitation of carbides and development of a hard brittle phase called 'sigma-phase' which is basically an iron chromium compound. Sigma tends to form within the temperature range of 500–900°C and the presence of other elements, for example, molybdenum, may encourage its formation.

- *Embrittlement of the heat-affected zone* — associated with solution of carbide-formers during welding, and subsequent preferential strain-induced precipitation — this was the case with the difficulties associated with the Type 347 (18% chromium, 12% nickel, 1% niobium) steel steam piping at certain generating stations. These problems have been resolved, the solution being to use a relatively soft weld metal and to apply a high temperature post-weld heat treatment (1050°C) to dissolve harmful precipitates, the operation being arranged to keep contractional strains on cooling to a minimum. For the Drakelow C super-critical units other materials have been used, in particular Type 316 (17% chromium, 11% nickel, 2.5% molybdenum) steel, and Esshete 1250 (16% chromium, 10% nickel, 6% manganese, 1% molybdenum, 1% niobium,

0.26% vanadium, 0.005% boron) steel. All weld joints were post-weld solution heat treated at 1050°C.

2.6 Cast irons

Cast irons, like steels, are alloys of iron and carbon but, being essentially unrefined iron from the blast furnace contain larger proportions of carbon, silicon, sulphur and phosphorus. A typical composition of cast iron is:

Total carbon	Silicon	Manganese	Sulphur	Phosphorus
3.5%	2.0%	0.8%	0.1%	0.35%

The carbon in cast iron can exist at room temperature as iron carbide, or as graphite which is the more stable form. Irons containing carbon as graphite are soft, easily machinable and are called 'grey irons'. Irons with carbon present as iron carbide are extremely hard, difficult to machine, and are called 'white' or 'chilled' irons (Fig 6.18). Irons with fairly equal proportions of graphite and iron carbide have intermediate hardness and are called 'mottled' irons. The factors mainly influencing the form of the carbon are the rate of cooling during solidification and the chemical composition. Rapid rates of cooling prevent the formation of graphite and result in a high proportion of iron carbide and hard white irons. Conversely, slow rates of cooling promote graphite formation and result in soft grey irons. This effect of cooling rate on the structure of cast iron is reflected in the precautions necessary for the satisfactory welding of the material. White irons can be regarded as unweldable. Grey iron can be welded but it is imperative to avoid the formation of a zone of white iron adjacent to the weld deposit as a result of the chilling effect of the parent metal. It is essential, therefore, that grey iron be preheated to retard the cooling rate in the weld metal and adjacent parent metal.

The rate of cooling of a casting is largely dependent on the mass of the casting and consequently castings of a relatively thin cross-section tend to be 'white' and castings of large cross-sections are always 'grey'. Castings with varying sections often contain a gradation of structure from grey to white.

With regard to chemical composition, elements such as manganese, nickel, copper, carbon and silicon promote graphite formation whereas chromium and other carbide formers promote carbide formation. Sulphur and phosphorus are present in larger quantities than in steels because the processes involved in the manufacture of cast iron do not include refining operations to remove these elements. Grey iron is extensively used in engineering because of five notable characteristics, these being:

- Cheapness.

- Low melting point and high fluidity making it suitable for castings of intricate shape.

- Relatively good erosion- and corrosion-resistance.

- High damping capacity with respect to vibration.

- Relatively good mechanical properties under compressive loading.

The graphite in grey irons exists in the form of flakes which act as stress-raisers under tensile loading and consequently grey irons have relatively low tensile strength and ductility.

White irons, because of their extreme hardness and brittleness are generally confined to applications demanding high abrasion resistance and where they can be fully supported by less brittle materials.

Small castings which tend to be white because of their small cross-sections can, however, be heat-treated to convert the iron carbide into graphite which results in a soft, machinable iron suitable for general engineering applications. This process, which is similar to the annealing of steel, is termed 'malleablising' and the final product is 'malleable iron'. The graphite in malleable iron is in the form of regular particles rather than flakes and as such has relatively little embrittling effect under tensile loading. The room temperature tensile properties of these irons approach those of mild steel.

Deliberate alloying additions are made to cast irons, in the same way as to steels, to enhance their mechanical properties and resistance to corrosion, abrasion and high temperature.

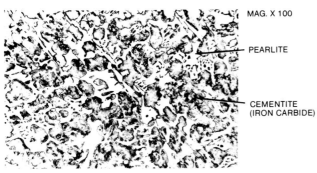

MAG. X 100

— PEARLITE

— CEMENTITE (IRON CARBIDE)

(a) White cast iron

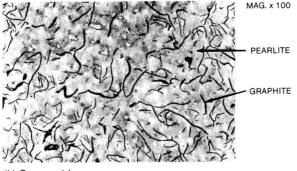

MAG. x 100

— PEARLITE

— GRAPHITE

(b) Grey cast iron

FIG. 6.18 'White' and 'grey' cast irons

2.7   Deformation and fracture

All engineering structures are required to sustain a load in one form or another, and a useful measure of a material for a given engineering purpose is its tensile strength. The design stresses imposed are generally within the elastic limit of the material, as defined by the tensile test, but a structure may be subject to local stresses above the yield point at certain locations due to discontinuities (such as stress raisers within the structure) and therefore must be capable of deformation without fracture. This requires that the material should have an adequate level of ductility in addition to strength.

In a metal the properties of individual grains are direction dependent according to the particular crystal form, but since they are generally randomly oriented and small in comparison with conventional material section sizes, the directional effect is usually eliminated. It is thus entirely feasible for the engineer to regard a piece of iron or steel as having uniform properties in all directions for design purposes. Certain exceptions to this generalisation include grain-oriented silicon-iron sheet for transformer cores, where a special effort is made to produce material in which the grains have the same crystalline directions in order to reduce electrical losses. It is important to consider the generally harmful effect of segregations of non-metallic inclusions, which are always present to some extent in commercial steels and which, if not properly controlled, may appear as laminations and strings of inclusions in plate and tube. Such inclusions impart directional mechanical properties by introducing weaker and/or more brittle material. Thus great care is taken in the working of important forgings such as those for turbine rotors so that any non-metallics are oriented in the least harmful direction.

Deformation of crystalline materials such as metals involves both 'elastic' and 'plastic' strain. The elastic portion is not permanent and disappears when the load is removed, but plastic strain involves the movement of atoms within individual crystals along certain planes by a shearing process known as 'slip' (Fig 6.16). It has been found in practice that metal crystals are much weaker than would be expected from theoretical predictions based on considerations of the inter-atomic forces which must be overcome to produce slip. This has now been explained by the concept of line imperfections within the crystal lattice, known as 'dislocations', which are able to move under relatively low stresses through the crystal lattice, producing slip of one part of the crystal with respect to the remainder. Modern methods of microscopy using high magnifications and resolutions now permit the observation of some of these phenomena, confirming the theories previously formulated.

If a specimen is subjected to an increasing uniaxial stress, as in a normal tensile test, it initially deforms in an elastic manner until a point is reached (the elastic limit or yield point), where the metal ceases to behave elastically and suffers permanent or plastic deformation. If the stress continues to increase the metal will extend a certain amount and then fracture, producing a crack which is usually trans-crystalline at ambient temperature. Under creep conditions (high temperature and relatively low stress values) movement can be concentrated at the grain boundaries resulting in an intergranular type of failure.

The well defined yield of mild steel is shown in Fig 6.19 (a) and the method of assessing proof stress, corresponding to 0.1% extension, in a steel not showing a sharp yield, is shown in Fig 6.19 (b). Design stresses for pressure parts not subject to creep are based upon arbitrarily selected fractions of one or another of these mechanical properties. The hardening effect of cold deformation is shown in Fig 6.19 (c).

Ductility is measured by noting the elongation undergone by a test specimen during straining to fracture. It may also be gauged by measuring the reduction of

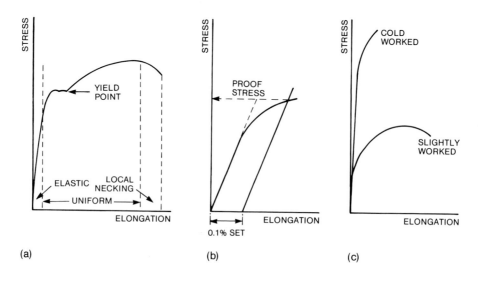

(a)   (b)   (c)

FIG. 6.19   Stress-elongation curves for mild steel

area at the point of fracture. If a metal exhibits low values of elongation and/or reduction of area, it is unlikely to be able to withstand stress concentrations or shock loading without premature failure.

The most common cause of failure of engineering components is fatigue under cyclic stresses. A tensile test gives an indication of how a steel will behave under static tensile loading, but it gives only an approximate indication of how a steel will behave under fluctuating stresses. Ferrous metals differ from many other metals in that they exhibit an endurance limit, which means that there is a maximum level of fluctuating stress which can be tolerated indefinitely. In most steels this stress is approximately 50% of the ultimate tensile strength and the endurance limit is defined as the stress which can be endured for ten million reversals of stress. The method of presentation of fatigue data as 'S-N' curves is shown in Fig 6.20, where the range of the cyclic stress (for a given mean stress) is plotted against the logarithm of the number of cycles. A definite endurance limit is only obtained in tests in air at temperatures below about 300°C. Tests at higher temperatures or in a corrosive environment do not display an endurance limit and the type of curve obtained under such conditions is also shown in the figure.

The presence of stress concentrations in the form of notches, holes, sharp changes in sections or re-entrant angles have a pronounced effect on the behaviour of a component under fluctuating stress conditions especially of those associated with crack-like defects. Such situations can now be adequately analysed by fracture mechanics methods (Chapter 12).

2.7.1 Impact toughness testing

The resistance of metals to crack propagation from defects at temperatures below the creep range (toughness) has traditionally been assessed by the impact test. For ferritic steels particularly, the measurement of brittleness has been assessed from the energy absorbed at impact by a blunt notched specimen (the Charpy Test).

A typical Charpy Impact curve is shown in Fig 6.21, where the energy absorbed and the percentage of brittle (crystalline) fracture on the fracture surface are plotted against test temperature. A smooth transition typically occurs from low to high energy fracture associated with a change from a brittle (lower shelf) to a ductile fibrous mode of failure (upper shelf). The crystalline appearance is due to trans or intergranular atomic cleavage and the dull fibrous appearance to localised internal plastic necking of the material between cavities nucleated primarily on non-metallic inclusions. Various transition temperatures can be defined, for example, the Fracture Appearance Transition Temperature (FATT) or the Ductile/Brittle Transition Temperature (DBTT).

An essential aspect of component design is therefore to ensure that the material is at all times in the 'upper shelf' region when subjected to significant stress during fabrication or operation. The micromechanisms of propagation and the controlling microstructural features, for example, grain size, inclusion content and residual element level, that affect both transition temperatures and upper shelf values have been identified in recent years and methods found to minimise their effects such that modern steels are now much more tolerant of defects.

Notched impact tests are useful for comparative purposes but data so obtained cannot be used for a quantitative assessment of the significance of cracks in a real structure. This is now achieved by the application of Fracture Mechanics and Fracture Toughness data (Chapter 12).

2.7.2 Elevated temperature properties

The tensile properties of most engineering materials at ambient temperature are independent of time for most practical purposes. At elevated temperature,

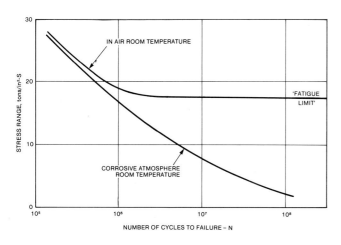

FIG. 6.20 Typical S-N (high cycle) fatigue curves

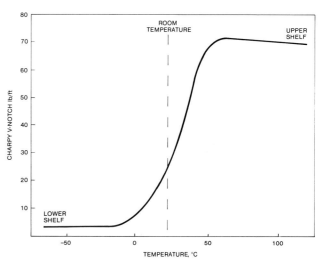

FIG. 6.21 Typical Charpy V-notch curve for a low carbon steel

however, the resistance to deformation becomes very dependent on both strain rate and time. Steels subjected to a constant tensile (or compressive) load at elevated temperature ($>300°C$) undergo time dependent deformation (creep) at stresses well below those required to induce plastic deformation at ambient temperature. Special tests are thus required to evaluate the performance of materials in different kinds of high temperature service. The creep test measures the dependence of dimensional changes (usually rate of increase of length) on stress and temperature while the stress rupture test measures the effect of temperature on the long-time load bearing characteristics. The engineering creep curve of a metal is determined by applying a load to a specimen (in tension or compression) at constant temperature and measuring the change in length as a function of time. A creep curve is shown in Fig 6.22 and typically exhibits three stages. The primary stage, where the initial rapid rate of elongation reduces as the material strain hardens. The secondary stage, where the rate of softening (recovery) due to heating balances the strain hardening results in a linear rate of deformation (steady state creep). The tertiary stage, where the material begins to decohere internally (by the formation of cavities or microcracks) and the strain rate accelerates rapidly to failure. The strain at failure is the creep ductility of the material. The basic creep data may be used to derive curves such a 'stress to produce a given creep rate or stress to give various amounts of plastic strain', for example, for high temperature rotors the design may be based on the stress to produce 0.1% strain after 100 000 hours' operation. Although simple in principle, the accurate measurement of creep extension at constant accurate temperature over periods of up to ten years requires a considerable investment in laboratory equipment. A log-log plot of stress against minimum creep rate frequently yields a straight line and is very useful for design purposes (Fig 6.23).

Creep tests are usually carried out at relatively low strains to avoid the tertiary stage. The design of many plant components however has historically been based on data obtained in the stress rupture test which is

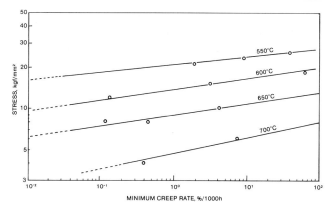

similar to the creep test except that the parameter of primary interest is the stress to cause failure in a given time (rupture time). Whereas the total strain in a creep test is often less than 0.5% the failure strain (rupture ductility) in a stress rupture test may be 50% or more. Simpler strain measuring devices are therefore used or, if strain rate data are not required, multiple specimens of varying cross-section to vary the stress can be tested in a single furnace.

Much of the high temperature data for steels is, because of its simplicity and lower cost, stress rupture data, but is generally only available for rupture times well short of the required operating life of the plant. The required rupture stress for design purposes must therefore be obtained by extrapolation. A typical design criterion is 60% of the stress that leads to failure in 100 000 h. The stress is plotted against the rupture time on a log-log scale which usually yields a straight line or smooth curve for each temperature, Fig 6.24. Sudden changes in slope are usually due to structural changes or changes in failure mode (e.g., transgranular to intergranular, carbide changes, etc.) and must be understood if serious errors in extrapolation are to be avoided. Much of the data for parent material is collected from laboratories in different countries from base materials whose composition varies within the specification limits. Together with minor variations in heat treatment and temperature measurement these factors result in a substantial scatter in the final plot

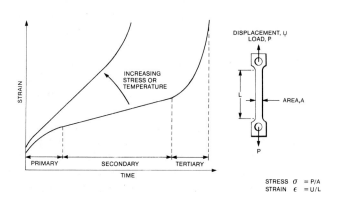

Fig. 6.22 Typical creep curves for steels showing primary, steady state secondary and tertiary stages

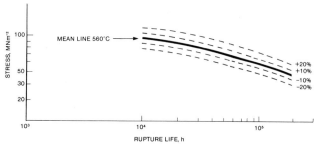

Fig. 6.24 Stress rupture data for $2\frac{1}{4}$ Cr1Mo at 560°C

for all data. The data base must then be analysed and fitted to a parametric relationship so that a mean line representing all data can be defined. The effect of deviations can then be covered by the use of upper and lower bound limits, normally quoted as 20% of the mean parametric stress value. Lower bound data are frequently used in design thus introducing a degree of conservatism. The relative 100 kh rupture strengths of the common boiler and turbine steels are shown in Fig 6.25.

2.7.3 Reference stress

The distribution of the time dependent stresses and strains throughout a complex component under load are governed by its creep properties and geometry and can be obtained by analysis. Such analyses were performed for simple geometries some fifty years ago. Since then, steady developments in analytical methods coupled with the advent of the digital computer now enable even the most complex shapes to be modelled and analysed under variable load and temperature histories. Computer analysis is very time consuming and costly and so parallel developments of simplified methods which avoid lengthy computation yet allow the designer to establish readily the rupture life have been developed. These enable simple determination of section thickness and identify necessary changes to improve performance. An important development has

been the definition of a 'reference stress' for a component such that the component lifetime is equal to the rupture time of a simple uniaxial specimen tested at the reference stress. Once the reference stress has been established the component lifetime can thus be obtained directly from the stress rupture curves for the component material. For thin-walled tubes the reference stress is the mean radius hoop stress but for thicker-walled tubes and complex components considerable variations of stress occur throughout the structure and the reference stress is less easy to establish. Theory and experiment have shown that the reference stress can frequently be based on the collapse load used in low temperature design, i.e., the maximum load a structure can carry without deforming excessively. Solutions are now available for a wide range of geometries.

2.7.4 Defect propagation data

A significant development over the last decade has been the evolution of analytical methods of assessing the significance of crack-like defects in structures subjected to combinations of static, dynamic and cyclic loading at both ambient and elevated temperatures. The general approach is similar in each case. The resistance of the material to defect propagation is characterised quantitatively in laboratory specimens containing defects of simple geometry. Specimens are loaded in special machines designed to simulate the conditions of loading, temperature and environment to which the defect will be subjected in the real structure. The application of analytical methods of failure analysis permits the rate of defect propagation in the laboratory specimens to be expressed in terms of a mathematical parameter defining the crack tip or ligament stress or strain field intensity driving the failure process.

The value of the parameter can then be calculated for defects occurring in the real structure and future propagation rates thereby anticipated. Propagation data and analytical methods are now available to cover many situations, for example, pressure loading (constant, cyclic or shock), effects of high cycle fatigue in rotating machinery, thermal fatigue during start-up and shut down, creep crack growth at elevated temperature. The relevant testing methods and analytical approach are indicated in more detail in Chapter 12.

Defect growth rates generally reflect the corresponding low temperature strength/toughness or high temperature creep/rupture properties. For example, material of low upper-shelf energy will generally exhibit poor fracture toughness and poor creep ductility will lead to high creep crack or thermal fatigue propagation rates. Defect growth rates therefore generally vary with changes in microstructure and/or composition in the same way as the low and high temperature material properties. Improvements in toughness and creep ductility through composition and microstructure control therefore lead to improved defect tolerance

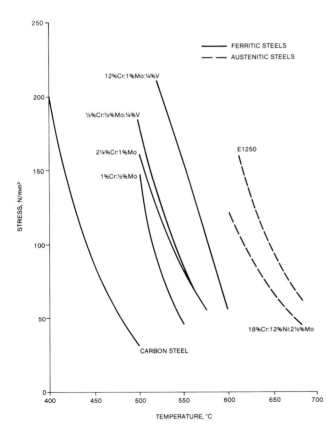

Fig. 6.25 Stress to fracture typical power plant steels in 100 000 h

and corresponding improvements in operating margins. Much of the current research effort is now devoted to demonstrating these effects and specifying means of avoiding situations which give rise to critical defect size situations and/or unacceptable defect propagation rates.

2.7.5 Stress redistribution in weldments

Component design for present plant based on stress rupture data is conservative and failure in parent material arising from life exhaustion in components operated within design conditions is very rare.

Effects of microstructure and composition known to reduce rupture lives of parent material (especially of ferritic steels) are recognised in the codes for plant design and construction which define the ranges of composition, heat treatment and fabrication processes for which the data apply. Such effects have been historically accommodated in weldments through the design safety factor and for many situations this has proved adequate. For materials which exhibit a strong microstructural dependence of creep properties, however, especially of rupture ductility, weldment performance can only be assessed by the development and validation of analytical methods of determining weldment lifetimes. This approach has been realised by the application of finite element stress analysis methods which make it possible to follow dynamically the changes in stress distribution throughout a welded component during post-weld stress relaxation and subsequent creep in service. The extent to which inhomogeneity in microstructure and associated elevated temperature creep and tensile properties affect the overall performance of the weld can thereby now be assessed.

At first sight it might appear that failure would occur in either the weakest or most creep brittle constituents. In fact this need not be so for the reason that, in order to maintain strain compatibility throughout the weldment, the weld metal, heat affected zone and parent must deform at similar rates leading to 'off-loading' of stress from the weaker on to the adjacent stronger constituents. For example, a weak weld metal will relax part of its load onto the adjacent heat affected zone, which then experiences a much higher stress than the weld metal during steady state deformation. The method requires the stress dependence of the minimum creep rate for each of the microstructural constituents of the weldment. For low alloy ferritic steels this involves fabrication of creep specimens of the simulated intercritical, fine-grained and coarse-grained bainitic micro-structures together with weld and parent metal specimens. The weldment lifetime can then be assessed by applying the steady state stress to the rupture life for each constituent or by defining a failure criterion, for example, time to onset of tertiary creep.

The model has been validated against the behaviour of full size weldments under realistic creep loading conditions. Long term experiments are being carried out on thick section pipe weldments and on more complex welded structures such as boiler headers in purpose built pressure testing facilities. The predictive capability of the finite element approach is good for deformation but the determination of failure criteria remains an area of active research for heavy section weldments. Much of the available data applies to low alloy ferritic steels ($\frac{1}{2}$ CrMoV, $2\frac{1}{4}$ Cr1Mo) but the method can in principle be applied to all alloy systems.

3 Determination of microstructure and distribution of composition within power plant steels by high resolution analytical microscopy

The understanding of the effects of bulk composition and microstructure necessary for interpreting and ultimately improving the mechanical properties of power plant materials, primarily steels but also nickel, copper and aluminium alloys, requires a detailed knowledge of the internal deformation mode of the material (distribution of internal defects and interfaces, for example, dislocation networks, grain and dendrite boundaries) together with the nature, distribution and composition of matrix and grain boundary particles and the nature and concentration of elements segregated to grain and particle interfaces. The last decade has seen significant advances in the provision of such information made possible through the development and application to metallurgical problems of advanced electron optical and micro-analytical methods, often coupled with automated image and data analysis facilities.

3.1 Optical microscopy

This relatively cheap method has long been used for the low magnification (1000 ×) general examination of polished and etched sections but resolves (i.e., spatially separates) features no smaller than 0.1–1 microns. It is the basic instrument for routine measurement of grain size and observation of phase and inclusion distributions and microstructural variation, especially across weldments. It is, however, severely limited in application by its low resolution and lack of analytical capability (phases can be recognised only by well catalogued characteristics of shape, colour, etc.).

3.2 Electron microscopy

The particle dispersions responsible for tensile and creep strength of ferritic steels are frequently too small to be resolved optically. The resolution of imaging systems is limited by the wavelength of the incident radiation and is improved from the micron level attainable with light to values as low as 0.1 nanometres (10^{-10} m) by using an electron beam emitted from a fine wire filament and accelerated under vacuum at high potential (typically in the range 50 kV–1000 kV). Instruments designed to optimise spatial resolution are now capable of routinely resolving atomic lattice planes.

3.2.1 Transmission electron microscopy (TEM)

The specimen is reduced to a thickness which is transparent to the defocussed electron beam, usually by chemical dissolution or ion bombardment of thin (0.2 mm) sheet. Perturbations in the atomic crystal structure of the material arising from the presence of dislocations, grain boundaries, second phases, carbides, etc., diffract the beam out of the primary image and the feature thereby comes into diffraction contrast (Fig 6.26 (a)). The angle through which the beam is diffracted (the Bragg angle, θ) is determined by the atomic crystal structure and atom spacing of the feature. If the diffracted components of the beam are focussed (by manipulation of the electromagnetic lense system) onto the recording screen, a pattern of diffracted spots characteristic of the feature results. Diffraction patterns for all the common phases in steels (carbides, nitrides, 66 oxides, sulphides, austenite or ferrite) and many other alloy systems have been compiled. It is thereby possible to identify rapidly the structure and composition by reference to the index.

3.2.2 Scanning transmission electron microscopy (STEM)

A more direct determination of particle composition is provided by the Scanning Transmission Electron Microscope (STEM). In addition to the normal transmission mode, the beam can be focussed to a very fine (<0.01 m dia.) spot (probe) and accurately located onto any feature or scanned in a continuous raster. Collision between a high energy beam of electrons and atoms stimulates the emission of X-rays of characteristic energy (wavelength/frequency) for each element. Analysis of the X-ray spectrum emitted by locating the electron probe on a particle, matrix or boundary region thus permits the identification of the nature and concentrations of elements present in the volume of thin foil material energised by the probe (Fig 6.26 (b)) Instruments are now available which permit the analysis of segregated boundary layers only a few atoms wide. For particle analysis the signal emitted from the foil matrix can be eliminated, and the accuracy of analysis thereby improved, by extracting the particles from a polished and etched surface using a carbon or plastic replica stripping method. The particles are then examined in isolation on the replica.

3.3 Fractography

An important aspect of modern analytical microscopy made possible by developments in electron analytical methods is the identification of the particulate phases and boundary segregants which control failure pro-

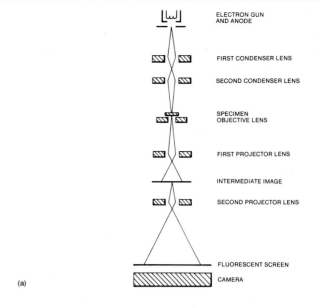

(a)

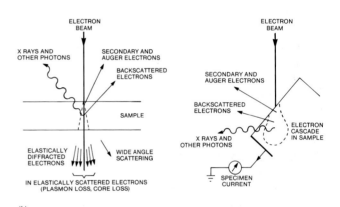

(b)

(c)

FIG. 6.26 Electron microscopy
(a) Transmission electron microscope (schematic).
(b) Detectable signals resulting from electron beam interactions with a metallic specimen.
The full lines represent electron trajectories, wavy lines photons. The dotted line represents the extent of the electron cascade in the sample.
(c) Simplified block diagram of a typical scanning electron microscope (SEM).

cesses. This is now routine practice where premature failure has occurred. Plant fracture surfaces are, however, generally too contaminated for direct analysis. The phases and segregants involved must therefore be exposed for identification by surface analytical methods by interrupting plant or laboratory specimens at a relevant stage of deformation or embrittlement and impacting under liquid nitrogen. The fracture path then follows the line of internal embrittlement or damage. Specimens for electron probe microanalysis may be fractured in air provided they are examined within a reasonable period. Specimens for surface analysis must be fractured in the analytical (high vacuum) chamber to avoid surface contamination.

3.3.1 Scanning electron microscopy (SEM) and the electron microprobe

Examination of fracture surfaces for contributory causes is facilitated if the highest and lowest parts of the specimen surface can be imaged simultaneously. This capability is referred to as the depth of focus of the instrument. Light optical instruments have a very small depth of focus (microns) and so are only suitable for examining the planar surfaces produced by sectioning the material. The electron optical characteristics of a scanning probe, which subtends a very small focussing angle from a fine final aperture, are such that the depth of field is over two orders of magnitude greater than for light at comparable magnifications. The high spatial resolution (to 10 nm) from modern filaments (LaB_6 or field emission tips) yields a magnification range from $50-250\,000 \times$ which, combined with the three dimensional nature of the image, provides great versatility for fracture surface examination.

The signals produced when a focussed electron beam impinges on a surface include secondary electrons and characteristic X-rays (Fig 6.26 (c)). In the SEM the surface is examined by scanning an electron probe in a raster to provide a visual image on a cathode ray monitor. The scan is modulated by the variation in secondary electron signal due to topographical changes as the beam is swept over the surface.

As in the STEM, the probe can be located in any position on the specimen using the cathode ray image and the characteristic X-rays emanating from the emission volume analysed for both qualitative and quantitative compositional information, in many cases with an accuracy of about 2%. An important feature of the scanning electron probe is the ability to image the distribution of elements on a fracture surface (from a depth of $1-10^{-1}$ m) by using the X-ray signal to modulate the scanning image. Typical applications of element mapping are the identification of fracture surface precipitates and the monitoring of compositional changes across sections through weldments and oxidised layers (Fig 6.27). Particle analysis, which

frequently entails the examination and identification of many hundreds of particles, has been facilitated in recent years by the development of computer controlled scanning microprobe analysers which stop the scanning beam when the signal indicates that a particle has been encountered, records the co-ordinates and initiates a step scan across the particle, counting the characteristic X-rays for each element present. The intensity of the substrate signal can be used to give an indication of particle thickness for sufficiently flat surfaces. Computer processing of the large amount of data gives particle size, shape and distribution in addition to composition for suitable specimens.

Similar computerised techniques are also now in routine use for processing light optical images (Quantimet Systems).

3.3.2 Surface analysis

The past decade has seen the rapid development of techniques capable of analysing the top few layers of atoms at a surface. This became possible only through advances in vacuum technology in the 1960s which

OPTICAL   Cr Ka

20μm

(a)

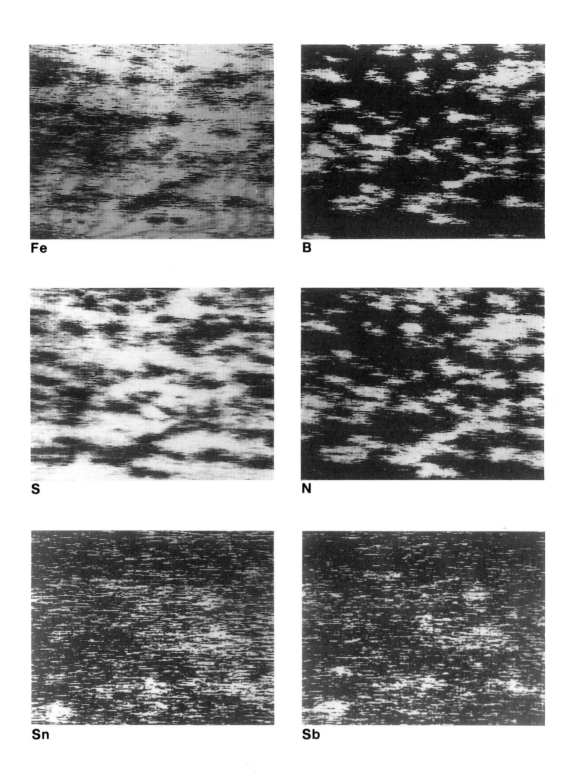

(b)

FIG. 6.27 Element mapping
(a) X-ray area scan of a section through an oxidised layer on the steam side of $2\frac{1}{4}$ Cr1Mo boiler tubing
reveals repeated laminations of Fe_3O_4 and $(FeCr)3O4$. Subsequent examination revealed cracks across the scale
by which oxygen is transported to the metal/oxide interface. Repeated laminate growth and crack propagation
produce the alternating structure.
(b) Distribution of iron, boron, sulphur, nitrogen, tin and antimony at the surface of a low alloy steel ($\frac{1}{2}$ Cr $\frac{1}{2}$ Mo $\frac{1}{4}$ V),
revealed by Auger electrons excited by scanning with an electron probe after a simulated weld thermal cycle in the
Auger vacuum chamber. The boron and nitrogen distributions coincide due to the formation of patches of the stable
compound of boron nitride which repel the sulphur. The tin and antimony also tend to cosegregate.

achieved routine pressures of ϱ 5.10^{-8} Torr and made possible the maintenance of a contamination-free surface for long periods (>1 hour). Surfaces can be bombarded by a variety of means (electrons, ions, X-rays, ultraviolet light) to yield a variety of characteristic emitted radiations, each with its own advantages for surface analysis. Of the many available techniques Auger Electron Spectroscopy (AES) has found widest acceptance in grain boundary and surface adsorption and segregation studies of power plant materials, largely due to its rapid progress in improving spatial resolution. Auger electrons are electrons of *characteristic* low energy emitted as a result of ionisation by bombarding the surface atoms with low incidence electrons, photons or light at ions (Fig 6.26 (c)). Their energy is such that they escape only from the topmost surface layers of atoms.

They can therefore be used, as are X-rays from an electron probe, to identify the nature and distribution of atoms at a surface (element mapping, Fig 6.27). Modern instruments tend to adopt a scanning electron system and probe facility, and frequently incorporate other surface analytical methods.

Many premature intercrystalline metallurgical failures are a consequence of the segregation of elements to grain boundaries. Auger analysis of boundary facets revealed by fracture of progressively embrittled specimens at liquid nitrogen temperature inside the vacuum chamber can be used to follow the segregation process (for example, temper and high temperature embrittlement, see Section 4.1 of this chapter). Depth profiling is also possible by stripping successive layers of atoms from the surface with a stream of argon ions, either continuously or in increments.

4   Failure mechanisms

Design of pressure components for power plant operation has historically been based on mechanical property data obtained from relatively simple (generally uniaxial) specimens taken from material representative of that of the body of the component. The data was then applied to the often complex configurations and service temperature and pressure environments of real components by incorporating an appropriate safety factor to accommodate the unknown effects of microstructure, composition and complex stressing situations.

Steels adopted for early ($\leqslant$100 MW) plant were generally tolerant of such effects. The more highly alloyed steels developed for improved tensile and creep rupture strengths to meet the exacting requirements of 500–660 MW plant have however proven to be more susceptible to toughness and ductility losses arising from variations in microstructure and composition, particularly at weldments. Whilst operational experience with such alloys has in general been very good, to the extent that the lives of many major components are being extended from 25 to 40 years or more, a number of situations have arisen where the

effects were not accommodated by the initial design assumptions and premature failure occurred. These have been relatively isolated incidents over long operational periods which stimulated considerable research effort to improve understanding of the failure processes involved. More rational bases for material specification and component design and repair necessary to avoid recurrence have consequently emerged.

The following sections describe some of the problems which have arisen from the adoption of the higher strength alloys, explaining the basic structural reason for their occurrence and indicating some of the practical means by which the problems have been overcome.

4.1   Intercrystalline failure of steels

The root cause of many plant failures has been found to lie in the susceptibility of commercial steels, primarily the ferritic rather than austenitic alloys, to premature intercrystalline failure. Basic studies have shown, however, that high purity experimental ferritic alloys are in fact generally resistant to intercrystalline failure at both low and high temperatures, and that the extreme cases of ductility loss which have led to plant problems are usually attributable to combinations of excessive levels of residual or impurity elements together with an adverse microstructure.

Where these have been identified, tighter specifications have resulted in significantly improved properties.

Problems arising from intercrystalline failure have been widespread and of such a generic nature that it is appropriate to summarise their more significant features before dealing with specific examples.

4.1.1   Low temperature intergranular embrittlement (temper embrittlement)

The segregation of impurity elements, for example, Sn, P, As, Sb, S to grain boundaries in general, but especially to the prior austenite grain boundaries of bainites and martensites, can occur from bulk concentration levels as low as 100 PPM, reducing boundary cohesion and promoting severe intergranular failure at low temperatures (below 100°C Fig 6.28). Segregation can occur during the quench from austenite or during tempering heat treatments in the range approximately 300–600°C and results in severe loss of material properties such as ductility, impact strength or fracture toughness. The loss of boundary cohesion is most easily measured as a shift in the transition temperature between ductile and brittle states in a Charpy Impact test (termed DBTT, Ductile/Brittle Transition Temperature or ΔFATT, Fracture Appearance Transition Temperature, Fig 6.29). The effect is of major concern in all equipment containing bainitic microstructures which must operate for long times in the critical temperature range or which are slowly cooled through this range during manufacture, for example, turbine

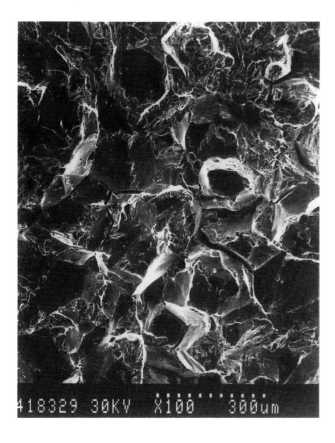

FIG. 6.28 Intergranular fracture by impact at 0°C in a 9Cr1Mo bainitic alloy induced by segregation of phosphorus to the prior austenite grain boundaries after ageing 4000 hours at 550°C.

rotors and pressure vessels. Moreover, boundary segregants also powerfully promote intergranular corrosion by inducing an electrochemical reaction between the grain boundary and the grain interior in the presence of corrosive environments. If the effect requires the presence of applied or residual stresses it is known as Stress Corrosion Cracking (SCC). The combined presence of boundary segregants, stress and corrosive environment can thus lead to a catastrophic situation where the boundary cohesion in the bulk of the material is so reduced that the critical defect size for fast fracture falls below the crack length of pre-existing defects (arising from weldment reheat cracks, unfused lands, slag inclusions, etc.) or cracks produced by corrosion at the surface. In severe cases critical crack lengths can fall to values below a millimetre. Such a situation can occur in the heat affected zones of heavy section weldments and in steam turbine rotors and discs. These latter are subjected to high and complicated stress states, including transient stresses from differential thermal expansion. Problems only arise in practice however, when fault conditions such as the existence of a corrosive product provides the necessary additional factor.

Whilst improvements in the quality and purity of rotor steels have reduced embrittlement susceptibility in recent years, some change in mechanical properties can be anticipated with increase in time and associated microstructural changes. Such aspects will affect the remaining life of the components. For future plant, materials and fabrication methods with reduced susceptibility to such time dependent changes remain a desirable target.

A quantitative method for characterising the degree of embrittlement in terms of the three principle controlling variables (degree of segregation, hardness and grain size) can be developed for a given alloy by using experimental steels to determine:

(a) The Charpy V-notch ductile-brittle transition temperature (ΔDBTT) dependence on boundary segregation level for individual elements (mainly P and Sn), grain size and hardness.

(b) The dependence of boundary segregation level on thermal history, impurity level and alloying element content.

The information in (a) and (b) can then be combined in an equation relating embrittlement severity (ΔDBTT) to boundary segregation level, grain size and hardness. The data from simple experimental melts is then extended to more complex laboratory alloys and finally to commercial alloys, thus establishing guidelines for control of composition and heat treatment. ΔDBTT can be related to changes in fracture toughness, enabling changes in critical defect size to be determined.

The characterisation of an alloy system for short time embrittlement by this approach is, because of the ease of Charpy testing, relatively rapid. The determination

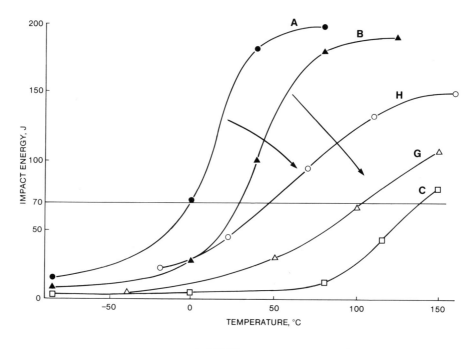

HEAT TREATMENT CODE	NORMALISED	TEMPERED	AGED
A ●	1h AT 1000°C AC	2h AT 750°C	NONE
H ○	1h AT 1000°C AC	2h AT 750°C	4kh AT 550°C
B ▲	1h AT 1000°C AC	½h AT 650°C	NONE
G △	1h AT 1225°C AC	½h AT 650°C	4kh AT 550°C
C □	1h AT 1000°C AC	½h AT 550°C	

FIG. 6.29 Variation of impact toughness of 9Cr1Mo steel with change in austenitising, tempering and ageing treatments

of boundary segregant levels, particularly of phosphorus, is now also a relatively straightforward procedure due to the rapid development and widespread adoption of the Auger method of quantitative surface analysis of boundary facets exposed by impact under vacuum.

The degree of embrittlement is directly dependent on the grain boundary concentration of impurities and typically exhibits *C* curve behaviour reflecting equilibrium segregation, with increasing concentrations at lower temperatures and longer times. The rate and amount of segregation and associated embrittlement depend not only on the bulk residual element levels but also on the total composition of the system. This arises because of strong interaction effects between individual residual elements and alloying elements which can complicate the extrapolation of data to long times. Element effects classify into three broad categories:

● Elements which enhance residual segregation but do not themselves segregate, e.g., chromium.

● Elements which enhance segregation by segregating with residuals, e.g., manganese and nickel.

● Elements which prevent segregation by chemically stabilising the residuals in the matrix, e.g., titanium, molybdenum and rare earths scavenge phosphorus.

To understand all combinations of possible effects between even the most common elements in steels would be prohibitively time consuming and current efforts are aimed at identifying systematic trends in existing data to allow segregation behaviour in complex systems to be predicted. Attempts have been made with varying degrees of success to model ternary systems by attributing interaction coefficients to solute elements; if the solute elements strongly attract they will precipitate in the matrix (for example, titanium or lanthanum can be used to chemically stabilise phosphorus). If their attraction is less strong they will *cosegregate* to the boundaries but the segregation of one will enhance the segregation of the other.

The situation can be further complicated by changes in the matrix and boundary carbides and associated levels of both embrittling and alloying elements in solution. This can lead to a complex pattern of behaviour in which embrittlement first increases, then decreases and finally increases again at very long times (thousands of hours).

4.1.2 High temperature intergranular embrittlement

The ductility of normalised steels at elevated temperature, in parallel with toughness at low temperatures, generally increases as the material softens with increasing degrees of tempering or service exposure. Failure of creep ductile steels occurs only after extensive plastic deformation by coalescence of voids nucleated within the grains on incoherent inclusions (sulphides and oxides in the size range 1–20 µm). Grain boundaries remain coherent despite the considerable elongation and shape changes of individual grains. Under certain conditions, however, compositional and microstructural factors can combine to induce separation of the grain boundaries after very low overall strains (< 1% elongation) and the material becomes creep brittle. The intercrystalline fracture surface then reveals the internal grain boundary structure in the same manner as does temper embrittlement.

Intercrystalline microcracks can propagate at temperature or during cooling to promote elevated temperature failure or generate defects which, if greater than the critical size, may subsequently initiate low temperature brittle failure.

Ferrite/ferrite or ferrite/pearlite grain boundaries are very resistant to both low and high temperature decohesion and intergranular cracking is usually only found after high strains in normalised ferritic steels transformed to a ferrite/pearlite aggregate. Bainitic microstructures developed at the high cooling rates experienced in weldments of low alloy steels, or at the slower cooling rates of heavy sections of more highly alloyed steels, are however susceptible to intercrystalline failure over a wide range of temperature and remain responsible for a high proportion of failures experienced on modern plant.

Bainitic microstructures are not however inherently susceptible to intercrystalline failure. Basic studies have shown that high purity bainites are in fact very resistant to boundary decohesion. This has led to the identification of the impurity and secondary elements responsible for the embrittlement and thereby to improved specifications for their control.

The fundamental cause of the embrittlement lies in the development of a weak zone some 1–2 µm wide at the PAGB (Fig 6.30). With increasing temperature above approximately 450°C, the ductility falls away dramatically as the boundary deformation zone becomes relatively weaker with respect to the matrix. Above 500–550°C virtually the whole of the overall deformation occurs in the boundary zones. Fracture can then occur in two distinct ways. If the PAGB itself remains coherent cavities nucleate, grow and coalesce within the zone, the overall elongation at failure (Ef) being given by:

$$Ef = \frac{\phi c}{\phi g}$$

where ϕc is the mean cavity diameter

   ϕg is the diameter of the prior austenite grains.

Failure is thus exacerbated by large grains as occur, for example, in weldment heat affected zones, but for

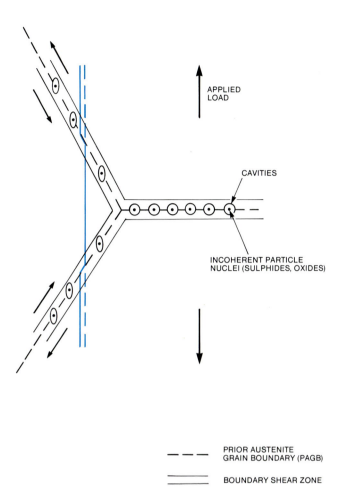

APPLIED LOAD

CAVITIES

INCOHERENT PARTICLE NUCLEI (SULPHIDES, OXIDES)

– – – – PRIOR AUSTENITE GRAIN BOUNDARY (PAGB)

————— BOUNDARY SHEAR ZONE

FIG. 6.30 Schematic of deformation processes operating during tensile deformation of low alloy bainitic microstructures at high temperatures

a given grain size the dominant factor in ductility is the cavity diameter at failure or, since cavities nucleate readily only on incoherent particles — particularly sulphides, the spacing of the boundary incoherent particle dispersion. Cavitation therefore, although leading to a brittle failure after overall uniaxial strains as low as 0.1%, is thus actually an example of ductile failure localised in the weak PAGB zones.

If the boundary plane becomes weakened however, for example, by segregation of impurity elements, then failure occurs by crack propagation along the boundary in a truly brittle manner known as high temperature brittle fracture. It is not uncommon to find a mixed mode of failure initiated by cavitation but changing at a later stage to brittle crack propagation.

Susceptibility to brittle failure, in common with cavitation, is also strongly dependent on both the base alloy composition and the levels of PAGB segregating trace and residual elements. Bainites of $\frac{1}{2}$CrMoV for example, generally fail by cavitation at all stress levels and exhibit brittle failure only at exceptionally high residual levels (> 0.10 wt%), whereas brittle failure is equally prevalent for $2\frac{1}{4}$Cr1Mo or MnMoNi (pressure vessel) bainites for residual contents within specification limits.

The segregation of elements to the PAGB responsible for the embrittling effect has been studied by Scanning Electron Auger analysis of embrittled fracture surfaces and free surfaces. Segregation of S, Sn and P occurs during the quench from austenite and the segregation sequence in ferrite follows the order CrN (300–450°C) replaced by P (450–550°C) replaced by SnSb (550–700°C). Embrittlement during service and preheat to 550°C is thus generally associated with sulphur, nitrogen and phosphorus whereas cracking at higher temperatures is exacerbated more by tin and antimony. A similar sequence is commonly found on cavity surfaces. This behaviour is reflected in the relative embrittlement severities of residuals in temper and stress relief embrittlement determined from alloys doped with individual residuals (see Table 6.1).

TABLE 6.1

Relative embrittlement severities for SRE and TE in MnMoNi steels

	P	As	Sn	Sb	Cu	S
SRE	1.0	0.81	1.18	1.49	0.12	0.19
TE*	1.0	0.25	0.08	1.09	0.014	0.056

*For equivalent weight fractions of each element

The tendency to high temperature brittle failure, as for that at low temperature (temper embrittlement), usually increases with increasing segregant level at the grain boundary, which in turn is related to the bulk level. The current approach is therefore to restrict residual levels to as low a value as is practicable for the particular application with anticipated improvements in both ambient and elevated temperature properties. Where large melts are involved, residual levels can be reduced to acceptable values by ore and scrap selection and by adoption of modern methods of steelmaking (see Section 1 of this chapter).

4.1.3 Significance of grain size in intercrystalline failure

While compositional control remains the primary means of reducing susceptibility to high temperature intercrystalline failure in transformable steels, for a given microstructural condition susceptibility is almost always reduced by a reduction in grain size. Since grain growth is driven by increasing the austenitising temperature this usually means restricting the final austenitising temperature seen by the component prior to service. In fact there are strong experimental indications that it is actually the restriction in the austenitising temperature rather than the grain size *per se* which achieves the benefit, although for most practical purposes they are synonymous.

Susceptibility to many forms of low temperature embrittlement (including temper embrittlement) also reduces with grain size. A uniform, fine grained microstructure, preferably resulting from a low austenitising temperature, therefore represents an ideal microstructural state for many plant applications. It can be achieved in most parent material by restricting the normalising temperature (in the absence of memory effects, see high temperature fasteners Section 4.5 of this chapter) and in weldments by adopting a welding process designed to produce continuous refinement (see Control of reheat cracking, Section 4.4.1 of this chapter) or through the use of a final renormalising heat treatment (Section 2.3).

4.2 Effect of chromium in ferritic steels and its significance for future plant designs

The family of low alloy chromium/molybdenum alloys was developed to provide an economic combination of high temperature strength, oxidation and corrosion resistance in power and chemical plant. Their strength however falls away rapidly above 570°C and cannot be further increased by dispersion methods without reducing weldability, since these strengthen the grain interiors of bainitic constituents more rapidly than the weaker grain boundary zones with drastic loss of ductility. A major benefit for power plant applications derives from the powerful effect conferred by chromium on bainite *ductilities*, additions above 5% Cr being sufficient to suppress (almost totally) cavitation and intercrystalline fracture in tempered bainites. Alloys of $2\frac{1}{4}$ Cr Mo are widely used for tubes and for heavier section components, but whilst generally exhibiting a significantly higher resistance to reheat and creep cracking than $\frac{1}{2}$ Cr MoV alloys, they do suffer these failures if the composition, especially with respect to residual elements, is not strictly controlled.

The beneficial effect of chromium on bainite ductility is readily explicable in terms of its effect on the prior austenite grain boundary deformation zone. Observations of the behaviour of surface fiducial scratches in experimental alloys reveal the progressive suppression of grain boundary zone shear (GBZS) and cavitation with increasing chromium content Fig 6.31.

The creep and rupture strength of 9Cr and 12Cr alloys can therefore be further improved by the addition of the carbide forming elements Nb, Ti, V and W without significant deterioration of hot ductility and therefore with good weldability and good resistance to thermal fatigue cracking.

Global trends in fossil-fuelled power plant are, because of steadily increasing fuel costs, moving towards more efficient 'advanced' steam conditions with temperatures to 650°C ultimately envisaged and pressures increased by a factor of about 2. This moves the steam into a 'supercritical' condition in which there is no phase boundary between the boiling water and steam.

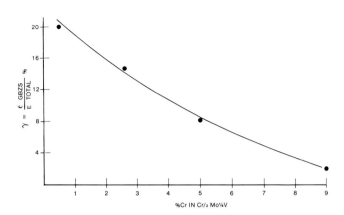

FIG. 6.31 Suppression of grain boundary zone shear during high temperature deformation of bainitic steels by progressively increased chromium content

Existing and newly developed alloys based on the 9Cr and 12Cr families are therefore being assessed for heavy section components (steam headers, pipes, valves, chests, casings) for supercritical plant operating at temperatures up to, or even exceeding, 600°C. Further increase in main steam temperature would necessitate the adoption of austenitic alloys which, because of their inherently greater coefficient of thermal expansion and poor thermal conductivity, are more susceptible to thermal fatigue cracking than the high chrome ferritic alloys.

4.3 Experience of low temperature cracking problems

4.3.1 Control of temper embrittlement and stress corrosion cracking in turbine discs

A classical example of low temperature brittle failure occurred in a CEGB Power Station in 1969, when an LP rotor of one of the 82 MW turbines failed catastrophically during routine overspeed testing. The cause of failure was shown to be the disintegration of the shrink-fitted turbine discs by brittle fracture originating from stress corrosion cracks less than 2 mm deep in the concentrated stress field at the keyway crown in the disc bore. Discs from the failed turbine were made of 3CrMo steel manufactured from both acid open hearth (AOH) and basic electric arc (BEA) processes, air melted using silicon deoxidation.

The three failed discs were of AOH material which subsequent testing showed to be more susceptible to stress corrosion and temper embrittlement than the BEA steels. Fourteen of the sixteen discs on the failed unit were of AOH origin and exhibited high FATT values, reflecting the higher sulphur and phosphorus levels and larger grain size of the AOH material. The embrittling effect was further exacerbated by a segregated band of sulphur and phosphorus compounds which met the bore in the region of crack initiation

in the keyway. All the AOH discs had FATT above 100°C and in some cases 200°C, i.e., well above their normal operating temperature at the time of failure of 57–90°C. Tests showed that the bursting behaviour of sharply notched experimental discs was related directly to FATT. Disc forgings were given a full dehydrogenation anneal prior to rough machining, followed by the quality heat treatment to develop mechanical properties of soaking for several hours at 900°C. Subsequent oil quenching to 300°C (to develop the bainitic microstructure) was followed by tempering at 600–650°C for several hours and slow cooling at 60°C/h to 500°C, declining exponentially to 25°C/h at 350°C. The combination of thermal history and high phosphorus level thus led to a high degree of prior austenite boundary segregation and associated temper embrittlement. Fracture toughness was determined using specimens of the compact tension type and it was shown that the toughness of segregated AOH material was so low that defects of the order of a few millimetres in highly stressed regions would be sufficient to initiate spontaneous brittle fracture.

The high density of keyway crown cracks, and much lower density of disc bore cracks, were shown by cyclic bend and crack propagation tests not to have initiated during start-up or overspeed, and therefore that the service environment was responsible for both initiation and propagation of the keyway and bore cracking. Subsequent tests showed that intercrystalline crack growth rates of pre-cracked specimens designed to eliminate the surface initiation stage in 3CrMo and $3\frac{1}{2}$NiCrMoV disc steels were adequate to account for the observed keyway crack lengths in simulated keyway environments or even pure water. Initiation was considered to arise from the concentration of liquid films along the keyway.

The conclusions of the investigation were that defect initiation and growth could be controlled by avoiding constructions which expose regions of high stress to wet steam environments and the replacement rotors were consequently designed for torque transmission without keys in the highly stressed bore region. It was also clear that defects could be better tolerated by improving fracture toughness and by reducing temper embrittlement susceptibility. This is now readily achieved by the adoption of modern steelmaking methods to reduce residual element and inclusion levels (see Section 1 of this chapter).

4.4 Experience of elevated temperature cracking problems in weldments

4.4.1 Improved welding procedures for control of reheat cracking of weldments

Cooling of heavy section weldments from the deposition temperature to ambient, induces large differential contractions across the heat affected zone and in the weld metal. This creates a system of residual stresses in the as-welded joint which can attain yield point magnitude. Post-weld heat treatment (PWHT) is specified for many applications for reasons including the avoidance of subsequent distortion, stress corrosion, accelerated creep, brittle fracture, etc. The process is generally also beneficial in reducing the net service stress and in increasing fracture toughness, provided that the microstructural constituents of the weldment can each accommodate the strain (approximately 0.3%) produced as the stress is relaxed by elevated temperature deformation. Unfortunately the alloy additions made to promote good high temperature strength (especially Mo, Ti, V in low alloy steels) can lead to a corresponding loss of high temperature ductility and relaxation may then occur by cracking.

During PWHT the weldment is heated slowly to a specified temperature (600–700°C for ferritic alloys), where it is held for several hours before being cooled. This is done slowly to minimise the generation of further residual stresses which might arise from differential thermal contraction strains between parent, HAZ and weld metal. Reheat cracking can occur at any time during the cycle when the conditions of residual stress, accumulated strain, microstructure and boundary segregation satisfy the conditions for cavitational or high temperature brittle failure. The most extensive experience with reheat cracking in the UK and Germany has occurred in recent years in the heat affected zones and weld metal of components fabricated in CrMoV parent metal (pipes, steam chests, valves, adaptors, etc.) welded using the Manual Metal Arc welding process and $2\frac{1}{4}$ Cr1Mo weld metal. Two characteristic types of failure have occurred.

4.4.2 Circumferential heat affected zone cracking

Cracking develops in the coarse grained annuli of the heat affected zone during post-weld heat treatment or early service, spreading circumferentially around the weldment and leading potentially to complete severance of the joint. Failure is usually by prior austenite boundary cavitation if residual elements are within specification limits, but exceptionally high levels of tin can induce the high temperature brittle mode. Repair rates have been found to increase with increasing wall thickness for steam pipes but unexpectedly high rates occurred at terminal and header welds. This was initially attributed to increased residual stress levels in the heavier section components, exacerbated by sectional changes and system stresses at terminal welds. Whilst this may be a contributory factor, subsequent investigation showed the principal cause to be the wide weld preparation angle of terminal joints. This allows weld beads to be deposited with a low degree of overlap which gives rise to a high proportion of coarse grained microstructure in the HAZ. The low creep ductility of this material means that it is unable to sustain the strains generated during stress relaxation.

Suitable welding techniques can be devised to overcome this problem.

An additional factor affecting susceptibility to cavitation of CrMoV alloys of commercial purity is the vanadium content. Turbine components frequently require a high degree of dimensional stability and, to compensate for the lack of control of cooling rate of large castings and forgings, vanadium contents up to 0.4wt% were originally permitted for 500 MW units in the early 1960s, subsequently restricted to 0.3wt%. Increasing vanadium content improves the parent metal rupture life but unfortunately correspondingly reduces the HAZ rupture life by strengthening the bainite matrix (through V_4C_3, precipitation), thereby increasing the concentration of the overall strain into the boundary deformation zones and reducing ductility values to $\leqslant 0.3\%$ elongation. A value of 0.27wt% vanadium, is the point at which the HAZ rupture strength falls below that of the parent. Alloys containing less than 0.27% vanadium appear resistant to reheat cracking for residual contents within current specification limits.

4.4.3 Control of transverse weld metal cracking (TWMC) of 2Cr1Mo welds between CrMoV components

Cracks develop by cavitation during both PWHT and service, normal to the welding direction. Macroscopic cracking during PWHT is rare but cavitation developed during PWHT can contribute significantly to subsequent cracking during service. Cracks rarely penetrate the weld metal into the bore but can propagate into the $\frac{1}{2}$ CrMoV heat affected zone. More recently it has become apparent that creep cracks can form in apparently sound metal after as long as 30 000 hours, indicating a potential mid and long term problem. The problem has been widespread and costly in terms of outage, NDT, repairs and reinspection; of the (approximately) 25 000 welds in service in the UK with an overall repair rate of 10% at least half were due to TWMC. Dominant factors influencing susceptibility of 2Cr1Mo weld deposits to TWMC are discussed in Chapter 10.

4.4.4 Type IV cracking

Circumferential cracking during early service or stress relief due to cavitation and brittle boundary failure in the coarse-grained regions of pipework heat-affected zones is now well understood and preventive measures have largely proved effective. More recently circumferential cracking has been experienced at $\frac{1}{2}$ CrMoV hot reheat pipework welds in middle or late design life. The problem is still incompletely recorded and understood but exhibits the following characteristics:

- Cracking occurs circumferentially, usually initiating as discrete cavities in outer regions of the refining annulus, but spreading later into the intercritically annealed regions of the heat affected zone (the annuli defined by the Ac_1 and Ac_3 isotherms and the overtempered parent). Less commonly, cavitation initiates in fully refined heat affected zones adjacent to the weld fusion boundary.

- The majority of affected welds are at terminal joints between hot reheat pipework and cast steam chests, manifolds or control valves and frequently involve section changes. Pipe-to-pipe welds in lines suffering sagging during operation are, however, also susceptible.

- Cracking is unlikely to develop before 35 kh operation, but crack growth, once initiated, is rapid and remanent life for propagation through the wall is likely to be between 5 kh and 10 kh.

Inadequate inspection procedures will thus result in steam leaks. If the trend for cracking to be confined to hot reheat pipe terminal welds continues, however, inspections can be limited to a relatively few critical welds.

Cracking appears to be a direct consequence of exhaustion of creep ductility with precise location depending on the relative creep rupture strengths of the sub-critically annealed, intercritical and refined microstructures. Operating temperatures have been ruled out as a primary cause since hot reheat pipework is much more susceptible than main steam pipework yet both operate at similar temperatures. Consideration of the components of axial stress arising from thermal expansion, residual welding strains, hanger lateral stiffness, inadequate deadweight support and internal pressure has concluded that primary pipework system stresses arising from inadequate pipework support and concentrated at terminal joints are the primary cause of Type IV accelerated creep rupture. The higher susceptibility of hot reheat pipes compared with main steam pipes is thought to arise from the increased tendency to overweight of the thinner reheat pipes, a 1% increase over design without increase in hanger support being sufficient to increase the terminal stress by 20%, while a 2% excess increases terminal stresses by 50%.

Cracking is currently contained by inspecting critical welds at 35 kh in addition to the current specification of inspections after 2–10 kh (depending on welding procedure), after 20 kh and after 50 kh. Magnetic particle and ultrasonic methods are supplemented by a metallurgical examination of one or both of the heat affected zones, depending on weld geometry and whether components are castings or forgings; the development of cavitation is revealed by taking replicas covering the parent/HAZ/fusion boundary of the polished surface at several locations around the circumferences.

Although only currently significant as a plant problem in CrMoV hot reheat weldments, Type IV failures are possible in all weldment heat affected zones con-

taining constituents of significantly reduced rupture strength.

4.4.5 Dissimilar metal welds

Low alloy ferritic steels have been widely adopted for tubes operating at moderate temperatures in the steam generators of modern coal-fired and nuclear power stations because of their attractive combination of high temperature strength and cost.

At superheat temperatures, however, the requirements for higher creep strength and oxidation resistance have dictated the use of austenitic stainless steels. Fabrication of the steam generator thus involves the manufacture of many thousands of welds between ferritic and austenitic tubes. These are known as dissimilar metal welds or transition joints. Some of these joints have exhibited cracking at times well short of design life. The reasons for this are (because of joint complexities) difficult to evaluate but the following factors are known to be important:

- Mismatch in the coefficients of thermal expansion of the ferritic steel, the weld metal and the austenitic steel generates high thermal stresses near the weld interface as the temperature of the weldment changes.

- Carbon atoms migrate from the ferritic steel to the weld metal and can produce a soft, coarse grained ferritic material adjacent to the weld interface.

- More rapid oxidation of the ferritic steel in comparison to that of the austenitic results in a step, or more frequently a notch, on the ferritic side of the interface.

- Markedly different creep strengths between the ferritic steel, the austenitic and the weld metal generate triaxial stresses in the vicinity of the interface.

- Post-weld heat treatment must be selected to stress relieve the joint but avoid temper embrittlement of the ferritic steel and sensitisation of the austenitic material.

- Excessive dilution of the weld pool, particularly in thin sections, can have a major influence on the performance of the weldments.

- Different thermal conductivities of the component materials can generate thermal stresses during operation.

- The ferritic and the austenitic components usually have different dimensions.

Transition joints are used mainly in thin section tubes in boilers but some thick section pipes also contain dissimilar metal welds. CEGB power plants have approximately 250 000 tubular welds in service located in superheater and reheater pendants (or plattens) in the furnace and in the dead space above the boiler roof where temperatures are expected to be more uniform across tube walls.

$1Cr\frac{1}{2}Mo$ steel was used in earlier stations operating at a nominal 540°C steam temperature. In newer stations designed for 565°C steam temperature, $2\frac{1}{4}Cr1Mo$ has been used, whereas the advanced gas-cooled reactor (AGR) steam generators employ 9Cr1Mo. These ferritic components are usually welded to a Type 300 series stainless steel. Tubular welds were initially made by flash butt welding, or with an austenitic filler metal, both of which result in a ferritic/austenitic interface. Carbon migration across this interface was initially thought to be responsible for premature failures. The austenitic filler was therefore replaced by a nickel-based filler which does not induce a carbon denuded zone and offers an additional advantage in that the thermal stresses generated due to the coefficient of thermal expansion mismatch are better distributed between the ferritic/weld metal and the weld metal austenitic interfaces.

Thick section pipes have been welded using austenitic nickel based filler. In order to reduce carbon migration, in one type of transition joint the ferritic side was 'buttered' with a stabilised ferritic weld metal (for example 2Cr1MoNb) prior to welding with the austenitic filler. Alternatively, a graded joint is fabricated such that the composition changes gradually from the ferritic to that of the austenitic steel.

The stresses on a dissimilar metal weld in service arise from different sources:

- Residual stresses at the operating temperature due to the welding procedure and due to the mismatch in the coefficients of thermal expansions of the components. These are expected to relax in the early stages of operation.

- The hoop, axial and radial stresses due to the internal pressures can be readily calculated.

- Thermal stresses due to the radial and the longitudinal temperature gradients can be calculated.

- The thermal stresses due to a change in the mean temperature are difficult to calculate although they are estimated to exceed the yield stresses during a shutdown.

- Stresses due to the dead weight, poor support structures resulting in sagging, vibrational stresses, and constrained thermal expansions/contractions of the system constitute the system stresses. These are difficult to measure or to calculate.

A large number of utilities operating dissimilar metal welds have experienced sporadic premature failures. Typically, low ductility circumferential cracks follow the weld bead contours, within about 50 μm of the fusion line. Such failures have caused numerous outages of boilers. Tubular dissimilar metal welds made with austenitic fillers begin to fail after approximately 20 000 h service and the failures become more frequent as the

plants get older. The weldments made with a nickel-based filler are superior to those made with austenitic fillers by a factor of about 4 in life.

Experience with large diameter $2\frac{1}{4}Cr1Mo$ steam pipe joints is also unsatisfactory; a number of such components have been taken out of service due to cracking through 10% of the pipe wall.

In general, welds made with nickel-based fillers are superior to those made with austenitic fillers but neither provides a trouble-free solution to the problem of premature failures. Available data suggests that a large number of joints may have to be replaced at least once during the 250 000 h life of a boiler.

Dissimilar metal welds may fail in service by cavitation along prior austenite grain boundaries, following the weld bead contours in the ferritic HAZ if coarse carbide stringers have not developed. These carbides form only in welds made with a nickel-based filler, after prolonged exposure, and then become the preferential sites for cavity nucleation. It is important to note that at the low stresses typical of operating conditions the rupture lives of weldments fabricated with nickel-based fillers are similar irrespective of whether cavity damage accumulation occurs at carbides or at the prior austenite grain boundaries. In other words, cavitation at either of these two different locations has approximately similar detrimental effect on the long term rupture strength of the joint.

Transition joints usually operate at higher metal temperatures than the nominal steam condition and sometimes even above the design temperature. A number of failures have been attributed to this overheating. Alternatively, the dissimilar metal welds can be subjected to high stresses. High bending stresses have been clearly shown to lead to early service failures. A combination of these factors can also expire the creep life of the joint at the operating metal temperature and stress.

For a given condition of temperature and stress, dissimilar metal welds fail prior to the expiry of the creep life of the ferritic steel because of the differences in the creep rates of the materials across the interface, which generates high triaxial stresses. Cavitational damage accumulation is further exacerbated by poor ductility of the HAZ near the interface where the triaxiality is very high. The ductility of the HAZ is very critically dependent on the local microstructures generated by the complex thermal cycles during welding. The lives of transition joints failing by cavitation in the ferritic HAZ can therefore be expected to be sensitive to the same compositional and microstructural factors affecting reheat and creep cavitation (Sections 4.1.2 and 3, 4.4.1 and 2 of this chapter).

A possible engineering solution is to reduce the stresses due to pressure, system loading and temperature changes by local thickening of the ferritic component in the interface region. A reduction in the system stresses can also be attained by careful relocation of the joints. It may also be possible to reduce the operating temperature of the joint by, for example, relocating the weldment.

More fundamental solutions may lie with an improved ductility of the HAZ, for example, by careful control of the residual elements or sulphide stabilisation (e.g., rare earths) in the low alloy steel. For nickel fillers it is also desirable to replace the sharp interface with a diffuse, continuous interface band by increasing the post-weld heat treatment temperature.

4.5 High temperature fasteners

High temperature fasteners operate under severe conditions of stress concentration in the engaged thread roots arising from tightening strains and the dimensional changes which occur in both fasteners and joined components during thermal cycling. A significant increase in failures on UK, American, Canadian and German plant in recent years has demonstrated that, while stresses can be reduced by stricter control of temperature gradients (through improved lagging and operational procedures), an important factor in failure is the low and variable tolerance of the material to thread root cracking. The two materials most commonly involved are Durehete 1055, a high strength, bainitic $1Cr1Mo\frac{3}{4}V(Ti,B)$ alloy and Nimonic 80A, a stronger nickel-based alloy originally developed for creep resistance at 700–800°C and adopted for situations where fasteners of reduced cross-section were required.

4.5.1 Improved performance of Durehete 1055 by residual element and grain size control

Creep ductility of bainites is sensitive to prior austenite grain size and a uniform fine grained structure is essential for adequate resistance to cavitation and associated intercrystalline crack propagation. Fine structures are usually achieved by austenitising at the lowest temperature consistent with the rate of dissolution of alloy carbides which will ensure adequate strengthening on subsequent cooling and tempering (980°C is specified for Durehete 1055 bolt stock). Several alloy bainites, however, are known to exhibit a powerful memory effect for the coarse grain structure established at a high austenitising temperature during subsequent austenitisation at a lower temperature. The effect is particularly strong in D1055 and has been shown to be due to untransformed austenite between bainite laths. These 'retained austenite' nuclei grow and coalesce on heating to reconstitute the coarse austenite grain structure immediately on retransforming to austenite at 900°C (Fig 6.32).

Industrial practice is for the final austenitisation treatment at 980°C to be carried out on bar material which has been previously hot rolled from 1300°C, i.e., in the grain growth range, leading to a mixture of fine and coarse grained microstructures, before air or pit cooling to bainite. The grain structure of the machined bolts thus ranges from uniformly fine (20–40 μm) through fine/coarse to uniformly coarse (100–150 μm). The memory retention can be destroyed by subcritically annealing at 660–700°C to fully transform the retained austenite to ferrite, followed by air cooling. The subsequent 980°C austenitising treatment then produces the desired uniform grain structure. A rapid heating rate between Ac_1 and Ac_3 also favours the formation of fine grains and these two factors are now recommended for inclusion in the Durehete 1055 bar production specification. The memory effect appears to be enhanced by the presence of residual elements, for reasons not yet fully understood.

A survey of failed D1055 bolts has also indicated a correlation with high levels of Sn, Sb, P and As. Scrap selection to lower the overall residual content is successful in raising properties to the top of the general D1055 stress rupture scatterband. A combination of scrap selection and grain size control yielding consistent properties in the top of the existing scatterband is therefore probably adequate for service requirements, assuming strict adherence to tightening and replacement schedules and adequate control of temperature gradients.

4.5.2 Nimonic 80A

Alloy steel bolts, tightened to a predetermined strain, undergo stress relaxation in service and must therefore be periodically retightened to secure the joint. Above 550°C, Nimonic 80A also exhibits normal relaxation behaviour, but at lower temperatures the stress in constant strain stress relaxation tests is found to increase with time due to matrix contraction, a phenomenon known as 'negative creep'. The stress on a tightened bolt thus increases in service. In association with grain boundary embrittlement and stress corrosion, negative creep has led to severe intercrystalline failures in the 320–520°C range. The contraction exhibits C-curve kinetics with a maximum rate at 450°C (−0.13% after 30 000 h) and is due to the time dependent rearrangement of the matrix chromium and nickel atoms into ordered domains of Ni_2Cr, this involving a decrease in the mean atomic separation.

A significant reduction of impact properties and lowered notch opening displacements in high temperature tensile tests of service exposed material, together with the detection by auger spectroscopy of phosphorous at the grain boundaries, suggests that a temper embrittlement type of phenomenon accompanies the ordering reaction. It is not yet known, however, whether the ordering *per se* induces the segregation process.

Cracking of this type is characterised by rapid propagation of single intergranular cracks. The corrosive environment is believed to result from the interaction of steam condensation with molybdenum disulphide anti-seizure compound during shutdown. Ex-service material has been shown to be susceptible to SCC in

411

FIG. 6.32 Reconstitution of the coarse austenite grain structure established by a high austenitising temperature during subsequent austenitising at a much lower temperature is a consequence of the retention of the high temperature austenite between the bainite laths during transformation. The process is visible directly in the photo emission electron microscope. (a) Initial austenite grain structure.

FIG. 6.32 Reconstitution of the coarse austenite grain structure established by a high austenitising temperature during subsequent austenitising at a much lower temperature is a consequence of the retention of the high temperature austenite between the bainite laths during transformation. The process is visible directly in the photo emission electron microscope. (b) Bainite needles with retained austenite after transformation.

FIG. 6.32 Reconstitution of the coarse austenite grain structure established by a high austenitising temperature during subsequent austenitising at a much lower temperature is a consequence of the retention of the high temperature austenite between the bainite laths during transformation. The process is visible directly in the photo emission electron microscope. (c) Retained austenite nuclei initiate transformation to austenite on subsequent heating.

FIG. 6.32 Reconstitution of the coarse austenite grain structure established by a high austenitising temperature during subsequent austenitising at a much lower temperature is a consequence of the retention of the high temperature austenite between the bainite laths during transformation. The process is visible directly in the photo emission electron microscope. (d) Original austenite grain structure as in (a) about to be re-established.

this type of environment, particularly when in a coarse grained condition. Reduced initial tightening strains, periodic slackening and retightening, exclusion of molybdenum disulphide and provision of drain holes have been recommended to overcome the problem. Generic studies to coarsen the strengthening precipitates and hence improve ductility at the expense of strength are in progress to provide a long term solution.

A general palliative for both Durehete 1055 and Nimonic 80A is to reduce first engaged thread root strain accumulation by waisting the bolts, combined with lower tightening strains. Work is in progress to demonstrate that adequate joint closure can be achieved and that waisted bolts can be incorporated into existing designs.

4.6 Improved steel for generator rotor end-rings

The most highly stressed component on large modern generators is the end-ring which retains the copper windings at each end of the generator rotor (Fig 6.33). These rings are shrunk onto the rotor and are stressed to about half their proof strength both when at rest and at operating speed. They are non-magnetic to reduce electrical losses and since the early 1960s an alloy steel of 18%Mn4%Cr has been used. Earlier generators used an alloy steel containing 8%Mn8%Ni4%Cr. Both these alloys are strengthened by cold forming and are known to be susceptible to cracking if subject to moist environments. Further, the hydrogen coolant used in modern generators can itself accelerate cracking in these end-ring steels. These cracking phenomena have led to over thirty failures of end-rings worldwide since the late 1940s. To contain the risk of failure in the UK, rigorous (but expensive) inspection procedures for end-rings were implemented.

A recently developed non-magnetic alloy steel containing 18%Mn18%Cr combines greatly improved resistance to stress corrosion and hydrogen assisted cracking with good toughness. For example, specimens of 18%Mn18%Cr steel have resisted cracking at over 30 000 hours when loaded up to their proof strength in either water or solutions of nitrate or chloride. By comparison, specimens of 18%Mn4%Cr steel would fail within a few hundred hours at stresses of 60% proof strength. The tolerance of 18%Mn18%Cr to possible existing cracks is also very high since its threshold stress level above which cracks can grow slowly in hydrogen is more than five times that for 18%Mn4%Cr steel. The new steel can also tolerate cracks four times deeper than the older steel at the same stress levels without fast fracture. The metallurgical reasons for the superior fracture resistance of the 18%Mn18%Cr steel are probably associated with the preferential formation of non-embrittling nitrides in the 18%Mn18%Cr steel, rather than the alloy carbides which tend to embrittle grain boundaries in the 18%Mn4%Cr steel.

On the basis of this excellent resistance to fracture and environmental cracking, 18%Mn18%Cr end-rings will be used whenever financially cost effective on all future CEGB generators. Also, wherever replacement end-rings are required in existing generators then 18/18 steel will be preferred. These changes both improve the integrity of operating generators and permit substantial financial savings on inspection costs.

4.7 Repair of steam chests

The low alloy CrMoV steel castings and forgings used for turbine steam chests are expensive and take a long time to manufacture. These components operate under demanding temperature and pressure fluctuation and often exhibit creep and thermal fatigue crack damage after extended periods of service. These cracks may grow as a result of localised creep at the crack tip.

Weld repairs can be contemplated in these conditions after consideration of the component's previous history and the results of detailed metallurgical investigations and non-destructive examinations. The extent of damage can be defined, establishing that these damaged regions can be safely removed leaving sound parent material capable of withstanding subsequent welding and service stresses.

Original mechanical features, such as stiffening webs, can be conducive to cracking and so each component is carefully measured and subjected to engineering scrutiny. Areas of design weakness or high stress are identified by photo-elastic studies of pressurised scale models or by numerical computer aided analysis (Fig 6.34). The same methods can be used to study the effects of design modifications which can be incorporated into the chest during the repair.

Whilst the weld processes employed are usually conventional manual metallic arc and gas shielded tungsten arc, very sophisticated weld reconstructions are now possible and regularly carried out. The general practice is to pre-heat the component and remove all defects. The welding techniques are optimised to give maximum refinement to the parent heat affected zone. Novel methods have been developed for the bridging of the large gaps caused by excavating damaged material using fitted backing plates and for the utilisation of new cast or wrought sections which are incorporated into the component by welding.

Each design of chest poses its own problems, but by the use of such methods the integrity of the chest fabric and its suitability for future service can be restored quickly and cheaply to very high standards. These techniques have enabled complete families of chests to be restored using spares from a single unit, thereby saving replacement costs and, more importantly, maximising generating plant availability.

The realisation of these techniques has been made possible only as a result of the background knowledge developed over the last 10 to 15 years into the control

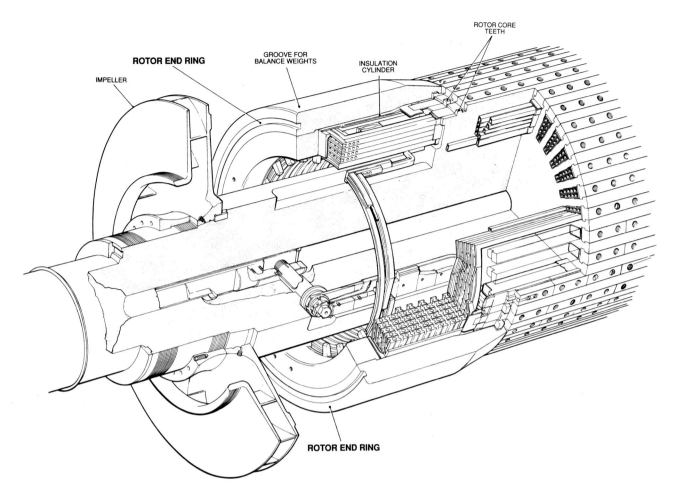

IMPELLER

ROTOR END RING

GROOVE FOR
BALANCE WEIGHTS

INSULATION
CYLINDER

ROTOR CORE
TEETH

ROTOR END RING

FIG. 6.33 Rotor end-ring assembly

of heat affected zone properties, residual stresses, weld metal chemistry, creep and thermal fatigue crack growth, oxide dating techniques and computer-based numerical methods of stress analysis.

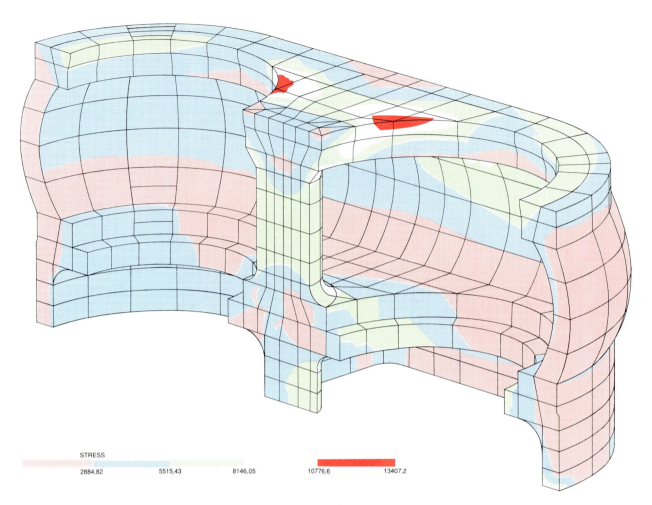

STRESS

2884,82   5515,43   8146,05   10776,6   13407,2

Fig. 6.34 Section of a numerical model for assessing areas of high stress or weakness in a steam chest

CHAPTER 7

Non-ferrous metals and alloys

Introduction

The extensive application of ferrous materials in power plant components arises from a combination of economic considerations, availability, relative ease of fabrication and a formidable range of properties that can be obtained by alloying additions and heat treatment. However, for some applications special characteristics and properties of other metals and their alloys may render them a more viable choice. This chapter considers the non-ferrous alloys that find usage within the power industry. It does not purport to be a complete review of non-ferrous metals for which there is extensive literature. Books by Rollason (1973)[1], John (1972)[2], Bailey (1973)[3], Higgins (1982)[4] and Metals Handbook (1979)[5] cover the subject more thoroughly.

1 Copper and its alloys

The main advantage of copper is that it offers a combination of high conductivity (electrical and thermal), ease of formability and good corrosion resistance. In terms of electrical conductivity it is second only to silver which is a more expensive material. Table 7.1 lists the electrical resistivity of some metals and indicates why copper finds many applications as an electrical conductor.

Since copper is malleable and ductile it is readily cold-worked and formed. Joining processes such as soldering, brazing and welding can be carried out easily. Copper also possesses good corrosion resistance particularly to atmospheric corrosion and aqueous corrosion in clean waters.

The effect of the addition of an alloying element to a metal can be conveniently represented by an equilibrium diagram. Such diagrams (Figs 7.1, 7.4, 7.5 and 7.6) can be viewed as a map that defines the regions

TABLE 7.1

Electrical resistivity of some metals

Metal	Electrical resistivity at 20°C, $\mu\Omega$m
Silver	1.50×10^{-2}
Copper	1.67×10^{-2}
Gold	2.07×10^{-2}
Aluminium	2.65×10^{-2}
Nickel	6.84×10^{-2}
Iron	8.85×10^{-2}
Tin	11.5×10^{-2}
Lead	20.6×10^{-2}
Titanium	42×10^{-2}

of stability of the various phases that can occur in an alloy system. With binary systems there is usually some degree of mutual solubility of the alloying element in the solvent metal and this primary phase is generally referred to as the α-solid solution (subsequent phases are usually referred to in order by the letters of the Greek alphabet). In the α-solid solution the atomic lattice is of a type characteristic of the solvent metal but the points on the lattice are occupied indiscriminately by the atoms of both elements. In some instances the solute atoms fit into the spaces between the solvent atoms. Solid solutions, therefore, do not have a fixed composition but extend over a range of compositions. If the lattice structure of the two metals and their atomic sizes are similar the metals are often completely miscible in the solid state, as in the case of copper-nickel alloys (see Fig 7.5). If the lattice structure or the atomic size of the two metals differ then, in addition to the primary solid solutions, other intermediate phases, mostly further solid solutions with a different atomic lattice, can be formed.

In the case of copper-zinc alloys (Fig 7.1) the zinc atoms at first replace copper atoms in the face-centred cubic lattice giving the α-solid solution. When the limit of solubility is reached a new phase, the β-solid solution, with a body-centred cubic lattice appears. This in turn is succeeded by another solid solution, the α phase, which has a complicated structure derived from the body-centred cubic lattice.

Commercial copper alloys vary widely in composition and offer considerable variation in mechanical properties and corrosion resistance. The alloys do not, however, offer satisfactory properties for prolonged service above 400°C. The most common alloying elements are zinc (brasses), tin (bronzes) and nickel (cupronickels) or combinations of these elements. When alloying elements are added in moderate amounts which do not alter the face-centred cubic atomic structure of pure copper, the resultant (α-phase) alloys are harder, stronger and maintain their ductility. They also remain readily amenable to conventional joining processes. However, the addition of alloying elements increases electrical resistivity. In some instances, e.g., the windings in hydrogen cooled rotors, relatively small amounts (about 0.1%) of silver are added to copper to increase strength without significantly increasing resistivity.

In the case of α-brasses (Fig 7.1) the addition of zinc up to a maximum of 30% actually increases the ductility of the alloys (Fig 7.2). The 70/30 alloy offers the optimum combination of strength ductility and hardness, and finds common usage.

As with most metals and alloys, cold working produces further increases in tensile strength dependent

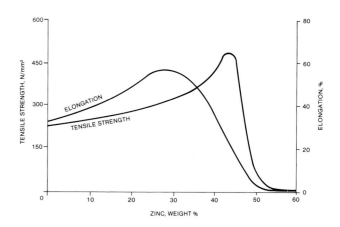

FIG. 7.2 Effect of zinc additions to copper on strength and ductility

on the extent of deformation. However, the increase in strength is achieved at the expense of ductility. In addition, work hardened material is susceptible to a form of stress corrosion known as season cracking. Annealing of cold worked material recovers the ductility but with a loss of strength. Figure 7.3 shows the effect of annealing on the tensile properties of a cold worked 70/30 brass.

It should be noted that brasses are susceptible to a form of selective corrosion known as dezincification. With this type of attack the original metal is replaced by a porous mass of copper which, though retaining the shape of the original article, has little or no strength. The addition of about 0.04% arsenic completely stifles dezincification and is generally included in α-brasses. With further increases in zinc content the ductility of the material decreases until at a composition of about 37% zinc, the limit of the phase, the ductility is similar to that of pure copper. Alloys containing between 37 and 46% zinc have a microstructure which contains α and second, β, phase.

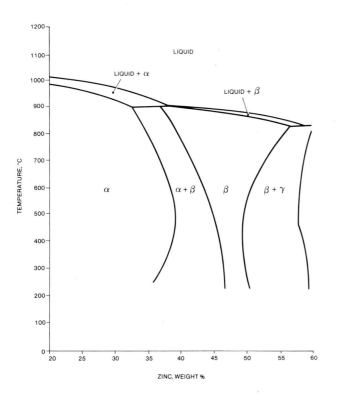

FIG. 7.1 Equilibrium diagram of copper-zinc alloys

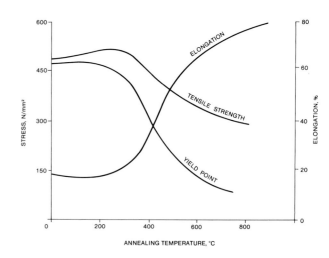

FIG. 7.3 Effect of annealing on cold-worked 70/30 brass

The β phase is associated with increased strength but this increase is at the expense of ductility. Alloys containing the β phase can only accommodate a small amount of cold deformation without rupture and are therefore generally hot worked. Additions of greater than 50% zinc result in the formation of brittle constituents and such alloys are of little commercial use. To improve the corrosion and erosion resistance of brasses, other alloying elements, in particular aluminium and/or nickel, may be added.

The most commercially used copper-tin (bronze) alloys are the α types with less than about 10% tin to prevent the formation of the brittle δ phase (Fig 7.4). These alloys find little usage in the power industry but are occasionally found in components such as lubrication pumps. It should be noted that some alloys with the hard second phase find application as bearing materials in the cast form (see Section 6 of this chapter).

Copper and nickel form a complete series of solid solutions (Fig 7.5). The cupro-nickels (the copper rich alloys) are characterised by toughness and malleability combined with high resistance to corrosion. In particular, the 70/30 alloys offer good corrosion resistance to marine environments. Further improvements to corrosion, especially impingement attack in fast moving sea water and abrasion by gritty or silty waters, can be obtained by increasing the iron and manganese contents to 2 per cent each. The 90/10 alloys are less susceptible to exfoliation and have better thermal conductivity than the 70/30 alloys and are used in feedheater tubes. For comparable technical require-

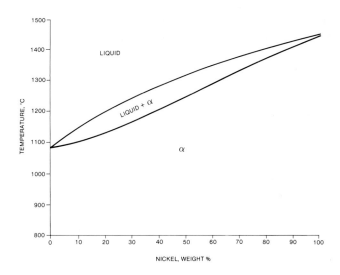

FIG. 7.5 Equilibrium diagram of copper-nickel alloys

ments they are more expensive than mild steel feedheater tubes. However, the tight specifications on mild steel tubes in terms of composition for easy weldability and defect size raise the costs of mild steel.

Although copper alloys have been used extensively in condenser components they have been replaced for tube applications in new power plant; this arises from the increasing availability and superior properties of titanium.

2 Aluminium and its alloys

One of the most important characteristics of aluminium is its lightness. It has a specific gravity of 2.7×10^3 kg/m^3 compared with iron or steel at 7.8×10^3 kg/m^3, copper at 8.9×10^3 kg/m^3 and titanium at 4.5×10^3 kg/m^3. Aluminium also offers very good corrosion resistance, ease of fabrication, high electrical conductivity and relatively low costs.

The pure metal has very low strength, but by alloying with other elements the strength can be considerably increased to give alloys with high strength to weight ratios. The electrical conductivity of aluminium is about 60% of that of copper (see Table 7.1) but, because of its low specific gravity, aluminium is superior on a weight for weight basis. For this reason, together with its resistance to atmospheric corrosion, aluminium has been extensively used in the past for overhead transmission conductors and, in some instances, busbars. To maintain its high electrical conductivity, impurities were controlled and 99.5% pure aluminium was used. Since the pure metal has low strength the cables were reinforced with steel wire. Recently, however, new or replacement conductors have utilised aluminium alloy conductors.

Although the alloys used principally contain 1% Si 0.5% Mg, they are of slightly lower conductivity than commercially pure aluminium but offer superior mechanical properties, reductions in weight and can

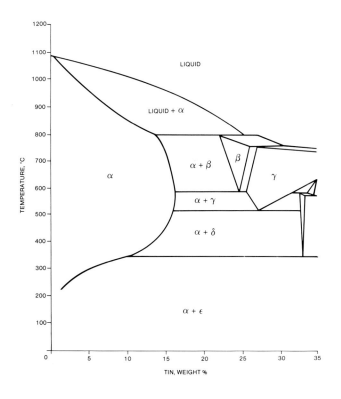

FIG. 7.4 Equilibrium diagram of copper-tin alloys

operate at higher temperatures. These properties enable uprating of lines using existing towers. The alloy strands are used either in aluminium conductor alloy reinforced (the 7 middle strands in aluminium alloy) or as all aluminium alloy conductors (all strands in aluminium alloy).

Aluminium is an active metal, i.e., it has a high affinity for oxygen and will rapidly oxidise. The oxide layer which forms is only a few atoms thick but is protective and resistant. In addition, if the oxide film is damaged it readily reforms and is self-healing. Further improvements in corrosion resistance can be achieved by the process of anodising, which increases the thickness of the protective aluminium oxide layer. Anodised aluminium sheets are used as cladding for buildings.

Apart from the applications as conductors and cladding, aluminium and its alloys are little used in the power industry. Some designs of generator rotors incorporate aluminium alloy slot wedges. The alloys used are those of the Al-Cu system with about 4% copper and small additions of magnesium and silicon. These alloys are non-magnetic, light and are capable of being hardened by and strengthened by a process known as age-hardening. The aluminium-rich portion of the aluminium copper alloy is shown in Fig 7.6. At room temperature the solubility of copper in aluminium forming the α-solid solution is less than 0.5%. The solubility increases with temperature reaching a maximum of 5.7% at 548°C. If an alloy containing 4% copper is cooled slowly from the α range, say 500°C,

coarse particles of $CuAl_2$ will be precipitated out. In this condition the alloy is relatively weak and brittle. However, if the alloy is quenched from 500°C, the rapid cooling causes the precipitation of $CuAl_2$ to be retarded and the structure consists of a super-saturated solution of copper in aluminium. Such a solution is unstable and there is a tendency for the copper to precipitate out as sub-microscopic $CuAl_2$ particles. The copper-rich areas are associated with a considerable increase in the amount of atomic lattice strain and increases in hardness and strength occur with time, even at room temperature. This natural ageing may be artificially accelerated by heating for a few hours at 100 to 150°C. Similar effects occur from the addition of magnesium and silicon when Mg_2Si is precipitated. Both the naturally aged and the artificially aged treatments are used for generator field winding wedges.

3　Titanium and its alloys

Titanium is a relatively low density material and also offers excellent corrosion resistance to a wide range of corrodents. The mechanical properties of titanium are affected by small amounts of oxygen and nitrogen in solid solution. By careful control of these elements it is possible to produce commercially pure titanium with a range of mechanical properties. Titanium is difficult to extract, refine and work when cold which formerly tended to make it expensive. However, the growing demand for titanium products, particularly in the aerospace industry and recently for power station condenser tubes, has had the general effect of cheapening its cost relative to its competitor materials especially copper and iron based alloys.

The high corrosion resistance of titanium to a wide range of corrodents is the prime reason for titanium having become the accepted material for condenser tubes in new power plant. Although titanium was known to have many benefits it was expensive in comparison to conventional condenser materials. The current price of titanium is such that it is a practicable proposition. This is partly because with no allowance for corrosion wastage, condenser tubes with a wall thickness of 0.7 mm or less can be installed. The first installations used seamless condenser tubes, but subsequent successful development of seam welded tube with its economic advantage led to widespread adoption of the latter.

Like aluminium, titanium is intrinsically very reactive, so that whenever it is exposed to air a film of oxide is formed. This oxide which is present on fabricated surfaces is thin, hard and tenacious, and acts as a protective layer. Titanium resists attack under conditions that corrode copper based alloys and stainless steels. On the waterside of condensers it is totally immune to salt, brackish and most industrially polluted waters. It does not pit under marine growths or deposits and is not susceptible to crevice corrosion nor stress corrosion. On the steamside of condensers tita-

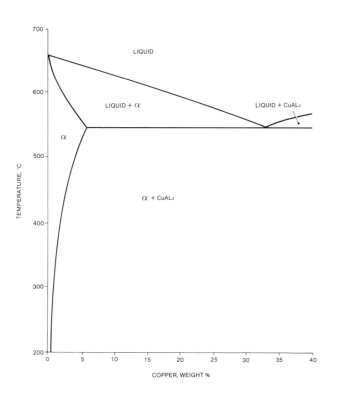

FIG. 7.6 Equilibrium diagram of
copper-aluminium alloys

nium is not attacked by the non-condensable gases, carbon dioxide, ammonia and oxygen. In terms of erosion-corrosion resistance, titanium is unaffected by excessive water velocity and silt and has good resistance to impingement attack. Since it does not suffer corrosion, a smooth surface is retained and fouling by deposits and marine life is avoided. The widespread application of titanium for replacement condenser tubes has to allow for the possibility of attack by galvanic corrosion on other components in the condenser. Such components would either have to be carefully protected by effective coatings or cathodic protection.

4 Nickel and its alloys

Nickel possesses a good combination of strength and corrosion resistance. A considerable number of nickel based alloys are available and these have excellent resistance to corrosion and oxidation, and high temperature strength. The principal alloys that find use in the power industry are 'Monel', 'Inconel', 'Incoloy' and 50Ni 50Cr alloys.

Monel metal contains about 67% of nickel, 28% of copper with appreciable amounts of iron and manganese. It possesses good general corrosion resistance and in particular, being virtually immune to chloride stress corrosion, has excellent formability and strength. Feedheater and condenser tubes have been fabricated in this material, but have not had a very wide application.

The Inconel and Incoloy alloys are based on the nickel-chromium-iron alloy system. Inconel nominally contains 15% chromium and 8% iron while the Incoloy alloys nominally contain 20% chromium, 30 to 45% iron with small additions of titanium aluminium and molybdenum. Both groups of alloys exhibit excellent corrosion resistance in particular they are resistant to pitting, intergranular attack and stress corrosion in the presence of chloride. They are easily formed and fabricated and also possess good high temperature strength. Mainly because of its resistance to stress corrosion cracking Inconel is used for steam generator tubes and other components in PWRs.

The high chromium content of the 50Ni 50Cr alloys provides them with excellent high temperature oxidation and corrosion resistance. Cast alloys are used for components that operate at high metal temperatures, e.g., burner components, superheater and reheater tube attachments, spacers and hangers. For superheater and reheater tubes which experience exceptionally high rates of fireside corrosion coextruded tubes, comprising an outer layer of corrosion-resistant 50Ni 50Cr alloy and an inner creep-resistant Incoloy layer, have given excellent service.

The Nimonic series of alloys are basically nickel-chromium alloys containing between 70 and 80% nickel and around 20% chromium with small additions of titanium and carbon. These alloys were developed for high temperature blading in gas turbines. As operating temperatures increased so titanium and aluminium levels were increased and cobalt and molybdenum were added. These modifications progressively improved the creep rupture strength of the alloys. The alloys derive their high temperature strength from a heat treatment which produces finely dispersed precipitates of intermetallic compounds of nickel and the alloying elements and also some metallic carbides. Apart from gas turbine applications, Nimonic 80A alloy is used for high temperature bolts in steam turbines.

Alloys based on the equiatomic nickel titanium system have a capacity to 'remember' or return to a previous shape if deformed at another, usually lower, temperature. At room temperature a 56% alloy consists of a stable austenitic phase but at lower temperatures, typically minus 120°C, it transforms to another phase, martensite, which is soft. The shape recovery (or memory) effect depends on the ability of the plastically deformed martensite to lose all evidence of its deformation on transforming back to the higher temperature austenite. This is providing that the amount of deformation does not exceed a critical maximum value of about 8%. Such an alloy has application as pipe-to-pipe couplings in cooling water, heating steam/water, lubricating and hydraulic oil circuits. It has also been used successfully to plug leaking condenser and inlet feedheater and economiser tubes. The major deficiency of the alloy which restricts wider usage is its poor creep strength at temperatures in excess of 350°C.

5 Magnesium and its alloys

The major power plant application for magnesium alloys is as a canning material for uranium fuel in Magnox nuclear reactors. The favourable properties of magnesium as a canning material include a very low neutron absorption cross-section, compatibility with the uranium fuel and carbon dioxide coolant and reasonable strength and ductility at service temperatures to accommodate changes in the dimensions of the fuel without failure. The alloys can be easily fabricated into finned cans.

The magnox alloys A.12 or AL80 contain 1.0% aluminium and a very small proportion of other elements. These alloys also have the advantage that plutonium transferred from the uranium to the can reacts with the aluminium to form $PuAl_3$ which does not diffuse through the magnesium and therefore remains at the inner surface.

Some reactors use a magnesium 0.6% zirconium alloy (ZA and ZR55) which offers higher strength with excellent ductility. However, these alloys suffer from the disadvantage that plutonium, formed during service in the reactor, is readily soluble in it and diffuses through. To overcome this problem a layer of graphite is incorporated to act as a barrier between the uranium fuel and the can.

It should be noted that the elements, aluminium, zirconium and beryllium which are alloyed with magnox have very low neutron capture cross-sections.

6 Bearing alloys

The requirements for bearing alloys are that they should be sufficiently hard to resist wear and abrasion, tough to withstand shock, plastic enough to shear when metal to metal contact occurs and eliminate high spots. They should have good melting and casting properties and readily bond to the backings of the bearings. A resistance to corrosion by lubricants is also necessary.

Bearing metals tend to be either tin based (white metal) or copper based alloys. The tin based alloys known as 'Babbit' metals contain up to 10 per cent antimony and 7 per cent copper. During freezing of tin-antimony alloys the hard compound SnSb crystallises in the form of cuboids. These cuboids are lighter than the melt and tend to float to the surface resulting in the bearing being too hard at the top and too soft at the bottom. However, by the addition of copper needles, a further phase Cu_6Sn_5 crystallise out prior to the production of SnSb cuboids. These needles form a network which prevents the relative movement of the SnSb cuboids and gives a uniform structure. Lead

may be added to tin based alloys for cheapness but the maximum pressure at which the bearing can be used is reduced.

Copper based bearing metals are usually bronzes. The tin bronzes with 10 to 15 per cent tin consist of a hard δ phase in a soft matrix α phase (see Fig 7.4). The addition of phosphorus to tin bronzes enhances their bearing properties. The phosphorus separates as hard particles of CU_3P associated with the δ phase. Some bearing bronzes contain up to 30 per cent lead; with these leaded bronzes the lead, which is insoluble in the copper, appears as globules in the microstructure and acts as a sort of metallic lubricant if the oil film breaks down.

7 References

[1] Rollason, E. C.: Metallurgy for Engineers: Edward Arnold, London: 1973

[2] John, V. B.: Introduction to Engineering Materials: MacMillan, London: 1972.

[3] Bailey, F. W. J.: Fundamentals of Engineering Metallurgy and Materials: Cassell, London: 1973

[4] Higgins, R. A.: Properties of Engineering Materials: Hodder and Stoughton, London: 1982

[5] Metals Handbook: Properties and Selection of Metals, Non Ferrous Alloys and Pure Metals: American Society for Metals, Ohio: 1979

CHAPTER 8

Non-metallic materials

---

1 The use of plastics and rubbers in power plant

1.1 Structure

Plastics and rubbers are, in general, organic polymers, which means they are composed of long-chain molecules having a carbon 'backbone'. The polymer chains are generally in a random, tangled form, with relatively weak, long-range attractive forces between them. This leads to properties quite different to those of metals, in which atoms are closely packed in an orderly form giving strong, short-range attractive forces.

Cross-links may be introduced between neighbouring polymer chains to create a strong interchain bond and it is also possible for crystalline regions to exist, in which chain molecules fold themselves in an orderly fashion with resultant close-range, strong chain-to-chain attraction. These variations in molecular structure can produce a wide range of mechanical properties in polymeric materials.

1.1.1 Thermoplastics

Thermoplastics are uncrosslinked polymers which may be amorphous or crystalline. The latter comprise an array of crystallites in an amorphous matrix. Figure 8.1 shows in schematic form the basic structure of an amorphous thermoplastic. Deformation under the action of stress occurs by the sliding of molecules over one another. The rate at which this can occur depends on the molecular flexibility of the particular polymer

and on the temperature. The properties of amorphous polymers therefore range from stiff, brittle materials at low temperatures to flexible, rubber-like materials at higher temperatures. Crystalline regions act as physical cross-links, which serve to increase stiffness while retaining toughness and give dimensional stability at elevated temperatures. The most significant property of thermoplastics is that they undergo a reversible change to a semi-liquid state when heated above a certain temperature. This phenomenon provides the basis for fabrication of thermoplastics components by various hot moulding methods (e.g., extrusion of pipe, injection moulding of pipe fittings).

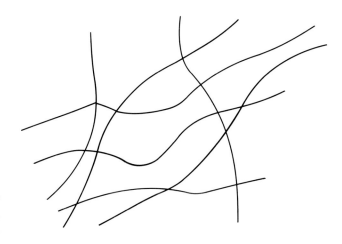

FIG. 8.1 Structure of thermoplastics

1.1.2 Thermosets

Thermosets are polymers with chemical cross-links between chain molecules, prepared by a reaction between at least two components (usually both liquid). A schematic diagram of the basic structure is shown in Fig 8.2. Since cross-linking is an irreversible chemical process, thermosets cannot be thermally formed in the same way as thermoplastics and different methods of manufacture are applicable. Under stress, the cross-links inhibit relative movement of polymer chains and this results in more rigid materials and much lower elongations-to-failure (typically 2%) than is the case for thermoplastics. In their basic form thermosets find only limited application because of their brittle nature. Major structural applications become practicable when these materials are combined with strong fibres such as glass or carbon to form reinforced plastics. Glass-fibre reinforced plastics (GRP) materials are widely used within the power industry.

1.1.3 Rubbers

Rubbers are lightly cross-linked polymers with particularly 'flexible' molecules in the chains which allow very high, reversible elastic deformations to occur in the short term under the action of stress. Elongations-to-failure of rubber can be considerably higher than with thermoplastics, and stiffness is much lower.

1.2 General properties and applications

The foremost properties of polymers relative to the more familiar traditional metals are:

- Lower strength and stiffness.

- Lower operating temperature limits.

- Electrically non-conducting.

- Free of corrosion.

This combination of properties results in extensive applications in two particular plant areas: electrical plant and cooling water systems. In addition, numerous applications can be found across the full range of plant.

1.2.1 Coatings and adhesives

All three types of polymer mentioned above are used in paints, protective coatings and linings, and adhesives, which find application in many areas of plant. These specialised uses of polymers are not dealt with in detail here.

1.2.2 Thermoplastics

A list of thermoplastics materials used in power plant is given in Table 8.1, together with an indication of properties and typical applications. Also listed are relevant British Standards. The table provides information on allowable service temperatures, but it

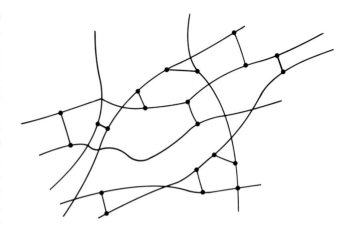

FIG. 8.2 Structure of thermosetting resin

must be noted that practical temperature limits may be lower or higher than the figures given, according to the level of mechanical stress present and the time for which maximum temperatures are operative.

Thermoplastics undergo time-dependent deformation under load (creep) even at ambient temperatures, and this phenomenon is taken into account in design of load-bearing components. Allowable design stress levels for pipe are defined in relevant Standards, and provide for a fifty-year service life with an additional superimposed safety factor.

The most widespread use of thermoplastics in non-electrical applications is probably in pipe form. In auxiliary cooling water systems, the use of thermoplastics pipe can offer substantial financial savings both in terms of first costs and more significantly in respect of lack of subsequent maintenance costs. Mild steel or cast iron pipe will corrode if unprotected, especially in sea or estuarine water, and may therefore require replacement three or four times in the life span of a power station. In addition, corrosion of mild steel causes the ingress of iron-oxide debris into the cooling water and hence can cause blockages in heat exchanger tubes and consequent expensive plant outages and replacement exercises. The use of polymers as protective coatings is, of course, an alternative to the use of plastics pipes, but most coatings require regular maintenance and repair to maintain their protective capacity and this must be borne in mind in making a materials choice.

Similar financial and maintenance benefits are available in auxiliary cooling water systems through the use of thermoplastics (or GRP) valves, of butterfly type, which are available in proprietary forms in sizes up to 0.6 metre and of pressure rating in some cases as high as 10 bar. Metallic valves, even when coated, can have particularly short useful lives in many salt water systems due to corrosion and erosion.

Other areas in which thermoplastics pipe finds wide usage are in water treatment plants for handling acidic or caustic solutions, and in effluent treatment plants (active or non-active) and discharge systems.

TABLE 8.1

Types of thermoplastic materials used in power plant

Name/s	Acetal	Acrylics	Acrylonitrile-butadiene-styrene (ABS)	Nylons, Polyamides	Polycarbonate	Polyethylenes (PE, Polythene)				Polypropylene (PP)	Polytetrafluoro-ethylene (PTFE)	Polyvinylidene Fluoride (PVDF, PVF$_2$)	Polyvinylchloride PVC	
						Low Density (LDPE)	Medium/High Density (MDPE/HDPE)	Ultra-High Molecular Weight (UHMWPE)	Cross-linked (XLPE)				Unplasticised uPVC	Plasticised
Tradenames	Delrin	Perspex			Lexan						Teflon, Fluon			
Strength†	4	4	3	3	4	1	1	2	1	2	2	2	3	1-2
Stiffness†	4	4	4	3	4	1	2	2	2	3	2	3	4/5	–
Toughness†	2	1	3	3	4	5	4	5	5	2	3	3	2	–
Acid Resistance††	Poor	Good	Moderate	Poor	Moderate	Moderate	Moderate	Moderate	Moderate	Moderate	Excellent	Excellent	Good	Moderate
Mineral Oil Resistance††	Good	Excellent	Good	Good*	Good	Good	Excellent	Excellent	Good	Good (but poor at 60°C)	Excellent	Good	Good	Moderate
Maximum long term service temperature	60-85°C*	50°C	70°C	60-80°C*	115°C	50°C	55°C	55°C	85°C*	55°C	180°C	135°C	50°C	<50°C
Short term or intermittent thermal limit	110°C	80°C	100°C	95-150°C*	140°C	70°C	80°C	80°C	110°C	110°C	260°C	150°C	80°C	80°C
Ionising radiation resistance†	1	2	3	3	3	3	3	3	3	1	1	3	3	3
Other properties	Good dimensional stability. Cannot readily be filled.	Transparent and optically good. Not solvent resistant.	Good mould finish and stability. Not solvent resistant.	Many types, can be filled. Large dimensional changes with moisture content, eg types 6 and 6.6.	Transparent grades. Low design strength. Poor hydrolysis resistance. Not solvent resistant.	All difficult to bond and rather poor dimensional stability. XLPE has a good creep resistance and UHMWPE good abrasion resistance.				Good fatigue resistance. Susceptible to oxidation.	Very low co-efficient of friction. High cost. Difficult to fabricate. Very difficult to bond.	Can be fabricated (unlike PTFE). High cost. Difficult to bond.	Inexpensive transparent grades. Tendency to brittle service failures. Not very solvent resistant.	Adaptable but creeps and may suffer loss or migration of plasticiser.
Applications	Gears, bearings, guide rollers	Glazing, light diffusers, filter housings, solvent cement	Pressure pipe (BS5391), safety helmets, instrument cases, CW screens, solvent cement	Condenser tube inserts, bandscreen rollers, valves, rope (BS4928), pneumatic hose (BS5409), gears, bearings, bushes, electrical components, coatings	Glazing, safety visors, electrical components	Cable insulation pipe (BS6234).	Pressure pipe (BS6437, CP312 Parts 1 & 2), coatings.	Coal bunker linings, valves.	Cable insulation, heat shrink products.	Pressure pipe (BS4991), rope (BS4928), valves, linings.	Generator hoses, seals, sealants, gaskets, piston rings, cable insulation, valve and pump components, low friction coatings. (BS6564).	Chemical pipe, linings for chemical pipes and vessels, valves.	Pressure pipe (BS3506 and CP312 Parts 1 & 3). Valves. Cooling tower packings. CW screens.	Cable insulation. conveyor belts. Anti-corrosion coatings.

†† Additives may stress crack plastics with poor solvent resistance
† Rating 1-5, 5 highest value and 1 lowest
* Depending on grade.

Care must be exercised when installing a thermoplastics pipe system to ensure its proper design in respect of pipe supports, accommodation of thermal expansion (greater than with metals) and external loadings. If there is concern regarding the possibility of damage due to mishandling or abuse after installation, then it is advisable to use a pipe of the highest available pressure rating (and hence of maximum wall thickness) even though the system pressure may be much lower. Several types of jointing method are available, the most appropriate method depending on factors such as pipe material and size.

1.2.3 Thermosets and reinforced thermosets

The two most widely used thermoset materials are polyester and epoxy resins, though phenolic resins are sometimes used where fire resistance is of special importance.

Applications for unfilled thermosetting resins are limited but nonetheless important. Epoxies, for example, are widely used as adhesives and protective coatings, as additives for cement (to improve mechanical properties) and for repairs to cracked structural concrete by the resin-injection method. Epoxies and polyesters are used in high pressure or vacuum impregnation techniques for sealing porous metal castings.

Particulate filled epoxies are often used for castings in various electrical applications; for example, for cable stop joints, insulating rings, stator cooler water boxes. Epoxies, with a wide range of metallic or ceramic fillers, are also extensively used as metal repair compounds (i.e., for rebuilding metallic surfaces wasted by erosion and/or corrosion) often in conjunction with a thermosetting resin coating material. Similar filled epoxies are used as a chocking or bedding medium for heavy machinery.

In reinforced form, both epoxies and polyesters are in common use. Hot-laminated reinforced epoxies (and phenolics) are available as proprietary products in different geometrical forms, with paper, cotton or glass-fibre reinforcement. Such materials are used in dry or wet bearing applications and for various electrical components, including slot liners and non-conducting bolts in generators.

As with thermoplastics in pipe form, there are particularly beneficial applications for reinforced thermosets in cooling water systems of coastal and estuarine power stations, where corrosion is a constant and costly problem with traditional structural metals. The much higher strengths and stiffnesses available with reinforced thermosets means that, unlike thermoplastics, they can be used in major structural applications in main turbine condenser cooling systems as well as in auxiliary systems. The major benefit of using reinforced plastics for structural components rather than a coated metal is the total lack of maintenance required. Protective coatings require regular, and sometimes costly, maintenance and repair. Furthermore, coating failure

can have serious consequences in some circumstances, especially where the presence of 'noble' metals (e.g., brass tube plates or stainless steel valve trim) can cause galvanic corrosion of ferrous metals at a high rate.

On the basis of cost and service requirements, the most appropriate reinforced plastics material for cooling water systems is GRP (glass-fibre reinforced plastics) usually based on a polyester resin of an isophthalic type. Components successfully installed in this material include: pipe (flanged or unflanged) in sizes ranging from 150 mm to 2700 mm diameter; main condenser inlet and outlet water boxes for a 500 MW unit and for auxiliary heat exchangers; butterfly valves in sizes ranging from 150 mm to 1800 mm diameter; bellows units as large as 1200 mm diameter; and penstock doors. Figure 8.3 shows an example of a large GRP component installed on a 500 MW station: a 2700 mm inside diameter, flanged pipe section. In most cases, fabrication of components is by a hand lay-up method because of geometrical complexities; but unflanged pipe components may be manufactured by filament winding, an automated technique in which continuous glass fibre strands are wound around a rotating mandrel at controlled angles.

Careful attention to design is necessary when fabricating components in GRP. The material is of much lower stiffness and higher thermal expansion coefficient than traditional structural metals. Most importantly, its allowable design stress is determined by considerations of time-dependent failure criteria as opposed to the plastic yield criterion applicable to mild steel. Hence replacement GRP components must be designed in their own right and on no account should be made as 'carbon copies' of existing metallic components. Design of GRP structures for cooling water systems should be to BS4994: 1986; and pipe to BS5480, Part 1: 1977 and BS5480, Part 2: 1982 or to BS6464: 1984.

Other applications of GRP in power plant include rigid linings for corrosion protection in cooling water systems and water treatment plans; over wrappings for temporary repair of corroded pressure pipework; storage tanks of various kinds; ballistic shielding; insulators; and cooling tower forced draught fan blades.

1.2.4 Rubbers

The most common usage of rubbers is in seals and gaskets, components of critical importance in many areas of power plant. Other important applications are in pressure hoses, conveyor belting, condenser tube vibration damping tubing, pipe bellows (including very large sizes for main condenser cooling systems), diaphragms for valves and protective coatings and linings.

Rubbers can be broadly classified into chemical types (e.g., nitrile or ethylene propylene) but their complexity is such that two rubbers within the same type classification can have very different properties, accord- to their precise chemical formulation. Correct choice of rubber type is important in many applications: for

FIG. 8.3 A 2700 mm GRP flanged pipe section installed in a 500 MW station

example, in respect of resistance to oil or to elevated temperature. A range of hardnesses is normally available for any particular rubber type; for sealing applications, the most suitable hardness will depend on the precise service conditions to be accommodated.

Table 8.2 gives properties and application areas of the main types of rubber.

Under long term stress, rubbers undergo irreversible deformation due to scission of cross-links (caused, for example, by oxidation). This phenomenon is of particular importance with seals, where 'compression set' is often the life-limiting factor.

Experience has shown that conditions of storage of rubber components can be of critical importance. It is possible, for example, for items such as O-ring seals to 'fail' effectively whilst on stores shelves. Guidance on this matter is available in BS3574: 1963.

1.3   Relevant standards

1.3.1   Thermoplastics

BS3506:1969	Unplasticised PVC pipe for industrial purposes.
BS4928:1985	Specification for man made fibre ropes.
Part 1:1973	Polypropylene ropes.
Part 2:1974	Polyamide (nylon), polyester and polyethylene filament ropes.
BS4991:1974 (1982)	Polypropylene copolymer pressure pipe.
BS5391:Part 1:1976	Specification for acrylonitrile-butadiene-styrene (ABS) pressure pipe: pipe for industrial uses.
BS5409:	Specification for nylon tubing.
Part 1:1976	Fully plasticised nylon tubing types 11 and 12 for use primarily in pneumatic installations.
Part 2:1978	Plasticised and unplasticised nylon tubing types 11 and 12 for use primarily in the automobile industry.
BS6234:1969	Polythene insulation and sheath of electric cables.
BS6437:1984	Specification for polyethylene pipes (type 50) in metric diameters for general purposes.

TABLE 8.2

Properties and applications of rubbers

Rubber:Common Name(s)	Natural	Neoprene* Chloroprene	Nitrile	Ethylene Propylene EP EPDM EPT	Silicone	Viton*
ASTM Designation	NR	CR	NBR	EPM[1] EPDM[2]	Si	FPM
Chemical Name	–	Polychloroprene	Acrylonitrile Butadiene Copolymer	Ethylene Propylene Copolymer[1] Ethylene Propylene Diene Terpolymer[2]	–	Fluorocarbon Polymer
British Standards and DTD Material Quality Standards	BS1154	BS2752	BS2751 & 3222	BS6014	DTD 5531A DTD 5582 DTD 818	DTD 5612
General Fluid Application Areas	Water; Sea Water; Most cold moderate chemical solutions	As for natural rubber plus mineral lubricating oils	As for neoprene plus hydrocarbons, greases; hydraulic fluids	Sea water; Hot water; Steam; Oxidising chemicals	Hot water; Low temp steam; Oxidising Chemicals	Hydrocarbons; Halogenated Hydrocarbons; Oils; Greases; Hot water; Low temp steam
Incompatible Fluids (swollen or attacked by)	Oils; Greases; Hydrocarbons; Strong acids	Strong oxidising acids; Ketones; Chlorinated Aromatic Hydrocarbons	As for neoprene	Mineral oils and solvents; Aromatic Hydrocarbons	Many solvents and oils; Concentrated acids	Ketones; Alcohols
Harmful Degradation Products (corrosive or toxic)†	–	HCl (combines with water to form hydro-chloric acid)	HCN (Hydrogen cyanide)	–	–	HF (combines with water to form hydrofluoric acid)
Environmental Resistance Ozone Ultra Violet	Poor Poor	Good Good	Poor Poor	Excellent Excellent	Excellent Excellent	Excellent Excellent
Maximum Long Term Operating Temperature** for 'O' Ring Type Seals	<60°C	90°C	90°C	100°C	130°C	150°C
Maximum Short Term or Intermittent Temperature*** for 'O' Ring Type Seals	<80°C	120°C	110 °C	130°C	250°C	250°C
Relative Radiation Resistance	Moderate	Poor	Moderate	Good	Poor	Poor

Notes: * DuPont trade name.

** Exposure for typically 20 000 – 30 000 hours. Lower temperatures will considerably extend life. In applications other than for seals, materials will often have a longer life; i.e. where seal force retention is unimportant.

*** Up to a few hours on a small number of occasions.

† Other than hydrocarbons and carbon monoxide, which are evolved in rapid degradation (i.e. fire) of all rubbers and plastics.

BS6564:Part 2:1985   Specification for fabricated unfilled polytetrafluoroethylene products.

CP312:   Plastics pipework (thermoplastics materials).

Part 1:1973   General principles and choice of material.

Part 2:1973   Unplasticised PVC pipework for the conveyance of liquids under pressure.

Part 3:1973   Polyethylene pipes for the conveyance of liquids under pressure.

1.3.2 Reinforced plastics

BS4994:1986   Vessels and tanks in reinforced plastics.

BS5480:Part 1:1977   Glass fibre reinforced plastics (GRP) pipes and fittings for use for water supply or sewerage. Part 1: Dimensions, materials and classification.

BS5480:Part 2:1982   Glass fibre reinforced plastics (GRP) pipes and fittings for use for water supply or sewerage. Part 2: design and performance requirements.

BS6464:1984   Reinforced plastics pipes, fittings and joints for process plants.

1.3.3 Rubbers

BS3574:1963   Recommendations for the storage of vulcanised rubber.

BS1154:1978   Specification for natural rubber compounds (high quality).

BS2751 + 3222:1982   Specification for acrylonitile-butadiene rubber components.

BS6014:1980   Specification for ethylene propylene rubber compounds.

BS2752:1975   Specification for vulcanised chloroprene rubbers.

DTD Specifications   (Procurement Executive, Ministry of Defence).

DTD 5612:   Vulcanised fluorocarbon rubber (low compression set) for aircraft.

DTD 5582::   Oil resistant vulcanised silicone rubbers for aircraft.

DTD 818:   Aircraft material specification; silicone rubber.

DTD 5531A: Silicone rubbers (general purpose).

2 Use of ceramics in the power industry

2.1 General applications

The bulk properties of ceramic materials (e.g., chemical inertness, thermal stability, high strength and low cost), make them ideal candidates for many applications in the hostile conditions in power generation. The application of ceramics falls into four broad categories:

- Electrical insulation — transmission.
- Thermal barriers — thermal insulation and oxidation prevention refractory applications.
- Mechanical barriers — wear.
- Nuclear — moderators, fuel containment.

These are in order of relative usage of ceramic material regardless of material type. With 150 000 tonnes of transmission line insulators in the UK electricity supply industry this is by far the most common application. Thermal insulation comprises the greatest volume of consumable ceramic material with the necessity to remove large quantities either as a result of degradation or to permit inspection of high temperature components. The nuclear application in the UK relates primarily to graphite, with well established performance as a moderator in the thermal reactor fission process as well as providing the containment sleeves for fuel in the later AGR design of reactor.

Ceramic fibre pressure vessel liner insulation is used in four AGRs with 35 tonnes/reactor. Ceramics for mechanical applications is an area of developing interest in the power industry. These are divided between bulk ceramics to combat wear in fuel (coal) handling plant and ceramic coatings to inhibit the combined effects of wear and oxidation on sliding surfaces.

A summary of the materials properties requirements for the major application areas are given below.

2.1.1 Thermal barriers

Thermal insulation

With fossil-fuelled boilers operating at gas temperatures up to 1500°C and the steam feed and turbine systems up to 568°C, thermal insulation is a vital element in the cycle efficiency. In addition, structural integrity calculations must take account of thermal effects such as creep behaviour and thermal fatigue which could be significantly affected by insulation performance. In the one extreme, inside a gas cooled reactor, insulation must contend with high pressure CO_2 (40 bar), a temperature range of 300–650°C and sound pressure levels up to 160 dB. At the other extreme cryogenic plant has to be effectively insulated.

The greatest use of ceramic materials for thermal barriers is in the fibrous form either as preformed sections, blanket or sprayed with a cementitious binder. The base material for these fibres is related to working temperature. Rock and slag based mineral wool is adequate for most applications up to 600°C. Above 600°C, and for extended lives, as in the nuclear environment, alumino silicate ceramic fibre is used. Fibrous insulations are formed by spinning or blowing processes (Fig 8.4) and contain a degree of non-fibrous material as well as lubricants.

In recent years the demands placed upon high temperature insulation materials have considerably increased due to changes in both technical and operational parameters. In specific terms most materials problems can be related to three important factors:

(a) Traditional insulants were formerly based upon asbestos formulations which exhibited good thermal, mechanical and chemical stability under most operating conditions. With the appreciation of the asbestos health hazard such materials are no longer used.

(b) Modern plant imposes much more severe conditions of higher operating temperature and mechanical cyclic loading on thermal insulation systems when compared with previous generation designs.

(c) The integrity of insulation materials and retention systems must be adequate to perform successfully over the period between plant outages which in British practice is typically 3 years.

To ensure adequate behaviour in service the suitability of commercially available materials needs to be continuously appraised. The currently employed insulation materials within CEGB plant are now briefly described in terms of their properties and intended duties.

2.1.2 Insulation materials

Thermal insulation materials used on CEGB plant are given in Table 8.3 and their general uses shown in Fig 8.5. The figure also includes information on the types of applications for which they should be used.

All rock-based man made mineral fibrous material is blown or spun giving a wide range of fibre sizes and some non-fibrous material. These are usually a rock diabase composition with additions of MgO and CaO.

The majority of ceramic fibre is manufactured from either alumino silicate or pure alumina.

The use of glass fibre insulation material is limited within the CEGB to low temperature plant. It is also found in a variety of ropes, sealants and gaskets. In bulk insulants, blown glass fibres are employed. The use of drawn fibres is generally limited to ropes, cloth and the reinforcement of other products.

Calcium silicate insulants, although not strictly ceramics, are based on reacted hydrous calcium silicate. A small percentage of glass fibre is added which imparts a degree of cold handling strength.

For the bulk fibrous insulating materials the density and heat stability are governed largely by the composition, the binder used and the manufacturing route. Standard methods are available for assessing thermal conductivity and for most materials these are sufficiently well characterised for our purposes.

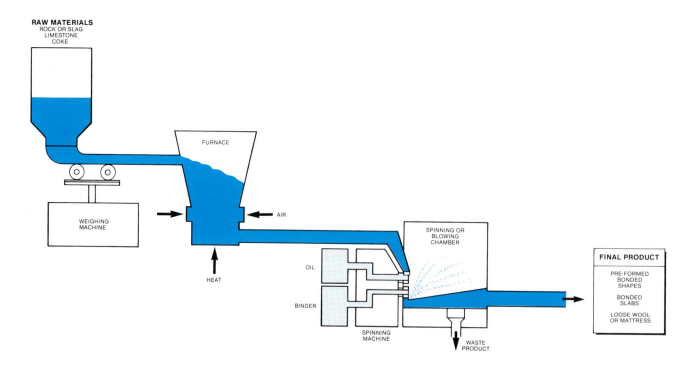

FIG. 8.4 Man made mineral fibre manufacturing route

TABLE 8.3

Major use of thermal insulation materials within the CEGB

Plant area	Materials employed	CEGB And GDCD Standard Number
Boilers internal ducts	Organic and inorganically-based MMMF slabs, calcium silicate slabs	CEGB 08991 Pt 1 and 089907 GDCD 14 Pt 1
Internal pipework	Organic and inorganically-bound MMMF and calcium silicate pipe sections/ radiused bevelled logs/moulded pipe sections	CEGB 08991 Pt 2 and 089907 GDCD 14 Pt 2
External ducting	Organically and inorganically-bounded slab	CEGB 08991 Pt 3 and 089907 GDCD 14 Pt 2
External pipework	Preformed calcium silicate, inorganically-bounded MMMF pipe sections	CEGB 08991 Pt 4 and 089907 GDCD 14 Pt 4
Storage tanks	Polyurethane and phenolic foams. MMMF slabs. Modular systems	CEGB 08991 Pt 5 and 089907 GDCD 14 Pt 5
Cryogenics	Cellular glass, phenolic foam	CEGB 08991 Pt 6 and 089907 GDCD 14 Pt 6
Turbines	Sprayed MMMF. MMMF mattress	CEGB 089903 Pts I and II and 089907 GDCD 15 Pts 1 and 2

Note CEGB 089907 refers to Materials Standards

Preformed man made mineral fibre (MMMF) (organically and inorganically bound)

The bulk fibre is processed in various ways to produce the following products:

- *Organically bound slabs and pipe sections* In this case the fibre is lightly compressed and held in a rigid predetermined shape by binders. This assists in the handling and fitting of the insulation and also provides restraint of the fibres.

  Organic binders (usually Phenolic resins) are widely used but although the fibres can withstand temperatures in excess of 600°C the organic binders will migrate and burn off above 250°C. This leaves the hot face of the insulation in a flexible form, similar to a mineral fibre mattress. This situation is considered to be acceptable because the resin migrates to the cold face but it can reduce the ability of the insulation to withstand compressive loads. Provided the section has a density greater than 120 kg/m^3 satisfactory service can be obtained up to 600°C under normal conditions.

  It should be noted that organically bound MMMF products with density below 120 kg/m^3 have inferior thermal conductivities and will tend to collapse when the binders are burnt off.

  Inorganically bound (usually a formulation containing bentonite clays) MMMF sections have recently become commercially available. Although their thermal resistance is not quite as good as the organically bound sections, the binding agent does not burn off. This significantly increases their ability to withstand compressive loads after exposure to temperature and therefore makes them an alternative to calcium silicate sections.

- *Ceramic fibre* Ceramic fibre is made from aluminosilicates and is generally white in appearance. The claimed maximum service temperatures range from 1000°C to 1500°C depending on the fibre composition. Ceramic fibre is more expensive than MMMF (rock or slag based) although the cost difference is now tending to reduce. There is generally no justification for using ceramic fibre below 600°C since cheaper alternatives are available as general insulant.

- *Calcium silicate (asbestos-free)* Calcium silicate sections are rigid but they effectively have no resilience. Incorrectly designed calcium silicate insulation systems can therefore suffer from thermal expansion cracking. However, calcium silicate is widely used since it has greater compressive strength than organically bonded insulants. In addition it is reasonably resistant to deterioration under alternate wetting and drying cycles.

- *Specialist applications* Some insulants use high temperature metal oxides of special particle size and distribution to reduce infrared radiation and radiation transmissions to very low levels. This produces an insulant with an extremely low thermal conductivity.

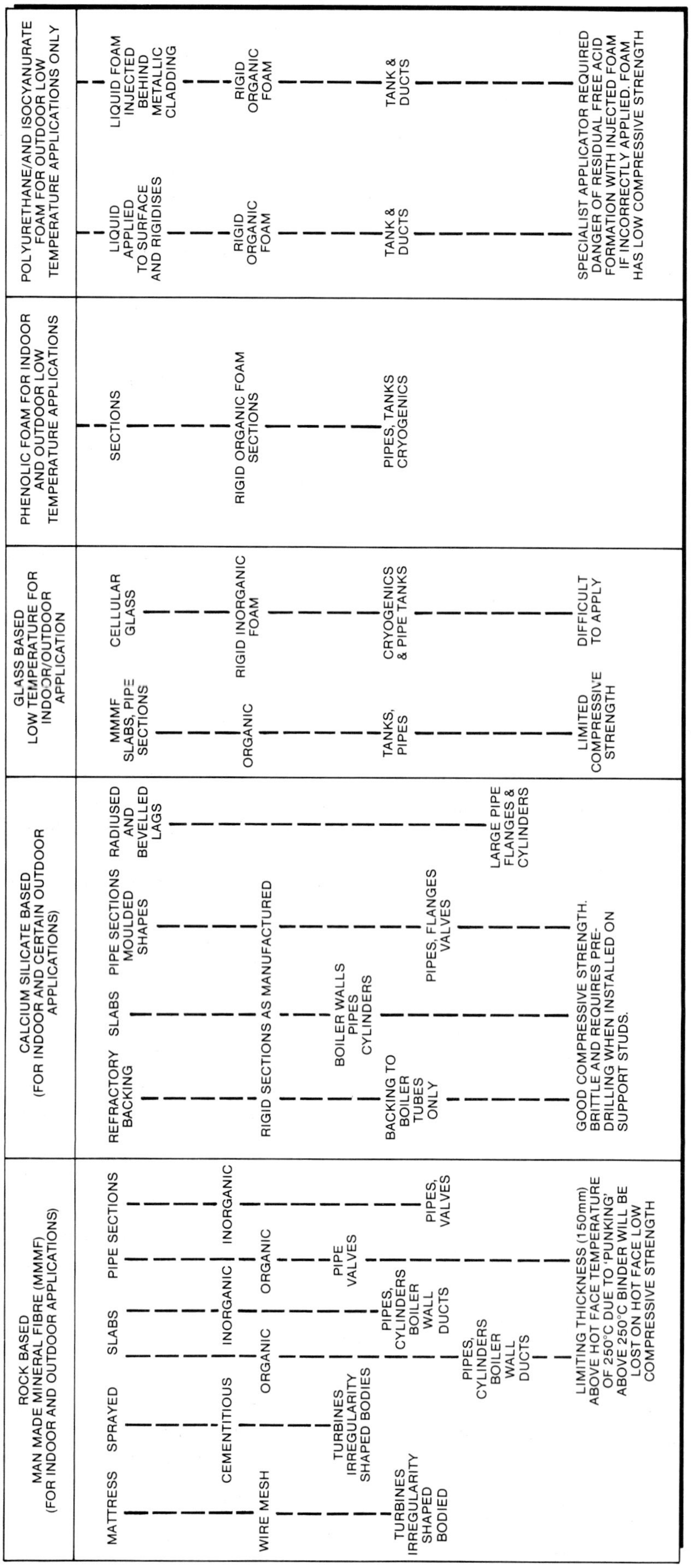

Fig. 8.5   Summary of approved thermal insulation materials and their applications

It is usually supplied as a compacted powder encapsulated in quilted or slatted glass fabric. The glass fabric normally has a limiting temperature of 550°C. However, the microporous insulation can withstand temperatures up to 950°C provided it can be retained.

2.2 The use of refractories in conventional boiler applications

The four principal reasons for installing boiler refractory materials are:

(a) To contain generated heat.

(b) To protect the working and structural parts of the boiler.

(c) To facilitate the correct distribution of the heat throughout the boiler.

(d) To prevent the egress of combustion gases and dust.

The purpose of refractory materials is therefore purely a sacrificial one, and hence the designers have over the years made considerable effort to reduce the quantity of refractory materials contained within each boiler. This reduction in refractory usage is immediately apparent if the refractory requirements for chain grate boilers are compared with those for a modern PF or oil-fired 500 MW boiler.

Most CEGB boilers now have water tube walls instead of radiating walls. The main refractory usage in these newer boilers is now limited to the burner quarls, ash hoppers, sections of the furnace roof, economiser and baffles. However, on some oil-fired boilers flame impingement has meant that designers have had to incorporate refractory belts adjacent to, or just above, the burners.

Shift working of power generation boilers is often an operational requirement. This, combined with high operating temperatures, has greatly increased the technical demands on the remaining areas of boiler refractory. These technical demands have been further exacerbated by the need for longer life materials due to the increased period between major boiler outages. For economic reasons it has also been desirable to reduce the work content and application time for the installation of refractory materials.

It has therefore been necessary to use high duty products, correctly installed, in order to avoid premature refractory failures. Premature failure can significantly effect boiler efficiency and in some cases plant availability. Furthermore, the consequential damage to other boiler components is usually extremely expensive to rectify.

The principal chemical and physical properties required from refractory materials and the major plant areas where they are employed are briefly reviewed below.

2.2.1 Refractory properties

The term *Refractory material* is usually defined as meaning a substance which is *hard to fuse*. In technology, however, the meaning of the term *Refractory* is broadened to include ceramic materials used in the construction of furnaces, steam raising plant, flues, heaters, etc. In these situations both high temperature stability and adequate environmental durability must be considered.

The principal properties required from a refractory material are as follows:

● Resistance to the temperature to which it is likely to be exposed in use.

● Resistance to any stress likely to be put upon it by adjacent material.

● Resistance to vibration such as may occur and any mechanical blows.

● Resistance to the slagging action from the fuel.

● Resistance to the erosion and abrasive action of flame from oil, or pulverised fuels, and flue dust.

● Expansion or contraction properties sufficiently uniform within conveniently narrow limits.

● Resistance to environmental attack associated with oxidising or reducing conditions.

Few refractory materials fulfil all these requirements. Therefore it is essential to consider whether pre-fired shapes or monoliths are the best material for each section of a boiler lining.

When making a decision on choice of materials and design it is equally important to bear in mind that the material is only as good as the method of application and of particular importance are sound and proven anchorage systems for good retention.

2.2.2 Chemical composition of refractories

The great diversity of purpose for which refractory materials are used has necessitated their classification according to their behaviour with various reagents, so that broadly speaking the materials recommended for steam raising plant falls into the following categories:

(a) Pre-fired shapes manufactured from fireclay, grog or from silicon-carbide based materials.

(b) *Clay-based mouldable or ramming materials* each being either an alumino silicate or silicon carbide chemical-bonded refractory.

(c) *Cement-based castable refractories and insulating materials* designed for use by casting or gunning; these using as a base the following types of cement:

● Portland cement

● High alumina cement

● Low iron high alumina cement.

435

Such materials could be composed of mainly:

- Refractory cement + aggregate
- Refractory cement + aggregate + x% of clay (for gunning)
- Refractory cement + aggregate + steel fibres.

2.2.3 Physical and mechanical properties of refractories

The physical and mechanical properties of refractory materials are as important as the chemical properties.

The most important physical properties may be listed as follows:

(a) Refractoriness (pyrometric cone equivalent) and maximum continuous operating temperature.

(b) Changes in dimension when heated:

- Reversible thermal expansion
- Permanent changes in volume (linear shrinkage) and permanent linear change PLC
- Thermal shock resistance.

(c) Strength of refractories:

- Modulus of rupture
- Cold crushing strength.

(d) Spalling resistance.

(e) Thermal conductivity.

2.2.4 Major plant areas utilising refractory materials

The plant areas where the majority of refractories are employed have been divided into the categories shown in Table 8.4. CEGB Standards have been written relating to the correct selection and installation of the refractory material for each application.

Electrical insulators

Porcelain and glass insulators are used extensively on transmission towers and in substations with the high purity porcelain insulators being the most frequently used. To provide the resistance path between conductor and support structure, insulators are strung together in

TABLE 8.4

Major use of refractory materials within the CEGB

Plant area	Materials requirements	Materials employed	CEGB Standard
Burner quarls	Temperature resistance. Dimension stability. High abrasion and corrosion resistance. Good thermal shock properties	Silicon carbides (coal-fired). Chrome concretes. Aluminosilicates (oil-fired) (Hydraulic and chemically bonded).	CEGB 078901 GDCD7
Ash hoppers	Dense castable refractory having high abrasion and impact resistance.	Abrasion resistant castable concretes.	CEGB 078902 GDCD8
Refractory belts	Reduce heat input to water-wall tubes. Good insulation properties.	Phosphate bonded mouldable materials containing 47–50% SiO_2 45–85% Al_2O_3.	Formerly CEGB 078903 GDCD9
Backing for tangential and finned tubes	To reduce boiler casing temperatures and to seal boiler.	Cast or gunned concretes.	CEGB 078905 GDCD10
Non-water-cooled walls	Capable of withstanding service temperatures of 1650°C. Generally confined to chain-grate boilers.	High temperature castables. Plastic or chemically-bonded mouldable grades. Typical composition Al_2O_3 43%, SiO_2 50% Fe_2O_3 1%.	CEGB 0789051 GDCD11
Backing for spaced tubes	Heat resistant furnace furnace gas seal. Adequate thin-section strength.	Fine-grained high-strength castables, reinforced with stainless steel fibres.	Formerly CEGB 078906 GDCD12
Renovation gunning	Used for renovation of partially-worn surfaces. Not suitable for ash hoppers.	Gun mixes	Formerly CEGB 078907 GDCD13

a variety of ways inevitably using mechanical links in the process.

The basic element in an insulator string for an overhead line is shown in Fig 8.6. The insulator itself is a glazed porcelain shed manufactured from Al_2O_3, SiO_2 and K_2O. High purity is essential for electrical resistance with a potential difference of 20 kV across each element. In addition, the string of insulators must support the conductor weight under all conditions, with wind and ice contributing considerably to the stresses. Small amplitude oscillations can arise from aerodynamic drag but occasionally large amplitude galloping of conductors will arise. The shed is designed to accept a metallic slotted cap and tie pin to link them together and accommodate wind loading effects. This requirement provides the designer with his greatest problems; to ensure compressive loading of the ceramic whilst enabling the transmission of tensile loads through the string.

In the broad range of wear problems in the power industry the application of ceramics is being developed in two particularly aggressive situations: particulate erosion in fuel handling plant; fretting wear in nuclear plant.

Within pulverised fuel transport systems there is a need for wear resistant liners in order to achieve satisfactory lives. Traditionally, cast basalt has been widely employed but in many cases its performance is not adequate and the use of high density alumina tiles (85–97.5% Al_2O_3) is becoming more usual. Typical wear rate data for a range of alumina containing materials versus impact angle is presented in Fig 8.7. In order to achieve the full benefit of these tiles it is important that they are properly installed with minimal gaps between tiles for the pulverised fuel to penetrate.

Silicon nitride bonded silicon carbide, whilst more expensive than alumina tiles, also has a role as it can be cast into the complex shapes required for such components as burner elbows and trifurcations. Here, the trend is to supplant the use of wear-resistant white cast irons.

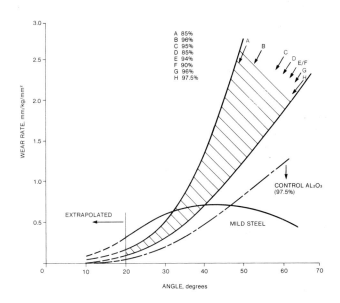

Fig. 8.7 Typical wear rate data for a range of materials plotted against impact angle

Fretting wear has been much neglected yet, whilst slip amplitudes are small (approximately 200 μm), sliding distances can be significant. In the power industry fretting is found in a wide plant range but is most difficult to contend with at elevated temperature in gas-cooled reactors (see Volume J). There have been major work programmes to study surface coatings in order to ensure design life because of anxiety over the performance of stainless steel interfaces. A limitation with most coatings was spalling due to thermal shock and interfacial oxidation. The techniques of detonating ceramic coatings (carbides and oxides) onto surfaces, however, overcame these problems and the coatings have been subjected to extensive fretting wear tests. Typically, wear rates are found to be moderate during the very early life but rapidly fall to low levels typical of a chrome carbide deposit.

2.2.5 Future requirements

Electricity transmission

All-ceramic insulators are perceived as promising in future to overcome the current problem of ensuring compatibility between the metal/ceramic interface. Already advances in technology have allowed large solid core all ceramic units to be developed. These are replacing troublesome multicone post insulators throughout the National Grid.

Thermal insulation

In the context of developing trends, where longer lifetime insulation systems are required, it is important to establish more definitively the kinetics and mechanisms of the degradation of materials. This has relevance to both environmental and economic factors associated with thermal insulation.

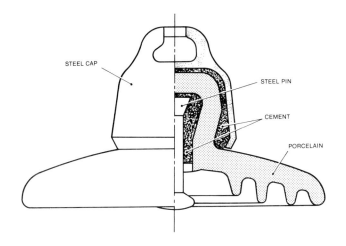

Fig. 8.6 Cap and pin transmission insulator element

Mechanical applications

The use of bulk ceramics for wear problems will grow to combat wear, but more work is required to relate material specifications to wear behaviour. Installation practice is of equal importance in achieving optimum lives and further development is required in this area.

Whilst plasma-sprayed coatings have been established in the fretting field where thin layers are adequate, alternative, high build, ceramic coatings would enable a much wider application in the wear field.

High temperature furnace applications

The replacement of metallic components by advanced ceramic for furnace burner tips could provide significant performance and maintenance advantages but as yet no material has been developed to resist the severe furnace gas environment.

The continuing development of the pressurised fluidised bed boiler is expected to demand high temperature corrosion and abrasive resistant materials and further development of advanced ceramics is expected to support this technology.

CHAPTER 9

Materials selection

1 Materials selection for design

The wide range of conditions under which power plant components operate demands knowledge of how the materials will behave under these conditions. In the selection of the most suitable design and material for a particular component a number of interacting factors will have to be considered. Design of a particular component must consider the range of operating temperatures, expected service life, stressing modes, acceptable deformation and resistance to the environment. Other aspects of relevance are cost and fabrication characteristics with respect to formability and weldability.

Design selection therefore requires a knowledge of manufacturing capability in addition to an understanding of the short term and long term behaviour of the component.

The prime property requirements for the great majority of power plant materials are adequate mechanical strength and ductility at the operating temperature. Other special requirements such as shock resistance, corrosion and oxidation resistance, fatigue resistance or combinations of these are also often necessary and have to be considered for specific application.

At ambient temperatures the stress and strain characteristics of most metal alloys are not time dependent. Thus the amount of strain produced by the application of a given load occurs instantaneously and does not increase with time. Many components are therefore designed on a time independent criterion such as yield point or proof stress, whereby the component does not permanently change its shape by plastic deformation. As operating temperatures increase there is a reduction in the yield point and tensile strength but

for moderate increases, e.g., up to about 340°C for mild steel, strain can still be considered to be independent of the time for which the load is applied. Again a yield point or proof stress criteria can be used for design. However at higher temperatures strain continues with time and deformation by creep occurs at loads considerably less than the yield point. At these elevated temperatures the component design may therefore be based on creep/stress rupture data.

It is apparent that materials selection should consider the duty required from the component and ensure that the manufacturing route and properties are compatible with these requirements. Potential failure mechanisms in service including those arising from fault conditions also have to be allowed for and particularly in the case of nuclear reactors, an awareness that access to components may be very difficult and remedial measures expensive.

It is convenient to examine materials in relation to the major items of plant. These are the heat producing sources; fossil-fuelled boilers or nuclear reactors, the power producing source; the turbine-generator, and ancillary equipment.

1.1 Boiler components

The most dramatic recent change in boiler design has been one of scale rather than increases in operating temperatures and/or pressures.

Typically a conventional boiler for a 660 MW set produces 2×10^6 kg/h of steam at 568°C and 166 bar (superheat) and 568°C and 41 bar (reheat). In the immediate future boiler designs for 900 MW output are under development, although steam conditions will be

439

at similar or slightly increased design steam temperatures and pressures.

In component detail a boiler remains an assemblage of tubes and thick walled cylinders providing heat transfer between the furnace gas at temperatures of up to 1600°C and 2000°C in coal and oil-fired boilers respectively and the water, water/steam and steam carried in the tubing.

The design and manufacture of water-tube steam generating plant subject to internal pressure are covered by BS1113 [1].

This document specifies design stress values for plates to BS1501 [2], sections and bars to BS1502 [3], forgings to BS1503 [4], castings to BS1504 [5], steel boiler and superheater tubes to BS3059 [6], and steel pipes and tubes for pressure purposes to BS3601 [7], BS3602 [8], BS3604 [9], and BS3605 [10]. BS1113 also specifies aspects of workmanship, inspection, testing, documentation and marking valves and gauges.

Most materials for pressure containing components of boilers fall within the following categories:

- Carbon steels.

- Chromium molybdenum alloy steels.

- Austenitic stainless steels and, to a lesser extent, ferritic stainless steels.

Table 9.1 lists steels that find common usage in existing boiler plant. The steels are listed in relation to their approximate maximum operating temperatures but the listing also reflects relative costs of the materials. It is evident that economics favour carbon steels for applications where metal temperatures are below about 450°C. Such steels therefore provide the bulk of material for boiler components in areas such as economiser, furnace wall evaporator, primary superheater and primary reheater tubes, headers and manifolds. Between 450°C and 580°C the low alloy ferritic steels find extensive usage. Above 580°C austenitic stainless steels have to be used. The temperature limitations primarily arise for two reasons, firstly the loss of mechanical strength with increasing temperature exemplified by loss of creep strength and secondly loss of the metal by oxidation and/or fireside corrosion.

Figure 9.1 shows a typical group of mean stress rupture curves for a low alloy steel which have been determined from laboratory tests. It illustrates how the stress to cause failure varies with time at different temperatures. Such data can be further grouped into a family of curves which give the rupture stresses of each material for a life of say 100 000 hours as shown in Fig 9.2. Historically, power plant has been designed on a 100 000 hour life basis but more recently this has been extended to 160 000 hours. The translation of these properties to a design stress is usually carried out by applying a 'factor of safety' to the mechanical property values of the steel. Figure 9.3 shows a typical design curve indicating how operating metal tempera-

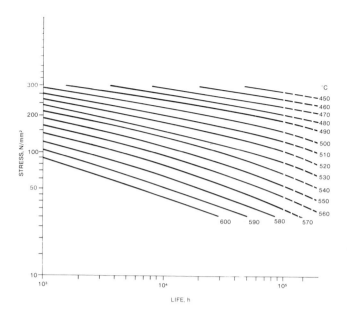

FIG. 9.1 Stress rupture properties of 1% Cr, $\frac{1}{4}$% Mo steel The stress rupture curves are derived from polynominal functions representing agreed ISO (International Standards Organisation) parametric curves. For lifetimes in excess of 100 000 h, the curves have been obtained for extended extrapolation. The curves can also be expressed conveniently as a polynominal function of log stress which can be readily computed to predict creep rupture lives of high temperature components. Such data are available for the common tube and pipe materials.

tures affect consideration of the mechanical properties for the design of a component.

The term carbon steel essentially applies to alloys of iron and carbon with varying amounts of other elements the principal ones being manganese with small amounts of silicon, sulphur, phosphorus and aluminium. These elements are derived from the raw materials used during steel making or are added deliberately to achieve certain properties. The range of carbon steels used for boiler pressure components is small compared with general engineering and structural steels. Typically a boiler carbon steel will contain only 0.05 to 0.25%C and 0.4 to 0.7% Mn, whereas carbon steels used in general engineering will contain up to 1.0%C and 1.0% Mn. The major considerations in the choice of these carbon and manganese levels is that they possess the required mechanical properties together with the ability to be fabricated readily into the desired component shape and size. In addition, sound welds can be easily made in these alloys without the need for expensive heat treatments. Under most conditions they also possess adequate resistance to the flue gas and water/steam environment. Their major limitation is that the maximum operating metal temperatures is about 450°C above which creep and scaling become excessive.

For operating metal temperatures above about 450°C there is a need for materials offering more creep resistance than carbon steel. In the case of steam power plant alloy ferritic steels containing additions of molyb-

TABLE 9.1

Materials for boiler components

	Typical Composition Weight %							Approximate maximum service temperature °C	Applications
	C	Mn	Cr	Mo	Ni	V	Others		
Mild Steel	0.12	0.6	–	–	–	–	–	450	Furnace, economiser tubes, feedwater heaters.
Carbon-manganese steels	0.15	1.0	–	–	–	–	–	450	Furnace, economiser, early stages of super-heater and reheater tubes. Boiler drums, headers, receivers, pipes.
	0.28	0.8	–	–	–	–	–	450	Feedwater pump, valve body castings.
	0.20	1.25	–	–	–	–	–	450	Boiler drums, headers, receivers.
Cr.Mn.Mo.V	0.15	1.25	0.55	0.24	0.7 max	0.1	–	450	Boiler drums, headers, by-pass vessels.
½% Mo (obsolescent)	0.12	0.60	–	0.50	–	–	–	510	Superheater tubes, pipes, headers.
1% Cr ½% Mo	0.12	0.50	0.9	0.50	–	–	–	550	Superheater, reheater tubes, pipes, headers and receivers.
2¼% Cr ½% Mo	0.12	0.50	2.25	1.0	–	–	–	580	Superheater, reheater tubes, pipes, headers and receivers.
	0.15	0.50	2.25	1.0				580	Valve body castings.
½% Cr ½% Mo ¼% V	0.12	0.50	0.40	0.60		0.25		580	Headers, pipes, valve bodies.
9% Cr 1% Mo	0.10	0.50	9.0	1.0	–	–	–	550	Evaporator, superheater tube (nuclear).
AISI 316	0.07	1.5	18.0	2.5	13.0	–	–	650	Superheater, reheater tubes. Headers, pipes.
AISI 321	0.07	1.5	18.0	–	11.0	–	0.35Ti	650	Superheater, reheater tubes.
AISI 347	0.07	1.5	18.0	–	12.0	–	0.7 Nb	650	Superheater, reheater tubes. Headers, pipes.
AISI 310		1.5	25	–	20	–		650	Outer layer of co-extruded furnace, superheater and reheater tubes.
ESSHETE 1250	0.10	6.0	15.0	1.0	10.5	0.25	1 Nb + B	670	Superheater, reheater tubes, pipes.

denum, chromium and in some alloys vanadium provide suitable properties for operating metal temperatures up to about 580°C.

Molybdenum has long been known to be a potent strengthener of steels at elevated temperatures acting both as a solid solution hardener and by the formation of molybdenum containing carbides which are more stable than those present in carbon steels. The simplest molybdenum containing steel is that containing 0.5% Mo. However, there have been instances of low ductility creep failures associated with intergranular fracture with this material and it is no longer used in modern

designs of power plant. Some older plant may still have headers and tubes fabricated from 0.5% Mo steel. Where, following temperature monitoring and dimensional checks, it is considered that such headers are at risk of failure they have been replaced with alternative low alloy steels such as 1% Cr $\frac{1}{2}$% Mo or $2\frac{1}{4}$% Cr 1% Mo steels.

The low alloy steels containing up to $2\frac{1}{2}$% chromium and 1% molybdenum provide superior creep properties to those of carbon and 0.5% Mo steels. These steels were initially developed in the oil refining industries for their corrosion resistance. Two steels,

441

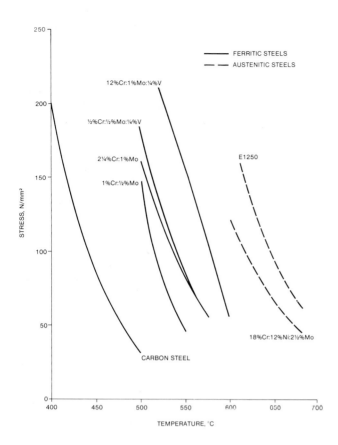

FIG. 9.2 100 000 hour rupture stresses for
boiler materials
At a given stress there is a temperature limit for each
steel and the creep resistance improves in the order:
carbon steel, low alloy ferritic steels, ferritic stainless
steels, austenitic stainless steels.

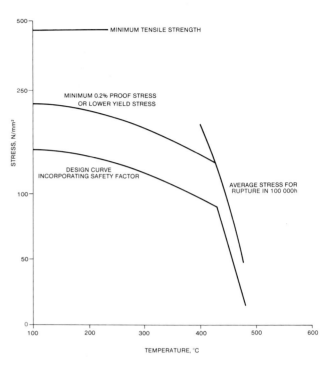

FIG. 9.3 Typical design curve for carbon steel
The design stress adopted is the lowest of the following
three values:

(a) $$\dfrac{\text{(Minimum room temperature)} - \text{(tensile strength)}}{2.7}$$
for carbon and low alloy steels

(b) $$\dfrac{\text{Minimum 1\% proof stress at temperature for austenitic steels}}{1.35}$$

(c) $$\dfrac{\text{Average stress for rupture in design lifetime at temperature}}{1.3}$$

Experience has shown that the two material properties
most likely to govern design of boiler plant are the
minimum proof stress and the average stress for rupture
in the design lifetime of the component.

in particular those nominally containing 1% chromium
with 0.5% molybdenum and $2\frac{1}{4}$% chromium with
1% molybdenum, were adopted for power plant because of their good creep strengths. The alloying elements chromium, molybdenum, and vanadium are all
strong carbide formers and improve the creep resistance
properties.

For components operating with metal temperatures
in excess of about 580°C, design with low alloy ferritic steels would require excessive wall thicknesses to
achieve an acceptable design stress and to accommodate
metal losses by oxidation. For such components, austenitic stainless steels offer the optimum material
properties.

The term stainless steels typically refers to steels
with in excess of 12% chromium; the chromium con-
fers excellent corrosion resistance due to the formation
of a protective surface layer of chromium oxide. The
stainless steels can be categorised either as ferritic
stainless steel or austenitic stainless steel. The ferritic
stainless steels, as the name implies, are ferritic in
nature (with a body-centred cubic structure) and are
basically iron-chromium alloys with minor additions of
other elements, often the strong carbide formers such

as molybdenum, vanadium, niobium which improve
creep resistance.

The austenitic stainless steels contain both chromium and significant levels of nickel. The addition
of nickel stabilises the austenitic (face-centred cubic)
structure of the steel. The most commonly used austenitic stainless steels in power plant are those based
on the 18% chromium 12% nickel alloys. Again minor
additions of other elements, such as molybdenum and
strong carbide formers such as niobium or titanium
(type 316, 347 and 321 steels respectively) may also be
present in the steels. A development of these austenitic
alloys which contains molybdenum, vanadium, niobium
and boron is Esshete 1250 steel. This steel offers superior creep strength to the Type 300 steels without the
penalty of loss of creep ductility. The superior creep
strength of this alloy allows components to be designed
with thinner wall thicknesses or extended life. It is
worth noting, however, that there are other limitations
on the minimum wall thickness of certain components.

For instance a minimum wall thickness of 3.7 mm is recommended for reheater tubes to avoid problems in tube bending and welding.

Compared with ferritic steels, austenitic stainless steels have high coefficients of thermal expansion, low thermal conductivities and relatively low yield strengths (Table 9.2). The above properties give rise to distortion problems in thick section and complex components, therefore austenitic stainless steels find usage principally for tubes and to a lesser extent pipes that carry steam at temperatures above 550°C. Although 'stainless steel' implies superior corrosion resistance, many of the austenitic stainless steels are prone to stress corrosion cracking, particularly in the presence of caustic soda or chloride containing solutions. Since both corrodants can be present in boiler water and saturated steam, austenitic stainless steels should be avoided in these parts of the water/steam circuit.

Experience with 9% Cr 1% Mo steel in fossil-fired plant is limited, it has found occasional application as a replacement material for $2\frac{1}{4}$% Cr 1% Mo or 1% Cr $\frac{1}{2}$% Mo steel superheater tubes which suffered fireside corrosion. For this purpose it has performed satisfactorily for service times in excess of 100 000 hours.

Its major application is in the boilers of advanced gas-cooled nuclear reactors as evaporator and superheater tubing with service life, to date, of up to 60 000 hours. The material offers good tolerance to fault conditions, with excellent high temperature ductility in both regimes of creep and creep fatigue. In terms of its performance under creep-fatigue conditions, hold times have no effect on the number of cycles to failure. This is attributed to the resistance of this material to grain boundary cavitation during creep which is in marked contrast to the behaviour of low alloy ferritic steels and Type 316 stainless steel.

In terms of waterside corrosion resistance the 9% Cr 1% Mo steel has a general superiority to $2\frac{1}{4}$% Cr 1% Mo particularly in respect of stress corrosion. It is, however, susceptible to stress corrosion cracking in the presence of caustic soda if left in the unheat-treated condition after welding and therefore requires post-weld heat treatment.

1.2 Reactor components

The three principal reactor systems in use are cooled by either CO_2, water or liquid sodium, and the constructional steels used depend primarily on the coolant and the operating temperature of the reactor.

There are two types of CO_2 cooled reactor, Magnox and AGR. In the Magnox type, which uses magnesium alloy fuel element cladding, the maximum operating temperature is about 360°C and the whole structure of the reactor and boilers can be fabricated from carbon steel. In the AGRs, which use 20% Cr 25% Ni Nb stainless steel fuel cladding, the maximum operating temperature is about 650°C. The materials of construction therefore range from carbon steel at low temperatures to austenitic stainless steels in the highest temperature regions, and include a large amount of 9% Cr steel in the middle temperature range evaporator section of the boilers. A list of many of the steels used in CO_2 cooled reactors is given in Table 9.3.

For water cooled reactors such as the PWR, the maximum operating temperature is about 325°C so that austenitic stainless steels with their high temperature properties are not required and most of the major structural items such as the pressure vessel, pressuriser and steam generators are manufactured from low alloy ferritic steels. Austenitic stainless steels are however employed in the primary circuit of PWRs to avoid build-up of corrosion products in the circuit.

Austenitic steels are therefore used for weld deposited cladding on the interior of the pressure vessel, for the core support structure, primary circuit coolant pipework, reactor coolant pumps and steam generator tubing. A list of typical constructional materials is given in Table 9.4.

In Fast Reactors where the coolant is liquid sodium, virtually the whole of the structure is manufactured from stainless material. There is an intent to reach international agreement on a common material for the primary circuit, which includes the main vessel, the primary coolant pipework, coolant pumps and intermediate heat exchangers. The most likely material is AISI 316L, although AISI 304 may be used in the cooler regions of the circuit. The most likely material for the secondary circuit is a 12% Cr ferritic stainless steel, and boiler tubing may be 9% Cr, 12% Cr ferritic of an iron-nickel-chromium alloy, Incalloy 800 (A800).

1.3 Turbine-generator components

As with boiler components, the most recent development in turbine-generator design has been one of scale rather than increases in operating pressures and temperatures. Operating conditions are not as extreme as those experienced on boiler plant components with typical maximum temperatures and pressures of 568°C and 169 bar, and high quality well-controlled steam and steam/water environments.

The materials adopted for turbine-generator components are listed in Table 9.5. For high temperature turbine casings and steam chests the material selection primarily involves the provision of adequate high temperature strength and ductility. These properties are required to accommodate the processes of creep and creep fatigue that arise naturally from the operation of such components.

These items of plant are designed to minimise pressure and thermal stresses by the avoidance of stress concentrating features and minimising section size. In the selection of materials for turbine casings there is also a need to use a material which can be fabricated into complex shapes with thick walls necessary

TABLE 9.2

Some differences in properties of ferritic and austenitic boiler tube steels

	Coefficient of thermal expansion, 20 to 500°C	Thermal Conductivity W/M°C	0.2% Proof stress for ferritics, 1.0% Proof stress for austenitics, N/mm^2					
			100°C	200°C	300°C	400°C	500°C	600°C
Low alloy ferritic steel	14.5×10^{-6}	44	250	230	200	185	175	150
Austenitic steel	18.5×10^{-6}	23	170	140	125	110	110	100

TABLE 9.3

Steels for CO_2-cooled reactors

	Typical Composition (Wt %)							Approximate maximum service temperature °C	Applications
	C	Mn	Si	Cr	Mo	Ni	Others		
Carbon Steel	0.1	1.0	0.1					360	Boiler tubing, structural steelwork
1 Cr Mo	0.15	0.5	0.3	0.9	0.5			360	Economiser tubing
1 Cr ½ Mo	0.35	0.6	0.3	1.0	0.6			360	Bolting
5 Cr	0.1	0.5	0.4	5.0	0.5			360	Tube transition pieces
9 Cr 1 Mo	0.1	0.5	0.6	9.0	1.0			550	Evaporator and super-heater tubing
12 Cr	0.1	0.6	0.2	12.0	1.7	2.5		360	Bolting
AISI 347	0.07	1.5	0.6	18.0		12.0	1.0 Nb	650	Structural steelwork
AISI 316	0.07	1.5	0.6	18.0	2.5	13.0		650	Superheater and reheater tubing, bolting, steelwork
AISI 321	0.07	1.5	0.6	18.0		11.0	0.5 Ti	650	Insulation, bolting, structural steelwork
AISI 304	0.03	1.5	0.6	18.0		11.0		650	Structural steelwork
AISI 310	0.1	1.5	0.6	25.0		20.0		650	Structural steelwork, insulation
E.1250	0.1	6.0	0.6	16.0	1.0	10.0		580	Structural steelwork
A.600	0.7	0.8	0.4	17.0		75.0		550	Tube transition pieces

to achieve acceptable stress levels. For casings operating up to about 450°C, carbon steel castings have been widely used with the low alloy ferritic steels being used for the higher temperature (up to 580°C) stages. Austenitic stainless steel castings have been used for operating temperatures in excess of 580°C; both Type 316 and Type 347 being utilised in such high temperature plant.

For high temperature rotors, material selection must take into account the high body stresses in the forging when rotating and the mechanical and thermal stability of the rotor to preserve running balance. There is a need to give attention to design details to ensure that there is no excessive creep deformation either of the rotor bore or at blade root fixings, and that stress concentrations at these positions are held to a minimum. For modern high temperature rotors 1% Cr 1% Mo $\frac{1}{4}$ V steel has become the accepted material due to its combination of adequate creep strength and ductility at the operating temperature together with satisfactory room temperature properties.

For low pressure and generator rotors the principal criteria for materials selection are adequate strength to maintain mechanical stability at running speeds and fracture toughness. The fracture toughness must be of a sufficient level to avoid any possibility of crack growth to a critical defect size during either normal or fault conditions that might occur in service. Indeed over recent years there has been a steady improvement in the fracture toughness properties (Table 9.6) due

TABLE 9.4
Steels for water-cooled reactors

	Typical Composition (Wt %)							Approximate maximum service temperature °C	Applications
	C	Mn	Si	Cr	Mo	Ni	Others		
SA 508	0.2	1.3	0.2	0.15	0.5	0.5		325	P.V., pressuriser and steam generator forgings
SA 533	0.2	1.3	0.2	–	0.5	0.5		325	P.V., pressuriser and steam generator plate
SA 540	0.4	0.8	0.2	0.8	0.3	0.8		325	Vessel closure studs and nuts
A.8	0.1	1.5	0.5	20.0	4.0	10.0		325	Weld deposited cladding
SB 166	0.1	0.7	0.3	16.0	–	75.0		295	Core support structure
AISI 304	0.07	1.5	0.6	18.0	–	11.0		295	Coolant pump forgings and castings, primary coolant pipes
AISI 316	0.07	1.5	1.0	18.0	2.5	13.0		295	Coolant pump forgings and castings, primary coolant pipes
A.600	0.07	0.8	0.3	17.0	–	75.0		325	Steam generator tubing
SA 516	0.25	1.0	0.2	–	–	–		295	P.V. and steam generator support structure
SA 193	0.4	0.9	0.2	1.0	0.2	–		295	Bolting
SA 453	0.05	1.5	0.6	15.0	1.2	25.0	2.0 Ti 0.3 Al	325	Bolting

to improvements in steelmaking technology and closer control of chemical composition. Most low pressure and generator rotors are forged in Ni Cr Mo V or Ni Cr Mo steels which can be heat treated to a minimum 0.2% proof stress of 555 MPa and ultimate tensile strength of 710 MPa.

Turbine blades are subjected to both dynamic tensile stresses from centrifugal force and variable stresses from bending forces arising from steam pressure drops and load variations. The important properties for blades are adequate mechanical strength to restrict deformation and maintain clearances at the operating temperature, good fatigue strength, good internal damping capacity (to reduce resonant frequency and vibrational effects) and corrosion resistance. The ferritic 12% Cr steels are standard materials for blading applications. For HP and IP blades operating at metal temperatures up to 480°C the basic 12% Cr alloy has given excellent service. For higher metal temperatures, 12% Cr steels with further alloying elements such as molybdenum, vanadium and niobium to enhance creep properties are selected. For LP blades which are subjected to larger centrifugal stresses because of their longer lengths, higher strength 12% Cr alloys with further alloying additions and a 14% Cr Ni Mo Nb Cu steel are generally used. For future designs, e.g., 900 MW units, there is an incentive to consider titanium alloy blading for the last row of

LP blades as an economic substitute for the well established 12% Cr alloys.

One of the most highly stressed components on large modern generators is the end-ring which retains the copper windings at each end of the generator. These rings are shrunk onto the generator rotor and can be stressed to about half their proof-strength, both when at rest and at operating speed. To reduce electrical losses they are fabricated from non-magnetic alloys, utilising high manganese steels such as 8% Mn 8% Ni 4% Cr up to the early 1970s and 18% Mn 4% Cr alloy on generators manufactured after that date. Both these alloys are strengthened by cold forming, but are susceptible to stress corrosion cracking if subjected to moist environments which can be encountered in the vicinity of the generator. In addition the hydrogen coolant used in modern generators can itself be a factor in promoting cracks.

To contain the risk of failure, rigorous and expensive inspection procedures have had to be adopted. However a more recently developed non-magnetic alloy steel containing 18% Mn 18% Cr and 0.4 to 0.7% N has been found to exhibit greatly improved resistance to stress corrosion (Fig 9.4) and hydrogen assisted cracking while retaining good toughness properties. Selection of this material removes the need for a significant programme of operational surveillance. The 18 Cr 18 Mn alloy will be used both on new generators

TABLE 9.5

Materials used in turbine-generators

	Typical Composition (Wt %)							Approximate maximum service temperature °C	Applications
	C	Mn	Cr	Mo	Ni	V	Others		
Carbon Steel (cast)	0.25	1.00	–	–	–	–	–	450	Steam valves, chests and turbine casings.
Carbon steel	0.40	1.00	–	–	–	–	–	450	LP rotors
½% Mo (cast) (obsolescent)	0.20	0.75	–	0.50	–	–	–	530	Steam valves, chests and turbine casings.
2¼% Cr 1% Mo (cast)	0.15	0.50	2.25	1.00	–	–	–	580	Steam valves, chests and turbine casings.
½% Cr ½% Mo ¼% V (cast)	0.15	0.50	0.50	0.50	–	0.25	–	580	Steam valves, chests and turbine casings.
1% Cr ½% Mo ¼% V (cast)	0.15	0.60	1.00	1.00	–	0.25	–	580	Steam chests and turbine casings.
1% Cr 1% Mo ¼% V	0.25	0.50	1.00	1.00	–	0.25	–	580	HP, IP rotors
3% Cr 1% Mo	0.20	1.00	3.00	1.00	–	–	–	580	HP, IP and generator rotors
3½% Ni, Cr Mo V	0.25	0.50	0.30	0.50	3.50	0.10			LP and generator rotors
2½% Ni Cr Mo	0.20	0.50	0.80	0.75	2.30	0.30			LP and generator rotors
1% Cr ½% Mo	0.40	0.60	1.25	0.60	–	–	–	480	Turbine bolts
1% Cr ¾% Mo ¼% V	0.40	0.60	1.00	0.75	–	0.25	–	510	Turbine bolts
½% Cr ¼% V	0.20	0.60	–	0.50	–	0.30	–	540	Turbine bolts
1% Cr 1% Mo V TiB	0.20	0.60	1.10	1.00	–	0.70	0.1 Ti + B	565	Turbine bolts
12% Cr	0.08	0.25	13.0	0.5	–	–	–	480	HP, LP blades
12% Cr Mo	0.10	0.3	12.5	0.75	–	–	–	550	HP, IP blades
12% Cr Mo V	0.10	0.60	12.0	0.60	0.80	0.20	–	565	HP, IP blades
12% Cr Mo V Nb	0.13	1.00	11.2	0.60	0.80	0.30	0.3 Nb	565	LP blades and bolts
12% Cr Ni Mo V Nb	0.15	0.80	11.0	0.60	1.00	0.30	0.4 Nb	565	LP blades
14% Cr Ni Mo Nb Cu	0.05	0.80	14.3	1.60	5.50	–	0.35 Nb 1.8%Cu	565	LP blades
8% Mn NI – Cr	0.50	8.0	4.0	–	8.0	–	–		Generator end rings
18% Cr 18% Mn	0.10	18.5	18.5	–	–	–	0.55 N		Generator end rings
AISI 316 (cast)	0.08	1.5	18.0	2.5	10.0	–	–	650	Steam chests
AISI 347 (cast)	0.08	1.5	18.0	–	11.0	–	0.5 Nb	650	Steam chests
ESSHETE 1250	0.10	6.0	15.0	1.0	10.5	0.25	1 Nb + B	650	Bolts
16% Cr Ni Mo Nb	0.08	1.0	11.5	1.4	16.0	–	1 Nb	650	Bolts
Nimonic 80A	0.06	0.30	20.0	–	Balance	–	2.4 Ti 1.4 AL	650	Bolts

and wherever replacement end-rings are required on existing generators.

Turbine cylinders and many valves are normally split horizontally for purposes of assembly and disassembly. The upper and lower halves of the assembly are separated by flanged joints held together by bolts. The joint tightness is maintained by tensile forces in the bolts and compressive stresses at the flange faces. The main consideration in design is the maintenance of steam tightness of the joint by means of

TABLE 9.6

*Reduction in FATTs for
3.5% NiCrMoV rotor steels*

Sampling period	Fracture appearance transition temperature °C (FATT)	
	Average	Range
1970–1975	+25	0 to +50
1976–1980	+16	−18 to +40
1981–1985	+10	−4 to +20

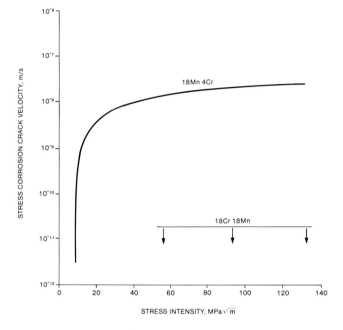

FIG. 9.4 Comparative stress corrosion resistance of end-ring materials.

the clamping force of the bolting without over-stressing the bolts. If the bolts operate at temperatures at which creep can occur the stress in the bolts is relaxed, together with the clamping force, over a period of time. In consequence, the bolts will require retightening after a period of time. The important factors in achieving safe and reliable performance of bolts are to select the correct material for the application, to lubricate with appropriate anti-seize compounds for ease of dismantling, to apply an accurate control of tightening and to replace before life expiry. For operating temperatures below about 370°C carbon steel bolts are satisfactory. However for higher operating metal temperatures two main types of bolting alloys are in common use; low alloy ferritic steels containing chromium and molybdenum, vanadium and Nimonic 80A — a nickel based alloy originally developed for gas turbine blades. A small amount of austenitic stainless steel is used for bolting but only in conjunction with austenitic components so as to retain matched thermal expansions. The characteristics of the low alloy steel bolts are quite

different from the Nimonic bolts. The low alloy steel bolts relax their clamping forces by creep and will require retightening, usually after about 30 000 hours operation. The Nimonic bolts have the unusual characteristic of exhibiting a time dependent volume contraction at operating temperatures. This results in an increase in bolt stress with time.

1.4 Piping and valves

Steam is conveyed between the boiler and turbine through steam pipes and valves. For modern conventional plant these components normally operate at temperatures of 540°C (oil-fired units) to 568°C (coal-fired units) and pressures of 169 bar (main steam) and 41 bar (hot reheat). With supercritical plant, operating conditions are up to 600°C with superheat pressures of 250 bar and reheat pressures of 50 bar. Other connecting pipework, e.g., cold reheat pipes which carry steam back from the turbine to the boiler to be reheated, operate at about 250°C.

The requirements for steam pipes are that they should carry the large volumes of steam with a limited drop in pressure along their length. They therefore have relatively large internal diameters, typically about 240 mm and 440 mm for main steam and hot reheat lines respectively. To achieve design lives in excess of 100 000 hours operation in the creep range, the wall thicknesses are typically 65 mm for main steam lines and 28 mm for hot reheat lines. Cold reheat pipework, which operates at lower temperatures, is designed on proof stress rather than rupture stress criterion.

The materials adopted for pipework have been included in Table 9.1. For older, lower outlet temperature and pressure plant, carbon-manganese steels $\frac{1}{2}$% molybdenum (now obsolescent) and 1% Cr $\frac{1}{2}$% Mo have been used, while in modern conventional plant the choice has been either $\frac{1}{2}$% Cr $\frac{1}{2}$% Mo $\frac{1}{4}$% V or $2\frac{1}{4}$% Cr 1% Mo steels. The higher design stress of $\frac{1}{2}$% Cr $\frac{1}{2}$% Mo $\frac{1}{4}$% V steel compared with $2\frac{1}{4}$% Cr 1% Mo steel and the higher cost of the austenitic steels makes the $\frac{1}{2}$% Cr $\frac{1}{2}$% Mo $\frac{1}{4}$% V steel the preferred material. Austenitic stainless steels Type 316 and Esshete 1250 have been installed in supercritical units and occasionally Type 347 has been used in conventional plant.

Other components associated with pipework, e.g., valves, are usually fabricated in the same, or some compatible, material as the pipe.

For plant conditions beyond 169 bar/580°C, e.g., 180 bar/580°C or 240 bar/580°C in possible 900 MW units there is an interest in using high alloy 12% Cr Mo V or 9% Cr Mo V Nb ferritic steels.

1.5 Gas turbines

Gas turbines have a range of uses in power generation, both producing power on their own via a generator — a 'simple' cycle — and to obtain higher efficiencies in conjunction with a steam cycle and a second alternator — a 'combined' cycle. In the former case with a clean distillate fuel, gas turbines are valuable for peak-lopping and as station auxiliaries particularly for 'black start' capability [11]. In some countries by using natural gas or impure oils as fuel, longer term operation can be economic, but as the hours/years of operation increase it rapidly becomes preferable to utilise the gas turbine waste heat in a steam cycle and operate a combined cycle. Looking to the future, coal-fired combined cycles are under active development [12].

For industrial use there are two distinct types of turbine available. Aircraft engines can be adapted for use as 'gas generators' to drive a separate power turbine connected to the alternator. Such aero-derivative engines typically have powers of 15–30 MWe, but several can be used to drive a single power turbine. A turbine designed solely for ground operation is built much more robustly with the power turbine in the same casing as the rest of the machine; indeed often a single shaft is used through the whole engine from compressor to alternator. Unit powers up to 150 MWe are available. The heavier construction means start-up times are considerably longer than for aero machines which can come up to full power from cold in 3 minutes. Reliability and time-between-overhauls should be better for the industrial machines. On the other hand, with aero-derivatives, if spare units are available a faulty unit can be replaced in a few hours. In Britain the vast majority of gas turbines used have been aero-derivatives, as exemplified by the Rolls Royce Avon (Fig 9.5).

Figure 9.5 identifies the basic components of a gas turbine: air is drawn into a compressor, generally radial, and compressed to typically 10–30 bar. The air is then used to burn the fuel in combustion chambers ('cans') designed to give an outlet gas temperature compatible with the rest of the machine. Air-fuel ratios are typi-

cally 50 : 1 by weight. The hot pressurised gases are then expanded through a series of turbine stages, which drive the compressor and alternator. About two-thirds of the power is used to drive the compressor.

Materials selection is an important aspect of the design of a turbine, particularly the compressor blades, combustion chambers, turbine blades and discs [13]. The aero engine requirements have generally led the developments; the industrial application frequently utilising the materials and other technological advances after they have been proven in aircraft use. The following paragraphs describe the materials requirements and properties for the above components and consider future development possibilities. The composition of some typical alloys for each application are listed in Table 9.7.

1.5.1 Compressor materials

Selection parameters include high levels of tensile and fatigue strength, with high Young's modulus and low density to avoid resonances. Compression of the gas increases its temperature and it is worth noting that the air temperature (approximately 600°C) entering the combustion chambers in a modern engine is similar to the combustor outlet temperatures in turbines of early design. This increasing temperature requirement has promoted the development of titanium alloys for the lower pressure stages of the turbine to replace aluminium alloys and of 12% Cr steels. In the final stages of a modern compressor, nickel base alloys are specified such as Inco 718 or Nimonic 90.

1.5.2 Combustion chambers

Combustion chambers are designed to introduce the secondary air in a controlled manner into the primary air-fuel flame. They comprise a cylindrical sheet metal construction containing a complex array of slots. The secondary air naturally cools the metal but as modern

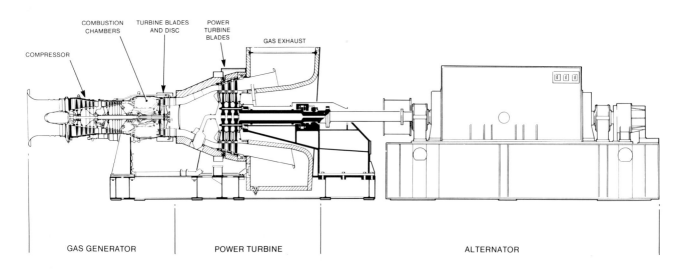

FIG. 9.5 A 30 MWe aero-derived gas turbine power generator set (Rolls Royce)

TABLE 9.7

Composition of typical gas turbine alloys

	Typical Composition Weight %											Applications
	C	Ni	Cr	Co	Mo	Nb	V	Ti	Al	Fe	Others	
FV 535	0.07	0.3	11	6	0.7	0.3	0.25	–	–	Balance	–	Compressor Blading
IMI 829	–	–	–	–	0.3	1	–	Balance	5.5	–	3.5 Sn 3 Zr	
Hastelloy X	0.1	Balance	22	15	9	–	–	–	–	19	0.6 W	
Nimonic 86	0.05	Balance	25	–	10	–	–	–	–	–	0.03 Ce	Combustion Chambers
Inconel 617	0.07	Balance	22	13	9	–	–	–	1	–	–	
IN 901	0.04	41	13	–	6	–	–	3	0.2	Balance	0.014 B	
IN 718	0.04	Balance	19	–	3	5	–	1	0.6	20	0.003 B	
Waspalloy	0.08	Balance	19	13	4	–	–	3	1.3	–	0.006 B	Turbine Discs
Astroloy	0.06	Balance	15	17	5	–	–	3	4	–	0.03 B	
RENE 95	0.06	Balance	13	8	4	3	–	3	3.5	–	3.5 N: 0.01B	
Nimonic 115	0.15	Balance	14	13	3	4	–	4	5	–	0.016 B	
IN 100	0.18	Balance	10	15	3	5	–	4.7	5.5	–	0.014 B	
IN 738 LC	0.11	Balance	16	8	2	3.5	1	3.5	3.5	–	2.5W: 0.008B	Turbine Blading
IN 939	0.15	Balance	23	19	–	4	–	3.7	1.9	–	2W: 0.009B	
FSX 414	0.25	10	30	Balance	–	–	–	–	–	–	7W: 0.012B	

gas outlet temperatures are in the range 1000–1500°C the extent of cooling is limited. As a consequence, oxidation resistant nickel base alloys such as Hastelloy X, Nimonic 80 and Inconel 617 (Table 9.7) are used. In addition, ceramic insulating coatings are often adopted to protect the inner surface of the metal. Such 'thermal barrier' coatings include stabilised zirconia applied by flame or plasma spraying.

1.5.3 Turbine discs

The turbine discs are protected by design from the direct impingement of the combustion gases but thermal conduction causes significant heat transfer to the discs. Thus they operate at elevated temperatures, the upper limit being a function of the allowable degree of air cooling. Up to about 1960, discs were manufactured from steels but the increase in temperature and stress in later designs led to the use of nickel base alloys such as IN 901, IN 718 and Waspalloy (Table 9.7). Such materials were developed against requirements of creep or low frequency cycle fatigue and this led to major advances in thermo mechanical processing routes and techniques of non-destructive examination to achieve the property and integrity targets. These approaches enabled metal temperatures up to approximately 700°C to be utilised. More recently, the need for even stronger alloys for aero engines has led to the use of pre-alloyed powder metallurgy disc materials, exemplified by Astroloy and Rene 95 (Table 9.7).

1.5.4 Turbine blading

As with steam turbines, the turbine component in a gas turbine consists of alternate sets of stationary and rotating blades. The latter are subject to the higher stresses and the first row of the former to the highest temperatures. The evolving designs have mainly used nickel based alloys but some cobalt based alloys have been used as stator blades. The achievement of enhanced properties in nickel based alloys has involved solid solution and precipitate strengthening mechanisms and a transition from forged to directly cast blades. This strength increase has been paralleled for aero engine applications by a decrease in chromium content of the alloys which reduces their level of corrosion resistance. Since the fuels used in gas turbines contain contaminants that can cause accelerated corrosion, it has become necessary to apply corrosion resistant coatings. Principal contaminants are sulphur, from the fuel, sodium, from water in the fuel or from air generally as chlorides, and vanadium from the fuel.

Even with nominally distillate fuel, sufficient sodium can be present to give corrosion problems. The mechanism for such corrosive attack is the reaction of sodium chloride and sulphur oxides to form sodium sulphate. This deposits on the turbine blades leading to corrosion of the metal by sulphidation. In certain instances for power generation turbines this has resulted in operational lifetimes of gas turbine blading being less than 1000 hours [11]. Corrosion can be minimised by control of the sodium level in the fuel: a limit of 0.5 PPM is generally stipulated, corresponding to a maximum concentration in the combustion gas of only 10 PPB.

Most practical experience of protective corrosion resistant coatings has been by forming a nickel aluminide surface layer on the nickel alloy blade, by diffusion of aluminium into the alloy surface. Whilst more sophisticated coatings are now available such aluminides remain very cost effective. More recent techniques of coating application involve deposition of a desired metallic composition on the blade surface by vapour evaporation or low pressure plasma spraying. A typical example would be MCrAlY where M is cobalt and/or nickel.

The requirement for high strength nickel based alloys with higher levels of chromium (>15%) for industrial use with less-pure fuels has led to the development of alloys such as IN 738 and IN 939 (Table 9.7). They are, in general, still coated but their composition is such that with less-pure fuels corrosion is not catastrophic once the coating is breached.

The general thrust of development in gas turbines remains focused on aero engines in the first instance. The viability of their present high turbine inlet temperatures depends upon sophisticated blade air cooling systems and the use of directionally solidified, and, more recently, single crystal blades [14]. One can expect some of these benefits to be translated in due course to industrial gas turbines.

2 Component fabrication

The importance of metals in modern technology is due partly to the ease with which they may be fabricated into useful shapes by a variety of processes. In power plant a large and complex range of shapes with widely differing properties are required and many fabrication routes may be involved.

This section deals with conventional fabrication techniques and also discusses developments which have improved product quality.

2.1 Production methods

Materials have to be processed into a great variety of shapes in order to make component parts. The shapes required vary enormously, both in size and complexity, ranging from bolts to large castings and forgings of perhaps several hundred tonnes mass. The properties of the material in the finished component are also influenced to a considerable extent by the type of shaping process employed and the various processes have advantages and limitations. Since a breakdown, because of component failure, of a 500 or 660 MW unit involves loss of generation and high costs, there is a demand to minimise such failures. This has necessitated a commensurate improvement in the general properties and integrity of power plant components, with stringent inspection at all stages of manufacture. Overall the strategy that has to be adopted is one making due allowance for the sometimes competing requirements of economy and safety. Most metallic components are initially produced from the liquid phase and are then cast into shapes, either to give a casting which may only require machining operations to obtain the final desired shape, or into ingots which can be further processed by manipulative techniques such as rolling, forging or extrusion.

2.1.1 Casting

Probably the easiest route for achieving a complex-shaped component is to directly pour molten metal into a preformed mould. Although there are many casting techniques, sand castings find the widest application in power plant components such as turbine casings and valve bodies.

The initial manufacturing stages in the production of a casting comprises pattern making, moulding and coremaking. At this stage the design optimises the feeding and running systems in order to account for shrinkage arising from contraction on cooling and solidification and hence produce a casting with minimum unsoundness. This objective is achieved by strategic positioning of adequate feeder heads and the use of taper sections and chills to increase feeding distances. Moulding techniques should avoid cracking that may result from the resistance to deformation of the mould as the casting contracts on cooling. In addition care is taken to ensure that large masses of still liquid metal are not retained between rigid sections which have already solidified. After casting the component is stripped from the mould and the headers and runners removed. The castings are then heat treated, usually by annealing, normalising and tempering to provide the optimum properties. An early assessment of the quality of the casting is then carried out. This involves shot blasting and perhaps light machining with visual and ultrasonic examinations to determine dimensions and detect defects. With all castings there is a possibility of shrinkage porosity and other defects. Should any specific defects be considered unacceptable, rectification by welding may be undertaken. If the defects are associated with internal porosity very large excavations may be necessary to provide sound foundations for welding. After any repairs the casting is subsequently stress relieved. The casting is then subjected to final inspection by non-destructive testing.

With the introduction of more stringent inspection techniques and as size and weight requirements for large turbine castings have increased with increased unit size, certain methods of casting have become less suitable for production as a single process. Instead it has become necessary to produce the component from a number of individual castings, each of which is processed and upgraded to the required standard. The individual castings are then welded together to produce the final assembly. Similarly, it has also been found necessary to apply the same philosophy to smaller types of components which previously were generally produced as single pieces. In these cases, size and shape preclude access for inspection and upgrading to the more stringent quality standards demanded by current requirements mean it is necessary to produce a component as a number of pieces which are subsequently welded together.

Few, if any, castings are perfect and often contain solidification/contraction defects together with brittle constituents at both grain and dendritic boundaries. The inhomogeneities can result in poor mechanical properties of the metal.

2.1.2 Forging

However, if plastic forming operations are carried out the redistribution of microconstituents, refinement of grain size and introduction of strain hardening can significantly improve the mechanical properties. By compressive deformation it is often possible to fragment a brittle microconstituent such that the ductile matrix flows into the spaces between fragments and welds together to leave a sound structure. Once the brittle constituent is broken up, its effect on mechanical properties is minor and ductility and strength are improved. Forging, extrusion and rolling processes are normally used to break down the cast structures of ingots. These manipulative techniques involve both hot working (above the recrystallisation temperature) and cold working (below the recrystallisation temperature). Both methods have advantages and disadvantages. The most important advantage of hot working is that when the material is at elevated temperature its resistance to deformation is lower whereas its plasticity is higher. The lower strength permits larger deformations with the same power. Another advantage is that the elevated temperatures and the deformation favour diffusion and consequently breaking down of cast structures. Although cold working requires more power than hot working it is extensively used because it offers closer dimensional tolerances, better surface finishes, grain refinement and controlled properties. In cold working it is not necessary to correct thickness measurements for thermal expansion and oxidation effects are negligible. With cold working, lattice strains can markedly change the material properties and by controlling the amount of deformation a wide range of properties can be obtained.

An alternative technique for the manufacture and fabrication of complex-shaped turbine castings and valves is the closed die forging process. For massive components, large (30 000 tonne) multiple-ram forging presses are utilised. Such presses offer the combined functions of extrusion and closed die multiple-ram forging (Fig 9.6). The resultant products are uniformly worked with the desired grain flow distribution and require very little machining. By using the multiple-ram forging process the dies are completely closed before they are filled. Complex shapes with multiple cavities may be produced, with one heating cycle and one closing of the die system. In addition to producing closed die forgings the large presses can be used for vertical extrusions (Fig 9.7) producing tubes up to 960 mm inside diameter and up to 12 m in length, with wall thicknesses varying from 19 mm to 203 mm depending on diameter and weight. Such tubes find particular application for cold reheat lines where because of previous limitation on solid drawn tube sizes they were manufactured from longitudinally welded plate.

For many years there has been a continued growth in the size of turbine-generators and a consequence was the problem of procuring large forgings of increased integrity. Although single piece rotors were always preferred, restrictions on the size of forgings available resulted in alternative methods of construction. The three common methods of construction are built-up, monobloc and welded rotors (Fig 9.8). It should be noted that single piece forgings are now readily available and suitable for all present sizes of design.

A typical manufacturing route for monobloc rotors is listed in Table 9.8. The forging process primarily improves the soundness and structure of the ingot and also achieves the final desired shape. Since the worst of the coarse structure of forgings tend to be at the core, many forgings in the past had a centre hole bored through them. This operation was carried out to eliminate centre line casting shrinkage and defects, and facilitate inspection of the interior of the rotor. However, improvements in steel making and forging techniques now render this operation unnecessary.

Prior to the investment into the manufacturing capacity to increase the size of rotor forgings large rotors were produced by welding together a series of fully heat-treated forgings and stress relieving after welding. Although some compromise in material properties was required to facilitate weldable materials the relatively small size of the forgings made it possible to ensure good properties throughout. The rotor design was also such that the weldments were positioned in areas of low stress.

Prior to the welded rotor design, the problem of the lack of availability of single piece forgings was overcome by the construction of built-up rotors involving shrunk-on discs. Since the construction depends on shrink-fits, built-up rotors have been confined to LP rotors which operate at temperatures below the creep range. This sort of construction gives a relatively flexi-

TABLE 9.8

Stages in a typical manufacturer's process for monobloc rotors

Melting

Ingot making — vacuum degassing

Hot working

Preliminary heat treatment (conditioning)

Visual inspection

Rough machining

Ultrasonic test (private test)

Quality heat treatment — vertically

Rough machining

Ultrasonic test (private test)

Mechanical test (private test)

Boring of centre hole

Ultrasonic test, bore inspection and other non-destructive tests

Stress relief heat treatment — vertically

Final mechanical test and other metallurgical tests

Machining prior to heat stability test

Heat stability test

Final machining

Dimension measurement, final inspection and marking

Packing and shipping

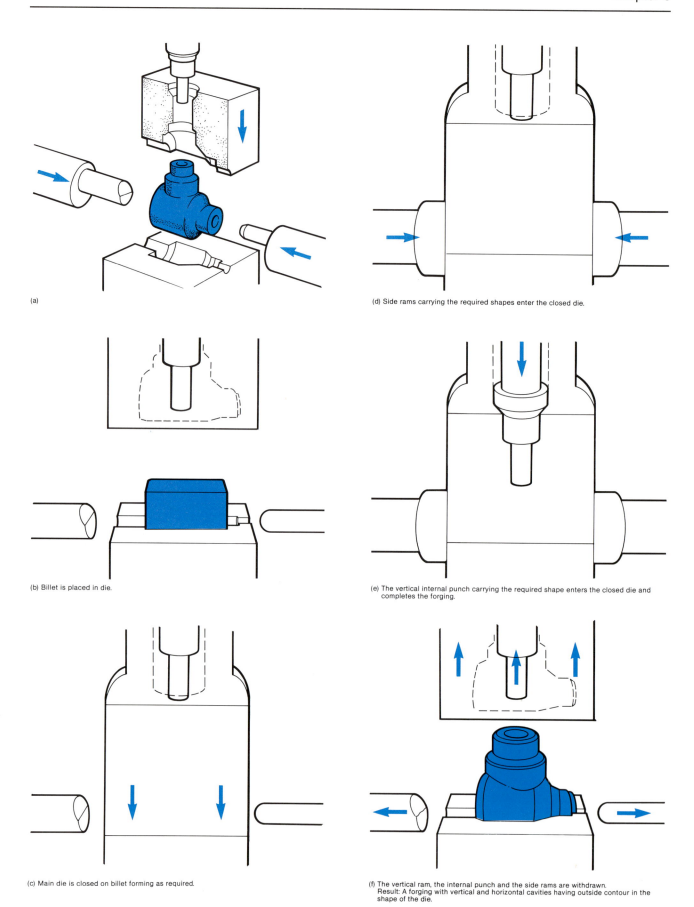

(a)

(b) Billet is placed in die.

(c) Main die is closed on billet forming as required.

(d) Side rams carrying the required shapes enter the closed die.

(e) The vertical internal punch carrying the required shape enters the closed die and completes the forging.

(f) The vertical ram, the internal punch and the side rams are withdrawn.
Result: A forging with vertical and horizontal cavities having outside contour in the shape of the die.

FIG. 9.6 Closed die forging process
Complex uniform forgings with multiple cavities are produced by activity from both vertical and side rams in conjunction with the shaped die.

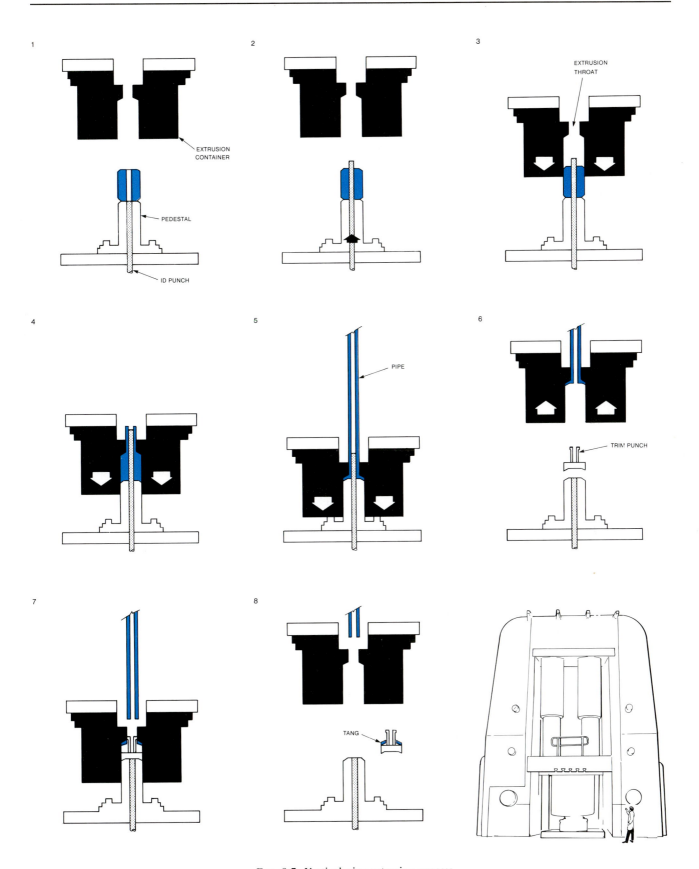

FIG. 9.7 Vertical pipe extrusion process
Initial ingot processing is carried out on a smaller press to produce a clean, hollow billet ready for final extrusion.
The deformation produced at this stage greatly improves the initial as-cast structure of the ingot.

453

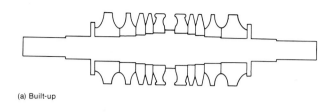

(a) Built-up

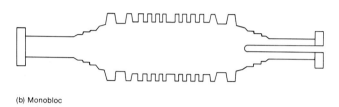

(b) Monobloc

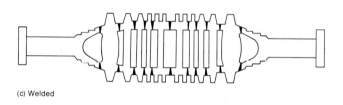

(c) Welded

FIG. 9.8 Methods of turbine rotor construction

ble rotor and hence any increase in disc weight requires a disproportionate increase in spindle diameter to attain the desired stiffness and hence a large bore in the disc. There is therefore a limit to the size for this type of construction at about 500 MW rotors.

Large turbine blades are manufactured by forging to produce a taper section with extra metal thickness where bosses and end fittings are required in the design. Finish forging is achieved by drop stamping or rolling in shaped rolls.

All sizes can be made as precision forgings but larger blades are usually machined all over by copy milling or, as with nozzle blocks, by electrochemical machining.

Another example of a component fabricated by forging techniques is the non-magnetic end-ring of generators. The end-ring is one of the most highly stressed components on a generator and is shrunk onto each end of the generator to retain the copper windings on the overhung portion at each end of the generator. For these components the steel ingot is pierced and forged into a ring approximately equal in length but smaller in diameter and larger in wall thickness than the finished component. After solution-treating the ring is expanded either by pushing a tapered mandrel through it at a temperature of 180°C or, alternatively, by explosive expansion. There may be several expan-

sions and intermediate heatings with a final stress relief heat treatment.

Early high pressure pipework systems were constructed by welding together sections machined from solid forgings. This process has also been used for some relatively small and thick walled headers but the metal wastage produced by boring and the limitation in length has made the process uneconomic for pipework. Most steampipes are therefore now made by piercing or extrusion.

2.1.3 Pipe bending

Pipe bending falls basically into two categories, namely cold bending and hot bending. The selection of the method of bending depends mainly on the material and size of the pipe related to the wall thickness and bend radius required.

For pipes from 12 mm to 300 mm outside diameter, bends can be produced by hydraulically-operated cold bending machines with a suitable mandrel and former. There are certain limitations on wall thickness related to pipe size since the cross-sectional area determines the amount of bending effort required.

For pipes with outside diameters up to 800 mm, hot bending is carried out. There is a limitation on the minimum bend radius, this being about three times the nominal bore but also depending on the outside diameter, wall thickness and material. The pipe is packed with sand, except where the outside diameter/ wall thickness ratio is such that the pipe wall will not distort when packed, then heated to the correct bending temperature. The pipe is then pegged at one end of a table while the other end is pulled by an electrically operated winch. A skilled bending operator, working to a curved template, controls the rate of bending and the contour by using stop pegs which are inserted as required into holes in the bending table during the pulling operation. When cool the pipe is emptied of sand, dressed and checked for dimensional accuracy.

As an alternative to the traditional flame bend process for pipework systems induction bending offers a high productivity method with better control over dimensional tolerances. The method involves pushing the pipe through an induction heating coil against the action of a pivot arm attached to the front end of the pipe. The length of the pivot arm determines the radius of the bend produced. The induction coil produces a narrow uniformly heated bend within which the deformation is concentrated. Water cooling from a ring of nozzles immediately beyond the coil minimises the width of the heated band.

For some steels, e.g., 12 Cr MoV, a waterspray cool is too fierce and may result in cracking. In these cases the pipe is air cooled, although this increases the width of the heated band and decreases the speed at which the pipe may be pushed through without impairing dimensional control.

Large thick-walled pressure vessels, e.g., boiler drums and reheater drums are generally fabricated by welding shaped rolled plate. For drums, the plate is bent into a cylindrical shape and joined with a single longitudinal weld usually made by electro-slag welding. The cylinders are then circumferentially welded together by submerged arc welding with the cylindrical welds being positioned to avoid cruciform welded joints.

2.1.4 Tube manufacture

Steel tubes are principally produced in two ways: either by piercing and elongating a solid billet to give seamless tube or by forming strip into tubular shape and welding the joint.

The manufacture of seamless tube takes many forms but is usually carried out as a continuous process (Fig 9.9). Typically a manufacturing route for carbon and low alloy steel tube would involve:

- Heating a billet to about 1250°C in a rotary hearth furnace; at high temperatures the steel is most amenable to extensive deformation.

- Converting the billet into a rough tube in a tapered roll piercing machine which forces the billet over a bullet-shaped plug.

- Elongating the pierced billet in a mandrel mill made up of pairs of rolls. Each pair of rolls forms a 'pass' which is partially oval. Each succeeding pass reduces in diameter so that the thickness of the workpiece is gradually reduced by being rolled down onto the internal mandrel bar. A final pair of rolls give the tube a circular cross-section. After extraction of the mandrel bar the ends of the tubes are trimmed to remove end scrap.

- Reheating the mandrel mill tube to 1000°C, descaling and then passing the tube through a stretch reducing mill to establish the final diameter and wall thickness. Stretch reducing involves passes of three grooved rolls which are machined to produce orifices which gradually reduce in diameter.

For seamless stainless steel, tube manufacture is generally by a direct extrusion process. Prior to extrusion a short round bar is pierced or bored to form a hollow billet. The billet is heated and inserted into the container of the hydraulically-operated extrusion press and a mandrel pushed through its axis to protrude on the far side. Elongation is achieved by squeezing the billet, at a temperature of about 1150°C, through the annular space formed by the die and the mandrel. Operations subsequent to the extrusion vary according to the type of product required, i.e., a hot finished tube or a cold finished tube. For hot finished tubes the tubes are cooled, either in air or by quenching, depending on the metallurgical factors involved. For instance, the non-transformable austenitic materials are water quenched, whereas the transformable martensitic grades are air cooled. Cold finished tubes offer better tolerances with cleaner and smoother surface finishes. Dependent upon the finished size requirements, the tubes may be cold reduced and subsequently reheat-treated.

A development from monobloc extruded tube is coextruded tube which finds application where a tube

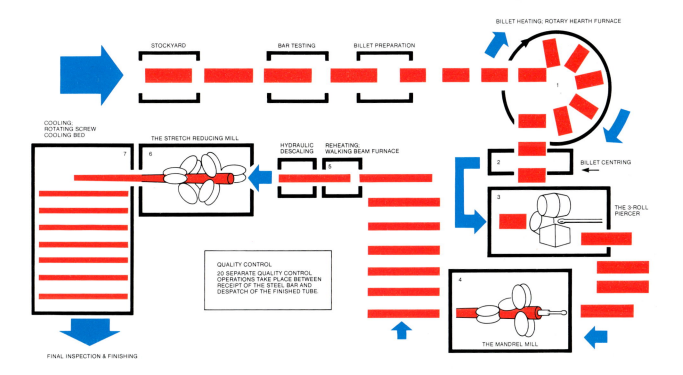

FIG. 9.9 Production route for seamless tube

is required to possess a combination of properties which are not readily obtainable in a single material. A sleeve of each of the component materials is machined to close tolerances from either a pierced billet or solid bar. The sleeves are then fitted together to form a composite billet (Fig 9.10). The billet is then heated to the extrusion temperature and processed by the nominal extrusion route which results in a metallurgical bond between the inner and outer materials.

An alternative production route for the manufacture of composite tube involves the use of a forged billet for the inner layer and a metal powder packed into a 'can' for the outer layer. The cans containing the centralised billet and powder are capped and vacuum sealed followed by sintering and hot extrusion and conventional cold reducing ky pilgering to final dimensions. The advantages of using powdered alloys include:

(a) The elimination of high machining costs associated with the preparation of outer layer cylinders from solid forged billets or castings.

(b) Flexibility of size range without stocking many billet sizes.

The powder metallurgy technique is used for the production of Incoclad tubes, e.g., 50% Ni 50% Cr alloys on the nickel based alloy 800. There are however, some additional process requirements primarily the need to remove, by grinding, the outer stainless steel can from the extruded shell.

Occasionally special tubular products may be incorporated into a boiler design. Two types of special tube, rifle bore and faceted tube (Fig 9.11), have been used to overcome problems associated with overheating/waterside corrosion in ash hopper slope tubes and fireside corrosion of furnace wall tubes.

For many applications such as superheater and reheater elements and sootblower parts, tubes will require further manipulation by bending. The choice of bending method is influenced by a number of factors including the tube material, the tube diameter/thickness ratio and the radius of the required bend. Whenever possible tubes are cold bent. Tubes with heavy walls generally present few problems, but as the diameter/thickness ratio of the tubes increases it becomes more difficult to produce bends free from buckling, particularly with close bending radii. The use of mandrel, or loaded bending techniques extends the limits of cold bending operations, but in some instances, particularly for low bend radii to tube outside diameter ratios, hot bending is specified. In any bending operation ovality and thinning should be controlled.

While most boiler tubes are seamless some boilers have mild steel or 1% Cr $\frac{1}{2}$% Mo steel electric resistance welded (ERW) tubes. These tubes are made by forming flat strip into tubular shape and welding the edges together in an electric resistance welding machine.

Such tubing offers smoother surfaces, more uniform wall thickness and less eccentricity.

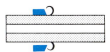

A SLEEVE OF EACH OF THE COMPONENT MATERIALS IS PREPARED FROM EITHER A PIERCED BILLET, SOLID BAR OR OUTER CAST CYLINDER.

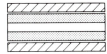

THE SLEEVES ARE MACHINED TO CLOSE TOLERANCES AND THE INNER FITTED TO THE OUTER TO FORM A COMPOSITE BILLET. IT IS IMPORTANT THAT THE GAP BETWEEN THE SLEEVES IS CONTROLLED WITHIN TIGHT LIMITS AND THAT NO CONTAMINATION EXISTS ON THE SURFACES THAT WILL SUBSEQUENTLY FORM THE METALLURGICAL BOND.

ALTERNATIVELY THE OUTER LAYER AS POWDER ALLOY IS ENCASED IN A WELDED STAINLESS STEEL CANISTER.

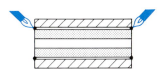

THE ENDS OF THE SLEEVES ARE WELDED TO PREVENT CONTAMINATION DURING SUBSEQUENT PROCESSING.

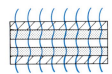

THE COMPOSITE BILLET IS HEATED TO EXTRUSION TEMPERATURE, CARE BEING TAKEN TO ALLOW FOR DIFFERENTIAL THERMAL EXPANSION. FOR POWDER ALLOY THE ASSEMBLY IS ALSO SINTERED.

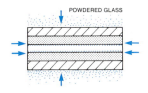

GLASS LUBRICANT IS APPLIED TO THE BILLET WHICH IS THEN TRANSFERRED TO THE 3000 TONNE EXTRUSION PRESS.

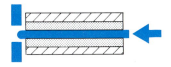

A MANDREL IS PUSHED THROUGH THE AXIS OF THE COMPOSITE BILLET AND ARRANGED CONCENTRICALLY WITH THE CIRCULAR DIE.

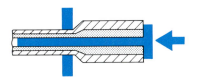

REDUCTION AND ELONGATION IS CARRIED OUT BY SQUEEZING THE BILLET THROUGH THE ANNULAR SPACE BETWEEN THE DIE AND THE MANDREL. THE RATIO OF MATERIAL DISTRIBUTION IS FIXED AND REMAINS UNCHANGED DURING FURTHER PROCESSING.

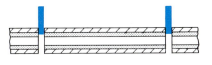

AFTER EXTRUSION THE CO-EXTRUDED HOLLOW IS COOLED, TRIMMED AT EACH END AND CLEANED.

DEPENDENT UPON THE FINISHED SIZE REQUIREMENTS OF THE CO-EXTRUDED TUBE, SUBSEQUENT WORKING OPERATIONS ARE CARRIED OUT BY COLD REDUCING.

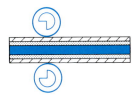

COLD REDUCING EMPLOYS SEMI-CIRCULAR, TAPER GROOVED DIES WHICH RECIPROCATE OVER THE TUBE AGAINST A MANDREL, THEREBY REDUCING THE WALL THICKNESS AND ELONGATING THE TUBE. THE TUBE IS SIMULTANEOUSLY FED INTO THE DIES AND TURNED ON ITS AXIS TO ENSURE EVEN DISTRIBUTION OF DEFORMATION.

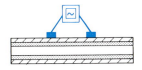

THE METALLURGICAL BOND IS CHECKED FOR INTEGRITY USING A SPECIALLY DEVELOPED ULTRASONIC TEST.

DESCALING
CO-EXTRUDED TUBES ARE DESCALED BY THE SODIUM HYDRIDE PROCESS. SODIUM HYDRIDE DOES NOT ATTACK THE PARENT METAL AND HENCE THE PROBLEM OF PREFERENTIAL ATTACK BY ACID PICKLING IS OVERCOME.

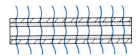

HEAT TREATMENT
CAREFUL TEMPERATURE SELECTION IS NEEDED WHEN HEAT TREATING CO-EXTRUDED TUBE. DEPENDENT ON THE MATERIALS UNDER CONSIDERATION, IT IS NECESSARY TO SELECT A HEAT TREATMENT TEMPERATURE THAT WILL GIVE THE REQUIRED PROPERTIES IN THE FINISHED TUBE.

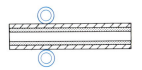

SURFACE FINISH
IN ADDITION TO DESCALING BY SODIUM HYDRIDE, ALL CO-EXTRUDED TUBES CAN BE POLISHED EXTERNALLY TO REMOVE ALL TRACES OF SCALE IF REQUIRED. POLISHING IS CARRIED OUT BY NORMAL POLISHING TECHNIQUES.
BORE FINISHING CAN ALSO BE PROVIDED USING A SPECIAL GRIT BLASTING TECHNIQUE.

MANIPULATION
THE CONDITION OF CO-EXTRUDED TUBE IS SUITABLE FOR MANIPULATION INTO SUPERHEATER AND REHEATER CONFIGURATIONS.

FIG. 9.10 Production route for co-extruded tube

457

FIG. 9.11 Examples of special tubular products

Rifle bore tube (shown on the right hand side of the photograph) was developed to increase turbulence and prevent overheating by steam blanketing in evaporator tubes. The tube is manufactured using a cold drawbench with tungsten carbide dies and spiral-grooved plugs, fabricated from hardened alloy-steel coated with hard chrome. The plug is allowed to rotate, as it forms the internal ribbing, by means of a thrust block inserted in line with plug anchor bar. After drawing, the tubes are normalised and rotary straightened in the conventional manner.

Faceted tube was originally fabricated by machining flats on opposite sides so that an increased wall thickness could be attained without reducing the tube bore in existing tangent-wall furnace areas. A thicker-walled tube was required to improve the life of tubes affected by fireside corrosion. Such tubes are produced by cold drawing through a special profiled carbide die and a standard carbide plug. After drawing, the tubes are normalised and straightened. The latter step departs from the normal procedure of rotary straightening due to the external geometry of the tube and a 7-roll single-plane straightening machine is used.

3   References

[1]   BS1113: Design and manufacture of water-tube steam generating plant (including superheaters, reheaters and steel tube economisers)

[2]   BS1501: Steels for fired and unfired pressure vessels: Plates

[3]   BS1502: Steels for fired and unfired pressure vessels: Sections and bars

[4]   BS1503: Steel forgings (including semi-finished forged products) for pressure purposes

[5]   BS1504: Steel castings for pressure purposes

[6]   BS3059: Steel boiler and superheater tubes

[7]   BS3601: Steel pipes and tubes for pressure purposes: carbon steel with specified room temperature properties

[8]   BS3602: Steel pipes and tubes for pressure purposes: carbon and carbon manganese steel with specified elevated temperature properties

[9]   BS3604: Steel pipes and tubes for pressure purposes: ferritic alloy steel with specified elevated temperature properties

[10]   BS3605: Seamless and welded austenitic steel pipes and tubes for pressure purposes

[11]   F. J. D. Tasker, C. W. Harris, N. S. Musgrave: 'Corrosion Experience in the CEGB Gas Generators' pp 385–408 in 'Deposition and Corrosion in Gas Turbines' ed A. B. Hart, A. J. B. Cutler (Applied Science, London): 1972

[12]   Davidson B. J., Meadowcroft D. B., Stringer J.: pp 219/244 in 'High Temperature Alloys for Gas Turbines and Other Applications' ed W. Betz et al (Reidel Holland): 1986

[13]   Meetham G. W. (ed): 'The Development of Gas Turbine Materials' (Applied Science, London): 1981

[14]   Meetham G. W.: (as ref [13]) pp 1–18

CHAPTER 10

Welding processes

1 Welding in power plant construction

The large number of welds required in power plant fabrication are a direct consequence of the size and complexity of efficient design. The Drax 2000 MW coal-fired station commissioned 1984/86 involved, in the boiler alone, 21 250 shop tube butt-welds and 13 350 site tube butt-welds. Experience has shown that welds are critical features in determining plant availability and reliability but there is only limited scope for rationalisation of design by reducing the total number of welds. However, there is a general emphasis on reducing site welding since more rigorous controls may be exercised over workshop fabrication. There is also an increasing trend towards the utilisation of automatic welding processes and implicit in this approach is that ways in which weld quality can be assured are identified and implemented by stringent process control.

Safeguards on weld quality via this route can come only from a thorough understanding of welding metallurgy, control of materials procedures and process vari-

ability and must be able to be implemented in a commercially feasible manner.

Modern power plant uses a range of materials from carbon steel for structural steelwork, through low alloy steels for boiler tubes, pipework and steam chests to austenitic stainless steels for superheater tubes and headers. Each of these materials poses different technical problems that have to be catered for in welding process selection and procedure development. The sizes of components encountered vary from, for example, approximately 150 mm thick steam drums to 2 mm attachment welds. In economic terms one of the most important factors in welding process selection is the volume of weld metal deposited. Labour costs and associated overheads account for the majority of total fabrication costs and so the incentives for developing high deposition rate, mechanised techniques is clear.

A technical/economic assessment of individual applications provides different solutions. For example in conventional plant, steam drum seam-welds are made using the electro-slag process while in the AGR, the

459

large number of boiler tube spacer attachments (approximately 8×10^6) utilises fully automatic robotic metal inert gas (MIG) welding. These examples represent the two ends of the spectrum, but by far the largest percentage of all power plant welds continue to be made by manual metal arc (MMA) welding as it has a sound record for flexibility, confirmable quality and cost effectiveness. However, commercial pressures are now fostering the increasing consideration of alternative processes with higher cost-benefits.

Future trends are difficult to predict because a great deal depends upon the level of industrial activity, but it is clear that while MMA welding will always be used to some extent its usage for carbon steel and low alloy steels will decline over the next decade. It is likely to be supplanted by the MIG process with the usage of flux-cored wires expanding faster than that of solid wire. The submerged arc process has probably reached saturation usage, while the tungsten inert gas (TIG) process has scope for increased utilisation with the development of orbital systems for tube welding and hot wire additions to increase deposition rates. Developments such as electron beam or laser welding will remain less attractive until the high capital investment they demand can be justified by high usage factors. Previously, technological innovation has often foundered because welding equipment has been unreliable but modern developments in solid state electronics have enabled very close control of the welding process to be obtained and the availability of robotic systems, seam sensing, feedback controls and on-line monitoring indicates immense potential in welding technology.

2 Weld quality and acceptance standards

There are many factors that have to be taken into account in deciding which welding process is the most suitable for a particular application. The manufacturer of power plant components requires flexibility of operation so that the technique may be applied as widely as possible; the designer of power plant requires a process that will deposit sound welds with acceptable mechanical properties.

Traditionally, both of these requirements have been adequately served by manual metal arc (MMA) welding. However, welding can contribute significantly to overall fabrication costs which both supplier and utility seek to reduce. Consequently there are clear economic incentives to replace manual welding techniques by higher deposition rate, mechanised or automatic equipment. Since the agreed standards for welding procedure acceptance are based on manual techniques this does present a problem in deciding the validation work that needs to be carried out before these more advanced techniques can be approved for future use. The economic penalties arising from unplanned plant outages impose a justifiably conservative approach on the

part of the utility towards novel developments. However, this does not imply a steadfast resistance to technological advances, rather an over-riding concern that innovation is properly researched and satisfactorily implemented.

The types of welding processes available for power plant construction fall into three categories:

(a) Well-developed processes that have an established history of plant application, e.g., MMA, manual TIG, submerged arc (SA).

(b) Modifications to these processes to increase efficiency and reduce costs such as narrow gap welding, the use of robotics and mechanical TIG.

(c) Processes or techniques that are technological innovations *per se* or have never before been used for power plant construction, e.g., pulsed MIG, plasma and hot wire TIG.

The current approval route for process or procedure approval is based upon the manufacturer carrying out a welding procedure test to demonstrate that the procedure is technically correct and that sound welds can be produced to a stringent acceptance standard.

Once this has been done successfully then, provided no further changes to essential variables in the welding procedure are made, the procedure test results are valid for an indefinite period.

A separate welder qualification test is also carried out to demonstrate that the welder is competent to carry out the procedure. These are generally specific contractual requirements and also necessary for statutory approval by inspection or insurance authorities. Thus, approval of procedures using processes in category (a) should be straightforward since probably the best proof of acceptability is the continued operation of plant manufactured according to a specific procedure. Therefore with such confidence in the suitability of a procedure based on category (a) it is only necessary to be satisfied that this will be maintained for the application under consideration. Thus, in fact, a new procedure test may not need to be carried out. Current UK and USA design codes specify BS4870 and ASMEIX respectively which define the requirements for welding procedure approval.

Categories (b) and (c) demand a different approach. Often extensive pre-production development work is required to devise satisfactory welding procedures and the emphasis in assessing and approving these procedures must be on ensuring that this has been adequately carried out. The difference between category (b) and (c) techniques, may only be in the type and extent of additional development work required to demonstrate satisfactory weld quality and properties. The customary standards for approval, BS4870 and ASME IX, are based on establishing that a particular process used in a particular procedure can produce a defect-free weld in which the room temperature tensile properties of weld metal and heat affected zone are at least as good as those of the parent metal. Some provision is made for toughness testing should this be part of the design requirement, but neither standard covers the testing of time dependent properties for high temperature applications. However, both standards are concerned essentially with manual welding while modern trends are towards automated techniques often with sophisticated control of welding parameters. This ostensibly introduces a large number of essential control variables each of which, it could be argued, require separate procedure testing if changed for any reason. The costs of such tests would be considerable and in real terms would not significantly increase the probability of guaranteeing a high level of weld quality in production. Since automated techniques can have a very high rate of weld production it is essential to ensure that the control systems designed to replace the skilled welder are capable of the same high standard of operation. If such a high standard can be attained then far higher levels of consistency of operation are possible. In order to achieve this, the variability in weld quality in the context of the specific design requirements must be examined systematically in the pre-production development work. In addition, sufficient in-line production monitoring of weld parameter variation, weld quality and sample testing, should be carried out as appropriate. Both of these aspects require the development of new techniques and methods of quantitative assessment, some of which are discussed in later sections. In summary, the approach to evaluating new techniques, whilst not losing sight of traditional methods for demonstrating the acceptability of a welding process, should place more emphasis on quantitative assessments of weld quality and variability. At the same time new technical acceptance criteria must be agreed which are not so restrictive as to frustrate technological innovation by nullifying the commercial incentives for change. In this way new manufacturing standards for automatic and mechanised processes can be developed.

3 Metallurgical effects of welding

Fusion-welded processes are by far the most important processes used in the fabrication of modern power plant. Components are joined by forming a pool of molten metal. For thin section components this may be comprised entirely from the parent material, i.e., autogenous welding. Thicker components require the addition of a filler wire into the welding arc to fill the weld preparation which has to be wide enough to accommodate the process in question. In either case peak temperatures in excess of the melting point are attained followed by very rapid cooling of the weld metal, which is effectively a miniature casting. The parent metal adjacent to the weld is also heated to somewhat lower temperatures and cooled very rapidly, giving rise to a heat affected zone (HAZ). The rate

of heating of this region depends upon the speed at which the welding arc travels and the amount of heat supplied. The rate of cooling depends upon the thermal conductivity of the material and the geometry of the component to be welded since heat is removed by conduction into the surrounding metal.

The welding arc can be considered theoretically as a moving point source of heat and the thermal cycles emanating from it, as welding progresses, can be described by classical heat flow equations. Consequently the heat flow characteristics associated with welding are, for most general applications, quite well understood and representative assessments of cooling rates after welding can be defined. From a metallurgical standpoint, cooling rate is particularly important since for a given material it determines the as-welded microstructure, hardness, residual stress level and hence the propensity to cracking.

For example, if data on the continuous cooling transformation (CCT) behaviour of a ferritic steel and the cooling rate under specific welding conditions can be combined then it is possible to predict the metallurgical condition of heat affected zones.

The choice of welding process also plays an important role in determining weld metal and HAZ structures and properties. Process selection for a given application depends upon a number of factors such as deposition rate, joint size and welding position. Usually the prevailing criteria are based on reducing economic penalties but in critical applications with materials of high 'hardenability', weld quality must also be guaranteed with a high degree of reliability.

The technique that provides the balance between economics and quality for the widest range of applications is manual metal arc (MMA) as it is a versatile all-positional technique suitable for a wide range of materials. Component thickness is generally the determining factor governing its use. At low thicknesses the heat input associated with MMA welding is too high and weld pool control too difficult to ensure satisfactory weld quality. Therefore it is supplanted by manual or mechanised tungsten inert gas (TIG) welding, which is a technique that provides such good weld pool control that it is consistently used in thicker joints for root passes. At very high thicknesses MMA welding becomes uneconomic and is replaced by or used in conjunction with submerged arc (SA) welding. These three processes are the foundation for the majority of the components fabricated for modern power plant. Electro-slag, flash butt and metal inert gas (MIG) welding are also used but to a much lesser extent.

However, the main feature that all welding processes have in common is that they create a discontinuity in the metallurgical structure and mechanical properties of the parent material. The metal adjacent to the molten pool is heated to a temperature which depends upon the distance from the fusion boundary. Melting and solidification occur rapidly during welding and as the thermal conductivity of metals is high the

461

heat affected zone extends over only a very small distance, typically a few millimetres.

In ferritic steels which are transformable on heating, all areas that become austenitic during the heating part of the cycle transform on cooling to give a range of structures which can be martensitic, bainitic or ferritic depending upon the alloy content and hardenability of the material. Those regions experiencing temperatures below the lower critical temperature for transformation to austenite show little visible microstructural change.

In austenitic stainless steels where there is no solid state transformation the weld metal is characterised by a coarse columnar grain solidification structure with heat affected zones exhibiting a gradation in grain size from the fusion boundary outwards. These features are shown in Fig 10.1.

The other main legacy conferred by the act of welding is to introduce high levels of residual stress into the weldment. These arise due to local straining which occurs because the metal that is heated on welding is inhibited from contraction during cooling. This produces tensile stresses in the vicinity of the weld. These stresses depend upon the component geometry but usually approach the material room temperature yield point. They are important because, if the local strain experienced during cooling exceeds the weld metal or HAZ ductility, then certain types of cracking occurs, e.g., stress relief cracking. Alternatively, the stresses can relax in service at high temperatures and may cause creep damage or cracking in unfavourable microstructures. Residual stresses are reduced or eliminated by application of a post-weld heat treatment which depends upon material thickness and material metallurgical characteristics.

4   Potential welding defects

The production of high quality welds with a high degree of consistency is readily achievable otherwise the economic manufacture of modern power plant with large numbers of welds would not be feasible. However, the occurrence of defects is more likely to be associated with welds than with wrought material. In addition, when welding defects do occur, they are often not isolated incidents and can cause considerable problems in maintaining construction programmes and in guaranteeing plant availability. Therefore it is necessary to devote effort to diagnosing the causes of the defects and defining remedial action.

4.1   Process generated welding defects

Welding defects can arise as a result of poor welding technique and process control. Within this category fall defects such as unacceptable weld profile, e.g., reinforcement or undercut, porosity and lack of inter-run or sidewall fusion. Factors responsible for these are, for example, inadequate welder training, incorrect cover gas or poor cleaning of weld preparations and insufficient welding current. The tendency for such defects to occur should be detected initially by standard welder or welding procedure testing which incorporates visual, NDT and bend tests. Remedial action lies in the revision of welding procedure details. These are important types of defect and a high standard of quality assurance is needed during fabrication to avoid unacceptably high repair rates. However, without underestimating the potential problems they can cause, it is intended in this section to concentrate upon the other source of welding defects namely those which are metallurgical in origin. Typical process defects are shown in Fig 10.2.

4.2   Metallurgical welding defects

4.2.1   Solidification cracking

During solidification, grains are nucleated at the fusion boundary and grow towards the centre of the weld pool. As this proceeds, the remaining liquid becomes enriched by a process of solute rejection and impurity elements are concentrated in the last liquid to solidify, which has a lower freezing point than the bulk composition. If the weld is subjected to tensile straining during the period when liquid films are present then solidification cracking may occur. This is a high temperature form of cracking and is typified by an intergranular crack morphology (see Fig 10.3). Cracks can form between individual solidifying dendrites and take the form of small microfissures. Alternatively, cracking can occur down the centre of the weld between the two opposing freezing fronts. This results in 'centreline cracking' which occurs longitudinal or parallel to the welding direction and can occupy a significant proportion of the weld length.

Solidification cracking is more likely to occur under conditions of high restraint and particularly if weld metal composition is incorrect. Thus nozzle welds and fillet welds are likely to be more susceptible, especially where fit-up is inadequately controlled. Weld pool geometry is also important and deeply penetrating narrow configurations are prone to centreline shrinkage cracks.

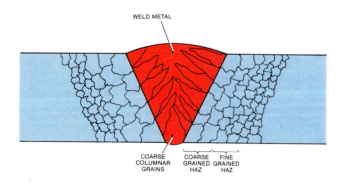

WELD METAL

COARSE COLUMNAR GRAINS   COARSE GRAINED HAZ   FINE GRAINED HAZ

FIG. 10.1   Weld metal and heat affected zone (HAZ) structures

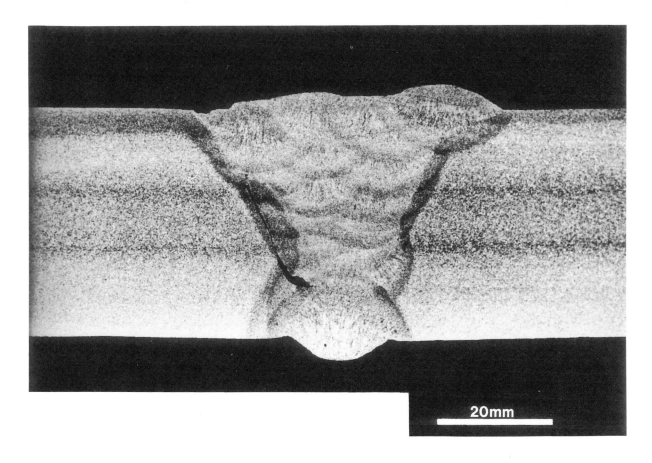

FIG. 10.2 Lack of fusion defect (The Welding Institute UK)

As the presence of impurities contributes significantly to the formation of low melting point films, it is important to restrict trace elements that contribute to them in both filler metal and parent metal. This should ensure freedom from cracking when autogenous welding or in applications where high dilution is possible.

Solidification cracking occurs in ferritic steels but fully austenitic welds, particularly Ni base alloys, are probably even more susceptible. In these materials compositional control is particularly important, although in austenitic stainless steels the problem is readily overcome by introducing 3–8% δ ferrite into the weld deposit. The δ ferrite has a higher solubility for sulphur and phosphorus and also dilutes the effect of these impurities by increasing the grain boundary area in the weld.

4.2.2 Liquation cracking

This form of cracking is also related to the appearance of low melting point films and can occur either in the weld metal or the heat affected zone (HAZ). It is caused by the reheating effect when one weld bead is deposited on top of another or on to the sidewall. It can be related to the grain boundary segregation of sulphur and phosphorus or sometimes to the presence of low melting point eutectics such as boro-

carbides. It is exacerbated by the use of a high heat input during welding, due to the longer time spent at high temperatures causing increased grain growth and also increased strain on cooling.

4.2.3 Ductility dip cracking

This form of cracking occurs at intermediate temperatures in the range 700–1200°C and is usually due to the formation of precipitates such as carbides which increase strength but reduce ductility. However, it can also be attributed to segregation effects causing grain boundary weakness.

4.2.4 Stress relief cracking

Upon completion of welding, cooling to ambient temperature introduces high levels of residual stress into the weldment that can approach yield point in magnitude in heavy section components. Residual stresses arise from the thermal contraction which occurs as the temperature decreases. In many applications it is desirable to eliminate residual stresses in order to reduce the risk of brittle fracture or stress corrosion cracking. This is achieved by application of a post-weld heat treatment in a temperature range where there is sufficient reduction in yield stress to allow residual stresses to be relieved by plastic deformation. During

463

FIG. 10.3 Centreline solidification cracking (The Welding Institute UK)

this process, strains are imposed upon the weldment which must possess adequate ductility to avoid cracking. Generally in low carbon steels there is very little risk of stress-relief cracking occurring, but this risk is considerably increased as material strength increases and certain creep resistant steels, notably the ferritic CrMoV steel and the austenitic type 347 steel, are susceptible to cracking in thick sections. This is primarily because they derive their properties by a precipitation hardening mechanism that increases strength at the expense of ductility. Thus, if the strain imposed during stress relief exceeds the ductility of either weld metal or HAZ, then cracking occurs (see Fig 10.4). Cracking does not arise solely during post-weld heat treatment. In-service cracking can be encountered during operation in the creep range if the post-weld heat treatment is inadequate or has been omitted.

A number of factors contribute to the appearance of cracks and these are briefly outlined below for both of these steels.

Chromium-Moly-Vanadium (CrMoV) steel

Welds in this steel are fabricated using 2Cr/Mo weld metal and in practice two types of stress relief or reheat cracking have been observed. Firstly circumferential cracking confined to the coarse grained region of

the HAZ. Secondly transverse weld metal cracks normal to the welding direction and occasionally extending into the HAZ.

Microstructure

The region of the HAZ adjacent to the fusion boundary experiences the highest peak temperature. At this temperature, thermally activated grain growth can occur which gives rise to a large austenite grain size that transforms on cooling to bainite. Coarse grained bainitic structures have very low ductility and crack during stress relief by creep cavitation of the prior austenite grain boundaries. However, the problem can be overcome by application of a suitable welding technique to restrict grain growth.

Transverse weld metal cracking is usually associated with coarse solidification structures in multi-pass welds. These arise from the nucleation of dendrite crystals which grow down the thermal gradients towards the weld bead centre. Each dendrite is initially a single crystal of austenite and the interdendritic boundaries are regions of high segregation due to solute rejection on cooling. This promotes cavitation and brittle boundary fracture which is invariably found to initiate on those boundaries untransformed by subsequent weld passes. The extent of these boundaries is a function

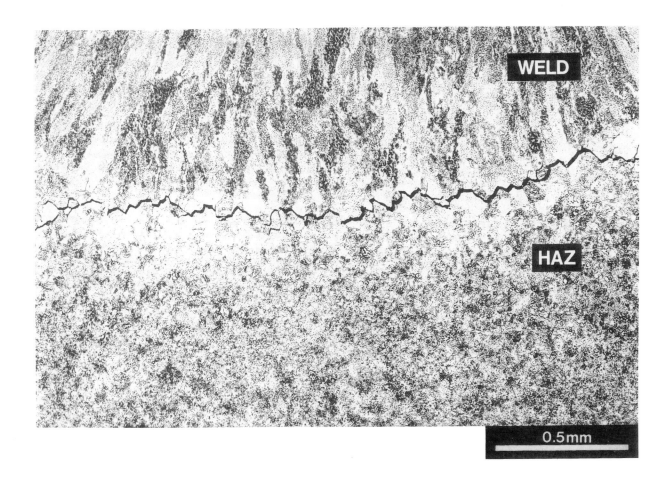

FIG. 10.4 Stress relief cracking (The Welding Institute UK)

of electrode gauge size and techniques such as TIG, which operate at a high heat input per unit volume of metal deposited, and produce large re-heated zones. Hence TIG welds are less susceptible to transverse weld metal cracking than, for example, submerged arc welds which tend to exhibit coarse untransformed weld metal microstructures.

Composition

Certain residual elements play an important part in HAZ stress relief cracking since they segregate to the prior austenite grain boundaries and promote cavity nucleation. Therefore tight specification of CrMoV parent material is required for elements such as Tin, Copper, Arsenic, Antimony, Sulphur and Phosphorus.

Vanadium content also needs to be limited (to ≯0.27% max) since although it improves the strength of the bainitic matrix, the HAZ grain boundaries are thereby relatively weakened and unable to sustain the strains imposed upon them during stress relaxation.

For transverse 2Cr/Mo weld metal cracking the element that has the most detrimental effect is copper and this should be restricted to 0.1–0.15 wt% in the weld deposit. Thus for submerged arc welds, uncoated consumable wire should be used with controls applied to packaging and storage to ensure that contamination does not occur.

Residual stress

Stress relief cracking is a function of the residual level and problems are generally more severe in welds in thick section components with unfavourable geometries compounded with service stresses. Inadequate stress relief also contributes to the incidence of in-service cracking since creep damage is introduced during post-weld heat treatment which may propagate during operation.

This has been particularly relevant for transverse weld metal cracking of submerged arc deposits which exhibit low ductilities at temperatures below the standard post-weld heat treatment temperature (690°C) and are unable to accommodate the strain required to relax yield point stresses.

The occurrence of stress relief cracking requires an unfavourable combination of microstructure, composition and stress. It can be avoided by controlling material specification and heat input during welding. However, there are instances, particularly in the repair

465

of old plant, where materials outside specification have to be welded. In such cases successful welds can be made using a two-layer HAZ refinement technique. This is dealt with more fully in Section 6.3 of this chapter.

Type 347 steel

Type 347 is a Niobium stabilised austenitic steel where Nb is present in order to prevent weld decay by forming NbC in preference to $Cr_{23}C_6$ carbides. However, NbC is a very strong precipitation hardening agent which considerably increases strength but decreases ductility. Stress relief in the temperature range 800–850°C reduces residual stresses but also promotes the formation of NbC on dislocation networks generated by the restraint imposed during welding. Thus the material grains are strengthened at the expense of the grain boundaries which are unable to withstand the off-loading effects occurring during stress relaxation. Solution treatment at 1050°C dissolves the NbC precipitates and avoids the problem. Type 316, which is a Molybdenum-bearing austenitic steel that gains its strength by solid solution hardening, is not susceptible to stress relief cracking.

4.2.5 Hydrogen cracking

Hydrogen cracking is a low temperature form of cracking that occurs in welds and heat affected zones in hardenable ferritic steels (see Fig 10.5). It occurs at temperatures below about 200°C since at temperatures greater than this hydrogen can diffuse rapidly away to reduce concentrations to safe levels away from the weld. Hydrogen is introduced into the molten weld pool either as an inherent consequence of the welding process selected, or by contamination caused by inadequate removal of impurities which decompose to produce hydrogen. The molten pool solidifies initially as austenite which has a relatively high solubility for hydrogen. Transformation of the austenite reduces the concentration of hydrogen in the matrix and the excess is precipitated as a gas in the voids and discontinuities present in the metal. This provides a source of hydrogen that can initiate cracking *per se*, or diffuse to pre-existing crack initiating sites either by thermal activation or under the influence of a stress gradient. Hydrogen lowers the energy for continued crack growth and consequently cracking occurs under conditions of static loading. The time dependence of the process is such that cracking may not occur immediately after welding, but may be delayed for up to a matter of days until a critical local hydrogen concentration is attained. Thus it is prudent to inspect for these defects some time after welding is complete and to use a combination of volumetric and surface inspection techniques.

Heat affected zone hydrogen cracks are usually associated with stress concentrating features present in the weld. Cracking often occurs at the weld toes or root and may run for some distance along the weld

seam if they are longitudinally orientated. Underbead heat affected zone cracks have also been observed but these are not as obviously related to stress concentrating features. Weld metal defects may be transverse to the welding direction, although longitudinal cracks can initiate from a region of stress concentration such as an unfused land. In general there is no characteristic identifying morphology for hydrogen cracks as they have been found to be intergranular, transgranular or a mixture of both depending upon the type of microstructure in which they appear.

Cracking in austenitic weld metals or heat affected zones is unknown due to the much higher solubility and much lower diffusivity of hydrogen in austenite. As a result, austenitic filler metals are sometimes used in repair situations where considerable difficulties are presented by the application of more conventional remedial measures.

The factors responsible for hydrogen cracking are described in the following paragraphs.

Hydrogen content

The hydrogen content of a process is expressed as ml H_2 at NTP per 100 gm deposited weld metal. Currently only MMA weld metal is covered by British Standards *viz* BS639, in which various electrode designations are specified. The 'hydrogen controlled' designation refers to electrodes that deposit < 15 ml/100 gm deposited weld metal. MMA welding introduces hydrogen from moisture and/or organic materials in the electrode coating. Cellulose and rutile coatings are only gently dried in manufacture and rely for their satisfactory operation on the generation of a reducing atmosphere containing hydrogen. These electrodes deposit typically > 20 ml/100 gm hydrogen and cannot be baked prior to use without destroying their properties. Thus in important plant applications where there is a need to control hydrogen content, basic coated electrodes are used. These work under an oxidising atmosphere of carbon dioxide gas formed by the decomposition of the coating constituents with simultaneous deoxidation of the weld pool using added deoxidants. This gives a low oxide inclusion deposit with excellent toughness. These electrodes are baked during manufacture and this gives a much lower weld metal hydrogen content typically < 15 ml/100 gm. This can be further reduced by baking during fabrication (generally between 350–450°C) to increase the control over potential cracking problems in critical plant components. However, baked electrodes absorb moisture from the atmosphere and so great care must be taken in their storage, distribution and re-issue during fabrication.

In the submerged arc process the fluxes are the major source of hydrogen. Fused fluxes which are heated to very high temperatures during manufacture are generally low in moisture and can be used to give weld metal hydrogen contents of < 5 ml/100 gm. Agglomerated fluxes, however, are only heated to intermediate temperatures (approximately 800°C) and contain signi-

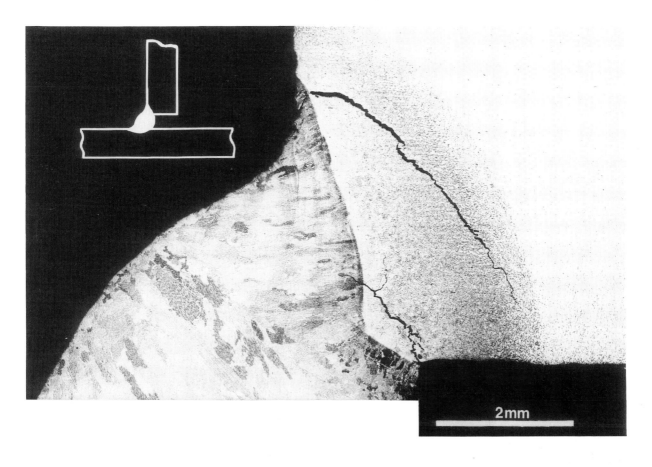

FIG. 10.5 Hydrogen cracking (The Welding Institute UK)

ficant quantities of moisture. Current practice during manufacture is to bake these fluxes at temperatures higher than MMA electrodes. Thus they can be rendered more resistant to moisture re-adsorption and can give hydrogen levels approaching those of fused fluxes.

Gas shielded processes such as TIG, and MIG are both potentially low hydrogen processes. The major source of hydrogen is the shielding gas. Argon can be very low in hydrogen but CO_2 (used in the Metal Active Gas (MAG) process) has a much higher inherent moisture content. For good hydrogen control the gas moisture content should be <15 VPM and is controlled by specifying its dewpoint temperature.

Gas shielded flux-cored arc welding has the possibility of a hydrogen contribution from the flux core. Little data is available on these consumables at present, but as flux-cored arc welding wires are generally baked during manufacture there is the possibility that low hydrogen levels can be achieved. However, the temperatures involved are lower than those for MMA electrodes and submerged arc fluxes and so hydrogen levels probably vary between wide limits from <5 ml/100 gm to 20 ml/100 gm.

Microstructure

The susceptibility of a given microstructure is related to its hardness, the harder the steel the more prone it is to cracking. The weld metal and HAZ structures which form as a result of rapid cooling from high temperatures can be very hard and particularly susceptible to cracking unless measures such as preheat or post-weld heat treatment are adopted. However, it should be noted that for a given hardness level, different microstructures will not have the same susceptibility to cracking. Thus any precautionary measures devised which are based on the precept of hardness control should only be applied to those classes of materials for which they have been specifically developed. For example, BS5135 — 'Metal arc welding of carbon and carbon manganese steels' — contains nomograms for deciding pre-heat levels for welding. These recommendations are based on hardness, which is considered to be a function of joint thickness and material carbon equivalent, and they do not necessarily apply to fully hardenable steels where pre-heat has no effect on hardness but serves merely to encourage hydrogen diffusion away from the weld.

Stress

Clearly, cracking will only be generated if stresses of sufficient magnitude are present. These are produced as a result of the restraint imposed by the surrounding material, the solidification stresses produced on

467

cooling and the yield point of the material which represents an upper limit of residual stress.

Generally restraint increases with section size and can be further exacerbated by poor weld design or local stress concentrating features such as slag inclusions or poor weld profile.

Material composition

Material composition is important as it determines the temperature range over which stresses may accumulate. This is the interval between the austenite transformation temperature and the pre-heat temperature. However, for highly alloyed materials, e.g., a 12CrMo ferritic steel with low austenite transformation temperature, there is a limit to the reduction in residual stress obtained by increasing pre-heat since, if welding takes place when austenite is present, the subsequent transformation on cooling generates a brittle untempered martensite.

The methods available for deriving welding procedures to avoid cracking are based on cracking test results that utilise single pass butt or fillet welds. Therefore the relevance of these data for thick section multipass welds depends critically on the experience gained in practice. However, the general guidelines for avoidance of hydrogen cracking should always be based on stringent control of welding consumables to restrict the source of hydrogen, application of a suitable preheat to slow down cooling rates after welding and finally incorporation of a post-weld anti-hydrogen heat treatment for critical applications.

4.2.6 Cast-to-cast variability

Marked variations in weld bead penetration and shape can occur in welding materials of the same specification using nominally identical welding conditions. This gives rise to the term 'cast-to-cast variability' and it is a problem that can cause serious manufacturing difficulties. In the power generating industry, most examples have been experienced with automatic TIG welding of austenitic stainless steel and nickel alloys, but it can occur in ferritic materials although it is much less likely.

The difference in welding response of individual casts is manifest by some casts displaying narrow deep weld pools while others exhibit wide shallow pools with poor penetration. In these casts there is an increased risk of welding defects due to incomplete fusion of the root area. However, a situation that has potentially worse consequences occurs when two casts with different welding responses are welded together. In this case there is an additional tendency for the weld pool to be displaced towards the low penetration cast giving rise to what has been termed 'weld pool wander'.

The problem has been well researched but to date no effective explanation has been found which is generally applicable to all situations. One explanation for variable penetration behaviour correlates weld bead

shape with surface tension effects. Surface tension gradients exist on a weld pool surface because surface tension is temperature dependent and there are significant temperature gradients associated with the molten weld metal. For pure metals and many alloys, surface tension is greatest in the coolest part of the pool, that is at the edges, and lowest in the hottest part of the pool, that is at the centre. Therefore there is fluid flow from regions of low surface tension to regions of high surface tension. This flow pattern transfers heat from the centre to the outside of the pool and produces a relatively wide shallow weld. Some of the materials present in commercial materials (often as impurities) are surface active and segregate preferentially to the surface of the molten pool. Small concentrations of these elements can also change the temperature dependence of surface tension so that for a limited temperature above the melting point the surface tension gradients are reversed. This produces fluid flow inwards along the weld surface and then downwards transferring heat to the bottom of the weld pool giving a narrow deep weld bead shape. Some correlations have been observed between cast-to-cast variability and sulphur, oxygen and aluminium contents. For TIG welding of stainless steel, very low sulphur contents cause poor weld penetration and as a precautionary measure minimum sulphur levels of 0.005–0.007 wt% should be specified. High oxygen contents of 0.08–0.1 wt% also alleviate the problem but this remedy could cause further difficulties since toughness properties may be impaired.

Another explanation for cast-to-cast variability argues that the shape and size of the anode spot controls the energy distribution in the workpiece and therefore will determine the fusion response of the material. The anode spot geometry is affected by the presence of impurities. Elements such as Ce, K, Ti, Ca widen the anode zone while elements such as O_2 and S have electrical characteristics that constrict the welding arc. Often poor penetration characteristics are associated with the occurrence of particles on the molten pool surface during welding. These are invariably oxide based inclusions containing those elements possessing a low ionisation potential which affect the formation of anode spots that amplify the spreading tendency of the arc.

Despite over twenty years of research into cast-to-cast variability, there is no conclusive agreement about the controlling mechanism. Since its occurrence is very difficult to predict it is often first identified during production welding when all contract materials have been purchased. Thus at this stage control via material specification is not feasible and successful fabrication depends upon being able to devise suitable welding procedures which can cater for this inherent variability. This situation should be avoided by correct initial materials specification which incorporates a weldability test with an acceptance criterion based on weld bead width to depth ratio.

4.2.7 Lamellar tearing

This is a form of weld-associated cracking that occurs in welded configurations in which a high level of restraint is present. Those configurations with the highest risk are T-fillet or T-butt welds in which the welding stresses are effectively applied across the thickness of at least one of the plates being welded. If the through-thickness ductility of the plate material is low then cracking occurs. Thus cracks may occur at some distance below the HAZ of the weld (see Fig 10.6).

Lamellar tears are due to the presence of planar inclusions that lie parallel to the plate surface along the rolling direction. Large inclusions or laminations of manganese sulphide are particularly detrimental but arrays of small inclusions with a degree of longitudinal alignment may also be responsible for cracking. Silicates, sulphides and alumina inclusions have all been implicated and lamellar tearing can occur in many different types of mild and low alloy steels. However, high strength, low ductility steels are less tolerant of inclusions and require tighter specification.

Generally, lamellar tearing can be avoided by using forged rather than plate components so that a more uniform distribution of inclusions is present due to the hot working process. However, if plates are to be used then the specification should impose a minimum requirement for through-thickness tensile ductility as well as stipulating a particular level of steel cleanliness by controlling the steelmaking route. Ultrasonic inspection cannot guarantee freedom from lamellar tearing but may be useful in identifying large laminations if repairs to old material are being undertaken.

Improved design can also be beneficial since many instances of lamellar tearing can be avoided by altering the weld geometry so as to avoid through-thickness stresses by ensuring that loads are not applied in the plate transverse direction.

A fairly effective solution is to apply a layer(s) of soft weld metal to the plate surface before making the fillet weld. It is an expensive remedy particularly if it is necessary to groove the plate as well, but as it effectively eliminates the susceptible material and reduces the applied through-thickness strain, it is usually effective.

Fortunately lamellar tearing is virtually a thing of the past. The advent of concast steels with low sulphur and low inclusion contents have eliminated the problem. However, caution must always be taken in design and checks applied to the steel type and on its properties.

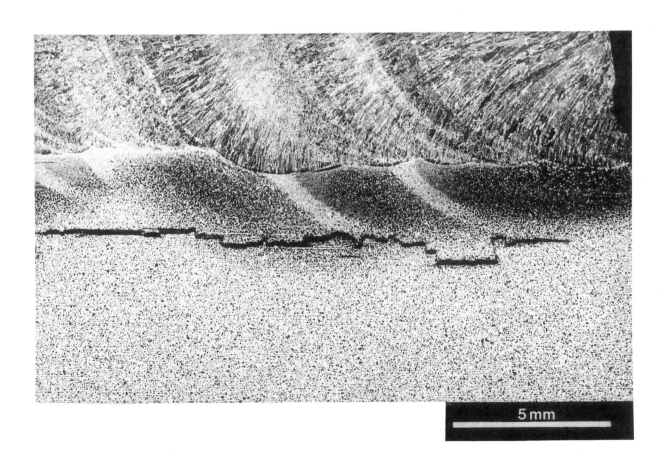

FIG. 10.6 Lamellar tearing (The Welding Institute UK)

5 Welding processes

5.1 Tungsten inert gas welding

Tungsten inert gas (TIG) welding is a process that relies upon the formation of an arc between a non-consumable tungsten electrode and the workpiece and in which the arc is generally initiated by a high frequency unit and protected by an inert gas shroud. The electrode tip angle determines the spread of the welding arc which is contained within an envelope of argon gas. The gas generates an arc plasma and also protects the molten pool from undesirable oxidation effects (see Fig 10.7).

Various shielding gases are available for different metals: argon for carbon steels; Ar/H_2 for austenitic steels. Alternatively Ar/He may be used for higher heat input and increased penetration. TIG welding is generally carried out under direct current conditions with electrode negative polarity and the process is capable of being operated within the current range 1–200 A. Alternating current is a less attractive option due to problems with arc re-ignition as the welding current polarity is reversed and overheating of the electrode tip when it is at positive polarity.

The conventional arrangement provides a small controllable intense source of heat suitable for autogenous welding or with filler wire additions into the arc underneath the cover gas shield. As the electrode is not consumed the current and voltage settings for a given heat input are independent of the melting characteristics of the added filler. Therefore, arc power and wire feed speed can be independently controlled which allows greater control of weld pool fusion. This in turn offers considerable potential for all-positional welding.

The TIG process is one of the most versatile of all welding processes but requires a high level of welder skill for manual applications. It can be used at low currents (less than 1 A) for very thin components (0.1 mm thick) but, although it is equally suitable for thick sections, the relatively low deposition rate makes it uneconomic. However, the use of hot wire techniques and narrow gap preparations can offset this deficiency and with these modifications TIG welding will have further potential in applications where there are high demands on weld integrity. Currently it is the usual practice in fabricating thick-section components to use manual TIG welding only for root passes where good control is essential. It is also widely used in power plant in mechanised form for welding thin tubes where high quality welds are needed.

5.1.1 Pulsing techniques

A variation of the TIG and microplasma processes is the use of a pulsed current that is varied from a high peak level to a low background level at a regular frequency that is related to travel speed.

The high peak current causes increased penetration to occur but burn-through is avoided by restricting the time for which it is applied and allowing the molten pool to solidify under the low background current. Thus the final weld is made up of a series of overlapping weld pools. A variety of waveforms are available and at first sight this introduces a large number of additional control parameters, particularly when gradual increase (slope-in) and decrease (slope-out) of welding current is employed at the start and end of each weld pass. However, this can be simplified since background current and time are not critical and the peak current and time are determined by material thermal diffusivity and thickness and can be readily established.

The particular advantage that the pulsing mode of operation possesses is that it provides far greater control of penetration in geometrical configurations with an asymmetrical heat sink. This makes it suitable for welding thick-to-thin components, and metals with dissimilar thermal diffusivity.

5.1.2 Plasma welding

The plasma welding process is very similar to and has been developed from TIG welding. The arc is again formed between a non-consumable tungsten electrode and the workpiece although it is deliberately constricted. However, the main difference is that the argon shielding gas is separated from the argon plasma forming gas and the electrode is situated in the body of the torch. The benefit of this arrangement is that the plasma can be collimated by a suitably designed nozzle geometry resulting in increased gas velocity and increased penetration compared with the TIG arc. This technique utilises a non-transferred arc and it has been

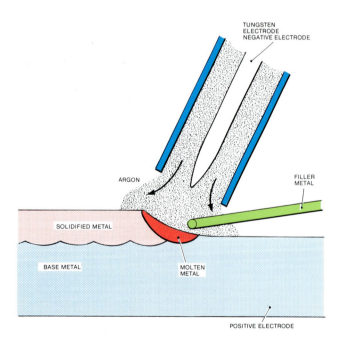

FIG. 10.7 Tungsten inert gas (TIG) welding

developed as a cutting technique as well as a welding technique.

Plasma techniques are utilised over three main current ranges:

(a) *Microplasma welding — 0.1 to 15 A*   This mode produces an arc that is stable over a wider range of arc lengths than is possible with TIG welding at low currents. It is particularly suitable for very thin components where it is important to minimise distortion and to maintain a high degree of dimensional stability.

(b) *Soft plasma — less than 100 A*   This mode competes directly with the TIG process. The main advantage it possesses is the increased penetration associated with the constricted arc which can make it suitable for sections that are thicker than those for which TIG welding would normally be used.

Initially an arc, called the pilot arc, is formed between the electrode and the plasma nozzle. This makes for more reliable starting following which the arc is transferred to the workpiece for welding. This arrangement gives rise to the term 'plasma transferred arc (PTA)' process.

(c) *Keyhole plasma — greater than 100 A*   This mode utilises the high energy density of the plasma arc to provide a deeply penetrating weld pool. The high current and plasma flow rate cause the jet to form a hole in the metal being welded and this hole travels along the joint as welding proceeds. It is this feature that gives rise to the term 'keyhole plasma welding'. In practice, control of the keyhole is difficult and defects such as porosity and craters can occur. Using this technique it is possible to weld relatively thick-section components in a single pass, e.g., seam welds. However, there are difficulties in closing the keyhole when the weld is complete which is particularly problematic with circumferential welds. This is a limiting feature of the keyhole plasma technique and demands very precise adjustment and control of welding parameters.

5.1.3 Summary

The advantages of the TIG process are:

- The process offers good control for root pass welding in both manual and mechanised versions.

- It can be used at low currents for thin component applications.

- Clean welds with low inclusion contents are produced.

- It can be mechanised for orbital tube welding and other applications.

The disadvantages of the TIG process are:

- It is a low deposition rate process.

- It has a limited range of application in terms of material thickness.

However, these disadvantages can be offset to some extent by the use of new techniques such as hot wire additions, narrow gap preparations and pulsing.

5.2   Manual metal arc

This is the most widely used method of welding in power plant construction combining, as it does, a flexibility of operation in all positions with a low capital cost and reasonably high deposition rates.

The process utilises an electrode which consists of a metal core wire covered with a mineral flux coating. The welder manipulates the electrode which is housed in a holder through which the welding current is introduced. For economic reasons it is desirable to use the largest diameter of electrode practicable but, for good control of the weld pool (for example in overhead welding) or restriction of heat input, smaller gauge sizes have a demonstrable technical benefit.

During welding the flux coating is fused, thereby forming a protective layer of slag over the weld bead. This must be completely removed, usually by chipping or brushing, before a second pass is deposited.

The stop/start position in manual metal arc welding can be a source of problems such as crater cracking inclusions or lack of fusion and these must be obviated by good welder technique. Metal is transferred to the molten pool from the electrode by the formation of a spray of small droplets (see Fig 10.8).

The heat of the welding current breaks down the electrode flux so that it provides a gas cover and a slag layer that protects the arc and molten pool from

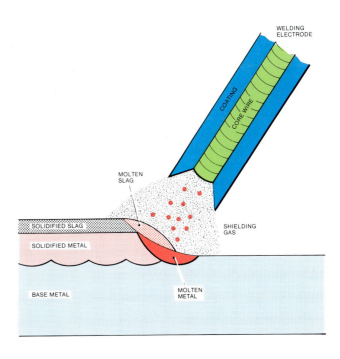

FIG. 10.8 Manual metal arc (MMA) welding

atmospheric contamination. However, it also allows alloying additions to be added to the molten metal so that weld metal composition and properties can be controlled. There is a wide range of coatings that can be used and the choice depends upon the specific application being considered. Coating technology can be complex involving the blending of a combination of mineral powders such as titanium oxide, metal silicates, cellulose and carbonates which are bound together with an agent such as sodium silicate. The coatings are extruded co-axially around the wire and the electrodes are oven-baked to harden and dry the coating and to reduce moisture content. Coating formulation critically affects electrode handleability which should be as easy as possible to reduce the skill factor necessary to produce defect free welds. However, this is not always compatible with the need to achieve satisfactory mechanical properties and a balance must be struck.

Electrodes vary in their shielding capacity since this is derived partially from the breakdown of the coating ingredients to form gases, e.g., carbonates which give carbon dioxide, and partially from the formation of the welding slag. Consequently control of the length of the welding arc is critical to avoid, for example, the ingress of nitrogen from the atmosphere giving rise to porosity defects. Finally coating constitution determines slag detachability which must be as easy as possible to minimise the risk of leaving inter-run slag inclusions. To facilitate this, some electrodes have been developed with self-detaching slags.

Control of weld metal composition can be achieved by addition of ferro alloys, e.g., ferro-silicon or ferro-boron, to the electrode coating. In this way it is possible to formulate a range of electrodes using a single composition core wire. For example, for low alloy ferritic steels, a rimming mild steel core wire can be used whilst stainless steel electrodes may be based on a standard 304 grade wire. However, for critical applications this may not always be acceptable since coating defects which cause spalling during welding could lead to grossly incorrect weld metal composition and inadequate properties.

Consequently it may be prudent to utilise core wires of near matching composition to reduce this risk. However, electrodes can generally be manufactured to very high quality standards and the problem lies in ensuring that appropriate quality systems are implemented so that incorrect electrodes are not used during manufacture. The various types of electrode coating are listed in Table 10.1 with a brief description of their operating characteristics.

5.2.1 Summary

The advantages of manual metal arc welding are:

- Low capital costs.
- Flexibility of operation. The technique is not restricted by component geometry and can be used in all-positions.

- Reasonably high deposition rates.
- Good metallurgical control of weld metal composition.

The disadvantages are:

- It is a discontinuous process with large numbers of stop/start positions that may be associated with the formation of cracks if welder technique is poor.
- Each weld run must be thoroughly de-slagged before a subsequent pass is made.
- Stringent control of electrode baking and storage procedures is necessary to avoid hydrogen cracking in hardenable materials.
- Depends heavily on welder training and qualification.

5.3 MIG/MAG welding

The MIG process possesses a number of features which should make it particularly attractive. It can be used as either a manual or mechanised technique in a number of operating modes which confer considerable flexibility of application in terms of material type, thickness, welding position and joint fit-up. Since the consumable electrode is fed continuously into the welding arc, high deposition rates can be achieved particularly when a mechanised operation is employed (Fig 10.9).

The original development of the process was for welding aluminium and utilised argon as the shielding gas. However for steels, the unfavourable economics and limited field of application led to the use of CO_2 and subsequently to the development of Ar/CO_2 mixtures for cover gas. Gas composition has a very significant effect upon process characteristics, weld bead shape spatter and weld metal mechanical properties (notably toughness). CO_2 welding is widely used for welding mild steel for general applications but where surface appearance, improved tolerance to welding parameter variation and good toughness properties are important then Ar/CO_2 or $Ar/CO_2/O_2$ mixtures are used — the increase in cost being offset by improved weld appearance.

The MIG process can be operated in a number of modes:

- *Dip transfer (short-circuiting)* — this mode uses CO_2 shielding gas and is applied primarily when welding steels positionally and involves the use of low currents (< 200 A) and voltages. The molten droplet forming on the end of the electrode transfers by touching (i.e., short-circuiting) on the surface of the weld pool. In this mode the arc characteristics are very sensitive to weld parameter variations and this can cause considerable difficulties in weld pool control.

- *Globular pulsed MIG* — utilises current pulsing (typically at frequencies of approximately 50 Hz) to provide the characteristics of spray transfer at low mean currents. The extent of weld spatter is reduced

TABLE 10.1

Various types of electrode coating
Electrodes for welding C and C/Mn steel are classified according to tensile strength and ductility, impact strength and type of covering. The choice of electrode is usually governed by the need to achieve good handleability to minimise the risk of introducing operator defects and to obtain high integrity welds with good mechanical strength. Electrode coating is a major factor in striking this balance. A complete classification of electrode characteristics may be found in BS639: 'Covered electrodes for the MMA welding of C and C/Mn steel'.

Type	Constitution	Characteristics
Acid	Iron and manganese oxides. Ferro-manganese and other deoxidants.	High fusion rate, good penetration, most suitable for welding in flat position.
Acid-rutile	Similar to acid but higher TiO_2 content (35% max)	More fluid slag than acid type.
Basic	Calcium carbonate and fluorspar	Suitable for positional welding. Resistant to cracking. Require stringent controls to avoid H_2 pick-up.
Cellulosic	Organic compounds	High penetration arc. Good gas protection in addition to slag cover.
Oxidising	Based on iron oxide	Easy slag removal. Used mostly where weld appearance is more important than strength.
Rutile	Based on TiO_2 (50% min)	Good handleability for positional welding and improved weld appearance

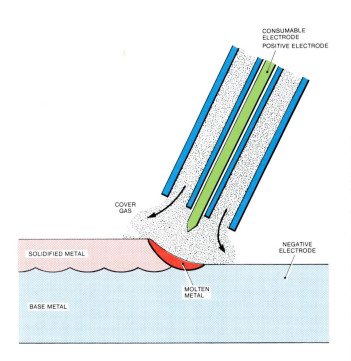

FIG. 10.9 Metal inert gas (MIG) or metal active gas (MAG) welding

in this mode but the technique is sensitive to variations in joint fit-up and pulse parameters. A number of additional parameters are introduced in using this method and these require careful optimisation to achieve stable conditions for a given arc length.

• *Spray transfer* — requires the use of Ar/CO_2 gas mixtures and relies upon high currents >250 A to detach metal droplets and transfer them across the arc. The method is feasible over only a narrow current range determined by a compromise of conditions to prevent excessive arc forces at high currents and the detachment of large droplets in an asymmetric fashion that makes positional welding impossible.

Dip transfer is the most commonly used mode but requires a moderately high level of skill for welding parameter optimisation and for manual welding. The risk of 'lack of fusion' defects prevents the process giving consistent Class 1 weld quality. In pulsed MIG welding the lack of tolerance to parameter variations during welding demands a power source stability that has only been provided by commercially available equipment in recent years. Therefore, for a number of reasons, of which the above are probably the most important, MIG welding has been beset by type defects, notably lack of sidewall fusion, spatter and porosity. Consequently, the perceived advantages over manual metal arc welding have never been realised in practice and MIG welding has not achieved the impact on industrial fabrication that was anticipated some twenty years ago.

5.3.1 Automatic pulsed MIG welding (synergic MIG)

A recent development in pulsed MIG welding has occurred with the advent of solid state power sources. This has provided the means for accurate control of current waveform parameters so that any practical frequency may be generated which can then be varied in a continuous manner. Consequently it is possible to link wire feed speed to the output of the power supply. In this way waveform parameters can be continuously altered in response to changes in wire feed speed while maintaining a steady welding arc. Having established the interaction between the physical and electrical characteristics of the welding technique it is then necessary to optimise the conditions of metal transfer. This requires a systematic selection of pulsing parameters to produce one droplet of molten metal transferring per pulse. The droplet size approximates to the wire diameter and under optimised conditions is detached at the end of the pulse peak. Once this has been achieved then, in theory, it is possible to control the entire welding operation using only a single control — the wire-feed speed — so that a range of conditions can easily be utilised for root, fill and capping passes.

5.3.2 Flux-cored wires

In an attempt to overcome some of the problems impeding the widespread acceptance of MIG welding, flux-cored MIG wire consumables were developed to attempt to reduce the level of operator skill required. Subsequent development has aimed to increase further weld deposition rates and to improve weld metal mechanical properties. In addition, weld pool control for positional welding has also been a benefit since readily ionisable elements in the flux core contribute to a stable arc and weld pool shape.

A logical progression has been the manufacture of self-shielded flux cored consumables. Shielding is provided by a combination of CO_2 generated by breakdown of carbonates and formation of metal vapour from elements such as Li and Mg contained in the flux. The latter cause problems of noxious welding fume which requires good ancillary extraction equipment for shop applications.

5.3.3 Summary

The advantages of MIG welding are:

- It can be a manual or a mechanised technique.
- It offers good flexibility of operation in terms of material type and welding position.
- High deposition rates can be achieved.
- Flux-cored consumables are available to improve weld pool shielding.
- Pulsed MIG offers the potential for easier control of the entire welding operation compared to conventional MIG.

The disadvantages of MIG welding are:

- The process has been prone to type defects such as lack of sidewall fusion, spatter and porosity which

have limited its impact since it is not possible consistently to get Class 1 quality welding.

- Precise control of welding parameters through good equipment design is necessary and so capital costs are higher.
- Flux-cored, especially self-shielded, welding can give rise to noxious welding fume that must be extracted from confined environments.

5.4 Submerged arc welding

Submerged arc welding involves introducing an uncovered electrode into the molten weld pool which is completely covered by a welding flux. The flux is delivered as a powder into the joint ahead of the advancing weld pool. Part of the flux is fused during welding and the remainder is returned to the storage hopper for re-use (Fig 10.10). The process usually operates at high currents (typically 400–1000 A) and, due to the insulating effect of the flux, has a high thermal efficiency. At very high current densities and travel speeds, deep penetration weld beads can occur which are more susceptible to centre line solidification cracking. Submerged arc welding is also characteristically a high dilution process with a ratio of approximately 2:1 parent metal to filler wire forming the as-deposited weld. It is a tolerant mechanised process that utilises continuous wire feed and flux supply and under optimised welding conditions offers good control of the weld pool shape and stability at high deposition rates. However, a potential disadvantage of the technique is that the arc is completely concealed from the operator who is unable to assess directly the progress of welding and apply suitable corrective action if required. In order to compensate for this, considerable care is necessary in preparation for welding and in weld bead placement. Joint cleanliness is particularly important and fit-up must be closely controlled so

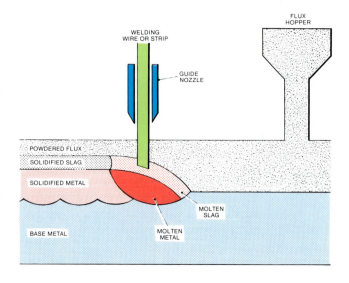

FIG. 10.10 Submerged arc welding

that a mechanical guiding device is able to locate accurately the position of the arc throughout the course of the weld. If these steps are taken then welds of a consistently high quality in terms of soundness and properties can be obtained in practice. It is probably the most widely used process for pressure vessel manufacture and gives consistent Class 1 quality welds.

Traditionally, wires for submerged welding have been copper coated, primarily to guard against rusting. However, this invariably leads to relatively high weld metal copper contents (approximately 0.3 wt%) which, although in many cases such as structural steels is not detrimental to performance, has been shown to promote transverse weld metal cracking in $2Cr\frac{1}{4}Mo$ deposits and to induce irradiation embrittlement in nuclear pressure vessel steels. Therefore for these applications, uncoated wires are used which with due care and attention paid to packaging and storage possess storage shelf lives at least comparable to coated wires. Wire composition has to be formulated specifically for the material and technique under consideration, since all alloying additions are added via the wire and not the flux. Stainless steels are, however, an exception to this and some compensation for element loss is necessary through additions to the flux.

Although the wire determines the overall weld metal composition, the constitution of the flux significantly influences the slag/metal reactions occurring during welding; consequently, the levels of elements such as manganese, silicon and oxygen occurring in the solidified deposit depend on flux formulation. High silica fluxes, described as 'acid', increase silicon and give a high oxygen and inclusion content while 'basic' fluxes have relatively little effect on carbon and silicon partitioning but reduce oxygen to low levels. Fluxes can be categorised in terms of basicity index, but although these provide an approximate means of classification they are largely empirical and have a limited range of applicability.

The most typical flux, the calcium silicate type, contains three major components. Lime (calcium oxide) is the basic oxide which is metallurgically beneficial. Silica which produces a glassy slag with attractive technological features affecting, for example, bead shape and slag detachability. Finally calcium fluoride which fluxes the lime and reduces the solidification temperature of the slag to below that of the metal. In other types of submerged arc flux the oxides of manganese or magnesium silicate are substituted for lime, alumina for silica and titania for calcium fluoride. Specific types may be selected for specialist applications such as single or two pass welding at high currents with enhanced productivity.

Fluxes are produced by fusion or agglomeration. Basic fluxes are made by agglomerating the finely crushed powder with a binder (such as sodium silicate) and drying at intermediate temperatures in the region of 700–800°C. Considerable care is necessary in the drying and storage of submerged arc fluxes to prevent moisture pick-up and possible risk of hydrogen cracking during welding. This is particularly true if agglomerated fluxes are to be recycled and re-used.

Choice of flux also affects weld deposit mechanical properties, notably toughness. Transfer of elements that increase hardenability, e.g., carbon, to the weld pool can result in coarse bainitic or martensitic structures and a reduction in toughness. Low carbon contents can give rise to pro-eutectoid ferrite structures which possess adequate toughness but optimum resistance to brittle fracture occurs with acicular ferrite or refined bainitic structures which are promoted by manganese or nickel. Weld metal cleanliness is also vital for good toughness. In order to achieve this, weld metal oxygen must be kept low, typically 0.02–0.03% and the sulphur and phosphorus levels limited to around 0.01%. This requires the use of basic fluxes with particular attention being paid to the parent material specifications.

5.4.1 Summary

The advantages of submerged arc welding are:

- High deposition rates can be readily achieved.

- Fused fluxes can be used to reduce hydrogen potentials and alleviate the risk of weld cracking.

- Good toughness properties can be obtained by correct flux formulation to produce low weld metal oxygen contents.

- Class 1 quality welding can be consistently achieved.

The disadvantages of submerged arc welding are:

- The welding arc is completely concealed and often a mechanical guiding device is necessary.

- Joint cleanliness and fit-up must be closely controlled.

- Flux drying and recirculation must be carefully controlled to avoid moisture pick-up, particularly if agglomerated fluxes are used.

5.5 Electro-slag welding

Electro-slag welding was developed from submerged arc welding to facilitate the economic fabrication of very thick-section components.

The technique utilises resistance heating to form a large molten pool and in order to maintain process control welding is carried out in the vertical position. Water-cooled copper shoes are placed on either side of the joint and these are raised as welding proceeds to prevent the molten metal and slag escaping.

The slag on top of the weld pool is electrically conducting and therefore acts as a source of heat as well as providing protection for the molten metal. The wire electrode is introduced into the pool through the slag and continuously oscillated to ensure uniform fusion at the faces of the weld preparation throughout the joint thickness. In the process of oscillation a slight dwell is introduced at the points of reversal where heat is extracted directly by the water cooled shoes. For joint thicknesses in excess of 100 mm, more than one wire may be used and extremely high deposition rates can be achieved. However, a significant proportion of the parent metal is melted and high dilution approaching 50% can be obtained.

The process is initiated by forming an arc between the electrode and a run-on plate connected to the workpiece while using only a small quantity of flux. The amount of flux is gradually increased until current conduction through the slag occurs and a stable welding condition exists without any arcing. Voltages of approximately 50 V are typical with the welding current being dependent on electrode feed rate which also determines the weld pool depth. Slag constitution must be carefully selected to ensure the correct electrical properties and a suitable viscosity so that reliable weld protection is provided.

The weld cross-section is usually barrel-shaped and the fusion boundary at mid-thickness can extend significantly beyond the original weld preparation. The heat input associated with electro-slag welds is very high. Which is due to the use of a very large and slow moving weld pool. Consequently weld metal structures with very large columnar grains and a coarse grained heat affected zone are typical (see Fig 10.11).

As-welded properties are inferior to those of the parent material and in order to refine the HAZ and weld metal microstructures in low alloy steels it is necessary to use a post-weld normalising heat treatment.

5.5.1 Summary

The advantages of electro-slag welding are:

- Extremely high deposition rates can be achieved enabling very thick components to be fabricated economically.

- Very little distortion.

The disadvantages of electro-slag welding are:

- High parent metal dilution is obtained and high heat inputs are inevitable, leading to unfavourable metallurgical structures and properties.

- Weld metal structures are very coarse-grained and post-weld normalising treatment is required to achieve the desired properties.

- It can be used only for longitudinal/straight seam welds.

5.6 Narrow gap welding

Narrow gap welding is a welding technique that involves the deposition of weld metal into a joint preparation that is generally parallel sided (see Fig 10.12). However, small groove angles are sometimes selected to compensate for distortion in large scale components

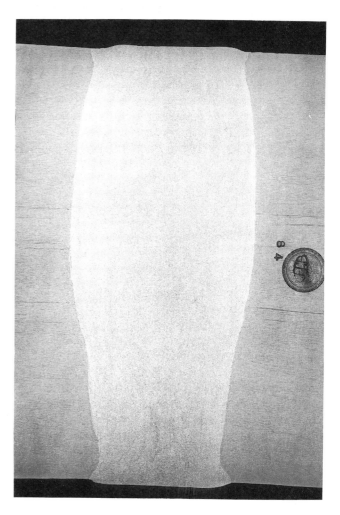

FIG. 10.11 Electro-slag weld

which causes joint closure that can trap the welding head. The weld is deposited with a constant number of beads per layer — usually one or two — and the narrow gap technique can be used with a variety of welding processes notably, submerged arc, TIG and MIG.

Narrow gap welding was developed in the 1960s primarily because of the economic advantages associated with the reduced weld metal volume required to fill parallel-sided joints compared to conventional weld preparations which are devised to provide good access during welding.

The increasing use of thick-section plate and forgings, for example, in nuclear pressure vessels, has meant that the greater welding time associated with open-angle joints contributes disproportionately to overall costs. Therefore, in the last decade, narrow gap techniques have been the subject of considerable research and development and narrow gap MIG welding has been widely used in industrial applications in the USA, Japan and the USSR.

From the technical standpoint, narrow gap welding offers the potential of producing high quality sound welds, once careful process optimisation and equipment

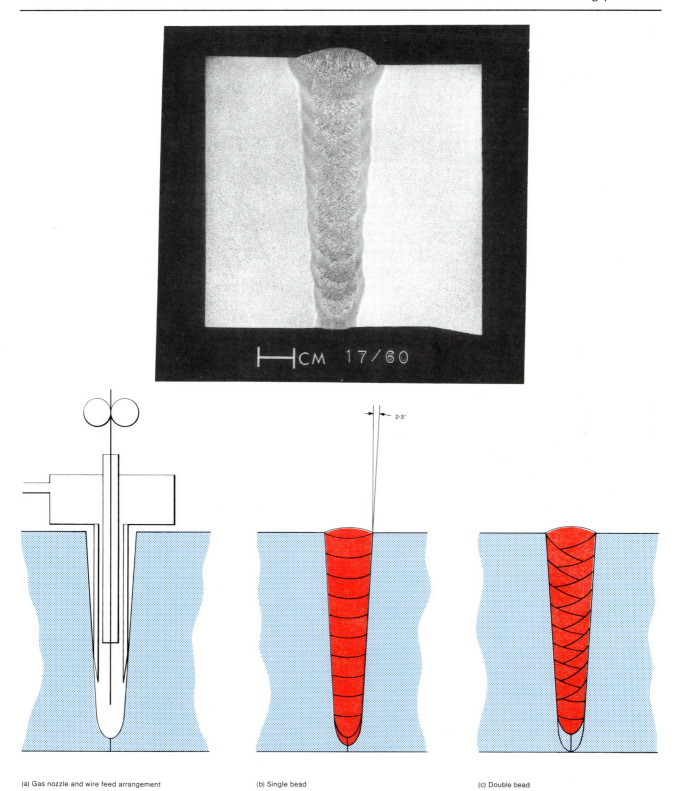

(a) Gas nozzle and wire feed arrangement

(b) Single bead

(c) Double bead

FIG. 10.12 Narrow gap welding (Babcock Power Ltd)

design has been completed. In addition, because the weld beads are stacked almost vertically on top of one another, the narrow gap technique produces a high degree of heat affected zone grain refinement. This is achieved because, apart from capping beads, there is an inherent high degree of overlap of successive thermal cycles which refines those regions of the heat affected zone that have experienced high peak tem-peratures. To achieve the same effect with conventional manual techniques requires a reduction in heat input that often results in an increase in welding time. Narrow gap welding has the advantage that high levels of HAZ grain refinement can be achieved without necessarily incurring any cost penalties. These heat affected zone structures possess particularly good toughness and creep properties and thus narrow gap welding is capable of realising economic and technical benefits that are not mutually exclusive.

The problems associated with narrow gap welding lie in process parameter optimisation and, in particular, guaranteeing reliable wire feeding into the narrow weld preparation so that sidewall defects can be eliminated. In order to alleviate this problem a variety of wire feed devices have been developed to ensure that, local to the sidewall, the wire impinges at a high angle of incidence. In this way greater arc energy is directed into fusion of the sidewall.

In gas shielded processes, good gas cover must be provided at all positions within the joint to ensure adequate protection of the molten pool. This can be particularly difficult for root passes in very thick plate where the geometry increases the susceptibility to air entrapment and the occurrence of porosity. This must be catered for by design of a satisfactory nozzle arrangement which generally requires insertion into the weld preparation.

The aspect ratio of the weld gap makes direct observation of the arc very difficult and this can be further impeded by the gas nozzle. Consequently accurate component alignment and reliable seam tracking devices are essential to ensure precise positioning of the welding arc to achieve good fusion. This is particularly important where a low heat input process with a small molten pool is being used since the likelihood of obtaining lack of sidewall fusion is considerably increased.

Weld bead shape must also be controlled to minimise lack of fusion defects and this can best be achieved by welding parameter optimisation. Thus, for narrow gap MIG welding the low voltage dip transfer mode is to be avoided, but equally, too high a voltage in the spray transfer mode can cause unwanted arcing and melt-back. Accordingly, a pulsed current is desirable to provide a stable arc and reduce spatter.

High currents are also undesirable since they result in a high weld bead depth-to-width ratio which can cause solidification cracking. This in turn can give rise to a plane of weakness down the weld centreline since individual weld bead effects are enhanced by the solidification pattern of narrow gap welds which is from the sidewalls to the weld centre.

Shielding gas is also important in controlling weld bead shape and for narrow gap MIG applications $AR + CO_2$ mixtures are suitable for most grades of steels.

Despite the preponderant commercial usage of narrow gap MIG welding, both TIG and submerged arc welding have been applied with narrow gap preparations. Narrow gap TIG welds are generally less susceptible to defects due to the improved arc stability of the process and they also possess very good mechanical properties. Deposition rates are lower but improvements may be made by using, for example, a hot wire TIG system.

Narrow gap submerged arc welding possesses all the advantages of submerged arc welding and gives high quality welds.

However, the fact that the arc is totally concealed places a high premium on good seam tracking equipment. Often, for high integrity applications, double-U preparations are adopted where the root passes are removed by grinding before welding from the second side. This, coupled with the need for wider weld grooves which require two beads per layer and inter-run slag removal, offsets some of the potential cost savings associated with the high deposition rates obtainable from this process.

5.6.1 Summary

The advantages of narrow gap technique are:

- Reduction in welding time and costs.
- Reduction in weld volume shrinkage and distortion on cooling.
- Improved HAZ properties due to increasing HAZ grain refinement.

The disadvantages of narrow gap welding are:

- Wire feeding must be carefully designed to ensure freedom from sidewall and inter-run defects.
- In gas shielded processes it can be difficult to ensure good gas cover of the root passes in thick-section joints.
- Component assembly time is increased.
- Repairs can be difficult to carry out.

5.7 Explosive welding

Explosive welding is a solid state method of joining that can be particularly useful for dissimilar metal combinations where pre- and post-weld heat treatment are not required. In power plant it is used primarily for tube plugging in heat exchangers and to a lesser extent in tube to tubeplate welding. The process relies upon the detonation of an explosive charge which accelerates one of the metal components (the tube or the plug) against the second component (the tubeplate) so that the collision occurs under high pressure and high velocity. The kinetic energy is converted into heat and a jet of metal is formed at the junction of the two components. The high rate of strain involved makes the metal behave in a viscous manner and under ideal process conditions this jet incorporates surface oxide and other contaminants and is expelled ahead of the weld. The 'liquid-like' jet generated during the explosion oscillates and creates a characteristic rippled interface indicative of a sound weld.

One of the key variables in the process is the stand-off distance which is the separation between components. In tube/tubeplate welding or plugging at the front face the requisite stand-off distance is achieved

478

by machining the tubeplate (see Fig 10.13(a)), so that sufficient relative velocity can be generated.

The explosive charge in the form of a detonator is positioned in a plastic insert in the base of the tube and arranged to produce a collision front which travels towards the face of the tubeplate. The tube is positioned to extend beyond the tubeplate so that any unbonded areas occurring as a result of the reduction in velocity as welding progresses can be removed. If it is required to position the weld below the front

face and within the body of the tubeplate then the stand-off effect can be achieved by modifying the design of the explosive charge holder to a tapered rather than parallel sided geometry (see Fig 10.13(b)).

In a successful explosive weld the tube/tubeplate interface behaves as a viscous liquid and forms a rippled structure. However, this is sensitive to collision velocity and pressure and in practice some re-solidified metal may be retained and appears as isolated pockets at the peaks and troughs of the ripples. If the explosive charge is too great then a continuous re-solidified phase occurs along the interface, while if the charge is insufficient an expanded or stuck weld occurs that is not properly bonded.

Under optimised welding conditions a metallurgical bond occurs between both components and the heat affected zone is restricted to a very narrow (1 mm) band. However, re-solidified molten weld metal is cooled extremely rapidly and in ferritic alloy materials has a very high hardness compared to either tube, tubeplate or HAZ. Consequently, if present in significant amounts and if there is a risk of stress corrosion cracking which could be enhanced by the presence of a crevice, post-weld heat treatment may need to be considered.

The process is sensitive to variations in a number of parameters notably surface finish, charge strength, detonator and tube position. Non-destructive examination by ultrasonics cannot provide an estimate of the quality of the interface in terms of ripple characteristics, although complete lack of bonding is detectable and the fused weld length can be measured. Therefore it is important to characterise accurately weld quality and its potential variability in pre-production trials. A further problem of ligament cracking can occur if tube pitching is small and insufficient tubeplate material is present between the holes. In this situation the impact forces generated during welding cannot be sustained and it may be necessary to support adjacent holes by mechanical or hydraulic means.

5.7.1 Summary

The advantages of explosive welding are:

- An attractive repair technique notably for tube plugging.

- Useful for welding dissimilar metal combinations.

The disadvantages of explosive welding are:

- It is sensitive to small variations in process parameters.

- Non-destructive examination by ultrasonics cannot provide an estimate of the quality of the welded interface although weld length can be measured.

- In tube/tubeplate applications ligament cracking in the tubeplate can occur during welding if tube pitching is small.

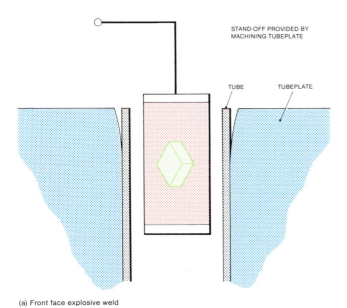

(a) Front face explosive weld

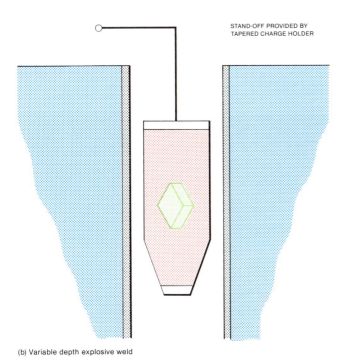

(b) Variable depth explosive weld

FIG. 10.13 Explosive welding

5.8 Flash butt welding

The flash butt welding process is an automatic technique that has traditionally had its main application in the power generation industry in welding boiler tubes. It is a very rapid, low cost process that involves the forcing together of the two components under the application of a high current and is a development of resistance welding. In flash butt welding one component is maintained in a fixed horizontal position while the other component is forced into contact.

Initially only a light contact is maintained between the workpieces while the current is applied. During this period the current flows entirely through the initial points of contact which melt and are ejected forming short-lived arcs. This is repeated as the tubes are gradually forced into more intimate contact and the 'flashing' stage of the operation persists until the surfaces of the tubes are uniformly heated or molten. At this juncture the moving component is rapidly advanced towards the fixed component under the action of a high force which forges together the ends of the workpiece and expels the molten metal on the tube surfaces. This metal forms a characteristic circumferential fin which should contain all the molten metal and any oxidation products leaving a sound weld that has much in common with a solid phase weld.

The important welding parameters that need specification and control are:

(a) *The speed of approach* If this is too high then the molten metal is quenched prematurely; if it is too low then excessive flash is generated or the tubes in the region of the weld are excessively deformed.

(b) *Upset length* This determines the length of the tubes involved in the production of the flash and in the forging process. It must be carefully controlled to maintain an acceptable weld profile and to ensure that all contaminated material is ejected.

(c) *Welding current and voltage* These electrical parameters determine the heat input to the weld. The voltage must be sufficient to maintain an arc during the flashing period while the current must be sufficient to raise the metal temperature to close to the melting point.

It is possible to incorporate a pre-heating stage into the procedure in order to shorten the flashing period. This is achieved by bringing the two components together for periods of 1 or 2 seconds with intermediate withdrawal to allow the temperature to build up by resistance heating.

The external flash is invariably removed after welding while the internal bore protrusion either has to be controlled to within agreed limits or otherwise removed by a reaming operation which of course increases fabrication costs.

480

The structure of a flash butt weld has much in common with a solid phase weld (see Fig 10.14). It should not contain any as-cast material since this is ejected during fabrication. Essentially the weld consists of two impinging heat affected zones with microstructures typical of those to be found in fusion welds. However, if the heat input is too high or upset force and length too low, then overheating of these structures can occur with a detrimental effect upon properties. Conversely if welding parameters are insufficient to heat the joint to the required temperature then lack of interfacial fusion can occur. It is also possible to obtain oxide or non-metallic inclusions concentrated at the interface due to retention of contaminated material during welding or due to dirty parent material. Such features are particularly troublesome if aligned normal to the tube surface where they may act as crack initiation sites.

Another source of concern with flash butt welding is the difficulty in reliable non-destructive examination. This arises because some process type defects that can occur, do so in an orientation which makes them difficult to detect using radiography. In addition, the presence of the internal flash confuses the interpretation of radiographs and ultrasonic traces unless it can be reliably removed. Consequently, as conventional NDT cannot be definitive, much greater emphasis must be placed upon strict monitoring and control of welding parameters and in-line production testing to reveal gross defects.

Modern equipment has the facility to control and display all the key welding parameters during the manufacture of a weld. This enables any perturbations in the process to be detected as they occur and suitable remedial action taken. Alternatively, these records are a useful diagnostic tool in the event of problems occurring. In-line testing as specified by BS4204 relies upon the production of test coupons at the beginning, middle and end of every shift which are subjected to

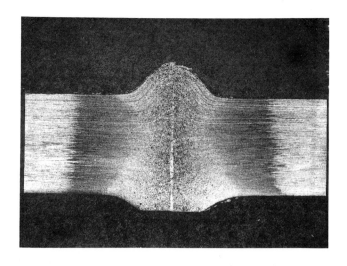

FIG. 10.14 Flash butt weld (NEI Ltd)

180° bend tests and examined for linear defects, misalignment and bore protrusion. Sufficient test data can be generated in this way to establish trends in weld quality during production.

5.8.1 Summary

The advantages of flash butt welding are:

- A cheap and rapid method of tube/tube welding.

The disadvantages of flash butt welding are:

- Any defects that do occur at the weld junction could be aligned normal to the tube surface.

- The interpretation of radiographic examination of these welds can be difficult due to the presence of the internal flash produced during welding.

- Inadequate removal of the internal flash by a reaming operation can create locations where enhanced in-service corrosion may occur.

6 Application of welding processes to power plant component manufacture

6.1 Steam drum

The steam drum of a conventional power station is a massive component typically 30 m long × 3 m diam × 150 mm thick. It is fabricated out of sections of rolled plate involving approximately ten longitudinal seam welds and eleven circumferential welds. For economic reasons electro-slag welding is used for the longitudinal welds and submerged arc for the circumferential welds which, due to their curvature, are unsuitable for electro-slag welding. Electro-slag welding is a very high deposition rate process that was developed over 40 years ago and service experience with steam drums has essentially been trouble-free. However, judged on the current knowledge, electro-slag welds have a number of metallurgical features that might be considered undesirable.

These are caused by the very high heat input which gives rise to a coarse weld metal structure and wide heat affected zone. Due to the large scale of the weld these structures are very similar to as-cast ingot structures, difficult to break down by heat treatment alone. Consequently, it is necessary to apply a double normalising post-weld heat treatment to achieve some measure of refinement followed by a tempering treatment to achieve the required properties. In this condition, electro-slag welds meet the elevated temperature design criteria and possess the satisfactory fracture toughness properties which are necessary to comply with hydraulic testing requirements.

Circumferential seams are welded using the submerged arc process and these require only a post-weld stress relief heat treatment in order to achieve toughness properties that satisfy the design criteria.

Consequently this component is an example where both technical and economic considerations have been optimised at an early phase in power plant construction and the need for further process development and revision has not arisen.

6.2 Furnace panels

Conventional boiler furnace panels are comprised of a series of tubes connected by fins made of thin strip (see Fig 10.15). The fin acts as a heat transfer path, a seal for boiler gases and a structural connection between tubes. The fin-to-tube welds must be of sufficient quality to satisfy these requirements. Single pass fillet welds are used and in a typical 2000 MW station approximately 30 000 metres of weld is involved and clearly only a mechanised welding process would be economically feasible.

The submerged arc process is ideally suited to this application and panels are made using multi-head equipment capable of simultaneous deposition of up to four welds. The welding arcs remain stationary and at a constant angle to the workpieces which are passed beneath them via sets of rollers that can be adjusted to cater for a variety of tube and fin dimensions.

Welding speed and welding current are two important variables that must be controlled to optimise

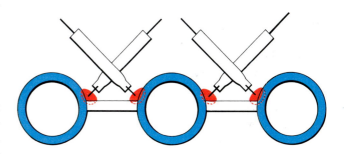

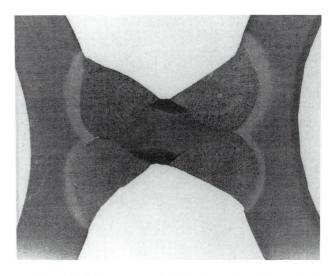

Fig. 10.15 Furnace panel fabrication using submerged arc welding

481

the fusion characteristics of the technique. Too high a welding speed and too low a current will give inadequate fusion, while too low a speed and too high a current could result in over-penetration and burn through into the tube bore. In this respect tubes with reduced wall thickness present a greater risk of burn-through and careful process parameter optimisation is necessary. Therefore systematic pre-production trials on contract material are essential prior to welding procedure approval so that there is a high level of confidence in the tolerance of the process to variations that may occur in practice. This involves producing unacceptable welds by using unsatisfactory welding parameters settings to establish the onset of over-penetration and lack of fusion. It is also necessary to produce a series of welds using the optimised welding procedure to demonstrate the consistency of the process. The results from these tests may then be analysed by a quantitative technique such as factorial analysis to give an estimate of reliability.

Welding procedure development must also contain an assessment of the likelihood of metallurgical defects to ensure that appropriate quality checks are imposed. For example, if there is a perceived risk of hydrogen cracking, then controls over material cleanliness and moisture pick-up that may occur with re-circulating submerged arc flux must be included. Finally, in-process surveillance is required to monitor welding parameters during production so that variations in, for example, welding current and welding speed, are recorded with a facility for detecting changes outside set limits so that the operator can be alerted and corrective action taken before there is a detrimental effect on weld quality. This can then be supported by selective sampling of welds for visual inspection and production of test samples at beginning and end of shifts for destructive assessment. A combination of all these factors enables a high level of confidence in weld quality, process efficiency and reductions in cost.

6.3 Boiler headers, pipework, valves and turbine steam chests

Welding is used extensively in the manufacture and repair of these components and considerable understanding of the metallurgical response to welding of heat affected zones in low alloy ferritic material has been obtained from close study of CrMoV steels. The underlying causes of HAZ stress relief cracking and potential remedies were described in Section 4 of this chapter. Control of material specification, notably vanadium content and residual elements P, Sn, Sb and As, is of considerable importance but, in addition, heat input during welding needs to be restricted to limit the extent of grain growth in the HAZ.

Adequate heat input control is usually achieved by restricting the gauge size of the electrode that can be used in contact with the sidewall. However, although this encourages low heat input it does not guarantee

it. This distinction can be particularly important in critical applications if welding is carried out in the downhand (flat) position. Other welding positions, such as vertical and to a lesser extent horizontal–vertical, are self-regulating in respect of low heat input by the need to maintain good control over the weld pool. A more precise method of controlling heat input is to specify the run out ratio for a given welding electrode. The run-out ratio is defined as the length of electrode burned off divided by the distance travelled and it is possible to correlate this to the area of weld bead deposited and also to the prior austenite grain size. Nomograms for the calculation of heat input from run-out ratio can be devised and the use of this ratio provides a convenient way of controlling heat input and hence metallurgical structure that can be implemented by adequate welder training, and monitored by simple measurement.

This method of heat input control can be used in the technique of two-layer HAZ grain refinement which is useful in situations where material specification is unfavourable or if previous attempts at welding have failed.

The main objective of this technique is to obtain a high level of grain refinement in the HAZ in order to improve mechanical properties such as toughness and creep crack growth resistance. This is achieved by the use of overlapping weld beads under conditions of controlled heat input. In the heat affected zones of adjacent weld beads, some areas of coarse grains in the first weld bead are reheated by the thermal cycle associated with the second bead so that some re-transformation to austenite will occur. In areas of the heat affected zones, where the temperature experienced is only slightly above the austenite transformation temperature, then grain growth will be limited and the microstructure will be refined (see Fig 10.16). However, the amount of refinement depends critically on the amount of bead overlap and, for example, for 50% overlap, approximately 75% refinement is obtained. However, the maximum HAZ grain size is not reduced, only the extent of the coarse grain structure. Consequently since single layers of weld beads leave residual areas of crack-susceptible microstructure it is sometimes necessary to refine these areas by superimposing the thermal cycles of a second layer of weld beads. In this way these regions are heated to temperatures just above the austenite transformation temperature without appreciable grain growth occurring. The exact welding conditions can be calculated if the appropriate material dependent bead shape data are available. It is then possible to specify electrode gauge sizes or a heat input for both layers.

The simpler method is to specify smaller electrodes for the first layer than for the second layer together with a fixed degree of overlap. The exact combination depends upon the specific application and needs to be determined by welding trials. The objective is to maintain a balance between failing to refine the existing

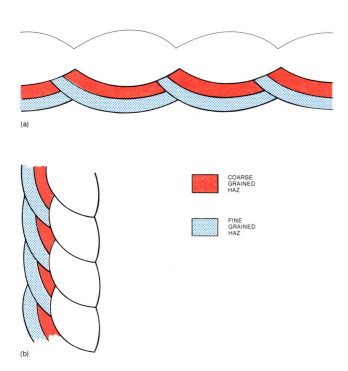

(a)

(b)

COARSE
GRAINED
HAZ

FINE
GRAINED
HAZ

FIG. 10.16 Refinement of heat affected zone
grain structures

coarse grain areas and avoiding burn-through which would introduce new coarse grained areas. Weld beads are deposited with the minimum of weaving to avoid partial solidification of the weld pool which invalidates the technique.

Since the extent of refinement is defined by the heat input during welding it is preferable to control this parameter directly. Optimum heat input ratios can be obtained by welding trials based on a series of bead-on-plate deposits using various heat inputs covering the anticipated range. From these tests the relationship between bead width and heat input can be measured and suitable welding conditions calculated for the first layer. A heat input ratio of about 2 kJ/mm should be aimed for and on this basis the welding conditions for the second layer can be derived. Alternatively, a series of two-layer deposits with different heat input ratios can be made until sufficient refinement is confirmed by sectioning. The heat input of the first layer should be as low as possible to minimise the peak HAZ grain size and the layer thickness. The success of this method depends to a large degree on careful control of welding technique, thorough welder training and strict supervision.

6.4 Tube-to-tubeplate fusion welding

In the design of heat exchangers such as feedheaters, condensers or nuclear steam generator units, one of the most critical aspects of fabrication is the connection between tube and the tubeplate. Difficulties arise because of the difference in thickness between the two components which causes an asymmetrical distribution of heat during welding and an overall weld geometry which can be difficult to inspect and post-weld heat treat.

Manual welding techniques are unsatisfactory for making this weld due to difficulties in access and control. In addition, large numbers of welds are usually involved which require very high consistent weld quality. Therefore in recent times the automatic orbital TIG process has been used almost exclusively. The fusion weld can be positioned at the front or back of the tubeplate and a number of different geometries have been devised (see Fig 10.17). In selecting one of these options all aspects of design, materials characteristics, inspectability and service performance must be carefully assessed.

Specific design issues will differ for each component. However, some typical factors that need to be considered include:

- The environment to which the joint is exposed since many weld configurations incorporate a crevice in which unfavourable impurity concentrating effects may occur.

- The ability of the joint to withstand thermal shocks.

- Cyclical loads and vibration effects.

- Thermal response between tube and tubeplate.

- Volumetric inspectability.

FRONT FACE WELDS

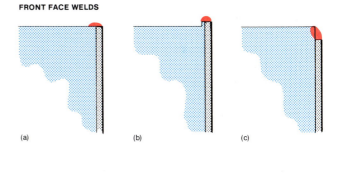

(a)   (b)   (c)

BACK FACE WELDS

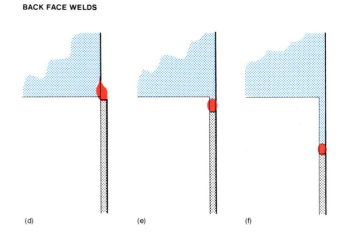

(d)   (e)   (f)

FIG. 10.17 Tube-to-tubeplate weld geometries

The simplest variant is the front-face seal weld (Fig 10.17(a)), but this may present problems of achieving uniform fusion of the two components due to the much larger heat sink provided by the tubeplate. Some compensation for this can be achieved by machining a castellation around the tubeplate hole to equalise these thermal effects (Fig 10.17(b)). Alternatively, a weld geometry which has the tube recessed in the tubeplate has been used to control the dilution in cases where dissimilar metals are being joined, e.g., clad tubeplates (Fig 10.17(c)).

In practice, hydraulic or explosive expansion of the tube into the tubeplate is used in conjunction with a front face seal weld in an attempt to close up the crevice. Autogenous TIG welding can be used but it is usually advisable to introduce a filler wire so that weld quality is less sensitive to the response of the tubeplate material, e.g., cast-to-cast variability. The front face weld is difficult to inspect volumetrically since there are problems of interpretation of results from X-ray and ultrasonics techniques. The only useful estimate of weld quality is via surface inspection and leak detection, which for many critical applications will be unsatisfactory.

Back face welds are much more attractive because they can be designed to alleviate these inspection problems and do not incorporate a crevice. A simple butt fillet geometry can be manufactured using an internal bore TIG technique involving autogenous welding (Fig 10.17(d)), but this is sensitive to the properties and cleanliness of the adjacent material and to the extent of tube insertion into the tubeplate. However, the volumetric inspectability of this weld is similar to that of the front face weld but this can be improved by incorporating an upstand which also alleviates the asymmetrical heat sink effects (Fig 10.17(e)). In fact this geometry has been introduced as standard for the manufacture of feedheaters in response to severe leakage problems associated some years ago with the seal weld and operational experience has been excellent.

A further modification of this latter design is to extend the length of the upstand to produce what is essentially a tube butt weld which can be made either by internal bore autogenous welding or external orbital TIG welding with filler wire providing the tube pitching is sufficiently large (Fig 10.17(f)). Clearly this is an expensive option but has many advantages over other types of weld for use in applications that are particularly onerous, for example, steam generator units operating in the creep range, and the material combinations require post-weld heat treatment to reduce residual welding stresses. The extra spigot length permits an effective local heat treatment to be carried out which is not possible with the conventional style where the entire assembly would have to be heat treated.

Heat exchangers may be required in a range of materials ranging from carbon steels through to stainless steels. For feedheaters, carbon steels or low alloy steels are adequate to minimise scantlings since re-

latively thick tubes are difficult to weld because a large weld pool is created, which is difficult to control due to gravitational effects. However, these steels contain a higher carbon content and some alloy additions which make them more sensitive to cracking during welding and under thermal shocks during operation. This is exacerbated by front face welds where the crevice acts as a stress raiser. Therefore, for this geometry, tubeplates made from lower carbon contents are used but clad with Inconel weld metal to improve erosion resistance. This presents a situation where a dissimilar metal weld has to be made and selection of compatible materials is necessary along with development of a welding technique to optimise dilution effects, weld bead profile and cracking resistance.

Because of the effect of defective heat exchangers on plant efficiency and availability only very small failure rates are acceptable and it is important to establish during pre-production work the likely variability in weld quality and the effects of any changes in control parameters that may occur during manufacture (see Section 7.3 of this chapter). Mechanical testing requirements will clearly depend upon the design chosen and the service conditions but should include pull-out, peel and fatigue tests in addition to the normal standard requirements.

Explosive welding can be used for tube/tubeplate connections and is dealt with in Section 5.7 of this chapter.

6.5 Arc stud welding

This is a well established process for attaching studs and is used extensively in the fabrication of plant. It provides a rapid and economical method of attachment which is especially advantageous when large numbers of studs are to be fixed such as the provision of pressure vessel insulation in advanced gas cooled reactors.

An arc is struck between the end of a stud and the plate to which it is to be welded. Currents of typically 1000 A are used and the arc is maintained for approximately 1/3 s, after which the molten-ended stud is plunged into the molten pool which has been created on the plate surface and a fusion weld is achieved. The process is capable of producing good quality welds but the quality is critically dependent on energy input.

The overall weld cross-section is greater than that of the stud and the weld metal, being fast-cooled, is harder and stronger than the stud material, therefore weld failure is only to be expected if the weld zone is grossly defective. Stud welding is an open-air technique, i.e., the arc is not shielded by gas or flux, hence the formation of pores may be expected in the molten pool due to oxidation. These will remain in the weld metal unless sufficient upset takes place during the plunging operation to sweep them out of the weld zone. Consequently weld quality is likely to improve as the upset increases and a minimum upset is required to give a serviceable joint. As well as porosity, lack of fusion defects can occur and impair weld integrity.

Traditionally, weld quality is assessed by methods such as hammer testing, tensile load testing, stud length measurement and visual inspection. These methods are generally satisfactory in cases where studs are used in lightly static loaded applications. However, in more critical applications some assessment of stud performance under service conditions is required, for example, fatigue testing.

Problems in stud welding can occur as a result of arc blow which causes flow of metal in a direction away from the welding return attachment, causing an uneven weld flash. Consistent alignment can also be difficult to guarantee with manual techniques where access may be difficult and good design involving lightweight equipment with inherent levelling facilities is required. Finally, surface preparation is critically important and in order to ensure satisfactory welds, all traces of surface contaminants such as paint, grease or oxide must be completely removed.

7 Future trends

7.1 Electron beam welding

In electron beam welding, electrons are emitted from a refractory metal cathode and travel through a high potential difference of between 30–150 kV. The electron beam is produced in a hard vacuum and finely focused by electromagnetic lenses. Sufficient heat is generated by impingement of the beam onto the workpiece to form a 'keyhole' containing metal vapour which is surrounded by molten metal.

Beam power determines the thickness of material that can be welded and in this respect the process is very flexible and can be adjusted to weld from 0.2 to 200 mm thick steel. In order to achieve good quality welds, power and welding speed must be balanced to ensure adequate penetration. Generally electron beam welding is an autogenous technique and requires very high standards of cleanliness, machining and fit-up. These standards need not be quite as stringent if filler wire is used but care must be exercised in wire positioning. Welding is usually carried out in vacuo for which a purpose built chamber is required and so for large scale components considerable capital investment is entailed. In order to reduce costs, portable vacuum chamber systems have been developed with sliding seals and out-of-vacuum systems can be used in conjunction with a cover gas to prevent contamination. However, this mode of operation has the drawback of being hazardous to personnel due to the production of X-radiation.

The main advantage of electron beam welding is its high welding speed, which can be used for welding thick section materials in a single pass. In addition, the heat affected zone is small and the associated workpiece distortion is low so that often finish machined components can be welded. Some control of material specification is required as high oxygen contents can give rise to porosity defects. However, in order to realise the potential of increased productivity, a very high throughput of manufacture is required to repay the initial capital outlay within a reasonable period. In addition, whole factory re-organisation will have to be undertaken to utilise fully the improvements in efficiency.

7.2 CO_2 laser welding

Laser welding uses a high voltage electrical discharge to excite a mixture of helium, nitrogen and carbon dioxide gases to produce light of a constant wavelength. This light is then amplified by repeated reflection between two parallel mirrors. One of these is partially transmitting and allows a portion of light to escape and form the laser beam. The gas mixture requires constant replenishment to avoid overheating. For industrial applications lasers with a power rating of up to 5 kW are required. The beam diameter on leaving the laser is large and the power density is too low for welding.

Therefore the beam is concentrated on to the workpiece by a focussing mirror and the power density and travel speed must be carefully matched to material type and thickness to achieve good weld quality. The focussed spot size is small and so accurate control over component alignment is necessary. Wire feeding can be used although it is possible to make autogenous welds in quite thick sections. Gas shielding must be used to protect the weld zone but only high ionisation potential gases such as helium will avoid the formation of a plasma cloud which could reduce the beam power density.

Laser welding has many features in common with electron beam welding including the high initial capital cost. However, the laser beam can be projected over several metres without loss of power and the process has a flexibility of operation that makes it very attractive for high throughput applications which can be fully automated.

Laser welding is likely to compete with TIG welding for components up to 10 mm thick and with electron beam welding up to 25 mm thick. However, it is less critical than electron beam welding as regards fit-up and it needs no vacuum.

7.3 Factorial analysis

Fully automatic welding procedures are usually adopted where a large throughput of welding operations is required. This often involves the specific control of many welding parameters in order to achieve high standards of weld quality. The consequence of this is that, in addition to demonstrating that the welding procedure is technically satisfactory by standard procedure tests, it becomes essential to establish that this can be consistently reproduced in production. The way in which this is done will depend upon the particular application being considered but will of necessity involve additional testing. Generally some form of consistency run is carried out where an agreed number of welds is made and assessed. However, while this may be a suitable approach for some applications, its chief deficiency is that no measure of the variability of weld quality is provided and hence no judgements can be made of the probability of maintaining weld quality in production. In order to do this it is necessary to carry out a statistical analysis of weld quality which will generate quantitative information about variability which can then be assessed against the engineering design acceptance criteria. Providing that these are adequately defined then decisions can be made about whether welding procedures will be tolerant to variations in welding parameters and what potential repair rates will be.

Factorial analysis is a method that can be used to carry out a statistical assessment of the variability of weld quality. It is essentially a planned and balanced experiment that enables the sensitivity of a chosen working point to be established as a function of changes in welding parameters. However, it is important to recognise that factorial analysis is not intended to indicate favourable parameters for weld procedure selection. The working point should be arrived at in the usual way on the basis of limited experimentation, experience and research data. Factorial analysis can then be used to test whether this working point is tolerant to variations and is acceptably reproducible. In addition, since rates of change of quality features may be estimated, it will be possible to indicate the probability of continuously satisfying acceptance criteria in practice. This can be done by carrying out relatively few experiments in which critical variables are varied systematically about their working point settings. In this way a quantitative assessment of variability of weld quality is obtained and interactive effects between variables can be detected. Overall this is a more cost effective and technically a more sound approach than performing consistency trials.

For example, it is well established that increasing current increases weld penetration. However, factorial analysis will provide not only a quantitative measure of their interrelationship but also an estimate of how penetration will be affected by changes in current, travel speed and/or wire feed speed. This enables an overall picture to be assembled of the implications that either planned or intrinsic changes in welding parameters will have on weld quality. Given this information the welding engineer can use it to make reasoned changes in the working point to increase his chances of success.

Clearly factorial analysis will not be suitable for all welding applications. It would not be feasible to apply it to the electro-slag welding of steam drums, but for tube/tubeplate welding using automatic orbital TIG

or attachment welding utilising robotic techniques it can be applied with tangible benefits.

However, factorial analysis is only suitable for assessing continuous trends in weld quality. Cracking, which is a discontinuous event, would not be predicted *per se* by using this technique unless it could be correlated with a continuous quality feature that can be measured.

In most welding situations the numbers of input parameters are large, e.g., current, voltage, travel speed, wire diameter, pulsing parameters. Likewise quality features, e.g., throat thickness, leg length, hardness, penetration. If each of these were to be varied systematically about the working point then an unreasonably large number of experiments would have to be carried out. Therefore the first step in any factorial analysis is to distinguish the critical variable from the rest, either by limited experimentation or by experienced judgement. This can often be done quite easily and the numbers of welds required to carry out a meaningful quantitative assessment is invariably less than would be done if more conventional methods of procedure development and consistency trials were adopted. In addition the results obtained are more useful.

Factorial analysis will undoubtedly become of more value and importance to the welding engineer as it replaces guesswork and broadly based procedure approval. Benefits will accrue in primary manufacture and be carried over into plant performance where the penalties of failure are large.

7.4 Expert systems

An expert system is essentially a computer programme that operates with a knowledge base which can be set up to incorporate the technical parameters relevant to making correct technical and economic decisions within a specific field. Some of these parameters will be known with a high degree of certainty and others may have to be estimated. In addition, rules must be defined by which data can be assessed and reasoned judgements made. In the simplest systems the data is contained within the rules and they operate as a checklist against which the individual steps comprising a decision making route can be assessed. However, greater potential usefulness will be obtained with systems that are capable of dealing with rules and data on a separate basis.

Although expert system design and usage are at an early stage of development they are being used successfully as diagnostic aids in medicine, agriculture, oil prospecting, fault diagnosis and financial analysis and it is clear that there is scope for their introduction into the field of welding technology. Currently it is envisaged that expert systems will either:

- Replace human experts (e.g., welding engineers), in which case the knowledge and decision making base will be directly accessible to, for example, design engineers who are non-expert in the field of welding. By making this information available at the conceptual stage of component design, then consideration of fabrication requirements could be integrated into the design route. If this can be successfully achieved, then the problems that often currently arise due to lack of welding expertise within an organisation or poor communication between engineers and welding technologists should be much less likely to occur in future, or

- Assist human experts (e.g., welding engineers), in which case the welding engineer would use the system to impose a higher level of consistency on his decision making by a process of iteration and interaction.

The expert system is a repository for all the up-to-date requirements of a particular field — including, for example, specifications, costings, welding procedures, research information — and should be serviced by all relevant experts within an organisation so that much greater efficiency of information exchange and usage is ensured. However, the expert system is not merely a data bank. It is capable of offering intelligent advice or of taking intelligent decisions by logical usage of the knowledge data base. Moreover, when required to do so, it must be capable of justifying the line of reasoning to the user. This is particularly important since the system is only as reliable as its input information and the advice given will have a degree of uncertainty associated with it. It is not a deficiency of the system because it is designed merely to replace human experts who are themselves often subject to uncertainty. Therefore the facility for continual justification of the advice given by an expert system is essential so that, if the line of reasoning is suspect, then a human expert can be consulted and may over-rule his electronic surrogate.

There would appear to be considerable scope for expert systems to be used in the field of welding technology. Some areas where benefits may be achieved are:

- Selection of welding process for fabricating specific components.

- Devising welding procedures for primary fabrication and in-service repairs.

- Approval of welding procedures.

- Materials selection and development.

- On-line control of welding processes by direct analysis and correction of parameter variation during the course of fabrication.

Figure 10.18 illustrates the component parts of an expert system with an explanation of some of the terms used and also some of the aspects that would need to be incorporated into an expert system de-

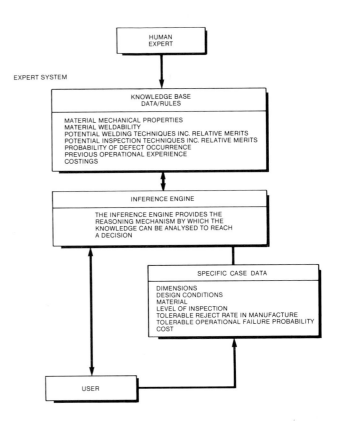

FIG. 10.18 Expert systems and their application

signed to give advice on a suitable technique for tube/tubeplate welding. It should be noted that an important feature of programs such as these will be the need to establish quantitative information on aspects which are currently considered on the basis of engineering judgement, e.g., tolerable in-service failure probabilities. Often it may not be possible to specify these figures exactly but a suitably devised expert system will enable a sensitivity analysis of these parameters to be carried out.

7.5 Adaptive welding systems

The previous sections of this chapter have illustrated the range of welding processes available for the fabrication and repair of power plant. Many factors have to be considered in process selection and for many applications there may be a choice of acceptable options. In other cases the correct decision will be clear cut. However, in all cases successful application depends upon correct procedure development and implementation. The standard methods for procedure development are well established but empirical, although more quantitative methods of assessing variability in weld quality do exist. Implementation depends upon quality control and implicit in this approach is firstly identifying what needs to be done and secondly devising a system that ensures that it is done. For manual welding, which is currently the most widely used technique, the traditional methods of supervision, inspection and repair are utilised. However, this will not be an efficient *modus operandi* with the increasing development and usage of fully automatic welding techniques to improve productivity.

In future much greater emphasis will need to be concentrated on ensuring that quality is engineered into the welding system. In this context there will be significant expansion in the area of adaptive control of welding where information obtained from the weld itself will be used to control the welding variables. Such systems will require sensor techniques to position accurately the welding torch at the point of initiation and seam tracking techniques to maintain the correct position during the course of the weld. Among current techniques are those based on mechanical devices which follow the weld preparation directly, or arc sensing systems that monitor changes in arc voltage as the arc gap between torch and workpiece varies during the course of a weld. However, both of these techniques rely on well defined weld preparations and may be inadequate for complex geometries. More sophisticated techniques based on laser range finders capable of providing three dimensional geometric information of the weld configuration are being developed and show promise for use in remote reactor repairs. In addition, by combining this information with computer models which predict the effects of welding variables on weld bead size, optimum joint filling conditions can be calculated. Thus it will be possible to find, measure and track a given welding preparation, accurately position the welding torch and place all the weld beads in the correct position.

Another aspect of adaptive control is the on-line monitoring and modification of welding parameters by direct computer control. This approach has been developed at the CEGB Marchwood Engineering Laboratories and successfully used in the modification of a heat exchanger unit in an advanced gas cooled reactor. This involved the removal of existing flow control devices and their replacement by extension tubes welded onto each hole of the tubeplate while maintaining a very high standard of surface finish and weld profile.

Metallurgical considerations prevented welding directly onto the tubeplate surface and it was necessary to build up the tubeplate with weld metal, machine a tube preparation in the built up region and weld on the extension tube using the internal bore TIG process. The entire operation which involved hundreds of welds was carried out *in situ* and, in order to ensure high quality and minimise operator error, a microprocessor based system was used to monitor and supervise each stage. By applying stringent controls during the manufacturing stage the amount of post-weld inspection was reduced.

The microprocessor control system was used for:

- Control of the sequence of operation and to maintain certain variables to within set reference limits.

- On-line monitoring of welding parameters.

- Modification of these parameters on the basis of information from sensing devices.

- Logging of data for subsequent analysis to confirm weld quality.

The face weld used to build up the tubeplate was made by the TIG process, with a filler wire addition and feedback control of the arc voltage to maintain the correct gap between the electrode and the weld surface. Other sensing devices monitored the wire feed speed shielding gas flow rate and electrical parameters. This information was fed into the microprocessor and was continuously monitored throughout the welding cycle. Any variations were assessed against reference parameter levels and, depending upon their extent, the system either activated an operator alarm or terminated the weld.

The internal bore TIG weld was made autogenously and had to satisfy stringent profile requirements. This was achieved by using a penetration controller based upon an optical radiation sensor facing the back of the weld. The output of this sensor was proportional to the molten weld pool size and, when the penetration obtained was satisfactory, the microcomputer terminated the first welding pulse. This was then repeated at $10°$ intervals around the tube until the tube was fully fused to the tubeplate. Other variables such as voltage, and pulsing parameters were monitored and compared against limits which had been experimentally derived and corresponded to a satisfactory weld. As a result the microprocessor was able to classify each weld into a category that required inspection and another that did not.

This example illustrates how all the aspects of a welding operation can be fully integrated and computer controlled. It is a complex process that requires a full understanding of the important controlling variables and considerable computer software development. However, in view of the potential technical benefits and cost savings it is an area of welding technology where future progress is likely to be rapid.

CHAPTER 11

Non-destructive testing

Introduction

For reasons of both economy and safety it is necessary to be assured of the integrity of modern power station plant. In many cases, particularly on nuclear plant, a demonstration of integrity is a formal requirement placed on the operator: for all plant, a reasonable level of assurance is called for by normal engineering judgement.

This assurance is obtained in two ways. For new or repaired plant, the materials and components are checked to ensure that the specified quality standard has been met. For operating plant, the standard is maintained by inspection, i.e., by ensuring that there has been no significant deterioration in service.

In some cases, destructive (or potentially destructive) tests may be called for — testing a sample of some component to destruction, for example, or carrying out a proof test on a pressure vessel. In most cases, however, there are sound economic reasons for using methods which will provide the required information on the condition of a component without rendering it unfit

for further service. This is referred to as non-destructive testing (NDT) and is the subject of this chapter.

First, the major NDT methods — visual, magnetic particle, liquid penetrant, radiographic, ultrasonic and eddy current inspection — are described and their use illustrated with some typical power station applications. The main advantages and disadvantages of each method are also summarised.

1　Visual inspection

A careful visual inspection is the first stage of any non-destructive test, whatever other methods are used. In some cases, indeed, it may provide the required information directly and avoid the time and expense of more complex methods. Very often, however, the object to be inspected is not directly accessible — it may be down a hole, round a corner, or in a hostile environment, for example. Here it is necessary to make use of the wide range of visual aids which are available, principally:

- Borescopes.

- Fibrescopes.

- Television cameras.

Borescopes are essentially special purpose telescopes fitted with a built-in light source, usually with either a forward or side-viewing head, although other viewing directions are possible. A wide range of sizes is available, although a typical example would be 10 mm diameter by about 1 m long. Typically this could be used for the inspection of the inside of a header after removal of an inspection nipple. For specialised purposes, borescopes are available with diameters as small as 2 mm, although these are fairly short and rather fragile. At the other extreme, sectional borescopes are available which can be extended up to 15 m (with a diameter up to about 30 mm). These could be used for the inspection of condenser tubes or rotor bores.

Borescopes produce an image of high quality but their rigidity may be a disadvantage and, if flexibility is required, a fibrescope is used. In these, the image is transmitted along a coherent bundle of optical fibres. The resolution is determined by the number of fibres and is generally lower than for a borescope. A typical general purpose instrument will have a diameter of about 8 mm and a length of about 1.5 m. Very thin (down to 2 mm) or very long (5 m or more) fibrescopes are available for special applications but they can be very expensive. In a typical application, a fibrescope could be fed into a complicated piece of machinery, a gearbox for example, to check for possible damage.

The use of closed circuit television equipment allows the range of visual inspection to be extended considerably and often into hostile environments such as the core of a nuclear reactor. Colour cameras are available but monochrome cameras are usually more compact and are adequate for most purposes. A typical commercial system has a waterproof camera 35 mm in diameter fitted with built-in lights and usable with a cable length up to about 70 m. Inspections can be recorded on videotape, providing a valuable permanent record. A typical application is the inspection of boiler waterwall tubes for internal pitting. Initially, an ultrasonic survey is usually carried out (as described later). Any sections showing severe damage will be cut out for repair and at this point it is useful to lower a television camera down the tubes to check for fine-scale attack which may be difficult to detect ultrasonically.

Visual inspection, usually with a suitable visual aid, has the advantages that:

- It is usually quick and flexible.

- In many cases, large coverage can be obtained in a reasonable time.

- The images are usually easy to interpret (although experience is still valuable).

But it has the disadvantages that:

- Not all defects are visible.

- In conditions of difficult access, the inspection may be severely limited by the quality of the image obtained.

2 Magnetic particle inspection

The previous section showed the power and versatility of visual inspection. But there are many cases where visual inspection, even with the best visual aids, cannot reveal fine surface discontinuities — either because they are too small or because there is too little contrast between them and the rest of the surface. There is a need to highlight these discontinuities and the next two sections describe the two most common ways of doing this.

This section describes magnetic particle inspection — usually abbreviated to 'MPI' in the United Kingdom. This works only on ferromagnetic materials but, as will be seen, that is not a great limitation for power station plant and is more than made up for by the advantages of the method.

Figure 11.1 shows the basic principle. The effect of magnetising a ferromagnetic material is visualised by drawing magnetic flux lines. These are continuous lines which are concentrated in the magnetic material but which spread out in air: the flux in air around a bar magnet will be familiar from school experiments with iron filings. The figure shows how surface-breaking discontinuities distort the regular pattern and produce a 'leakage flux' in the air next to the discontinuity. Finely divided magnetic particles applied to the surface, either as a powder or suspended in a liquid, will be attracted to the position of leakage and will reveal the discontinuity.

The figure also shows that sub-surface defects can be detected by MPI. The reliability of detection falls off very rapidly with depth, however, and for most practical purposes MPI has to be considered as a surface inspection method.

A practical magnetic particle inspection consists of the following main stages:

- Surface preparation.

- Magnetisation.

- Application of the magnetic particles (the magnetisation is usually maintained during this stage).

- Inspection.

- Cleaning and, if necessary, de-magnetisation.

The main points will now be considered in more detail.

2.1 Surface preparation

It will be seen that MPI is inherently capable of a very high sensitivity of inspection. That capability can

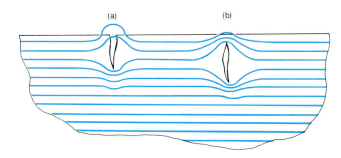

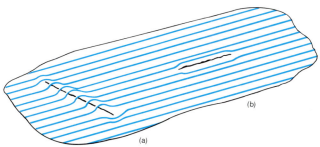

FIG. 11.1 The principle of magnetic particle inspection
shown here in cross-section
Defects in the ferromagnetic material distort the
magnetic flux lines, shown coloured. At surface-
breaking defects (a) a leakage flux is produced in the
air adjacent to the defect and this attracts the finely
divided magnetic particles to reveal the presence of
the defect. A sub-surface defect (b) may also disturb
the flux but this is a weaker and more diffuse effect
than for the surface-breaking defect. The reliability
of detection of sub-surface defects falls off very
rapidly with increasing depth.

FIG. 11.2 The effect of defect orientation
This perspective view shows two surface-breaking
defects and (in colour) the family of flux lines
immediately below the surface. For the defect at
right angles to the direction of magnetisation (a) the
flux lines are forced to jump through the air to give a
leakage flux. The defect lying parallel to the direction
of magnetisation (b) causes almost no distortion to
the flux lines and will not be detected.

only be realised if it is matched by a high standard of surface preparation. In particular, surface irregularities such as weld ripple or undercut must be ground to a smooth profile to avoid the formation of non-relevant indications which might mask real defects. The layer of magnetic iron oxide which forms on boiler components such as tubes and headers operating at high temperatures also presents problems, since the thickness and relatively high permeability of the layer reduce considerably the sensitivity of the test. Such components must be ground to bright metal before inspection.

In general, grinding is needed to prepare the surface of power station components for MPI, although wire brushing may be adequate for smooth surfaces which have not seen operation at high temperatures. The more violent methods such as shot blasting must be avoided because of the risk of peening over surface-breaking defects.

If a lower standard of test can be tolerated, then the surface preparation requirement can be relaxed. An example of this is the inspection of large lengths of weld in structural steelwork, where the critical defect size is usually large and where the cost of surface grinding would be prohibitive. But this course of action is rarely advisable on power station plant and must be used with extreme caution.

2.2 Magnetisation

The crucial factor which has to be considered here is the angle between the discontinuity and the magnetic field; this is shown in Fig 11.2. A discontinuity at right angles to the applied field effectively 'cuts' the flux lines, forcing them to jump through the air to give the required leakage flux. The discontinuity lying parallel to the field, however, causes almost no

disturbance to the flux lines and is unlikely to be detected. In practice, linear discontinuities lying up to 45° either side of the perpendicular direction can usually be detected and jagged discontinuities lying generally more nearly parallel to the field may also be detected. The objective, therefore, is to produce in the component under test a magnetic field of the correct strength, lying in the correct direction. In some cases the orientation of expected defects is known and the magnetisation technique can be selected accordingly. In general, however, it will be necessary to carry out two tests, with the surface of the component magnetised in perpendicular directions.

There are three main ways of magnetising a component:

- An electric conductor carrying a heavy current is arranged so as to produce the required magnetic field in the component — 'induced magnetisation'.

- The magnetic field is produced by passing the current through the component itself — 'current flow'.

- The component is made to complete a magnetic circuit — 'magnetic flow'.

The details of these techniques — the exact geometry, the current required, whether to use AC or DC — are complex and cannot be dealt with here. But it is possible to illustrate the general considerations by describing the techniques found most often on power station plant.

The 'coil technique', shown in Fig 11.3(a), is the most common example of *induced magnetisation*. The coil acts as a solenoid, magnetising the component along the axis of the coil and therefore giving the greatest sensitivity to defects lying perpendicular to the axis. A common use is the inspection for cracks along the toes of stub to header welds, with the coil wound around the stub close to the weld. The 'threader

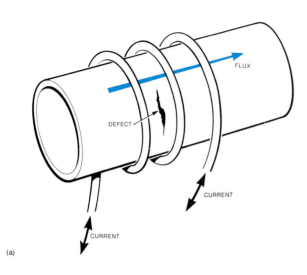

(a)

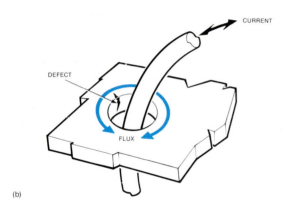

(b)

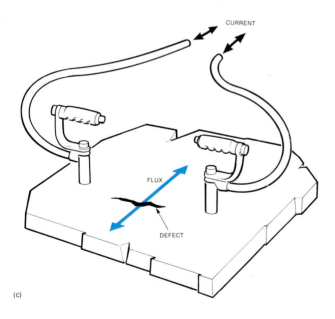

(c)

FIG. 11.3 Methods of magnetisation
The methods of induced magnetisation most often
used on site are the coil technique (a), commonly
used for pipes, and the threader bar technique (b),
used for inspecting the bores of holes.
The current flow technique is also widely used,
almost invariably with the current being applied
by hand-held prods (c).

bar' is another induced magnetisation technique, illustrated in Fig 11.3(b). This is used to magnetise the bores of holes to detect defects lying inside, along the axis of the hole. Other arrangements are also found: various types of coil may be used, both rigid and flexible, and one or more conductors may be laid adjacent to the area to be inspected.

The almost universal example of current flow found on power station plant is the 'prod technique', shown in Fig 11.3(c). A magnetic field is produced in the surface perpendicular to the direction of current flow and the technique is thus most sensitive to defects lying along the line joining the two contacts. The technique is simple to apply but currents of several hundred amps are needed and this can easily result in arcing and overheating which are hazardous to the operator and may be harmful to the component.

The *magnetic flow* techniques can be understood most easily by visualising a traditional horse-shoe magnet with its poles placed on the surface of a ferromagnetic component. The component completes the magnetic circuit and flux lines flow through it from one pole to the other. The technique is thus most sensitive to defects lying between the poles at right angles to the line joining them (confusion between this and the prod technique is common).

Permanent magnets provide a simple illustration but they are not particularly effective in carrying out MPI; the field they produce is small and many defects may be missed. Their use is confined to areas where an electricity supply is not available or not permitted. A much more attractive option is the use of a portable electromagnet yoke operating on 110 V AC. These yokes are very popular because they are quick and easy to use but some caution is needed. A good magnetic contact is essential and if the yoke is not positioned carefully, or if the soft iron feet become rounded, the sensitivity may be reduced.

2.3 Magnetic particles

The magnetic particles are usually applied as a suspension in a liquid — often referred to as a 'magnetic ink'. These magnetic inks can be water-based or kerosene-based. Water-based inks (which also contain wetting agents and corrosion inhibitors) are cheap and safe but their use may be prohibited: in the circuits of gas-cooled nuclear reactors, for example, or near electrical machinery. The kerosene-based inks avoid these problems and are generally considered to be rather more sensitive but they present problems of flammability, odour and possible damage to the operator's skin.

Magnetic inks are available in aerosol cans or they can be sprayed onto the surface with a pressure sprayer or flooded on from a plastic wash bottle. In all cases, the container has to be kept thoroughly agitated to maintain the magnetic particles in suspension.

The build-up of particles at a position of leakage flux can be made visible in two ways. Most commonly the particles have a distinctive colour — usually black — and here it is usual to apply a thin coat of quick-drying white contrast paint before the inspection. Alternatively, the particles may be made fluorescent and the component examined under ultra-violet light, as described below. This offers a very sensitive technique giving high contrast indications and it is widely used for critical applications or where contrast paint cannot be used.

2.4 The use of ultra-violet light

It has been mentioned that fluorescent inks have to be inspected under ultra-violet (UV) light. This will be met again in discussing penetrant testing and this is a convenient point to digress to describe the use of UV light in non-destructive testing.

The visible spectrum extends from red light at a wavelength of about 700 nm (10^{-9} m) down to violet at about 400 nm. Below this, from 400 nm down to about 200 nm, is the UV part of the spectrum. The radiation at the shortest wavelengths from 200 nm to about 320 nm, called 'hard UV', is very active; it causes sunburn, damages the eyes and kills bacteria. In contrast, the UV radiation between 320 nm and 400 nm — the 'near UV' — is not damaging and it has the additional useful effect of causing many substances such as minerals and dyes to fluoresce in the visible part of the spectrum. The fluorescent inks and penetrants used in non-destructive testing exploit this effect: when illuminated by near-UV they fluoresce with a yellow-green light, to which the human eye is most sensitive. So long as the ambient light is sufficiently subdued, small indications are very readily revealed.

The usual source of near-UV for non-destructive testing applications is a mercury vapour lamp. This produces a wide spectrum from visible light down to hard-UV and a filter is therefore fitted to remove all the harmful UV below about 300 nm and almost all the visible light, except for some deep violet. It is very important that this filter is securely fitted and not damaged: leakage of visible light decreases the contrast of any indications and leakage of hard UV is a serious hazard to the operator.

2.5 Summary

MPI is a powerful method for surface inspection. It is quick and fairly easy to apply and it is capable of high sensitivity if used correctly. It only works on ferromagnetic materials but these make up a very large proportion of the components of interest on a power station. To give some practical examples:

- Welds on thin-walled vessels such as tanks and de-aerators are usually inspected very quickly and simply using a portable electromagnet with a white contrast paint and black ink.

- Larger items such as steam chests and valve bodies are inspected most easily in service by the prod technique, again with white contrast paint and black ink. For alloy steel components, it is necessary to grind off the arc marks after testing and to retest the areas using an electromagnet.

- Turbine blades often need to be tested while still on the rotor. The usual way is to arrange a flexible coil so as to magnetise groups of blades in turn and to use fluorescent ink for high sensitivity and to avoid the use of contrast paint. For large components like this it may be difficult to arrange sufficiently subdued lighting and it is usually necessary to erect a temporary tarpaulin tent on a scaffolding framework.

MPI is a very natural technique for detecting service-induced cracking, since this almost invariably occurs at the surface of components. A common example in power station plant is the thermal fatigue cracking which occurs as a result of thermal expansion stresses at stress raisers, such as changes of section inside steam chests and valve bodies. The usual remedy — while sufficient thickness remains — is to grind out the cracks during overhauls and here the speed of MPI offers another advantage, since the excavation can be checked very rapidly to determine if the crack has been removed.

In brief, the advantages of MPI are that:

- It is a rapid and simple test.

- It produces direct indications.

- It works even if the cracks are plugged with debris or oxide.

- It works through a surface coating — although sensitivity may be considerably reduced.

- The equipment is rugged and relatively simple.

The disadvantages are that:

- It works only on ferromagnetic materials.

- It is unreliable for sub-surface defects.

- A sample must be magnetised in the correct direction.

- Odd shapes may be difficult to magnetise correctly.

- Heavy currents may be required, with problems of arcing and overheating.

3 Liquid penetrant testing

Magnetic particle inspection works only on ferromagnetic materials. Penetrant testing, which depends on the capillary effect, does not have this limitation. The basic principle is shown in Fig 11.4. The penetrant, which is either a fluorescent liquid or a coloured dye, is spread over the surface of the component, from where it is drawn into surface-breaking discontinuities by capillary action. Excess penetrant is

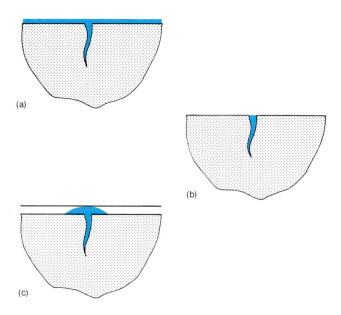

FIG. 11.4 The principle of penetrant testing
The penetrant is spread over the surface of the test
piece and is drawn into open defects by capillary
action (a). Excess penetrant is wiped off, taking care
not to remove penetrant from defects (b). An absorbent
developer is then applied to the surface and the
penetrant is drawn out of the defect, again by capillary
action, thus revealing its position (c).

then removed from the surface and a developer in
the form of an absorbent white powder is applied.
The penetrant is now drawn out of the discontinuities
into the developer, again by capillary action, and the
results are inspected either directly, in the case of
coloured dye penetrants, or under ultra-violet light, in
the case of fluorescent penetrants.

This simple description immediately illustrates both
the major advantage and the major disadvantage of
penetrant testing. The method does not depend on
the properties of the material, as in the case of magnetic testing: it works on magnetic and non-magnetic
materials, on metals and non-metals. However, defects
will only be detected if they communicate directly with
the surface and are not blocked with grease, oxide,
wear debris and so on.

There are six major stages in a penetrant inspection:

- Surface preparation.
- Application of the penetrant.
- Removal of excess penetrant.
- Development, to reveal any indications.
- Inspection.
- Final cleaning.

These points are discussed in more detail below.

3.1 Surface preparation

Poor preparation is the most common cause of test
failure. Contaminants such as oxide scale, paint, oil,

494

grease or wear debris can seal off surface openings
and prevent entry of the penetrant, or they can trap
excess penetrant on the surface and give rise to false
indications. Some contaminants, particularly oil and
grease, can degrade or mask the fluorescence or the
colour of the penetrant.

Mechanical cleaning may be necessary to remove
loose contamination but it must be used as sparingly
as possible to avoid closing up surface discontinuities.
Gentle wire brushing is usually the most that should
be allowed.

Wherever possible, cleaning should be done chemically, using paint stripper, if required, and a degreasing
agent. In critical applications, a metallurgical etch may
also be used. It must be remembered, however, that for
some materials and some applications there are restrictions on the use of chemical cleaning agents. The use of
an incorrect substance could result in the stress corrosion of austenitic steels, for example, and there are
very stringent requirements on the use of the materials
within the circuits of nuclear reactors. Specialist advice
must be sought in any case of doubt.

3.2 Penetrant application

Penetrant can be applied by dipping, brushing or
spraying. For inspections carried out on power station
plant, spraying from an aerosol can is used almost
exclusively, with brushing being used in a few specialised cases or where there are difficulties of access.

The penetrant must be allowed time to seep into
the surface discontinuities. This is usually called the
'dwell time' and is determined by experiment and past
experience. Times of up to 30 minutes are typical but
dwell times of up to an hour or more may be needed
for the detection of very fine cracking on austenitic
generator end-rings. For these longer dwell times it is
usually necessary to re-wet the penetrant periodically.

3.3 Removal of excess penetrant

The aim here is to remove the excess penetrant from
the surface without removing it from the surface discontinuities. There are three main types of penetrant
system, which are characterised by their method of
removal:

- *Water-washable* penetrants have an oily base but
  contain an emulsifier which allows the excess penetrant to be removed by a water spray. This has the
  advantages of speed, simplicity and cheapness and is
  widely used for testing large components during
  manufacture. It has the disadvantage of being the
  least sensitive system.

- *Post-emulsifiable* penetrants are not themselves
  water-washable but they can be made so by adding
  an emulsifier as an extra step after penetration. This
  blends with the excess penetrant on the surface
  and the mixture is then removed with a water spray.
  So long as the test is carried out correctly, the emul-

sifier penetrates the surface discontinuities only to a very limited extent and there is much less chance of penetrant being washed out of shallow defects. This has all the advantages of the water-washable system — apart from needing an extra process step — and the additional advantages of being more controllable and more sensitive. It is widely used for the critical testing of large numbers of components during manufacture where a high throughput is required.

- *Solvent-removable* penetrants, as their name suggests, are removed with a special solvent. This is the most sensitive system but it is not suitable for large scale applications. It is, however, very well suited for the relatively small scale, local inspections usually required on power station plant.

The water-washable and post-emulsifiable systems will be met only rarely in power station use and are not discussed any further here. Details can be found in text books, standards and the manufacturer's literature.

The solvent removable system is used extensively, almost always in the form of aerosol cans. These will include a 'remover' or a 'remover/cleaner'. As much of the excess penetrant as possible is first wiped off with a dry lint-free cloth and the surface is then cleaned using a cloth moistened with penetrant remover. The remover must not be sprayed directly onto the surface and the amount of cleaning should be kept to a minimum to avoid drawing penetrant out of surface discontinuities. If a fluorescent system is being used, the surface can be examined under ultra-violet light to check for complete removal.

3.4 Development

The absorbent developer powder can be applied in a number of ways, depending on the penetrant system being used. It can be a dry powder, or it can be held in suspension or solution in water or an organic solvent. For the solvent-removable system, the developer is usually suspended in a solvent and supplied in the form of an aerosol can. After thorough agitation of the can, this should be applied to give a light uniform coating.

The time required for development is again determined by experiment and previous experience. It must be long enough for the indications to develop but not so long that they become blurred as the penetrant spreads out through the developer. Times of 10 to 30 minutes are typical but longer times may be required for very fine cracks. It is usually helpful to inspect the component at intervals during the development.

3.5 Inspection

The success of the penetrant method depends on two factors:

- The enhanced contrast provided by the fluorescent or coloured indication.

- The sideways spread of the penetrant which produces an indication wider than the actual defect.

When applied carefully under favourable conditions, penetrant testing can reveal very small indications — certainly of the order of 1 mm or even smaller — and care must be taken to ensure that these are detected visually. For coloured penetrants, good lighting is essential while, for fluorescent penetrants, it is necessary to subdue the ambient lighting to increase the contrast of the fluorescent indications. For critical work, the use of a hand lens is recommended.

Defects such as porosity, inclusions or blow holes are revealed readily as fine dots and the size, shape and the extent to which the penetrant spreads can give an idea of the size and depth of the defect. Some very porous materials can 'bleed' penetrant so profusely that there is a risk that small but potentially more damaging defects, such as fine cracks, may be obscured. It is sometimes possible to detect these by cleaning the surfaces and re-applying developer but, as a general rule, the penetrant method is not suitable for the detection of very fine defects in porous materials.

Linear defects such as cracks will usually be revealed as continuous linear indications, although they may appear intermittent if the defect is fine or slightly closed.

In general, fine cracks produce faint indications while large cracks produce clear, rapidly spreading indications. The characterisation of crack type and depth from the appearance of penetrant indications is not, however, a very reliable process.

False indications are common and they almost always arise from incomplete removal of penetrant, especially where it is trapped in loose oxide, keyways, threads and so on. These are fairly obvious but the main risk is that they may mask real indications. If they cannot be avoided, further information may be obtained by watching the indications carefully as they develop.

3.6 Final cleaning

It is good practice to clean components after penetrant inspection and this is essential if the residues could damage the material — by causing general corrosion or stress-corrosion, for example. In some cases the components may need to be re-inspected — to monitor a defect to check for possible growth, for example — and here it is essential to remove penetrant from the defect before it dries out. Cleaning is done by similar methods to those used for surface preparation, making use of the remover-cleaner in the solvent-removable system.

3.7 Health precautions

A word of warning. The chemicals used in penetrant systems are often both flammable and toxic and can present a considerable safety hazard to operators. In

the large scale fixed penetrant testing installations used by manufacturers, considerable care is taken to cover tanks and to provide adequate ventilation. This is not usually possible under site conditions and the operator must be aware of the danger and arrange his work accordingly. This will involve, for example, taking fire precautions, providing ventilation in confined spaces and minimising the use of penetrant and the time of any exposure.

3.8 Summary

To summarise, penetrant testing is used widely for the surface inspection of non-magnetic components in power station plant, almost always in the form of the solvent-removable system although engineers should be aware of the existence of other systems. Typical applications include the testing of austenitic welds on superheaters and reheaters, austenitic-end rings on generators, copper conductor bars, aluminium generator wedges, brazed-on erosion shields and hard faced components such as discs and seats in valves.

The advantages of penetrant testing are:

- Versatility — it works on any material.

- Simplicity and economy — little equipment is needed.

- Direct indications.

- High sensitivity (under favourable conditions).

The disadvantages are:

- The discontinuities must be open to the surface and not blocked or contaminated.

- Porous materials or rough surfaces may be impossible to test.

- Penetrant materials can be harmful both to the operator and to the material being tested.

- An individual test is very slow compared to a magnetic test.

4   Radiography

The methods described so far are confined to surface inspection. Their usefulness should not be underestimated because of this — many defects, especially those arising in service, are surface breaking — but there will be many occasions where the inspection needs to be carried into the interior of a specimen, or maybe to surfaces to which there is no immediate access. In radiography, this is done by the use of penetrating radiation.

The basic principle is shown in Fig 11.5. The contrast is provided by differential absorption, with thinner or less dense areas on the specimen giving rise to a darker image on the film. The various components will be considered in more detail.

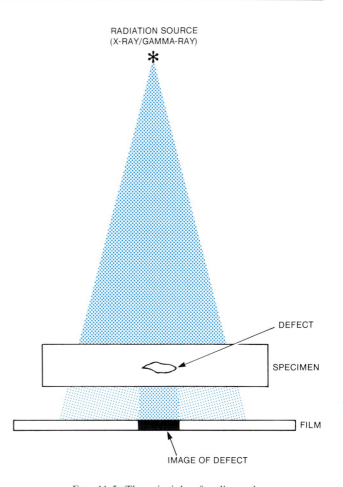

FIG. 11.5 The principle of radiography
The penetrating radiation (X-rays or gamma rays) passes through the specimen and exposes the film. Some of the radiation is absorbed by the specimen, the amount absorbed increasing with increasing thickness and density. Differences in absorption across the specimen give rise to the radiographic image. In this case, an internal pore absorbs less radiation and appears as a darker image on the film.

4.1   Sources of radiation

There are two sources of penetrating radiation suitable for carrying out radiography under site conditions:

- X-rays, which are produced when electrons are decelerated by colliding with atoms.

- Gamma-rays, which are produced during the radioactive decay of some isotopes.

Although an X-ray photon is indistinguishable from a gamma-ray photon of the same energy, there are considerable differences in the method of production and the practical details of using the two sources.

A typical X-ray source is shown in Fig 11.6. The maximum energy produced is determined by the accelerating voltage which can be up to 400 kV for the transportable equipment used on power station sites.

A gamma-ray source is much simpler, consisting merely of a small (a few mm) pellet of radioactive material, sealed in a capsule to prevent loss or damage. The complexity arises here from the need to store the

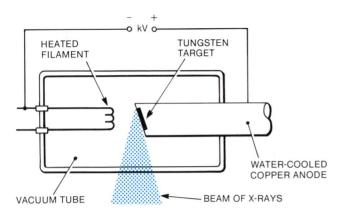

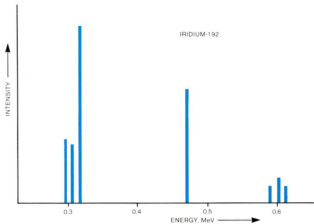

FIG. 11.7 Gamma-ray energy spectrum produced
by iridium-192
Like most gamma-ray sources, this consists of a
family of discrete lines.

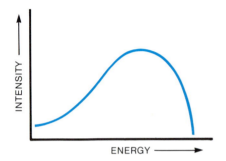

FIG. 11.6 Schematic view of an X-ray source
showing a typical energy spectrum
Electrons are produced from the heated filament and
accelerated across the evacuated tube by a high
voltage to strike a tungsten target attached to a
water-cooled copper anode. The sudden deceleration of
the electrons produces a beam of X-rays. The spectrum
shows a high energy cut-off which occurs when an
electron gives up all its energy in one collision. Not all
electrons do this and so a continuous spectrum is
produced extending to lower energies.

source in a shielded container when it is not in use
and to provide a mechanism to expose the source when
required with the minimum hazard to the operator.
The simplest containers have a removable cap or an
internal mechanism which allows the source to be
moved from the safe storage position to an 'expose'
position where radiation is emitted in a defined direction. Many other arrangements are possible. The most
common for power station work is one in which a
flexible cable is operated remotely to push the source
out of its shielded container and along a flexible tube
into the required position.

Several isotopes are used as gamma-ray sources,
most commonly cobalt-60 and iridium-192. The
gamma-ray energy spectrum produced by iridium-192
is shown in Fig 11.7 and it will be seen that, unlike
the X-ray spectrum, it consists of a family of discrete
lines.

A gamma-ray source is thus considerably more
portable than an X-ray tube and it can be used in
conditions of difficult access. The X-ray source has
the advantage, however, that it can be easily turned
on or off, while the gamma-ray source must be moved

out of or into its container. X-ray equipment can also
generate, within limits, a range of energies by varying
the accelerating voltage: gamma-ray energy is fixed
for a particular type of source.

There is a third, less welcome, form of penetrating
radiation. When X-rays or gamma-rays strike other
materials, secondary electrons are produced and these
are capable of exposing photographic film. Secondary
electrons produced within the specimen, surrounding
objects or the film itself, usually called 'scattered radiation', can significantly reduce the image quality.

4.2   Image recording

In principle, anything which suffers a physical or
chemical change on exposure to radiation could be
used to record a radiograph: fluorescent screens, now
being coupled with television recording and image
processing equipment, are used to carry out 'real time'
radiography and arrays of solid state detectors are used
in computerised medical scanners. Further development is likely to result in these techniques being used
more widely on site but, at present, the vast majority
of site work is carried out using photographic film
which produces a visible image after suitable processing.

Like ordinary photographic film, radiographic film
consists of a sensitive emulsion containing silver halide
grains on a transparent base. In order to absorb the
radiation as efficiently as possible, however, the emulsion is thicker than for conventional photography and
there is a coating on either side of the base.

In use, the film is usually sealed in a light-tight
cassette which can be placed in a suitable position
against the specimen. The cassette usually incorporates thin (0.02–0.2 mm) lead 'intensifying screens'
on either side of the film. Secondary electrons produced by the X- or gamma-rays in the front screen
assist in the formation of the image, while both screens

reduce the amount of low energy scattered radiation reaching the film.

The emulsion contains silver halide grains which are too small to be visible to the naked eye. Secondary electrons produced within the emulsion can, however, result in several grains being made developable for each quantum of radiation absorbed. This results in a blurring of the image, usually termed the 'film unsharpness' (sometimes 'inherent unsharpness'), U_f. As might be expected, U_f increases with the energy of the incident radiation. Typical values are 0.05 mm for 100 kV X-rays, 0.15 mm for 400 kV X-rays and 0.35 mm for cobalt-60 gamma-rays.

4.3   Factors affecting image quality

Image quality is characterised by two main factors:

● Contrast: the ability to detect small differences in absorption.

● Definition: the ability to detect fine detail.

Practically, it is also necessary to consider penetration: the ability to produce an image in a realistic time.

The variables in a radiographic examination affect these factors in a complex, interlinked way, as follows:

(a) *Energy of the radiation* Clearly, higher energy gives greater penetration. This can be expressed as the maximum thickness of material which can be inspected for a given energy. For a high sensitivity inspection of steel, for example, typical maximum thicknesses are 10 mm for 100 kV X-rays, 25 mm for 200 kV X-rays and 60 mm for iridium-192 gamma rays. If a lower sensitivity can be tolerated the values are correspondingly greater: the practical maximum for site work is about 150 mm of steel using cobalt-60 gamma radiation.

On the other hand, a higher energy of radiation results in lower image contrast since, if a large proportion of the radiation passes straight through the specimen, small differences in absorption will be more difficult to detect. For X-rays, the solution is to use the lowest accelerating voltage possible, compatible with a reasonable exposure time. The energy of gamma-ray sources is fixed, however, and this sets a minimum thickness which can be inspected for any given isotope. For example, the minimum thickness of steel which can be inspected at high sensitivity using iridium-192 radiation is about 18 mm, dropping to about 6 mm if a lower sensitivity can be tolerated. The corresponding figures for cobalt-60 are 50 and 30 mm and it will thus be seen that the commonly available gamma-sources are unsuitable for thinner-section steel components. Even above the recommended minimum thicknesses, the contrast obtained using the common gamma-sources is usually inferior to that obtained from X-rays. Lower energy sources such

as Ytterbium-169 and Thulium-170 overcome these difficulties to some extent and are coming slowly into more common use.

As mentioned above, increasing radiation energy also has a detrimental effect on definition, as the film unsharpness increases with increased scattering.

(b) *Geometry* Radiation sources have a finite size, usually of the order of a few mm, and this results in a blurring of the edges of the image, as shown in Fig 11.8. It will be seen that this blurring — or 'geometric unsharpness', U_g — can be minimised by:

● Minimising the source size.

● Minimising the specimen to film distance.

● Maximising the source to film distance.

The first two of these are, in general, outside the control of the operator. Source sizes are set by the equipment commercially available and for conventional equipment are probably already as small

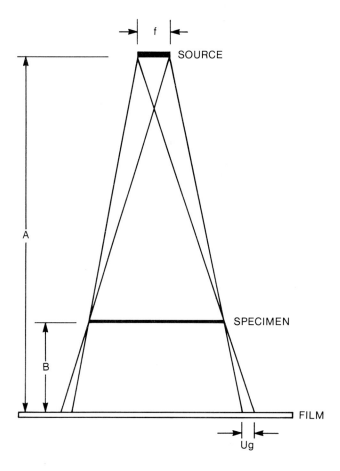

FIG. 11.8 The effect of geometry
The finite size of the radiation source produces a blurring, or penumbra, at the specimen edges — the 'geometric unsharpness', U_g. This can be calculated from the source size, f, the source to film distance, A, and the specimen to film distance, B, by similar triangles: $U_g = fB/(A - B)$.

as possible. 'Microfocus' X-ray tubes, in which the electrons are focused on the target to give a very small spot, are now being used more widely but are not yet regarded as general purpose tools.

The specimen to film distance can be minimised by placing the film as close as possible to the specimen but, even in the best case, the specimen to film distance cannot be less than the specimen thickness.

For a given specimen and equipment, then, the geometric unsharpness can be influenced only by the source to film distance. The criteria for setting this will be discussed in the next section.

(c) *Film type and processing* A range of film types is available, characterised by the grain size. Finer grain films give improved definition but at the cost of a longer exposure to achieve a given density. The density itself is important since, for X-ray films, contrast increases with increasing density. Radiographic standards usually specify a minimum average density.

For various types of film, processing conditions are defined in terms of times, temperatures and the freshness of the chemicals. In general, any departure from the specified conditions is likely to have adverse effects on both contrast and definition.

(d) *Scattering* The scattering of secondary electrons, as mentioned above, is a particularly potent source of loss of contrast. It can be limited by the use of screens and by collimating the beam, to restrict it to the area under inspection. For specimens of complex shape, it may also be necessary to add shielding around the edges of the specimen or within holes in the specimen.

4.4 Practical radiography

Practical radiography involves a compromise to achieve an acceptable balance of the variables referred to in the previous section It is possible to reduce geometric unsharpness, for example, by increasing the source to film distance but this will very rapidly increase the exposure required to achieve the specified film density, since the intensity of the radiation falls off as the inverse square of distance. A lower film density cannot be tolerated, since that will reduce the contrast, and the energy of the radiation cannot be increased, since that will both reduce contrast and increase film unsharpness. A faster film could be used but this would reduce definition.

A good radiographic procedure aims to achieve a balance of all these variables. Typically, the considerations might be:

- The film type is selected to give the required definition.

- The specimen thickness determines the energy to be used.

- The film unsharpness, U_f, is then derived from reference tables.

- The source to film distance is selected to give a value of geometric unsharpness, U_g, about equal to U_f.

- The exposure is calculated to give a suitable value of film density.

If the exposure is unrealistically long, further compromise will be necessary. This is usually done by a calculated reduction in the source to film distance: because of the inverse square law the benefit to exposure will be more rapid than the deterioration in U_g.

The use of an image quality indicator (IQI) is usually specified. This is placed on the specimen surface away from the film during exposure and consists of a series of wires of varying diameter or a series of metal steps each containing one or more holes. Other designs are also found. The 'IQI sensitivity' is calculated from the finest wire or smallest hole visible. This value cannot be related directly to the sensitivity to defects and it is best regarded as a check that the procedure has been correctly applied.

In general, X-rays produce higher quality radiographs than do gamma-rays. This arises partly because of the relatively high energy of the common gamma-ray sources, resulting in lower contrast, greater scattering and increased film unsharpness. In addition, the lower energies present in the continuous X-ray spectrum improve the contrast of the X-ray image. In spite of this, gamma-radiography is adequate for many purposes and is widely used for its simplicity and cheapness.

Radiography is readily applicable to butt welds where there is access to both sides. It can also be applied to butt welds in small bore pipes using a double-wall technique, as shown in Fig 11.9, although the sensitivity is reduced because of the double thickness of material. This situation is encountered very commonly on power stations in the repair or replacement of boiler tubes. Radiography is also widely applied to more complex geometries: tee-butt and nozzle welds, castings, etc. The coverage may be limited in such cases and each one needs to be assessed separately.

4.5 Radiographic interpretation

It is clear that a great deal of care and attention is required to produce a radiograph: equal care and attention must be applied to extracting the maximum information from it. The viewing area should be darkened and the interpreter should allow his eyes to become dark-adapted before starting work. A suitable viewer must be provided and the edges of the film must be masked to avoid glare.

Before commencing interpretation, it is necessary to check that the radiograph has been correctly identified (this is done by placing small lead letters on the specimen before exposure) and that the specified

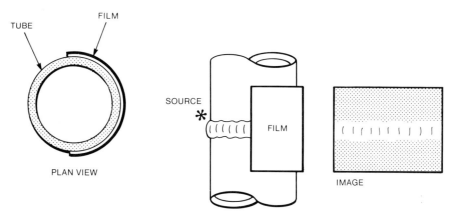

(a) Single image technique

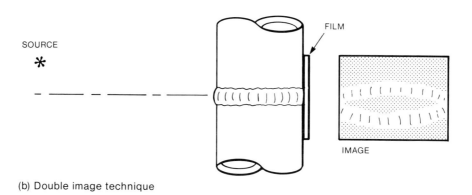

(b) Double image technique

FIG. 11.9 Double wall radiography of small bore tubes where the interior of the tube is inaccessible to both the source and the film
In the single image technique (a) only the wall adjacent to the film is imaged and three or more exposures are needed for complete coverage. Both sides are imaged in the double image technique (b) and the source is usually offset to separate the two images, giving an elliptical appearance. Because of distortion at the edges of the image, three exposures of 120° are needed for complete coverage. The exposures will be longer than for the single image technique because of the greater source-to-film distance.

density and IQI sensitivity have been achieved. The radiograph is also checked for artefacts such as pressure marks, scratches or dust spots.

When all these factors have been satisfied, the interpretation proper can begin. This can be illustrated with the most common application — the inspection of welds — although radiography is used widely for other applications.

The key point which must be borne in mind is that contrast depends on differential absorption. The amount of blackening at any point on the film depends on the amount of material through which the radiation has passed: the less material, the less attenuation and the more blackening of the film. The radiographic image is thus analogous to a 'shadow picture' projected onto the film. This is shown in Fig 11.10, which shows the effect of different weld root conditions.

Figure 11.11 shows a range of other weld defects. In general, it will be seen that three-dimensional de-

fects such as slag and porosity are readily detected and usually easily recognisable. The detection of planar defects such as cracks or lack of side wall fusion, however, is less certain and depends critically on the angle between the defect and the beam of radiation.

4.6 Radiological safety

If misused, the penetrating radiation used in industrial radiography can present a serious health hazard, both to individuals and possibly even to their children as a result of genetic damage.

It is not possible to discuss this here: the subject is too complex and the regulations change from time to time. It is essential, however, that anyone involved with radiography — whether the operator carrying out the radiography or the station engineer authorising the work — is properly trained, understands the dangers involved and applies the latest regulations diligently.

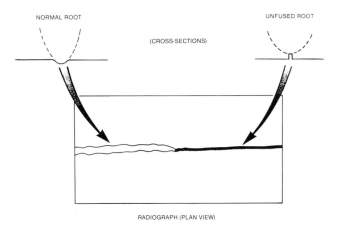

FIG. 11.10 Effect of different weld root conditions
on the radiographic image
A normal root bead appears as a slightly irregular
broad band at the centre of the weld image, lighter
than the general density of the radiograph because
of the thicker material. The unfused root appears
darker, because there is less material, and the image
is a sharp line resulting from the machined edges of
the weld preparation.

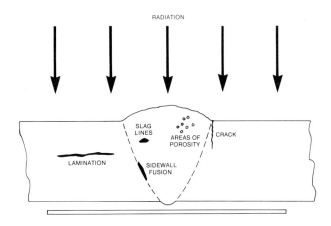

FIG. 11.11 Detection of typical weld defects
by radiography
Three-dimensional defects such as slag lines and areas
of porosity are readily detected and (usually) easily
recognised. Planar defects lying parallel to the direction
of the radiation beam, such as the crack, produce a
narrow line on the radiograph which may be detected if
the definition is sufficiently good. Planar defects at
other angles such as the lack of sidewall fusion and the
lamination are unlikely to be detected.

4.7 Summary

The most common application of radiography is undoubtedly to check the quality of new or repaired plant.
Most station staff will be familiar with the radiography
usually carried out on boiler tube repairs. The types
of defect found most readily by radiography — slag
in welds, shrinkage and porosity in castings, for example — are good indicators of the quality of workmanship and many product acceptance standards are
written in terms of radiographic inspection — the

maximum area of film over which porosity can occur,
for example, or the maximum length of a slag line. In
addition, radiography produces a visible record, which
can form a useful basis for contractual discussions.

Radiography is, however, also capable of more sensitive diagnostic inspections. Cracking can be detected,
if the direction of the beam is carefully chosen, and
it is also possible to detect, for example, corrosion
pitting at the bore of a pipe. Radiography also has
more general uses: a valve actuator may be radio-
graphed to check that its internal components are
working correctly, or boiler tubes may be checked for
the presence of blockages. These applications are not
routine ones and they will usually call for an experienced radiographer, careful preparation and a number
of trial exposures.

The advantages of radiography are thus:

- A large volume of material can be covered easily.

- A permanent record is produced.

- The sensitivity to three-dimensional defects is high.

- Sizes in the plane of the film (length of slag lines,
  area of porosity, etc.) are easily measured.

The disadvantages are:

- The sensitivity to planar defects, in particular cracks,
  is low.

- Sizes perpendicular to the film (for example, the
  through-thickness extent of a defect) are difficult
  to measure.

- Some geometries (tee-joints, nozzles, etc.) may not
  be inspectable.

- Safety considerations make the work difficult and
  may hold up other work nearby.

5 Ultrasonic testing

Like radiography, ultrasonics is able to inspect the
interior of a material, in this case by the use of elastic
waves, which are transmitted through materials and
reflected or absorbed by discontinuities. The principle
is the same as that used in echo-location (or 'sonar')
for the detection of submarines.

The velocity of elastic waves in materials of interest ranges between about 3000 and 6000 ms^{-1}. The
defects of interest are typically a few mm in size and
so the wavelength of the elastic waves must be of that
order, if not shorter. Using the relationship between
velocity, frequency and wavelength:

$$\text{velocity} = \text{frequency} \times \text{wavelength}$$

it can be seen that frequencies of a few MHz are
needed to give a wavelength of about 1 mm. Higher
frequencies would give shorter wavelengths and hence
more resolution, but the attenuation of the waves
increases rapidly at higher frequencies and so for normal applications (although there are special exceptions)

the frequency used for testing lies between 1 and 5 MHz. This is well above the audible frequency, hence the term 'ultrasonic'.

The following sections will consider the production and detection of ultrasonic waves and their use in non-destructive testing.

5.1 Basic equipment

The most common way of producing pulses of ultrasonic waves is by applying a short electrical pulse to a piezoelectric crystal of suitable thickness, as shown schematically in Fig 11.12. Such a device is known as an ultrasonic probe or, sometimes, 'search unit'. This makes use of the property of piezoelectric materials whereby an electrical voltage is converted into a mechanical deformation. Because the effect is reversible — mechanical vibrations being converted into an electrical signal — the device can also operate as a receiver of ultrasonic waves.

Naturally occurring piezoelectric single crystals can be used, of which quartz is the best known example, but much better performance can be obtained from synthetic ceramics such as barium titanate or lead metaniobate. These are polycrystalline and are made to display the piezoelectric effect by a process of 'polarisation' during manufacture — that is, cooling in an electric field from above a critical temperature known as the Curie temperature.

The instrument which is used to produce the electrical pulses and to detect and display the results is known as a flaw detector. This is shown schematically in Fig 11.13, which also shows the result obtained from a probe coupled to a simple test piece. This type of display, where the signal is plotted against

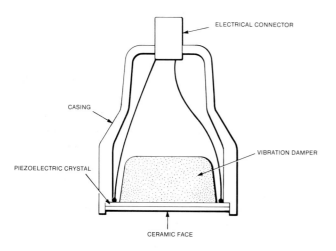

Fig. 11.12 Basic equipment — schematic view of an ultrasonic probe
The piezoelectric crystal is made to vibrate by the application of an electrical pulse. The vibration is damped to give a short pulse (a few cycles only) of ultrasonic energy which passes out through a protective ceramic face. The casing carries a connector for the electrical signals.

time (which is equivalent to range) is known as an A-scan display.

It will be seen that even this simplest arrangement is capable of measuring thickness and of detecting defects such as porosity or laminations in simple geometries, such as plate material. But this simple arrangement is not capable of handling the wide range of geometries and defect types found in practice and to see how this is done it is necessary to discuss ultrasonic wave propagation in more detail.

5.2 Ultrasonic wave propagation

The simple probe described above generates what is known as a compression wave, in which the atoms vibrate in the direction of propagation. The other important type of wave is the shear wave, where the atoms vibrate at right angles to the direction of propagation. These two types are illustrated in Fig 11.14. There are other types of wave motion, some of which are used in specialised tests, but they are not discussed further here.

The velocity of the waves depends, among other things, on the modulus of the material — the Young's modulus for compression waves and the shear modulus for shear waves. The stiffer the material, the higher the velocity. This has two important practical results. First, fluids (that is, liquids and gases) have very small shear moduli — they cannot support a shear force. This means that for all practical purposes, shear waves can propagate only in solids. Second, the shear modulus of solids is always lower than the Young's modulus. This means that shear waves travel more slowly than compression waves and hence have a shorter wavelength at any given frequency. At 4 MHz in steel, for example, shear waves have a wavelength of 0.8 mm, compared to 1.5 mm for compression waves. Shear waves thus have a higher resolution than compression waves of the same frequency.

Consider now the general case of a compression wave incident on a boundary between two different materials, as shown in Fig 11.15. The first point to note is that the waves are both reflected and transmitted — like light falling on a glass surface. The relative proportions reflected and transmitted depend on the acoustic impedances of the two materials, where acoustic impedance is the product of density and ultrasonic velocity. The greater the difference in acoustic impedance, the higher the proportion of energy reflected. Because of its low density, air has a very low acoustic impedance compared to all other materials of interest in ultrasonic testing and thus ultrasonic waves are almost completely reflected on meeting any interface with air. In all ultrasonic tests, a liquid or gel couplant is needed between the probe and the test surface in order to exclude all air. Without it, no energy would pass into the specimen.

The second point is that the reflected and transmitted components contain, in general, both shear and

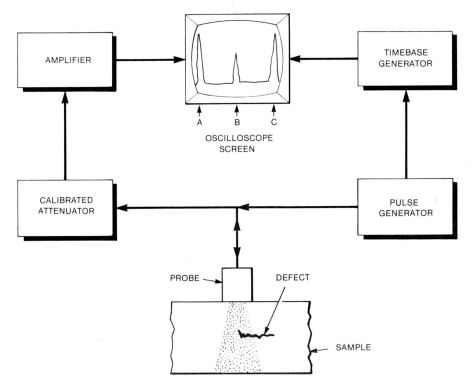

FIG. 11.13 Basic equipment — schematic diagram of a flaw detector
The pulse generator produces electrical pulses at a rate of typically 1000 per second. These are applied to the probe and simultaneously trigger the timebase generator, the output of which produces a horizontal deflection of the oscilloscope trace. The horizontal scale thus represents time, which is equivalent to distance through the sample under test. The output from the probe passes through a calibrated attenuator and an amplifier, which also rectifies the signal. This signal then produces a vertical deflection of the trace. The results show signals from (A) the initial pulse, (C) the back wall of the sample and (B) a defect at about mid-thickness.

compression beams. This phenomenon is known as mode conversion. The angles of reflection and — for the transmitted beams — refraction are related to the appropriate velocity by Snell's law, which states that sin θ/V is a constant — where θ is the angle and V the velocity. Since shear wave velocity is the lower, the shear wave direction in each material is nearer to the normal.

The case shown in Fig 11.15 is of an ultrasonic beam passing from lower to higher velocity material. The incident and refracted angles are related by:

$$\frac{\sin \theta_{1c}}{V_{1c}} = \frac{\sin \theta_{2c}}{V_{2c}} = \frac{\sin \theta_{2s}}{V_{2s}}$$

where the subscripts 1 and 2 relate to the materials and c and s to the compression and shear waves respectively. It will be seen that in this case the refracted compression wave is at a higher angle to the normal than the incident one ($\theta_{2c} > \theta_{1c}$, since $V_{2c} > V_{1c}$).

If the incident angle is now increased, the point will be reached at which $\theta_{2c} = 90°$, where the refracted compression wave is said to go critical. If the incident angle is increased still further, the refracted compression wave disappears, leaving only the shear wave. The shear wave angle at the critical point can be cal-

culated by putting $\theta_{2c} = 90$ in the relationship above. For steel, the compression and shear wave velocities are 3250 and 5920 ms^{-1}, respectively, and hence the shear wave angle at the critical point is sin^{-1} (3250/5920), or about 33°.

This provides a simple and effective means of producing a shear wave probe, as shown schematically in Fig 11.16. The piezoelectric crystal produces compression waves in a perspex wedge and these are converted to shear waves in the steel specimen. The compression velocity in perspex is 2730 ms^{-1} and substitution in the above relationship shows that a wedge angle of about 27° produces critical compression waves and, as shown above, shear waves at about 33°. Higher wedge angles produce higher angled shear waves up to a practical maximum of about 80°. Wedge angles below 27° will produce a mixture of compression and shear waves. These may be used in special applications but careful interpretation will be needed to avoid confusion between the different wave modes.

This ability to produce shear waves at various angles is of considerable importance in practical applications, as will be seen later.

Finally, it should be noted that ultrasonic beams are not parallel but that they diverge by an amount which depends on the ratio of the wavelength to the

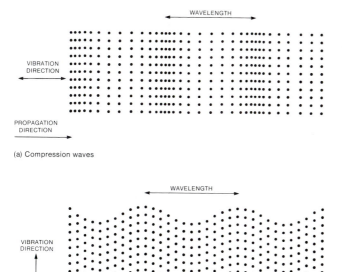

(a) Compression waves

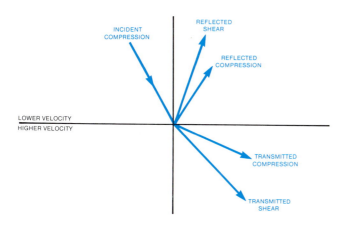

(b) Shear waves

FIG. 11.14 Schematic representation of (a) compression waves and (b) shear waves
For compression waves the atoms vibrate in a direction parallel to the direction of propagation; for shear waves they vibrate at right angles.

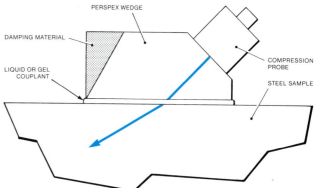

FIG. 11.16 Schematic view of a shear wave probe
A compression probe produces compression waves in the perspex wedge. These are mode-converted to shear waves as they pass into the steel sample. A thin layer of liquid or gel couplant excludes air from between the probe and the sample under test. Damping material traps reflected shear and compression waves within the wedge.

FIG. 11.15 General case of a compression wave incident on a boundary between two different materials
The wave is both reflected and transmitted and the reflected and transmitted waves contain both compression and shear components.

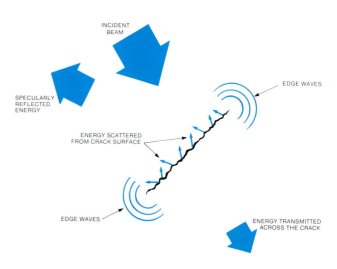

FIG. 11.17 Interaction of an ultrasonic beam with a crack
When the incident beam strikes the crack the majority of the energy is specularly reflected. Some of the energy is diffracted at the crack edges — the 'edge waves' — and some is scattered from the surface, if the crack is rough. The edge waves and the scattered waves can be detected over a wide range of angles but they are much weaker than the specular reflection. If the crack is under compressive stress some of the energy may be transmitted across it.

crystal size. Lower frequencies (that is, longer wavelengths) and smaller crystals both lead to more divergent beams. A typical value for a general purpose probe is 10° either side of the main beam direction.

5.3 Interaction of ultrasound with defects

Figure 11.17 illustrates the important practical case of the interaction of an ultrasonic wave with a planar

defect such as a crack. Of the variety of echoes produced, the strongest is usually the specular reflection — where the angle of reflection equals the angle of incidence — and this is usually the one relied upon for defect detection. Some typical testing arrangements are shown in Fig 11.18.

In general manual testing, it is simplest to use the same probe as both receiver and transmitter and in this case it is thus necessary to achieve normal incidence on the defect. To inspect a weld for lack of

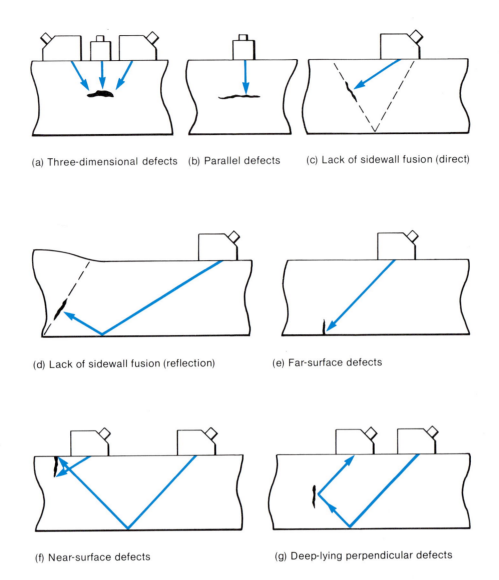

(a) Three-dimensional defects   (b) Parallel defects   (c) Lack of sidewall fusion (direct)

(d) Lack of sidewall fusion (reflection)   (e) Far-surface defects

(f) Near-surface defects   (g) Deep-lying perpendicular defects

FIG. 11.18 Detection of defects

Three-dimensional defects such as slag or porosity (a) can usually be detected from any direction. Planar defects, however, reflect most of the energy specularly and so, if the same probe is used both as transmitter and receiver, it is usually necessary to achieve normal incidence on the defect. Defects parallel to the testing surface can be detected with a normal compression probe (b) but for defects in other orientations, a shear wave probe of the appropriate angle is needed. To detect lack of sidewall fusion in a weld, for example, the probe angle is matched to the weld preparation angle and the sidewall is scanned either directly (c) or, if the weld cap is in the way, after reflection from the far surface (d). Defects breaking the far surface can be detected easily by the 'corner effect' (e), while defects at the near surface can be detected either by a high angle probe or by the corner effect after reflection from the far surface (f). Using a single probe, it is impossible to obtain normal incidence on defects perpendicular to the testing surface lying deep within the sample. In this case, a 'tandem arrangement' may be used (g).

sidewall fusion, for example, it is necessary to match the beam angle to the angle of the weld preparation, as shown in the figure. If the likely orientation of defects is not known, then it will be necessary to inspect the component with a variety of angles. In practice, the effect of beam spread means that a signal is obtained even if the defect is not exactly normal to the beam and a comprehensive inspection can usually be carried out using a 0° compression probe and 45°, 60° and 70° shear probes.

For some defects, especially those with their plane at right angles to the testing surface, it may be difficult to obtain normal incidence. If the defect is close to a surface, a strong signal may be obtained because of the 'corner effect', in which a beam striking two perpendicular reflecting surfaces is reflected directly back towards the transmitter. For defects near mid-section, however, it may be necessary to use two probes in a tandem arrangement, as shown in Fig 11.18(g).

It will be clear that amplitude alone is not a good guide to defect size, since the amplitude depends also on factors such as defect orientation and roughness. A large defect at an unfavourable orientation, for example, would give rise to a much smaller signal than a smaller defect normal to the beam. Another example

is shown in Fig 11.19. Signal amplitude is thus not used to assess size, except in special applications. Instead, most defects are sized by probe movement techniques, where the operator moves a variety of different probes over the test surface, observing the ultrasonic response in order to locate the edges of the defect. The details of this are too complex to describe here and, in fact, a number of different techniques are used, depending on the type of defect. The accuracy of sizing depends on many factors, but in a typical high quality test it is of the order of 2 or 3 mm.

Ultrasonics thus has the potential to detect planar defects, particularly cracks, throughout the thickness of a component, to locate their position and to measure their size. These are exactly the parameters needed for a fracture mechanics analysis, in which the integrity of a component is assessed by considering the stress-concentrating effect of crack-like defects. This match between the capabilities of ultrasonics and the needs of structural integrity is one of the most important reasons for the widespread interest in ultrasonic testing.

5.4 Practical considerations

Given the complexity of ultrasonic testing, it is not surprising that there are a large number of factors which can conspire to degrade the quality of the test. The most important of these concern access and surface preparation (see Fig 11.20).

An unground weld cap will limit probe movement and can give rise to confusing echoes which may mask real indications. Simple grinding to a smooth profile may not be sufficient, since long wavelength undulations form 'acoustic lenses' beneath the probe and distort the beam — rather like viewing a scene through hammered glass. Because of this, it is usual to specify a surface flatness of 1.5 mm deviation in any 50 mm, or 0.5 mm for critical applications. Fine scale roughness also needs to be carefully controlled to avoid serious attenuation of the signal, particularly at higher frequencies.

Ultrasonic testing depends to a large extent on probe movement and the quality of the test will be impaired if that movement is limited by, for example, adjacent bends, tapers, flanges or other attachments.

Many of these problems can be avoided by careful design of components or care in preparation. If problems remain, as is often the case for older welds which may not have been made with ultrasonic inspection in mind, the operator has to identify and report carefully any limitations to his test.

5.5 Display of ultrasonic results

The A-scan data presentation, as described above, is used widely and effectively in ultrasonic testing. But there are disadvantages:

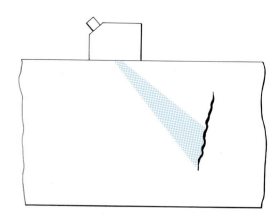

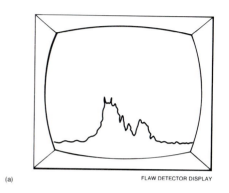

(a)   FLAW DETECTOR DISPLAY

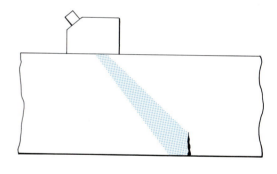

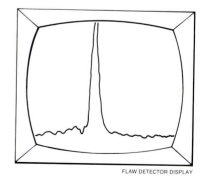

(b)   FLAW DETECTOR DISPLAY

FIG. 11.19 Signal amplitude as an indication of defect size Signal amplitude is not always a good indication of defect size. Both tests are carried out at the same sensitivity. The rough crack (a) away from normal incidence gives a much lower signal than the smaller surface-breaking crack (b) which gives rise to a strong corner effect.

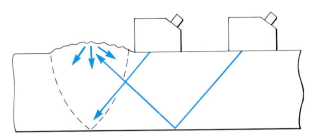

(a) Unground weldcap

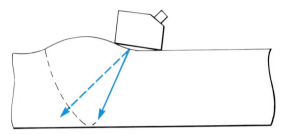

(b) Weldcap ground smooth but not flat

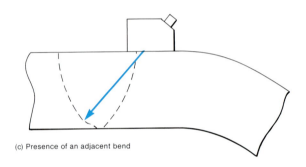

(c) Presence of an adjacent bend

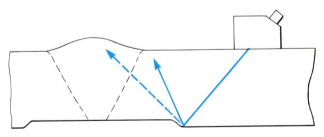

(d) Presence of a counterbore

FIG. 11.20 Factors which can degrade the quality
of an ultrasonic test
The most important of these concern surface
preparation and access. An unground weld cap (a),
for example, limits probe movement towards the weld
and can give rise to spurious echoes which may mask
real defects. Grinding to a smooth, but not flat, profile
can give rise to an 'acoustic lens' (b) which distorts the
beam. The presence of an adjacent bend (c) may limit
backward probe movement, with the result that the
weld is not fully tested from one side. Tests which rely
on reflection from the inside surface of a pipe depend
on the condition of that surface; the presence of a
counterbore (d), for example, can confuse the test.

- It relies on careful measurements by the operator of probe position and this is a possible source of error.

- The operator must transfer his results to a scale drawing of the component under test. Again, this is a source of error and it can become very tedious for complex geometries or extensive defects.

- There is no permanent record for later reference.

Because of this, there has long been pressure to improve the display of ultrasonic results and one of the first ways of doing this was the B-scan, as shown in Fig 11.21. This displays a cross-sectional view of the component, as seen by ultrasound, and may be more familiar as a medical diagnostic tool. The position and orientation of the ultrasonic probe are determined by an indexing device and the position of any ultrasonic echoes exceeding a predetermined threshold are plotted onto a storage oscilloscope screen. This can then be photographed to provide a permanent record.

It is also possible to produce views in the other two perpendicular directions: the C-scan or plan view and the D-scan or end elevation. Portable devices are now available which allow any of these views to be selected.

As described, the B-, C- and D-scans only show where the threshold has been exceeded: they give no further information on the amplitude of the signal. There are developments, however, which can record the signal and display the image with all the amplitude information — as a colour-coded image, or a contour plot, for example. Such devices are not yet widely used as site tools, although this is likely to change.

5.6 Automated testing

To a growing extent, large inspections are becoming completely automated. Motorised devices scan not one but a large battery of probes, the signals are mutliplexed into the flaw detector and are recorded, along with the positional information. The results are then analysed, usually by computer, to give B-, C- or D-scans as required. This offers several advantages:

- Inspections can be carried out in areas inaccessible to a human operator. For example, a small carriage has been devised which is able to move down the narrow space between a generator rotor and stator to inspect pole teeth without the need to remove the rotor.

- Inspections can be carried out in areas which it would be unsafe for an operator to enter. A wide variety of devices has been developed to inspect components within nuclear reactors.

- Perhaps the greatest advantage is the ability to inspect large areas quickly and reproducibly, with a comprehensive display of the results. An outstanding example of this is the inspection of the high pressure components at the CEGB's Dinorwig pumped storage station, as shown in Fig 11.22.

507

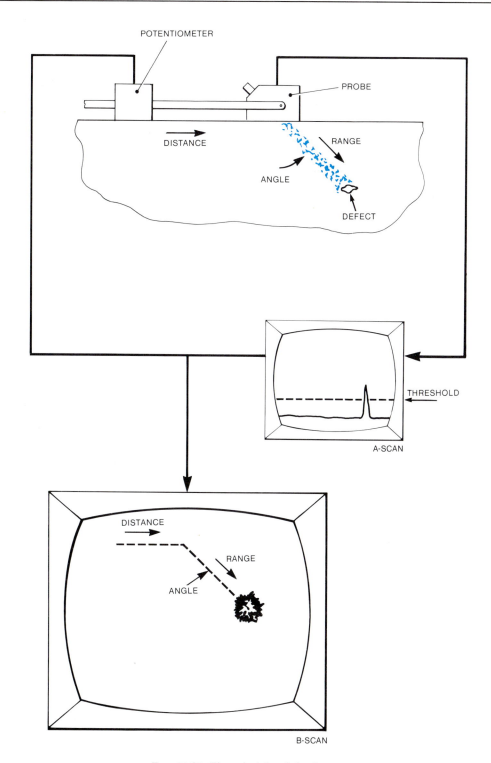

FIG. 11.21 The principle of the B-scan
The ultrasonic probe is connected mechanically to a potentiometer which gives an electrical signal
corresponding to the probe position. The B-scan equipment combines: • Distance, obtained from the potentiometer.
• Range, obtained from the A-scan. • Probe angle, preset on the equipment.
This produces a bright spot on a storage oscilloscope screen whenever the signal breaks a predetermined threshold on
the A-scan trace. The effect is to produce a cross-sectional view of the component as seen by ultrasound. Scanning
frames are also available to record the probe position in both the X and Y directions on the specimen surface and,
with more complex equipment, views can be produced as plans or elevations.

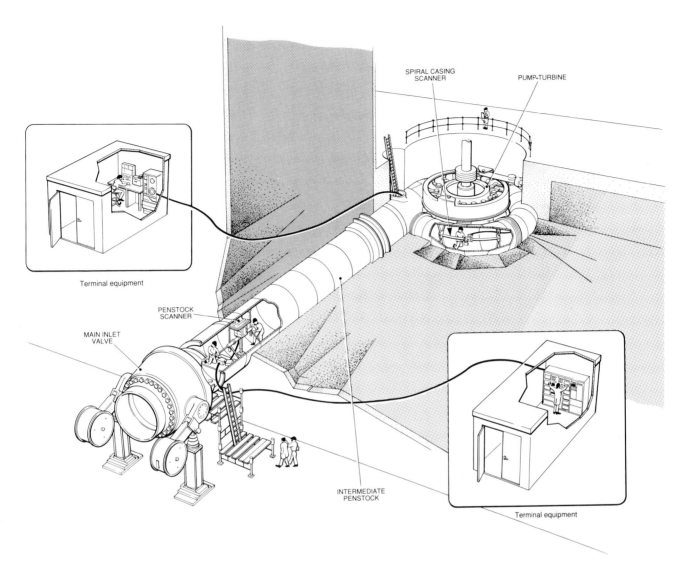

FIG. 11.22 Ultrasonic testing of penstocks at Dinorwig

The Dinorwig pumped-storage station makes use of two enlarged natural lakes with a difference in elevation of 517 m. Water is pumped to the higher lake at times of low electricity demand and then released back to the lower lake, the pumps now acting as turbines, to meet peaks in demand and to help stabilise the grid frequency. The steel intermediate penstocks and spiral casings see frequent pressure variations corresponding to the full head of water as the main inlet valves open and close, leading to the possibility of fatigue damage. The design and quality control applied during manufacture are such that no defect growth is considered possible, but as an extra safeguard, in-service inspections are carried out at planned intervals. The challenge is to inspect a large length of weld in a short outage period, with a high degree of reproducibility both in sensitivity and positioning. This is only feasible with automated equipment. With the main inlet valve shut, the system is drained and specially designed scanners are passed through manholes and assembled within the penstocks and spiral casings. Arrays of ultrasonic probes are automatically scanned along the welds and the results — both positions and ultrasonic responses — are passed out for recording and subsequent analysis.

Although such automated systems are not used widely at present, this is an area which is sure to grow in importance, helped by continuing developments in electronics and computing technology.

5.7 Summary

Ultrasonic testing has several important advantages:

- It can detect defects throughout the thickness of a component.

- It is sensitive to planar defects.

- It can readily provide information on the size and position of defects.

- It poses no health hazard and can thus be carried out without interruption to neighbouring work.

But there are disadvantages:

- The operator needs a high level of skill, both in carrying out the test and in interpreting the results.

- In many cases no permanent record is produced.

These disadvantages are being overcome by more advanced developments but this is often at the expense of increased cost and complexity.

6 Eddy current testing

The final method to be considered here is eddy current testing. This uses electric currents to probe the material under test, which must be a conductor of electricity. Although the method can reveal sub-surface defects — just as radiography and ultrasonics can — the depth of testing is limited by the material and the test parameters and the method is best described in most cases as a 'near-surface' one.

6.1 The basic principle

The basic principle is illustrated in Fig 11.23. An alternating current is passed through a coil adjacent to the sample under test, the frequency ranging between about 5 kHz and 1 MHz, depending on the application. This induces circulating electric currents in the sample, called 'eddy currents' from their resemblance to eddies in a fluid.

The induced currents set up their own alternating flux in opposition to that of the exciting coil and this couples the exciting coil electromagnetically to the sample, effectively forming a transformer with a single turn secondary coil. This 'transformer' will have a certain impedance, measured at the exciting coil, and this impedance will vary with any variation in the eddy currents. Since the system includes both resistive and reactive components, the impedance variation will in general display both amplitude and phase changes.

This then offers the possibility of detecting any phenomena which alter the distribution of eddy currents. The ones of most practical importance are:

- Geometrical variations in the sample, such as cracks, pitting or areas of thinning. This is the effect of most interest for power station plant use.

- Changes in the electrical or magnetic properties of the material under test. Eddy currents can be used, for example, to detect small variations in electrical conductivity resulting from the presence of impurities in the manufacture of copper wire. The method can also be used to check that steel components — valves for car engines, for example — have received the correct heat treatment, since both magnetic and electrical properties are sensitive to microstructure.

- The spacing of the coil from the specimen surface, termed 'lift-off' in eddy current testing. This is usually an undesirable complication to be avoided or minimised but it can be utilised to measure clearances in rotating equipment, for example, or the thickness of insulating films (usually paints) on metal substrates.

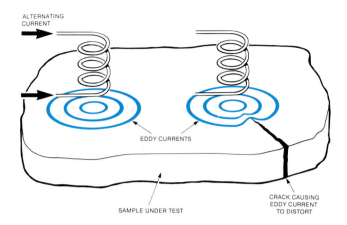

FIG. 11.23 The basic principle of eddy current testing
An alternating current is passed through a coil adjacent to the sample under test. Circulating electrical currents —'eddy currents' — are induced in the sample, coupling the exciting coil and sample together electromagnetically and effectively forming a transformer with a single-turn secondary coil. Anything which distorts the eddy current distribution, such as the crack, will alter the impedance of this transformer and can thus be detected.

6.2 Practical considerations

Although the basic principle of the method is fairly simple the practical application can be complicated, mainly because of the large number of effects which can cause changes in the eddy current distribution. The following factors need to be considered:

(a) *Skin depth* Because eddy currents are alternating, they exhibit the familiar 'skin effect' — that is, they are concentrated at the surface of the test specimen. The depth to which they penetrate — the 'skin depth' — depends on the test frequency and the electrical conductivity and magnetic permeability of the test specimen.

Increasing frequency, conductivity or permeability all decrease the skin depth. For non-ferromagnetic materials — copper, aluminium, austenitic stainless steel, for example — the relative magnetic permeability is 1 and so the skin depth at a given frequency is determined by the electrical conductivity. Typical values are of the order of 1 mm. For ferromagnetic materials, however, the relative permeability is about 1000, resulting in a considerably smaller skin depth.

Because of the skin effect, the test frequency will usually need to be selected to suit the particular application. A frequency of about 2 kHz, for example, would be suitable for the detection of corrosion thinning on the hidden side of aluminium panels a few mm thick — a frequent problem in the aerospace field. But such a test would be very insensitive to shallow surface breaking cracks, since these would result in a very small disturbance to the total eddy current distribution. The detection of such cracks in aluminium alloy components (another common task in aerospace inspection) requires a much higher test frequency — typically 100 kHz.

The very small skin depth which results from the high relative permeability of ferromagnetic materials means that for ferritic steels — which form the majority of power station plant — eddy current testing must be considered a surface inspection method. It is possible to increase the depth of penetration by magnetising the steel to saturation, thus reducing the slope of the magnetisation curve — and hence the relative permeability — to near unity but this requires complex equipment and is not well suited for site applications.

(b) *Test coils* The single coil shown in Fig 11.23 — known as an absolute coil — may be used in some applications but this arrangement is very sensitive to spurious effects such as changes in temperature or small changes in the electrical conductivity or magnetic permeability of the material under test. These effects can be minimised by the use of a differential arrangement, as shown in Fig 11.24. Two coils are used, forming two arms of a bridge circuit. A small difference in impedance between the two coils is revealed as an out of balance signal across the bridge. The two coils may be arranged in various ways and some of these are shown in the figure, which takes as an example the testing of tubes using external circumferentially-wound coils. This is an arrangement commonly employed for the continuous inspection of tubes during manufacture.

Many other arrangements of coil are possible, depending on the application. Fairly large, flat 'pancake' coils may be used for testing plate where a large coverage is required at modest sensitivity. For higher sensitivity tests, or for more precise location of the defects, a 'focused' coil wound on a small diameter ferrite core may be more suitable, although the coverage will be much smaller. Test coils may also be produced to suit particular geometries: for the detection of cracks at thread roots, for example, the end of the coil may be profiled to fit into the thread.

(c) *Test equipment* There are many effects which produce a change in the eddy current distribution and the parameters of a practical test must be selected to maximise the effect being sought and to minimise all others. The test equipment must then produce an easily understandable indication of the required effect.

Because the system has both resistive and reactive components it is necessary to consider both the amplitude and the phase of the eddy current signal and these are visualised most easily on a vector diagram, as shown in Fig 11.25. In fact, modern general purpose eddy current instruments

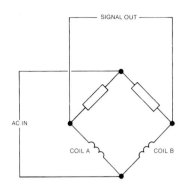

(a) The bridge circuit

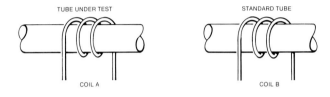

(b) Practical example of bridge application

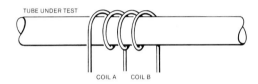

(c) Alternative arrangement for small defects

FIG. 11.24 The use of a bridge circuit
Small differences in the impedance of the two coils in the bridge circuit (a) are revealed as an out-of-balance signal. Changes in temperature affect both coils equally and thus do not give rise to spurious signals. A typical practical example is the testing of tubes using external circumferentially-wound coils. The tube under test may be passed through one coil, with the other placed in a standard, defect-free tube (b). Alternatively, the tube may be passed through the two coils wound close together (c). This arrangement is suitable for detecting defects smaller than the coil spacing, such as small holes or pits, or circumferential cracks. Defects with a longer axial extent, such as gradual changes in conductivity or tube dimensions, are not detected.

usually include such a display, usually in the form of a storage oscilloscope.

An example is given in Fig 11.25 and this shows how, with a suitable choice of test parameters, the effect of interest can be made to give a signal 90° out of phase with a confusing, unwanted effect. The desired signal can then be picked off using a phase sensitive detector or, on the more general equipment, the display can be rotated and the output taken along one of the axes.

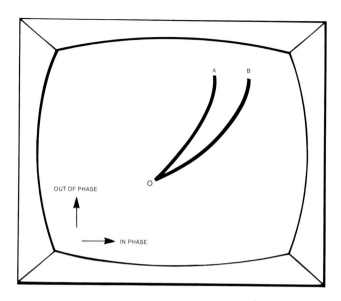

FIG. 11.25 Eddy current vector display

An eddy current signal contains both phase and amplitude information and this can be illustrated on a vector display, where the horizontal axis represents the in-phase component of the signal and the vertical axis the out-of-phase component. On such a display, the length of a vector represents the amplitude of a signal and the angle it makes with the in-phase direction represents the phase. Modern general purpose eddy current equipment usually includes such a display in the form of a storage oscilloscope. On the example shown, O is the signal with the coil in air, remote from the specimen. As the coil is brought into contact with the specimen, the signal follows the line from O to A — termed the 'lift-off line'. If the coil is now moved similarly on to a specimen with a slightly different conductivity, say, it may follow the lift-off line OB. The vector AB then represents the difference in conductivity between the two specimens — between a standard and an unknown, say. The signal can be displayed using a phase sensitive detector. Note that, in this case, the test frequency and coil design have been chosen so that AB is nearly perpendicular to the lift-off lines for both specimens. This means that the signal representing conductivity is little affected by small changes in lift-off — the ideal situation for a conductivity meter. Alternatively, the signal at right angles could be detected. This would register lift-off and be insensitive to small changes in conductivity, and would be suitable for a paint thickness meter. In general, eddy current test parameters are chosen to maximise the effect being sought and to minimise all others.

For many specific applications, modern electronics hides most of these details from the operator. Battery powered instruments are available, for example, to detect surface cracking using a 'pencil' probe. In such a piece of equipment the test parameters may be selected automatically to minimise the lift-off signal. A crack will then give a simple indication on a meter and an audible alarm if some preset threshold is exceeded.

(d) *Calibration* In most eddy current tests, defects are classified by signal amplitude and so the correct and reproducible setting of sensitivity is an important part of any inspection. For on-site test-

ing, the sensitivity is almost invariably set by reference to a standard specimen containing artificial defects — holes or slots in a tube, for example, or fine saw cuts in a plate.

This is usually satisfactory so long as the reference 'defects' are a reasonably good representation of the real ones. Small drilled holes, for example, are often used very successfully to calibrate equipment for the detection of corrosion pitting in tubes. But a test carried out at this sensitivity would probably fail to detect fine scale cracking, if this was present.

6.3 Practical applications

By far the most common application of eddy current testing on power station plant is the inspection of condenser tubes, usually to detect incipient pitting attack. Access is gained through the tubeplate from a water box after draining the cooling water and a pair of internal, differential, circumferentially-wound coils is used as shown in Fig 11.26. The coils are blown down each tube in turn and then winched back using the connecting cable.

With this coil arrangement, a pit or hole gives a bipolar signal as the first then the second coil passes over the defect. The figure shows the appearance on the vector display. The phase is selected to maximise the defect signal and this is then plotted on a chart recorder against time — which corresponds to distance along the tube, if the coils are withdrawn at a constant speed. A typical chart record is shown on the figure. The sensitivity is set by pulling the coils through a calibration tube containing a graded series of artificial defects — usually holes — and the defects are categorised according to their amplitude. Experience, based on metallurgical examination of typical tubes with a range of defect severities withdrawn from a condenser, suggests what severity of defect is likely to result in a leaking tube in the near future and tubes containing such defects are usually plugged.

Since modern condensers contain several thousand tubes, the emphasis is on speed of testing and typical rates of 40 tubes per hour, depending on tube length, are usually achieved. In a more modern development, the results are analysed on-line by a computer which gives an immediate indication of the condition of each tube and stores all the results for further analysis.

The other applications of eddy current testing on power station plant tend to be specialised and far less common than condenser tube testing. Examples are the detection of cracking in ventilation holes on older types of austenitic generator end-rings and checking for damage on the end of graphite sleeves in nuclear fuel element assemblies. However, the method has considerable potential and the number of applications is likely to grow considerably.

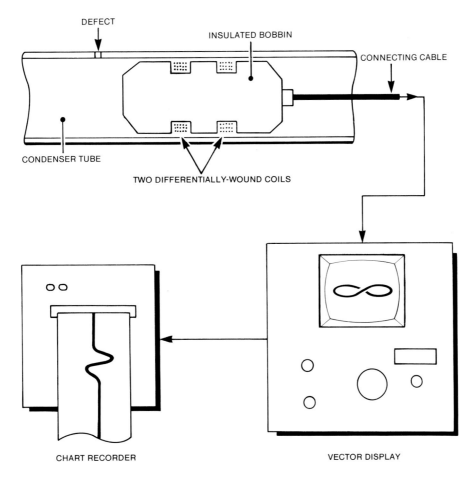

FIG. 11.26 Condenser tube testing

An insulated bobbin, carrying two differentially wound coils, is blown by compressed air down each tube in turn and winched back by means of the connecting cable. A defect, such as a small hole, gives rise to a bipolar signal on the vector display. An output signal with a phase selected to maximise the defect response is then displayed against time on a chart recorder. If the coils are withdrawn at a constant rate, the position along the chart can be related to the distance along the tube. The amplitude of the defect signal is categorised by comparison with the signals obtained from a calibration tube containing a series of defects of known size.

6.4 Summary

The eddy current method has the advantages:

- The probe does not need to be in direct contact with the testpiece.

- It works on a wide number of materials of importance to power station plant.

- It is very sensitive to near-surface defects.

- In favourable applications, high rates of testing can be achieved and a high degree of automation may be possible.

But the method has the disadvantages:

- A large number of effects give rise to eddy current signals and the interpretation may thus be difficult.

- The depth of penetration may be limited and the sensitivity and resolution may be limited for sub-surface defects.

- For other than relatively simple geometries, it is usually necessary to design and manufacture specialised probes.

7 NDT in practice

The previous sections have shown the great range and complexity of modern NDT methods. The understanding of these methods and the ability to apply them are of great importance, but of equal importance are the knowledge and experience of using the methods to solve practical problems: analysing a requirement for inspection, selecting the appropriate method (or methods) and ensuring that the inspection is carried out in the correct way.

To understand this properly, it is necessary to look beyond the technical details of the methods in order to appreciate the interaction of NDT with other disciplines such as design, structural analysis and metallurgy. The subject is complex and cannot be covered in detail here. The essential features can, however, be illustrated under three main headings:

- Specification of the problem.

- Design of the inspection.

- Control of the inspection.

7.1   Specification of the problem

Because NDT is so complex, with several major methods and many ways of applying each method, it can only be used efficiently if the problem which it is called upon to solve is sufficiently well defined.

This is encountered first in the interaction between the designer and the NDT specialist. Any practical component contains defects — either present in the original material or introduced during fabrication. To produce a 'perfect' component would probably be impossible and certainly uneconomic, and the aim of NDT during manufacture is partly to demonstrate that these inevitable defects are kept at an acceptable level. In designing a component, then, the designer must be aware of the capabilities of the NDT methods available, to ensure that unreasonable demands are not placed on NDT. The designer must also be aware of factors within his control which can limit the quality of an inspection: the presence of bends or flanges adjacent to a pipe-to-pipe weld which would limit access for ultrasonic inspection, for example, or unnecessarily complex geometries which can hinder radiography. The choice of materials is also relevant. In some older plant, components were fabricated from steel plate containing a large number of laminations and inclusions. The mechanical properties of this material may have met the original specification but if, at a later date, an ultrasonic inspection is called for to demonstrate the continued integrity of any welds it may be found that the 'dirty' plate completely hides the weld from the ultrasonic beam. Improved steel-making practices and better quality control have removed this problem but other materials present different problems. Austenitic steel castings or welds contain large, aligned, columnar grains. This gives rise to a highly anisotropic structure which can distort and skew the ultrasonic beam, producing in extreme cases a completely uninspectable structure. A thorough understanding of the physics of wave propagation in such materials, coupled with close collaboration with metallurgists and welding engineers, is needed in order to produce inspectable components in such materials. The problem of inspecting austenitic materials is aggravated still further since, because they are not magnetic, surface inspection must usually be carried out by penetrant inspection, rather than the more sensitive magnetic particle method.

Inspections on new or repaired plant are usually carried out to satisfy a quality standard which will specify acceptable defect sizes at a level intended to reflect good workmanship. These are invariably much smaller than the size of defect which could threaten the integrity of the structure. Radiography is often specified in such cases since the defects to which it is most sensitive — slag, porosity and so on — are good indicators of quality.

But defects may also be found on plant which has been in service. They may have formed in service — by fatigue or stress-corrosion, perhaps — or they may be original manufacturing defects revealed for the first time by an escalation of the inspection requirements. These defects are usually assessed on a fitness for purpose basis — that is, demonstration by a combination of inspection and structural analysis that the component is safe to operate. In this case, ultrasonics is almost always used, as it is particularly sensitive to cracks, which are of most concern to structural integrity, and because it measures those parameters — the through-thickness size and position of defects — which are needed by the structural analyst.

Different requirements for NDT may thus dictate the use of different methods. Even if the same method is used, it may need to be applied in different ways: ultrasonic inspection carried out to satisfy a fitness-for-purpose requirement will need to pay careful attention to accurate sizing and location of any defects, particularly as there is often a further requirement for periodic monitoring of some defects to check for possible growth. If an inspection is to be carried out properly, therefore, it is essential to understand what it is trying to achieve.

7.2   Design of the inspection

With the requirements of the inspection clearly identified, the NDT engineer is able to design a detailed inspection to meet that specification. This involves the choice of an NDT method — or methods — and the technique by which that method is applied. In many cases, established techniques are readily available but for novel geometries or critical applications extensive development work may be needed. An example of this is the development of manipulators to allow ultrasonic inspection to be carried out on components not accessible to manual operators.

An essential component in the development of any technique is the validation — the demonstration that the proposed inspection is capable of meeting the stated specification. This can be done in a variety of ways:

- At the simplest level, inspections are validated on a basis of experience and established practice. This is the approach followed, perhaps implicitly, in many 'standard' inspections where an established technique is applied to a familiar problem.

- Non-standard problems are often dealt with by straightforward demonstration. In developing an ultrasonic test to detect a particular defect in a particular component, for example, the inspection is often validated by demonstration on a mock-up containing a representative real defect, or a realistic artificial one.

- A similar type of 'test block' approach may be used in a more general way to demonstrate the capability of NDT methods. When the possibility of introducing a pressurised water reactor into the UK was first discussed, there was some uncertainty over the capability of ultrasonic inspection to detect defects of concern in the thick-walled pressure vessel. This led to a series of test block trials in which full size welds were made containing a wide range of deliberately implanted defects. These were inspected by a number of teams who had no prior knowledge of the defects.

- Such test block exercises are a very direct way of validating an inspection but they can be expensive and time consuming and it may not be possible to cover all possible geometries or defect types. Validation may also depend, therefore, on a sound physical understanding of the processes involved and, where appropriate, on the development of mathematical models of them. This allows the range of the validation to be extended and permits an assessment of the sensitivity of the inspection to the various parameters.

- In very critical applications, the final stage may be a completely independent validation of the whole inspection by an organisation not involved in its development.

The level of validation adopted will depend on the complexity of the problem and the importance which is attached to the results. One of the most demanding applications is the inspection of the pressure vessel of the proposed UK pressurised water reactor. All of the options described above are being used, in close consultation with the designers, in order to guarantee the integrity of this critical component.

7.3 Control of the inspection

However good the inspection which is developed, it can only be of value if it is applied correctly and a significant amount of effort is used to control the inspection. This is embodied in a written procedure — a document defining exactly how the inspection is to be carried out. It is essential that this defines unambiguously all the factors which may affect the outcome of the inspection since the job may be done by a number of different operators, or a component may need to be reinspected in subsequent years. The principal factors to be included in the procedure are:

- A careful definition of the scope, to ensure that the procedure is used only in the circumstances for which it was designed.

- A specification of the equipment and consumables to be used — the type of film for radiography, the type of ink for magnetic particle inspection, the calibration requirements for ultrasonic equip-

ment, and so on. In many cases recognised national standards are available to cover such items and should be invoked where appropriate.

- The requirements for the NDT operator. Many inspections depend critically on the skill, experience and integrity of the operator. A number of nationally-accepted certification schemes exist and an appropriate level of operator approval should be specified in the procedure.

- A clear and unambiguous description of how the inspection is to be carried out.

- The way in which the results are to be presented including, if possible, a standard reporting form.

In some cases it may be desirable to carry out an independent audit of the work. This may be a simple check that the correct procedure is being used, that the equipment is correctly calibrated and that properly certified operators are being used. It may also include an independent reinspection of a sample of the work, typically 5 or 10%.

8 Summary

This chapter has described the major NDT methods and has shown how they can be used to demonstrate the quality of new or repaired plant and to check for deterioration in operating plant.

The first and most basic method described was visual inspection. This has many advantages and should not be overlooked but it has obvious limitations even when assisted by such devices as borescopes or closed circuit television.

For surface-breaking defects, the limitations of visual inspection can be overcome to a large extent by the methods of penetrant or magnetic testing, which work by highlighting the defects. The advantages and limitations of the two methods were described and it was noted that the magnetic method is inherently more sensitive and more reliable and should be used in preference to penetrant testing wherever tests are carried out on ferromagnetic components. It was also noted that the characteristics of magnetic testing are well suited to the detection of service-induced defects, such as those arising from creep or fatigue, and this accounts for the extensive use of this method on operating power station plant.

Methods were then described which can also detect defects hidden within components. The first was radiography, familiar from its medical uses. This is a well established method which provides a permanent record and can readily detect three-dimensional welding defects such as slag or porosity. Such defects are good indicators of weld quality and radiography is thus used widely to monitor the quality of new welds. But radiography has its disadvantages: there is a health hazard and its use can disrupt work on surrounding plant. More importantly, radiography is usually rather insensitive to the planar defects, particularly cracks, which are of greatest structural concern.

The next method described was ultrasonics which, in contrast to radiography, is very sensitive to planar defects — so long as the test is carried out correctly. The method is also capable of measuring the position and size of defects and these parameters are exactly those required for structural integrity assessments. This explains to a large extent the considerable interest in ultrasonics. Even the brief description given shows the high level of understanding and expertise needed to apply the method correctly. The potential for further development was also noted, involving data storage and display, the use of automated systems and so on. Such developments are driven by the increasing demands placed on inspection and are fuelled by advances in electronics and computing.

The last method to be described was eddy current testing. The full potential of this method for power station plant has not yet been realised and it is used at present in only a few specialised applications. But, like ultrasonics, the method has considerable development potential and the number of applications is sure to grow.

Finally, the practical application of NDT was discussed. This is in some ways the most important section, although it was not possible to do it justice in this short chapter. The main point is that NDT does not exist in isolation, but needs to be integrated with design, structural analysis, welding, metallurgy and so on. There is a need to specify clearly the aims of any inspection and detailed technical considerations need to be taken into account in designing the inspection. Finally, close control is needed to ensure that the inspection is carried out correctly.

Non-destructive testing is an essential component in the task of assuring and demonstrating the integrity of power station plant. Its use is certain to grow with growing demands for efficiency and safety.

CHAPTER 12

Defect analysis and life assessment

1 Defect analysis within design life

1.1 Low temperature behaviour

Many engineering components are designed according to criteria which do not take into account defects such as cracks, porosity and inclusions, which may be present in the component. To overcome this problem design criteria often specify some 'allowable stress' to avoid failure by modes such as excessive plastic deformation (general yielding), buckling or fatigue while at the same time making economic use of the engineering material. However, engineering components often do contain cracks or crack-like defects which may have arisen during manufacture or from service exposure. The assessment of the significance of defects on the integrity of a component has very important implications with regard to both safety and economics. Therefore an understanding of defect behaviour can allow acceptance standards to be formulated, avoid the unnecessary rejection of components and enable decisions to be made on whether repairs are required immediately or can be deferred to a more convenient time, or indeed whether a repair

is necessary at all. An understanding of the behaviour of cracks in engineering components is therefore essential for the safe and economic operation of power plant. This section considers the analysis of the effect of defects at low temperature, i.e., at such temperatures that there is no significant time dependent material behaviour such as creep. The assessment of defect behaviour in the creep range under steady and cyclic (thermally induced) stresses is considered in Sections 1.2 and 1.3 of this chapter.

The failure of an engineering component by a fracture process can generally be considered to take place in three stages:

- Crack initiation.

- Crack growth.

- Final fracture.

Final fracture normally takes place very rapidly by, for example, a brittle crack running at approximately the speed of sound in the material. The initiation and growth stages effectively determine the life of the component and either may dominate.

For example, a smooth metal component essentially free of significant stress concentration which is loaded cyclically slightly above its fatigue strength could spend the vast majority of its life in the crack initiation stage. During this period very small defects in the metal surface develop into cracks which are generally contained within a single grain of the material. Very late on into the fatigue life such a crack begins to penetrate more deeply into the material. The growth of this particular crack then dominates the failure process as it progresses across the section. At the other extreme, the initiation stage may be non-existent, i.e., a defect may already be present, for example, a large inclusion or a weld defect, and under service duty this defect is capable of growing.

Some of the important crack initiation mechanisms are fatigue, fretting-fatigue, stress corrosion cracking and pitting corrosion (see Section 2.3 of this chapter). A well designed and manufactured component subjected to its normal design duty should not generate cracks by 'pure' fatigue. The fact that fatigue failures do occur is often attributable to a weakness in design and particularly to sudden changes in section which give rise to stress concentrating features. Fretting fatigue is an important consideration in shrink fit assemblies. In essence, microscopic movement between parts causes surface damage and the shrink fit stresses promote the development of fatigue cracking. A very large reduction in the fatigue strength of a material can be brought about by such fretting and can contribute to the failure of turbine rotors beneath shrunk-on thrust collars (see Section 2.3 of this chapter).

Defect assessments involve determining whether a defect at its current size can be expected to cause failure under the envisaged operating conditions and whether subsequent defect growth during service can be expected eventually to reach critical defect size and cause fast fracture. The assessment of the significance of defects is based on analysis by fracture mechanics which is briefly considered below.

1.1.1 Fracture mechanics

Much of modern fracture mechanics has developed from Linear Elastic Fracture Mechanics (LEFM) in which material behaviour is taken to be perfectly elastic. It can be shown that the stress field ahead of an embedded crack tip in an infinite body loaded in tension can be characterised by a single parameter — the stress intensity factor, K, given by:

$$K = Y\sigma \sqrt{(\pi a)}$$

where σ is the tensile stress and a the defect semi-length. This equation relates a crack tip parameter (K) to the crack length (a), the externally applied stress (σ) and a factor embodying the geometry of the body (Y). When the externally applied stress is sufficiently high, the crack may become unstable and extend causing failure of the body at a critical stress given by:

$$\sigma_f = K_c/[Y\sqrt{(\pi a)}] \qquad (12.1)$$

The value of stress intensity at the failure stress is a measure of the fracture toughness, i.e., the resistance of the material to unstable crack propagation. Within the normal range of experimental scatter of results expected in materials testing this fracture, toughness is independent of the geometry of the test piece (as long as elastic conditions prevail) and can therefore be considered as a material property at the particular temperature.

In its simplest form a defect assessment based on LEFM requires:

● Measuring or estimating the fracture toughness, K_c.

● Determining all relevant stresses contributing to crack opening.

● Establishing the geometric factor (Y) — solutions for which exist for many different components and loading circumstances.

Equation (12.1) can then be used to calculate the critical crack length at prevailing loading conditions. Defect assessments using such techniques have been successfully used for cracked welds in steam drums and cracked turbine discs.

1.1.2 Plastic behaviour

Although finding wide application, LEFM does have limitations. This can be appreciated by inspection of Equation (12.1) which predicts that the failure stress for a crack-free body would be infinitely large; this is clearly unrealistic, and arises from the assumption of elastic behaviour. The stresses near the tip of a crack loaded in tension are, according to LEFM, proportional to $K/\sqrt{(2\pi r)}$ where r is the distance ahead of the crack tip. Sufficiently close to the crack tip, therefore, the yield stress (σ_y) of the material will always be exceeded and a plastic zone of a size proportional to $(K/\sigma_y)^2$ will exist. If this plastic zone size is small, then the body behaves essentially in an elastic manner and LEFM can be applied. However, when the plastic zone is of a size comparable with the typical dimensions (the crack length or the remaining uncracked ligament) of the body then the stress intensity factor, K, can no longer be taken to be representative of the stress field ahead of the crack. It follows, therefore, that K_c is no longer truly indicative of the fracture toughness under conditions of large scale plasticity. This is illustrated in Fig 12.1 where a normalised fracture stress is plotted against a normalised crack length.

At the opposite extreme of elastic failure behaviour is 'fully plastic'. This mode of failure, plastic collapse, occurs when the shear stress across the net section is everywhere equal to the material's yield stress (in non-work hardening materials). This can occur, not only as in Fig 12.1 at the short crack lengths and high

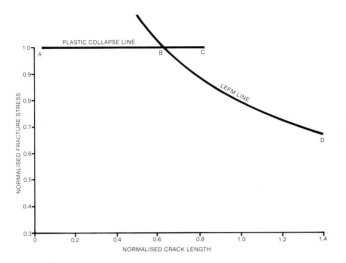

FIG. 12.1 Fracture stresses as a function of crack size
The fracture stresses are normalised by the material's
ultimate tensile stress σ_u, and the crack lengths by
$(K_c/\sigma_u)^2$. For longer cracks, the fracture stress is in
good agreement with LEFM predictions. For shorter
cracks, the LEFM line becomes increasingly inaccurate
as the plastic zone size at failure becomes comparable
with the crack length. At short crack lengths, the
failure stress is high, approaching the ultimate tensile
strength of the material.

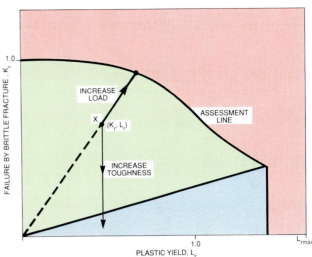

FIG. 12.2 The failure assessment diagram
The procedures outlined in Fig 12.1 allow an assessment
point to be plotted on a *Failure Assessment Diagram
(FAD)*. The ordinate (K_r) represents the proximity to
failure by brittle fracture, while the abscisus (L_r)
represents failure by plastic yield/collapse (L_{rmax}). The
assessment line interpolates between these two extreme
modes of failure. Points lying outside the assessment
line therefore represent defects that would cause failure.

failure loads, but at long crack lengths when the net
section stress is high. As with stress intensity factors,
many plastic collapse solutions are available in com-
pilations for various structural geometries and loading
circumstances.

1.1.3 The failure assessment diagram (R6)

In the preceding paragraphs it has been shown how
fracture can occur by distinct modes — brittle fracture, which can be treated by LEFM or fully plastic
fracture. These two modes represent two extremes of
failure behaviour, brittle fracture being governed by
crack tip events and fracture toughness, and plastic
collapse essentially being controlled by the net section
(the remaining uncracked ligament) and the material's
yield stress. In practice fracture behaviour is bounded
by these two extremes and an interpolation between
them (based on the J-integral or Crack Tip Opening
Displacement methods) forms the basis of the defect
assessment procedure widely adopted within the CEGB
— the so called R6. The failure curve is shown on
Fig 12.2, in the form of a failure assessment diagram (FAD) which reduces greatly the complexities of
elastic-plastic failure analyses. In effect two relatively
simple and independent calculations are performed,
the elastic and plastic, as summarised in the flow
diagram of Fig 12.3, and an assessment point (X) of
the co-ordinates K_r (crack tip failure), L_r (plastic yield)
is plotted on the FAD. If K_r, L_r lies within the red
sector, failure is conceded. From point X an increase
in load is required for this, so the structure is safe.

An increase in toughness would lower the point of
K_r, L_r, indicative of an increase in margins. If point
X fell into the blue sector, failure could only occur
by plastic collapse, at a loading of L_{rmax}. The simplest
FAD requires that $K_r < 1$ and $L_r < L_{rmax}$; this ignores
any interaction between the two failure mechanisms.

The assessment of the significance of a flaw in a
structure requires careful consideration of a number
of factors as the formalised procedure shown by Fig
12.3 indicates:

- The defects must first be characterised, i.e., their
  shape, size, orientation and location within the
  structure must be established.

- The material properties at the appropriate temperature and environmental conditions must be
  determined; these are principally measures of fracture toughness and plastic flow properties.

- All important loadings of the component must be
  known. These may give rise to primary stresses (σ^p)
  such as pressure stresses, and secondary stresses (σ^s)
  such as residual stresses arising from welding.

- An assessment point may then be plotted on the
  failure assessment diagram to determine whether or
  not failure is avoided.

- If failure is avoided, consideration may need to
  be given to the possibility of defect growth during
  service and another assessment point plotted.

- Finally, some investigation of the sensitivity of
  the defect analysis to variations in the input data
  should be made.

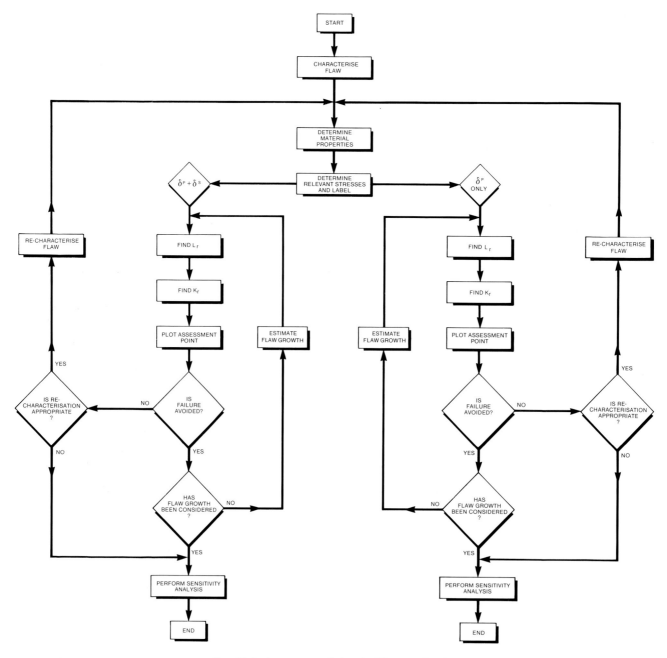

Fig. 12.3 Assessment of the significance of a flaw

In carrying out assessments of this kind, it must be emphasised that care is taken to consider all potentially significant stresses (for example, residual, thermally induced or system stresses) under all operating conditions (steady state, start-up and faulted). Care must also be taken that the materials property data used are suitably lower band since there is no explicit safety factor incorporated into the R6 procedure.

An uncertainty often encountered in these analyses is in quantifying defect sizes. Thoughtful consideration must therefore be given to the sensitivity and reliability of any non-destructure test method used for defect sizing.

It is rare, if ever, that during a defect assessment attention is confined to a single (or postulated) defect of fixed size. Almost invariably careful consideration must be given to the possibility of defect growth, either to show that a defect will not reduce the effective life of a component or to determine future inspection intervals or replacement dates.

1.1.4 Defect growth by fatigue and environmental attack

By far the most common defect growth mechanism arises from fatigue caused by fluctuating or cyclic loading. Under static conditions, however, aggressive environments may also promote defect growth by stress corrosion cracking (SCC) mechanisms.

Cyclic loading is a common operational feature, for example, in pressure vessels arising from pressurisation/depressurisation cycles or in rotors where self-weight bending occurs at the frequency of rotation. In some cases the amplitude of the load fluctuation may be relatively constant throughout defect growth as in the case of cycles between zero and normal operating pressure for pressure vessels, or extremely variable, perhaps near random, as in the case of vibration induced stresses in turbine blades. Fracture mechanics concepts can be applied to both variable and constant amplitude loading but attention will be restricted to the latter case.

Just as the stress intensity factor, K, has been applied to fast fracture, fatigue crack growth can be described in terms of the range of stress intensity factors (ΔK) during cyclic loading. A fatigue crack growth test may be conducted by subjecting a fatigue-cracked specimen to constant amplitude load fluctuations and periodically measuring the crack length, a, and the number of elapsed loading cycles, N. The measured fatigue crack growth rate ($\delta a/\delta N$) may then be plotted against the stress intensity range, ΔK, as shown in Fig 12.4. For many materials the crack growth rate dependence on ΔK falls into three regions. In the first, a fatigue threshold ΔK_{th} is, exhibited, i.e., a value of cyclic stress intensity range exists below which the fatigue crack growth rate is zero. Above the near-threshold region, the linear portion of the curve can be represented by:

$$\frac{\delta a}{\delta N} = A(\Delta K)^n$$

where A and n are material constants. In the final region III the crack growth rate accelerates rapidly as the maximum value of the stress intensity range during the fatigue loading cycle approaches the material's fracture toughness, K_c.

A further important factor in determining fatigue crack growth rate, in addition to the cyclic stress intensity range, is the mean stress ratio, R, which is K_{min}/K_{max}. The growth rates in region II are little affected by the stress ratio, but in both regions I and III the growth rates tend to be higher for higher values of R.

The preceding comments on fatigue crack growth are fairly generally applicable but ignore a major factor influencing crack growth and threshold, namely the environment. Power plant items can experience a wide range of environments, from high purity water and/or steam in the boiler circuit and turbines to hydrogen gas in the generator. The environment need not necessarily be regarded as aggressive towards a particular material but may still significantly affect its fatigue behaviour. For example, Fig 12.5 shows the effect of hydrogen gas on the fatigue crack growth rates of a generator rotor steel. Unlike many high strength steels, the rotor steel shows no tendency to hydrogen-induced cracking under static loading, but under fatigue loading hydrogen can greatly enhance

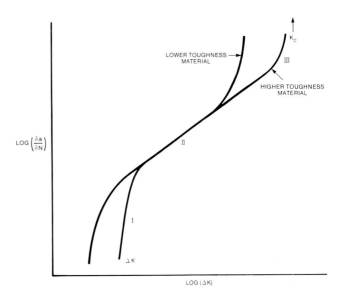

FIG. 12.4 Fatigue crack growth rate dependence on ΔK The fatigue crack growth rate often shows three regions of dependence on the stress intensity range. In region I the growth rate is particularly sensitive to mean stress, microstructure and environment. Region II is relatively insensitive to these parameters.

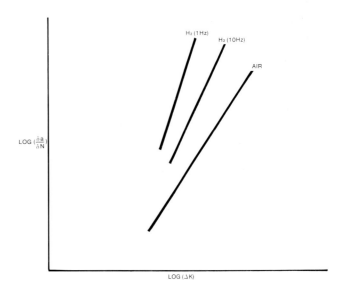

FIG. 12.5 Effect of hydrogen on fatigue crack growth rates of generator rotor steel

the crack growth rate compared with that measured in air.

A notable feature of this enhancement of growth rates is its dependence on frequency. Similar behaviour has been observed in water environments where the growth rates at high frequencies are very similar to those in air but faster growth is observed at lower frequencies.

1.1.5 Proof testing

Some components, particularly pressure vessels, are sujbected to a proof test to give some assurance that

521

the component will be fit for future service. An overpressurisation with a load considerably in excess of its normal service loading is normally used. This procedure can enable some inferences to be made with regard to the size and type of defects present in the vessel at the time of testing. Great care is required, however, in interpreting the result of such a simple test in terms of the future integrity of the component under normal service conditions. This is due to both the possibility of defect growth during service and the nature of the test itself.

From what has already been discussed above, the principal parameters determining defect tolerance are the imposed stresses and the material properties (toughness and plastic flow) and both these groups of parameters may be significantly different under the proof test and the normal operating conditions.

It is likely that the only significant stresses during a hydraulic proof test are those due to the internal pressure with perhaps some contribution from residual welding stresses. In service however, thermal stresses may contribute and their significance will depend on the position and orientation of any defects with respect to the induced tensile stresses. The fracture toughness and yield stresses are temperature dependent and proof testing is often carried out at a temperature other than the operating temperature and this must clearly be taken into account.

In order to make a conservative calculation of 'safe life' it is necessary to estimate the maximum size of defects which may survive the test and the minimum likely critical defect size under the most onerous operational conditions. Any assumptions made in the assessment calculation must therefore be optimistic (in terms of tolerable defect size) for the proof test part and pessimistic for the operational part. The estimation of crack growth rate between these two crack lengths during service should also be an upper bound consideration.

When all these factors are taken into account it is quite possible to discover that a proof test might give little or no assurance of any future safe operating period. In this circumstance it may be necessary to refine the assessment calculations by making more detailed stress analyses or materials property measurements. Stress analysis may highlight parts of the plant which give cause for concern and these can then be non-destructively examined if there is a suitable method with an adequate resolution.

1.2 High temperature behaviour (creep crack growth)

Defect behaviour at temperatures within the creep range has close parallels with that at lower temperatures where failure is bounded by fully plastic and brittle modes. The corresponding cracking modes at creep temperatures are usually referred to as creep ductile and creep brittle.

Creep ductile cracking behaviour is exhibited by materials of creep ductility typically greater than around 10% and is analogous to plastic collapse at low temperatures. The effect of the crack is principally to reduce the section thickness with failure controlled by the average stresses across the section ahead of the crack rather than those at the crack tip.

The failure mode is illustrated schematically and microscopically in Fig 12.6. Highly stressed material ahead of the crack tip is the first to enter tertiary creep and develops a high concentration of cavities and/or microcracks. The damaged material can no longer sustain the applied load which transfers to adjacent material of the structure (or continuum). The increased stress level then accelerates this material into tertiary creep when it is itself heavily damaged. The process continues with time and can be envisaged as the progression of a front enclosing the increasing volume of damaged material. The observed macroscopic cracking occurs by coalescence of cavities or microcracks behind the damage front and does not itself affect failure, which occurs when a sufficient proportion of the component section is so damaged that the remaining volume is unable to support the load.

Failure in the creep ductile mode is often referred to as *Continuum Damage Mechanics*. The development and propagation of damage zones in complex structures is amenable to analysis by finite element methods, from which it emerges that the failure time of structures by creep ductile cracking approximates to the rupture time at the reference stress. The reference stress (which is related to the low temperature plastic

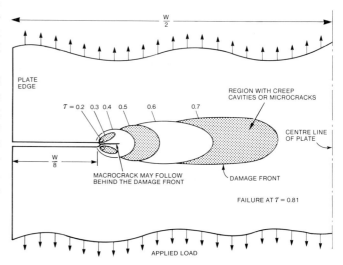

FIG. 12.6 Crack front development due to creep
Finite element computer solution of the development of a blunt damage front with microcrack, following creep ductile cracking for a notched plate held under instant load at high temperature. The illustration shows one half of the plate which has two symmetrically-placed cracks, one at each edge. Failure occurs when the damaged fronts meet at the centre.

collapse load) can thus be used to estimate failure times of cracked structures from the uniaxial creep properties of the material obtained from smooth (uncracked) specimens. This is seen to be valid because the initial crack is effectively blunted by the progress of the damage front.

Creep ductile cracking in microstructurally homogeneous material under well defined steady stresses is however a relatively uncommon occurrence on power plant. The majority of creep cracking problems are found to be associated with creep brittle microstructures, especially those occurring at weldments. The situation is then frequently further complicated by the interaction of system and or residual stresses with the steady state pressure stresses.

Creep brittle cracking behaviour occurs in materials exhibiting creep ductilities within and below the 5–10% elongation range. Cracks in such microstructures effectively remain sharp and initiation and growth are relatively rapid. The lifetime of the component is therefore governed by the stresses and strains local to the crack tip in parallel with the behaviour of brittle cracks at low temperature. The period under load at the crack tip required before propagation can begin is known as the incubation time (t_i). Incubation times can be characterised in the laboratory in terms of the displacement recorded between the crack surfaces as the crack begins to propagate (Fig 12.7). The crack opening displacement (δ_i) is a material property and relates to the incubation time by an expression of the form:

$$t_i/t_r(\sigma_{ref}) = F_{MAT} (\delta_i/R)$$

where $t_r(\sigma_{ref})$ is the uniaxial creep rupture time at the reference stress. F_{MAT} is a function determined from the constants of the uniaxial creep law of the material obtained from creep tests and R is a characteristic length obtained from tables of standard solutions which relates the stress/strain conditions local to the crack tip to the nominal conditions in the structure.

The main problem in treating creep brittle crack growth is, as fracture mechanics has successfully achieved at low temperatures, to identify a mathematical parameter describing the stress/strain field at the crack tip which adequately relates to the observed crack growth rates. Correlations have been found with stress intensity, crack opening displacement and reference stress but the most generally applicable appears to be with a quantity commonly referred to as C^*, related to the uniaxial creep strain of the material at the relevant stress applied.

Having established the crack growth rate dependence of the material in the laboratory on, for example, C^*, growth rates in cracked structures can be anticipated by calculating the prevailing C^*. These methods are currently at the validation stage on full-size components and existing data as yet remains comparative. Assessments of creep crack initiation and growth be-

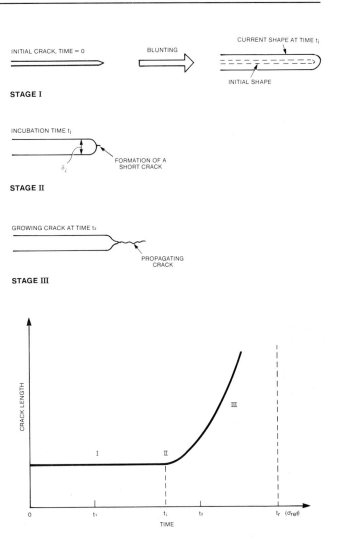

FIG. 12.7 Creep crack propagation
Stages in the development of a creep brittle crack at elevated temperature.

haviour are however proving of value in assessing the significance of plant defects. The following examples give an indication of the range of application.

1.2.1 Assessment of solidification crack behaviour in AGR transition joint weld metals

Solidification cracks extending across a significant proportion of the section of the 9Cr1Mo weld metal were suspected in the 9Cr1Mo to austenitic AISI Type 316 superheater transition joints early in the operation of an Advanced Gas-cooled Reactor (AGR). The effects of the worst possible operating conditions, represented by maintaining the peak temperature and maximum system stresses seen by joints during operation, were therefore assessed by testing internally pressurised and end-loaded tubes containing real joints for the equivalent of 25 000 hours steady creep operation and 1500 start-up/shutdown thermal cycles, the latter representing five times the number of thermal cycles expected during the plant life. Creep crack propagation over the expected plant life was shown to be less than 0.5 mm, an effect attributed to weld metal strengthening during service. Crack extension from thermal fatigue cycling was similarly shown not to present a risk of failure, even over five times the anticipated life cycles.

1.2.2 Defect assessment in Type 316 austenitic stainless steel welds

Austenitic weld metals, used as fillers for austenitic boiler tubing, inherently suffer from solidification cracking. A successful palliative is deliberately to raise the chromium equivalent of the metal and thereby introduce 3–5% volume fraction of delta ferrite into the otherwise purely austenitic microstructures. Unfortunately, creep failure then occurs more readily along the delta ferrite/austenite interface and the weldment becomes susceptible to failure by creep crack growth from pre-existing defects. Manufacturing defects are more likely in welds than tubes or plates and are difficult to assess in austenitic steels because of poor transmission of ultrasonic signals.

An assessment of creep crack growth was performed for an AGR boiler component for which, despite careful design for integrity and inspection, it was impossible to prove the absence of crack-like defects in a small portion of weld. Stress relief heat treatment of the welds was impractical, hence it was necessary to assume that residual stresses in the weld might extend defects within the uninspectable region. Procedural welds were made and pre-cracked samples were tested under load at 538°C in conditions which modelled the loading of the component. The results proved that even the maximum size of uninspectable defect would not cause failure within the next 25 full power years.

1.2.3 Assessment of transition joint behaviour

The premature failure of existing designs of transition joints is attributable to the presence of a brittle zone or interface between the ferritic tube and filler material. Nickel based fillers show a significant improvement in performance over the original austenitic material but still fail to meet design requirements. Current efforts are therefore aimed at producing improved designs by variations of post-weld heat treatment and welding procedure. The resulting changes in joint performance have traditionally been assessed by cross-weld rupture testing, which is relatively insensitive and requires tests to long times to yield significant indications of improvement. Deformation in creep crack growth tests by contrast is concentrated in the weak interface region to which cracking initiation and propagation are confined. Creep crack growth testing is thus proving to be a sensitive and rapid means of assessing changes in interface and therefore joint integrity. For example, a change from austenitic to nickel filler material leads to an improvement in cross-

rupture life by only a factor of two whereas creep crack growth rates are increased by one or two orders of magnitude.

1.2.4 Cracking in CrMoV weldments

The occurrence of circumferential heat affected zone and transverse weld metal cracking in main and reheat steam lines, of $\frac{1}{2}$CrMoV steel welded with 2CrMo weld metal, is described elsewhere. A comparison of growth rates has revealed the critical dependence of weldment failure in $\frac{1}{2}$CrMoV weldments on the local microstructure of the heat affected zone and weld metal. Cracks have been grown through structures simulating the coarse, fine and intercritical annuli of the heat affected zone, through real heat affected zones, along the weld/parent interface and through various weld metal microstructures. Growth through coarse grained bainite occurs by cavitation on the prior austenite grain boundaries at rates more than an order of magnitude greater than through refined bainites or the parent material, thereby validating the case for weld repair by continuous refinement methods.

Crack growth studies also provide a sensitive means of assessing the effects of various compositional variations (for example, of residual elements) and the effects of inadequate post-weld heat treatment, both of which can result in dramatic increases in growth rates. Routes for assessing the significance of defects on plant, where the distribution of microstructure in each individual weldment is usually unknown, have for this reason been based on a statistical evaluation of the cracking behaviour of the total weldment population.

1.3 Thermal fatigue (transient behaviour)

Power plant components experiencing a heating and cooling cycle during routine operation vary over three orders of magnitude of thickness from fuel element cladding tubes in nuclear reactors (0.5 mm) to vanes and blades of gas turbines (5 mm); ending with cast valve chests, rotor casings and forgings in conventional turbines of 100–1200 mm thickness. As the proportion of total electrical output from nuclear plant (which operates only at base load) has increased, so fossil-fuelled plant has increasingly been required to two-shift (16 hours on, 8 hours off). Thermal fatigue behaviour frequently becomes life limiting for heavy section components (chests, casings, rotors) operating under these conditions. Whatever the component size, however, if successive thermal cycles produce reverse plastic yielding at the surface, or at the root of a stress raising notch or groove, a small thermal fatigue crack will initiate after a characteristic number of cycles. The crack then propagates through a surface plastic zone the dimensions of which are defined by the notch or groove geometry and the surface temperature gradient. This is the High Strain Fatigue (HSF) reversed plasticity or Low Cycle Fatigue (LCF) growth stage.

Ultimately the crack emerges from the area of local surface plasticity into a region experiencing only reversing elastic strains and its progress can then be described by Linear Elastic Fracture Mechanics (LEFM). Beyond a certain crack depth the cyclically induced elastic stresses decay to a minimal value. Thermal fatigue cracking then ceases but the crack may subsequently propagate under the action of high frequency fatigue, in rotating machinery, or pressure stresses (creep crack growth) in high temperature components.

Suppose a thick turbine casing is at a uniform temperature of say 200°C and that steam at 550°C suddenly impinges at A (see Fig 12.8). The surface of A goes into compression as it attempts to expand against the surrounding cooler material, yielding along OQ. As heat flows into the casing the whole system expands taking the surface at A into tension approximately isothermally along QR to R, i.e., the original strain. During the steady running period the residual tension stress relaxes along RR′. On cooling the process is reversed; the surface layers at A now go into tension during attempted contraction, yielding along R′S. Later, as the whole structure cools, the yielded section A is forced into compression by the surrounding metal. Heating and cooling is then repeated from point P and the system shakes down to a closed hysteresis loop PQRS. The driving force for crack initiation and propagation in HSF is thus the degree of reversed plasticity ($\Delta\epsilon_p$) at the surface arising from the transient surface temperature gradients during heating and cooling. Principal factors in limiting the thermal fatigue life of plant components are therefore material properties, plant start-up and shutdown rates, and the presence of surface stress concentrating features (notches, grooves, profiles).

1.3.1 Laboratory simulation of the service cycle

To determine materials behaviour in the laboratory and provide initiation and growth data for multivariable conditions, the thermal expansion strains in the component are calculated and reproduced by mechanical strain introduced and controlled, usually isothermally, by the laboratory test machine. The thermal strain cycle seen by the component is replaced by an accelerated fatigue cycle lasting only a few minutes. For simulation of two-shift conditions, laboratory dwell times are superimposed on the accelerated fatigue cycle (0.5–16 hours or more). Tests simulating HSF in the laboratory are usually carried out between constant total strain limits and the plastic strain (ϵ_p) per cycle is measured. The peak stresses are left to adopt their own values, and the tension and compression values are usually similar. Although HSF in practice occurs over a range of temperature, a frequent simplification is to test at the maximum temperature of exposure, assuming this is the most damaging.

During the steady running period the residual tension stress relaxes from R to R′ (Fig 12.8) by creep,

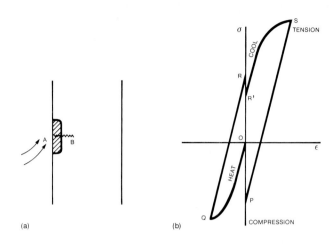

FIG. 12.8 Heating and cooling the surface of a structure (a) leads to the compressive/tensile hysteresis loop (b)

so that the material becomes damaged to a degree determined by the peak stress and hold time. The loop width (thermally induced strain range) attenuates with increasing distance into the yield zone, ultimately becoming elastic beyond the yield zone boundary. Initiation and propagation in the laboratory are therefore generally simulated and measured as follows:

1.3.2 Initiation — the endurance test

A smooth specimen representing the surface at A is strain cycled in push-pull or reverse bend. In analogous manner to the more familiar high frequency fatigue test, where stress amplitude S is plotted against total endurance N_T, total strain range ϵ_T (peak tension to peak compression) is plotted against cycles to failure to yield an 'endurance curve' (Fig 12.9). In fact for some materials the HSF curve blends in well with the high frequency data.

Provided the practical situation can be reproduced accurately in terms of strain range and peak temperature, the number of cycles to initiate a crack can be taken from the endurance curve, assuming rapid crack growth. Alternatively, initiation can be taken from cycles to a specific fractional load drop or cycles to an observed crack length.

1.3.3 Propagation in the yield zone (short cracks)

A shallow starter notch is introduced and testing continued under fully reversed net section plasticity, usually in push–pull. This is the High Strain Fatigue test simulating short crack growth in the yield zone to a depth usually of only a few millimetres. The rate of crack growth $\delta a/\delta n$ is monitored as a function of the total reversing strain range (ϵ_p) and can usually be expressed in the form:

$$\delta a/\delta n = \beta a^Q = C\epsilon_p n_a^Q$$

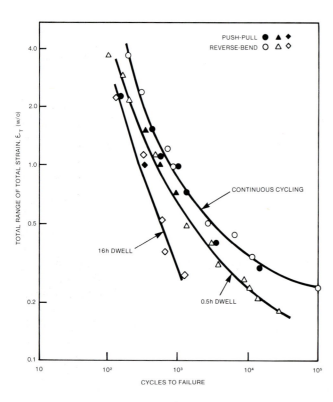

FIG. 12.9 Failure data for $\frac{1}{2}$ CrMoV steel obtained at 550°C in reverse-bend push–pull tests

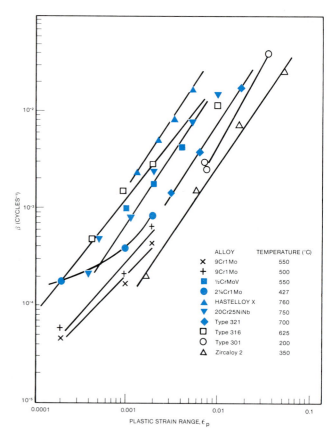

FIG. 12.10 Variation in growth rate with plastic strain for a range of power plant alloys

where a is the crack length. Values of β have been collected for many alloys and are summarised in Fig 12.10.

1.3.4 Propagation in the reversing elastic field (long cracks)

Given that the variation of stress intensity arising from the surface fatigue cycle is known across the component section, the growth of long cracks (i.e., in Region B, Fig 12.8) can be simulated and measured on standard edge notched specimens by visual, striation or electrical methods. This is the LEFM test and growth rates can be expressed in the form:

$$\delta a / \delta n = C \Delta K^m$$

where ΔK is the range of applied stress intensity.

1.3.5 The thermal shock test — crack arrest

In large static components, significant economic savings can result if it can be shown that relatively deep cracks discovered during operation will become dormant, thus allowing the component to continue in service.

Reproduction of the entire cracking sequence, from initiation to the limit of the thermal fatigue influence deep into the structure, can be simulated on thin or thick sections by using a thermal shock test (Fig 12.11). A hollow cylinder is cycled either by quenching the inside of the bore or by subjecting the outer surface

to an up-shock by radio frequency heating. The latter has the advantage of better simulating component surface conditions and the capability for introducing dwell effects. Once the crack has emerged from the region of reverse plasticity, growth becomes retarded as the elastic stress gradient decays. The test therefore permits the possibility of crack arrest to be examined as a function of strain range, temperature and dwell period.

1.3.6 Creep-fatigue interactions

The combined effects of the relaxation of the cyclically-induced stress and the creep generated in pressurised components under steady running conditions are simulated by introducing a dwell in the Endurance, HSF or LEFM tests, usually at peak tension for simplicity and to ensure conservative data. Much of the required data necessitates long term tests (for durations of several years). Collaborative programmes in the UK have however generated extensive data banks for the creep resisting steels of importance to power plant.

The effects of strain range and hold time are illustrated in Fig 12.9 for $\frac{1}{2}$ CrMoV at 550°C. As the strain range is reduced from 4% to 0.2% the endurance increases from 100 to 10 000 cycles. If, however, hold periods of 0.5 h and 16 h are introduced, the endurance reduces to less than 10% of the continuous

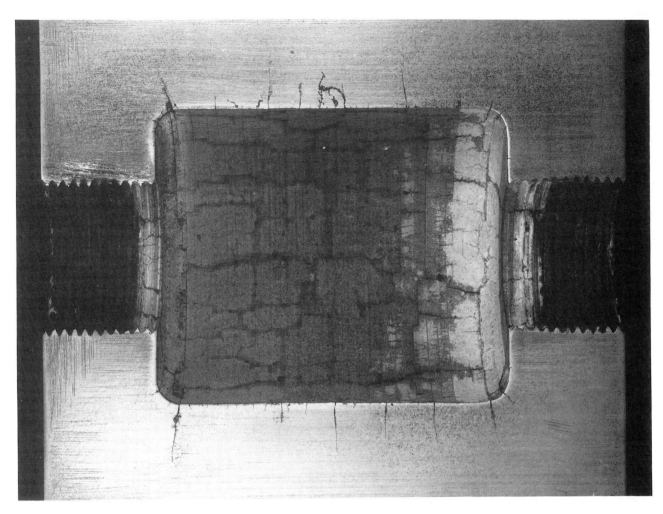

(a)

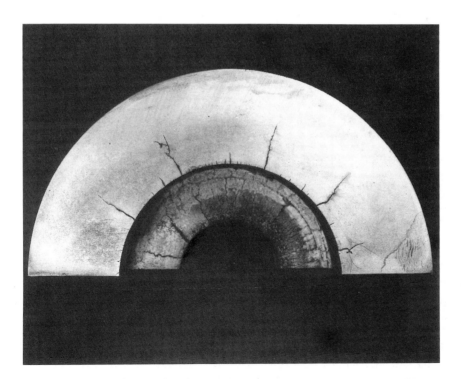

(b)

FIG. 12.11 Thermal shock test
Circumferential (a) and longitudinal (b) cracking in a specimen simulating low cycle thermal fatigue
in heavy section components.

527

value at the low strain range. The effect becomes more pronounced with increasing temperature as the effect of creep damage becomes more significant.

The effects of temperature, strain range and dwell period are associated with changes in crack growth mechanisms. At high strain range and lower temperatures in continuously cycled or short tensile dwell period tests, cracks propagating from the surface of reverse bend tests tend to be transgranular and macroscopically straight, showing all the characteristics of low temperature high strain fatigue cracks. At lower strain ranges and higher temperatures the cracking mode becomes more intergranular and with increasing dwell period the intergranular crack front is preceded by general intergranular creep damage, in analogous fashion to the behaviour of creep crack fronts. The effect is generally associated with increases in crack growth rates of two orders of magnitude in LEFM and a similar, though less pronounced effect in HSF. At the dwell periods (16–100 h) and strain ranges of major interest in component design therefore, fracture generally occurs in low alloy steels by intergranular cracking with associated creep damage at the higher temperatures. The thermal fatigue resistance of the alloy therefore usually reflects its creep ductility.

The parallel contributions of creep and fatigue are recognised in the life fraction rule, used to describe damage accumulation by creep-fatigue

$$\sum N/N_f + \sum t/t_R = 1$$

where N_f = cycles to failure by fatigue

t_R = time to failure by creep

Results from 0.5 h dwell tests are used to define the fatigue damage component (approximately 50% of stress relaxation occurs in the first five minutes) while the creep damage component is defined by the time at stress during the dwell.

1.4 Defect assessment routes

In practice the life of many components of high temperature plant, especially turbines, is often limited by the combined effects of thermal fatigue and creep damage experienced during start-up, two-shifting and steady state operation. Assessment routes have therefore been devised which address the various requirements arising from thermal fatigue/creep interactions. These typically split into two main decision paths; those dealing with damage accumulation and defect initiation (defect-free route) and those assessing crack growth (pre-existing defect route). An example is given in Fig 12.12.

The defect-free route splits into damage accumulation and lifetime assessment paths, the former to demonstrate safe operation over a specified lifetime or to judge the effects on damage accumulation of variations in the input data at the design stage. The

lifetime assessment path further divides to allow for the possibilities of either a remanent life or design life calculation. The assessment of damage or lifetime refers to the failure of a particular region of the component by bulk damage mechanisms. This may be taken as failure of the component or, alternatively, as crack initiation in, for example, HP and IP rotors.

The pre-existing defect route splits into remaining life and tolerable defect size paths. The former determines the lifetime of a cracked component of known initial defect size while the latter further divides to enable calculation of either the initial tolerable defect size (for design purposes) or the final defect size after some growth (for service assessments).

The two main routes are linked to allow for the possibility that an assessment starts from the assumption of a defect-free component, but subsequently involves crack initiation and propagation.

1.4.1 Calculation of damage

Components are, in general, subjected to a number of different cycle types, each containing transient and steady state components. The calculation of either total damage or lifetime therefore requires determination of both fatigue and creep damage accumulated per cycle for each cycle type.

The fatigue damage rate is calculated from the representative total strain range. Prediction of the fatigue damage rate leading to crack initiation is achieved by using an established model for creep/fatigue interaction. The model may then be used to obtain the number of cycles N_i required to cause failure or to produce an initial crack of size a_i.

Calculation of the creep damage rate requires summation of the local values of primary and secondary stress attained during the steady state total time (Δt) of each cycle. This creep damage does not include that previously accounted for in the fatigue damage calculation. Standard creep rupture data are used to obtain an appropriate rupture time (t_r). The damage rate is then the fraction of the creep rupture time spent during the steady state dwell period ($\Delta t/t_r$). Creep rupture pertains to failure or initiation of a defect of size a_i.

The total damage D is obtained by summing the creep and fatigue damage component for all cycle types over the specified lifetime. The current damage D_c is the creep and fatigue damage component summed over the known load history. Life is assessed from the creep and fatigue damage D^* accumulated per year calculated from the respective damage rates per cycle and future loading. Total life is then given by $1/D^*$ and remaining life by $(1 - D_c)/D^*$.

The pre-existing defect route addresses the problem of crack growth rates between an initial crack length (a_o) and an acceptable or critical value of crack length (a_c). The initial defect size a_o might be related to fabrication flaws (e.g., casting defects), the limit of

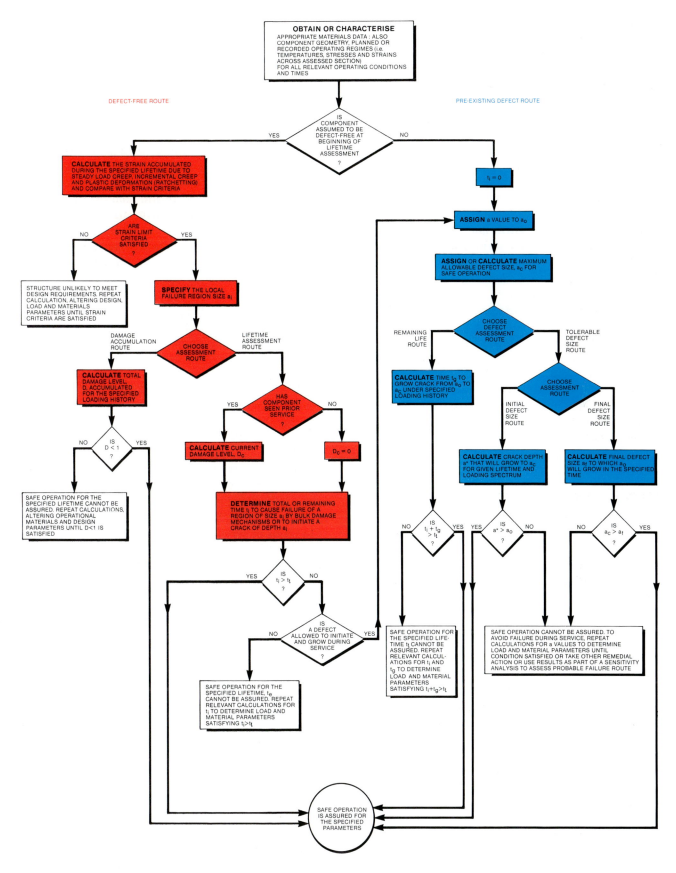

FIG. 12.12 Flow chart for thermal fatigue assessment procedure

NDT resolution at the design/manufacturing stage, crack depth discovered in service or the maximum defect size which would have just avoided failure in an overload test, estimated by the defect assessment procedure. The acceptable final defect depth (a_c) might be defined from the load bearing capacity of the uncracked section (determined by creep rupture or plastic collapse considerations) or from the most onerous loading conditions which could arise in service, including postulated overload conditions estimated by R6 (see Section 1.1 of this chapter).

The prediction of crack growth requires an established law describing the cyclic propagation rate. If the representative stress distribution in the component is known, the growth law can be integrated to yield the time t_g (remaining life) to grow the crack from a_o to a_c. If t_g plus t_i, the time to initiation (if applicable), exceeds the desired lifetime, then safe operation is assumed.

In general the law should encompass both the high stress condition, required for short cracks emanating from surfaces where plastic deformation occurs due to severe thermal cycling, and the low strain region of elastic deformation described by LEFM.

The tolerable defect size a^* is that which would not grow to an unacceptable size during the desired remaining lifetime of the component, or over a given time interval. The final defect size a_f is that to which an initial defect of size a_o will grow in a specified time.

If the safe operation of the component cannot be assured after assessment then the remaining life may have to be reduced, the damaged region repaired or the input parameters reassessed. If the latter, then a sensitivity analysis will identify those parameters (material, environmental or operational) having the dominant effect on lifetime or tolerable defect size.

2 Remaining life assessment beyond design life

2.1 Remaining life assessment of boilers

Boilers and generating units have generally been built to a range of design codes incorporating a nominal life which may be up to 140 000 operating hours. Current strategy however is to operate some of the 500 MW units for up to at least 300 000 hours. Since many components will have shorter lives than this because of progressive deterioration in service there are requirements to allow extension of life for major stations by developing a life prediction capability, and thereby:

- Avoid costly unscheduled outages and underwrite safe operation by preventing high temperature failure and where necessary obtain insurance certification.

- Optimise the planning, replacement and installation of components.

- Ensure that replacements are technically acceptable for the application and, if feasible, provide replacements which will last the total anticipated lifetime of the unit.

The requirements impose a need to examine systematically and monitor plant during and after service to assess its future reliability and safety. The procedures have potential for avoiding replacement generation costs but the most important potential benefit arises from the possibility of substantially extending the life of existing stations, thereby reducing the capital investment associated with construction of new plant.

In developing a methodology, consideration must be given to the types of components and materials most likely to require accurate life prediction. In general these will depend on plant design and operational practice. Other criteria, such as the severity of the operating regime, current failure statistics, the difficulty of repair and the cost of associated outages, the cause of failure and the potential benefits to be gained by developing an accurate life prediction capability may also have to be considered. Components likely to be affected and their probable life-limiting failure modes are shown in Table 12.1.

Boiler components tend to show a more pressing need than turbine components. This is because turbine components, although operating under extremely onerous conditions, nevertheless generally operate within design parameters for temperature, pressure and cyclic loading. Boiler components, however, frequently operate outside these parameters, particularly with respect to temperature and corrosive environment.

In terms of service failures, CEGB statistics indicate a high failure incidence in superheater and reheater tubing and at first sight the development of a predictive capability could reduce these. However, boiler tubing can be replaced relatively easily and the outage time for individual replacements is low at correspond-

TABLE 12.1

Components and their probable life limiting failure modes

Component	Failure mode
Furnace wall tubes	Fireside corrosion
Superheater/reheater tubes	Creep, fireside corrosion
High temperature pressure vessels	Creep, thermal fatigue
Steam/pipe systems	Weldment integrity
Turbine valves/chests	Thermal fatigue, creep
Turbine HP casings	Thermal fatigue, creep
Turbine HP rotors	Creep, fatigue
High temperature bolts	Creep, stress corrosion cracking
Turbine LP blades	Fatigue, erosion
Turbine LP rotors	Fretting fatigue, corrosion fatigue
Generator end-rings	Stress corrosion cracking
Condenser tubes	Stress corrosion cracking
Feed systems	Corrosion fatigue

ingly modest cost. Furthermore, it is doubtful whether high accuracy can be attained in a remaining life assessment in view of the variation in operating parameters seen by tubing in different locations within the boiler. In these circumstances it is by no means assured that the life controlling mechanisms in one region correspond to those pertaining in another. Techniques of remaining life prediction therefore have limited applicability in the case of boiler tubing and in general are expected to give poor accuracy. A preferred route is often to treat this problem by means of a statistical analysis of boiler tube failures and take remedial action on the basis of operating experience with individual boilers, with particular emphasis on those factors giving rise to localised fireside corrosion. In contrast, the development of life predictive techniques for steam headers is considered a worthwhile undertaking. These are components:

- Which are extremely expensive and difficult to replace requiring extensive outage and have long delivery times.

- For which creep is the principal mode of damage under normal operating conditions.

- Which suffer from a known operational problem of localised overheating.

- Which are of sufficient section to avoid the complications due to environmental interaction.

- Which are complex vessels containing multiple penetrations which cause localised variation in stress and temperature.

Given this last consideration, a predictive capability of failure at least 3–4 years before the event is required so that replacements can be ordered and manufactured. For these reasons, therefore, steam headers were selected for initial attention and improved techniques of creep life prediction have been developed. Many of these are however applicable, with some modification, to other components.

To date there have been very few failures of high temperature headers resulting in major steam leakages. However, in one instance in 1969 a catastrophic rupture did occur. This incident was in part caused by operation at excessive steam, and hence metal, temperatures during the hours preceding the failure. Since then considerable caution has been applied to the assessment of remaining creep life and safety legislation has been enforced.

In some instances the temperature distribution across a boiler may give rise to a characteristic M-shaped distribution with peak temperatures occurring at the outside quadrants. In addition, the temperature distribution through an element may be such that some inlet stub tubes operate at temperatures in excess of design temperatures. The regions of the header surrounding these stubs therefore experience a higher rate of exhaustion of creep life.

A survey of all CEGB fossil-fired boilers indicated that nearly all header replacements had been made on the basis of creep life exhaustion. In about 25% of the cases investigated direct evidence of creep damage was identified in the form of inlet stub tube weld cracking or measurable swelling of the header body. The need for replacement of the remainder was indicated by a record of excess temperatures and the determination of safe operational life on the basis of the life fraction rule. Post-service testing and examination of some headers in this latter category showed that many of these had been replaced prematurely indicating considerable conservatism in the procedures then used to assess remaining life. Since, over the next 25 years, the CEGB is expected to have to replace several hundred headers at a high cost, there is considerable incentive to improve the accuracy of life predictions to prevent further premature renewals and, if possible, extend useful life.

Most of the headers of interest are constructed in $1Cr\frac{1}{2}Mo$, $2\frac{1}{4}Cr1Mo$ or $\frac{1}{2}CrMoV$ steel. However, it was not immediately clear whether the effort should be concentrated on weldments or parent metal. The most common signs of distress in service are associated with cracking at inlet stub tube welds and if this is allowed to develop there is a risk of unscheduled outage due to steam leakage. It would seem therefore, that the main thrust to improve life prediction should be concerned with the weld regions. However, cracking of this form is readily detected and can be repaired, albeit less readily, at convenient outages. A crucial consideration for such repairs is the extent of creep damage in the main vessel body. If this is extensive then welding followed by post-weld heat treatment may not be possible. Thus for the purposes of remaining life prediction and life extension it is considered that deterioration of the base material by creep is the overriding consideration. Nevertheless, some work was required on weld structures because these are important in the following contexts:

- The number and frequency of weld repairs may become prohibitive and require a decision to be taken to replace the component.

- In certain cases weld failure could cause failure of the complete header due to, for example, a steam cutting action, even though life usage in the header body is insignificant.

- Damage accumulation at welds can be used to give an indication of life usage in regions remote from the weld.

- The welds may be inaccessible for easy repair.

For these reasons the integrity assessment of weldments is an important aspect of the overall assessment methodology to determine header lives.

2.1.1 CEGB procedure for header life prediction

Methods of remaining creep life prediction are considered under two broad categories:

(a) Methods involving the acquisition and monitoring of operational parameters, the use of standard material data and the life fraction rule.

(b) Methods based on post-service examination and/or testing which require direct access to the component for sampling and measurements.

Both categories have been incorporated into a procedure which involves three stages of assessment, each stage being more refined and less conservative than the preceding one. In this way effort is concentrated on the headers of major concern and minimum costs are incurred:

Stage I — based on steam temperature data

Stage II — based on measured metal temperatures

Stage III — based on newly developed methods of assessing expired life.

The purpose of Stage I assessment is to classify the headers as those where a detailed analysis is necessary and those where, because remaining lives are significantly greater than planned operational lives, no further work is required. Steam temperature records are necessary to enable this analysis to be carried out. When these data are not available the more rigorous approach of Stage II is adopted, but this requires a substantial resource in terms of thermocouple attachment and monitoring equipment.

In making a quantitative assessment of the header life, arithmetic averaging of temperatures and pressures is not acceptable because temperature and stress are not linearly related to life. The usual approach is the adoption of methods using Robinson's Life Fraction Rule. In this method it is assumed that when a material is subjected to a stress and a temperature in the creep range for a time interval t, then the fraction of life usage during the interval is t/t_r, where t_r is the rupture life at that stress and temperature. When the stress or temperature changes in operation then failure will occur when the sum of life fractions equals unity, i.e.,

$$\sum t/t_r = 1$$

2.1.2 Stages I and II

Stages I and II involve the incorporation of the operating stress and temperature history data, together with the lower bound creep rupture properties, into the life fraction relation. The operating stress at this stage is generally taken from the mean diameter formula:

$$\sigma = Pd/2tl$$

P = operating pressure, d = mean diameter, t = minimum thickness l = ligament efficiency as per BS1113 (1969).

Estimates of life must, for considerations of safety, involve adopting the most conservative values from the ranges of input parameters. Life estimates obtained in this way will therefore be minimum values. Stages I and II are essentially an exercise in sorting headers into the categories of those whose estimated lives substantially exceed design and those which require the more detailed assessment of Stage III and closer monitoring during the planned life of the station.

The overall accuracy of the method is not very high, however, since considerable uncertainties are associated with each element of the assessment. These are now considered briefly, emphasising the more important sources of error and indicating where emphasis has been placed to improve accuracy.

2.1.3 Applicability of the life fraction rule

A recent study of the validity of the life fraction concept in low alloy steels, aimed at establishing the effects of fluctuating stress and temperature by performing temperature and stress accelerated rupture tests, has concluded that life fraction summations are realistic for temperature changes but not for changes in stress. Residual life predictions made from increased stress tests were found to fall short of the true residual life, the inaccuracy increasing with the degree of acceleration for even small increases in stress over the service level. Tests involving increased temperatures, however, were found to provide reasonable estimates of residual life, both when life fraction techniques were used and when straightforward extrapolations of higher temperature tests were made to the service temperature (Fig 12.13).

The reason for the divergence is that the stress rupture properties of normalised and tempered low alloy steels are controlled by the mean separation of the creep strengthening carbide dispersion and hence by exposure to temperature excursions. Fluctuations in stress at constant temperature have little effect on the carbide dispersion and the material life is little changed.

The establishment of the validity of the life fraction rule for changes in temperature has been an important step in assessing the lives of steam headers, which are likely to experience marked variations in temperature rather than pressure during service. The application of the life fraction rule to headers is therefore seen to be limited more by uncertainties in the input data rather than by the validity of the rule itself:

2.1.4 Stress rupture data

Two factors here have an important bearing. Data often does not exist for an adequate duration at the design stress and must therefore be extrapolated. Potential structural degradation at long times makes this a hazardous procedure, however. In addition, the scatter of $\pm 20\%$ in stress means that, if the position of the material in the scatterband cannot be accurately located (see Stage III), the assumption of minimum

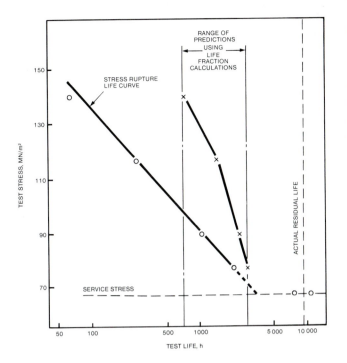

(a) Increasing stress at constant temperature (557°C)

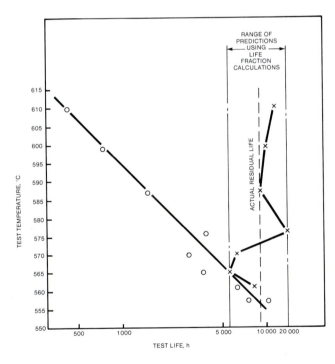

(b) Increasing temperature at constant stress (66.26MN/m²)

FIG. 12.13 Remanent life test procedures
(a) Test acceleration by increasing stress at constant temperature for $\frac{1}{2}$CrMoV specimens leads to a considerable underestimate of the actual remaining life, either by simple extrapolation of the stress rupture life curve or by using the life fraction calculations.
(b) Acceleration by increasing the test temperature at constant stress shows good agreement with the actual residual life both by rupture life curve extrapolation and life fraction calculations.

rupture properties dictated by safety considerations can be extremely pessimistic (by up to a factor of five in time).

2.1.5 Oxidation effects during stress rupture testing

Standard materials data are obtained by testing relatively small diameter specimens ($\leqslant 12$ mm) in air at or above the plant operating temperature. Stresses are calculated on the basis of the initial gauge cross-sectional area, with no compensation for the progressive reduction in area arising from oxidation during the test. Oxidation effects, even at relatively low temperatures, have now been shown to increase creep rates significantly and reduce rupture lives such that the application of standard materials data may significantly underestimate the life of a thick-walled component, where oxidation effects are insignificant (Fig 12.14). The effect is of particular concern in rupture tests accelerated by temperature (Stage III) which should be performed in a non-oxidising atmosphere when thick section components are being assessed.

2.1.6 Temperature measurement

Rupture life for low alloy steels in the range 540°C to 600°C is very sensitive to small changes in temperature, with a reduction by a factor of $\geqslant 1.5$ for each 10°C rise in temperature. The temperature profile along and around a header can exhibit variations of typically 40°C which leads to gross pessimism if metal temperatures are estimated from the steam temperature since, for conservatism in the life estimate calculation, it is necessary to assume a generous steam to metal temperature differential.

2.1.7 Stage III

Techniques developed for Stage III assessments require direct access to the component for sampling and measurement.

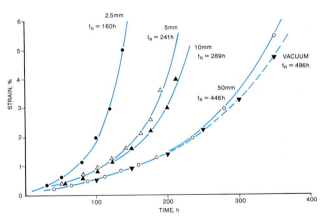

FIG. 12.14 The effect of test piece oxidation on rupture life resulting from progressive loss of section, is revealed by testing specimens of varying diameters in air and comparing with tests in vacuum or argon.

533

A variety of models have been developed which relate creep damage induced by the service exposure to remaining life. The relevant creep damage feature or property must then be measured on the component. Ideally this should be related to a strain or strain rate measured on the plant and related to the life fraction consumed (Fig 12.15). If the strain or strain rate can be assessed directly then the creep model alone can be used to estimate the life. Alternatively, post-exposure accelerated creep rupture tests can negate the need for both stages. A knowledge of the creep damage processes is, however, still needed to give confidence in extrapolating from accelerated to service conditions.

Mechanistic models have been developed which attempt to identify structural changes which can be measured and correlated with exposed life. Two distinct creep damaging processes usually occur simultaneously under service conditions in low alloy steels; structural degradation from precipitate coarsening or dislocation mesh growth leads to continuous reduction in creep strength, and creep cavitation nucleated on boundary particles leads to a progressive increase in boundary damage. The prevalence of either is determined by the material microstructure, purity, temperature and stress state, with structural degradation dominating in base material (normalised and tempered ferrite, ductility usually >10%) and cavitation in bainitic materials of ductility <5% (heat affected zones in particular).

The problems involved in developing and validating structural models (for example, the creep rate dependence on particle composition or spacing, the variation of particle composition and spacing with time and the solid solution strength contribution) are complex and not yet fully developed. Parametric models describing the evolution of strain and strain rate over the component lifetime have, however, been used for some time. A model using a generalised damage parameter W has been shown to correlate well with creep data for low alloy steels. By integrating the parametric relationships:

$$E = A\sigma^n/(1 - W)^n$$
$$\dot{E} = B\sigma^v/(1 - W)^\eta$$

(where A, B, n, v and η are material constants, W = 0 when the material is in the undamaged state and W = 1 at rupture) the strain (E) and strain rate ($\dot{E}$) can be related to time throughout the life of the material. Materials data are required only in the form of the minimum creep rate/rupture life product which is found to be approximately constant for low stresses (known as the Monkman Grant constant). The specific creep response of the material is not therefore needed and the rupture life can be related to a single strain and strain rate measurement on plant without a knowledge of the rupture strain (for ductile materials).

2.1.8 Measurement of creep damage

Creep damage must usually be measured in the most critical area of the component and considerations of accessibility and sample size therefore frequently dictate the choice of technique. At present most forms of post-service mechanical testing require relatively large specimens involving weld repairs to the component. There is a strong incentive therefore to develop techniques which require only very small samples, thus obviating the need for weld repair.

Methods of mechanical testing of miniature (3 mm dia.) specimens are currently under development. Measurements of changes in physical or mechanical properties are relatively easy to perform and many have been investigated (e.g., Charpy V-notch, hardness, density, resistivity), but the sensitivity tends to be low and none can at present be related satisfactorily to remaining life.

Micro samples (boats or trepanned cylinders or cones) enable detailed optical and electron microscopy to be carried out. A possible application of metallographic techniques is to use the degree of carbide coarsening or changes in carbide composition to measure the effective operating temperature. Life estimates can then be made from the life fraction rule. Although the accuracy is less than that from thermocouples, it is rare for a component to have been fully instrumented through life. The method is therefore particularly useful where marked changes in boiler operation have occurred. Modern image analysis techniques are essential in quantifying the distributions of particle sizes and spacings since the structures of low alloy steels are highly inhomogeneous.

The volume fraction of cavities or fraction of cavitated boundaries are promising parameters for assessing the lives of weldments and critical areas of plant. The number fraction of grain boundaries containing creep cavities can be related to the life fraction con-

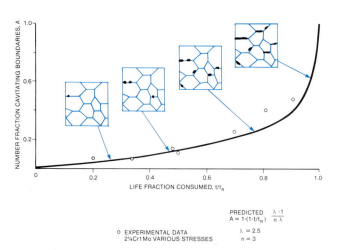

FIG. 12.15 Development of cavitation
Fraction of cavitating grain boundaries increases as the life of the material is consumed.

sumed in simulated weld heat affected zone specimens and the experimental data is found to be in reasonable agreement with predictive models of cavity growth. The approach thus offers a quantitative non-destructive route applied directly by taking plastic replicas.

Since measurements of strain or strain rate can be related parametrically to life consumed, direct measurements of strain or strain rate on plant components provide an obvious route to life prediction, and are used extensively. Methods used include the bow gauge micrometer, strain gauges and photographic techniques.

Where the availability of specimens permits, techniques based on accelerated creep or rupture testing are now generally accepted as reliable methods for providing accurate assessments of remaining life. Stress rupture tests are cheaper to perform in protective atmospheres. Creep tests however, have the advantage of assessing strain tolerance and enable estimates of life at times well before rupture, thereby reducing the test time and required degree of test acceleration.

Life can be predicted from post-exposure tests either by parametric extrapolation or by application of the life fraction rule. Parametric extrapolation involves obtaining data under accelerated conditions on used material and extrapolating to service conditions. In choosing test conditions, however, it is important to perform post-exposure tests in the same creep damage regime as that pertaining under service conditions. Stress acceleration results in rapid refinement of the dislocation mesh size controlling the creep strength. The deformation rate is thus very sensitive to stress, such that extrapolations from high to low stress can be subject to considerable error. Temperature extrapolation at constant low stress generally results in precipitate and dislocation mesh coarsening which more closely simulates that under service conditions. Accordingly, provided no major changes occur in precipitate type and associated coarsening constants over the extrapolative range, then the rupture life will be less sensitive to temperature changes, allowing more reliable extrapolations than for stress changes. The sensitivity of the cavitational damage process to stress and temperature changes follows a similar pattern, albeit for different reasons. Relatively short (10^3–10^4 hours) post-exposure tests, extrapolated in temperature, can thus give reliable life estimates.

Test acceleration by increase in temperature rather than stress also yields more accurate life prediction using the life fraction rule. A disadvantage with using the life fraction rule as a basis of a post-exposure test programme is the need for virgin material rupture data, which are not generally available. If material is removed from service at two or more service times, however, the life fraction rule can be used without the need for virgin material properties. Furthermore, this procedure requires, in principle, only a single post-exposure test on each occasion and hence has the advantage over the parametric approach where more tests are required to ensure confident extrapolation.

There are advantages in using all the above techniques in a complementary manner. Thus, although an accurate method, post-exposure creep rupture testing is restricted not only by the number of samples which can be taken but also by the test duration before a life estimate can be attempted. Similarly, strain measurement on plant will generally require the time of operation between outages before a life estimate can be made using this method. In contrast the metallographic techniques and the more promising physical/mechanical property measurements, although unlikely to yield the same degree of accuracy in the life estimate, might provide a useful means of conducting an initial assessment of life and might also allow more samples to be taken from complex components such as headers, giving greater assurance of the total reliability in the overall life estimate for the component.

2.2 Remaining life assessment of HP and IP rotors

The development of methods of plant life assessment concentrated initially on components such as boiler headers and pipework for which the costs of unplanned outage and replacement times were highest. A similar approach for key turbine components is currently in the process of development. Many of the larger 500 MW machines constructed in Britain in the 1960s are now approaching 100 000 hours' operation and, in common with boiler headers, there is a considerable economic incentive to extend the operating lives to 200 000 hours or beyond to obviate the need to build new units.

The two major reasons for retirement of rotors are unacceptable dimensional changes due to creep and the development of cavitational damage or cracking from creep and/or thermal fatigue. A flow chart summarising the various input parameters and assessment procedures required to avoid these two basic problems is illustrated in the life assessment route shown in Fig 12.16. Dimensional and metallographic changes together with long term mechanical property data are incorporated with non-destructive defect examination and stress analyses to optimise the evaluation. It is intended that a procedure such as that outlined in the flow chart should eventually form the basis of a CEGB Generation Operation Memorandum (GOM) for HP and IP rotors.

The positions of fastest strain accumulation in HP and IP rotors are at the rotor bore and blade root fixings. For the lead rotor families the degree of bore creep strain accumulation is considered to be the principal life limiting factor but for some designs blade root fixing strains are higher. Rotors manufactured from 1CrMoV have a bainitic creep resisting microstructure and exhibit low and variable creep ductilities which tend to decrease to values as low as 2% at 10^5 hours or beyond. Failure is by prior austenite boundary cavitation. Boundaries can therefore become virtually

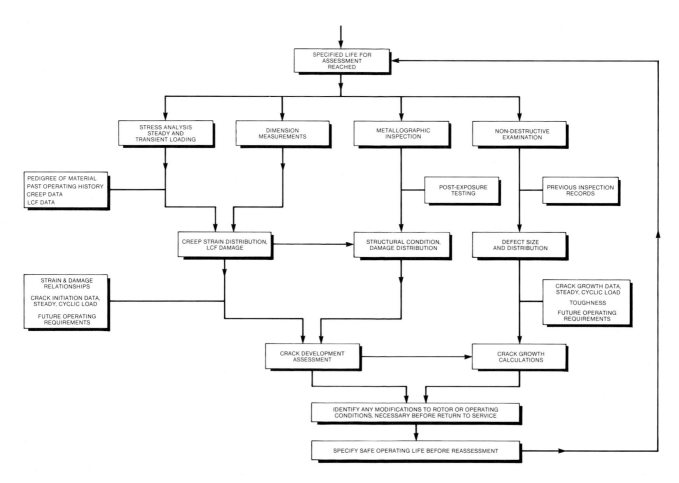

FIG. 12.16 Metallurgical aspects of remanent life assessment for high temperature rotor forgings

decohered in the more brittle alloys after strains of only 0.5–1%, leading to initiation of discrete microcracks.

The identification of a realistic life limiting strain criterion at which detectable creep cavitation and incipient cracking occurs and the way this varies with temperature and stress in the differing rotor alloys is therefore a basic requirement. Extensive long term creep and rupture data are now available from laboratory programmes which facilitate refinement of the original design considerations and provide the information defining the essential relationship between creep strain and damage development (at times up to and beyond 150 000 hours). The adoption of 0.5% accumulated bore strain is consistent with the need to ensure adequate margins of creep ductility, but in view of the variation of long term creep ductility among rotor alloys, it is more appropriate to take this as an assessment point rather than an absolute limit on life. This is an important requirement since an increase from 0.5% to 0.6% can correspond to an increase in life expectancy of several years. Times to a given strain can be calculated from a knowledge of the prevailing steady state stresses, the operating history and the long term creep data. Extrapolation of the data to longer times is being explored using temperature acce-

lerated creep tests of 2–3 mm diameter specimens taken from the rotor bore (post-exposure testing). The degree of damage in the surface region can be assessed directly by the use of surface replication techniques, given adequate accessibility, and compared with the damage levels expected from the known behaviour of the material in laboratory tests.

Dimensional measurements (rotor diametral strains converted to bore strains by calculation) provide an important means of confirming creep deformation calculations and are becoming an integral part of life assessment. Many rotors on large units in the UK have been installed with provision for creep deformation measurement at scheduled outages. These indicate that the bore creep strain generally approaches 0.3% in 100 000 hours and the trend line based on uniaxial data suggests the average bore strain will reach about 0.6% in 300 000 hours.

Particular rotor forgings demonstrate a range of creep and rupture behaviour which is consistently above or below the mean, generally as a result of within specification variations in manufacturing route. Such variations are important for life assessment since they can easily result in differences by up to a factor of three in calculations of accumulated strain and thereby classify rotors into families of similar expected life.

The above considerations refer primarily to the development of cavitational damage by creep processes. Cracking associated with thermal transients during start-up and shutdown is also a major concern. The low cycle (high strain) fatigue properties of rotor and casing materials have therefore been defined with hold periods up to 16 hours, representing daily stop/start cycles, and durations exceeding five years. These data have been used to define the allowable strain ranges and safe operating conditions at steam inlet regions where transients are most severe. This is discussed in greater detail in Section 1.3 of this chapter.

If the avoidance of crack initiation from creep, thermal fatigue or at steelmaking defects (oxides, silicates or sulphide inclusions) cannot be demonstrated then it may be necessary to calculate the extent of crack growth during subsequent operation. As a continuation of the trend to move away from strength oriented data towards assessments of ductility and crack formation, considerable effort has been devoted to generating crack growth data relevant to start-up and shutdown cyclic operations in rotor forging material. The interacting effects of crack tip deformation and surrounding ligament deformation has been found to define the growth rates, such that simple fracture mechanics parameters are invalid and methods of creep crack growth analysis must be applied (see Section 1.2 of this chapter).

A large body of information already exists to form the basis of remanent life assessments of rotors at a significant proportion of their design times. Additional information is however required to define crack growth rates under relevant conditions and establish the influence of complex loading on damage development. Information from in service measurements and destructive evaluation of retired rotors is also required to establish relationships between laboratory data and service behaviour. Representative rotors with appropriate operating histories are currently being selected for destructive evaluation as part of the life assessment and extension programme for UK units.

2.3 Life assessment of LP rotors

The operating temperature range of LP rotors falls below that at which creep damaging processes operate. Failure processes are therefore driven by high cycle fatigue at both on-load and barring frequencies, usually initiating at stress concentrating features such as keyways and stress relief grooves. Since the initial elastic stress analysis incorporated during design ensures that the fatigue strength is not exceeded in such regions, failure can only initiate from surface defects, either pre-existing or created during service, of a critical depth to exceed the threshold stress intensity for propagation (ΔK_{th}, see Section 1.1 of this chapter).

Defects can arise from a number of sources, including those resulting from the steelmaking process, but the most common derive from pitting corrosion, stress corrosion cracking and fretting.

Pitting corrosion occurs during off-load periods when the rotor surface is wetted by the steam condensate. Electrochemical interaction with inclusions intersecting the surface leads to the formation of surface crevices or 'pits' which can grow at rapid rates. Fatigue cracks initiate at pits and, for situations where crack growth cannot be allowed, the time to a critical pit depth is taken as the rotor lifetime, beyond which the pits must be skimmed out. Application of laboratory and plant pit growth kinetics has led to operating limits on the number of allowed cold starts between inspections. Growth kinetics are however found to be very variable, specific to individual rotors and depend on operating practice, for example, nitrogen blanketing, degree of condensate polishing or forced air cooling. Rates do however decrease with time and tend to a limiting value. Where a fracture mechanics assessment shows this to lie below the critical depth for propagation and pit growth can be shown to be controlling, then safe future operation is assured.

Stress corrosion cracking (SCC) is usually intercrystalline in nature and tends to initiate at corrosion pits, although in severe cases cracking begins at the surface. The classical example of failure from SCC occurred when a disc of a 60 MW rotor disintegrated during routine overspin testing (see Chapter 6). SCC was found to occur in this case even in pure water. The turbine was non-reheat and so ran wet; modern 500 MW rotors tend to incorporate a reheat cycle and therefore run much drier so that SCC problems are less likely. Design details that involve significantly lower levels of stress concentration are also incorporated.

Crack growth rates have been determined for a range of stress corroding conditions and can be used to determine inspection intervals required to avoid growth to the critical size for catastrophic fracture in service. Current effort is concentrated on assessing the effects of two-shifting, especially during off-load periods when the combination of temperature and moisture could produce conditions which may enhance stress corrosion attack.

Fretting is the term used to describe the generation of surface damage when components in contact under pressure experience a low amplitude oscillatory sliding motion between their contacting surfaces.

The associated wear processes rapidly generate oxide debris and surface cracks to depths of up to 0.25 mm. The enhanced crack initiation rates, together with the additional driving force for propagation which the fretting action also provides, result in significant reductions in fatigue life (Fig 12.17).

Fretting has been a major source of fatigue crack initiation and propagation in modern LP and generator rotors where large contacting surface areas exist in the many slots and keyways which are essential features of the basic design.

The assessment of rotor life due to fretting fatigue is complex since the process can occur at many locations and the fretting conditions are specific to each.

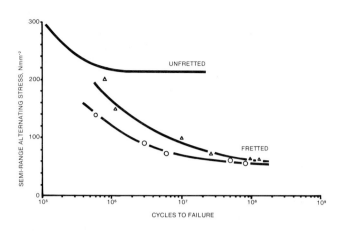

FIG. 12.17 Fretting fatigue data for $3\frac{1}{2}$ NiCrMoV steel in contact with 1CrMo steel with a mean stress of 300 Nmm^{-2}

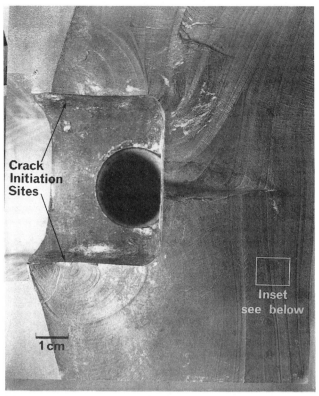

(a)

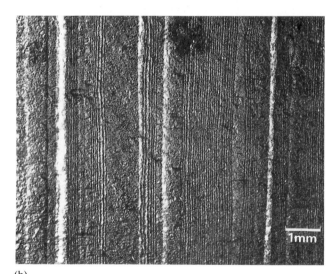

(b)

FIG. 12.18 Progress of:
(a) Fretting fatigue crack, initiated at a rotor disc keyway, is revealed by beach marks on the fracture surface. (b) Detail of beach marks in (a).

Assessment methods are therefore based on predicting the accelerated crack propagation rates due to fretting rather than on initiation, since this latter occurs so readily. Recent research has shown that both crack initiation and growth are largely controlled by the contact stresses between fretting components, surface wear processes being of secondary importance.

It is possible to calculate crack tip stress intensity factors for fretting cracks from a knowledge of surface contact stresses.

These can then be compared with threshold stress intensities (ΔK_{th}) and the size of defects which might grow under fretting conditions at a particular location thereby determined.

2.3.1 Life assessment from observation of beach mark patterns

The considerations outlined above relate primarily to life defined by the onset of crack initiation. Situations may arise however where families of rotors are found to be susceptible to fatigue failure but under conditions where high cycle fatigue crack growth occurs slowly, perhaps at the rate of only a few millimetres per month or year. Any change in operational conditions, especially during start-up and shutdown, is recorded on the propagating crack surface as a curve delineating the instantaneous crack front position (Fig 12.18). The beach mark pattern obtained by destructive examination of a rotor fatigue crack thus permits the crack growth rate to be related fractographically to the operational history thereby defining a safe operational framework for other rotors of the same family. Situations can arise, for example, where the pattern indicates that the crack grew relatively rapidly in the few weeks

following initiation but thereafter slowed and ultimately arrested.

The following case histories are offered as examples of the successful application of fracture mechanics to the assessment of defects in LP and Generator rotors.

2.3.2 Centre collar stress relief cracking in LP rotors

Cracking in stress relief grooves of 3CrMo steel LP rotors have been found to initiate from corrosion pits growing from machining score marks (Fig 12.19). A combined depth of 0.25 mm was sufficient to initiate fatigue crack growth during barring, verified by experimental tests on smooth push–pull specimens pre-pitted to various depths by exposure to condensing steam; severe reductions in fatigue strength were found and fracture mechanics analysis showed that corrosion pits could act as even more severe defects than sharp pre-existing cracks.

The dominant stresses in the stress relief groove were the alternating self-weight bending stress (± 14 MPa), amplified by a stress concentration factor associated with the groove, and a high tensile mean shrink stress from the disc assembly of approximately 500% MPa at rest or barring speed, reducing to 250 MPa at 3000 r/min (running). During cold starts therefore, the centre collar groove is wet for about 2 hours and pitting is at its most severe. At a pit depth of 0.25 mm fatigue cracks initiate and grow rapidly during barring, driven by the self-weight bend stresses within the high tensile shrink stress field. As the defect penetrates the rotor during barring it can eventually begin to grow (at 1–2 mm) during normal running when the mean stresses are lower. All rotors of this design are susceptible to this failure mode but the analysis outlined above has enabled further failures to be avoided by reprofiling the groove to reduce the stress concentration factor and by reducing the barring speed from 30 r/min to 2 r/min.

2.4 Generator rotors

The difference in flexural stiffness about the winding and pole axes of generator rotors, which otherwise leads to unacceptable vibration levels, can be compensated by machining either transverse (inertia) or longitudinal slots to reduce flexural stiffness. Both of these methods have, however, led to fatigue cracking problems.

2.4.1 Inertia slot cracking

A classical example of the application of fracture mechanics to enable plant containing a sizeable defect to remain safely in operation, with substantial consequent savings in replacement generation costs, was provided when the inertia slot ends of a 500 MW rotor were found during a routine outage to have been overheated to about 800°C, Fig 12.20. The overheating was traced to a severe negative phase sequence electrical overload during recommissioning trials six years earlier when the rotor had seen only 6 months' service. Material at the slot ends had expanded under the heat, eventually yielding in compression at temperature. Subsequent cooling then resulted in large tensile residual stresses, confirmed as of yield point level by centre hole drilling methods. Cracks 180 mm deep were present at most slot ends, exhibited similar profiles and had initiated to a depth of 8 mm by ductile tearing under the action of the tensile residual stresses. The

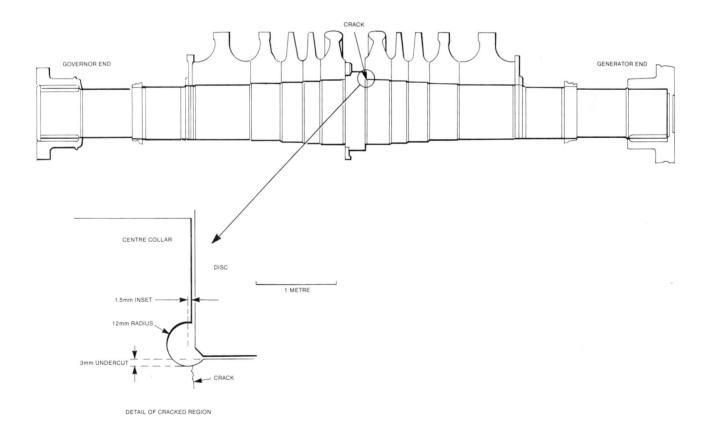

FIG. 12.19 Location of centre collar stress relief grooves in LP rotor

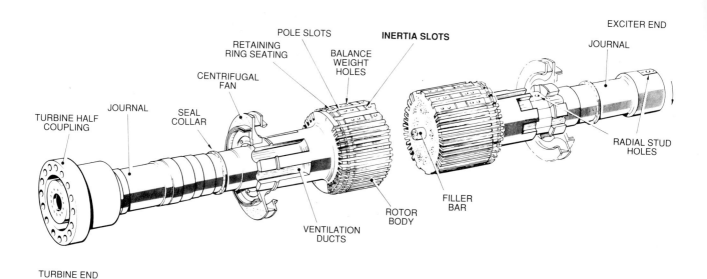

FIG. 12.20 Location of inertia slots in a generator rotor

remaining propagation had occurred by high cycle fatigue under self-weight bending. Comparison of the fracture surface beach mark spacing with the outage and operational records indicated that most of the crack growth had occurred in a period of 1500 hours' service following the overheating incident. The combination of the facts that the cracks were all of similar length and had stopped early in the rotor life indicated that the driving force for propagation was the mean stress, dominated by the short range residual tensile field, rather than the self-weight bending stresses which increase considerably towards the centre of the rotor. The residual stress distribution was such that it fell away rapidly below the surface of the rotor. As the fatigue cracks propagated therefore, they entered material of progressively lower mean stress and hence of higher ΔK_{th}. Ultimately the applied stress intensity (K) equalled the local threshold value and the crack arrested. Since both the metallurgical and fracture mechanics analyses showed that the cracks had arrested and a spare rotor was not available, the cracked rotor was returned to service where it operated satisfactorily for two years to the next planned outage.

2.4.2 Fretting fatigue in longitudinally slotted rotors

Longitudinal slots reduce the flexural stiffness but, since the pole is subjected to high magnetic flux densities, it becomes necessary to replace most of the lost steel to minimise magnetic saturation. Steel filler bars and retaining wedges were initially used in 500 MW sets, loaded in short lengths to reduce stiffness. When adopted for 660 MW rotors, however, this led to the initiation and propagation of fatigue cracks driven by the relative movement between the ends of the retaining wedges and the land against which these impinged on the pole teeth of the rotor under the action of the self-weight bending stresses. At operating speeds the contact pressure between wedge and tooth land is so high that the wedge is effectively coupled to the rotor. The wedge gap therefore acts to extend the cracks initiated by fretting (Fig 12.21), and this facilitates initial growth to the tooth peak fillet radius. Fracture mechanics calculations showed that subsequent propagation was dependent on the presence of high tensile residual stresses (>210 N/mm^2) induced by the slot machining process. Further failures have been successfully avoided by replacing the short steel wedge bars with continuous wedge bars of high strength aluminium.

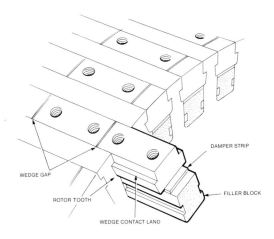

(a) Detail of pole face slot

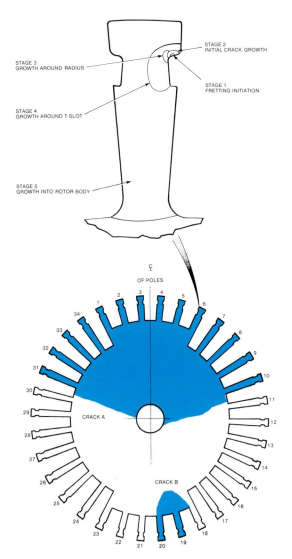

(b) Crack development stages

FIG. 12.21 Fretting fatigue in longitudinally-
slotted rotors:
(a) Detail of pole face slot
(b) Crack development stages in a 660 MW
longitudinally-slotted generator rotor.

3 Additional references

B. J. Cane, R. D. Townsend: Prediction of Remaining Life in Low Alloy Steels: TPRD/L/2674/N84: 1984

R. V. Hart: Assessment of Remaining Life using Accelerated Stress Rupture Tests: Metals Technology: January 1976

B. J. Cane: Remaining Creep Life Estimation by Strain Assessment on Plant: Int. J. Pressure Vessels and Piping, *10*, p 11: 1982

CHAPTER 13

Environmental effects

Introduction

With the exception of some of the noble metals which may occur naturally in the elementary state, metals are usually found as compounds (ores) principally oxides and sulphides. To extract metals from their ores therefore requires a supply of energy. Conversely most metals will react spontaneously with their environment and tend to revert to compounds. The reactions between a metal and its environment to form compounds constitutes corrosion and in many cases are predictable from thermodynamic considerations. However, since corrosion processes occur at surfaces the properties of the corrosion product can control the rate of corrosion. In particular the corrosion product may act as a barrier between the environment and the metal. If the effectiveness of the barrier increases with time the rate of corrosion reduces with time: this is the frequently observed behaviour of metals in gaseous atmospheres. If, however, the corrosion product can be removed from the site of reaction, as in the case of a soluble corrosion product in an acqueous solution, the rate of metal loss cannot be expected to diminish with time. Similarly if the corrosion product is defective, e.g., cracked or porous, localised attack of the metal may ensue.

In practice the methods used to combat corrosion include:

- *Modification of the corrosion properties of the material* For steels and nickel alloys alloying with other elements, particularly chromium, is often beneficial in promoting a more protective oxide. The metallurgical condition may also influence the mechanism or rate of corrosion and improvements in corrosion resistance may be achieved by variations in heat treatments or fabrication routes.

- *Modification of the environment* Clearly if a readily identifiable contaminant is responsible for governing rates of corrosion, the removal or neutralisation of this contaminant is beneficial. Similarly if critical concentrations of certain reactants are responsible, care should be taken to avoid such concentrations.

- *Use of protective coatings* The principal function of such coatings is to exclude the corrosive media. Organic coatings, e.g., paints, are probably the most familiar and used for the greatest variety of applications but are limited to maximum temperatures of up to about 200°C, above which they break down. Ceramic and metallic spray coatings nominally offer superior elevated temperature properties but have proved to be of limited value because of spalling/separation due to poor bond strength, particularly under thermal cycling conditions, and lack of barrier effectiveness due to porosity.

1 High temperature oxidation and fuel ash corrosion

Before examining in detail the importance of fuel impurities in fireside metal corrosion, mention should be made of the oxidation of the metal surfaces by oxygen and carbon dioxide in the combustion gases or high temperature steam to form a protective (or sometimes non-protective) oxide scale. Mild steel or low-chromium steels generally have sufficient oxidation resistance to serve as satisfactory and economic materials for power plant components operating at relatively low metal temperatures.

Above about 600°C it is necessary to adopt alloys containing major additions of chromium and nickel, which have an 'austenitic' as opposed to a 'ferritic' lattice structure. The chromium, in concentrations greater than 15%, provides better oxidation resistance and the austenitic structure provides greater creep strength. Other elements such as aluminium and silicon confer improvements in oxidation resistance but are limited in the quantity which can be added because they are detrimental to mechanical properties. All these alloying elements produce more protective oxides and since they have a higher affinity for oxygen they also tend to oxidise preferentially. For the most severe temperatures (on tube supports and burner components) the oxidation resistance may become the dominant design requirements and alloys containing chromium concentrations of up to 50% may then be preferred.

At high temperatures, the oxidation rate is controlled by atomic diffusion processes which allow transport of metal and oxygen atoms through the layer of oxide formed by the reaction on the surface of the metal. The rate of oxidation thus decreases as the thickness of the oxide layer increases and can be described by a parabolic law. Good oxidation resistance requires that the protective oxide layer should retain its integrity and remain adherent to the surface of the metal. Factors such as scale separation arising from differential expansion between the metal and the oxide or the presence of porosity or voidage particularly near the metal/scale interface can significantly affect rates of oxidation.

It is worth noting that corrosion processes, which are time and temperature dependent, can provide a technique for service failure investigations (Pinder, 1981 [1]) and the determination of service histories. For instance, measurement of the variation of oxide thickness down the length of a crack can, when used in conjunction with detailed oxidation kinetic data, provide a reliable estimate of the time of initiation and the rate of propagation of a defect. An examination of free surface oxide thickness, morphology and, in some cases, the lattice parameter of certain oxide phases can provide an estimate of the component exposure temperature.

The oxidation behaviour of boiler component materials can be significantly changed when they are in contact with flue gas. The flue gases contain carbon dioxide and water from the combustion of the coal or oil together with a small excess of oxygen. In such an environment, steel tubes operating at their design metal temperatures would be expected to behave as in oxygen and exhibit sufficient oxidation resistance to achieve their design lives. In addition to carbon and hydrogen both coal and oil contain a wide variety of organic and mineral impurities. Although in the case of coal most of these impurities are alumino-silicates, minerals which are considered inert; other impurities such as sulphur, chlorine, sodium and potassium can be present. In the case of oil, vanadium is also present. These impurities can considerably influence the high temperature oxidation behaviour of alloys used for boiler tubes and result in a significant increase in the rate of oxidation and hence metal loss. A comprehensive documentation of the aspects of fuel ash corrosion is given by Laxton et al [2].

In practice, two distinct forms of fuel ash corrosion are encountered; furnace wall (evaporator) tube corrosion and superheater/reheater corrosion.

1.1 Furnace wall corrosion

The fireside metal temperature of mild-steel evaporator tubes on the furnace wall is about 450°C. Under normal oxidising flue gas conditions this produces a protective oxide scale and acceptably low corrosion rates. However, the dynamic nature of the combustion process can vary considerably with coal-fired burners, and a complex range of non-equilibrium chemical conditions can exist if the combustion departs locally from the ideal oxidising condition (with approximately four per cent excess oxygen) towards more-reducing conditions (no excess oxygen).

Reducing conditions promote aggressive fireside corrosion due to the complex gaseous interaction between sulphur, chlorine, carbon, and oxygen in the presence of high heat fluxes, to yield a wide range of different and relatively non-protective corrosion scales. Depending on the type of scale formed, corrosion rates of between two and six millimetres per year have been measured on sidewall tubes in the burner belt over an area as much as 100 tubes wide. This rate is well in excess of the corrosion rate of 0.3 mm per year which would allow the design life of the 8 mm wall tubing to be achieved.

In coal-fired boilers (Fig 13.1), the temperature at the outer surface of the ash deposit on furnace tubes can be as high as 1250°C. Sintering and partial melting of the ash, commonly referred to as 'slagging', can then occur. This does not cause direct attack on the protective oxide, but can prevent the access of oxygen to the oxide scale and so encourage reducing conditions which allow formation of iron sulphide. Such conditions render the scale markedly less protective and allows rapid diffusion through the scale because of its defective atomic structure. The presence of sulphide within the oxide also causes mechanical disruption associated with its greater molar volume.

The most damaging of the reducing conditions occur when incompletely burnt coal particles impinge upon and stick to the sidewall tubes. Combustion of these coal particles is retarded on the relatively cooler ash surface and a smouldering action slowly releases high local concentrations of carbon monoxide together with the mineral impurities — including volatile compounds of sulphur, sodium, potassium and chlorine. The high carbon monoxide concentration assists sulphidation and the gaseous hydrogen chloride which is released also enhances corrosion.

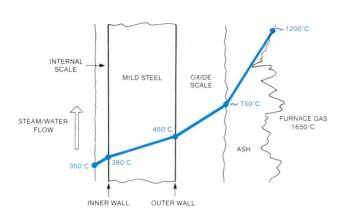

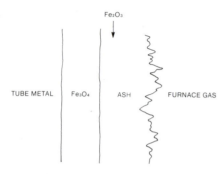

PROTECTIVE-OXIDE SCALE FORMED IN
OXIDISING CONDITIONS

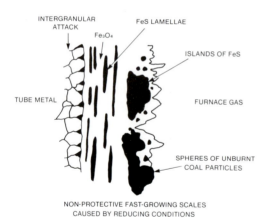

NON-PROTECTIVE FAST-GROWING SCALES
CAUSED BY REDUCING CONDITIONS

FIG. 13.1 Oxide films: coal-fired boilers
Although the metal temperature of mild-steel furnace
wall tubes is relatively low, the high heat flux gives a
steep temperature gradient through the oxide scale and
deposited ash (upper diagram). The form of the scale
grown under normal oxidising conditions is an inner
magnetite (Fe_3O_4) layer with a covering of haematite
(Fe_2O_3). Under reducing conditions, generally caused
by a deficiency of oxygen relative to the fuel, a rapidly
growing magnetite layer forms, containing lamellae of
iron sulphide (FeS) with (in a coal-fired boiler)
incompletely burnt coal particles near its surface. These
smouldering particles release volatile compounds
(particularly hydrogen chloride) which promote
intergranular attack of the tube metal.

Hydrogen chloride diffuses rapidly through the oxide
layer to the metal surface, where volatile iron chloride
is formed and can diffuse back to the scale/gas interface. The metal-chloride formation is related to severe
intergranular attack of the metal surface, and it may
also promote cracking in the oxide scale. At certain
power stations, consideration has been given to the
possibility of bleeding air through the furnace walls
to avoid the local reducing conditions which permit
this kind of attack.

It is notable that the corrosion of evaporator tubes
is not generally a severe problem in oil-fired boilers.
The combustion of fuel oil is more rapid and efficient so that, although local reducing conditions may
still occur, they are not sufficiently severe to cause
excessive corrosion through sulphide formation. There
is also considerably less chlorine in fuel oil than in
coal and chloride attack on the metal does not occur.

Furnace wall corrosion is primarily a combustion-
related problem exacerbated by chlorine in coal. In
the majority of cases it can be minimised by careful attention to those combustion parameters, e.g.,
abnormalities in burner characteristics, milling and
classification, or fuel distribution which give rise to
localised reducing conditions adjacent to furnace wall
tubes.

Where reducing conditions are not amenable to combustion control, remedial actions involving materials
options will be required. Various options are available
which can be ranked in order of increasing lifetime
as follows:

- Regular replacement in the same mild or low alloy
  steel tube material. This would be appropriate if
  rates of metal loss are relatively low, i.e., up to
  40 nm/h which is equivalent 4 mm in 100 000 hours.

- Replacement with faceted tube (Fig 13.2). The concept of faceted tube is to use a tube with an eccen-
  trically shaped profile such that it is thicker at both
  front and rear while maintaining the original bore
  and tube thickness on the sides. This enables installation of the tube within existing wall clearances but
  can provide up to a factor of two improvement in
  tube life.

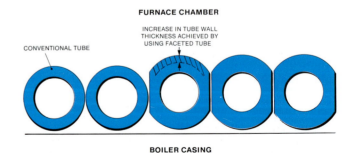

FIG. 13.2 Faceted boiler tubing to achieve increased
wall thickness for tangent furnace wall tubes

- Replacement with composite coextruded tube comprising an inner layer of mild steel and outer layer of corrosion-resistant austenitic stainless steel, usually Type 310 Nb — a 25% Cr 20% Ni steel stabilised with 1% niobium. The installation of monobloc austenitic stainless steel tube is precluded because of the risk of stress corrosion from the waterside of the tubes. Coextruded tubing is generally considered to offer a factor of about 3 improvement in corrosion resistance and in some instances the improvement factor can be an order of magnitude. Although relative material costs are about 8:1 for coextruded tube: mild steel tube installation costs reduce this ratio to less than 3:1 and a coextruded tube installation becomes economically viable for rates of metal loss >150 nm/h.

1.2 Superheater and reheater fireside corrosion

Fuel ash corrosion of superheater and reheater tubes generally manifests itself by producing two distinct wastage flats on the hottest exposed faces of tubes (Fig 13.3). On occasions, at geometrical/aerodynamic features such as support lugs, fillet welds beads and tube butt welds, local deep isolated pits can occur.

The worst fuel ash corrosion occurs on aerodynamically exposed tubes, i.e., the first tube of a major bank or the first tube within a bank following a major ($\geqslant$200 mm deep) cavity. Contra-flow superheater banks which combine the hottest flue gases with the hottest steam are the most vulnerable; conversely, those designs which are parallel flow are generally less susceptible.

In coal-fired plant the major coal impurities relevant to fuel ash corrosion are sulphur (typically 1 to 3% as received in British coals), sodium (0.05 to 0.4%), potassium (0.3% to 0.6%) and chlorine (0.02 to 0.75%).

When the coal is burnt these impurities give rise to sodium and potassium sulphates which gradually accumulate by condensation at the base of the porous ash deposit on the tube (Fig 13.4). Such mixed sulphate deposits are molten at surface metal temperatures above about 580°C and form a thin layer of highly corrosive liquid next to the protective oxide scale. The molten sulphate layer contains free sulphur trioxide which stabilises the sulphate and attacks the scale by dissolving the normally protective oxide to form iron, chromium and nickel sulphates.

The stability of the iron and chromium sulphates decreases as the temperature increases, and the temperature gradient through the deposit causes a concentration gradient for the dissolved metal ions. These factors ensure that the dissolution is effectively a continuous process, with iron, chromium and nickel diffusing outward through the molten sulphate layer. At the hotter outer surface, iron/chromium oxides are re-precipitated within the inner ash, but they no longer form a protective layer. This dissolution process results in linear corrosion kinetics, as against parabolic kinetics for protective oxidation.

The rate of this continuous transport process determines the rate of corrosion, which is therefore highly dependent upon the temperature gradient between the hot gas and the metal surface, i.e., the heat flux established through the tube. Consequently, for a given

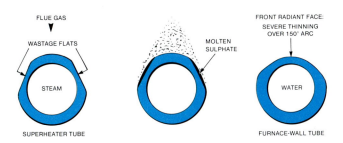

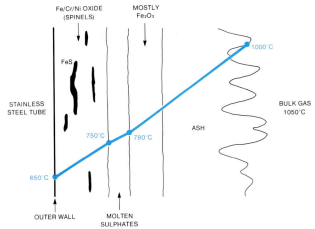

FIG. 13.4 Superheater tubing — oxide scale characteristics

FIG. 13.3 Schematic of wastage profiles on superheater/reheater tubes and furnace wall tubes
Superheater and reheater corrosion profiles typically occur as two wastage flats at the two o'clock and ten o'clock positions with leading face (at twelve o'clock) largely uncorroded. On the front face, the deposited ash provides sufficient insulation to keep the metal surface temperature quite low. The ash is much thinner around the side of the tube and the high heat flux allows the formation of a molten sulphate layer, resulting in severe wastage. Furnace wall tube corrosion profiles are quite different and occur on the radiant (i.e., hottest) face with thinning of the tube wall over a considerable arc.

The structure of scale and deposits on superheater tubing and the temperature gradients within them may be compared with those for furnace-wall tubing shown in Fig 13.1. The overall mechanism of superheater tube corrosion may be viewed as a continuous process of metal oxidation at the tube surface, dissolution of this oxide in the molten sulphate, diffusion across this layer and precipitation as porous haematite (Fe_2O_3) beneath the outer covering layer of ash. The thickness of the protective oxide layer on the surface of the metal is thus reduced and the oxidation of the metal can proceed more rapidly.

metal temperature, higher corrosion rates are observed at higher gas temperatures. As metal temperatures increase so do reaction (corrosion) rates (Fig 13.5). Above a peak temperature of 650–700°C, molten sulphate solutions become thermodynamically unstable and this causes reaction rates to fall off at higher temperatures, eventually leading to normal high temperature sulphidation/oxidation. This results in the classical 'bell-shaped curve' shown in the inset of Fig 13.5.

Operational experience has shown that there is a good correlation between chlorine content and the rate of corrosion of austenitic superheaters and reheaters. In the past this had been explained by assuming that sodium and chlorine co-existed in coal as sodium chloride and that chlorine was therefore a guide to sodium content. In addition, sodium chloride was thought to promote the release of naturally involatile potassium from mineral matter in the coal.

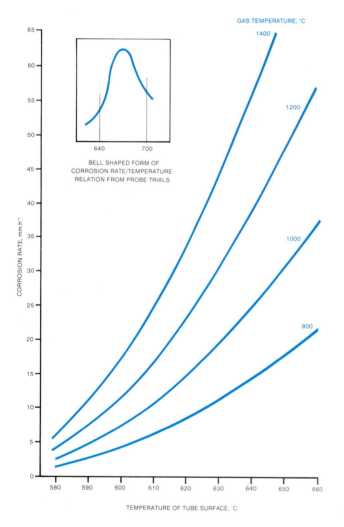

FIG. 13.5 Corrosion rates of austenitic tubes in coal-fired boilers for non-leading tubes and a coal chlorine content of 0.25%. CEGB research into the mechanism of sulphate attack has enabled theoretical calculations of the expected corrosion rate to be made which are in agreement with measurements on superheaters and reheaters in boilers.

Research work has suggested that chlorine is not directly related to alkali release, for the following reasons. Firstly chlorine does not occur in coal primarily as sodium chloride, but rather as ions weakly bonded to the organic coal matter. The majority of the sodium in coal may be present in a similar manner, but the ratio between sodium and chlorine in coals is variable and chlorine is not therefore a reliable guide to sodium content. Secondly, studies of the release of metals from coal during combustion have shown that potassium release is not related to sodium content, but rather to chlorine content. Chlorine promotes the release of both sodium and potassium and it is this fact that helps to explain the correlation with corrosion rate. However, this correlation is better than the experimentally derived relationship between alkali metal release and chlorine content, it is therefore possible that chlorine may play an additional, as yet unknown, direct role in the corrosion mechanism. Alkali metal release is also influenced by ash content. The amount of ash present will affect the partition of alkali metals between the ash melt and the vapour phase, i.e., the higher the ash content of a coal the greater the alkali metal sulphate rendered innocuous.

Sulphur occurs in coal predominantly as either sulphur or pyrites, FeS_2, and is released mainly as SO_2 during combustion. Any alkali metal chloride vapour will rapidly react with SO_2 and O_2 to form alkali metal sulphates. However, as all British coals contain an excess of sulphur compared with the possible alkali release, sulphur content is not a significant factor in considering the corrosion potential of fossil fuel burnt in CEGB boilers.

In modern oil-fired boilers fireside corrosion of superheater and reheater tubes has not been a serious problem. This absence of fireside corrosion primarily arises because steam outlet temperatures have been limited to 540°C, which restricts surface metal temperatures to less than 580°C and hence minimises metal losses by fireside corrosion.

Residual fuel oils generally contain less than 200 WPPM each of sodium and vanadium, which in practice are the only impurities relevant to the mechanism of high temperature corrosion. Typical values for Middle East oil supplied to the UK are about 50–100 WPPM of each. Unlike coal, which typically produces 20% ash, the residual fuel oils produce only about 0.2% ash on combustion. Thus even lower concentrations of these elements, down to about 5 WPPM, can cause corrosive deposits on superheaters and reheaters because there is very little dilution by other fuel impurities as occurs with coal firing. In some boilers when both coal and oil are burned simultaneously, there is good evidence that coal ash can adequately dilute the vanadium content of the deposits, thus rendering the deposits harmless.

Corrosion is not sensitive to the absolute amount of sodium and vanadium when they are both present, but the ratio between them (i.e., the V/Na weight ratio in the fuel) can be a useful guide to the corrosiveness of deposits on metal. At the higher metal temperatures (above 590–610°C) corrosion rates can increase by a factor of >4 when the V/Na ratio moves from the lowest (0.2) to the highest values. It should be noted that most CEGB boiler experience is restricted to oils with V/Na <2. The V/Na ratio in the bulk deposits is usually very similar to that in the oil. However, with age, the deposits may segregate into Na and V rich bands with the highest concentrations of Na at the metal oxide interface.

Sulphur contents in the fuel up to 4.5% do not appear to affect corrosion rates below metal surface temperatures of 620°C. Above this temperature an increase in sulphur from 0.5% to 4.5% might double the corrosion rate.

A series of corrosion-rig tests using an oil-fired burner, and plant trials using operational boilers, have demonstrated the wisdom of the decision to apply this limit to steam outlet temperature. The primary objective of the tests was to establish the corrosion rates of ferritic and austenitic steels over the typical operating temperature range of superheater and reheater tubes, for different flue gas temperatures. For austenitic steels, corrosion rates up to two millimetres per year (equivalent to a tube life of two years) were measured with a metal temperature of 600°C and a flue gas temperature of 1150°C. Acceptable corrosion rates (0.2 mm per year) were obtained only below normal operating temperature. Ferritic steels displayed much better corrosion resistance (Fig 13.6).

The most likely explanation for the poor performance of austenitic steels is a combination of their susceptibility to oxide cracking and spalling and attack of the thin protective oxide scale by vanadium compounds. The spalling is exacerbated by thermal expansion stresses between the metal and oxide. On a second exposure to the hot flue gases, faster oxidation occurs at the exposed metal surface during re-growth of the protective oxide. With ferritic tubes, on the other hand, spalling is rare because the expansion coefficients of the metal and the oxide are more closely matched and most of the oxide remains as an adherent layer throughout the life of the tube.

The important influence of vanadium compounds on the melting behaviour of sulphates in ash deposits has already been mentioned. They may also influence the corrosion directly. Depending on the relative concentrations of the compounds and the effective pressure of oxygen, a wide range of vanadate compounds based on vanadium oxides and sodium sulphate can exist. If the molten vanadates reach the protective oxide layer it may be dissolved as iron and chromium vanadates, thus diminishing the protection which it affords.

Since metal losses by fireside corrosion can be life limiting, it is important to be able to predict the rate of deterioration of boiler tubing to determine its original design or residual life. This will also enable the

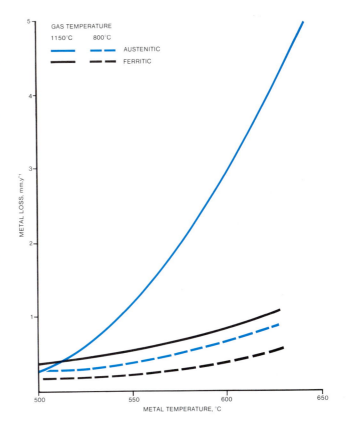

FIG. 13.6 Corrosion rates of austenitic tubes in oil-fired boilers
In oil-fired boilers, austenitic steels are particularly susceptible to corrosion at high flue gas temperatures, largely because they are more prone to oxide cracking and spalling, than ferritic steels. The higher rates of corrosion are caused by the repeated re-growth of new protective oxide following each spalling incident.

most economical replacement of corroded tubes to be planned and provide data for design modifications.

The exact calculation of residual life is precluded by the imprecise nature of many of the factors involved. Nonetheless useful estimates can be made. The mechanism of failure depends largely upon the operating temperature of the component involved, i.e., superheater and reheater tubes often operate in the temperature range where time dependent creep deformation and rupture are relevant, while furnace wall tubes operate at lower metal temperatures where creep effects can be neglected and the main consideration is whether the tube is of sufficient thickness to contain the internal pressure and are considered on a proof stress basis. Some stages of superheater/reheater may operate at temperatures below the creep range and are also dealt with using a proof stress related criterion.

The two failure mechanisms involve different approaches or life assessment; full details are given in Laxton *et al* [2]. Briefly the proof stress regime is based on the assumption that failure occurs when the hoop stress in the tube attains a certain critical level related to the ultimate tensile strength of the tube material at its operating temperature. From a knowledge of the tube dimensions, operating conditions and corrosion rate the operating time to reach this critical value can be readily estimated. In the case of tubes operating at metal temperatures in the creep regime, tube lives are estimated using a life fraction rule which summates the fraction of creep life consumed in defined temperature bands and time intervals. These estimates can be calculated either manually or with the aid of computer programs.

2 Stress corrosion

Stress corrosion is a phenomenon associated with cracking resulting from the conjoint effects of corrosion and stress, usually a tensile stress. The magnitude of the stress at which failure occurs is generally well below that required for failure in the absence of corrosion, and similarly, the environment alone is insufficiently corrosive to cause a stress-free assembly to fail.

For almost every alloy there appear to be particular environments which render it susceptible to stress corrosion cracking; these 'classical' systems are listed in Table 13.1. In practice, however, failures have occasionally occurred in structural materials in which the classical species were either absent or existed only in very low (PPM) concentration in the bulk solution. An example of such behaviour was the catastrophic failure of a low pressure steam turbine disc at a power station in 1969 while undergoing routine overspeed testing. Although the low to medium strength (<900 MN/m^2 UTS) low alloy steels used for discs were known to suffer stress corrosion in hot nitrate, hydroxide, phosphate and carbonate/bicarbonate solutions, they were not considered to be susceptible to stress corrosion in the high purity wet steam environment of an LP steam turbine. However, subsequent tests on similar steels showed that stress corrosion cracks can initiate and propagate in wet steam with no chemical contamination or impurity concentration. Service experience with other alloys and environments has enabled the 'classical' systems to be extended as indicated in Table 13.2.

TABLE 13.1

'Classical' stress-corrosion systems

Alloy	Damaging environments (Aqueous solutions)
Mild steel (low strength ferritic steels)	OH^{-*}, NO_3^{-*}, CN^-, PO_4^{3-}
Austenitic stainless steels	Cl^{-*}, OH^{-*}, H_2O^* (oxygenated)
High strength steels	H_2, H_2O, CN^-, Cl^-
Copper base	NH_4^+, moist NH_3 vapours
Aluminium base	Cl^-
Titanium base	Cl^-, HNO_3^*
Nickel base	OH^{-*}

* Boiling or hot ($>50°C$)

TABLE 13.2

Some recorded stress corrosion systems

Alloy	Damaging environments
Mild steel (low strength ferritic steels)	OH^{-*}, NO_3^{-*}, CN^-, $NH_4^+{}_{(an)}$, H_2O, moist $CO^†CO_2$ gas, $CO_3^{2-†}$ HCO_3^{-*}, molybdates*, acetates*, propionates, phosphates, saturated steam, acid SO_4^{2-*}, $SO_4^{2-}+H_2S^*$, polythionic acids.
Austenitic stainless steels	Cl^{-*}, OH^{-*}, $H_2O(O_2)^{*†}$, $SO_4^{2-†}$ Cl^-, $Na_{(liq)}$ H_2O with Pb contaminant*, SO_4^{2-}, saturated steam, polythionic acids, $F^{-†}$, sea water†
High strength steels	H_2, $H_2O_{(l)}$, CN^-, Cl^-, $HCl_{(g)}$, $HBr_{(g)}$, $H_2S_{(g)}$, $HCl_{(liq)}$, H_2+O_2, saturated steam, $NH_{3(g)}$, acid SO_4^{2-}, $NH_3^†O_2^†CO_2$, $Cl_{2(g)}$
Copper base	NH_4^+, citrates, tartrates, $FeCl_3$, $H_2O_{(g)}$, moist $SO_{2(g)}$, moist NO_x, moist NH_3 vapours, $H_2O_{(l)}$
Aluminium base	Cl^-, Br^-, I^-, $H_2O_{(l)}$, organics containing H_2O, moist H_2, $H_2O_{(g)}$
Titanium base	Cl^-, HNO_3^*, fused salts, $H_2O(O_2)$, CH_3OH, CCl_4, $Br_{2(l)}$, $H_{2(g)}$, $N_2O_{4(l)}$, $H_2O_{(g)}$, distilled water.
Nickel base	$H_2O^*{}_{(l,g)}$, OH^{-*}, polythionic acids, HF acid vapour

* Boiling or hot ($>50°C$)   † Sensitised condition only

Although it is commonly accepted that any given alloy is only susceptible to stress corrosion cracking in the presence of specific environments, the occasional failures in environments in which the required species were either absent or existed only in extremely low concentrations suggest that they can be associated with time dependent changes in either the solution chemistry or the metal. The former usually involves a concentration mechanism for the damaging species either by chemical reaction or by physical means. Metallurgical influences can involve pre-existing defects, changes in alloy surface composition by leaching or oxidation, compositional changes at grain boundaries by elemental segregation or second phase precipitation. These initiation mechanisms which can lead to the establishment of conditions for stable stress corrosion crack propagation are especially insidious because they may be associated with no apparent deviation in environmental material specification or plant operation.

Generally most engineering materials operate outside the relatively narrow regimes of intolerance, either fortuitously or through application of a fundamental understanding of the problem. Nevertheless, random failures still occur, many result from adventitious events involving residual stresses but others are the consequence of lack of knowledge regarding:

- The role of the environment.
- The importance of crack tip stress intensity factor as opposed to general (net section) stress.
- The influence of changes in microstructure brought about by heat treatment, operating temperatures, mechanical working or contamination.

Other unanticipated failures are the direct result of the trend towards designing both for higher operating stresses, which demand more advanced high strength alloys, and for longer operating lives.

The stresses necessary for the attack may arise from internally or externally applied loads or both. Internal stresses can result from cold working, cold forming, welding, thermal gradients or even from metallurgical transformations involving volume changes. External stresses arise as a result of loading or external pressures.

The visible manifestations of stress corrosion are cracks that create the impression of inherent brittleness in the material since the cracks propagate with little attendant deformation and only a very small amount of corrosion.

Initiation sites for stress corrosion cracking divide broadly into those which are either pre-existing or those which are developed in service. The sites may also be associated with a highly localised mechanical effect which acts as a stress concentrator or chemical effects associated with the concentration of ionic species or the development of local acidity.

Examples of stress concentrators are:

(a) Geometrical features such as sharp keyways or inadequately radiused changes in section.

(b) Manufacturing defects such as laps and emergent laminations. There is a strong association between stress corrosion crack initiation and the presence of non-metallic inclusions. The inclusion may locally alter the chemistry of the environment by being attacked and its removal may leave a stress concentrating surface flaw or generate pit-like chemical conditions.

(c) Localised pitting or crevice corrosion occurring during storage, installation or service.

The formation of corrosion pits would be expected to influence stress corrosion behaviour by producing a stress concentrator, enabling departure from the bulk environment composition within the pit and changing the local electrochemical corrosion processes.

Clear evidence of stress-corrosion cracks initiating from pits has been found in LP turbine steels both in the UK and the USA. In the previously mentioned turbine disc failure, stagnant conditions within the wet keyway promoted pitting at (Mn,Fe)S inclusions. Under the shrink fit stress of the disc on the rotor, these pits sharpened and developed into intergranular stress corrosion cracks, eventually leading to brittle fracture. In laboratory and power station rig tests, no stress corrosion cracks were found unassociated with pits. Again in the UK, pits growing in the circumferential interdisc grooves of a 500 MW LP turbine rotor shaft under operational conditions where condensate was present have developed into stress corrosion cracks. Major fatigue cracks which have propagated in the centre collar stress relief groove under load are believed to have initiated from these blunt stress corrosion cracks.

(d) Processing or fabrication damage such as machining marks, hammer blow marks, identification stamps or welding arc strikes (Fig 13.7). These forms of damage also often induce tensile residual stresses which may be of yield magnitude and have a detrimental effect on the susceptibility to stress corrosion.

After initiation, the growth of the stress corrosion cracking can be transgranular and/or intergranular (Fig 13.8). It is also possible for the same alloy in a given environment to show transitions from one mode to the other as a result of changes in heat treatment, applied stress intensity, cold work, or chemical heterogenity.

Optical microscopy combined with scanning electron microscopy may reveal features which enable stress corrosion cracking to be distinguished from brittle, transgranular cleavage or fibrous, ductile rupture. Intergranular fracture is readily identified by the presence of smooth grain facets in the scanning electron microscope, but it can be difficult to deduce an environmental influence if the solution is passivating or if hydrogen diffusion to the maximum stress region ahead of the crack tip is the mechanism of propagation. Where final unstable fracture in the specimen or component has not occurred it is often revealing to break open the remaining ligament in liquid nitrogen to identify the stress corrosion cracking/cleavage interface where, generally, there will be relatively little masking of detail by corrosion product if the crack was active.

Examples of stress corrosion cracking systems exhibiting transgranular cracking are:

- Austenitic stainless steels in acidic chloride solutions.
- Low strength ferritic steels in acidic media, often referred to as hydrogen cracking.
- Ferritic steels in phosphate solutions.

The fracture surfaces of stainless steels which have suffered transgranular stress corrosion cracking exhibit characteristics of cleavage, being relatively flat and exhibiting feathery features superficially resembling cleavage river patterns but possessing a greater degree of coalescence of surface contours, giving a fan-like effect. Similar behaviour has been observed in low

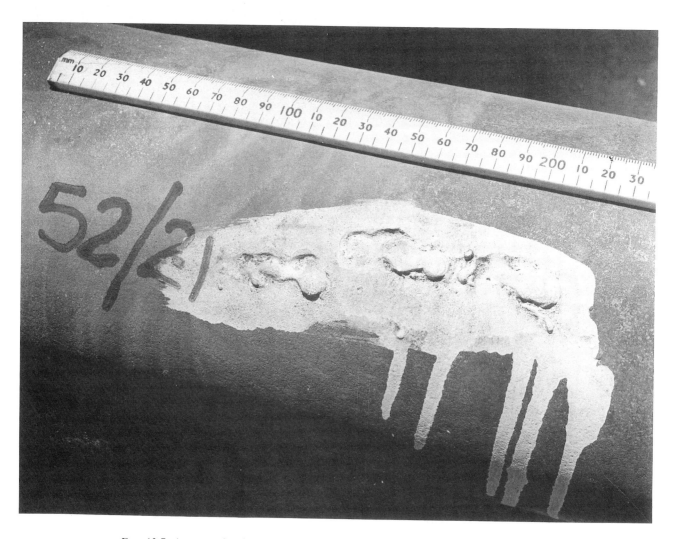

FIG. 13.7 An example of welding electrode arc strike damage on C-Mn steel boiler pipe
Axial cracking, which exhibited features characteristic of stress corrosion cracking, has been found to have grown from similar damage. Other crack origins have been associated with either hammer indentations or impressions made during erection, most being less than 0.5 mm deep.

alloy steels when the crack is propagating normally to the rolling direction.

Examples of stress corrosion systems exhibiting intergranular corrosion are:

- Sensitised austenitic stainless steels in oxygenated water and water containing sodium sulphate.
- Carbon steels in caustic solutions.
- Low alloy steels in pure water.
- Nickel based alloys in high purity water and caustic solutions.
- Tempered martensitic stainless steels in chloride solutions.

In these systems the grain boundary is a preferred path owing to chemical heterogeneity resulting from solute segregation or precipitation of a second phase. The fracture surfaces are characterised by clearly defined grain facets (Fig 13.9).

With low and medium strength ferritic steels, fracture can occur along prior austenite grain bounda-

ries when temper embrittled by heating and slowly cooling in the range 350–550°C. Various damaging environments have been identified including pure water, saturated oxygenated steam and boiling caustic solution. The grain boundary segregation of impurity elements such as Sb, P, Sn, S, Se, Ge and Te is believed to augment a hydrogen damage mechanism by retarding a recombination reaction of hydrogen atoms to gas molecules, thereby increasing the hydrogen flux into the metal from the crack tip.

Austenitic stainless steels and nickel-base alloys are subject to intergranular failure associated with a phenomenon known as 'sensitisation', which, like temper embrittlement, involves a critical departure from bulk alloy composition. Chromium carbides ($Cr_{23}C_6$) precipitate at the grain boundaries when the component is slowly cooled from the solution annealing temperature of 1050 to 1100°C, or operates in the temperature range of 500 to 750°C.

Precipitation of the chromium carbides at the grain boundaries causes depletion of chromium in the regions

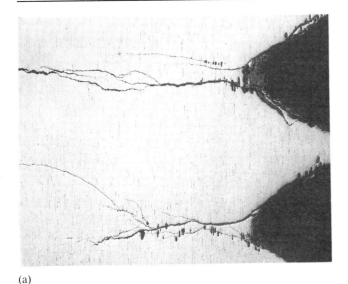

(a)

(b)

FIG. 13.8 Examples of stress corrosion cracking:
(a) Forked transgranular crack propagation
(b) Intergranular crack propagation

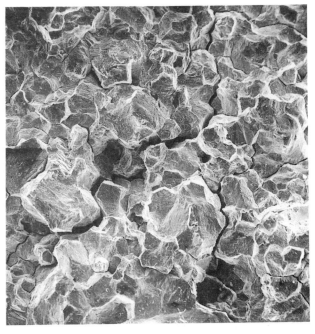

FIG. 13.9 Scanning electron macrograph showing
intergranular stress corrosion cracking with clearly
defined grain facets

adjacent to the grain boundaries and these areas then exhibit reduced corrosion resistance. Any stress corrosion attack in this instance is intergranular.

Many austenitic stainless steel superheater and reheater tubes operate at metal temperatures within the range 500–750° and would therefore be expected to become sensitised with service exposure. However, stress corrosion arising from sensitisation has rarely been encountered in service due to the physical distance of the austenitic steel from the evaporation zone and the high degree of superheat associated with superheater and reheater stages fabricated in austenitic stainless steel. It should also be noted that the heating introduced during a welding cycle can result in bands of sensitised material a short distance on each side of the weld. This may lead to localised attack, known as weld decay, in suitable environments, and is one reason why austenitic stainless steels are not used in the water circuits of boilers. In some instances the problem of sensitisation can be overcome by the use

of stabilised grades of austenitic stainless steel, i.e., steel with additions of alloying elements such as titanium or niobium which have a greater affinity for carbon than chromium. Alternatively, the risk of intergranular corrosion arising from sensitisation can be considerably reduced by using low carbon ($<0.03\%$) grades.

3 Corrosion fatigue

When a metal is subjected to an alternating or fluctuating stress it is liable to develop cracks that gradually propagate through the material and cause failure by fatigue; such failures are extremely common. The fatigue cracks usually start from some stress concentration, e.g., undercut at an undressed weld, a keyway or a scratch on the metal surface.

Figure 13.10 shows a plot of stress versus the number of cycles to failure (a S–N curve) for ferritic steels below about 200°C subjected to cyclical stress. A number of cycles at the corresponding stress to the right of the upper solid line results in failure, but no failure occurs for an infinite number of cycles at or below the fatigue limit. With most non-ferrous and austenitic steels no definite fatigue is found. In a corrosive environment, failure at a given stress level usually occurs within fewer cycles and in the case of ferritic steels a true fatigue limit is no longer attained, i.e., failure occurs at any applied stress if the number of cycles is sufficiently large. Cracking of metal resulting from the combined action of a corrosive environment and repeated or alternate stress is known as

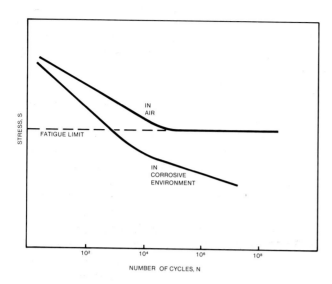

FIG. 13.10 S–N curve for steels subjected to
cyclical stress

corrosion fatigue. It should be noted that the damage
is usually greater than the sum of damage by corrosion
and fatigue acting separately.

Corrosion fatigue cracks are typically transgranular
(Fig 13.11), often multiple parallel cracks and usually
initiate from corrosion pits formed at the surface. The
environments in which corrosion fatigue can occur
are numerous and are not specific, contrary to the
situation for stress corrosion cracking for which only
certain ion-metal combinations result in damage. Steel
for instance can suffer corrosion fatigue in relatively
pure boiler water, fresh waters, sea water and combustion product condensates.

In boiler plant components, thermal cycling and
expansion constraints are the primary stressing mode
with high temperature oxidation the primary corrosion
process. Examples of where corrosion fatigue has been
a problem are from heater to stub tube welds, tube
attachments where a rigid connection has been made
or binding has occurred, at joints in tube finning and
in economiser tubes. Most of these problems are associated with differential thermal expansion arising from
temperature differences. If the temperature differential
imposes sufficient strain on the tubes, failure occurs
after a number of start-up/shutdown cycles. Lower
merit units are therefore more prone to suffer such
failures, and the number of failures usually correlates
with cyclic operation of the boiler.

A feature of corrosion fatigue is that it usually
develops simultaneously in other similar positions, so
that when a failure occurs it is likely that cracks are
present in other tubes at similar positions. Non-destructive testing of other similar tubes is then required
to determine the extent of damage and repair.

During start-up and shutdown many components
will be subject to transient temperature gradients across
their wall thickness. For thick walled components such
as the high and intermediate pressure pipework, valve

FIG. 13.11 Transgranular corrosion fatigue cracking,
initiating from corrosion pits at the bore of a steam
generator tube

assemblies, casings and rotors, such transient temperature gradients will give rise to thermal expansion
stresses. If during start-up the temperature gradient
is excessive, the stresses may be sufficient to result
in compressive plastic yielding at stress-concentrating
features and then cause tensile stresses at the surface
when the temperature gradient decays. Although creep
may cause this stress to relax during operation, each
shutdown and start-up cycle regenerates the stress and
results in low frequency cyclic deformation. Under such
conditions long term operation may result in cracking
by a mechanism known as thermal fatigue.

Careful consideration to design detail to avoid stress
concentrations and compliance with correct start-up
procedures to avoid high transient temperature gradients are expected to avoid thermal fatigue cracking
during component life.

4 CO₂ corrosion

The two designs of nuclear reactor operated in Britain,
magnox and AGR, both employ CO_2 as the coolant
or heat transfer medium. Operating conditions for
the AGRs with higher temperatures and pressures are
much more severe than for magnox and the corrosion
behaviour of steels in the two reactor systems will be
dealt with separately.

4.1 Magnox reactor design

The design operating conditions for the various magnox reactors covered maximum gas temperatures from 350°C to 400°C and gas pressures from 9.3 bar to 27.6 bar. Under these conditions it was considered that mild steel was a suitable material for all the structural steelwork within the pressure circuit, including boiler tubing.

In all these reactors one can anticipate some oxidation of the steelwork within the pressure circuit according to the equation:

$$3Fe + 4CO_2 \rightarrow Fe_3O_4 + 4CO$$

Experimental work has shown that the parameters which influence the corrosion behaviour of mild steel in CO$_2$ are temperature, moisture content of the gas and composition of the steel.

Increasing the temperature or moisture content increases the corrosion rate, as might be expected.

The element which contributes most to the corrosion behaviour of mild steel is silicon which varies with the type of steel from around 0.005% in a rimming steel to a maximum of 0.35% in a fully-killed steel, although for most fully-killed steels the range is 0.1–0.2% silicon. The benefit of high silicon can be judged from the fact that a low silicon rimming steel corrodes at three to four times the rate of a fully-killed steel.

Initially the oxide formed on the steel is protective and the corrosion rate steadily decreases with time. This period of protective oxidation may continue for a time varying from tens of hours to thousands of hours, depending on the operating conditions and the type of steel, but eventually there is a change to a non-protective or breakaway oxidation which proceeds at a steady linear rate (Fig 13.12). Essentially this change is brought about by the formation of cracks in the original oxide, which permit ready gas access to the underlying metal, and the fact that slight chemical changes in the metal catalyse the formation of carbon according to the equation:

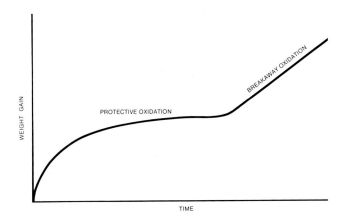

FIG. 13.12 Oxidation behaviour of mild steel in CO$_2$ coolant gas

$$2CO \leftrightarrow CO_2 + C$$

This carbon deposits within the newly growing oxide and produces a porous oxide which permits continuing access of CO$_2$ to the metal, thus enabling the breakaway oxidation process to continue.

The oxidation rates which occur on large components are generally not sufficient to cause concern about any weakening of the structure through loss of section but a problem can arise where bolted or welded interfaces occur. The volume of oxide produced is about 2.5 times the volume of metal consumed and even with a simple bolted joint comprising two plates, bolt, washer and nut there are eight surfaces contributing an oxide growth which can cause jacking apart of the surfaces and induce strain in the bolt. If the washer is rimming steel, as was often the case in the magnox reactors, then the jacking is further enhanced.

The failure of bolts in a reactor due to oxide jacking was first observed at the Bradwell magnox power station in 1968, where several bolts on a steel irradiation specimen sample basket were found to have broken. As a result of this discovery, engineering assessments were carried out on all reactors to determine the possibility of further failures occurring and the decision was eventually made to lower the operating temperatures of the reactors to a bulk gas outlet temperature of 360°C, thus obtaining a marked reduction in the oxidation rate at most stations. Some further reduction was obtained by strict attention to the maintenance of very low moisture levels in the coolant gas.

Since the problem of the jacking which occurs in high pressure CO$_2$ was not appreciated when most of the magnox reactors were being designed and built there was no opportunity to make any provision for it except in the case of Wylfa, the last of the magnox stations. Here an oxidation allowance of 0.006 inches per surface was introduced and modifications were made to allow the use of waisted bolts. These gave longer effective gauge lengths than the standard shanked bolts and enhanced the strain absorbing ability of the bolted joints.

As a result of these various actions, the magnox reactors have continued to run very successfully and CO$_2$ oxidation has been a contained problem.

4.2 Advanced gas-cooled reactor (AGR) design

In the AGR design the maximum gas temperature is approximately 650°C and the gas pressure 41.4 bar, except at Dungeness *B* where the gas pressure is 31 bar. The gas also has an appreciably higher moisture content, around 400 VPM as opposed to about 5 VPM in the magnox system. This moisture arises from the methane additions made to the coolant to reduce the radiolytic corrosion of the graphite core, which is much more severe in the AGR since the power density is four to five times higher. In the AGR therefore the gas is much hotter, wetter and at a higher pressure than in the magnox stations and consequently more-corrosion-resistant materials have to be used in the upper temperature ranges.

The use of mild steel is generally restricted to a maximum temperature of 350°C and only fully-silicon-killed material is used. Even so, the steel will go into breakaway oxidation and the pessimistic assumption is made that it goes into breakaway from the beginning of service life. Fortunately the amount of oxidation is not great, generally less than 250 μm mean metal loss over the reactor lifetime of 250 000 hours and can be accommodated easily within the design allowances.

In the evaporator section of the boiler at temperatures between 350°C and 520°C, a 9% Cr Mo steel was chosen to provide improved CO_2 oxidation resistance whilst at the same time giving good resistance to steamside/waterside stress corrosion which could occur if austenitic stainless steel tubing was used in this temperature range. Above 520°C, austenitic stainless steel is used as it gives excellent CO_2 oxidation resistance and there is sufficient superheat margin to minimise the risk of stress corrosion cracking.

The original oxidation tests on 9% Cr steels showed good behaviour under the proposed service conditions, but at temperatures above 550°C they showed a tendency to go into a breakaway oxidation mode similar to that observed in mild steel, and at these higher temperatures the oxidation rates were very rapid. Since the indications were that breakaway oxidation was likely to start at some time after a weight gain of 20 mg/cm^2 had been reached, it was considered possible that after very long exposures breakaway oxidation might also occur at the lower temperature seen in reactor operation.

Further research on 9% Cr Mo steels was devoted to the influence of temperature, steel composition, coolant gas composition and component geometry. It has shown the 9% Cr steel behaviour to be similar to mild steel in that high silicon steels are more resistant to oxidation, and that higher temperature and moisture levels favour breakaway nucleation and growth. It also showed a geometric influence on time to breakaway, in that the shortest time to breakaway was seen at the corners of specimens, with edges taking longer to go into breakaway, while plain surfaces lasted longest of all. Thus in boilers where finned 9% Cr tubing is used to give improved heat transfer, the presence of fin edges means that, all else being equal, there is a greater likelihood of the material going into breakaway oxidation than is the case with a boiler using plain tubing.

Since the breakaway oxidation rate at the higher 9% Cr steel temperatures is sufficient to lead to penetration of a tube within a few years, the prime requirement is to avoid the onset of breakaway. Assessment of the data requires a sophisticated statistical approach which considers the probability of boiler tubes going into breakaway. This has shown that for all likely operating regimes the bulk of the 9% Cr steel will behave in a satisfactory manner although there is a small probability that a few tubes may go into breakaway towards the end of life of a station, which could give rise to some economic penalty.

Austenitic stainless steels (18 Cr/10 Ni/2 Mo) are used in the superheater and reheater regions of the boilers, and for all high temperature structural steelwork. Their oxidation behaviour is much more amenable to normal design practice where acceptable metal loss is the main criterion. The initial oxidation behaviour over the first few thousand hours can be variable, depending on surface condition and grain size but settles down to give a tolerable metal loss over the reactor lifetime. This value is generally less than 250 μm, although for large grain material (>80 μm) and welds a value of 450 μm should be taken over the range 550–700°C. Oxide jacking is not a problem with austenitic stainless steels as the protective oxide seals interfacial gaps at an early stage and does not produce a continuous jacking action.

Where very thin section material such as insulation foil is employed at temperatures above 400°C, a more highly alloyed stainless steel (25 Cr/20 Ni) is used, and the metal loss for this material is taken as 30 μm in 250 000 hours at 700°C.

To check that the in-service behaviour of the various materials of construction is as predicted, each AGR station is provided with an Oxidation Monitoring Scheme. In this a range of materials and components is exposed to AGR coolant gas at realistic temperatures, both within the pressure circuit and in autoclaves, and can be removed periodically for examination.

5 References

[1] Pinder, L. W.: Corrosion Science 21, 11, 749–763: 1981

[2] Laxton, J. W., Meadowcroft, D. B., Clarke, F., Flatley, T., King, C. W. and Morris, C. W.: CEGB Report No TPRD/L/2915/R85: 1986

Subject Index

strongly basic
  deionisation, 231
  deionisation units, 248
weakly basic
  deionisation systems, 251
Anions, 158
Annealing
  steelmaking, 385
Anodising
  aluminium, 422
Anthracites
  filtration
    water treatment, 225
  formation, 5
Argon
  welding, 470
Argon oxygen decarburisation
  steelmaking, 375, 376
Armohib 28
  cleaning
    boilers, 339
    inhibition, condensers, 348
Arsenic
  coal
    analysis, 60
Asbestos
  analysis, 5
  control, 145
  cooling towers
    packings, 309
  fibres
    counting, 146
  microscopy, 144
  monitoring, 144
Ascorbic acid
  feedwater
    deaeration, 177
Ash (see also Fine ash)
  adventitious
    coal, analysis, 50
  analysis, 5
  associated
    coal, analysis, 50
  cleaning
    boilers, 78
  coal
    analysis, 46, 50, 56, 68–78
  coal classification, 18
  commercial uses, 74
  disposal, 69, 76
    analysis, 60
    heavy fuel oils, 110
  fusion test, 70
  inherent
    coal, analysis, 50
  loss on ignition, 69
  marketing, 75
  sales, 69
  sources, 74
  transport, 77
Ash content
  fuel oil, 104
Ash lagoons
  cleaning effluents, 328
Asphaltenes
  combustion, 98
  fuel oil, 104
Astroloy
  turbine discs, 449
Atomic absorption spectrometers, 4
Atomic number, 158
Atoms
  models, 158
Attrition test
  ion-exchange resins, 242
Austenite
  structure, 380
Austenitic stainless steels

boilers, 440, 442
composition, 391
fracture
  prevention, 550
magnetic properties, 392
superheaters, 554
type 316
  welds, defect assessment, 524
  welding, 462
Automated testing, 507
Aviation fuel
  oil refining, 95

Babbit metals
  metallurgy, 424
Back-end corrosion
  fuel oils, 108
Bacteria
  control, 153
Bainites
  chromium
    ductility, 406
  crystallography, 386
  intergranular cracking, 405
  steelmaking, 385
Barium titanate
  ultrasonic wave production, 502
Basic oxygen steel process, 373
Baum washer, 23, 29
Beach mark patterns
  life assessment, 538
Bearing alloys
  metallurgy, 424
Bending
  pipes, 454
Bentonite
  greases, 125
Bessemer converter process
  steelmaking, 373
Best operational practices, 4
Bimetallic couples
  polarity
    electrode potentials, 164–165
Biofouling
  cooling towers, 310
Bituminous coal
  float/sink test, 20
Bituminous coals
  formation, 5
Blast furnaces
  steelmaking, 373
Block greases, 125
Blow holes
  liquid penetrant testing, 495
Blowdown
  evaporators
    water treatment, 279
Boiler drums
  fabrication, 455
Boiler headers
  welding, 482
Boiler tubes
  corrosion
    drum boilers, 181–186
    once-through boilers, 186–187
  debris deposition, 187
  deposition
    once-through boilers, 186–187
  manufacture, 456
  storage, 325
Boiler water
  alkalinity, 170–172
  chemical monitoring, 203
  chemical treatment
    recirculating steam generators, 187–188
  corrosion
    chemistry, 157–217